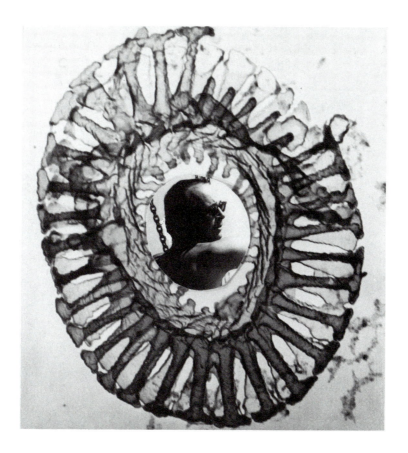

CESARE EMILIANI holds a doctoral degree from the University of Bologna, Italy, and a Ph.D. from the University of Chicago. At Chicago he pioneered the isotopic analysis of deep-sea sediments as a way to study the Earth's past climates. He then moved to the University of Miami where he continued his isotopic studies and led several expeditions at sea. His work has revolutionized our understanding of climate dynamics and the ice ages. He was instrumental in initiating the Deep-Sea Drilling Project, now in its 25th year of operation, a project that finally revealed how the Earth works. He is the recipient of the Vega Medal from Sweden and the Agassiz medal from the National Academy of Sciences of the United States. Cesare Emiliani is the author of several books and well over one hundred research papers.

The photograph shows the author encapsulated in a coccolith produced by *Emiliania huxleyi*, a coccolithophorid widespread in all oceans and named after the author.

Planet Earth

PLANET EARTH

Cosmology, Geology, and the Evolution of Life and Environment

Cesare Emiliani

Department of Geological Sciences,
The University of Miami

CAMBRIDGE
UNIVERSITY PRESS

Published by the Press Syndicate of the University of Cambridge
The Pitt Building, Trumpington Street, Cambridge CB2 1RP
40 West 20th Street, New York, NY 10011, USA
10 Stamford Road, Oakleigh, Melbourne 3166, Australia

First published 1992

Printed in the United States of America

Library of Congress Cataloging-in-Publication Data
Emiliani, Cesare.
    Planet Earth : cosmology, geology, and the evolution of life and environment / Cesare Emiliani.
        p.   cm.
    Includes bibliographical references and index.
    ISBN 0-521-40123-2 (hc). — ISBN 0-521-40949-7 (pb)
    1. Earth.   2. Geology.   3. Cosmology.   I. Title.
    QB631.E55   1992   92-5757
    550—dc20   CIP

A catalog record for this book is available from the British Library

ISBN   0-521-40949-7 (paper)   0-521-40123-2 (hardback)

CONTENTS

PREFACE

Our planet is at risk. The current explosion in the human population is forcing us to a simple but pivotal choice: stabilize the planet or perish. The first few decades of the next century will bring grave crises—environmental, economic, human. Only if we achieve a close understanding of the system of which we are part—how it came about and how it works—will the next generation be positioned to cope with the emerging problems.

Fortunately, the spectacular advances that have been made in many fields of science since World War II—in particle physics (quantum electrodynamics, the quark theory), in cosmology (the Big Bang, element formation, quasars, pulsars, black holes), in biology (the genetic code), and in geology (plate tectonics)—now make it possible, for the first time in history, to construct a model of the events from the Big Bang to today, a model that explains quite clearly how the Earth and life work. This achievement culminates 25 centuries of groping by the best thinkers and experimenters that humanity has produced.

Having a model does not mean that we know the ultimate truth—it only means that things *could be* that way or *could have gone* that way. Half a century ago, even such simple questions as "Why does the Sun shine?", "Why do children resemble their parents?", "How come are there so many volcanoes and earthquakes around the Pacific?" could not be answered. Today we either know the answer or have well-founded theories to explain the observations.

Our model demonstrates the unity of science. Cosmology is intimately connected to physics, and so are many areas of geology. Biology is related to paleontology, which documents the evolution of life on Earth. The motions of the Earth in space have timed the ice ages, which in turn have timed the evolution of the genus *Homo*. The examples that could be cited to demonstrate the close relationships of the various areas of science are truly innumerable. The common thread is, of course, mathematics— logarithmic expressions describe the timing of radioactive decay, which is the heart of the geological time scale; conic sections describe the paths of celestial bodies; forces, ubiquitous in the universe, relate to the second derivative of position with respect to time; the number e is at the core of growth and decay functions and of population dynamics.

There is much talk nowadays about the poor educational preparation of the present generation of students and the alarming level of science illiteracy at all levels of society. Much of the blame lies with the fragmentary way science is taught in the schools and is presented to the public in print and on television. It is as though an art lover had to be content to examine the Mona Lisa square inch by square inch, in no particular order, and across many months or years. How much would that person grasp of the meaning and aesthetic unity of the painting? How much does the average person grasp of the meaning and aesthetic unity of the universe? Very little, at best.

The media bear some blame, too, seeing that only one percent of the books reviewed by the leading newspapers and magazines are science books. Even though the venerable *National Geographic* prints dazzling pictures of natural landscapes, most of the time it does not bother saying what all those rocks are or what they mean.

This book is an attempt to present a global picture of modern science within the framework of the origin and evolution of the world in which we live. It provides the background necessary to understand why our planet is at risk. It also provides the background necessary for devising ways to stabilize the planet. The treatment is quantitative and rigorous throughout, but no expertise other than a knowledge of English and arithmetic is needed. The emphasis is on facts, figures, quantities, and interrelationships, without which no true understanding of how anything works can be achieved. The book is divided into eight parts:

I. Prolegomena: science and religion

II. Matter and Energy

III. Cosmology

IV. Geology

V. Evolution of Life and Evnvironment

VI. The Historical Perspective

VII. Appendixes

VIII. References and Subject Index

Part I includes a discourse on how religion views the cosmos, a brief discussion of the twelve most imortant scientific discoveries that have brought about our modern understanding of the physical world, and a presentation of the units of measurement, which form the basic vocabulary of science.

Part II discusses in some detail the structure of matter, without which radioactivity, radiometric dating methods, and the geological time scale cannot be understood. States of matter and photons are included, and so is the Uncertainty Principle, one of the cornerstones of our world. Part II provides the reader with the background that is essential to understanding the rest of the material presented in the book.

Part III deals with cosmology, without which it is impossible to understand how Planet Earth became the way it is and works the way it does. This part begins with the birth of the universe and traces its subsequent evolution. The basic trigonometric fuctions, which are essential for describing celestial motions and many other things, are reviewed, and the various methods used to measure astronomical distances are discussed in some detail. An analysis of the different types of objects in the sky and their evolutionary histories are presented, followed by a discussion of the origin, evolution, geochemical differentiation, and present status of the solar system and its components. A discussion of the motions of the Earth in space and of the calendar and seasons concludes Part III.

At this point the reader is prepared to tackle that highly complex geophysical, geochemical, and biochemical system that we call Earth, the subject matter of Parts IV and V. Starting with the planet's geochemical composition, crystallography and mineralogy, petrology, geophysics, plate tectonics, atmosphere and ocean, weathering and soils, and terminating with sedimentation, fossil fuels, diagenesis and metamorphism, rock deformation, and stratigraphy, Part IV is, in fact a brief treatise on physical geology.

Parts V deals with the evolution of life and environment. Following an indispensable introduction to biochemistry and molecular biology, Part V discusses in some detail the origin of life on Earth and the subsequent biological and environmental evolution. The various taxa are introduced in the order in which they appeared in the fossil record. This will give the reader a direct appreciation for the immensity of time and the workings of evolution, which in 4.5 billion years produced *Homo sapiens* from scratch.

Part VI is a fairly extensive list of the men and women who contributed to scientific progress through the ages, with brief discussions of their major contributions. An alphabetical list of names is included.

Part VII includes a set of tables of physical chemical data, arranged in alphabetical order, and a Chemical Formulary. Part VIII gives a complete list of cited books and papers and a detailed subject index.

This is a multipurpose book in the sense that it can benefit a wide variety of people in different walks of life:

1. The casual lay reader will learn quite a bit about our planet, even if the math is ignored.

2. People in public life, in the legal system, or in business, who may be called on to take action or pass judgment on issues that will affect the environmental stability of the globe, will learn much of what they need to know to make wise decisions and take appropriate actions.

3. College students interested in learning how our world works should dig into this book with some aggressiveness. In fact, the subject matter in this book could provide the basis for a pair of highly relevant, college-

level science courses (first course, Part I–IV; second course, Part V) that would benefit both science and nonscience majors.

4. Science teachers in high schools, colleges, and universities will find that this book can lead to a better understanding of how the various areas of science integrate into a harmonious ensemble that explains how our world originated, evolved, and works; this understanding will make it easier for teachers to demonstrate to students at all levels the relevance of science to their lives.

The wealth of physical and chemical data and the Chemical Formulary in the appendixes are for reference and to help solve the Think questions that dot the book.

One who reads this book with any kind of attention will come to the inescapable conclusion that Planet Earth and the life on it form a truly remarkable system—a system that began with the Big Bang and culminated with the evolution of humans and their diversification into a rich variety of races, languages, and cultures. It is unfortunate that this diversification has led to chronic physical and cultural tribalism that, fanned by political and religious shamans of all shapes and colors, still rules the world. The failure of political and religious systems to evolve as rapidly as science has is now placing the entire planet at risk.

To deal with the serious problems that the world is now facing, we should follow the advice of Confucius—to expand knowledge as much as possible. From an informed citizenry new ideas can emerge that may provide the solutions that are desperately needed. There is no time to waste: All indications are that the next few decades will be quite difficult for us (see Chapter 24). If the political and religious philosophies that have made the twentieth

century the bloodiest in the long and troubled history of humankind should carry over into the next, then the twenty-first century will be much, much worse. If, on the other hand, humans can succeed in burying the tribalism of the twentieth century and helping each other through the harsh times ahead, a new world should be waiting: better informed and better organized, with new machines doing much of the routine work, with enough food, shelter, health care, education, and leisure for everybody. Whether we get there or not will be determined by the children who are being born today. They will reach adulthood just as the Earth's economic and environmental crises force a decision: Change or perish. They will have to bury ancient philosophies, doctrines, and symbols; they will have to join hands with their brothers and sisters across continents and oceans; and they will have to come up with new ideas and new practices. To them, this book is dedicated.

ACKNOWLEDGMENTS

I am grateful to Steve Bruenn, of Florida Atlantic University, for reviewing portions of this book; to Peter-John Leone and Lauren Cowles of Cambridge University Press for much valuable advice; to Andrew Alden, of Oakland, California, and to Alan Gold, of New York, for skillful editing; to Charles Messing, of Nova University, who did the original illustrations appearing in this book; and to the authors and publishers listed in Part VIII, who gave their permission to reproduce illustrations and tables from their works.

Cesare Emiliani
Coral Gables, Florida, April 1992

Greek alphabet

Latin	Greek lowercase	Greek uppercase	Latin	Greek lowercase	Greek uppercase
a	α	A	o	o	O
b	β	B	p	π	Π
c	γ	Γ	r, rh	ρ	P
d	δ	Δ	s	σ, ς	Σ
e	ε, ϵ	E	t	τ	T
z	ζ	Z	y	υ	Υ
e	η	H	f	φ, ϕ	Φ
th	ϑ, θ	Θ	ch	χ	X
i	ι	I	ps	ψ	Ψ
k	$\varkappa, \kappa$	K	o	ω	Ω
l	λ	Λ	ng	$\gamma\gamma$	$\Gamma\Gamma$
m	μ	M	nk	$.\gamma\varkappa$	ΓK
n	ν	N	nx	$\gamma\xi$	$\Gamma\Xi$
x	ξ	Ξ	nch	$\gamma\chi$	ΓX

I

Prolegomena: Science and Religion

MORE than ten billion years have passed since the Big Bang. The universe began expanding at that time and is still expanding. Meanwhile, stars are born, mature, and die, and new stars are formed. The original flash of unimaginably intense radiation that accompanied the Big Bang has cooled to a mere 3 degrees above absolute zero, and its wavelength has incresed to 2 millimeters. The original photons, although immensely lengthened, are still around. You can see them on your TV screen by tuning to a blank channel and turning down the brightness untill the screen is almost dark: One in a hundred of the bright specks you see is a primordial photon (the rest is microwave communication noise). You can catch the moment of creation in your living room.

What followed is astonishing in its orderly development and apparent inevitability. The formation of hydrogen and helium in the observed proportions is an inescapable consequence of the initial blast. Also inescapable are the formation of stars, the formation of heavier elements inside stars, their dispersion to outer space during supernova explosions, the formation of more stars, and the formation of planetary systems, including planets like the Earth. Given a planet like the Earth as a substrate, it is virtually impossible to prevent the evolution of life. And so, through a series of steps not only probable but also logical, we end up with intelligent life.

Those who are familiar only with the final product—Earth and life as they appear today—find this product so astonishing in its complexity that they have to invoke a Higher Power to have made it and to run it. Out of this primordial need to explain the existence of our world has come religion, in all of its multifaceted, often contradictory, and even absurd manifestations. Yet, if we look at our world through the telescope of time, we see that, given the Big Bang, everything seems to have evolved necessarily and inevitably, from simple to complex, from the primaeval fireball to our modern civilization. Aside from a few setbacks here and there, the process seems to have been quite smooth and reasonable. A truly remarkable story.

Unfortunately, there was nobody to watch and take notes as the story unfolded. Everything had to be reconstructed from human observation. Humans began observing in prehistory. In antiquity, the Babylonians and the Egyptians were particularly good at observing and recording.

One of the first questions that even the most primitive society probably

1

would ask is "How did this world come about?" The earliest tradition on origins arose in India around 2000 B.C.E. and was codified in the *Rig-veda* (1200 B.C.E.). According to this tradition, in the beginning there was the *One*, who breathed by its own energy. Then desire entered the One and Thought was created. From that came light, and then all the rest.

According to the biblical tradition, as written in Genesis (ca. 1250 B.C.E.), "In the beginning God made heaven and earth, but the earth was invisible and featureless and darkness [was] on the abyss, and the breath of God was carried over the waters; and God said *let there be light* and light came into being". The creation of light made it possible to distinguish between night and day and established the first formal day. During the next five days, God made everything else. In summary form:

Day 1: Light (day and night)
Day 2: The sky, separating the water "below" from that "above"
Day 3: Water under the sky gathered into a basin (sea), letting dry land emerge (earth); trees on land, with fruits and seeds
Day 4: Sun, moon, stars
Day 5: Marine animals and birds
Day 6: Wild animals, cattle, reptiles, man and woman
Day 7: Day of rest

As one can see, the Jews make the world very simply, in six days and in (almost) logical order. The story makes rather good sense. Now let us see how the Greeks treated the same subject.

Genesis According to Hesiod (750 B.C.E)

In the beginning there was Chaos (empty space).
Next came Gaia (Earth), Tartarus (the underworld), and Eros (love).
Gaia was lonely and produced Uranus (heaven).
In union with Uranus, Gaia gave birth to 12 Titans (including Oceanus, Rhea, and Cronos, the youngest child), 3 Cyclopes, and 3 Hekatoncheirontes (monsters with 100 arms and 50 heads each).
Uranus, upon seeing these monsters, got mad and tried to stuff them back into Gaia.
Gaia got upset, gave Cronos a flint knife, and told him to castrate Uranus.
Cronos obliged and threw Uranus' testicles into the Aegean Sea.
The sea foamed, and out of the foam rose the island of Cyprus, with Aphrodite (Venus) nude on the beach (this, in my opinion, is a vastly superior origin for Cyprus, as compared with the tectonic push from the African plate advocated by geologists, who can offer nothing to rival a nude goddess on the beach).
Uranus predicted to Cronos that his own children would unseat him.
Cronos forced his sister Rhea to bear him Hestia (Vesta), Hera (Juno), Poseidon (Neptune), and Hades (Pluto), but each time a child was born, Cronos sneaked into the nursery, snatched the child from the crib, and swallowed it.

Table I.1. The Olympian gods and their functions

Name (Greek/Roman)	Function
Zeus/Jupiter	Ruler of the Olympian gods; uses lightning to keep order
Hera/Juno	Wife of Zeus; goddess of marriage and childbirth
Poseidon/Neptune	God of the sea
Hades/Pluto	God of the underworld ($\pi\lambda o\tilde{v}\tau o\varsigma$ means *wealth*, because gold comes from the underworld)
Athena/Minerva	Goddess of knowledge
Apollo	God of prophecy and sun god
Artemis/Diana	Virgin goddess of hunting
Aphrodite/Venus	Goddess of love
Hermes/Mercury	Messenger of the gods and god of travelers (see Hermes' head on American Express cards)
Ares/Mars	God of war
Hephaistos/Vulcan	God of fire and metalworking
Hestia/Vesta	Goddess of the hearth

When the last child, Zeus, was born, Rhea hid him in a cave on Mount Ida in Crete and put a stone in the crib. Cronos sneaked into the nursery and swallowed the stone. That made him vomit, and he vomited not only the stone but also all the children he had previously swallowed.

Meanwhile, in order to hide the crying of baby Zeus, Rhea hired a bunch of Korybantes to make a lot of noise. (The Korybantes were whirling dervishes dedicated to the cult of Cybele, the goddess of nature in a region in west-central Asia Minor called Phrygia.)

Zeus grew up, drove the Titans out of heaven into Tartarus, and proceeded to father Apollo and Artemis (Diana) in union with Leto (daughter of two Titans); Hermes (Mercury) in union with Maia (daughter of two other Titans); and Ares (Mars) and Hephaistos (Vulcan) in union with Hera (Juno).

Zeus also produced Athena (Minerva) out of his head when Hephaistos split it with an ax to cure him of a headache.

One must admit that this story is much more interesting than that chronicled in the Bible: At the end of creation, the Jews still had the same old god, but the Greeks had 12 major gods to choose from, plus a vast number of minor ones. The 12 major gods (called the Olympian gods) and their functions are shown in Table I.1, together with their Greek and Roman names. The Olympian gods reside on Mount Olympus, a mountain 2,910 m high, 260 km northwest of Athens.

Zeus made the present human race. It is a race of evil people who despise the good and praise the bad. Zeus eventually destroyed it with a flood, except for Deucalion (who happened to be his cousin) and Deucalion's wife Pyrrha. The couple built a boat, floated around for nine days and nine nights, and landed on Mount Parnassos (a peak 2,500 m high in central Greece). The two repopulated the Earth by throwing stones behind their backs—Deucalion's stones became men and Pyrrha's stones became women (fortunately, Greece is very stony—we would not be here if they had landed in Louisiana).

At the time of Homer (900 B.C.E.) and Hesiod (750 B.C.E.), the Greeks had a god, a goddss, or at least a nymph for practically everything. The sun was dragged along by Apollo's chariot. Lightning, of course, was thrown around by Jupiter when he got mad. And when Vulcan or Neptune got mad, there were volcanic eruptions or storms at sea. Even minor natural phenomena were the works of divinities. For instance, when the Greeks landed in Sicily, they found a spring on a hillside. A geologist would say that was because there was an impermeable layer of roch beneath permeable strata. According to the Greeks, however, it was because the nymph Arethusa went skinny-dipping in the river Alpheius in Greece. The river fell instantly in love with her, took human form, and chased her all the way to Sicily, where she hid inside a hill and transformed herself into a spring. Alpheius, presumably heartbroken, went back to Greece and, doubtlessly with a sigh, resumed his duties as a river.

During the sixth and fifth centuries B.C.E., Asia was shaken by a ferment of religious innovation. The *Rig-veda* evolved into Hinduism, and great religious leaders appeared—Zarathustra in Media (fl. 628 B.C.E.), Buddha in India (ca. 563–483 B.C.E.), and Lao-tzu (ca. 604–531 B.C.E.) and Confucius (ca. 551–479 B.C.E.) in China. Later, Jesus (4 B.C. to April 3, 33 C.E.) and Muhammad (ca. 570–632 C.E.) gave rise to two great religions based on the biblical tradition: Christianity and Islam. None of these Asiatic religions showed much interest in natural phenomena.

The Greeks, on the other hand, not only created a religion that attempted to explain natural phenomena but also went on to interpret and hypothesize. The Greeks are very curious people, a curiosity that stems from the geology of their land: There are mountains, rivers, caverns, a fretted coastline, and scores of islands, large and small. Just looking around there makes one curious, and curious people ask question—lots of questions. In antiquity, most of the questions could not be answered, and so the Greeks invented divinities—lots of them, big and small, in charge of all sorts of natural phenomena, and doing all kinds of extraordinary things, even changing shape. Greek religion soon grew into such a bizarre web of tales that at least some Greeks came to regard it with a healthy dose of skepticism. Skeptics tend to question, think, and probe. Questioning, thinking, and probing led the Greeks to kick off the evolution of Western thought. And so, after 2,500 years, with many stops and starts and numerous false turns, here we are today, finally, with a reasonably coherent story about our world.

1
MONUMENTAL SCIENTIFIC DISCOVERIES,
UNITS OF MEASUREMENT, AND
A WARNING ABOUT THIS BOOK

1.1. MONUMENTAL SCIENTIFIC DISCOVERIES

Scientists often are portrayed as superhumans, operating on a higher plane, in some sort of intellectual empyreum. In fact, scientists are human, richly endowed with human frailties, often involved in polemics, and often gifted with an excellent sense of humor. The number of scientists has increased in proportion to the growth of the human population, which is now exponential. There are more scientists alive today than ever existed in the past. The *Concise Dictionary of Scientists*, published by Chambers and the Cambridge University Press in 1989, lists the top 1,000 scientists the world has produced. All of these scientists, and some who are not listed, have made important contributions. But among all those contributions, which have truly revolutionized human thinking? It is not easy to come up with a short list, until we notice that the truly monumental scientific discoveries share one characteristic: They were so contrary to common sense and experience that everybody, including other scientists, was shocked into incredulity.

Few scientific discoveries rank as monumental. As a result, there are few monumental scientists. Here is my preferred list of the top twelve scientists of all time:

1. Thales of Miletus (ca. 640–548 B.C.E.) figured that a point has no dimensions, a line has length but no width, and a surface has length and width but no thickness. (You don't say! Here is a point: · , and here is a point with no dimensions: . See anything? No? That's because a point with no dimensions obviously does not exist! Care to try for a line?)

2. Anaximander (ca. 610–540 B.C.E.), a student of Thales at Miletus, claimed that the Earth was free in space. We do not know his line of reasoning, but he probably thought that because the Sun, the Moon, and the stars go down in the west and reappear next morning in the east, they must pass *under* the Earth. He thought that the Earth was shaped like a cylinder, with a flat top surface on which the known world was resting. All this is very nice, but obviously anything free in space just *has* to fall down. The Earth is not falling down, so it cannot be free in space.

3. Pythagoras (ca. 582–507 B.C.E.) claimed that the Earth was a sphere, seeing that the shadow of the Earth on the face of the Moon during a lunar eclipse is definitely curved. That is flatly against common sense, for anyone can plainly see, just by looking at the ocean or at a vast plain, that the surface of the Earth is flat. Pythagoras also maintained that the Earth was orbiting a "central fire" perennially located above the south pole (and therefore invisible from the Northern Hemisphere), with the Moon, the Sun, and the planets in the opposite direction (and therefore visible from the Northern Hemisphere). Pythagoras was the first to remove the Earth from the center of the universe and put it in orbit. He came to his proposed structure for the solar system in order to explain the fact that lunar eclipses were more frequent than solar eclipses. Pythagoras' scheme is highly ingenious, but if the Earth really were a sphere, everybody, except perhaps the Eskimos, who sit practically on top of it, would slide off the sides and plunge into empty space.

4. Democritos (ca. 460–360 B.C.E.) inferred the existence of atoms, probably by the simple argument that if one keeps cutting

5

a piece of matter in half, then after an *infinite* number of cuts one would end up with *nothing*. Because an infinite number of nothings is still nothing, there must be one smallest, indivisible particle, or else there would be no matter.

5. Aristarchos (ca. 310–230 B.C.E.) placed the Sun at the center of the solar system and ranked the planets around it in the proper order. By maintaining that the Earth orbited the Sun, not vice versa, he, like Pythagoras before him, ran flatly against common experience, which clearly shows that the Sun, the Moon, and the planets move across the sky and therefore orbit the Earth.

6. Galileo Galilei (1564–1642) claimed that *all* bodies fall at the same speed. (Fat chance, I say! Just toss a feather and your TV set out of the window and see which hits the ground first.)

7. Isaac Newton (1642–1727) may have come to the idea of gravity not because of a falling apple but because he noticed that when he as a kid threw a stone, the stone went farther the harder he threw it. If one were to throw a stone hard enough, the stone would circle the Earth and hit the thrower in the back—in other words, it would go into orbit. Newton reasoned that it was the same for satellites orbiting planets, and for planets orbiting the Sun. (Gee, why didn't I think of that!)

8. James Clerk Maxwell (1831–1879) claimed that electricity and magnetism are two manifestation of the same phenomenon. (Really! Just try to light a bulb with a magnet, or pilot a ship with a flashlight.)

9. Max Planck (1858–1947) claimed that even though a radiating body emits a continuous spectrum (as any fool can plainly see), the energy is not continuous, but is emitted in discrete amounts called *quanta*. He further claimed that energy is proportional to frequency. The proportionality constant, called *Planck's constant* (h), is how recognized as a fundamental natural constant. (How come, then, when we turn on a hot plate, the plate heats up gradually?)

10. Albert Einstein (1879–1955) discovered relativity, perhaps the most astonishing discovery of all time. He *may* have concluded that, because the electrostatic and electromagnetic systems of cgs units are related to each other by a constant that is the speed of light, then the speed of light must be a constant! What follows is the relativity of time. (This means, of course, that a car that has put on 100,000 miles should look a lot younger and be better preserved than it would if it had been kept in a garage all along. Do you really believe that? Albert does.)

11. Werner Heisenberg (1901–1976) formulated his famous *uncertainty principle*, which says that it is impossible to simultaneously measure two conjugated quantities, such as position and momentum, or energy and time, with an error smaller than Planck's constant (h) divided by 2π (i.e., with an error smaller than $1.05 \cdot 10^{-34}$ J/Hz). Cause and effect, which appear to follow each other so orderly in our world, break down at the atomic level: If the position and momentum of a particle cannot be precisely determined because of the uncertainty principle, then, clearly, particle interaction cannot be predicted. But there is more: The uncertainty principle also predicts that there are two worlds, the *virtual world* in which the product energy × time is less than $h/2\pi$, and the *real world* in which it is greater. (Really! Has anybody seen a virtual particle yet?)

12. Murray Gell-Mann (b. 1929) came up with the astonishing notion that protons and neutrons, and other particles as well, consist of point masses that have a *fractional* electrical charge! (Some nerve. We all know that one cannot split an elementary electrical charge, not even with an ax.)

This list of twelve scientists is my own preference. Others may come up with different lists (a greatly expanded list of men and women who have contributed to scientific progress through the centuries can be found in Part VI). The majority in my list (5 out of 12) are ancient Greeks, followed by 3 Germans, 2 English, 1 Italian, and 1 American. As mentioned earlier, the highly varied land of Greece made the Greeks curious and inquisitive. Compare with Egypt, a land that in antiquity was much wealthier than Greece: There is the flat Nile delta, bordered on two sides by desert and on the third by an empty sea—and that's it. If the land of Egypt was rich, it was also boring. Even the floods were regular yearly affairs occurring on a fixed schedule and therefore holding no promise of excitement.

Unfortunately, after they had launched Western civilization surely on its track, the Greeks were conquered by the Romans, who had little sense of humor but delivered Roman law and Roman order (in exchange for taxes, naturally). The spirit of Greece died. And then came the barbarians, who destroyed Rome and everything else, followed by Christianity, which offered no incentive for inquiry and innovation. When Dante Alighieri put three popes in Hell, the Renaissance was born. Those 400 years in ancient Greece and the most recent 400 years in Europe have largely made modern Western civilization as we know it.

1.2. UNITS OF MEASUREMENT

Omnia in mensura et numero: Everything rests on measurement and number, we may say. There is no science without measurement, and there is no measurement without units.

Units of length common in Greece were the foot (31.6 cm) and the stadion (600 ft or 189.6 m). The Romans adopted those units and added the mile (1,000 steps or 1.475 km); the word "mile" derives from the Latin *mille*, which means one thousand. During the Middle Ages and the Renaissance, Europe was fragmented into a number of states and provinces, each of which developed its own system of measurement. Even when the names were the same, the units were different. The foot, for instance, was 30.48 cm in England, 32.5 cm in France, 27.9 cm in Spain, 33 cm in Portugal, 30 cm in Belgium, 28 cm in Holland, 31.4 cm in Denmark, and 29.7 cm in Sweden and Norway. In addition, there were local units of length, mass, and volume with a variety of names. The units of mass and volume were even more numerous and varied. All those disparate systems of measurement created big problems, especially for commerce.

The French Revolution brought sweeping changes in France. Having dispatched king, nobility, and clergy, the French Assembly set out to develop a new system of measurement that would be "rational," not tied to old traditions. In 1791, they established the *meter* as a unit of length. The meter was defined as 1/10,000,000 of the meridional quadrant of the Earth. A meridional quadrant is one-fourth of the circumference of the Earth passing through the poles. The meter was adopted as a unit of length in France in 1801, together with its multiples (decameter = 10 m; hectometer = 100 m; kilometer = 1,000 m) and fractions (decimeter = 0.1 m; centimeter = 0.01 m; millimeter = 0.001 m).

A cubic centimeter of water was used to define the unit of mass, which was called a gram (g). Its multiples (decagram = 10 g; hectogram = 100 g; kilogram = 1,000 g; myriagram = 10 kg; quintal = 100 kg; ton = 1,000 kg) and its fractions (decigram = 0.1 g; centigram = 0.01 g; milligram = 0.001 g) were also established. As for time, the French Assembly retained the second (s) as a unit.

And then came Napoleon. Napoleon dragged his armies all over Europe, and in so doing he spread the metric system. The English detested Napoleon and refused to adopt the metric system. So did the Americans, even though the Americans rather liked the French, because of Lafayette and other things. The metric system, however, has one big advantage over the English system—it works by powers of 10, and so it is extremely easy to use.

In order to establish a standard for the meter, the French determined as accurately as they could the length of the quadrant of the circle passing through Dunkirk and Montjiuch, near Barcelona. One ten-millionth of the length of that quadrant was reproduced as a bar of platinum called the Archives Meter. On May 20, 1875, an international treaty was signed in France creating the International Bureau of Weights and Measures. The Bureau was built in Sèvres, near Paris, on land ceded by the French government and declared to be international. The Bureau is still run by an international body. In 1889 the Archives Meter was replaced by a platinum-iridium bar (90% Pt, 10% Ir) with three thin lines near one end and three thin lines near the other end. The meter was defined as the distance between the two middle lines, which reproduced, of course, the length of the Archives Meter. This became the Standard Meter. The International Bureau of Weights and Measures assumed the task of producing duplicate Pt-Ir standards for countries that wanted them. This was an inconvenient and expensive process, and so the meter was judiciously redefined in 1961 as being equal to 1,650,763.73 wavelengths of the $2p^{10}$–$5d^5$ transition (see Chapter 2) of the krypton-86 (^{86}Kr) atom. Because there are 1,650,763.73 waves in 1 m, the wavelength is $1/1,650,763.73 = 0.6057802106 \cdot 10^{-6}$ m = 0.6057802106 μm (i.e., micrometer, or millionth of a meter). Light of this wavelength is orange in color. Anyone with a krypton lamp and an interferometer can define a standard meter.

The meter was redefined once more in 1983 on the basis of the speed of light in vacuo. This speed is a universal constant. Because the speed is 299,792,458 m/s, the meter was redefined as the distance covered by light in 1/299,792,458 second *exactly*.

The unit of mass, the gram, which had been defined as the mass of 1 cm^3 of pure water at 3.98°C (the temperature at which water achieves its maximum density), had to be abandoned when it was realized that even triple-distilled water contains impurities; in addition, there are different isotopes of hydrogen and oxygen that make water molecules with different masses. In its place, a cylinder of platinum-iridium was manufactured, with mass closely similar to that of 1 dm^3 of pure water. This standard was called the kilogram (kg). The kilogram is equal to 1,000 g (*kilo-* derives from the Greek χιλιάς, which means 1,000; actually, it should be spelled *chilo-*). That cylinder is still the only standard of mass we have.

A most useful unit of mass, when dealing with atoms and molecules, is the *atomic mass unit* (symbol u), defined as $\frac{1}{12}$ of the rest mass of a neutral atom of carbon-12 (i.e., a carbon atom with its 6 electrons) in its nuclear and electronic ground state. We specify rest mass (i.e., the mass of a stationary ^{12}C atom) and the nuclear and electronic ground state because if the atom moves or the nucleus or the electrons are in an excited energy state, the energy increases the mass by $e = mc^2$.

The atomic mass unit (u) approximates the average mass of a nucleon (i.e., a proton or

a neutron) when bound inside the atomic nucleus (free protons and free neutrons have atomic masses about 0.7–0.8% greater). Because $e = mc^2$, the atomic mass unit can also be expressed in terms of energy. It is equal to 931.49432 MeV (million electron volts). There are $6.0221367(\pm 36) \cdot 10^{23}$ u in 1 g (the digits in parentheses refer to the standard deviation of the last two digits). This number, equal to g/u (the number of atomic mass units in one gram), is called the *Avogadro number*. I have recommended this number, with symbol Av, as a unit of numerical quantity.

[The Avogadro number is here expressed in exponential notation, where the exponent represents the number of zeros that follow the number 1. Thus,

$6.0221367 \cdot 10^{23}$
$= 6.0221367 \times 100,000,000,000,000,000,000,000$
$= 602,213,670,000,000,000,000,000$

It is clear that exponential notation is much easier to read and understand when dealing with very large numbers. Very small numbers are also expressed with exponential notation, but with the exponent negative. The atomic mass unit, u, for instance, is equal to $1/6.0221367 \cdot 10^{23}$ g, which is equal to $0.1660540 \cdot 10^{-23}$ g. The exponential notation is summarized as follows:

one trillion	= 1,000,000,000,000	= 10^{12}
one billion	= 1,000,000,000	= 10^9
one million	= 1,000,000	= 10^6
one thousand	= 1,000	= 10^3
one hundred	= 100	= 10^2
ten	= 10	= 10^1
one	= 1	= 10^0
one-tenth	= 1/10	= 10^{-1}
one-hundredth	= 1/100	= 10^{-2}
one-thousandth	= 1/1,000	= 10^{-3}
one-millionth	= 1/1,000,000	= 10^{-6}
one-billionth	= 1/1,000,000,000	= 10^{-9}
one-trillionth	= 1/1,000,000,000,000	= 10^{-12}

The exponents are called *logarithms* of the corresponding numbers *on base 10* (because the numbers are expressed as powers of 10). Any number can be expressed by another number (called the *base*) raised to the appropriate power. For instance, 4 is the logarithm of 16 on base 2 (because $2^4 = 16$). Logarithms on base 10 are called *common logarithms*; those on base e ($=2.718...$) are called *natural logarithms*. Fractional exponents represent roots. For instance, $4^{1/2}=\sqrt{4}=2$; $9^{1/2}=\sqrt{9}=3$. Fractional *negative* exponents represent the *inverse* of roots. For instance, $4^{-1/2}=1/\sqrt{4}=\frac{1}{2}$; $9^{-1/2}=1/\sqrt{9}=\frac{1}{3}$.]

Because the Avogadro number is a mass (g) divided by a mass (u), it is a *dimensionless number*—it does not represent an extensive property like mass, length, or time. There are dozens of dimensionless numbers of physics, many bearing the name of their proponents. Perhaps the most commonly known is the Mach number, which is the ratio of the speed of an object through a fluid to the speed of sound in that fluid. (Notice that one does not say "Mach's number," but "Mach number"; similarly, one should say "Avogadro number," not "Avogadro's number" as is commonly done.) The Avogadro number is a unit of amount, like *dozen* ($=12$) or *score* ($=20$). One can, therefore, speak of one Avogadro number of atoms, molecules, sand grains, people, or stars. The official symbol for the Avogadro number is N_A (or Av, as I have proposed elsewhere). A mole (symbol mol) is the *mass* of 1 Av of items. Thus, 1 mole of ^{1}H atoms = 1.008 g; 1 mole of ^{12}C atoms = 12 g; 1 mole of H_2O molecules = 18 g; 1 mole of ^{238}U = 238 g; 1 mole of hemoglobin molecules = 65.3 kg; 1 mole of sand grains = $1.8 \cdot 10^{15}$ tons (t). The names *Avogadro number* and *mole* are often used interchangeably, creating a lot of confusion. I repeat:

Avogadro number (Av) = g/u
$$= 6.0221367(\pm 36) \cdot 10^{23}$$
mole (mol) = mass of 1 Av of items

We will use this nomenclature in this book.

The most obvious unit of time is the day. The ancient Egyptians divided the day in 12 daytime hours and 12 nighttime hours. The day hours were longer than the night hours during the summer, and vice versa during the winter. To eliminate that inconvenience, the day came to be divided into 24 equal hours. There are 60 minutes (m) in 1 hour (h) and 60 seconds (s) in 1 m. These subdivisions, and the similar angular subdivisions, derive from the Babylonians, who used to count by sixties.

The modern unit of time is the second (s), which until 1968 was defined as 1/86,400 of the mean solar day [1 day (d) is equal to 24 h × 60 m × 60 s = 86,400 s], or, more precisely, 1/31,556,925.9747 of the tropical year 1900. The second so defined is called *ephemeris second* ("ephemeris" derives from the Greek ἐπί, which means *upon*, and ἡμέρα, which means *day*; all together it means *daily*). Since 1968, the second has been defined as being equal to the duration of 9,192,631,770 periods of the hyperfine transition of the cesium-133 (^{133}Cs) atom. The corresponding radiation has a wavelength of 3.26 cm. The second so defined is called *atomic second*.

[The hyperfine transition in an atom is the flip-flop of the magnetic field of the nucleus with respect to the magnetic field of the electron cloud. When the two are parallel, the atom has 1.9 electron volts (eV) more energy than when they are antiparallel. The ^{86}Kr atom normally occupies the lowest energy level, with the two magnetic fields antiparallel. In order for ^{86}Kr to emit radiation, it must be excited by an external radiation of *approximately* the same frequency as that of the hyperfine radiation. What the ^{86}Kr atom does is to make the frequency absolutely constant.]

Length, mass, and time are fundamental quantities; that is, they are quantities that cannot be expressed in terms of other quantities. All other quantities listed in Table 1.1 are derived from the fundamental ones. Consider, for instance, velocity and acceleration. Velocity v is the change in distance r from a reference point 0 with respect to time t. We can write $v = dr/dt$, where dr is the change in distance (it does *not* mean $d \times r$), and dt is the change in time. Velocity is expressed in m/s (or km/h, or even mph). Acceleration is a change in velocity v with respect to time t. Therefore, acceleration is equal to dv/dt. Because velocity is the change in distance with respect to time, acceleration is a change in the change of distance with respect to time. The expression $a = d^2r/dt^2$ means that acceleration (a) is a change in the change of distance with respect to change in time. The two expressions dr/dt and d^2r/dt^2 are called the first and second *derivatives* of distance with respect to time. Notice the position of the exponent 2 in the second derivative; it is applied to the letter d (which in fact means *derivative*) in the numerator, but to the letter t (for time) in the denominator.

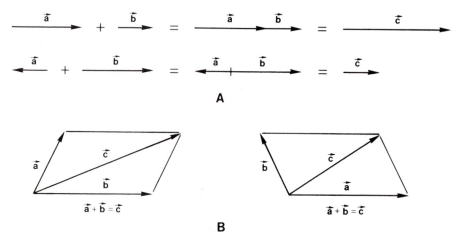

Figure 1.1. (A) Addition and subtraction of parallel vectors. (B) Addition of vectors at an angle.

There can be higher derivatives. The third derivative of distance with respect to time is called *jerk*. A jerk is a change in acceleration, which is a change in the change of velocity, which in turn is a change in the change of change in position. Jerk is equal to d^3r/dt^3. If the acceleration is smooth (no jerks), the third derivative is zero. If the body moves at constant velocity, the second derivative is also zero (there is no acceleration). If the body does not move, the first derivative too is zero (there is no velocity).

When a body is moving, distance is a function of time. Distance is called the *function,* and time is called the *variable*—as time changes, so does distance. Differential calculus is the study of derivatives. Integral calculus is used to reconstruct the original relationship when the derivative is known. For instance, knowing that a velocity is 10 m/s, one obtains $r = 10t$ if the body started moving away from the reference point 0. But if the body was already at a distance D from the reference point 0, the relationship becomes $r = 10t + D$. D is called the *constant of integration.* If all we know is that a car is moving at 50 mph, we do not know how far it is from the starting point. In this case, the distance from the starting point is the constant of integration.

Velocity, acceleration, and jerk are *vectors;* that is, they are quantities that have both magnitude and direction. Vectors are represented by arrows, with the length of the arrow being proportional to the magnitude of the vector (the scale, which must be specified, is up to the draftsman). In contrast, quantities that have only magnitude, such as mass, are called *scalars.*

[Vectors can be added or multiplied. Parallel vectors can simply be added (Fig. 1.1A). The addition is algebraic if they have opposite directions. Two vectors that form an angle other than 0° or 180° are added by completing a parallelogram with the two vectors forming two adjacent sides (Fig. 1.1B): The sum is the diagonal of the parallelogram. Three or more vectors can be added by forming a parallelogram between any two of them, adding the third vector to the vector representing the sum of the first two, and so on. If two vectors to be added do not originate from the same point, one can be moved, maintaining its direction in space, until its origin coincides with the origin of the other vector.

Two vectors can be multiplied by each other by again forming a parallelogram. The surface of the parallelogram gives the *magnitude* of the product. The *direction* of the product is at 90° from the plane of the parallelogram, and its sense is that of advancement of an imaginary, standard (right-handed) screw whose rotation would superimpose the multiplicand on the multiplier (Fig. 1.2).]

A vector is represented in print by a boldface letter or by a letter with a little arrow above it.

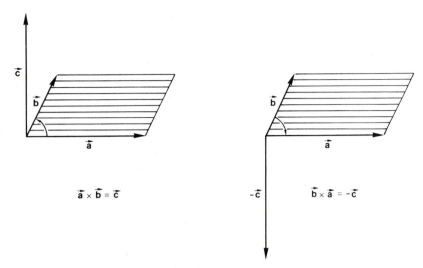

Figure 1.2. Multiplication of vectors. The product **a** × **b** of two vectors **a** and **b** is obtained by completing the parallelogram and by rotating the multiplicand (**a**) on the multiplier (**b**) via the smaller angle. The surface of the parallelogram gives the value of the product. The direction of advancement of a standard screw whose rotation would superimpose **a** over **b** via the smaller angle gives the direction of the product vector. Notice that **a** × **b** = **c**, but **b** × **a** = −**c**. The commutative law does not apply to the multiplication of vectors. Furthermore, two parallel, antiparallel, or identical vectors cannot be multiplied by each other, because no parallelogram can be formed (magnitude = 0).

Vectorial products are indicated by the sign "×" and are also called *cross products*. The sign "·" denotes scalar multiplication and the product is called a *dot product*.

A force is obviously a vector. The unit of force is the newton (N), defined as the force necessary to accelerate a mass of 1 kg by 1 m per second each second, or $1 \, \mathrm{kg \, m \, s^{-2}}$; in other words, $\mathbf{F} = m\mathbf{a}$ (where $\mathbf{F}$ = force, m = mass, $\mathbf{a}$ = acceleration). This is Newton's second law. Notice that $\mathbf{F}$ is a vector because $\mathbf{a}$ is a vector (m is a scalar). One newton is about equal to the weight of a lemon, not a large force. *Energy* is force times distance. Both force and distance are vectors, but they point in the same direction. Their vectorial product is therefore zero, and so energy is not a vector but a scalar. The unit of energy is the joule (J). One joule of energy is spent in lifting one lemon by 1 meter (or two lemons by $\frac{1}{2}$ meter). Notice that it is not specified how long the lifting takes or whether it is straight up or not. Lifting one lemon straight up by one meter in one second takes the same energy as lifting the same lemon by one meter along any path over any length of time.

Power takes time into consideration. Lifting one lemon by one meter per second requires 1 joule per second or 1 watt. The watt (W) is the unit of power. The human body can deliver 1 kilowatt (kW) for a short period of time, like lifting 10 kg by 1 m each second. By comparison, a medium-size car may deliver 100 kW of power.

During the nineteenth century, a system of units based on the centimeter, the gram, and the second (and therefore called the cgs system) became popular. In this system, the unit of force is the dyne (dyn), defined as the force necessary to accelerate a mass of 1 g by 1 cm per second each second ($1 \, \mathrm{g \, cm \, s^{-2}}$). Because 1 g = 1/1,000 of 1 kg, and 1 cm = 1/100 of 1 m, 1 dyn = 1/100,000 of a newton, or 10^{-5} N.

Following the construction of the first electrical generator by Faraday in 1831, it became necessary to formally define electric and magnetic units. The unit of charge was defined as that charge that, when placed at a distance of 1 cm from another similar charge, would exert on it a force of 1 dyne (attractive if the two charges are opposite; repulsive if they are the same). That unit was called the statcoulomb (statC). It is equal to the charge of $2.08 \cdot 10^{9}$ electrons. The unit of electrical

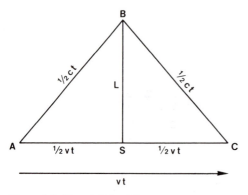

Figure 1.3. The relativity of time.

current, called statampere (statA), was then defined as being equal to 1 statcoulomb per second (1 statC/s). It was discovered, however, that a current of 1 statampere flowing through a conductor 1 cm long and 1 cm distant from another similar, parallel conductor in which an equal current flowed did not exert on it a force of 1 dyn/cm of length, but rather a force of 1/29,979,245,800 dyn/cm. The number 29,979,245,800 happens to be the speed of light in centimeters per second. Thus, 29,979,245,800 more electrons are needed to produce a unit current that will have the same effect on another unit current as a unit charge has on another unit charge. This unit of moving charge is called abcoulomb (aC) and is equal to c statcoulombs (where c is the speed of light in centimeters per second). A current of 1 aC/s is called abampere (aA). So the cgs system had *two* systems of electric units, the cgs electrostatic system of units (cgs_{esu}) and the cgs electromagnetic system of units (cgs_{emu}), related to each other by a constant equal to the speed of light raised to a positive or negative power (see Table 1.1). For instance:

$$1\,aC = c\,statC \qquad (=6.23 \cdot 10^{19}\ \text{electron charges})$$
$$1\,aA = c\,statA$$

where c is the speed of light. Other cgs_{emu} units are similarly related to the cgs_{esu} units. For instance: abfarad $= c^2$ statfarad; abhenry $= c^{-2}$ stathenry; abohm $= c^{-2}$ statohm; abvolt $= c^{-1}$ statvolt; and so forth. Clearly, cgs_{emu} is related to cgs_{esu} via the speed of light.

Although there is no clear record, it seems likely that Einstein concluded that the speed of light in vacuo must be a constant because

it relates two systems of measurement to each other. From that, special relativity (relative rectilinear motion, 1905) and general relativity (relative accelerated motion, 1916) followed.

[Let us see how special relativity works. Consider a streetcar in uniform, rectilinear motion in which a passenger flashes a beam of light at a mirror across from him. Fig. 1.3, looking down on the streetcar from above, shows what happens. According to the moving frame (the streetcar), the path of the light is from the source S to the mirror at B and back to S. The speed of light c is equal to $2L/t'$, where $t' =$ time in the moving frame. Therefore,

$$t' = 2L/c$$

According to the stationary frame (an observer on the sidewalk, for instance), the light beam has traveled the distance ABC, having left the source S when S was at A, hitting the mirror when the mirror had moved to B, and returning to the source when the source had reached C. Therefore,

$$(\tfrac{1}{2}ct)^2 = L^2 + (\tfrac{1}{2}vt)^2$$
$$\tfrac{1}{4}c^2t^2 = L^2 + \tfrac{1}{4}v^2t^2$$
$$(\tfrac{1}{4}c^2 - \tfrac{1}{4}v^2)t^2 = L^2$$
$$t^2 = L^2/\tfrac{1}{4}(c^2 - v^2)$$
$$\quad = 4L^2/(c^2 - v^2)$$
$$t = 2L/(c^2 - v^2)^{1/2}$$
$$\quad = 2L/[c^2(c^2 - v^2)/c^2]^{1/2}$$
$$\quad = 2L/[c^2(1 - v^2/c^2)]^{1/2}$$
$$\quad = 2L/c(1 - v^2/c^2)^{1/2}$$
$$\quad = (2L/c)(1 - v^2/c^2)^{-1/2}$$
$$\quad = (2L/c) \cdot \gamma$$
$$\quad = t'\gamma$$

where t is the time in the stationary frame, t' is the time in the moving frame, and γ is equal to $(1 - v^2/c^2)^{-1/2}$. Because $(1 - v^2/c^2)$ is always less than 1, γ is always greater than 1. Therefore t is always $> t'$. For instance, on hour on Earth is equal to 0.87 hours on a space ship traveling at 150,000 km/s (half the speed of light). For nonrelativistic speeds, the distance AC is negligible compared with the distance $2L$. For instance, if L were 5 m, $2L$ would be 10 m, a distance that light would cover in $3.3 \cdot 10^{-8}$ seconds. In that time, the fastest jet flying at the record speed of 2193.167 mph (980 m/s) would cover only 0.03 mm. Our streetcar, even on a fast run (30 mph), would cover only 0.000045 mm.]

In order to combine the electric and electromagnetic units, the Italian physicist G. Giorgi, in the early part of this century, proposed a system based on the meter, the kilogram, the second, and the ohm (the unit of electrical resistance). This system evolved into the International System (*Système International*, in French, abbreviated SI), which was adopted in 1960. In this system there are nine fundamental units: the meter, the kilogram, the second, the ampere, the radian, the steradian, the mole, the candela, and the kelvin.

The ampere (A) is defined as that current that when flowing in two parallel wires 1 m apart in vacuo will exert a force of 10^{-7} N per meter of length (exactly).

The radian (rad) is the unit of plane angle and is defined as being equal to $r/2\pi r = 1/2\pi = 1/6.283\ldots = 0.159\ldots$, from which we obtain 1 rad $= 0.159 \times 360° = 57.29\ldots°$ (notice that the radian, being a length divided by a length, is a dimensionless number). The steradian (sr) is the unit of solid angle and is defined as that solid angle with vertex at the center of a sphere that subtends on the surface of the sphere an area equal to the square of the radius; in other words, 1 steradian $= r^2/4\pi r^2 = 1/4\pi = 0.079\ldots$ (notice that the steradian, too, is a dimensionless number).

The candela (cd) is the unit of luminous flux. It is equal to the luminous flux of $1.464\cdot10^{-3}$ watts/steradian emitted by a source radiating at a frequency of $540\cdot10^{12}$ hertz/second. Because light covers 299,792,458 m in one second, the wavelength of this radiation is $299,792,458/5.4\cdot10^{14} = 0.555\,\mu$m, which is in the middle of the spectrum of visible light.

The kelvin (K) is the unit of temperature. Temperature is the *level* of heat energy. One kilogram and one gram of boiling water have the same temperature (100°C), although the kilogram has 1,000 times the heat energy of the gram. Although the larger mass contains more heat energy, it cannot transfer it to the smaller mass because they are both at the same temperature. The *triple point* of a substance is the temperature at which the solid, liquid, and gaseous phases are in thermal equilibrium with one another in an enclosed vessel from which air has been removed. The triple point of water, when ice, liquid water, and water vapor are in equilibrium with one another in the absence of air, is 273.16 K $= 0.01°$C. At the triple point,

water vapor pressure is 4.57 mm of mercury (4.57 mmHg). In the presence of air at a pressure of one atmosphere (atm), ice, liquid water, and water vapor are in equilibrium with one another at the temperature of 273.15 K $= 0.00°$C. This temperature is called the *freezing point* (or *melting point*) of water. The kelvin (K) is defined as 1/273.16 of the absolute temperature of the triple point of pure water. The kelvin temperature scale starts at absolute zero: 0 K $= -273.15°$C, and, conversely, 0°C $= 273.15$ K (notice that 0°C is defined on the freezing point of water, i.e., the temperature at which the three phases of water are in equilibrium with one another in the presence of air at a pressure of 1 atmosphere).

A mass creates a gravitational field, and an electric charge creates an electrical field. These fields are maintained by *virtual particles* (see Chapter 2). The position of a mass in a gravitational field with respect to a reference level or point, or the position of a charge in an electrical field, again with respect to a reference level or point, is associated with *potential energy* (or simply *potential*). If a mass or charge moves downfield (i.e., from higher to lower potential), energy is released; if the mass or charge moves upfield, energy is absorbed. Energy is neither delivered nor consumed if the motion is sideways on a line or surface along which the potential remains constant. The line or surface along which the potential remains constant is termed *equipotential*. The unit of electric potential gradient (or simply *potential*) is the volt (V). One joule of energy is absorbed or delivered in moving a charge of 1 coulomb across a potential difference of 1 volt. Therefore, V = J/C.

If the two terminals of a 1-V battery are connected through a light-bulb filament having a resistance of 1 ohm (Ω) (see below), one coulomb of electrons flows through the filament each second, developing a current of 1 ampere and delivering the power of 1 watt as visible light and heat. The flow continues as long as the chemical reactions inside the battery continue, acting as an electron pump.

The electrons are led in and out of the filament by wires. At room temperature (25°C $= 298$ K) the speed of the free electrons in the wire due to thermal agitation is about 1,200 km/s. The path, however, is not straight, but is a random zigzag in which the electrons bounce against the fixed ions that form the metal structure. The distance between one bump and the next at room temperature is about 10^{-8} m, or 200 ionic diameters for conducting metals (silver, copper, etc.). The bouncing increases with temperature, because although the speed of the electrons does not increase with temperature, the ions become more agitated and therefore disturb the electron flow. Indeed, the resistance to the movement of electrons increases in proportion to the absolute temperature over a vast temperature range. Resistance is expressed in ohms (Ω). One ohm is equal to 1 volt/coulomb, which means that a resistance of $1\,\Omega$ will allow $1\,C$ of electricity to flow through if the potential difference is $1\,V$.

Superimposed upon the bouncing around of the electrons is their net flow along the wire from the negative terminal to the positive terminal of the battery. The speed of this flow, called *drift*, is a fraction of a millimeter per second.

[**Caution!** It is important to note that *electron flow* is in the direction opposite that of the corresponding *electric current*. By convention, electric current is considered to flow from positive to negative, as though it consisted of positive particles flowing from the positive pole of a battery to the negative pole. The positive pole is considered to be at a higher potential than the negative pole. There are historical reasons for this convention. The ancient Greeks observed that when amber (a fossil resin) was rubbed with wool cloth, a spark was generated between the two. The spark was thought to be a sort of fluid. Benjamin Franklin (1706–1790) postulated that lightning was nothing but a big spark. He proved it by flying a kite in a thunderstorm and capturing electricity (1752). He also hypothesized that electricity, being a fluid, would run, like all fluids, from high (positive) to low (negative). We now know that when amber is rubbed with wool, electrons are stripped from the wool and collect on the amber. The Greek name for amber is ἤλεκτρον, and the electron became the elementary unit of negative charge. Although the electricity carriers, the electrons, flow from negative to positive, in practical work, such as wiring a house, the electric *current* is taken to flow from positive to negative.]

Magnetic fields are created by electric charges in motion. In a magnetized piece of iron, the iron atoms are aligned parallel to each other, and so are their net magnetic fields. In a horseshoe magnet, the iron atoms on the opposite poles of the magnet point in the same direction. An electric charge moving between the two poles is deflected by the magnetic field. The deflecting force is given by

$$\mathbf{F} = q\mathbf{v} \times \mathbf{B} \tag{1}$$

where q = electric charge, $\mathbf{v}$ = velocity vector of the charge, and $\mathbf{B}$ = magnetic-field vector (intensity and direction of the magnetic field). The force, as indicated by the vectorial product, is normal to the plane defined by the vectors $\mathbf{v}$ and $\mathbf{B}$. This means that $\mathbf{F}$ is 0 when $\mathbf{v}$ is parallel to $\mathbf{B}$, and it reaches a maximum when $\mathbf{v}$ is normal to $\mathbf{B}$.

From equation (1) we derive

$$\mathbf{B} = \mathbf{F}/q\mathbf{v} \tag{2}$$

For F = 1 newton, q = 1 coulomb, and v = 1 m/s, we obtain 1 newton/(coulomb·meter/second) as the unit for B. But coulomb/second = ampere; therefore, the unit of B is equal to 1 newton/ampere·meter or $NA^{-1}m^{-1}$. This unit is called tesla (T). The magnetic field between the poles of a small horseshoe magnet may be 0.1 tesla. The cgs$_{emu}$ unit of magnetic field intensity is the gauss (G), equal to 10^{-4} tesla. The magnetic field of the Earth is about 0.5 gauss. A further fraction, used in geophysics, is the gamma (γ), which is equal to $10^{-5}\,G = 10^{-9}\,T = 1$ nanotesla (nT).

Table 1.1 lists the basic units with their names, symbols, and definitions. More extensive lists of units appear in Appendix A (Tables A15 and A16). Appendix A also contains a list of symbols and abbreviations (Table A13) and an extensive table of conversion factors (Table A2).

1.3. A WARNING ABOUT THIS BOOK

This book is about science. Science rests on three legs: facts, figures, and theory. It is obvious that there cannot be any science without facts. But it should be equally obvious that there cannot be any science, nor any understanding of science, without figures (i.e., without math). This book is filled with numbers, equations, and formulas, which, together with the facts, form the very core of science. Although the math nowhere reaches a level higher than what is required in high school, there is quite a bit of it. Unfortunately, math is a subject that seems to instill terror in the most intrepid mind. It is indeed an alien subject, for two excellent reasons. First, as we saw in the Prolegomena, in six days God made Heaven and Earth, light, land and sea, plants, Sun and Moon, fish and birds, land animals, and, finally, man and woman. In six days God made everything—everything, that is, except math. If God did not make math, math must have been made by the Devil, which gives us the right to hate it. The second reason is that the human brain did not evolve to do math—there was no need to do math in the jungle. The human brain became instead a prodigious memory bank, with an uncanny ability for pattern recognition. This is, of course, essential for survival in the jungle—to remember which way to go, what to eat and what not to eat, how to distinguish friend from foe, how not to step on a snake or a scorpion, and so forth. As a result of this evolutionary course, normal humans rightly feel that math does not fit: our brain cannot tell the square root of 3, whereas a two-bit pocket computer can do that in a split of a second. There are some unusual humans, however, who really dig math and end up earning a Ph.D. in math. Unfortunately, by that time, their brains have become so exalted that they are unable to communicate with poor mortals like you and me. Even more unfortunately, these people end up teaching math, thus perpetuating the spreading of mathematical terror in the unwary minds of young students. The result is generation after generation of math haters. In my modest opinion, this could be avoided if math were to be taught as part of the natural sciences—little by little and as the need arises—instead of a separate subject.

But relax. Even if you have been terrorized as a kid, there is no need to panic or freeze at the mere sight of an equation. When you see an equation coming your way, the secret is to grab it and rip it apart: consider the separate pieces and what they mean and, step by step, reassemble the equation, observing the relationships between the various pieces. As an example, consider our discussion of relativity in the preceding section. You probably started reading it as one reads a novel, i.e., at a reasonable clip. When you hit the math you had two choices. Your first choice should have been to stop cold, slowly peruse the first line of the mathematical development, and only after you had convinced yourself that the equation was correct [i.e., that $(\frac{1}{2}ct)^2$ is indeed equal to $L^2 + (\frac{1}{2}vt)^2$], you should have proceeded to the second line. At the end of the development, you would have been *mathematically* sure that time is indeed relative. You do not have to remember the development—only the conclusion. The huge difference is that you would have proved to yourself that the conclusion is correct instead of relying on somebody else's word. Your other alternative would have been to skip the math and trust the conclusion that time is relative. This is far, far less satisfactory.

The third leg on which science rests is theory, an attempt to explain facts and figures in an intellectually rewarding and aesthetically pleasing way. This book contains lots and lots of facts and figures, but it also contains theories. Whereas facts and figures, once established, are not likely to change, theories are likely to change. Believing that this should not discourage the construction of new theories, I have included in this book some pretty wild theories, on the invention of math, the launching of the Renaissance, the wonderful world of virtual particles (which may or may not be what I say they are or aren't), the origin of the Moon, the dynamics of the mid-ocean ridges, the formation of geoclines, the role of viruses in evolution, the evolution of birds and horses, and more. The reader will have to exert utmost judgment in assessing these brilliant theories, keeping in mind that, like most theories, they are likely to have an ephemeral life, in some cases even as short as Planck's time ($5.390 \cdot 10^{-44}$ s). Nevertheless, for the time being, debating these theories, even in the solitude of one's mind, will serve the important purpose of bringing some reviving sparkle to the reader's mind, which is likely to have been anesthetized by watching too many vapid TV programs during the wonder years.

A final warning. This book necessarily deals, here and there, with nomenclature—the language of science—and definitions of terms. This is, of course, a crashing bore. For instance,

Table 1.1. Commonly used units of measurement (an asterisk identifies a fundamental SI unit)

Quantity	Unit	Symbol	Definition and remarks
length	meter*	m	distance in vacuo covered by light in 1/299,792,458 s
	astronomical unit	AU	mean Earth–Sun distance = 149,597,870.7 km = 8.3 light-minutes = 499.0 light-seconds
	light-year	—	distance in vacuo covered by light in 1 tropical year = $9.46 \cdot 10^{12}$ km
mass	kilogram*	kg	mass of Pt(90%)–Ir(10%) cylinder (diameter and height = 3.9 cm each) kept at Sèvres, France $\simeq$ mass of 1 liter of pure water at 3.98°C (the temperature at which water has its maximum density)
	gram	g	1/1,000 kg
	atomic mass unit	u	1/12 of mass of neutral ^{12}C atom = $1.66 \cdot 10^{-24}$ g = $1.66 \cdot 10^{-27}$ kg = 931.49 MeV
time	second*	s	duration of 9,192,631,770 reversals of the nuclear magnetic field with respect to the net magnetic field of the electron cloud in the ^{133}Cs atom (the second thus defined, also called *atomic second*, is equal to the ephemeris second); there are 31,556,926 s in a tropical year
	ephemeris second	s_E	1/86,400 of mean solar day
	ephemeris day	d_E	time interval equal to 86,400 s_E
	sidereal year	—	time required for the Earth to return to the same position with respect to the Sun and a distant star
	tropical year	y	time interval between successive vernal (spring) equinoxes
velocity	—	**v**	dr/dt, measured in m/s
acceleration	—	**a**	d^2r/dt^2, measured in m/s²; the gravitational acceleration at the Earth's surface ranges from 9.78 m/s² at the equator to 9.83 m/s² at the poles; the average is 9.81 m/s²
	galileo	gal	cm/s²
momentum	—	**p**	*m***v**
angle			
plane	radian*	rad	$1/2\pi$ of 360° = 57°18′
solid	steradian*	sr	solid angle, with vertex at center of sphere, subtending an area on surface of sphere equal to the square of the radius
angular momentum	—	**L**	**L** = **r** × **p** (where **r** = shortest distance of revolving particle from axis of revolution, with origin of **r** vector on axis; **p** = momentum of particle); direction of **L** vector = direction of advancement of standard screw whose rotation would superimpose **r** over **p** via the smallest angle

Quantity	Unit	Symbol	Definition and remarks
numerical quantity	avogadro	Av	$6.022 \cdot 10^{23}$
amount of substance	mole*	mol	mass of 1 Av of items (often considered synonymous with avogadro)
force	newton	N	from $\mathbf{f} = m\mathbf{a}$, where $m = 1$ kg. $\mathbf{a} = 1$ m/s; because $\mathbf{f} = m\mathbf{a}$, the force with which a 1-kg mass is attracted by the Earth's mass at the Earth's surface (= weight) is 9.81 N at 45° north latitude
energy	joule	J	$N \cdot m$
	electron volt	eV	energy of 1 electron accelerated through the potential difference of 1 volt $= 1.6 \cdot 10^{-19}$ J $= 1.073 \cdot 10^{-9}$ u
	million electron volts	MeV	10^6 eV
power	watt	W	J/s. For an average person (80 kg), 1 push-up/second may deliver the power of 200 W; $W = V \cdot A$. An ordinary flashlight has two 1.5-volt batteries in series, for a total voltage of 3 V. The light bulb rates 1 watt. The current is therefore $\frac{1}{3}$C/s $= 2 \cdot 10^{18}$ electrons/second. Each electron has the energy of 3 eV.
luminous intensity	candela*	cd	luminous intensity of 1/683 W/sr emitted by a monochromatic source radiating at the frequency of $5.40 \cdot 10^{14}$ Hz
frequency	hertz	Hz	1 cycle/second; $\approx$ frequency of human heart
charge	statcoulomb	statC	$2.1 \cdot 10^9$ e (where e = charge of the electron)
	abcoulomb	aC	c statC
	coulomb	C	$6.2 \cdot 10^{18}$ e $= 0.1$ aC
	faraday	—	1 avogadro of electrons = 96,485.3 C
current	statampere	statA	1 statC/s
	abampere	aA	c statA
	ampere*	A	1 C/s $= 0.1$ aA
potential	volt	V	J/C
magnetic-field intensity	tesla	T	$N/A \cdot m$
	gauss	G	10^{-4} T. The average intensity of the magnetic field at the Earth's surface $\simeq 0.5$ G.
	gamma	γ	$= 10^{-5}$ gauss $= 10^{-9}$ tesla $= 1$ nanotesla
magnetic moment	—	—	J/T. The magnetic moment of spinning proton $= 1.4 \cdot 10^{-26}$ J/T, with S-to-N direction of resulting magnetic field in the sense of advancement of a standard screw whose turning represents the sense of spin; magnetic moment of spinning electron $= 0.93 \cdot 10^{-23}$ J/T, with S-to-N direction opposite that of the proton for the same sense of spinning.
temperature	kelvin*	K	1/273.16 of the absolute temperature of the triple point of pure water

p. 313 contains the astonishing revelation that *mudflow* actually means a flow of mud. You, the reader, are invited to skip whatever you find boring (even the whole book, if you are so inclined). If after some skipping you should encounter an unfamiliar term, you should go to the subject index (which is quite detailed) and from there to the page where the term is defined. Upon discovering, for instance, that *mudflow* means a flow of mud, you will realize that the money you spent on this book was not wasted.

THINK

* Anthropologists say that the brain of *Homo sapiens* reached its present size and complexity perhaps as early as 125,000 years ago. If so, why did it take such a long time to develop modern civilization?

* Can you name an organized church or a cult based on the pursuance of scientific research? If there aren't any, how come?

* The meter, the kilogram, and the second—the fundamental units of length, mass, and time, respectively—are all three artificial. Can you come up with a set of *natural* units? If you fail, explain why. If you succeed, write to the Bureau International des Poids et Mesures, Sèvres, France. If they accept your suggestion, you will become instantly famous.

* Some proposals have been made to apply the decimal system to the measure of time. None has succeeded. Try again and see if you can come up with a practical way to do it (another way to become famous).

* Your car battery is powerful enough to start the engine, and yet if you place one hand on the positive pole and the other hand on the negative pole, you do not get a shock. How come?

II

Matter and Energy

THE substrate of matter and energy was a subject of much speculation by the ancient Greeks. Thales thought it was water. Anaximander thought it was an undefined state, which he called ἄπειρον. Anaximenes (ca. 570–500 B.C.E.) thought it was air.

Part II offers a reasonably good foundation for understanding the structure of matter, electronic systems and bonds, electromagnetic energy, radioactivity, and radiometric dating methods. The subject matter is presented in two chapters:

Chapter 2   Atoms and Photons
Chapter 3   The Atomic Nucleus

Chapter 2 deals with the states of matter, the electron structure of the atom, quantum numbers, chemical bonds, entropy, and the interaction between matter and energy. Included are such fundamental constants as the Avogadro, Boltzmann, and Planck constants. This is followed by an analysis of the uncertainty principle that shows how quantum reality differs from, and complements, common reality.

Elementary particles and the four forces of nature are treated rather extensively in Chapter 3. It is my strong opinion that this part of physical science is so fundamental to everything we know that it should be an intrinsic part of any introduction to science, however elementary it might be. Anyone dealing with biology or geology without a knowledge of elementary particles is like a writer who does not know the alphabet. Because all elementary particles spin, some attention is given to the concept of angular momentum. This concept is relevant to a broad range of subjects, from particle physics to cosmology, and yet it seems to be confusing to many.

Chapter 3 also deals with the formation of atomic nuclei and the nuclear binding energy, which is basic to an understanding of radioactivity, radioactive dating methods, and the geological time scale. Besides, this information is essential to anyone living in the nuclear age. The basic equation of radioactivity, $N = N_0 e^{-\lambda T}$, is introduced term by term, so that the reader is led step by step to clearly understand what it means and how it is used. The real meaning of the number e is revealed using poultry science.

Part II provides the reader with the basic principles needed to understand our megascopic world, namely, the objects that form our universe, their origin and evolution, their dynamics, the physics and chemistry of the solar system and of planet Earth, and the properties of the materials that form our planet. These are the subjects covered in Part III (Cosmology) and Part IV (Geology).

2.1. ATOMS

Everybody knows that matter consists of atoms and that atoms are very small. What is difficult to grasp is how small they really are and how many there are. Let us try anyway. Table 2.1 shows the composition of the human body in terms of number of atoms. The average atomic mass is 7.9 u. Suppose that your body weighs 70 kg or 70,000 grams. If the average mass of the atoms in your body were 1 u, you would have one Avogadro number (1 Av) per gram of mass, or $6.021 \cdot 10^{23} \times 70{,}000 = 4.2 \cdot 10^{28}$ atoms in your body. But because the average atomic mass is 7.9 u, there will be 7.9 times fewer atoms, or a total of $5.3 \cdot 10^{27}$ atoms. Now, the diameter of an average atom is 10^{-10} m, which means that if you took all the atoms in your body and lined them up in a row, they would form a line $5.3 \cdot 10^{27} \times 10^{-10}$ m $= 5.3 \cdot 10^{17}$ m long. This is equal to $5.3 \cdot 10^{14}$ km or 56.0 light-years—more than enough for three round trips to Sirius, the brightest star in the sky.

An atom consists of a *nucleus* surrounded by one or more electrons. The simplest atom, that of hydrogen-1 (^{1}H), consists of a single proton orbited by a single electron. Its radius, when the electron is closest to the nucleus, is $0.529 \cdot 10^{-10}$ m. The nucleus is much smaller than the atom, only 10^{-15} m, or 100,000 times smaller than the atom. This means that if the nucleus of a hydrogen atom were the size of a pinhead (radius = 1 mm), its electron would be 100,000 mm = 100 m away. The atom, therefore, is mainly empty space. The density of the nucleus is 10^{14} g/cm^3, but that of the atom as a whole is the same as the density of common solid matter, because atoms are packed in contact with one another in the solid state. Density in solids ranges from 0.07 g/cm^3 for

Table 2.1. Chemical composition of the human body

Element	Atomic mass (u)	Percent of atoms
H	1.008	63.070
O	15.999	25.606
C	12.011	9.460
N	14.007	1.324
Ca	40.078	0.233
P	30.974	0.202
K	39.098	0.065
S	32.066	0.049
Na	22.990	0.040
Cl	35.453	0.026
Mg	24.305	0.013
Fe	55.847	0.00045

solid hydrogen to 22.4 g/cm^3 for iridium (the densest metal).

Atoms heavier than ^{1}H also contain neutrons in their nuclei. Table 2.2 lists the properties of the three particles that form the atoms. These three particles form all the elements and isotopes in nature. An *element* is the set of atoms having the same number of *protons* in their nuclei. An *isotope* is the set of atoms having the same number of *protons and neutrons* in their nuclei.

Table 2.3 shows the composition of some of the many isotopes that occur in nature. As the table suggests, the lighter elements have equal numbers of protons and neutrons in their nuclei. The heavier elements, on the other hand, have an increasing excess of neutrons. Notice that the mass number, which identifies the isotope, is equal to the number of protons plus the number of neutrons. Normally, the number of protons in the nucleus is equal to the number of electrons that surround the nu-

Table 2.2. Atomic components

Particle	Mass (u)	Charge (e = electron charge)
proton	1.0072764	$+1\,e$
neutron	1.0086649	0
electron	0.0005486	$-1\,e$

Table 2.3. Examples of isotopes

Isotope	No. of protons	No. of neutrons
^{1}H	1	0
^{2}H	1	1
^{4}He	2	2
^{12}C	6	6
^{14}N	7	7
^{16}O	8	8
^{40}Ca	20	20
^{45}Sc	21	23
^{56}Fe	26	30
^{133}Cs	55	78
^{197}Au	79	118
^{235}U	92	143
^{238}U	92	146

cleus. The charges are thus balanced, and the atom is neutral. The electrons are distributed in concentric shells, forming an electron cloud. Atoms, especially those that have 1–2 or 6–7 electrons in their outermost shell, tend (respectively) to lose or gain electrons. If they lose electrons, they become positively charged; if they gain electrons, they become negatively charged. Charged atoms are called *ions*. Positively charged ions are called *cations*, and negatively charged ions are called *anions*. These two names derive from the fact that in electrolytic cells (like those used for electroplating) the negative pole is called the *cathode*, and it is toward it that the cations move; the positive pole is called the *anode* and it is toward it that the anions move. The term *cathode* generally refers to the negative electrode. In storage batteries, however, it is the positive terminal that is called the cathode.

The mass of an atomic particle is expressed in *atomic mass units*. Before 1960 there were two definitions of the atomic mass unit (abbreviated *amu*). There was a physical definition saying that 1 amu was equal to $\frac{1}{16}$ of the mass of the

neutral atom of ^{16}O. And then there was a chemical definition saying that 1 amu was equal to $\frac{1}{16}$ of the average isotopic mass of oxygen. The chemical amu was 1.000275 times the physical amu, because there are three isotopes of oxygen in nature, two of which—^{17}O and ^{18}O—are heavier (and rarer) than ^{16}O. The atomic mass of an element was taken to be the weighted average of the masses of the naturally occurring isotopes of that element in their natural proportions on Earth.

In 1960 the atomic mass unit was redefined as $\frac{1}{12}$ of the mass of the neutral atom of carbon 12, and its symbol was changed to *u*. The atomic mass unit is similar to the average mass of a *nucleon* (a proton or a neutron) when bound in an atomic nucleus. This is most convenient, because the number identifying a given isotope (238 for uranium-238, for instance), which gives the total number of nucleons, is also very close to the mass of that isotope (238.051 u in this case). The masses are not exact integers, because different elements and isotopes have different nuclear binding energies (see Chapter 3).

The electrons in atoms form an electron cloud around the nucleus. Within this cloud the electrons are distributed in *shells* that are identified, from inside out, by the letters K, L, M, N, O, P, and Q, or by the corresponding integers 1 through 7. These integers are called *principal quantum numbers* (symbol *n*). The principal quantum number determines the size of the atom. However, the size of the neutral atom does not change that much. In fact, the radius of the uranium atom is only 2.5 times that of the hydrogen atom. The reason is that the increasing numbers of protons in the nuclei of the heavier elements "pull in" the inner electron shells.

The radius of the hydrogen atom in the ground state (electron in K shell, with principal quantum number $n = 1$) is $0.5 \cdot 10^{-10}$ m. If the electron is in a shell n farther out, its radius r_n is given by

$$r_n = r_1 \times n^2$$

where r_n is the radius of the shell with principal quantum number n, r_1 is the radius of the hydrogen atom in the ground state, and n is the principal quantum number. The principal

Table 2.4.   Electron structure of the uranium atom in the ground state

	K	L		M			N				O				P				Q			
Shell (n)	1	2		3			4				5				6				7			
Subshell (l)	0	0	1	0	1	2	0	1	2	3	0	1	2	3	0	1	2	3	0	1	2	3
Orbital (m_l)	1	1	3	1	3	5	1	3	5	7	1	3	5	7	1	3	5	7	1	3	5	7
Number of electrons (max.)	2	2	6	2	6	10	2	6	10	14	2	6	10	14	2	6	10	14	2	6	10	14
Number of electrons in U	2	2	6	2	6	10	2	6	10	14	2	6	10	3	2	6	1	0	2	0	0	0

quantum number of the L shell is 2, and so its radius is

$$r_2 = 0.5 \cdot 10^{-10} \times 4$$
$$= 2 \cdot 10^{-10} \, \text{m}$$

When $n = 7$, the radius of the hydrogen atom is $24.5 \cdot 10^{-10}$ m. The electrons of uranium, which has the largest number of electrons (92) among all naturally occurring elements, occupy shells K through Q when the atom is in its ground state. If the atom is hit with radiation of sufficient energy, one of the two electrons in the Q shell can be moved to a shell with higher energy (R, S, T, etc.). If the radiation has an energy of 6.05 eV, one electron is removed and the uranium atom becomes an ion.

Within each shell, the electrons are distributed in *subshells*, each characterized by another number called *orbital quantum number* (symbolized by the letter l). The quantum number l can assume the values $0, 1, 2, 3, \ldots, n - 1$, where n is the principal quantum number. The orbital quantum number determines the shape of the volume within which a particular electron is largely confined. In the ground state, even elements with a large number of electrons (cerium and heavier elements) can accommodate their electrons within the first four values of l (0 to 3). By tradition, these are assigned the letters s, p, d, and f (meaning, respectively, *sharp*, *principal*, *diffuse*, and *fundamental*, which are old terms referring to spectroscopic lines).

Electrons are charged particles that orbit the nucleus. The orientation and intensity of the orbital magnetic field created by an atomic electron are defined by the *orbital magnetic quantum number* (symbol m_l); m_l can assume only integral values ranging from -1 through

0 to $+1$. For any given l, the number of the values is $2l + 1$. The energy level of an electron in a given shell and subshell, characterized by a given value of m_l, is called an *orbital*. An electronic orbital is then characterized by three quantum numbers: n, l, and m_l.

Electrons also spin, and the vector representing the spin may point "up" or "down" (with respect to an arbitrary frame of reference, such as an external magnetic field). Spin is given by the *spin magnetic quantum number* (symbol m_s), which can only have values of $\pm \frac{1}{2}$ (in units of $h/2\pi$). A maximum of two electrons, provided they have opposite spins, can occupy a given orbital. Therefore, the four quantum numbers n, l, m_l, and m_s characterize an individual electron. The four quantum numbers are sufficient to describe the properties of all 4,278 electrons occurring in the natural elements.

The distribution of electrons in atoms and molecules is governed by the *Pauli exclusion principle*, which says that *no two fermions of the same kind and belonging to the same system (atom, molecule, or a larger bound system) can exist in the same quantum state as specified by the set of quantum numbers, but must differ in at least one quantum number*. (Fermions are particles, such as the electron, that have spin $\frac{1}{2}$, see Chapter 3.) The Pauli exclusion principle means that an atomic orbital may contain only one or two electrons, and if it contains two electrons, they must have opposite spins (and therefore a different m_s quantum number). The electron structures of the elements are shown in Table A6, Appendix A; that of uranium is shown in Table 2.4.

Chemists express the electronic structure of

atoms by writing first the principal quantum number, followed by the l number, carrying as an exponent the number of electrons with that value of l. For example, the notation for sodium (in the ground state) is

$$1s^2 2s^2 2p^6 3s^1$$

meaning that the first shell (1) has one subshell (s) with two electrons; the second shell (2) has two subshells (s and p) with, respectively, two and six electrons; and the third shell (3) has one subshell (s) with one electron. In terms of orbitals, the 1s and 2s subshells have one orbital each (with two electrons); the 2p subshell has three orbitals (with two electrons each); and the 3s subshell has one orbital (with one electron).

The distribution of electrons in the uranium atom in the ground state is shown in Table 2.4. The corresponding chemical notation is

$$1s^2 2s^2 2p^6 3s^2 3p^6 3d^{10} 4s^2 4p^6 4d^{10} 4f^{14}-$$
$$5s^2 5p^6 5d^{10} 5f^3 6s^2 6p^6 6d^1 7s^2$$

The energy of an electron generally increases with shell number. There are many exceptions, however, because the energy depends on both the principal quantum number and the orbital quantum number. The 3d subshell, for instance, has more energy than the 4s subshell.

An electron can jump from one energy level to a higher or a lower level by, respectively, absorbing or emitting a photon of the appropriate energy. If the light emitted by an excited element is dispersed through a prism, a series of bright lines on a dark background appear (*emission spectrum*). The position of the lines in the spectrum is characteristic of the specific energy transitions of the specific electrons belonging to the atoms of that specific element. The spectrum is therefore characteristic of the element.

The shells of the hydrogen atom have only one subshell (s), because the hydrogen atom has only one electron. The energy E_n of an electron bound in a hydrogen atom depends only on the shell it occupies and therefore only on the principal quantum number n:

$$E_n = -e^4 m_e / n^2 h^2 8\varepsilon_0^2 \text{ J}$$
$$= -13.6/n^2 \text{ eV} \qquad \text{(for } n=1\text{)}$$

where e is electron charge, m_e is electron mass, n is the principal quantum number, h is Planck's constant ($6.626 \cdot 10^{-34}$ J/Hz, the energy

of a photon with a period of 1 Hz, and thus a wavelength equal to the distance covered by light in 1 second), and ε_0 is the permittivity constant ($8.85 \cdot 10^{-12}$ farad/meter), needed to express E_n in SI units.

Here, zero energy is the energy of a proton-electron system in which the proton and the electron are at infinite distance from each other. Energy totaling 13.6 eV is released by the system when the electron falls from infinity to ground level (i.e., to the K shell). The same amount of energy must be injected into the system (H atom) to ionize the atom when the electron is in the K shell. Correspondingly smaller amounts of energy are needed to ionize the H atom when the electron is already at a higher energy level. The energy levels of the hydrogen atom are shown in Fig. 2.1.

The energy levels of free electrons within a larger confined system (a plasma or a metal bar, for instance) are too numerous to be assigned letters or numbers. They form a practically continuous *energy band*. A copper penny, for instance, contains about $1.56 \cdot 10^{22}$ atoms. In forming the metallic bond, 1 electron per atom is turned loose and is free to roam through the entire penny. Because only two electrons (with opposite spins) can occupy a given energy level, the number of energy levels (orbitals) in a copper penny is $1.56 \cdot 10^{22}/2 = 7.8 \cdot 10^{21}$. Would you care to calculate how many energy levels there are in a length of railroad track?

Because metals have an extremely large number of free electrons (1–2 Av of free electrons per mole), and because only two of these electrons can be at the same energy level, free electrons in metals have a practically infinite number of energy states. As a result, hot metals emit a continuous spectrum rather than a specific number of well-defined spectral lines as electrons bound in atoms do.

If the continuous spectrum from a hot surface is passed through a cooler gas, the atoms of the gas will absorb the same frequencies they would emit and reradiate them in all directions. From any given direction, the spectrum appears as a series of dark lines on a luminous background (*absorption spectrum*).

A proton is a particle charged with one unit of positive electricity. Like the electron, it spins. Because the proton is charged and because it spins, it creates a magnetic field whose axis coincides with the spin axis and whose direction is

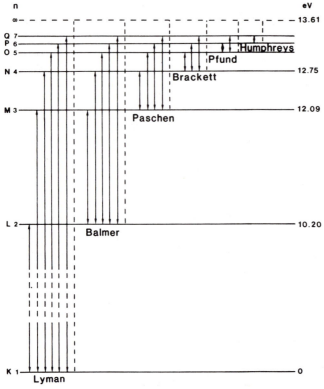

Figure 2.1. Energy levels of the hydrogen atom (energy-level letters and numbers to the left; energy in electron volts to the right). The single electron of the hydrogen atom occupies orbital 1, subshell 0, shell K ($n = 1$) when it has the least possible energy (ground state). It can be raised to higher energy levels by absorption of appropriate energy (i.e., light of the appropriate wavelength). The electron remains at the higher energy level for about 10^{-8} s, and then it drops down to a lower energy level, emitting light of the appropriate wavelength. Transitions between the K shell ($n = 1$) and higher shells involve ultraviolet radiation and form the Lyman series of spectral lines. Transitions between the L shell ($n = 2$) and higher shells involve visible light (Balmer series). Transitions between the M shell ($n = 3$) and higher shells involve infrared light (Paschen series), as do transitions among the N, O, P, and Q shells (Brackett, Pfund, and Humphreys series).

the same as the direction of advancement of an imaginary, right-handed screw threaded through the proton along its spin axis. By convention, the direction of the field is from magnetic south (the head of the screw) to magnetic north (the tip of the screw) (Fig. 2.2, left). For a spinning electron, the direction of the magnetic field is opposite, because the electron is negatively charged (Fig. 2.2, right). Both a spinning proton and a spinning electron produce a *dipole* field (i.e., a field with two poles, S and N).

The *magnetic moment* **μ** of a magnetic dipole is a measure of the strength of the dipole, as evidenced by the torque **τ** exerted on the dipole by a homogeneous external magnetic field **B** of 1 tesla: $\tau = \mu \times B$, from which $\mu = \tau/B$. Magnetic moment is expressed in joules/tesla or equivalent units.

[*Moment* and *torque* are synonymous. The moment (or torque) of the force **F** on a point *P* is equal to the product **r** × **F**, where **r** is a vector originating at *P* and terminating at the point of application of **F** (Fig. 2.3). The vector **F** can be moved, maintaining it parallel to itself, until its point of application comes to coincide with that of **r**. The product **r** × **F**

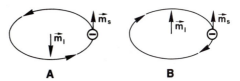

Figure 2.4. Angular-momentum vectors of a rotating and revolving electron. The magnetic spin angular-momentum vector **m**ₛ may be antiparallel (A) or parallel (B) to the magnetic orbital angular-momentum vector **m**ₗ (whose point of application is the center of the orbit).

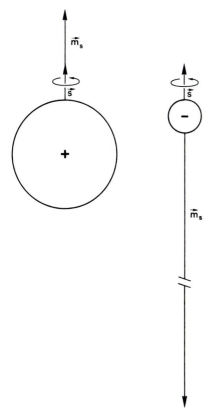

Figure 2.2. A proton and an electron rotating in the same sense have equal spin vectors **s** but different and opposite spin magnetic-moment vectors **m**ₛ. That of the electron is 658 times larger than that of the proton.

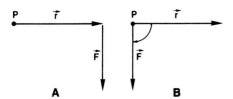

Figure 2.3. (A) The torque of force **F** (also called the *moment* of force **F**) on a point *P* is the vectorial product **r × F**, where **r** is the distance from *P* to the point of application of the force **F**. (B) The product is obtained by moving **F** to the point of application of **r** (while keeping it parallel to itself) and superimposing **r** on **F** through the smaller angle. The direction of the torque vector is that of advancement of a standard screw whose rotation superimposes **r** over **F**. In this case, the torque vector is directed into the page, making *P* turn clockwise.

is then a vector directed into the page at 90° from the plane defined by the vectors **r** and **F**. Being the product of force times distance, moment (or torque) is energy and therefore is expressed in joules.]

Electrons in the electron cloud surrounding the nucleus not only spin but also revolve around the nucleus. Their orbiting produces an additional magnetic field, which has a direction opposite that of a normal screw whose rotation would be in the same sense as the orbital motion (Fig. 2.4). The various spinning and orbital magnetic moments combine to give the electron cloud a net magnetic moment. Usually, this magnetic moment is very small. In some elements, however, most notably iron, this moment is not negligible. The net magnetic moment of the electron cloud can be either parallel or antiparallel to the magnetic moment of the nucleus.

In iron, cobalt, nickel, and a number of alloys, adjacent atoms line up parallel to each other, forming *domains* of about 10^{18} atoms in which the magnetic moments point in the same direction. The cubic root of 10^{18} is 10^6, and the diameter of an atom is about 10^{-10} m, so that the length of the edge of a domain is about $10^6 \times 10^{-10} = 10^{-4}\,\text{m} = 0.1\,\text{mm}$. Domains are visible with a microscope.

Adjacent domains have different magnetic orientations. However, if a piece of iron is placed in an external magnetic field of increasing intensity, the number of domains having the same orientation increases. When all domains have reached the same orientation, magnetic saturation is attained—all magnetic domains are oriented in the same direction. A magnetized piece of iron can be demagnetized by in-

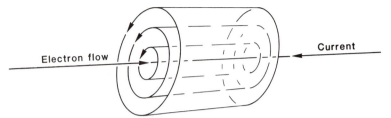

Figure 2.5. Magnetic field around a conductor carrying a constant current. Notice that electron flow and electric current are in opposite directions. The field is coaxial with the wire, and the equipotential surfaces are cylindrical. The sense of the field is counterclockwise when seen from the direction of flow of the electrons, and clockwise when seen from the direction of flow of the current.

creasing its temperature. At 770°C the thermal energy is sufficiently high to randomize the domains in iron, and iron becomes demagnetized. This temperature is called the *Curie point*. The Curie point is 1131°C for cobalt, 358°C for nickel, 578°C for magnetite (Fe_3O_4), and 675°C for hematite (Fe_2O_3).

Substances like iron, nickel, cobalt, and alloys that when solid have a domain structure are called *ferromagnetic*. In other substances, the atoms may become oriented in the same direction when placed within an external magnetic field, only to return to a random orientation when the external field is removed. These substances are called *paramagnetic*. Most substances are paramagnetic. In a few substances, such as bismuth, the atoms become weakly oriented opposite an external magnetic field when placed in it. These substances are called *diamagnetic*.

If the motion of a charged particle is uniform, the magnetic field that is produced is stationary. Consider a straight wire through which a current of 1 ampere flows. An axial magnetic field is created, with equipotential surfaces of cylindrical shape, with radii increasing away from the wire (Fig. 2.5). If the speed of the current changes, the magnetic field also changes. The change in the magnetic field radiates away as an electromagnetic wave traveling with the speed of light. In the antennae of radio transmitters, the electrons flow up and down, reversing their direction 500,000 to 1,500,000 times each second. As a result, electromagnetic waves with frequencies of 500,000 to 1,500,000 Hz radiate away.

THINK

* Consider the following:

  • A 12-ounce can of beer contains 355 g of beer, 90% of which is water. The molecular mass of water is 18.01 u, which means that 18.01 g of water contains 1 Av of molecules, or $6.022 \cdot 10^{23}$ molecules.

  • The volume of the oceans is $1,350 \cdot 10^6 \, km^3$, 96.5% of which is water and 3.5% salts. The volume of water in salty inland lakes (Caspian Sea, Dead Sea, Aral Sea, Great Salt Lake) is 390,000 km³, of which 320,000 km³ is water (for an average salinity of 1.8%). The volume of fresh water on the continents (freshwater lakes, streams, groundwater) is $4.325 \cdot 10^6 \, km^3$. The volume of water in the atmosphere is 13,000 km³.

  • $3.6 \cdot 10^{14}$ kg of water are processed each year through the biosphere, where the water molecules are broken up and rearranged to form organic compounds. Organic respiration forms new water molecules in approximately the same amount that was consumed.

  • A person excretes each day an average of 1 liter of urine (which is 98% water).

  • The mixing time for water in the mobile reservoir (oceans, surface water and groundwater on the continents, atmosphere) is 1,500 years, which means that water is homogenized within 1,500 years.

  • Julius Caesar lived 56 years, from 100 to 44 B.C.E., which is more than 1,500 years ago.

Now calculate the following:

- The total amount of water in the mobile reservoir on Earth (in grams), and from that calculate (a) the number moles and (b) the total number of molecules,

- the total number of water molecules produced by Julius Caesar during his lifetime,

- the number of water molecules in one can of beer that passed through Julius Caesar.

Repeat this calculation for any person of your choice (but remember that he or she must have lived more that 1,500 years ago for the water molecules to have homogenized in the mobile reservoir). If you wish, you can tell your bartender.

* Density is mass/volume. The mass of the Earth is $5.97 \cdot 10^{24}$ kg and its volume is $1.08 \cdot 10^{21}$ m^3. Calculate the density of the Earth in grams per cubic centimeter. What would be the diameter of the Earth if the density of the Earth were equal to that of the atomic nucleus (taken to be $2.8 \cdot 10^{14}$ g/cm^3)? How far would you have to be above the surface of such an Earth if you wanted to maintain your present weight?

* The electronic structures of the 92 elements are shown in Table A6, Appendix A. Can you spot three elements from which it should be easy to yank out an electron? What criterion did you use?

* A piece of copper weighing 63.5 g contains 1 Av ($6.022 \cdot 10^{23}$) of atoms. Because each atom releases one electron to form the metallic bond, there are also $6.022 \cdot 10^{23}$ free electrons. This amount of electricity is called a faraday (F) and is equal to 96,485.3 coulombs, one coulomb being equal to $6.24 \cdot 10^{18}$ electron charges. One ampere is equal to 1 coulomb/second. A 100-watt light bulb needs a current of 1 ampere (at a voltage of 100 V). If you could convince all those free electrons to flow through the filament of a light bulb, how long would the light bulb stay lit? What do you think would happen to that piece of copper if all free electrons were gone?

2.2. STATES OF MATTER

Except for a black hole, which is described only by its mass (see Chapter 6), the simplest state of matter is the *gaseous state*. In this state, the particles (atoms or molecules) are not linked to each other—they approximate point masses that interact with each other only by perfectly elastic collisions. At absolute zero (0 K), the particles either condense into the solid state or, if they are too far apart in space, cease to move. In either case, there is a residual, irreducible vibration called *zero-point energy*. If it were not for this residual energy, a particle at absolute zero would be immobile. We would then know at the same time both its position and its momentum (which would be equal to zero). That would violate the uncertainty principle.

However, for the moment let us suspend the uncertainty principle and observe one particle with mass $= 1$ u inside a box in an inertial frame at 0 K (an inertial frame is a frame not acted upon by external forces). Let us assume that at that temperature the walls of the box are immobile and that the particle is stationary in the middle of the box. Now we add energy. The particle begins to move. Its kinetic energy is a function of the absolute temperature. In order to raise the temperature by 1 K, we need to add $1.5k$ of energy, where $k = 1.38 \cdot 1^{-23}$ J/K is a constant called the *Boltzmann constant*. The amount of energy *per Avogadro number* of particles ($6.022 \cdot 10^{23}$) is $1.5R$, where R is called the *gas constant*. Therefore, $R = 6.022 \cdot 10^{23}k$.

[Notice that in the standard formula $pV = RT$ the term pV is *energy*. In fact p = pressure = force (F) per unit area $= F/l^2$ (where $l =$ length); V = volume $= l^3$. Therefore $pV = Fl^3/l^2 = F \times l =$ energy. The formula simply states that energy (actually $\frac{2}{3}$ of the energy, because $E = 1.5R = \frac{3}{2}R$, or $\frac{2}{3}E = R$) is proportional to temperature, with R = proportionality constant.]

Let us now attach a second box to the first one, with only a flexible membrane separating the two, and let us place one particle with mass $= 1$ u in each box. At any given temperature for the entire system the two particles will have the same energy and therefore will hit the membrane the same number of times per second (on the average). This means that the pressure on the two sides of the membrane is the same.

If we add a second particle to one of the two boxes and maintain the same temperature, each of the three particles will hit the membrane the same number of times per

Table 2.5. Mass of 1 Av of different substances

Substance	Particle mass (u)	Mass of 1 Av (grams)
^{1}H	1.007825	1.007825
^{2}H	2.014102	2.014102
^{12}C	12.000000	12.00000
^{16}O	15.994915	15.994915
^{56}Fe	55.934939	55.934939
^{238}U	238.050785	238.050785
H_2O	18.015	18.015
Hemoglobin	65,332	65,332 ($= 65.332$ kg)
Cellulose	500,000	500,000 ($= 500$ kg)
Starch	10,000,000	10,000,000 ($= 10$ tons)

second (on the average), which means that the pressure on the membrane from the side where there are two particles is now *double* the pressure from the other side. In order to have the *same* pressure, we must have the *same* number of particles on both sides. This is true even if the particles on one side have masses different from those of the particles on the other side. Let us assume that on one side there is 1 particle of mass 1 u and on the other side 1 particle of mass 2 u. If the temperature is the same, the energy will be the same. That is, $\frac{1}{2}m_1v_1{}^2 = \frac{1}{2}m_2v_2{}^2$. The more massive particle will have to move more slowly so that the product mv^2 will be the same on both sides. This simple reasoning leads directly to Avogadro's law, formulated in 1811 by the Italian chemist Amedeo Avogadro, count of Quaregna (1776–1856): "Equal volumes of gases at the same temperature and pressure contain the same number of particles".

Avogadro did not know how many particles there would be in, say, 1 cm^3 of H gas at room temperature and pressure. As mentioned in Chapter 1, we now define the Avogadro number as follows:

The Avogadro number is the number of atomic mass units (u) in 1 gram.

This quantity turns out to be 6.0221367 (± 36)$\cdot 10^{23}$, where the error refers to the last two digits of the figure (see Section 1.2). Different atoms and molecules have different masses. Therefore, the same number of different atoms or molecules will have different masses. Table 2.5 shows the masses of 1 Av of different substances.

Returning now to our original box, let us place in it two particles (not necessarily of the same mass) and set the temperature ot 0 K. At 0 K the two particles will not move. If we raise the temperature to a finite value, the two particles will have exactly the same energy. They will hit each other occasionally, but, having the same energy, neither can accelerate or decelerate the other. However, if we place in the box three or more particles, things will be different. Two or more particles may ac-

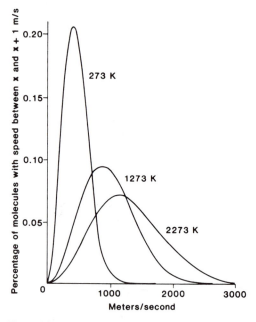

Figure 2.6. Maxwell distributions of atomic or molecular speeds (meters per second) for nitrogen gas at different temperatures. (From Emiliani, 1988, p. 37, Fig. 3.5; adapted from Moore, 1950, p. 185, Fig. 7.12.)

cidentally gang up on a single particle and either accelerate or decelerate it. If the box contains a large number of particles, at any one instant some will be practically at a standstill, some will have high velocities, and most will have velocities around some mean value. The problem of the velocity distribution of n particles in thermal equilibrium was solved in 1860 by the Scottish mathematician and physicist James Clerk Maxwell (1831–1879). He showed that the velocity distribution of n particles is given by

$$dN/N = 4\pi v^2 (m/2\pi kT)^{3/2} e^{-mv^2/2kT} dv \qquad (1)$$

where dN/N is the fraction of particles of mass m having speeds between v and $v + dv$. (The only reason I am flashing this equation is to show you how lucky you are that you do not have to derive it.)

Fig. 2.6 shows the distributions of molecular speeds for nitrogen gas at 0°C, 100°C, and 2,000°C. The curve becomes narrower the lower the temperature, until at 0 K all particles have the same energy—the zero-point energy. All three curves are characteristically skewed (Fig. 2.7 and 2.8), with a long tail toward the high-energy side, because energy has a limit on the low-energy side but not (neglecting relativity) on the high-energy side. This is very important because it enables a chemical reaction to proceed even when the average energy of the system (which is related to the average velocity of particles) is insufficient for a given reaction to occur. Remember, atoms and molecules are surrounded by electron clouds that repel each other. A reaction, therefore, can occur only when atoms or molecules rush at each other with sufficient energy to overcome their mutual repulsion.

By the way, the reason that the absolute zero cannot be reached is that in order to slow down a particle to zero velocity we would have to transfer *all* of its momentum to some other particle. The momentum, no matter how small (i.e., how close to 0 K the particle is), is always positive. A particle could reduce its momentum to zero only by transferring it to an infinite number of particles, each with zero momentum, or to a single particle with equal but *negative* momentum (i.e., with negative mass or negative velocity). Both contingencies are clearly impossible, and so the absolute zero cannot be reached.

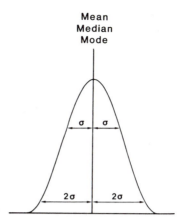

Figure 2.7. Gaussian curves represent the normal distribution of a set of values, Gaussian curves are bell-shaped (i.e., symmetrical about the mean of the values) and are given by

$$y = [1/\sigma(2\pi)^{1/2}]e^{-(1/2)[(x-\mu)/\sigma]^2}$$

where y = ordinate, σ = standard deviation; and μ = mean of the distribution. The *standard deviation* is a measure of the dispersion of the x values about the mean. It is given by $\sigma = (\sum x_i^2/N)^{1/2}$, where x_i = deviation of the value i from the mean, and N = number of values. The larger σ is, the broader the curve. Notice that the standard deviation is in the units that measure x. If x are the heights in meters in a population of elephants, σ is given in meters. If x are the lengths in millimeters in a population of mosquitoes, σ is given in millimeters. Two gaussian curves with the same σ will have the same shape. *Kurtosis* is a measure of the peakedness of a set of data normally distributed about a mean: $K = (\sum fd^4/N\sigma^4) - 3$ where K = kurtosis, f = frequency of values in class interval x, $d = x - \bar{x}$ = distance of class interval x from mean $\bar{x}$ (below, negative, or above, positive), N = number of values, and σ = standard deviation. Positive kurtosis indicates a more peaked, or *leptokurtic*, distribution than the normal curve; negative kurtosis indicates a flatter, or *platykurtic*, distribution than the normal curve. For normal (gaussian) distributions, the percentages of the values falling within given σ intervals from the mean are as follows: $\pm 1\sigma = 68.27\%$; $\pm 2\sigma = \pm 95.45\%$; $\pm 3\sigma = 99.73\%$; $\pm 4\sigma = 99.994\%$; $\pm 5\sigma = 99.99994\%$. For explanations of *mean, median,* and *mode,* see Fig. 2.8.

A formula much simpler than equation (1) gives the mean velocity of a particle as a function of temperature and mass:

$$v_{\text{mean}} = (8kT/\pi m)^{1/2} \qquad (2)$$

Table 2.6 shows characteristic mean velocities for selected atoms and molecules. The square of the velocity of an atom or a molecule in a

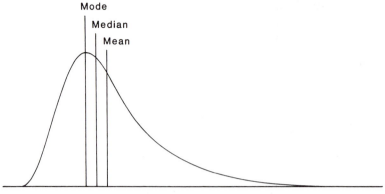

Figure 2.8. A skewed curve. A curve is said to be skewed when the distribution of values is not normal. In skewed curves, the most common value (called *mode*) and the value that divides the total number of values into two equal parts (called *median*) do not coincide with each other, nor do they coincide with the *mean.* In gaussian curves, the mean, median, and mode coincide. *Skewness* is given by $Sk = \sum fd^3/N\sigma^3$, where Sk = skewness, $d = x - \bar{x}$ = distance of class interval x from mean $\bar{x}$, N = number of values, and σ = standard deviation.

gas is proportional to the temperature and inversely proportional to the mass of the particle. In fact, we can rewrite equation (2) as follows:

$$v^2_{mean} = 8kT/\pi m$$

The *mean free path* is the distance a particle travels before hitting another particle. For identical particles (treated as perfectly elastic little balls), the number C of collisions per second experienced by a particle is equal to

$$C = 2^{1/2}\pi n d^2 v$$

where n is the number of particles per cubic meter, d is the diameter of a particle, and v is the average particle speed. Because in one second a particle travels a distance expressed by v, the mean free path l is therefore v/C or

$$l = 1/2^{1/2}\pi n d^2$$

Notice that whereas the number of collisions per second depends on the speed of the particles, and hence on temperature, the mean free path does not.

The mean free path of air molecules at sea level is about $6.6 \cdot 10^{-8}$ m, or about 200 molecular diameters. At a mean speed of 440 m/s (Table 2.6), the number of collisions per second will be $440/6.6 \cdot 10^{-8} = 6.6$ billion. Lots of bumps. On the other hand, the mean free path in intergalactic space, where the

Table 2.6. Mean atomic or molecular velocities at 0°C

Substance	Velocity (m/s)
^{1}H	2,395
^{4}He	1,202
N_2	454
O_2	425
CO_2	362
^{222}Rn	161

average density of matter (mainly hydrogen) is 1 atom/m^3, is $2 \cdot 10^{19}$ m, or about 2,100 light-years. The intergalactic temperature is 2.75 K, and so the average speed, calculated from equation (2), is 220 m/s. At that speed it takes an average of 2.9 billon years for a hydrogen atom to meet another atom. A truly lonely journey.

2.3. ENTROPY

An *ideal gas* is a system of particles that do not interact with one another other than by bouncing around like perfectly elastic balls. Consider a system consisting of 5 billiard balls on a table, with lines dividing its surface into 5 equal partitions. If all the balls are placed within a single partition, there will be 5

Table 2.7. Number of possible states of a system of 5 particles on a table with 5 partitions

System	Distribution	Number of states	Probability
A	5 0 0 0 0	5	0.038
B	4 1 0 0 0	20	0.154
C	3 2 0 0 0	20	0.154
D	3 1 1 0 0	30	0.231
E	2 2 1 0 0	30	0.231
F	2 1 1 1 0	20	0.154
G	1 1 1 1 1	5	0.038
Total		130	1.000

choices (because there are 5 partitions), and the system will have 5 possible states. If 4 balls are placed in one partition and 1 ball in one of the 4 remaining partitions, there will be $5 \times 4 = 20$ choices, and the system will have 20 possible states. If 3 balls are placed in one partition and 2 balls in another, there will again be 20 possibilities. However, if 3 balls are placed in one partition, the 4th ball in another partition, and the 5th ball in still another, there will not be $5 \times 4 \times 3 = 60$ possibilities, because the 4th and 5th balls can be interchanged without making any difference—the two systems will be identical. Two identical systems with 60 possibilities each reduce to one system with 30 possibilities (system *D* in Table 2.7). System *F* has 120 possibilities, but 6 are identical, because the 3 balls in individual partitions can be interchanged, and the number of permutations (see Section 19.3) of 3 balls is $3! = 1 \times 2 \times 3 = 6$ (the symbol *n!* reads *n factorial* and means $1 \times 2 \times 3 \times \cdots \times n$). System *G* has 120 possibilities, of which $4! = 1 \times 2 \times 3 \times 4 = 24$ are identical, and so there are $120/24 = 5$ states. The different distributions are shown in Table 2.7.

There are 130 possible states. Each state is equally probable, but distributions *D* and *E* are more probable simply because there are more of such distributions, and therefore the system has more states. The natural logarithm, of the number of possible states *W* (meaning *Wahrscheinlichkeiten* = probabilities), multiplied by the Boltzmann constant *k*, is called *entropy* (symbol *S*):

$$S = k \ln W$$

Entropy is therefore proportional to the number of different states that a system can

assume. In nature, one normally deals with a large number of possible states, and therefore *W* is generally a large number.

If the 5 balls are placed within one partition (a distribution that has 5 possible states) and energy is added (by shaking the table, for example), the balls will scatter and end up in a distribution that will have a greater number of states. Entropy will increase. In a system consisting of a few particles, entropy can accidentally decrease: Five balls scattered on the table might accidentally be grouped within a single partition by shaking. This becomes less and less likely the greater the number of particles.

Mixing also increases entropy. Consider a table with two partitions, with 5 black balls in one partition and 5 white balls in an adjacent one. There is only one distribution in which all black balls are on one side and all white balls on the other. Diffusion starts when a black ball moves into the white side and a white ball moves into the black side. The black ball has 5 choices as to which white ball to replace, for a total of $5 \times 5 = 25$ possibilities.

Table 2.8. Diffusion of particles

Partition 1	Partition 2	Number of states (*W*)	Probability
5 B	5 W	1	0.004
4 B, 1 W	4 W, 1 B	25	0.099
3 B, 2 W	2 W, 3 B	100	0.397
2 B, 3 B	3 W, 2 B	100	0.397
1 B, 4 W	4 W, 1 B	25	0.099
5 W	5 B	1	0.004
Total		252	1.000

The same for the white ball. For the next 2 balls to move, the choices are 4 each, for a total of 16. The two moves combined give a total of $25 \times 16 = 400$ possibilities. However, looking at the system after 2 balls have moved from each side, we cannot tell which ball moved first and which moved second: The system has simply changed to a state in which there are 2 black balls and 3 white balls on one side and vice versa on the other. The 2 black balls can replace any 2 white balls, and vice versa. There are 10 possible pairings of white balls and 10 of black balls (1 and 2, 1 and 3, 1 and 4, 1 and 5; 2 and 3, 2 and 4, 2 and 5; 3 and 4, 3 and 5; 4 and 5; these are the 10 combinations of 5 elements taken 2 at a time—see Section 19.3). There are, therefore, 10 possibilities for the black balls and 10 possibilities for the white balls, giving a total of $10 \times 10 = 100$ possibilities. If 3 balls are exchanged, the possibilities again total 100. If 4 balls are exchanged, we are back to the first case with 25 possibilities. The total number of possibilities is 252.

Both shaking and mixing will increase entropy, as well as the disorder of the system. In fact, entropy is a measure of the internal disorder of a *closed* system (i.e., a system that does not exchange matter or energy with another system). Both disorder and entropy can decrease if the system is open. A crystal or a living organism, which arrange matter in an orderly fashion while growing, are examples of open systems.

Consider now a system consisting of 100 g of hydrogen-1 gas at 0°C ($= 273.15$ K). This system contains $100 \times 6.022 \cdot 10^{23} = 6.022 \cdot 10^{25}$ particles. If we add 1 joule of energy per gram, the change in the energy of the system divided by the temperature of 273.15 K gives the change in entropy:

$$dS = dE/T$$

In this case, the change in entropy is $100/273.15 = 0.366$ J/K. If we add 1 joule per gram to a second system consisting of 100 g of the same gas, but at a temperature of 100°C ($= 373.15$ K), the change in entropy will be $100/373.15 = 0.268$ J/K. For the same amount of added energy, the change in entropy decreases with increasing temperature. Clearly, if temperature were increased to infinity, the decrease in entropy caused by the addition of 1 J/g would reduce to zero.

Thermodynamics is the study of energy transfer between systems. There are three laws of thermodynamics.

First law: In a closed system, energy is conserved.
Second law: A spontaneous change within a closed system always increases its entropy; an interaction between systems always increases the total entropy.
Third law: The entropy of a perfect crystal of a single element and a single isotope at 0 K is zero (because there is only one possible state, and $\ln 1 = 0$).

2.4. THE CHEMICAL BOND

In the liquid state, particles are weakly bonded to each other. Bonding is due to electric polarity generated by the uneven distribution of the electrons in the electron clouds of the particles. Some molecules, like H_2O and NH_3, have permanent polarity. They can form bonds of appreciable strength with each other and with other polar molecules. A water molecule at 25°C, for instance, needs an average of 0.45 eV of energy to free itself from other water molecules and evaporate. Polar molecules can induce transient polarity on adjacent, nonpolar molecules. Nonpolar molecules can also induce transient polarity on each other, resulting in mutual attraction. These kinds of bonds are weak (a fraction of an electron volt) and are generally referred to as *van der Waals forces*.

In the solid state, particles are linked to each other by bonds of significant strength (a few electron volts). If they are arranged in a regular array that is repetitive in space, the structure is called *crystalline*. If the array is irregular, the structure is called *amorphous*. There are four types of bonds.

1. The *ionic bond* is formed by electron transfer from one atom to another, resulting in oppositely charged ions that attract each other (e.g., Na^+Cl^-). The energy of ionic bonds is on the order of 3 to 8 eV (1 eV $= 1.602 \cdot 10^{-19}$ J).
2. The *covalent bond* is formed by atoms sharing a pair of electrons of equivalent energies and opposite spins (one from each atom); examples are H_2, N_2, O_2, and

diamond. The energy of a covalent bond is similar to that of ionic bonds, about 3 to 8 eV.

3. The *metallic bond* is formed by valence electrons released by a metallic atom (1–2 electrons per atom) and moving freely through the metal lattice. One mole of Cu atoms ($= 63.5$ g of Cu) has 1 Av of free electrons $= 6.0221367 \cdot 10^{23}$ free electrons. In 1 coulomb of electricity there are $6.2415064 \cdot 10^{18}$ electrons. Therefore, in 63.5 g of copper there are $6.0221367 \cdot 10^{23}/6.2415064 \cdot 10^{18} = 96,485.309$ C of electricity. This unit—1 Av of electrons—is called a faraday (do not confuse faraday with farad, which is the SI unit of capacitance). The energy of the metallic bond ranges around 1 to 3 eV.

4. The *hydrogen bond* is formed by a hydrogen atom, already part of a molecule, with a pair of unshared electrons in an electronegative atom in the same molecule or an adjacent molecule. Examples include ice, where the hydrogen in one molecule is attracted to the oxygen of the next molecule, and an immense variety of organic tissues and fibers. The energy of the hydrogen bond ranges from 0.2 to 1 eV.

A molecule continually exchanges energy with the environment. The following types of molecular energies (all quantified) are involved:

Translational: for a molecule of H_2 at $0°C$, $E = \frac{1}{2}mv^2 = 9.5 \cdot 10^{-22}$ J, corresponding, by $E = h\nu$, to the energy of a photon with frequency $\nu = 1.44 \cdot 10^{12}$ and wavelength $\lambda = 0.2$ mm

Electronic: by interaction with photons with wavelengths λ ranging from 0.1 to 1 μm, corresponding to the ultraviolet-to-red range of light

Vibrational: by interaction with photons with λ ranging from 3 to 30 μm (near infrared) (see Fig. 14.1C)

Rotational: by interaction with photons, with λ ranging from 0.1 to 1 mm (far infrared)

Fine structure: resulting from interaction between the spin and the orbital angular momentum of the electrons, with λ ranging from 3 cm for the light elements to 1 μm for the heavy elements

Hyperfine structure: resulting from interaction of the nuclear magnetic moment with that of the electron cloud, with λ ranging from 3.26 cm for cesium to 21.1 cm for hydrogen

In addition to the familiar solid, liquid, and gaseous states of matter, there are other states that are less familiar but equally important:

Plasma is an ionized gas with equal numbers of positive ions and negative electrons. Examples include neon lights, lightning, auroras, and the solar photosphere (the outer layer of the sun that emits the light we see).

Degenerate is a state of matter in which free electrons occupy all the lower energy levels available and cannot interact in any physical process until they are excited above those levels. Examples include metals at very low temperatures and the mixture of mostly protons and free electrons that forms the high-density (10^5 g/cm^3), high-temperature core of white dwarf stars.

The *neutron* state consists of solidly packed neutrons. It has the density of neutron matter (10^{14} g/cm^3), and it exists only in neutron stars.

Black hole is the state in which matter has infinite density, a result of the explosion of massive stars (see Chapter 6)

Ylem (or ἄπειρον) is the undefined state of matter-energy in existence immediately after time zero (see Chapter 4).

We have counted eight states of matter: solid, liquid, gas, plasma, degenerate, neutron, black hole, and ylem. These states are interchangeable. If energy is added, the solid becomes a liquid, and then a gas. If the gas is ionized, it forms a plasma. If it is squeezed by the pressures available in the cores of stars, it becomes degenerate. Some more squeezing and we have the neutron state and eventually the black-hole state. On the other hand, one can say that energy comes in a single state—a low-energy photon and a high-energy photon differ only in their wavelengths.

THINK

* A particle of mass m traveling at a velocity **v** has a momentum equal to m**v**. Because **v** is a vector, m**v** is also a vector. The kinetic energy of

the particle is $\frac{1}{2}mv^2$. Is energy also a vector? If not, why not?

* At a pressure of 1 atmosphere ($= 760$ mm of mercury) there are $2.55 \cdot 10^{25}$ molecules of gases per cubic meter of air. The mean free path (the average distance a molecule can travel without hitting another molecule) is $6.63 \cdot 10^{-8}$ m. A mass-spectrometer tube may be 1 m long. The atoms or molecules to be analyzed in a mass spectrometer need to travel freely, without bumping against air molecules. This means that the mean path must be at least 1 m, which is obtained by pumping the air out of the tube. To which value must the atmospheric pressure inside the tube be reduced for the mass spectrometer to work?

* What would be the change in entropy if 1 joule of energy were added to 100 g of hydrogen gas at the temperature of the solar photosphere (6,000 K)?

* Suppose you have a closed system, consisting of a number of separate cells, each containing different quantities of an ideal gas at different temperatures (your choice of number of cells, quantities of gas, and temperatures). Can you find a practical (as opposed to theoretical) way to decrease the entropy of the total system by removing or adding partitions?

* Your mother is cooking spaghetti on the stove while you are sitting at the kitchen table reading the newspaper. If the human eye could see infrared radiation, would it be easier or more difficult for you to read the newspaper?

2.5. THE UNCERTAINTY PRINCIPLE

It would be very nice if we could take a close look at a proton, a neutron, or an electron. Unfortunately, this is not possible. It is not for lack of instrumentation or ingenuity, but because of nature itself. All particles move. They move even at absolute zero, because they still have the zero-point energy. In order to examine a particle, we have to know where it is and how fast it moves, so that we can catch it. But there is a problem. The problem is the famous *uncertainty principle* enunciated in 1927 by the German physicist Werner Heisenberg (1901–1976). This principle says that it is not possible to precisely determine the position and momentum of a particle *at the same time*. We can determine position as accurately as we want, but then we do not know anything about the momentum; or we can determine the momentum accurately, but then we do not know anything about the position. Absurd? Not at all. Consider a frictionless billiard table with perfectly elastic sides and with a single, perfectly elastic billiard ball in the center. The rules of relativity and quantum mechanics can be ignored here because the ball is very big compared with a neutron or a proton and because its speed is very small relative to the speed of light. In particle physics, we can only make one measurement, to determine either the position or momentum of a particle, because the measurement itself inevitably scatters the particle—we do not know where it went and therefore cannot make a second observation. Similarly, we are allowed only one experiment to study the billiard ball. Let us choose a fast camera and see when we can do.

The ball, representing a particle, moves with a certain velocity. If we take a snapshot, we freeze out the motion, and we can tell exactly where the ball was at the moment the snapshot was taken. Unfortunately, we will not know anything about its velocity because the snapshot has frozen out the motion. If, instead, we take a time exposure, the picture shows a streak. By knowing the duration of the time exposure and measuring the length of the streak, we can determine the velocity of the ball. However, we do not know at which end of the streak the ball was when we opened and closed our shutter, which means we do not know the position.

A similar argument applies to particles, with the added complication that, whereas our photographic efforts do no disturb the billiard ball, anything we do to detect a particle will scatter the particle in an unpredictable way. The point is that even if we found a way not to disturb the particle, we still would not be able to determine its position and momentum at the same instant.

Pairs of quantities, such as position and momentum, that cannot be precisely determined together are called *conjugate variables*. Other examples are angle and angular momentum, and energy and time. Their common characteristic is that one of the pair must be zero in order to determine the other—velocity must be zero in order to determine position; angular momentum must be zero in order to determine the angle; and time must be zero in order to determine energy.

Having ascertained that conjugate properties cannot be determined simultaneously with arbitrary precision, the question arises how close can we come to doing so. As shown by Heisenberg in 1927, we can get no closer than Planck's constant divided by 2π:

$$\Delta x\Delta p \geqslant h/2\pi \qquad (3)$$

where Δx is the uncertainty in position, Δp is the uncertainty in momentum, and h is Planck's constant. For the pair energy-time, we have

$$\Delta E\Delta t \geqslant h/2\pi \qquad (4)$$

where ΔE is the uncertainty in energy (and therefore in mass), and Δt is the uncertainty in time.

Equations (3) and (4) have the same dimensions. In fact, $x = l$ and $p = mv = m{\cdot}l/t$, and therefore $\Delta x\Delta p = ml^2/t$; $E = mc^2 = m{\cdot}l^2/t^2$, and therefore $\Delta E\Delta t = m(l^2/t^2){\cdot}t = ml^2/t$. (In the preceding, l = length, m = mass, v = velocity, t = time, c = speed of light.)

From (4) we derive

$$\Delta t \geqslant (h/2\pi)/\Delta E \qquad (5)$$

Relationship (5) indicates that a particle of mass $m = E/c^2$ can materialize from nothing provided that it returns to nothing within the time $\Delta t \leqslant (h/2\pi)/mc^2$. Particles that last their pertinent Δt's or more are called *real*; particles that last less than their pertinent Δt's are called *virtual*. For example, the mass of the charged pion (π^+) is $0.15\,\mathrm{u} = 0.25{\cdot}10^{-27}\,\mathrm{kg}$. Therefore,

$$\Delta t = (6.62{\cdot}10^{-34}/2\pi)/(0.25{\cdot}10^{-27} \times 9{\cdot}10^{16})$$
$$= 1.05{\cdot}10^{-34}/2.25{\cdot}10^{-11}$$
$$= 0.47{\cdot}10^{-23}\,\mathrm{s}$$

Pions that last $0.47{\cdot}10^{-23}\,\mathrm{s}$ or more are *real*. Pions that last less than $0.47{\cdot}10^{-23}\,\mathrm{s}$ are *virtual*.

Not just pions, but all particles can be virtual. These virtual particles fill up all empty space, coming and going all the time. In fact, truly empty space apparently does not exist. Because of the uncertainty principle, the lighter or less energetic particles last longer as virtual

particles than do the more massive or more energetic particles.

Virtual particles cannot be detected directly. Their effects, however, can be readily seen. The British physicist P. A. M. Dirac (1902–1984) showed that the magnetic moment of a free, spinning proton should be equal to $eh/4\pi m_p c$ [where e = electron charge ($+1$ for the proton); h = Planck's constant; m_p = mass of the proton; c = speed of light]. All these terms are constant, which means that the free proton always spins at the same rate. The quantity $eh/4\pi m_p c$ is called the *nuclear magneton* (symbol μ_N). The magnetic moment of the spinning proton (symbol μ_p), however, is not equal to μ_N but to $2.7926\,\mu_N$ because of the effect of the accompanying virtual pions. The proton, in fact, polarizes the space around itself by attracting the negative virtual pions and repelling the positive ones. These form a positive shell farther out around the proton that increases μ_p to $2.7926\,\mu_N$. Dirac also showed that the magnetic moment of a free, spinning electron should be equal to $eh/4\pi m_e c$ (where e = electron charge; h = Planck's constant; m_e = mass of electron; c = speed of light). All these terms are constant, which means that the electron, too, spins at a constant rate. The quantity $eh/4\pi m_e c$ is called the *Bohr magneton* (symbol μ_B). The magnetic moment of the free electron, however, is not equal to μ_B, but to $1.0011\,\mu_B$, again because of the effect of accompanying virtual pions (in this case, the positive pions are close to the electron, and the negative ones farther put).

Because in the magneton formulas the mass term appears at the denominator and the mass of the proton is 1,836 times larger than that of the electron, the magnetic moment of the spinning proton should be 1,836 times smaller than that of the spinning electron. It is, in fact, only 658 times smaller. Again, this is because the proton is surrounded by a cloud of virtual pions, which increase its magnetic moment to the amount indicated.

THINK

* Estimate the mass of your body (in kilograms), and calculate how long you would last if you were a virtual person.

2.6. PHOTONS

Stationary charges arrange *virtual* photons so as to form an electrostatic field. Electrostatic fields do not radiate energy away from the charges that create them. Charges moving with uniform velocity arrange virtual photons around themselves and create a stationary magnetic field, which also does not radiate energy away.

On the other hand, accelerated charges produce *real* photons that radiate energy away. Real photons are as strange as their virtual cousins. For a start, they behave with equal aplomb as particles or as waves, *depending on what they interact with*. A photon of the right energy (a few electron volts) may interact with an electron bound in an atom and jack it up to a higher energy level, or it may interact with an electron in a metal and knock it out of the metal surface. In these cases, a photon acts like a bullet.

In the phenomenon of interference, a photon can cancel out another photon. This can happen only if a photon acts like a wave. Two wave trains of equal amplitude and frequency can cancel out if the two wave trains are 180° out of phase, in which case crest meets trough, and vice versa. If they are in phase, they will reinforce each other. Interference of two light sources produces alternating bands of light and darkness on a screen (Fig. 2.9).

The question then arises, What are photons *before* they interact? Are they particles or waves? They are neither. Consider, in analogy, water in a garden hose that can become a jet or a spray depending on how it interacts with the nozzle. Obviously the water in the hose is neither jet nor spray, but it has the potential to become one or the other at the moment it interacts. We can state with confidence only what a photon is *not*. For instance, a photon is not a tomato, because a tomato cannot interact as a wave. Two tomatoes thrown at a speaker from two different directions do not cancel out.

[Although a tomato is not a wave, it has, like any other material particle or object, an associated wave according to the de Broglie equation:

$$\lambda = h/p \qquad (6)$$

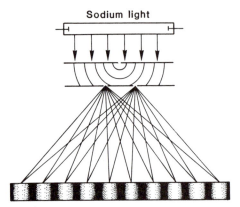

Sodium light

Figure 2.9. Interference. Light from a monochromatic light source passes through a pinhole in a screen and then through two narrow slits (0.01 mm wide, 0.1 mm apart for visible light) on a second screen. The two identical light beams emerging from the slits reinforce each other on a third screen (placed some 50 cm away) if the two paths are of equal lengths or if they differ by an integral number of wavelengths. They cancel out if the two paths differ by half a wavelength. Intermediate path differences produce intermediate light intensities. The result is a series of light bands grading into intervening dark bands. This experiment, performed in 1803 by the English physician and physicist Thomas Young (1773–1829), conclusively demonstrated that light can behave as a wave.

where λ is wavelength, h is Planck's constant, and p is momentum. If somebody throws a tomato with mass = 0.1 kg and speed = 10 m/s, the momentum of the flying tomato will be 1 kg·m/s (it will make a good splash), but the associated wavelength will be only $h/p = 6.62 \cdot 10^{-34}/1 = 6.62 \cdot 10^{-34}$ m, and the targeted person will have no way of noticing it. Momentum is in the denominator in equation (6), and so bodies with greater momenta will have even shorter wavelengths. For instance, if you were picnicking on top of a mountain overlooking the Mojave desert and, because of some slight pilot error, you found yourself suddenly scooped up and splattered across the nose of a Boeing B52-H Stratobomber (mass = 220 tons; speed = 1,100 km per hour), you might notice the momentum ($6.7 \cdot 10^7$ kg·m/s) but not the wavelength ($9.8 \cdot 10^{-42}$ m). At the other extreme, the gliding bacterium *Myxococcus*, capable of reaching the vertiginous speed of 0.12 mm per hour, has a momentum of only $4 \cdot 10^{-21}$ kg·m/s. Even so, its associated wavelength is only $1.6 \cdot 10^{-13}$ m, or about one-thousandth the size of an atom.]

You may have resigned yourself, by now, to the notion that it is not possible to know what a photon is. But there is more bad news. Not only do we not know what a photon is, we don't even know where it is.

Consider, for instance, a hydrogen atom in intergalactic space (if you wish, you may consider the same hydrogen atom discussed in Section 2.2, the one that spends 2.9 billion years looking for another hydrogen atom to bump into). Suppose that this hydrogen atom has its electron sitting on the M shell, and the electron is desirous of jumping down to the L shell. When it does, it loses 1.89 eV of energy ($= 3.0 \cdot 10^{-19}$ J). This energy appears as a photon with a wavelength of 0.62 μm, which corresponds to the red-orange color. The moment the photon appears, it is forced to move at the speed of light, 299,792,458 m/s $\simeq$ $3 \cdot 10^{8}$ m/s. We do not know which way it went. All we know is that, as it moves, it is somewhere along the surface of an imaginary sphere called a *probability wave*. The probability that our photon will be somewhere on this spherical surface is 1 (i.e., 100%). But the probability of the photon being at one specific place on the surface of the sphere is proportional to the ratio of the cross section of the photon [in this case, about $(0.62 \mu m)^2 = 3.8 \cdot 10^{-13}$ m^2] to the surface of the probability wave. After 1 second, this surface is equal to $4\pi(3 \cdot 10^8)^2 = 1.1 \cdot 10^{18}$ m^2, so that the probability of the photon being at a specific place on this surface is $3.8 \cdot 10^{-13}/$ $1.1 \cdot 10^{18} = 3.4 \cdot 10^{-31}$. Until the photon interacts with an atom, we do not know where the photon is, other than that it is somewhere along the probability wave. When it interacts, it becomes localized, and its probability of being somewhere else along the probability wave suddenly drops to zero. In a way, this is similar to you knowing that there is an enemy tiptoeing at night through your carefully laid mine field—you do not know where he is until he steps on a mine and blows himself up. Only at that moment does he become localized and his probability of being somewhere else suddenly drops to zero (unfortunately for him).

In conclusion, we do not know what a free photon looks like, and we do not know where

it is. If we knew where it originated, all we could say about it is that it is at distance $c \cdot t$ from its source (c = speed of light, t = seconds).

Photons are produced not only by jumping electrons but also by any accelerated charge, including protons, ions, and charged bodies. An electron oscillating up and down a ratio antenna once per second emits one photon per second, with frequency of 1 Hz, wavelength of 299,792,458 m, energy of $6.6 \cdot 10^{-34}$ J, and power of $6.6 \cdot 10^{-34}$ W. The energy of an electromagnetic wave with a frequency of 1 Hz is Planck's constant. The expression

$$E = h\nu \tag{7}$$

simply says that energy (E) is proportional to frequency (ν), the proportionality factor (h) being Planck's constant. From (7) we obtain

$$h = E/\nu \quad \text{(joules/hertz)}$$
$$= E \cdot s \quad \text{(joules} \times \text{second)} \tag{8}$$

because frequency is the inverse of time. Equation (8) shows that Planck's constant is the energy of a photon that has a frequency of 1 Hz

If the electron were moving up and down the antenna 1 million times per second (i.e., at a frequency of 10^6 Hz), the wavelength would be 299,792,458/1,000,000 = 299.792458 m, which is in the middle of the broadcast range.

The complete electromagnetic spectrum is shown in Table 2.9. Notice that frequency varies continuously from zero to about 10^{23} Hz (the frequency of the most energetic photons thus far detected), and so the boundaries are purely artificial. Different energy bands have been given different names for convenience and also because they are produced by different mechanisms.

Given an appropriate surface, medium, or obstruction, electromagnetic waves *reflect, refract,* or *diffract.*

In *reflection*, the angle of incidence is equal to the angle of reflection (Fig. 2.10):

$$\theta_i = \theta_r$$

where θ_i = incident angle and θ_r = reflected angle.

Refraction occurs when a wave strikes at an angle the interface between two media through which the wave propagates at different velocities. The angle of incidence θ_i and the angle of refraction θ_r (Fig. 2.11) define the *index of*

Table 2.9. Electromagnetic spectrum

Frequency (Hz)	Wavelength	Name	Typical source
10^{23}	3×10^{-13} cm	cosmic gamma rays	supernovae
10^{22}	3×10^{-12} cm	gamma rays	unstable atomic nuclei
10^{21}	3×10^{-11} cm	gamma rays, hard x-rays	unstable atomic nuclei
10^{20}	3×10^{-10} cm	hard x-rays	inner atomic shell
10^{19}	3×10^{-9} cm	x-rays	electron impact on solids
10^{18}	3×10^{-8} cm	soft x-rays	electron impact on solids
10^{17}	3×10^{-7} cm	ultraviolet	atoms in discharges
10^{16}	3×10^{-6} cm	ultraviolet	atoms in discharges
10^{15}	$0.3 \mu m$	visible spectrum	atoms, molecules, hot bodies
10^{14}	$3 \mu m$	infrared	molecules, hot bodies
10^{13}	$30 \mu m$	infrared	molecules, hot bodies
10^{12}	0.3 mm	far infrared	molecules, hot bodies
10^{11}	3 mm	microwaves	communication devices
10^{10}	3 cm	microwave, radar	communication and detection devices
10^{9}	30 cm	radar	communication and detection devices
10^{8}	3 m	video, FM	television, FM radio
10^{7}	30 m	shortwave	shortwave radio
10^{6}	300 m	AM	AM radio
10^{5}	3 km	longwave	longwave radio
10^{4}	30 km		induction heating
10^{3}	300 km		induction heating
10^{2}	$3{,}000$ km		rotating electromagnets
10	$30{,}000$ km		rotating electromagnets
1	$300{,}000$ km		rotating electromagnets
0	infinite	direct current (dc)	batteries

Note: The electromagnetic spectrum is continuous, and as a result the different types of electromagnetic radiation grade into one another.

refraction of medium 2 with respect to that of medium 1 (Snell's law):

$$\sin \theta_i / \sin \theta_r = V_i / V_r$$
$$= n_{21}$$

where V_i = velocity of the incident beam, V_r = velocity of the refracted beam, and n_{21} = index of refraction of medium 2 with respect to medium 1. If a ray (a ray is the normal to the wave front) strikes the interface perpendicularly, the angle of incidence and the angle of refraction are both equal to 90°; that is, the wave is not deflected. If $\sin \theta_r = 1$ (angle of refraction = 90°), we have

$$\sin \theta_i = V_i / V_r$$

and θ_i is called the *critical angle*.

Diffraction is a change in either the amplitude or phase of a wave resulting from a partial obstruction with dimensions of the same order as the wavelength of the incident wave.

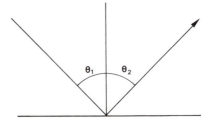

Figure 2.10. Reflection: The angle of incidence (ϑ_1) is equal to the angle of reflection (ϑ_2).

Monochromatic, parallel light rays passing through a pinhole, for instance, generate a new set of rays emerging from the pinhole, as if the pinhole itself were a source of light. These rays diverge, so that the distances between different points on the surface of the pinhole and any given point on a distant screen will be different. If the difference in length between two rays

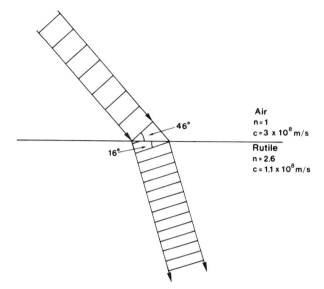

Figure 2.11. Refraction: When a ray of light passes from a medium with a lower index of refraction to a medium with a higher index, the speed of that ray of light decreases. If the ray hits the interface straight on, the ray will continue in the same direction. If the ray hits the interface at an angle, the ray will be deflected toward the normal to the interface, because one side of the wave front will hit the interface before the other. If the ray passes from the medium with a higher index of refraction to that with a lower index, and if it hits the interface at an angle, the ray will be deflected away from the normal to the interface. The illustration shows a ray of monochromatic light (sodium D line, with $\lambda = 589.3$ nm) being refracted when passing from air into rutile ($n =$ index of refraction; $c =$ speed of light in the media). (Emiliani, 1988, p. 21, Fig. 2.4.)

reaching the same spot on the screen is an even number of half wavelengths, the two rays will reinforce each other, and the spot on the screen will be bright; if the two paths differ by an odd number of half waves, the two rays will cancel each other, and the point on the screen will be dark. A pinhole will produce a concentric series of light and dark bands, one grading into the next. Similar effects are produced when light passes through a narrow slit or grazes a sharp edge perpendicular to the light: The space within the slit or immediately adjacent to the edge functions as a source of diverging light rays.

THINK

* Is there any connection between induction heating and microwave cooking? If so, why does food get hot in a microwave oven, seeing that it does not contain free electrons?

* Nerve conduction in humans and animals involves the movement of potassium and sodium ions in and out of nerve fibers. A nerve fiber at rest is enriched in potassium ions inside, while the fluid outside is enriched in sodium ions (each carrying a single positive charge). Their concentrations are different, however, so that the interior of the fiber has a negative potential (-60 millivolts) with respect to the outside. The fiber responds to an external stress by rushing Na^+ ions in, which causes the internal potential to become $+50$ mV. The K^+ ions rush out and the potential returns to normal. This potential spike travels down the fiber at a speed that ranges from 1–2 m/s in fibers a few micrometers in diameter to 20 m/s in giant fibers 1 mm in diameter. The ions only go in and out, across a distance of 1 μm or so. Only a few ions need to move to provide the potential pulse. The pulse lasts only 0.5 milliseconds. The pulse height remains the same, but the pulsing frequency depends on the magnitude of the external stress. Under high stress, the frequency may reach 2,000 Hz. Because the particles involved (K^+ and Na^+) are charged, electromagnetic waves are emitted. What is the energy and wavelength of the photons emitted? What kind of equipment would be needed to detect them? What part of your body should emit the most? What will happen to these photons as they propagate in space at the speed of light? Will they be absorbed, and if so, by what? Or will they last forever, a kind of cosmic karma summarizing your entire life? If so, would you expect your karma to be able to interact with the karmas of people who died before you? And what about your doggie? It's got a karma too, you know.

* Einstein says that $E = mc^2$, and Max Planck says that $E = h\nu = h(c/\lambda)$, where E is energy, m is mass, c is the speed of light, h is Planck's

constant, v is frequency, and λ is wavelength. Using Einstein's equation and setting $m = 1$ kg, calculate the energy. Next, using Planck's equation, calculate the wavelength of a photon having the same energy. De Broglie says that $\lambda = h/p$, where λ is wavelength, h is Planck's constant, and p is momentum ($= mv$). So, setting $m = 1$ kg and $v = 1$ m/s, calculate the de Broglie wavelength. Now calculate the ratio of the Planck wavelength to the de Broglie wavelength. Surprise? Explain. (For maximum effect, use at least seven decimal places in all your calculations.)

3

THE ATOMIC NUCLEUS

3.1. ELEMENTARY PARTICLES

Elementary particles are the ultimate constituents of matter and energy. They are particles that cannot be broken down into smaller particles. In this sense the proton, neutron, and electron are elementary particles and are listed as such in Table A4. But there are problems. To begin with, the neutron, when free, has a *mean life* of only 15.1 minutes (meaning that it lasts, on the average, only that long), after which it decays into a proton and an electron. Second, both the proton and the neutron are believed to consist of a set of three smaller particles called *quarks*. Physicists have not been able to break up the proton and neutron into quarks. Third, there are transient particles, called mesons, that decay, directly or indirectly, into electrons, neutrinos, and radiation. Mesons consist of a quark-antiquark pair, an antiquark being a quark with reversed electric charge and other properties.

All these particles can be grouped into three families:

1. The **hadrons** include all particles consisting of quarks. This family includes two subfamilies: the subfamily of the *baryons*, which consist of 3 quarks and include the proton, the neutron, and heavier, unstable particles called *hyperons*; and the subfamily of the *mesons*, which consist of quark-antiquark pairs.

2. The **leptons** include the electron, the muon, the tauon, and their respective neutrinos. The electron is the fundamental carrier of negative electric charge. The muon (symbol μ) is a lepton produced by the decay of mesons. As shown in Table A4, it decays into an electron and a neutrino (or antineutrino) 98.6% of the time, or into an elec-

tron and a pair of high-energy photons 1.4% of the time. The neutrino is a neutral particle that either has no mass at all or has an exceedingly small mass ($1.8 \cdot 10^{-5}$ u?). It spins, however, and so it can carry energy. It has *negative helicity* (i.e., it rotates counter-clockwise along the direction of flight). Its antiparticle, the antineutrino, has *positive helicity* (i.e., like a standard screw or a corkscrew, it spins clockwise in the direction of advancement).

3. The **field particles** are the virtual particles that create fields. Included are the gluon (strong force, holding quarks within hadrons, and its residual, the *nuclear force*, holding nucleons within atomic nuclei), the photon (electromagnetic force, creating a repulsion between like charges and an attraction between opposite charges), the $W^{\pm}$ and the Z^0 particles (weak force, interacting between hadrons and leptons), and the graviton (gravitational force, interacting between masses).

A *field* is a region of space where a force exists. The force is created by virtual particles. Interaction between real particles in a field is caused by the exchange of the pertinent force particle. Consider, for example, an electron stationary in empty space. Its presence causes virtual photons to appear and disappear. The maximum lifetime t_{max} of a virtual particle is (from the uncertainty principle)

$$t_{max} = (h/2\pi)/E$$

where h is Planck's constant, and E is the energy of the particle. For a photon, $E = h\nu$, where ν is the frequency, and so

$$t_{max} = 1/2\pi\nu$$

An electron continually emits and reabsorbs

virtual photons of all energies and therefore wavelengths. The distance r that a virtual photon can travel is

$$r = c \cdot t_{max}$$
$$= c/2\pi v$$

where c is the speed of light. This means that the weaker a virtual photon is, the longer it lasts and the farther it can travel.

Consider now a second electron and place it at a distance of your choice from the first electron. When the first electron emits a virtual photon, it suffers a recoil, because the virtual photon (like all photons) carries a momentum $p = h/\lambda$ (where λ = wavelength). When the virtual photon reaches the second electron, it gives it a push because of its own momentum. It is this recoil–push combination that makes the two electrons repel each other. If one is a positive electron, instead of a negative electron, the momentum vector is reversed, and attraction results. Of course, the farther apart the two particles are, the weaker the interaction because the virtual photons capable of bridging the distance will be proportionally weaker. Momentum transfer by virtual photons is what causes the interaction between charges.

Virtual gluons, $W^\pm$ and Z^0 particles, and gravitons cause interactions in the other three fields. The strength of the interaction by virtual photons and gravitons changes in inverse proportion to the square of the distance.

The quarks, which are the basic constituents of hadrons, carry an electric charge that is equal to either $+\frac{2}{3}$ or $-\frac{1}{3}$ that of the electron. In addition, they are characterized by properties called *flavor* and *color*. There are 6 flavors [up (u), down (d), strange (s), charm (c), bottom (b), and top (t)], each of which comes in 3 colors [red (R), green (G), and blue (B)]. These names only refer to quark properties and have no relationship to their common meaning. Because there are 6 flavors, each with 3 different colors, there is a total of 18 different quarks. If we add the 6 leptons and the 4 force-field particles, we have a total of 28 truly(?) elementary particles. Table 3.1 lists the quark flavors, electrical charges, and masses.

A quark of a given flavor but with opposite charge is called an *antiquark*. The symbols for antiquarks are the same as those for the quarks, but with a line above the letter. As an example, ū is the symbol for the up antiquark.

Table 3.1. Quark flavors, charges, and masses

Flavor	Symbol	Charge (e)	Mass (u)
up	u	$+\frac{2}{3}$	0.36
down	d	$-\frac{1}{3}$	0.36
strange	s	$-\frac{1}{3}$	0.46
charm	c	$+\frac{2}{3}$	1.67
bottom	b	$-\frac{1}{3}$	4.6
top	t	$+\frac{2}{3}$	4.5

Quarks have color, and antiquarks have anticolor.

Table 3.2 shows the quark structure of baryons (proton, neutron, and hyperons). Notice that the charges of the baryons ($+1$, 0, or -1) result from appropriate combinations of the fractional quark charges. The three quarks that form a baryon must have three *different* colors. The proton, for instance, consists of $u_R u_G d_B$ meaning a red up quark, a green up quark, and a blue down quark. The antiproton, on the other hand, consists of $\bar{u}_R \bar{u}_G \bar{d}_B$, meaning an antired up antiquark, an antigreen up antiquark, and an antiblue down antiquark. In mesons, which consist of a quark-antiquark pair, the antiquark must have the anticolor of the quark. In mesons, too, the charges combine so as to give one net positive, negative, or zero charge. The triplets of different colors or the color-anticolor pairs form *colorless* (or *white*) combinations, so that hadrons as a whole do not carry color. Table 3.2 shows that some baryons have the same quark composition as far as flavors are concerned: Λ and Σ^0, for instance, both consist of uds quark triplets. They differ, however, in other properties.

Ordinary hadrons consist of quarks with the charges indicated in Table 3.1, whereas ordinary electrons are negatively charged. Ordinary matter, therefore, consists of atoms that have a positive nucleus and negative electrons. Matter with reversed charges is called *antimatter*. Whenever energy is transformed into matter, equal amounts of matter and antimatter are formed. If a gamma ray of frequency greater than $2 \cdot 10^{20}$ Hz grazes an atomic nucleus, the gamma ray disappears, and an electron-antielectron (positron) pair

Table 3.2. Quark structures of baryons

Name	Mass (u)	Quark structure	Charge (e)
nucleons			
proton	1.00727	uud	$+\frac{2}{3}+\frac{2}{3}-\frac{1}{3}=+1$
neutron	1.00866	uud	$+\frac{2}{3}-\frac{1}{3}-\frac{1}{3}=0$
hyperons			
Λ	1.19764	uds	$+\frac{2}{3}-\frac{1}{3}-\frac{1}{3}=0$
Σ^+	1.27681	uus	$+\frac{2}{3}+\frac{2}{3}-\frac{1}{3}=+1$
Σ^0	1.28015	uds	$+\frac{2}{3}-\frac{1}{3}-\frac{1}{3}=0$
Σ^-	1.28539	dds	$-\frac{1}{3}-\frac{1}{3}-\frac{1}{3}=-1$
Ξ^0	1.41159	uss	$+\frac{2}{3}-\frac{1}{3}-\frac{1}{3}=0$
Ξ^-	1.41848	dss	$-\frac{1}{3}-\frac{1}{3}-\frac{1}{3}=-1$
Ω^-	1.79543	sss	$-\frac{1}{3}-\frac{1}{3}-\frac{1}{3}=-1$

appears. Countless experiments have proved that no net charge can be created—if a positive particle is created from pure energy, its antiparticle also appears. There is no way to make only one appear.

An antiparticle may be born, but it has no chance at all in our world because it soon hits its corresponding particle (of which our world is made), and the two annihilate each other and revert to pure energy. This transformation of energy into mass and of mass into energy is expressed in the most famous physical equation of all: $e = mc^2$.

When the universe was created, particles and antiparticles must have come into existence in equal amounts. In the first few instants, the fundamental particles (quarks, leptons, and photons) formed a kind of undifferentiated soup [the ἄπειρον of Anaximander called *ylem* by the Russian-American physicist George Gamow (1904–1968), who first proposed the Big Bang theory in 1948]. Under those conditions, not only was matter (quarks and leptons) continuously transforming into energy (photons), and vice versa, but also quarks were transforming into leptons, and vice versa, by such reactions as the following:

quark ⇄ X particle ⇄ antielectron (a positive electron or positron) + antiquark

antiquark ⇄ X particle ⇄ electron + quark

The X particle (of which there are 12 different types) is a virtual particle, like the $W^\pm$ and the Z^0 particles in weak interaction. It is, however,

much more massive, at least 10^{15} u, and carries a charge of $-\frac{1}{3}e$. If its mass is 10^{16} u, it can last, at most, $2\cdot10^{-39}$ s (by the uncertainty principle) and can travel, at most, 10^{-31} m. Time is reversible for reactions involving photons, because time does not pass for photons, which, being massless, must travel at the speed of light. Time is also reversible for processes involving material particles, except for the X interaction, which slightly favors the transformation of antimatter into matter over the reverse. This slight difference is responsible for the very existence of matter in our universe (see Chapter 4).

Mesons are unstable particles, each consisting of a quark and an antiquark of a different flavor (Table 3.3). They have masses that range from 0.15 to 0.5 u if they do not contain the heavier charm or bottom quarks, or 2 to 5 u if they do (Table A4). Mesons participate in many nuclear reactions. Table 3.3 shows the quark structure of mesons. When they are free, mesons decay, either directly or through other mesons, into electrons, neutrinos, and/or gamma radiation. The charges of the quark-antiquark pairs that form mesons are such that mesons, like baryons, exhibit either unitary charge (+ or −) or zero charge—no free fractional charge has ever been observed.

Particles interact with each other by means

Table 3.3. Quark structure of mesons

Name	Mass (u)	Quark structure	Charge (e)
π^+	0.14983	ud	$+\frac{2}{3}+\frac{1}{3}=+1$
π^0	0.14489	uu/dd	$+\frac{2}{3}-\frac{2}{3}/-\frac{1}{3}+\frac{1}{3}=0$
π^-	0.14983	du	$-\frac{1}{3}-\frac{2}{3}=-1$
K^+	0.52997	us	$+\frac{2}{3}+\frac{1}{3}=+1$
K^0	0.53427	ds	$-\frac{1}{3}+\frac{1}{3}=0$
K^-	0.52997	su	$-\frac{1}{3}-\frac{2}{3}=-1$
D^+	2.00687	cd	$+\frac{2}{3}+\frac{1}{3}=+1$
D^0	2.00687	cu	$+\frac{2}{3}-\frac{2}{3}=0$
D^-	2.00687	cd	$-\frac{2}{3}-\frac{1}{3}=-1$
F^+	2.11594	cs	$+\frac{2}{3}+\frac{1}{3}=+1$
F^-	2.11594	cs	$-\frac{2}{3}-\frac{1}{3}=-1$
B^+	5.6584	bu	$+\frac{1}{3}+\frac{2}{3}=+1$
B^0	5.6620	bd	$-\frac{1}{3}+\frac{1}{3}=0$
B^-	5.6584	bu	$-\frac{1}{3}-\frac{2}{3}=-1$

of field forces. Currently, four different types of forces are known to exist: the color or strong force, the electromagnetic force, the weak force, and the gravitational force. Their field particles are, respectively, the gluon, the photon, the $W^{\pm}$ and Z^0 particles, and the graviton (Table 3.4).

1. The **strong** (or **color**) **force** is the force that holds quarks within hadrons. The residual strong force between quarks belonging to adjacent nucleons holds the nucleons within nuclei and is called *nuclear force*. The strong force results from the fact that quarks, in addition to flavor, have color. Color is a property analogous to charge, but whereas there are two charges ($+$ and $-$), there are three colors and three anticolors. Charged particles interact with each other by means of virtual photons, which create an electrostatic field, are not themselves charged, and do not consume energy. Hadrons, on the other hand, interact with each other by means of virtual gluons, which carry one color and one anticolor and can also interact among themselves. The color force appears to be like a rubber band—it becomes stronger the farther apart the quarks are from each other. This explains why quarks are confined within the radius of a hadron. A hadron is not a little ball of matter, but a small region of space where two quarks (in mesons) or three quarks (in baryons) dance around each other without ever being able to break away. The dance is so fast that, from the outside, a hadron looks like a fuzzy ball. The gluon, which carries the strong force, is massless and colorless.

2. The **electromagnetic force** is caused by the exchange of virtual photons between charges. This exchange produces attraction between opposite charges and opposite magnetic poles, and repulsion between like charges or magnetic poles. The electromagnetic force has practically an infinite range, because it is possible to create virtual photons with practically zero energy and therefore with practically infinite lifetime. The electromagnetic force is 100 to 1,000 times weaker than the nuclear force.

3. The **weak force** is caused by the exchange

of virtual W^+, W^-, or Z^0 particles between hadrons and leptons. It has an extremely short range, because these particles are very massive (86.7 u for the W^+ and W^- particles; 99.7 u for the Z^0 particle) and therefore have a lot of energy ($8.0 \cdot 10^{10}$ eV for the $W^{\pm}$ particle; $9.2 \cdot 10^{10}$ eV for the Z^0 particle). Because of the uncertainty principle, which says that $\Delta E \Delta t \geqslant h/2\pi$, the $W^{\pm}$ particle can last a maximum of $0.82 \cdot 10^{-26}$ s, and the Z^0 particle a maximum of $0.71 \cdot 10^{-26}$ s. During that time they can reach only to a distance of 10^{-18} m. The weak force is about 10^6 times weaker than the nuclear force, and 10^4 times weaker than the electromagnetic force.

4. The **gravitational force** is caused by the exchange of virtual gravitons between masses. It has infinite range and is 10^{39} times weaker than the strong force.

An example of weak interaction is the *beta decay* process, by which a neutron (udd) becomes a proton (uud), or a proton becomes a neutron. In the β^+ decay process, a u quark is transformed into a d quark, and a W^+ particle is emitted that immediately decays into a positive electron (e^+) and a neutrino:

$$p\,(udd) \rightarrow n\,(udd) + W^+$$
$$W^+ \rightarrow e^+ + v_e$$

In the β^- decay process, a d quark is transformed into a u quark, and a W^- particle is emitted that immediately decays into an ordinary electron (e^-) and an antineutrino.

$$n\,(udd) \rightarrow p\,(udd) + W^-$$
$$W^- \rightarrow e^- + \bar{v}_e$$

The β^+ decay process occurs only in atomic nuclei that have too many protons. The β^- decay process occurs both in free neutrons or in nuclei that have too many neutrons.

Another example of weak interaction is the decay of the positive or negative pion:

$$\pi^+\,(ud) \rightarrow W^+ \rightarrow \mu^+ + v_\mu$$
$$\pi^-\,(du) \rightarrow W^- \rightarrow \mu^- + \bar{v}_\mu$$

The positive muon (μ^+) then decays into a positive electron and a neutrino-antineutrino pair, and the negative muon (μ^-) decays into an ordinary electron and a neutrino-antineutrino pair.

Neutrinos are scattered by hadrons and

Table 3.4. Field particles

Name	Field	Interaction	Mass (u)	Range (m)
gluon	strong force	between quarks	0	10^{-15}
photon	electromagnetic force	between charges	0	∞
$W^{\pm}, Z^0$	weak force	between hadrons and leptons	86.7, 99.7	10^{-18}
graviton	gravitational force	between masses	0	∞

leptons. Scattering occurs through the exchange of Z^0 particles, which, being very massive (99.7 u) and therefore very energetic, have extremely short range ($\langle 10^{-18}$ m). The field particles are summarized in Table 3.4.

The Inverse Square Law

The strength of the electromagnetic field and that of the gravitational field obey the inverse square law in empty, isotropic space (i.e., space that has the same properties in all directions).

The inverse square law works as follows. Energy is force times distance:

$$E = F \cdot d$$

from which obtain

$$F = E/d$$

where E is energy, F is force, and d is distance.

Consider an electron emitting a virtual photon of energy 1 (in arbitrary units). Another electron at distance 1 m receives the virtual photon and experiences a force $F = E/d = 1$. Now double the distance. The virtual photon, in order to reach the receiving electron, has to have half the energy, or 0.5 (in the same arbitrary units). But the distance is now 2 m, and so the force is $0.5/2 = 0.25$. If the distance increases to 3 m, the energy of the virtual photon must be $\frac{1}{3}$, making the force equal to $\frac{1}{3}/3 = \frac{1}{9}$. The numerator decreases and the denominator increases, both in direct proportion to the distance. As a result, the quotient (i.e., the force) changes in inverse proportion to the square of the distance. The same is true for the gravitational force.

Power, transmitted by real particles, also obeys the inverse square law. This derives from the fact that the surface of a sphere is equal to $4\pi r^2$.

[The volume of a sphere is $\frac{4}{3}\pi r^3$. The surface is simply the derivative of the volume with respect to the radius: $dV/dr = 4\pi r^2$.]

Consider, for instance, a constant point source of light in space surrounded by a concentric series of perfectly transparent spherical surfaces. The total power crossing each spherical surface remains constant and equal to the power emitted. However, the formula for the surface of a sphere, $4\pi r^2$, shows that the surface changes in *direct* proportion to the square of the radius. This means that the power *per unit surface* changes in *inverse* proportion to the square of the radius. Notice that the total power received by each surface remains constant, because as the radius increases, the surface increases in proportion to the square of the radius, but the power per unit surface decreases in the same proportion, and so the two cancel out. This consideration was used by Heinrich Olbers (1758–1840) to enunciate his famous paradox (see Chapter 5).

Spin and Angular Momentum

Elementary particles have intrinsic spin. The fundamental unit of spin energy is $\frac{1}{2}h/2\pi = 5.27 \cdot 10^{-35}$ J/Hz, where h is Planck's constant $= 6.626 \cdot 10^{-34}$ J/Hz. Notice that $\frac{1}{2}h$ is the energy *per turn*, while $\frac{1}{2}h/2\pi$ is the energy *per radian* (the symbol for $h/2\pi$ is $\hbar$). Quarks, leptons, and baryons have spin of $\frac{1}{2}\hbar$; mesons have spin $0\hbar$; the photon and the strong (gluon) and weak ($W^{\pm}$ and Z^0) force particles have spin $1\hbar$; and the graviton has spin $2\hbar$. Nuclei with odd numbers of nucleons have spins of $\frac{1}{2}\hbar$ or odd multiples of $\frac{1}{2}\hbar$ (up to $\frac{9}{2}\hbar$). Nuclei with even numbers of nucleons have spins of $0\hbar$, $1\hbar$, or multiples of $1\hbar$ (up to $5\hbar$).

A spinning body has spin angular momen-

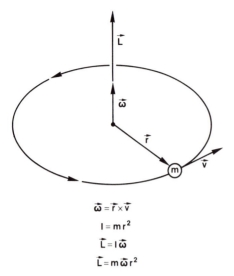

$$\vec{\omega} = \vec{r} \times \vec{v}$$
$$I = m r^2$$
$$\vec{L} = I \vec{\omega}$$
$$\vec{L} = m \vec{\omega} r^2$$

Figure 3.1. Angular-velocity vector ω, rotational inertia I (a scalar), and angular-momentum vector **L** of a mass m revolving about an axis. The product $\mathbf{r} \times \mathbf{v}$ is obtained by transporting **v**, while maintaining it parallel to itself, until its point of application coincides with that of **r**. The product $\mathbf{r} \times \mathbf{v}$ yields the vector ω, with the direction indicated.

tum. In addition to spinning, the electron also orbits. An orbiting body has orbital angular momentum. Angular momenta, like linear momenta, are vectors.

Angular momentum is a property of all spinning and orbiting bodies, from quarks to quasars. Consider a point mass revolving around an axis at a distance r from it (Fig. 3.1). The point mass has a rotational inertia I equal to mr^2. The orbital angular momentum is equal to the rotational inertia multiplied by the angular velocity. Angular velocities (symbol ω) are expressed in radians/second. One radian (rad) is the angle at the center of a circle subtended by a segment of circumference equal in length to the radius. Therefore, a full circle, or $360°$, has 2π rad or 6.28 rad. The angular momentum of a point mass revolving around an axis at a distance r is then equal to $mr^2\omega$. If $m = 1$ kg, $r = 1$ m, and $\omega =$ one revolution per second (rps) $= 6.28$ rad/second, then the angular momentum (L) is

$$
\begin{aligned}
L &= mr^2\omega \\
&= 1 \, \text{kg} \times 1 \, \text{m}^2 \times 6.28 \, \text{rad/s} \\
&= 6.28 \, \text{kg·m}^2\text{/s} \quad\quad\quad\quad\quad (1)
\end{aligned}
$$

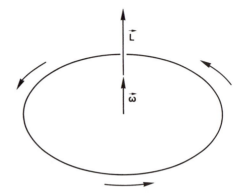

Figure 3.2. Angular-velocity vector ω and angular-momentum vector **L** for a rotating bicycle rim.

(the radian does not appear in the result because it has no dimensions).

Vectorially, the angular momentum of a body revolving around an axis of rotation at a distance r is the product of two vectors, one representing **r** (considered as originating at the axis of rotation and directed outward toward the revolving body), and the other representing the momentum **p** of the revolving body (Fig. 3.1). The two vectors are at $90°$ to each other. A mass of 1 kg making 1 rps at a distance of 1 m from the axis of rotation has an angular momentum **L** equal to

$$
\begin{aligned}
\mathbf{L} &= \mathbf{r} \times \mathbf{p} \\
&= 1 \, \text{m} \times 1 \, \text{kg·} 2\pi \, \text{m/s} \\
&= 6.28 \, \text{kg·m}^2\text{/s} \quad\quad\quad\quad\quad (2)
\end{aligned}
$$

If the sense of revolution is as shown in Fig. 3.1, the direction of the momentum vector **L** is up.

The angular momentum of the rim of a bicycle wheel (Fig. 3.2) is the same as if the mass of the rim were concentrated into a single lump at a distance from the center equal to the radius of the wheel. If the mass of the rim where 1 kg, the radius 0.35 m, and the angular velocity 1 rps, then its angular momentum **L** would be $\mathbf{L} = \mathbf{r} \times \mathbf{p} = 0.35$ m $\times$ 1 kg·6.28·0.35 = 0.77 kg·m^2/s (or 0.77 kg/m^2 s^{-1}). The angular-momentum vector is normal to the plane of the wheel rim and applied at the center of the wheel. Its sense is that of advancement of a normal screw, threaded through the hub, whose rotation would make the wheel turn (Fig. 3.2).

A solid sphere of uniform density rotating on its axis (Fig. 3.3) has an inertia $I = \frac{2}{5}mr^2$

Figure 3.3. Angular-velocity vector ω and angular-momentum vector **L** for a rotating sphere.

and an angular momentum $L = I\omega$, where ω is the angular velocity. The earth turns 6.28 rad in 24 h or 86,400 s. Its angular velocity is therefore $6.28/86,400 = 7.27 \cdot 10^{-5}$ rad/s. If it were a sphere of uniform density, its angular momentum would be $L = I\omega = 6.9 \cdot 10^{33}$ kg m^2 s^{-1}. Its actual angular momentum is $5.86 \cdot 10^{33}$ kg m^2 s^{-1}, which is 15% smaller because the Earth's density is not uniform: The core, which represents 30% of the mass of the Earth, consists of an iron–nickel alloy that has a density much higher than that of the mantle rocks above.

The angular momenta of the elementary particles, spin as well as orbital, are *quantified*, which means that they can assume only specific values. Particles with spin angular momentum equal to a half-integer times $h/2\pi$ are called *fermions*; particles with spin angular momentum equal to an integer (including 0) times $h/2\pi$ are called *bosons*. Quarks, leptons, baryons, and all atomic nuclei with an odd number of nucleons are fermions; the photon, the weak- and strong-force particles, mesons, and all atomic nuclei with an even number of nucleons are bosons.

THINK

* Guess (or find out) why physicists chose red, green, and blue (and not, for instance, orange,

yellow, and purple) as the three colors for quarks.

* Mars is 1.5 times more distant from the Sun than the Earth is. How much weaker is solar light reaching the surface of that planet?

* If you are standing on the equator, you are moving eastward with the Earth at a speed of 1,670 km/h = 463.88 m/s. This is your velocity vector. If you weigh 80 kg, your momentum is $80 \times 463.88 = 37,110$ kg m^2 s^{-1}. Because the equatorial radius of the earth is 6,378 km, your angular momentum will be $6,378,000 \times 37,110 = 2.37 \cdot 10^{11}$ kg m^2 s^{-1}. The vector **L** representing this angular momentum is applied at the center of the Earth and is directed *northward*. What would your angular momentum be if you were standing at the North Pole? Which direction would the angular-momentum vector have? What about if you were standing at the South Pole?

* Physicists, even with the most powerful machines, have not been able to overcome the strong force and take apart the three quarks that form a proton or a neutron or the two quarks that form a meson. How strong is the strong force? To find out, consider the two u quarks inside a proton. They are confined within the diameter (d) of the proton, which is $2.8 \cdot 10^{-15}$ m. They are each positive and carry a charge equal to two-thirds that of the electron. The force F between two charges q_1 and q_2 is equal to the constant $1/4\pi\varepsilon_0 = 8.99 \cdot 10^9$ N m^2 C^{-2} (where ε_0 is the permittivity constant $= 8.854 \cdot 10^{-12}$ C^2 N^{-1} m^{-2}) times the product of the two charges divided by the square of the distance: $F = (1/4\pi\varepsilon_0)(q_1 q_2/r^2)$. Using this equation, and choosing d as the distance between the two quarks, calculate the electromagnetic force between them (in this case, $q_1 = q_2$). The strong force is some 137 times stronger than the electromagnetic force (which means that the strength of the electromagnetic force is around 1/137 that of the strong force; this quantity, 1/137, is a pure number called the *fine-structure constant*). Calculate the value of the strong force between the two quarks and compare it to that of the gravitational force $F = G(m_1 m_2)/r^2$ (where G is the gravitational constant, r is the diameter of a proton $= 2.8 \cdot 10^{-15}$ m, and, in this case, $m_1 = m_2 =$ mass of each quark $= 0.36$ u). *Note:* By using the electromagnetic force between quarks to obtain the value of the strong force, we have tacitly assumed that, like the electromagnetic force, the strong force decreases in

inverse proportion to the distance. That may not be correct. There are theories saying that, like a rubber band, the strong force *increases* with distance to about $2.8 \cdot 10^{-15}$ m (the diameter of a nucleon) and that for the residual quark–quark interaction between adjacent nucleons, it rapidly decreases toward zero beyond that distance.

3.2. FORMATION OF ATOMIC NUCLEI

The atomic nucleus consists of protons and neutrons (with the exception of ^{1}H, whose nucleus consists of a single proton). As discussed in the preceding section, the nuclear force—the force holding protons and neutrons together within a nucleus—is due to the strong interaction between quarks belonging to adjacent nucleons. The nuclear force does not obey the inverse square law: Beyond a distance of about $3 \cdot 10^{-15}$ m it rapidly decreases toward zero. Let us see how strong the nuclear force is.

The disassembling of an atomic nucleus into its component nucleons requires energy. Conversely, when protons and neutrons are brought together with sufficient force to overcome the electromagnetic repulsion among the protons, the strong force takes over and glues these particles together. The nucleus as a whole has *less* energy than the individual nucleons when free. This energy, called *binding energy*, is the energy *lost* by the nucleus in the process of its formation and radiated away. Because $e = mc^2$, the bound nucleons, having lost energy, have less mass than they had as free nucleons (Table 3.5). As a result of the energy lost when forming nuclei, the atomic mass unit (u) approximates the *average* mass

of a *bound* nucleon (Table 3.6). The *binding-energy curve* (Fig. 3.4) shows the binding energy per nucleon as a function of nuclear mass. Binding energy per nucleon increases from H to Fe and then decreases from Fe to U. This means that fusion of light elements or fission of heavy elements yields products with *greater* binding energy per nucleon. The energy lost by the system is retrieved as nuclear energy.

Binding energy is large when viewed on the atomic scale, but on the human scale it is very small, because $1 \text{ MeV} = 1.60 \cdot 10^{-13}$ J. Thus, the energy needed to disassemble completely even the complex uranium nucleus is only $1,753.46 \times 1.60 \cdot 10^{-13}$ J $= 2.80 \cdot 10^{-10}$ J, which is about the energy expended by a small ant crawling at a speed of 1 mm/s. However, because force = energy/distance, and the distances are very short (on the order of $3 \cdot 10^{-15}$ m), the forces needed to split the nucleus are very large.

In theory, all isotopes can be constructed from ^{1}H by adding neutrons, because when an isotope contains too many neutrons in its nucleus, one neutron decays into a proton by β^- decay and forms the next element up. ^{3}H, for instance, decays into ^{3}He, and ^{14}C decays into ^{14}N. Notice that the transformation of a neutron into a proton does not change the mass number. George Gamow suggested (in 1948) that all elements were made in this simple way in the first half hour of the life of our universe. Unfortunately, adding neutrons works only up to ^{4}He. If a neutron is added to ^{4}He, ^{5}He is made, but ^{5}He is extremely unstable and immediately breaks down to ^{4}He and a neutron. In other words, the neutron bounces

Table 3.5.   Binding energy of selected nuclei

Nucleus	Nuclear structure	Calculated mass (from free nucleons) (u)	Measured mass (u)	Binding energy (total) (MeV)	Binding energy per nucleon (MeV)
^{2}H	1p 1n	2.01593	2.01300	1.70	0.85
^{4}He	2p 2n	4.03186	4.02967	27.25	6.81
^{12}C	6p 6n	12.09558	11.99671	89.03	7.41
^{40}Ca	20p 20n	40.3186	39.9516	331.61	8.29
^{56}Fe	26p, 30n	56.4488	55.9206	478.72	8.55
^{238}U	92p, 146n	239.9332	238.0004	1,753.46	7.37

Note: 1 u = 931.49432 MeV.

Table 3.6. **Masses of free and bound nucleons**

mass of free proton	= 1.00727 u
mass of free neutron	= 1.00866 u
average mass of free nucleon	= 1.00796 u
average mass of nucleon bound in ^{12}C	= 0.99972 u
average mass of nucleon bound in ^{40}Ca	= 0.99878 u
average mass of nucleon bound in ^{238}U	= 1.00000 u

right back. We shall see in Chapter 4 how the elements were made.

Table A9 in Appendix A shows a plot of all natural isotopes in terms of the number of protons and neutrons in their nuclei. The plot begins with ^{1}H and the neutron and starts with a 45° trend, indicating that the number of protons tends to increase *pari passu* with the number of neutrons. As the elements become heavier, however, the number of neutrons increases faster than the number of protons to keep the strong force from being overcome by the electromagnetic force among protons. The ^{238}U nucleus, for instance, contains 92 protons and 146 neutrons.

Of the 284 stable or long-lived radioactive isotopes occurring in nature, 166 have even numbers of protons (p) and neutrons (n); 56 have even p and odd n; 53 have odd n and even p; only 4 have odd p and odd n (^{2}H, ^{6}Li, ^{10}Be, and ^{14}N). It is evident that both neutron excess and nucleon pairing stabilize the nucleus.

Table A9 in Appendix A shows that the majority of the elements have two or more isotopes. The record is held by tin, with 10 stable isotopes, followed by xenon with 9. Twenty-one elements, however, have only one stable or long-lived isotope. Elements with an odd number of protons in their nuclei tend to have fewer isotopes than those with an even number of protons. Two elements of moderate mass, technetium (no. 43) and promethium (no. 61), and seven heavy elements between bismuth and uranium, have no stable isotopes. Technetium and promethium do not exist on Earth, but they have been observed in remnants of supernova explosions and have been made artificially in laboratories. The seven unstable elements between bismuth and uranium occur on Earth because they are

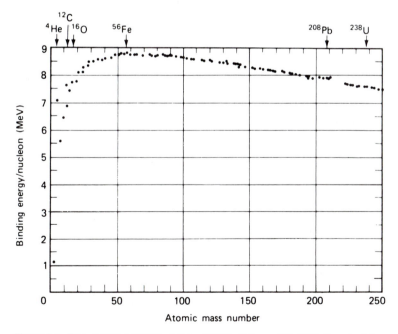

Figure 3.4. The binding-energy curve shows the nuclear binding energy per nucleon (in million electron volts) as a function of the atomic mass number for the most abundant (or most stable) isotope of each element. (Emiliani, 1987, p. 25.)

continuously produced by decay of the long-lived isotopes of uranium and thorium.

THINK

* Suppose the trend of the binding-energy curve were upside down, with iron at the bottom and both lighter and heavier nuclei higher up. By which process or processes could you get nuclear energy?

* In the preceding section we mentioned that nucleons forming nuclei are held together by the nuclear force, the strong force interacting between quarks that belong to adjacent nucleons. The strong-force particle is the gluon, and the interaction is maintained by virtual gluons. Internucleon distances in atomic nuclei are about $3 \cdot 10^{-15}$ m. Table 3.1 shows that the binding energy of ^{2}H is 1.70 MeV; convert that to joules (Table A2 in Appendix A) and, noticing that energy = force × distance, and, therefore, force = energy/distance, calculate the force necessary to separate the two nucleons that form ^{2}H. Calculate also to what weight on the Earth's surface that force would correspond, knowing that gravitational acceleration at the Earth's surface is 9.8 m/s². (Knowing that $F = ma$, where $m =$ mass and $a =$ acceleration, this means that a mass of 1 kg is attracted by the Earth with a force of 9.8 N.)

3.3. RADIOACTIVITY

Radioactivity is the processs by which unstable isotopes decay. The principal modes of decay are the following:

β^+ decay: the emission of a β^+ particle ($\equiv e^+$), transforming a proton into a neutron [e.g., ^{26}Al $(\beta^+) \rightarrow {}^{26}$Mg].

K capture (symbol ε): the capture by the nucleus of an electron from the K shell, transforming a proton into a neutron (same effect as in β^+ decay [e.g., ^{40}K $(\varepsilon) \rightarrow {}^{40}$Ar].

β^- decay: the emission of a β^- particle ($\equiv e^-$), transforming a neutron into a proton [e.g., ^{14}C $(\beta^-) \rightarrow {}^{14}$N].

α decay: the emission of an α particle ($\equiv$ nucleus of ^{4}He) [e.g., ^{238}U $(\alpha) \rightarrow {}^{234}$Th].

fission: the splitting of a nucleus into two major fragments plus 2–3 free neutrons (e.g., ^{235}U $\rightarrow {}^{143}$Nd $+ {}^{90}$Sr $+$ 2 neutrons). One in

2.3 million atoms of ^{238}U fissions naturally; the nucleus of U is shaped like a bowling pin rotating about its axis; fission occurs at the constriction; ^{235}U does not fission naturally, but it will fission if hit by a thermal neutron, i.e., a neutron traveling at relatively slow speed (< 20,000 km/s).

spallation: the breakup of a nucleus into three or more fragments plus free neutrons.

Beta decay and K capture involve interactions between hadrons and leptons, resulting in the emission or absorption of electrons by hadrons. Alpha decay, on the other hand, involves interaction among hadrons, resulting in the direct emission of an alpha particle. The color force that binds quarks within nucleons and the nucleons to each other within nuclei causes, in heavy nuclei, the grouping of two protons and two neutrons into alpha particles (and their breakup and regrouping). An alpha particle just formed inside, say, a nucleus of ^{238}U is kept there by the internucleon strong force acting between the alpha particle and other nucleons or alpha particles. For the alpha particle, the energy is 8.8 MeV, whereas the energy that it needs to emerge from the nucleus is 13.0 MeV. So how does the alpha particle escape? It escapes because the alpha particle can acquire the additional 4.2 MeV needed ($= 4.2 \times 1.6 \cdot 10^{-13} = 6.7 \cdot 10^{-13}$ J) thanks to the uncertainty principle. This energy is virtual energy that can be acquired by the alpha particle only for the duration Δt permitted by the relationship

$$\Delta E \Delta t \geqslant h/2\pi$$

The duration must then be less than $(h/2\pi)/6.7 \cdot 10^{-13} = 1.6 \cdot 10^{-22}$ s. Within that time, however, an alpha particle suitably located close to the surface of the nucleus can travel out of the nucleus before having to return to the virtual world the virtual energy that it had acquired. This process is called *tunneling*.

An unstable isotope may decay into a stable isotope in one step or through one or more intermediate unstable isotopes. ^{14}C, for instance, decays directly into ^{14}N by β^- decay, while ^{238}U decays to ^{206}Pb through 14 intermediate steps that include the emission of 8 successive alpha particles with intermediate β^- decays. Some isotopes can decay in two or even three different ways, a process called *branching*. ^{40}K, for instance, decays 89.3% of the time into ^{40}Ca by β^- emission and 11.7% of the time into ^{40}Ar by K capture (the absorption by the nucleus of an electron from the K shell); ^{226}Ac decays 92.79% of the time into ^{226}Th by β^- emission, 17.20% of the time into ^{226}Ra by β^+ emission, and 0.01% of the time into ^{222}Fr by α emission.

The *activity* (A) of a radioactive substance is the number of atoms decaying per unit of time:

$$A = -dN/dt$$

where A is the activity, N is the number of atoms decaying, and t is time (the negative sign signifies disappearance). Activity is the same as gross mortality. Mortality *rate* is usually expressed as the number of deaths per year per 1,000 people. The present mortality rate in the United States is 8.9/1,000 per year, or 0.0089 per year. Consider carefully these two expressions:

(8.9/1,000)/year

which means that 8.9 people die per year for each 1,000 persons present, and

0.0089/year

which means that 0.0089 person dies per year for each person present. The two expressions are obviously identical. The number 0.0089 is the mortality rate and is called the *decay constant* (symbol λ). We have, therefore,

$$\lambda = (dN/dt)/N$$
$$= 0.0089$$

If we take the second as a unit of time, the decay constant for humans in the United States becomes

$$\lambda = 0.0089/31.5569 \cdot 10^6$$
$$= 2.82 \cdot 10^{-10}$$

because there are $31.5569 \cdot 10^6$ seconds in one year. This means that for each person alive in the United States, $2.82 \cdot 10^{-10}$ person dies each second. Except for very unstable isotopes, the decay constant is always a very small number, because it is normalized to a single item (atom, person, etc.).

Life expectancy in the United States is now 75 years. This does not mean that everybody lives until the 75th birthday, only to die on that day. Obviously some people die sooner and others later. The *mean life* is 75 years. *Half-life* is the time is takes for a population with no new births to be reduced to one-half. In radioactivity, half-life (symbol $t_{1/2}$) is defined as follows:

Half-life is the time it takes for a given quantity of a radioactive substance to decrease to $\frac{1}{2}$ of its original amount.

The half-lives of unstable isotopes range from more than 10^{20} years (e.g., tellurium-130) to 10^{-20} seconds (e.g., helium-5). The half-life of the free neutron is 10.5 minutes, while that of the mesons ranges from 10^{-8} second to 10^{-18} second (see Table A4 in Appendix A). Half-life is 0.693 of mean life. To understand clearly the concepts of mean life and half-life, let us consider growth instead of decay, because growth is a phenomenon with which we all are familiar.

Consider a chicken that weighs 1 kg on January 1, and let it grow for one year, while regulating its food and exercise in such a way that at any time during the year its growth rate matches its actual weight. This means that the chicken grows at a rate of 1 kg/year on January 1. On April 1 the chicken will weigh 1.28 kg and will grow at a rate of 1.28 kg/year. On July 1 it will weigh 1.65 kg and will grow at a rate of 1.65 kg/year. On October 1 the chicken will weigh 2.12 kg and will grow at a rate of 2.12 kg/year. Its final weight, after 1 year, will be 2.72 kg. The growth function, when starting with one unit (1 kg in this case) and allowing it to grow proportionally to itself through a unit of time (1 year in this case), is given by the number 2.718281828450..., whose symbol is e. At the end of the first year we have

$$W = 2.718... = e$$

where W is the weight (in kilograms).

Suppose we decide to let the chicken grow through two years instead of one. During the second year, each of the 2.72 kilograms that we start with will grow to 2.72 kg, and so at the end of the second year the weight will be

$$W = e^2 = 7.4 \text{ kg} \quad \text{(nice chicken)}$$

But suppose we find a better chicken that will grow at *twice* the rate. It will reach a weight of 2.72 kg in 6 months. During the next six months, each of the 2.72 kilograms will itself grow to 2.72 kg, and so at the end of the year the weight will be

$$W = e^2 = 7.4 \, kg \quad \text{(still a nice chicken)}$$

Generally, if the growth rate is k, the final weight after one unit of time will be e^k.

If we now let our second chicken grow for 2 years, its weight at the end of the second year will be

$$W = e^{2 \times 2} = 54.8 \, kg \quad \text{(terrific chicken!)}$$

Generally, if the growth rate k continues for a period of time t, the final weight will be

$$W = e^{kt}$$

All this for a chicken that weighs 1 kg at the beginning. If we start with a chicken with a weight W_0 different from 1 kg, our chicken, growing at the rate k through time t, will end up weighing

$$W = W_0 e^{kt}$$

This is the general growth function, where W is the final weight, W_0 is the initial weight, $e = 2.718\ldots$, k is the growth constant, and t is time. In the case of decay, all we have to do is place a negative sign in front of the exponent

$$W = W_0 e^{-kt}$$

Decay functions are common in many areas of science and engineering. Attenuation of light through a semitransparent, homogeneous medium, for instance, is given by

$$Q = Q_0 e^{-\alpha x}$$

where Q is the light received by a detector inside the medium at distance x from the surface of the medium, Q_0 is the light hitting the surface of the medium, and α is the attenuation coefficient (which depends on the medium).

The attenuation of voltage or current along a transmission line is given by similar equations:

$$V = V_0 e^{-\alpha x}$$
$$I = I_0 e^{-\alpha x}$$

where V is voltage and I is current, V_0 and I_0 are the voltage and current at the source, α is the attenuation coefficient (which depends on the type of line), and x is the distance from the source.

Because power = current × voltage, we also have

$$P = P_0 e^{-2\alpha x}$$

In the case of radioactive isotopes, we use the letter N to represent the number of atoms now present, and N_0 to represent the number of atoms originally present:

$$N = N_0 e^{-\lambda t} \tag{3}$$

where N is the number of radioactive atoms still present, N_0 is the original number of radioactive atoms, λ is the decay constant, and t is time. Equation (3) is the general equation for radioactive decay. It is widely used to date rocks, minerals, and other natural and artificial substances. If N, N_0, and λ are determined, one can solve the equation for t (time).

From equation (3) we have

$$N/N_0 = e^{-\lambda t} \tag{4}$$

The definition of half-life says that after one half-life, half of the original atoms are gone. Therefore, at the end of one half-life we have

$$N = \tfrac{1}{2} N_0$$

or

$$N/N_0 = \tfrac{1}{2} \tag{5}$$

From equations (4) and (5) we derive the following:

$$e^{-\lambda t} = \tfrac{1}{2}$$
$$-\lambda t \ln e = \ln \tfrac{1}{2}$$
$$-\lambda t = \ln \tfrac{1}{2} = -\ln 2$$
$$\lambda t = \ln 2 = 0.693$$
$$t = 0.693/\lambda = t_{1/2} = \textit{half-life (symbol } t_{1/2})$$
$$\lambda = \text{decay constant} = 0.693/t_{1/2}$$
$$1/\lambda = t_{1/2}/0.693 = 1.44 t_{1/2} = \textit{mean life (symbol } \tau)$$

[Recall from Chapter 1 that the symbol "ln" means *natural logarithm*. The natural logarithm of a given number is the exponent that we have to give to the number $e (= 2.718\ldots)$ in order to obtain that number.]

Mean life is life expectancy. Mean life in the United States is 75 y, and therefore half-life is $75/1.44 = 52$ y. This should mean that if we take a random sample of 1,000 Americans (including children), we should have 500 left after 52 y, 250 after 104 y, 125 after 156 y, and so on. Can you explain why this does not seem to happen? Why does equation (3) not apply?

Equation (3) does apply, however, to radioactive decay. In radioactive decay, the decay is purely random. If we have 1,000 atoms of

^{238}U, some will decay right away, and others will decay billions and billions of years later. The half-life of ^{238}U is $4.468 \cdot 10^9$ y. The mean life, representing the life expectancy of a ^{238}U atom, is therefore $4.468 \times 1.443 = 6.445 \cdot 10^9$ y. Activity decreases by $1/e = 0.368$ with each mean life (and by 0.5 with each half-life).

Measurement of Half-lives

The half-life $(t_{1/2})$ is obtained directly from the decay constant λ:

$$t_{1/2} = 0.693/\lambda$$

The decay constant λ is determined by measuring the number of atoms decaying per second with respect to the number of atoms present:

$$\lambda = -(dN/dt)/N$$

For example, for 1 Av of ^{238}U atoms, it is found that

$$
\begin{aligned}
-(dN/dt)/N &= (2.960 \cdot 10^6 \text{ } \alpha \text{ emissions/s})/ \\
&\quad 6.022 \cdot 10^{23} \text{ atoms} \\
&= 4.915 \cdot 10^{-18} \text{ s}^{-1} = \lambda
\end{aligned}
$$

Therefore,

$$
\begin{aligned}
t_{1/2} &= 0.693/4.915 \cdot 10^{-18} \text{ s} \\
&= (0.693/4.915) \cdot 10^{18} \text{ s} \\
&= 0.141 \cdot 10^{18} \text{ s} \\
&= 4.468 \cdot 10^9 \text{ y}
\end{aligned}
$$

For 1 Av of ^{226}Ra it is found that

$$
\begin{aligned}
-(dN/dt)/N &= (8.265 \cdot 10^{12} \text{ } \alpha \text{ emissions/s})/ \\
&\quad 6.022 \cdot 10^{23} \text{ atoms} \\
&= 1.372 \cdot 10^{-11} \text{ s}^{-1} = \lambda
\end{aligned}
$$

Therefore,

$$
\begin{aligned}
t_{1/2} &= 0.693/1.372 \cdot 10^{-11} \text{ s} \\
&= 0.505 \cdot 10^{11} \text{ s} \\
&= 1,600 \text{ y}
\end{aligned}
$$

The trend of the proton-versus-neutron plot in the Isotope Chart (Table A9 in Appendix A) indicates that nuclear stability is achieved when nuclei of the light elements contain equal numbers of p and n and when nuclei of the heavier elements contain increasing excesses of n.

If the ratio of protons to neutrons is too far from the equilibrium ratio, the nucleus will undergo radioactive decay by throwing off the excess positive (β^+ decay) or negative (β^- decay) charges or by neutralizing an excess

positive charge by K capture. If the nucleus is too big (e.g., ^{238}U), it will decay by α emission, with intervening β^- emissions, or it will break into two major fragments plus 2 or 3 free neutrons (fission). If a nucleus is hit by a highly energetic (> 100 MeV) particle, it may break up in pieces (spallation).

The heaviest stable isotope is ^{209}Bi. Above it there is a series of elements that are represented only by unstable isotopes (Po, At, Rn, Fr, Ra, Ac, and Pa). These isotopes exist in nature, however, because they are continuously produced by the radioactive decay chains that start with ^{238}U, ^{235}U, and ^{232}Th and end with ^{206}Pb, ^{207}Pb, and ^{208}Pb, respectively. ^{238}U, ^{235}U, and ^{232}Th are also unstable, but their half-lives are sufficiently long for them to still exist in nature in significant amounts:

^{238}U decays to ^{206}Pb by emitting 8α and $6\beta^-$ $(t_{1/2} = 4.468 \cdot 10^9$ y).
^{235}U decays to ^{207}Pb by emitting 7α and $3\beta^-$ $(t_{1/2} = 0.704 \cdot 10^9$ y).
^{232}Th decays to ^{208}Pb by emitting 6α and $4\beta^-$ $(t_{1/2} = 14.05 \cdot 10^9$ y).

A uranium-containing rock that has not been geochemically altered will contain all isotopes of the decay chains of ^{238}U and ^{235}U in *radioactive equilibrium*, which means that while U decreases and Pb increases, the short-lived, intermediate radioisotopes will remain present in constant proportions. The relative abundances of these isotopes are given by the ratios of their half-lives. For instance, the number of ^{226}Ra atoms in equilibrium with ^{238}U is equal to $t_{1/2}(^{238}\text{U})/t_{1/2}(^{235}\text{U}) = 4.468 \cdot 10^9$ y$/1,600$ y $= 2.79 \cdot 10^6$. This means that there will be 1 atom of ^{226}Ra for $2.79 \cdot 10^6$ atoms of ^{238}U.

Radioactivity is the foundation of radiometric dating of all kinds of materials, from lunar rocks and meteorites to prehistoric human remains. Radiometric dating has yielded a reliable time scale for the evolution of the solar system, of planet Earth, and of the life on it.

THINK

* The mortality rate in Hungary is 13.8/1,000. Is the decay constant there larger or smaller than in the United States, where the mortality rate is 8.9/1,000?

* Considering how water molecules evaporate from water droplets at different temperatures, and therefore with different internal energies, would you expect that the alpha particles emitted by shorter-lived nuclei would be more or less energetic than those emitted by nuclei with longer half-lives?

* Radioactive decay involves energies ranging from 100,000 eV to millions of electron volts. By comparison, chemical energies range from a fraction of an electron volt to several electron volts and are therefore 100,000 to 1,000,000 times smaller. As a result, in the terrestrial environment, where temperatures are comparatively low (even in the Earth's core), radioactive decay proceeds at a steady rate. Of the various processes of radioactive decay, excluding spallation, which one would you expect to be the first to accelerate if temperature were raised sufficiently high?

* Considering that α emission from the nucleus of a heavy element occurs because the α particle can borrow the energy needed to leave without violating the uncertainty principle, how would you explain the existence of stable nuclei?

3.4. RADIOMETRIC DATING METHODS

Radioactive processes involve nuclear forces, which are a million times stronger than the forces involved in chemical processes. The energies commonly available on the surface and interior of planetary bodies are insufficient to affect nuclear processes. As a result, nuclear decay proceeds at a steady, unchanging rate in all planetary materials and provides a way to date these materials reliably.

In order to date a sample, the fundamental equation for radioactive decay

$$N = N_0 e^{-\lambda t}$$

must be solved for t (time). The decay constant is readily available from the half-life (which is measured in the laboratory):

$$\lambda = 0.693/t_{1/2}$$

N, the present number of atoms of the radioactive isotope of choice, is determined in

the laboratory. N_0, the original number of atoms when the sample was formed, is equal to the present number plus the number of atoms lost to radioactive decay. The latter is also determined in the laboratory. The accuracy of radiometric dating is within a few percent of the value obtained.

[*Accuracy* is a measure of how close a measurement approaches the true value; *precision* is the amount of consistency in repeated measurements of the same quantity, expressed by the number of decimal places. Thus, a set of high-precision measurements made with a poorly calibrated instrument will yield high precision but poor accuracy, whereas a set of sloppy measurements made with a properly calibrated instrument will yield low precision but high accuracy.]

The principal methods for radioactive dating are discussed next, together with the pertinent instrumentation.

Uranium-Thorium/Lead Methods

These methods are based on the decay of ^{238}U to ^{206}Pb, ^{235}U to ^{207}Pb, and ^{232}Th to ^{208}Pb. Uranium and thorium are measured by *flame photometry* or *atomic absorption*, giving the total U and Th. The relative abundances of ^{238}U and ^{235}U can be read directly from the Isotope Chart (Table A9). There is only one long-lived isotope of thorium (^{232}Th), and, therefore, the concentration of Th is also the concentration of ^{232}Th.

There are four stable isotopes of Pb: ^{204}Pb (1.4%), ^{206}Pb (24.1%), ^{207}Pb (22.1%), and ^{208}Pb (52.4%). These are the percentages today. All four isotopes were formed when the elements were last formed in the region of the solar system about $4.7 \cdot 10^9$ y ago. Since then, the amounts of ^{206}Pb, ^{207}Pb, and ^{208}Pb have been increasing because of the decay of ^{238}U, ^{235}U, and ^{232}Th, respectively, while that of ^{204}Pb has remained constant. A *mass spectrometer* is used to determine the concentrations of the lead isotopes in a rock sample to be dated.

We know that the ^{204}Pb in the rock was there to begin with when the rock formed. The relative amounts of ^{206}Pb, ^{207}Pb, and ^{208}Pb have slowly changed with time, because the half-lives of the parent isotopes, ^{232}Th, ^{235}U,

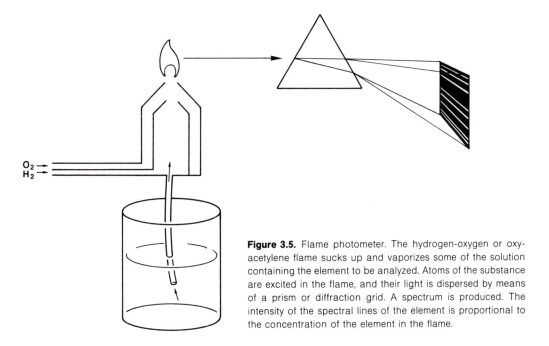

Figure 3.5. Flame photometer. The hydrogen-oxygen or oxy-acetylene flame sucks up and vaporizes some of the solution containing the element to be analyzed. Atoms of the substance are excited in the flame, and their light is dispersed by means of a prism or diffraction grid. A spectrum is produced. The intensity of the spectral lines of the element is proportional to the concentration of the element in the flame.

and ^{238}U, are different. ^{207}Pb increases fastest because ^{235}U has the shortest half-life, while ^{208}Pb increases slowest because ^{232}Th has the longest half-life. The approximate age of the rock sample can be determined from the ratios of the lead isotopes. The proportionate amounts of ^{206}Pb, ^{207}Pb, and ^{208}Pb that entered the rock together with ^{204}Pb when the rock was formed are calculated and subtracted. The rest is the *radiogenic* portion, which was added by the decay of ^{232}Th, ^{235}U, and ^{238}U during the lifetime of the rock. The radiogenic portion of ^{206}Pb, identified by an asterisk ($*^{206}Pb$), added to the amount of ^{238}U present in the sample (N), gives N_0, the original amount of ^{238}U in the sample. Knowing N and N_0, we can solve the radioactive decay equation (3) for time (t). We can do the same with $*^{207}Pb$ and ^{235}U, and with $*^{208}Pb$ and ^{232}Th.

With modern instrumentation it is possible to date rocks and minerals in which uranium, thorium, and lead occur only in minute concentrations. The major problem is the contamination that may have occurred since the rock was formed. Natural percolating solutions, for instance, can add or remove either some uranium or some lead or both; if that happened,

spurious ages will be obtained. In order to test for that possibility, a given sample is often analyzed by both U/Pb and Th/Pb. If the sample has not been geochemically altered, the two methods will give concordant ages.

[1. In *flame photometry* (Fig. 3.5), the concentration of an element in a substance is determined by dissolving the substance in a solution and then vaporizing the solution in a gas flame. If the flame is hydrogen-oxygen, the temperature of the flame is between 2,550°C and 2,700°C; if the flame is acetylene(C_2H_2)-oxygen, the temperature is between 3,050°C and 3,130°C. These temperatures are sufficient to energize some of the electrons in the atoms that have been vaporized. The atoms of the elements become excited in the flame and emit light of characteristic frequencies, producing an *emission spectrum*. The higher the concentration of the element, the stronger the emission lines. The instrument is calibrated by running solutions with known concentrations of the elements to be studied. Flame photometry is particularly useful for analyzing alkaline elements (group 1 elements, including Li,

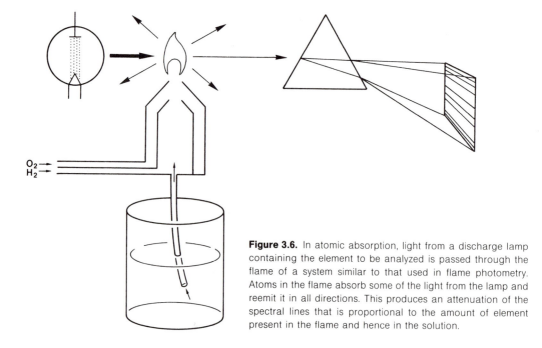

Figure 3.6. In atomic absorption, light from a discharge lamp containing the element to be analyzed is passed through the flame of a system similar to that used in flame photometry. Atoms in the flame absorb some of the light from the lamp and reemit it in all directions. This produces an attenuation of the spectral lines that is proportional to the amount of element present in the flame and hence in the solution.

Na, K, Rb, Cs) and alkaline-earth elements (group 2, including Be, Mg, Ca, Sr, Ba).

2. In *atomic absorption* (Fig. 3.6), a discharge lamp producing light from the vapors of an element to be analyzed is beamed through the flame in which the sample is dispersed, as in flame photometry. The light of the lamp will be absorbed by the atoms of the element and reradiated (within 10^{-8} s) on a 4π geometry (i.e., in all directions), thereby greatly attenuating transmission in the direction of the observer. An *absorption spectrum* is thus produced, the intensity of which is proportional to the concentration of the element in the flame. Atomic absorption is particularly useful for substances whose atoms do not become readily excited at the flame temperatures (group 11, including Cu, Ag, and Au; group 12, including Zn, Cd, and Hg; part of group 13, including Ga, In, and Tl; and such other elements as Mn, Ni, Sb, Pb, and Bi). The instrument is calibrated by running artificial samples, as in flame photometry.

3. *Mass spectrometry* (Fig. 3.7) is used to determine the relative abundances of different isotopes of the same element in a sample.

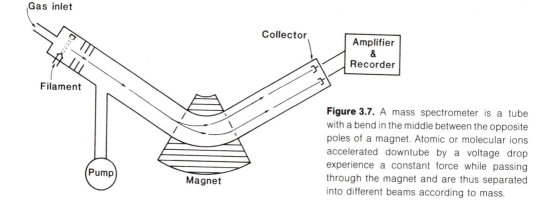

Figure 3.7. A mass spectrometer is a tube with a bend in the middle between the opposite poles of a magnet. Atomic or molecular ions accelerated downtube by a voltage drop experience a constant force while passing through the magnet and are thus separated into different beams according to mass.

The sample is vaporized and introduced into a metal tube with a 60° or 90° bend in the middle, across which a strong magnetic field (4,000–6,000 gauss) is maintained. The gas is introduced at one end of the tube, where a hot tungsten filament emits electrons that ionize the gas atoms or molecules (most commonly creating single-charge ions). The ions are accelerated downtube by a strong voltage drop (1,000–2,000 V positive to ground). Upon entering the magnetic field, the isotopes are deflected by the force $\mathbf{F} = \mathbf{I} \times \mathbf{B}$, where $\mathbf{I}$ is current (which is a positive ion flow) and $\mathbf{B}$ is the magnetic-field vector (which is directed from S to N). The same force is applied to different isotopes, with the result that the heavier isotopes are deflected less than the lighter ones. The mixture of isotopes is thus separated into different beams that are collected on electrodes at the other end of the tube. The ionic currents, on the order of 10^{-11} amperes, are led into high-resistance resistors ($10^{12}\,\Omega$), and the voltage drop (10–20 V) developed across them is amplified. The voltage drop is proportional to the current, which is proportional to the relative abundance of the ions producing it. Elements that cannot be made into gaseous form or gaseous compounds at room temperature (e.g., Ca, Sr, Ba, etc.) are placed into solution, and the solution is coated directly on the mass spectrometer's filament. When the filament is turned on, the atoms of the element are vaporized and ionized and shoot downtube as a gas.]

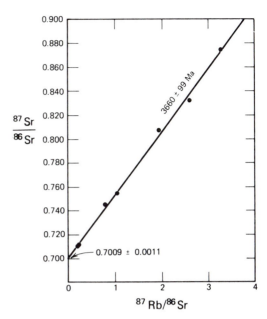

Figure 3.8. ^{87}Rb/^{87}Sr age of the Amitsoq gneiss, one of the oldest rocks on Earth (southwestern Greenland). Different minerals in the rock had different concentrations of ^{87}Rb and therefore have produced different amounts of ^{87}Sr since the rock formed. The slope of the line (isochron) gives the age of the rock – the older the rock, the steeper the slope. A modern rock would have a horizontal isochron intersecting the ordinate at the same point. (From Faure, 1986, p. 128, Fig. 8.7.)

^{87}Rb/^{87}Sr Method

Rubidium and strontium are less subject to geochemical alteration than are lead and uranium. Rubidium has two naturally occurring isotopes: ^{85}Rb (72.17%) and ^{87}Rb (27.83%). The first is stable, and the second has a very long half-life, $4.8 \cdot 10^{10}$ y, which accounts for the fact that it is still around. ^{87}Rb decays to ^{87}Sr by β^- decay. Rubidium does not form minerals of its own, but it takes the place of potassium or cesium in a variety of minerals. Because different minerals in the same rock usually contain different amounts of Rb, a rock to be dated is separated into its different minerals, and the concentration of Rb in each mineral is determined by flame photometry or atomic absorption. The proportion of ^{87}Rb is obtained from the Isotope Chart (Table A9). There are four stable isotopes of Sr: ^{84}Sr, ^{86}Sr, ^{87}Sr, and ^{88}Sr. Their relative abundances are obtained by mass spectrometry. The ratios ^{87}Rb/^{86}Sr and ^{87}Sr/^{86}Sr are plotted versus each other for the different minerals. The slope of the line that is obtained is a function of the age of the rock. The line is called an *isochron*, because all points on it represent the same age. Fig. 3.8 shows the isochron for the Amitsoq gneiss of southwestern Greenland, one of the oldest rocks on Earth. For a rock formed today, a similar graph would show a horizontal line starting at the same point. Such a rock would have minerals with different concentrations of rubidium, but still no increase in the ^{87}Sr/^{86}Sr ratio because ^{87}Rb would not have had time to produce any appreciable amount of ^{87}Sr. The intercept of the isochron with the ordinate represents the modern value of the $^{87}Sr/^{86}Sr$ ratio, as can be verified from the relative abundances of ^{87}Sr and ^{86}Sr given in Table A9.

$^{40}K/^{40}Ar$ Method

Potassium has three natural isotopes: ^{39}K (93.2581%), ^{40}K (0.0117%), and ^{41}K (6.7302%). ^{40}K decays either by β^- into ^{40}Ca (89.3% of the time) or by K capture into ^{40}Ar (10.7% of the time). Its half-life is $1.277 \cdot 10^9$ y. Argon consists of three stable isotopes: ^{36}Ar (0.337%), ^{38}Ar (0.063), and ^{40}Ar (99.600%). The reason ^{40}Ar is so preponderant today is in part because the original amount that formed at the time of element formation has been increased by the decay of ^{40}K during geologic time.

Argon is a noble gas, and therefore it does not form bonds with other atoms. One can assume, therefore, that a rock or mineral that formed slowly inside the earth did not contain any argon at the beginning. Only rocks chilled rapidly, such as lavas erupting on the ocean floor, may trap some argon. Otherwise, the argon contained in rocks and minerals today results essentially from the decay of ^{40}K.

In practice, the concentration of K is determined by flame photometry or atomic absorption. The rock is melted, and the Ar released is passed through a mass spectrometer to determine its isotopic composition. The reason for this is that atmospheric argon (which contains all three isotopes) always contaminates the rock or mineral sample in the field as well as in the lab. Mass-spectrometer analysis will reveal the relative abundances of the three isotopes. They will be in the ratio indicated earlier, plus the radiogenic ^{40}Ar derived from the sample ($*^{40}Ar$). This excess is the argon produced inside the mineral or rock since formation, by the decay of ^{40}K. The amount of ^{40}K now present in the sample (N) plus $*^{40}Ar$ gives the original amount of ^{40}K in the sample (N_0), making it possible to solve the radioactive decay equation (3) for time (t).

$^{40}Ar/^{39}Ar$ Method

Although argon does not form bonds with other atoms, as previously noted, some argon may be absorbed from a rock melt when a rock solidifies and crystals are formed. This argon will be found mainly along crystal boundaries. In addition, argon formed from the decay of a ^{40}K atom near the surface of a crystal in the rock may diffuse into crystalline interfaces and possibly escape. To overcome these problems, a rock sample to be dated is crushed and reduced to its individual crystals. The crystals are cleaned in an acid an irradiated with neutrons in a reactor. The reaction $^{39}K(n,p)^{39}Ar$ is produced, where (n,p) means "neutron in, proton out". In other words, the reaction changes a neutron into a proton and transforms the stable ^{39}K atom into the unstable ^{39}Ar atom. ^{39}Ar decays back to ^{39}K by β^- emission, with a half-life of 269 y. The irradiated sample is heated stepwise in a furnace, and the $^{40}Ar/^{39}Ar$ ratio is measured with a mass spectrometer as temperature rises. The ^{40}Ar released at the highest temperature is derived from the inner portion of the crystals and thus represents the true concentration of the radiogenic ^{40}Ar formed from the decay of ^{40}K since the rock began crystallizing. Because the concentration of ^{39}Ar is related to that of ^{39}K, which in turn is related to the concentration of ^{40}K, the ratio $^{40}Ar/^{39}Ar$ yields concentrations of both ^{40}K and its daughter product ^{40}Ar, which can then be used to solve the radioactive decay equation and determine the age of the rock.

Uranium Disequilibrium Methods

^{238}U and ^{235}U are the two primary, long-lived isotopes of uranium, with half-lives of $4.468 \cdot 10^9$ y and $7.04 \cdot 10^8$ y, respectively. A uranium-containing mineral or rock, if undisturbed, will contain not only the two primary uranium isotopes in their natural relative abundance (137.9 ^{238}U to 1 ^{235}U) but also all of their decay products in isotopic equilibrium with the parent isotopes and with each other. The equilibrium amount of each decay product with respect to the parent isotope or any other decay product in the mineral is given by the ratio of the two half-lives. For instance, the immediate daughter product of ^{238}U is ^{234}Th, which has a half-life of 24.10 days (d). The concentration of ^{234}Th in the mineral, with respect to that of ^{238}U, will therefore be 24.10 d/$4.468 \cdot 10^9$ y, or $1.48 \cdot 10^{-11} = 1.48 \cdot 10^{-9}\%$. The concentration of ^{234}U, with a half-life of 245,000 y, will be much greater, amounting to 245,000/$4.468 \cdot 10^9$ y $= 0.000055 = 0.0055\%$.

Seawater contains uranium at a concentration of 3.3 parts per billion (ppb). If the daughters of uranium were in isotopic equilibrium, seawater should also contain $5.5 \cdot 10^{-5}$ ppb of ^{230}Th and $3.4 \cdot 10^{-7}$ ppb of ^{231}Pa. Instead, the concentrations of these two isotopes are less than 10^{-9} ppb. The reason is that thorium and protactinium are highly insoluble elements— as soon as ^{230}Th and ^{231}Pa are formed in the ocean by the decay of uranium isotopes, they adsorb on clay particles and drop to the sea floor as part of fecal pellets. All daughter products of ^{230}Th and ^{231}Pa are also absent.

In addition to ^{238}U and ^{235}U in their natural proportions, the ocean contains a 15% excess of ^{234}U above the amount in equilibrium with ^{238}U (0.0063% instead of 0.0055%) because ^{234}U is more readily leached from continental rocks by percolating waters than is ^{238}U. ^{234}U is then transported to the ocean by rivers. The excess ^{234}U has no chance of decaying along the way, because the hydrologic cycle is much shorter than the half-life of ^{234}U. If rivers ceased to flow, this excess of ^{234}U in the ocean would practically disappear in a few half-lives. But the rivers keep flowing, creating a steady-state situation—the amount of ^{234}U reaching the ocean is equal to the amount that decays in the ocean.

Reef-building corals deposit skeleta of calcium carbonate ($CaCO_3$) in the form of aragonite. During the precipitation of aragonite from seawater, some uranium substitutes for calcium in the aragonite lattice because of their similar ionic radii. The concentration of uranium in corals amounts to about 3 ppm. After two very short decays (to ^{234}Th and ^{234}Pa—see Table A9), ^{238}U decays to ^{234}U, which has a half-life of 245,000 y. ^{234}U decays directly to ^{230}Th, which has a half-life of 75,380 y. The other primary uranium isotope, ^{235}U, decays to ^{231}Pa (half-life-32,760 y) through the short-lived ^{231}Th. Neither ^{230}Th nor ^{231}Pa is incorporated into the aragonitic structure of the corals, because they are practically absent from seawater and, in addition, their ionic radii would not fit. The coral structure, therefore, incorporates only uranium when it forms.

The 3 ppm of uranium in aragonite precipitating to form a coral skeleton consist of 99.273681% ^{238}U, 0.719994% ^{235}U, and 0.006324% ^{234}U, but no thorium or protactinium. As time goes by, ^{238}U starts decaying,

adding to the ^{234}U present; the excess ^{234}U decreases and eventually disappears; ^{230}Th, the daughter product of ^{234}U, begins to build up, and eventually both ^{234}U and ^{230}Th come to equilibrium with ^{238}U. At the same time, ^{235}U decays, through the short-lived ^{231}Th, to the longer-lived ^{231}Pa (half-life = 32,760 y). ^{231}Pa builds up until it comes to equilibrium with ^{235}U. The rate at which ^{230}Th and ^{231}Pa build up depends on their half-lives. Starting from zero (no ^{230}Th or ^{231}Pa), it takes one half-life to reach 50% of equilibrium, another half-life to reach 75% of equilibrium, still another half-life to reach 87.5% of equilibrium, and so forth. The result of all this is that ratios ^{234}U/^{238}U, ^{230}Th/^{234}U, ^{230}Th/^{238}U, ^{231}Pa/^{235}U, ^{231}Pa/^{230}Th, and so forth, keep changing during the first few hundred thousand years, until isotopic equilibrium is reached. These changing ratios are functions of time alone, affording an excellent method for dating young marine carbonates or sediments. In particular, the ratio ^{230}Th/^{234}U has been used to date corals, and the ratio ^{231}Pa/^{230}Th has been used to date deep-sea sediments (see Chapter 24).

²¹⁰Pb

^{222}Rn is part of the decay chain of ^{238}U. It is produced by the decay of ^{226}Ra and is a noble gas. As such it tends to escape into the atmosphere from surface sources. ^{222}Rn decays, through various short-lived isotopes (see Table A9), to ^{210}Pb, which has a half-life of 22.3 y. ^{210}Pb is rapidly removed from the atmosphere by precipitation and accumulates in snow and ice, as well as in marine and freshwater sediments. These deposits can be dated from the amount of ^{210}Pb they contain. As a deposit ages, its content of ^{210}Pb decreases by the amount $e^{-\lambda t}$, where λ is the decay constant of ^{210}Pb. This method applies only if the rate of accumulation of snow, ice, or sediment remains constant. Clearly, if snowfall were to suddenly double, the concentration of ^{210}Pb would be halved, not because one half-life elapsed, but because the ^{210}Pb would have been diluted.

Radiocarbon Method

The Earth is continuously bombarded by cosmic rays. Cosmic rays consist mainly of solar and galactic protons and ^{4}He nuclei

Table 3.7. Radionuclides induced by cosmic rays

| Radioisotope Produced | Source Nuclide | Half-life | Production Rate (atoms cm^{-2} s^{-1}) | | Global Inventory |
			Troposphere	Total Atmosphere	
^{3}H	^{14}N, ^{16}O	12.33 y	8.4·10^{-2}	0.25	3.5 kg
^{7}Be	^{14}N, ^{16}O	53.3 d	2.7·10^{-2}	8.1·10^{-2}	3.2 g
^{10}Be	^{14}N, ^{16}O	1.5·10^6 y	1.5·10^{-2}	4.5·10^{-2}	430 tons
^{14}C	^{14}N	5715 y	1.1	2.5	63 tons
^{22}Na	^{40}Ar	2.609 y	2.6·10^{-5}	8.7·10^{-5}	1.8 g
^{24}Na	^{40}Ar	14.959 h	—	—	—
^{26}Al	^{40}Ar	740,000 y	3.8·10^{-5}	1.4·10^{-4}	1.1 tons
^{28}Mg	^{40}Ar	20.90 h	—	—	—
^{31}Si	^{40}Ar	2.62 h	—	—	—
^{32}Si	^{40}Ar	172 y	5.4·10^{-5}	1.6·10^{-4}	1.4 kg
^{32}P	^{40}Ar	14.26 d	2.7·10^{-4}	8.1·10^{-4}	0.4 g
^{33}P	^{40}Ar	25.3 d	2.2·10^{-4}	6.8·10^{-4}	0.6 g
^{35}S	^{40}Ar	87.5 d	4.9·10^{-4}	1.4·10^{-3}	4.5 g
^{38}S	^{40}Ar	2.84 h	—	—	—
^{34}Cl	^{40}Ar	32.23 m	—	—	—
^{36}Cl	^{40}Ar	301,000 y	4.0·10^{-4}	1.1·10^{-3}	15 tons[a]
^{39}Cl	^{40}Ar	55.6 m	—	—	—
^{37}Ar	^{40}Ar	35.04 d	—	—	—
^{39}Ar	^{40}Ar	269 y	—	23	—
^{81}Kr	^{80}Kr	213,000 y	—	16.2	—

[a] Including estimate of ^{36}Cl produced by n capture at the Earth's surface.

Sources: Lal D. (1974). Radionuclides: cosmic-ray produced. In Fairbridge R.W. (ed.), *Encyclopedia of Geochemistry and Environmental Sciences*, Van Nostrand Reinhold, New York, pp. 996–1066; Kathren, R.L. (1984). *Radioactivity in the Environment: Sources, Distribution, and Surveillance.* Harwood Academic Publishers, Chur, Switzerland, 397 p.; as well as other sources.

(which are alpha particles, sometimes called *helions*). The speed of the solar particles is about 300 to 700 km/s. The speed of galactic protons and helions is much higher, ranging up to several tens of thousands of kilometers per second. At such speeds the particles are energetic enough to trigger a variety of nuclear reactions when hitting the nuclei of atoms in our atmosphere. Spallation of nitrogen and oxygen nuclei produces ^{3}H (tritium), and spallation of argon nuclei forms a whole series of nuclei ranging from sodium to chlorine (Table 3.7). Spallation also produces free neutrons, because heavier elements contain an excess of neutrons compared to the lighter elements. If a free neutron thus produced has the right energy, it may hit a ^{14}N nucleus and be absorbed. The excited ^{14}N nucleus will emit a proton and become the isotope ^{14}C by the reaction ^{14}N(n,p)^{14}C.

^{14}C is unstable. It has a half-life of 5,715 y, and it decays back to ^{14}N by β^-. Once formed in the atmosphere, ^{14}C quickly oxidizes to ^{14}CO$_2$, which is breathed by plants and incorporated into living matter or diffuses into the ocean and is precipitated as CaCO$_3$ (mainly by marine organisms, but also inorganically). When an organism dies, its matter ceases to absorb new ^{14}CO$_2$, and the ^{14}C concentration in the dead matter decreases by one-half each 5,715 y.

Carbon from a substance to be dated is extracted and oxidized to CO$_2$. The concentration of ^{14}C in the CO$_2$ is measured in a Geiger counter.

[The *Geiger counter* (Fig. 3.9) consists of a metal tube, 10 cm to 1 m in length, with an insulated wire running through its axis. The two ends of the tube are closed by windows

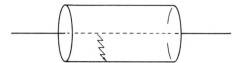

Figure 3.9. The Geiger counter is basically a metal tube (cathode) with a wire along its axis (anode). The two ends of the tube are closed by windows of insulating material, and the tube is filled with a gas containing the particles to be counted, or with an inert gas to count particles from an external source.

of insulating material. The tube is filled with gas, at 1 atm of pressure, that contains the particles to be counted. A potential difference of about 1,200 volts is maintained between tube and wire and is adjusted so that, with the gas inside, a spark *almost* develops between the two. When a ^{14}C atom decays in the gas, the β^- particle emitted is sufficient to ionize the surrounding gas, and a spark is produced. Because each spark represents one β^- decay, the activity $(-dN/dt)$ of the substance can be determined simply by counting the sparks. Geiger counters filled with inert gases at a fraction of one atmosphere of pressure are used to count particles entering the counter from such external sources as radioactive minerals and rocks or cosmic rays.]

Wood from trees grown before 1950, when nuclear-bomb testing began adding large amounts of ^{14}C to the atmosphere, produces about 16 counts per minute (cpm) per gram of carbon. Wood 5,715 y old produces 8 cpm, wood 11,460 y old produces 4 cpm, and so on.

Another method for measuring ^{14}C is *scintillation counting.* Carbon from the substance is made into benzene (C_6H_6), to which is added a special organic substance (e.g., 2,5-diphenyloxazole) that emits a tiny flash of light when energized by a β^- particle. This tiny flash is amplified by means of a *photomultiplier* so that it can actually be registered.

[The *photomultiplier* (Fig. 3.10) uses the photoelectric effect, by which free electrons can be removed from metal surfaces by photons of appropriate energy. The energies needed range from 2.14 eV (cesium) to 5.9 eV (selenium), which are within the energy range of visible light. In a photomultiplier tube, 6

to 16 successive plates (called *dynodes*) are maintained at a 75–150 volt difference from each other (increasingly positive). Photons hitting the first dynode release 2 to 5 electrons, which hit the second dynode, releasing more electrons. A cascade in thus formed. For a production of 4 electrons per stage and a total of 10 stages, the number of electrons reaching the anode will be $4^{10} = \sim 10^6$. The voltage developed by this electron flow across a high resistance (100 kΩ to 100 MΩ) is amplified and registered as a pulse.]

Because the half-life of ^{14}C is only 5,715 y, the radiocarbon method cannot be used to date materials older than about 40,000 y, which is about 7 half-lives. In 7 half-lives, the original amout of ^{14}C in a substance has been reduced to $1/2^7 = 1/128$, or less than one-hundredth. Conversely, a sample younger than 200 y will be indistinguishable from a modern sample.

Photon in

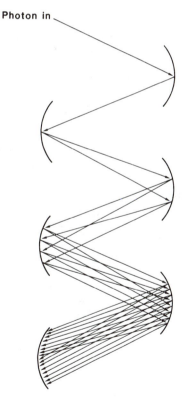

Figure 3.10. A photomultiplier consists of a set of metallic plates (dynodes) kept at progressively more positive voltages, through which a cascade of electrons is formed. The current thus generated is amplified and registered as a pulse.

The radiocarbon dating method is applicable to a variety of carbon-containing substances, including wood, charcoal, organic and inorganic carbonates, CO_2 dissolved in water or ice, hydrocarbons, and all kinds of organic matter.

There are two stable isotopes of carbon: ^{12}C (98.90%) and ^{13}C (1.10%). ^{14}C is greatly diluted by these two abundant isotopes, but its total amount in nature can be easily calculated. The decay constant of ^{14}C, calculated from its half-life, is $3.8 \cdot 10^{-12}$. The activity of modern carbon, because of its content of ^{14}C, is, as mentioned earlier, 16 counts (or decays) per minute per gram of carbon, which is equal to 0.27 decay per second per gram of carbon. In 1 g of carbon there are $5 \cdot 10^{22}$ atoms. There are, therefore, $0.27/5 \cdot 10^{22} = 5.4 \cdot 10^{-24}$ decays of ^{14}C per second per atom of total carbon. The concentration of ^{14}C in modern carbon is then $5.4 \cdot 10^{-24}/3.8 \cdot 10^{-12} = 1.4 \cdot 10^{-12}$. The mobile carbon reservoir on Earth, which includes the carbon in the hydrosphere and biosphere, amouts to $4.5 \cdot 10^{16}$ tons. The total amount of ^{14}C in this reservoir is therefore $4.5 \cdot 10^{16} \times 1.4 \cdot 10^{-12} = 63$ tons.

Table 3.7 lists the radioactive isotopes produced by cosmic rays in the Earth's atmosphere. As may be seen, some of them are exceedingly scarce. The record is held by two isotopes of phosphorus, ^{32}P and ^{33}P, for each of which there is a total of only half a gram in the entire world. But half a gram of radioactive phosphorus is still 10^{22} atoms. Most of radioactive phosphorus ends up in the surface water of the ocean, where it rapidly decays to stable ^{32}S and ^{33}S. Such minute quantities can be detected only because ^{32}P and ^{33}P are radioactive and emit β^- particles of reasonable energy (1.7 and 0.25 MeV, respectively). Each counter click represents *one* atom. Radiochemistry, therefore, can "see" individual atoms. Standard wet chemistry is lucky if it can detect a concentration of 10^{-10}, which still means something like 10^{11} to 10^{13} atoms!

[The ability of nuclei to capture neutrons produced by cosmic-ray particles is measured in *barns*. A barn (b) is a unit of surface, equal to $(10^{-14} \text{ m})^2$ or 10^{-28} m^2. It is the apparent cross section of a nucleus seen from the point of view of an incoming "thermal" neutron (a thermal neutron is a neutron at room-temperature, traveling at a speed of a few kilometers per second). If a thermal neutron zips by too fast, it may not even "see" a nucleus. But if it moves more slowly, it has a better chance to be captured. Some nuclei are excellent neutron captors. The record is held by ^{135}Xe. which has a capture cross section of 2,600,000 barns equivalent to an apparent diameter of $(2,600,000 \cdot 10^{-28})^{1/2} = 1,612 \cdot 10^{-14}$ m, or about 10,000 times larger than its actual physical diameter. Apparently, a thermal neutron passing in the neighborhood of a ^{135}Xe nucleus feels an irresistible attraction and rapidly falls in. Cadmium-113, which represents 12.22% of the naturally occurring isotopes of cadmium, also has a respectable capture cross section (19,900 barns), and so it is a good neutron absorber. It is for this reason that cadmium rods are used in nuclear reactors to absorb neutrons in case the fission rate should become too high.]

Tritium Method

Tritium is hydrogen-3 (^3H). It has two neutrons and one proton in its nucleus. It decays by β^- into ^3He with a half-life of 12.33 years. Tritium is produced in the atmosphere by spallation of nitrogen and oxygen nuclei. Tritium becomes rapidly oxidized to water. Because hydrogen has two stable isotopes (^1H and ^2H) and oxygen has three stable isotopes (^{16}O, ^{17}O, and ^{18}O), there are nine isotopic species of water molecules (Table 3.8).

Before nations began to test fusion bombs, which produce a lot of tritium by the reaction $^2H + {}^1H = {}^3H$, the concentration of tritium in rainwater was about 10^{-18}. This concentration, 1 atom of tritium per 10^{18} atoms of ^1H, is called a *tritium unit* (T.U.). The concentration of tritium in the atmosphere increased to 2,200 T.U. in 1964, at the peak of hydrogen-bomb testing. Because tritium has a relatively short half-life (12.33 years), atmospheric tritium is rapidly returning to its natural equilibrium.

Tritium has been used to date groundwater and to study the vertical mixing of the ocean. The measurements are done by converting water to hydrogen by electrolysis and by determining the concentration of ^3H with a Geiger counter or by scintillation counting.

Table 3.8. Isotopic species of water molecules

Composition	Natural abundance (%)
$^1H^1H^{16}O$	99.732
$^1H^1H^{18}O$	0.200
$^1H^1H^{17}O$	0.038
$^1H^2H^{16}O$	0.030
$^2H^2H^{16}O$	0.020
$^1H^2H^{18}O$	0.006
$^1H^2H^{17}O$	0.001
$^2H^2H^{18}O$	0.0004
$^2H^2H^{17}O$	0.00008

The application of radiometric dating methods to a variety of materials (meteorites and lunar rocks, terrestrial rocks of all ages, and recent carbonates and deep-sea muds) has made it possible to construct a rather accurate geological time scale (Table A8) that traces the entire history of our planet. This led to determining the rates of the major geophysical phenomena, the major geochemical cycles, climate change, and the evolution of life and environment on Earth. The remainder of this book concerns those topics.

THINK

* You find that a rock has equal amounts of ^{238}U and radiogenic ^{206}Pb. How old is the rock?

* When metallic atoms join to form a solid, each atom releases one or two electrons to form the metallic bond. Generally, the lower the ionization energy of the free atom, the lower is the *work function* (the energy required for an electron to be photoemitted from the metal surface). Iron, cobalt, and nickel are metals widely used in industry. Do their electronic structures (Table A6 in Appendix A) suggest that they could be used for dynodes in photomultiplier tubes?

* If you want to drink water that contains as little tritium as possible, what kind of water should you drink?

* In radiocarbon dating, modern wood gives 16 cpm. The Shroud of Turin, which many believe was used to wrap the body of Jesus Christ after he was taken down from the cross, was radiocarbon-dated and found to be only 660 years old. From the radioactive-decay equation $N = N_0 e^{-\lambda t}$, where, in this case, $N_0 = 16$ and $t = 660$, calculate N, the number of counts per second that the Shroud gave.

III Cosmology

C OSMOLOGY is the oldest science. All ancient civilizations made observations, often with antonishing accuracy, on the objects in the sky and their behavior. The positions of the planets were recorded with respect to the backdrop of the fixed stars, which, fortunately, never seemed to change position and therefore provided a perfect grid for plotting planetary positions. That keen interest in the motions of the celestial bodies was prompted by the common belief that these bodies ruled human events—ancient astronomy was in fact astrology. It remained astrology through the Renaissance. Kepler himself believed in astrology and drew up horoscopes for his sponsor, the emperor Rudolf, and for potentates at Rudolf's court.

Cosmology is treated in five chapters:

Chapter 4   Birth of the Universe
Chapter 5   Measuring the Universe
Chapter 6   The Objects in the Sky
Chapter 7   The Solar System
Chapter 8   Earth in Space

Chapter 4 describes the Big Bang events, including inflation and the stabilization of matter as temperature decreased during the first 10 seconds. Chapter 5 starts with trigonometry and trigonometric functions, because angular parameters and measurements are fundamental in astronomy. Chapter 5 provides a discussion of the various ways in which celestial distances are measured and how the size of the visible universe is assessed. The basic principles of lens and mirror optics and the way these principles are applied to the construction of refracting and reflecting telescopes are also discussed. Chapter 6 deals with the structure and dynamics of the universe as we know it today, and Chapter 7 deals with the solar system and the recent discoveries by unmanned satellites. Chapter 8 deals with the motions of the Earth in space and explains the calendar. Knowledge of these topics is essential to understand both time and the seasons, which intimately affect our daily lives.

Part III should convince the reader that, given the initial conditions, the Big Bang and the subsequent evolution of the universe was an inevitable development and a foregone conclusion. Like Parts I and II, Part III is essential to understanding how Planet Earth works, which is the subject matter of Parts IV and V.

4

BIRTH OF THE UNIVERSE

How and when did the universe come into existence? We do have an idea of the when (between 10 and 20 billion years ago), but the how is another story. The universe is now expanding. If it is "closed," that is, if its density is greater than about $6.5 \cdot 10^{-30}$ g/cm^3 (which we do not know yet), it will slow down, stop, and then collapse over itself in a giant cosmocrunch, some 100 billion years in the future. All particles will collapse into a "singularity", a point of extremely high (perhaps infinite) density and extremely high (perhaps infinite) temperature. Space itself may collapse into that point. Inside that point, matter and energy will be indistinguishable—the ἄπειρον of Anaximander or the ylem of Gamow. It is a truly unimaginable situation: a point containing all the mass and energy of the universe, around which there may have been no space and for which no time passed. How long will this situation last? The question is meaningless: if time does not flow, an eternity, or no time at all, is all the same. What seems certain, however, is that this state is totally unstable and the ἄπειρον has to blow up. This may have been the situation 10–20 billion years ago—a point consisting of ἄπειρον. The ἄπειρον blew up, the rest is predictable.

For the first $3 \cdot 10^{-10}$ s, temperature was too high for matter to be stable—only radiation would be stable. During this time, temperature is given by

$$T = 1.5 \cdot 10^{-10} \times t^{-1/2} \qquad (1)$$

where T is the temperature is kelvins, and t is the time in seconds since the beginning. Equation (1) shows that at the beginning ($t = 0$), temperature was infinite. The events immediately after the ἄπειρον blew up proceeded at such a furious pace that they are best summarized in a table (Table 4.1). Is this all

true? Well, most of it may be true, but we have a big problem with the ἄπειρον and the data on the first line of this table, because nothing is known about the Planckian. The Planckian, the first aeon of cosmic time, which ranged from 0 to $5.390 \cdot 10^{-44}$ s, is defined as the time it takes for light to cross the *Planck length*, a fundamental unit of length equal to $(Gh/2\pi c^3)^{1/2} = 1.616 \cdot 10^{-35}$ m (G = gravitational constant, c = speed of light). In fact, we do not know if the fundamental constants that work in our world (c, h, k, G, etc.) even worked during that time. So things could have gone differently.

The Planckian was followed by the Gamowian, the second aeon of cosmic time, about which we know quite a bit. The Gamowian ranges from $5.390 \cdot 10^{-44}$ second after time zero to 4.6 billion years ago, when the solar system formed. It is the longest aeon of cosmic time. During this time, the universe has been expanding. The rate of expansion has been either increasing, constant, or decreasing. An increasing or constant rate leads to an open universe—a universe that will keep expanding forever. A decreasing rate leads to a closed universe, a universe that will fall back on itself and eventually end in a cosmocrunch.

[Expansion involves the continuous creation of space between celestial objects that are not sufficiently bound to each other gravitationally, namely galactic clusters. It is the space between galactic clusters that increases with time, not the distance between one galaxy and the next or the distance between stars within a galaxy. The present rate of expansion is given by the *Hubble constant*, which appears to be in the neighborhood of 18 km/s/10^6 light-years at the present time. For an open universe, the Hubble constant either increases with time or

Table 4.1. **The Big Bang events**

Time after beginning	Radius of universe (m)	Temperature (K)	Event
0	0	Infinite	Appearance of space, time, and energy
$5.390 \cdot 10^{-44}$ s	$1.6 \cdot 10^{-35}$	10^{32}	End of the Planckian
10^{-43} s	$3 \cdot 10^{-35}$	10^{31}	Gravity separates
10^{-35} s	$3 \cdot 10^{-27}$	10^{28}	Strong and electroweak forces separate
10^{-33}–10^{-32} s	$3 \cdot 10^{-27} \to 0.1$	$10^{22} \to 10^{27}$	Inflation
10^{-10} s	0.13	10^{15}	Electromagnetic and weak forces separate
10^{-9} s	0.4	$7.5 \cdot 10^{14}$	t quarks (mass ~ 50 u) stabilize
		$7.5 \cdot 10^{13}$	b quarks (mass $= 5$ u) stabilize
10^{-6} s	300	$1.3 \cdot 10^{13}$	c quarks (mass $= 1.8$ u) stabilize
		$3.3 \cdot 10^{12}$	s, d, and u quarks (masses 0.5–0.4 u) stabilize
		$\sim 10^{12}$	Protons and neutrons stabilize
10^{-3} s	300,000	$1.4 \cdot 10^{10}$	^{2}H nuclei stabilize (binding energy $=$ 1.7 MeV $= 0.002$ u)
10 s	$3 \cdot 10^{9}$	$4.1 \cdot 10^{9}$	Electrons (mass 0.00055 u) stabilize
100 s	$3 \cdot 10^{10}$	$1.5 \cdot 10^{9}$	Stabilization of ^{3}He and ^{4}He nuclei
800,000 y	$6.6 \cdot 10^{21}$	3,000	Electrons captured by nuclei; formation of H and He atoms and of H_2 molecules. The universe becomes transparent to light. Relative abundances 74% H, 26% He by mass or 92% H, 8% He by number of atoms. Small amounts of Li and Be.

remains constant; for a closed universe, the Hubble constant decreases with time, reaches zero, and then becomes negative.]

During the earliest Gamowian, the four forces of nature were part of a single "superforce". If order to see how that could be, consider gravitation, the weakest of the four forces of nature. The mass of a particle increases with increasing temperature because the speed of the particle increases. In fact, the mass of even the smallest particle would reach infinity if the particle could be accelerated to the speed of light. At the very high temperatures ($> 10^{31}$ K) prevailing at time $t = 10^{-43}$ s, the mass of the particles was so large (because of their speed) that the gravitational force between the particles was as strong as the strong force. The other two forces, the electromagnetic and weak force, which are intermediate in strength between gravity and the strong force, follow suit. As a result, above 10^{31} K, the four forces are indistinguishable.

As the universe expanded, temperature decreased, which caused the mass of the particles

to decrease and gravitational interaction to weaken. Gravity, therefore, was the first force to appear as a separate force. Next to separate was the strong force, followed by the electroweak force (which soon split into the electromagnetic force and the weak force).

According to current theory, a period of *inflation* occurred between 10^{-33} and 10^{-32} (Table 4.1), during which time the radius of the universe grew from 10^{-33} light-seconds ($= 3 \cdot 10^{-25}$ m) to 10 cm. The increase in radius from practically zero to 10 cm in $9 \cdot 10^{-33}$ s resulted in a speed of radial expansion of 10^{32} m/s. This is $3 \cdot 10^{23}$ times faster than the speed of light. Relativity was not violated, however, because inflation did not involve transmission of signals or information, only an increase in space.

Notice that the greatest distance at which a point inside a 10-cm sphere could see would be only $3 \cdot 10^{-24}$ m, the maximum distance that light could travel within the 10^{-32} seconds since the beginning. The volume of a spherule with a radius of $3 \cdot 10^{-24}$ m is $1.1 \cdot 10^{-70}$ m^3, while the volume of the 10-cm sphere is $4.2 \cdot 10^{-3}$ m. This sphere, therefore, contains $4.2 \cdot 10^{-3}/1.1 \cdot 10^{-70} = 3.8 \cdot 10^{67}$ spherules. According to the inflation theory, the larger sphere, and all the spherules inside it, have since expanded, until our spherule has reached its present radius of about 15 billion light-years. This is what we call the *visible universe*. All the other spherules have also expanded and have become universes of their own, within which objects are in mutual, visible contact. Our universe would then be only one among $3.8 \cdot 10^{67}$ ($= 6.3 \cdot 10^{43}$ Av) other universes that we will never be able to see, or know anything about, because they started out too far apart from us, within the original sphere for their light to ever reach us. As the story goes, after 10^{-32} s, the inflated universe resumed its expansion at the rate given by the Hubble constant.

Inflation may seem an unnecessary complication in an otherwise smooth scheme, but it is invoked by astrophysicists to explain several things that the standard model (i.e., the model shown in Table 4.1, minus inflation) finds difficult to explain:

1. The fact that the spectrum of the microwave background radiation (the leftover of the Big Bang that still pervades the universe today—see Chapter 6) is the same in all directions in space if we neglect the small red shift–blue shift in opposite directions that is due to the motion of the solar system in space

2. The fact that the average amount of matter in space is the same everywhere on a large enough scale (a cubic billion light-years or so) but is irregularly distributed on a smaller scale (forming stars, galaxies, galactic clusters, and filaments)

3. The fact that certain massive particles called *monopoles* seem to be missing

4. The fact that the average density of matter in the universe ($\sim 0.9 \cdot 10^{-30}$ g/cm^3) seems rather close to the limit between an open universe (a universe that keeps expanding forever) and a closed universe (a universe whose self-gravitation is sufficient to stop the expansion, reverse it, and produce a final cosmocrunch)

5. The fact that the universe contains matter, not just radiation

   [The density limit between an open universe and a closed universe is called the *critical density* (symbol ρ_c) and is given by

   $$\rho_c = 3H_0^2/8\pi G$$

   where H_0 = Hubble constant, and G = gravitational constant. The critical density, therefore, depends only on the rate of expansion—the higher the rate of expansion, the greater the critical density. For $H_0 = 18$ km/s/10^6 l.y., the critical density is $6.5 \cdot 10^{-30}$ g/cm^3. The observed average density of the universe appears to be $0.9 \cdot 10^{-30}$ g/cm^3, which would indicate an open universe.]

Let us now consider, one by one, the five points listed above. The uniformity of the spectrum of the microwave radiation (see Fig 6.1) in all directions in space implies that the earliest universe had the same temperature everywhere. Opposite, distant regions of our present, visible universe, however, whose light is just now reaching us, have been cut off from each other since the beginning, because, according to the standard model, the universe started expanding at the speed of light immediately, at time $t = 0$; there was no time to reach thermal equilibrium. In contrast, the inflation model says that between time $t = 0$ and time $t = 10^{-35}$ s, the universe was small enough for light to reach across it and thus ensure thermal equilibrium.

Thermal equilibrium across the earliest universe suggests that radiation density was also uniform, except for exceedingly minute random fluctuations. As a result, when matter later condensed out of the radiation, its distribution on a large scale in space was also uniform. The minute, random density fluctuations were magnified by inflation to form small-scale fluctuations in the density of matter, which resulted in the accumulation of stars and galaxies. In the standard model, magnification would be far too slow to account for the stars and galaxies that we see all around us.

Monopoles are massive particles (about 10^6 u) consisting of individual N or S magnetic poles. According to theory, monopoles should have formed in great numbers in the early universe, but they are not seen today, even though they should be readily detectable because of their large mass. The inflation model says that there may be as many as 10^{67} different universes, so that any one universe may contain only a few monopoles. This would make their discovery very difficult, if not practically impossible.

The inflation model explains the "flatness" of the universe, that is, the fact that the average density of the universe is close to the critical density, as a result of the process of inflation itself. The universe before inflation was ultradense, and therefore its space-time was highly curved. The enormous inflation that took place "flattened out" space-time (just as inflating a balloon flattens out its surface), bringing the average density of the universe close to the critical value.

The inflation model also explains the existence of matter in the universe, by means of the X interaction (Chapter 3). Between the end of the Planckian (time $t = 5.390 \cdot 10^{-43}$ s, $T = 10^{33}$ K) and the beginning of inflation ($t = 10^{-33}$ s, $T = 10^{28}$ K), energy was transforming into matter + antimatter, and vice versa, because neither matter nor antimatter was stable at those high temperatures. When the universe inflated, temperature dropped, facilitating the transformation of energy into matter + antimatter. In addition, matter can transform into antimatter, and antimatter into matter, by the X interaction. As we saw in Chapter 3, the transformation of antimatter into matter is slightly favored over its reverse, so that $10^9 + 1$ quarks were created for each 10^9 antiquarks. Quarks and antiquarks did not stabilize until time $t = 5 \cdot 10^6$ s, when temperature dropped below $3 \cdot 10^{12}$ K. At that time, nearly all matter and antimatter had disappeared back into radiation, with only one particle of matter in a billion surviving (this explains the observation that the ratio of material particles to photons in the visible universe is $1/10^9$). The matter that survived is the matter that forms all the objects that we see in the visible universe.

Matter is stable when temperature is below its energy equivalent divided by $1.5\,k$ ($k =$ Boltzmann constant). Following inflation, the first to stabilize were the quarks. The masses of the quarks (referred to the free state) are shown in Table 3.1. The first to stabilize were the top quarks, which have the greatest mass and therefore the greatest energy:

mass of top quark = 50 u
$$= 50 \times 1.66 \cdot 10^{-27}$$
$$= 8.3 \cdot 10^{-26}\ \text{kg}$$
$$= 8.3 \cdot 10^{-26} \times c^2$$
$$= 7.5 \cdot 10^{-9}\ \text{J}$$

The temperature at which this energy transforms into mass, and vice versa, is

$$T = E/1.5\,k$$
$$= 7.5 \cdot 10^{-9}/2.1 \cdot 10^{-23}$$
$$= 3.6 \cdot 10^{14}\ \text{K}$$

Next to stabilize was the bottom quark:

mass of bottom quark = 5 u
$$= 5 \times 1.66 \cdot 10^{-27}$$
$$= 8.3 \cdot 10^{-27}\ \text{kg}$$
$$= 8.3 \cdot 10^{-27} \times c^2$$
$$= 7.5 \cdot 10^{-10}\ \text{J}$$
$$T = 7.5 \cdot 10^{-10}/2.1 \cdot 10^{-23}$$
$$= 3.6 \cdot 10^{13}\ \text{K}$$

Similarly, the strange quark stabilized at $3.7 \cdot 10^{12}$ K, and the u and d quarks stabilized at $3 \cdot 10^{12}$ K.

The quarks became confined and formed protons, neutrons, and mesons when the temperature dropped to about 10^{12} K. The binding energy of ^{2}H is 1.7 MeV, which indicates stabilization at a temperature of about 10^{10} K. That was followed by the stabilization of ^{3}H and ^{4}He nuclei and of the electron. And that was the end of particle formation, because by then the particles were too far apart and the energy too low to form other nuclei. Besides, as we know from Part II, ^{5}He is so unstable that it blocked the formation of heavier elements.

It is important to realize that the foregoing scenario is not the result of guesses (except for the Planckian), but the result of calculations. The fact that the element abundances that result (74% H and 26% He) are precisely those observed in stars and the interstellar matter gives credence to the story.

When the temperature had decreased to about 3,000 K, 800,000 y after the Big Bang, the energy was low enough for atoms to become stable. Until then the electrons had remained free, giving the universe the appearance of a luminous fog (like the gas discharge inside a neon light). When the electrons were captured, the universe became transparent to radiation. There was nothing to see, however, because density had already decreased to

$1.6 \cdot 10^{-17}$ g/cm^3, stars had not yet formed, and radiation had a wavelength around $3.2\,\mu$m, which is in the region of the infrared (not visible).

As the universe continued expanding, irregularities in the distribution of energy in the preinflation universe led to the formation of gravity centers. A gravity center attracts mass, and therefore it strengthens itself (positive feedback), leading to the formation of stars and the grouping of stars into galaxies and of galaxies into galactic clusters. Many details remain to be worked out, of course, but it seems that given the initial conditions (the ἄπειρον), the evolution of the universe as an expanding system of galaxies, stars, and planets was inevitable. Also inevitable is the conclusion that given the Titus-Bode law (which suggests that planetary distances from the central body are not random—see Chapter 7), a planet like the Earth is likely to exist in most, if not all, planetary systems.

This means that humans may not be alone in the universe. In fact, consider the following:

number of galaxies in the visible universe = 10^{11}
number of stars in an average galaxy = 10^{11}

Therefore,

number of stars in the visible universe = 10^{22}

of which 16% are spectrally similar to the Sun. Therefore,

number of stars similar to the Sun = $0.16 \cdot 10^{22}$
$= 1.6 \cdot 10^{21}$

Of these, 50% are double stars and the other 50% are single stars, most of which probably have a planetary system. Therefore,

number of planets like Earth in the visible
universe = $1.6 \cdot 10^{21} \times 0.5$
$= 0.8 \cdot 10^{21}$

As noted above, each of these planetary systems should have a planet like the Earth. The average age of a star like the Sun in the visible universe is probably similar to that of the Sun ($5 \cdot 10^9$ y). The human brain achieved its present size about 125,000 years ago. Assuming that *Homo sapiens* will soon become extinct (see Chapter 24), we have:

average time window for intelligent planetary
life = $125{,}000/5 \cdot 10^9$ y
$= 2.5 \cdot 10^{-5}$

Assuming that each planet has the same population as the Earth ($5 \cdot 10^9$ people), we have:

number of people in the visible
universe = $0.8 \cdot 10^{21} \times 5 \cdot 10^9 \times 2.5 \cdot 10^{-5}$
$= 1 \cdot 10^{26}$

No need to feel lonely.

THINK

* Calculate the frequency and wavelength of a photon whose energy would be the equivalent of the mass of your brain.

* Using your phone book, estimate the percentage of lawyers in your community. Assuming that the same percentage holds for the entire "human" population in the visible universe (see above) calculate how many avogadros of lawyers there are in the universe.

MEASURING THE UNIVERSE

5.1. TRIGONOMETRIC FUNCTIONS

Now that we know how the universe came about (or at least have a model to describe what may have happened), can we determine how large the universe is? how far away are the Moon, the Sun, the planets, the stars, the galaxies? Before tackling the measure of the universe (Section 5.2), we must develop a basic understanding of trigonometric functions. *Trigonometry* means *the measure of the triangle*. It is the most basic method for measuring distances, other than using a rod.

Trigonometry is based on the Pythagorean theorem, formulated by Pythagoras (ca. 582–507 B.C.E.). This theorem says, as everybody knows, that the sum of the squares of the sides of a right triangle is equal to the square of the hypotenuse (Fig. 5.1A). In fact (Fig. 5.1B), the area of the outer square $(a + b)^2$ is equal to the area of the inner square (c^2) plus the areas of the four triangles between the outer and inner squares, each of which is equal to $\frac{1}{2}ab$. Therefore,

$$(a + b)^2 = c^2 + (\tfrac{1}{2}ab) = c^2 + 2ab$$

but

$$(a + b)^2 = a^2 + 2ab + b^2$$

Therefore,

$$a^2 + 2ab + b^2 = c^2 + 2ab$$

Subtracting $2ab$ from both sides, we have

$$a^2 + b^2 = c^2$$

Q.E.D. (meaning *quod erat demostrandum*, Latin for *which was to be demonstrated*).

Consider now a circle. The equation of the circle is

$$x^2 + y^2 = r^2 \tag{1}$$

or

$$y^2 = r^2 - x^2 \tag{2}$$

where r is the radius of the circle. We can verify this by choosing an arbitrary value for r (e.g., $r = 1$) and solving for y using different values of x. For $r = 1$, equation (2) becomes.

$$x^2 + y^2 = 1$$

from which we obtain

$$y^2 = 1 - x^2$$

If we replace x with successive values, such as 0.1, 0.2, 0.3, and so forth, we can find the corresponding values of y by solving the equation. A plot of these values for x and y on graph paper will roughly form a circle. The circle can be better approximated if the number of values for x is increased. The vertical and horizontal axes passing through the center of the circle are called *Cartesian axes*, after the French philosopher and mathematician René Descartes (1596–1650).

Because we have worked out this graph by varying the value of x and calculating the value of y, x is called the *variable* and y is called the *function*. The value of the function depends on the value of the variable. In this case, we could also have done the opposite; that is, we could have chosen different values for y and calculated the corresponding values of x. If so, y would be the variable, and x would be the function.

Often, such exchange is not possible. In the inverse square law, for instance, the power received from a source (a light, for instance) is the function, and the distance is the variable. If we change the distance, the power received also will change. The law tells us *how* the function changes (inversely to the square of the distance).

The equation for the circle is based on the Pythagorean theorem (Fig. 5.2). For each point on the circle there is a corresponding value of

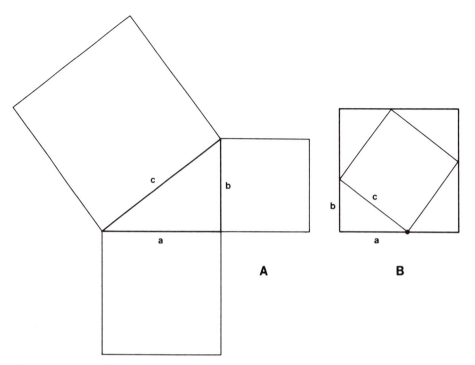

Figure 5.1. (A) The Pythagorean theorem: The sum of the squares of the sides of a right triangle is equal to the square of the hypotenuse $(a^2 + b^2 = c^2)$. (B) Demonstration of the Pythagorean theorem (see text).

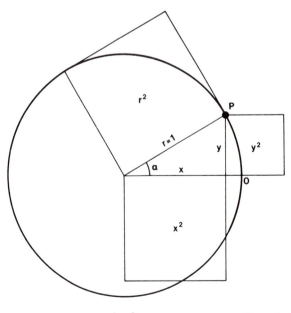

$x^2 + y^2 = r$

$x = \cos \alpha$

$y = \sin \alpha$

Figure 5.2. The Pythagorean theorem applied to a unitary circle (radius = 1): $y = \sin \alpha$; $x = \cos \alpha$. Therefore, $\sin^2 \alpha + \cos^2 \alpha = 1$. (Emiliani, 1988, p. 111, Fig. 8.2.)

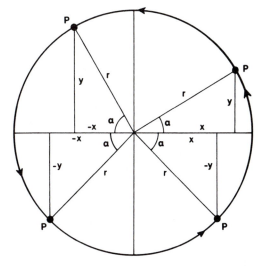

Figure 5.3. As point *P* rotates counterclockwise around the circle, sin α assumes values ranging from 0 (α = 0) to + 1 (α = 90°), back to 0 (α = 180°), on to − 1 (α = 270°), and then again back to 0 (α = 360° = 0°). At the same time, cos α ranges from + 1 to 0, to − 1, back to 0, and on to + 1. (Emiliani, 1988, p. 112, Fig. 8.3.)

coordinate *x* is called the cosine (abbreviated cos) of the angle formed by + x and *r* (Fig. 5.3). The ratio sine/cosine is called the *tangent* (abbreviated tan) of the given angle. It gives the slope of the radius.

There are additional trigonometric functions, as shown in Fig. 5.4 and more completely in Table 5.1. If we plot the values of the sine as the point runs around the circle, we generate a *sinusoidal curve* that starts at 0, increases to + 1, decreases back to 0, further decreases to − 1, increases back to 0, and keeps doing the same over and over as the point keeps running around the circle (Fig. 5.5). Each trip around the circle is a *cycle*, and the number of cycles per second is the *frequency* of the revolving point. If we do the same for the cosine, we obtain a similar curve, which starts at the value of 1 because cos 0° = 1 (Fig. 5.3). This, too, is a sinusoidal curve.

the coordinates *x* and *y*. Let us consider again a circle with radius = 1. If we take a point with coordinates *x* = 1 and *y* = 0 and make it go around the circle *counterclockwise* (Fig. 5.3), we see that the coordinates *x* and *y* form an angle that increases from 0 (*x* = 1, *y* = 0) to 90° (*x* = 0, *y* = 1), to 180° (*x* = − 1, *y* = 0), to 270° (*x* = 0, *y* = − 1) to 360° = 0° (*x* = 1, *y* = 0). Coordinate *y* is called the sine (abbreviated sin) and

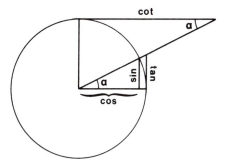

Figure 5.4. Trigonometric functions.

Table 5.1. **Trigonometric functions**

Name	Symbol	Definition	Formula (Fig. 5.2)
sine	sin	the ordinate of the endpoint of an arc of a unit circle starting at 3 o'clock and drawn counterclockwise	y/r
cosine	cos	the absicissa of the endpoint of an arc of a unit circle starting at 3 o'clock and drawn counterclockwise	x/r
tangent	tan	the ratio sine/cosine	y/x
cotangent	cot	the ratio cosine/sine	x/y
secant	sec	the inverse of the cosine (1/cos)	r/x
cosecant	csc	the inverse of the sine (1/sin)	r/y
versine	vers	the difference 1 − cos	$1 - x/r$
coversine	covers	the difference 1 − sin	$1 - y/r$

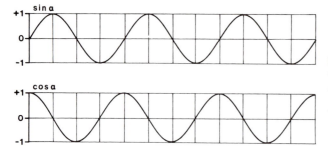

Figure 5.5. The upper curve represents the values of sin α as point P revolves counterclockwise around the circle, starting from the 3 o'clock position (Fig. 5.3). The lower curve represents the corresponding values for cos α. (Emiliani, 1988, p. 113, Fig. 8.4.)

Sinusoidal curves can be used to describe many cyclic phenomena. Consider, for example, two identical masses free in space and connected by a spring. If the spring is compressed and then released, the two masses will oscillate toward and away from each other. This type of oscillatory motion is what molecules perform if they consist of two identical atoms, such as $^1H_2$, $^{14}N_2$, and $^{16}O_2$. It is called *harmonic motion* and is represented by the equation

$$x = A \cos \omega t$$

where x is the displacement, A is the amplitude of the displacement, ω is the *angular frequency* (number of cycles per second times 2π), and t is time. The motion is represented by a curve similar to that shown in Fig. 5.5.

The circle is part of a family of curves called *conic sections* because each of the conic sections is outlined by the intersection of a plane cutting across a cone at a suitable angle (Fig. 5.6). The other members of the family are the ellipse, the parabola, and the hyperbola. If the angle of intersection is 90° to the axis of the cone, the conic section will be a circle (Fig. 5.6). If the plane cutting the cone is tilted, a series of ellipses will be cut by the plane. When the tilt becomes parallel to the side of the cone, the plane will cut an open curve called a *parabola*. Increasing the tilt of the plane will produce a series of more open curves called *hyperbolae*.

The ellipse is an elongated circle, with two centers called *foci* (singular *focus*). The ellipse is described by the equation

$$x^2/a^2 + y^2/b^2 = 1$$

where a is the semimajor axis and b is the semiminor axis (Fig. 5.7). If $a = b = 1$, that is, if the two axes are equal, this equation will reduce to that of the circle.

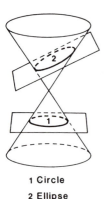

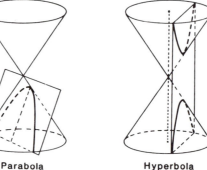

1 Circle

2 Ellipse

Parabola

Hyperbola

Figure 5.6. Conic sections. A cone is an open geometric surface that consists of two *nappes*, with a common vertex. The intersection of a plane with the cone normal to the cone's axis (other than through the vertex) forms a circle. At increasing inclinations of the plane, the intersections form a series of ellipses. When the plane is parallel to the side of a nappe, the intersection forms a parabola. At greater inclinations, the intersections form a series of hyperbolae. The circle, ellipse, parabola, and hyperbola are called *conic sections*. (Emiliani, 1988, p. 115, Fig. 8.5.)

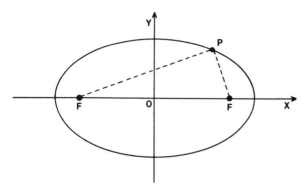

Figure 5.7. In an ellipse, the sum of the distances of any point *P* on the ellipse from the two foci (*F*) is constant. *Ellipticity* is the difference between semimajor and semiminor axes divided by the semimajor axis. The farther apart the two foci of an ellipse, the greater its ellipticity. (Emiliani, 1988, p. 115, Fig. 8.6.)

The parabola is a curve that may be described by the equation

$$x^2 = Ay$$

where *A* is a constant that determines how open the parabola is. This equation gives

$$y = x^2/A$$

If we assign to *A* any value (0.5, 2, or whatever) and then calculate *y* for a series of positive and negative values of *x*, we get a parabola (Fig. 5.8).

The hyperbola (Fig. 5.9) is a double, open curve that may be described by the equation

$$xy = A$$

or

$$y = A/x$$

where *A* is a constant that determines how open the hyperbola is. In the ellipse, the *sum* of the distances of a point on the ellipse from the two foci is constant; in the hyperbola, the *difference* remains constant.

Distances can be measured by triangulation. The distance of a point *C* from a point *A* can be determined by measuring the length *b* of a *baseline* between points *A* and *B* and by measuring the angles α and β (Fig. 5.10).

Surveyors use theodolites to range features on the surface of the Earth to be mapped. A theodolite consists of a small telescope mounted on a vertical graduated circle that rotates on a horizontal graduated circle (Fig. 5.11).

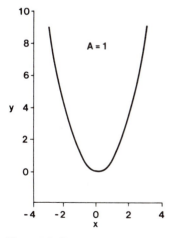
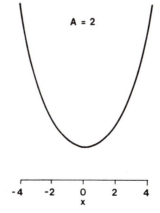

Figure 5.8. Two parabolas representing the equation $y = x^2/A$, one with $A = 1$ and the other with $A = 2$. The greater *A* is, the more open the parabola is. (Emiliani, 1988, p. 115, Fig. 8.7.)

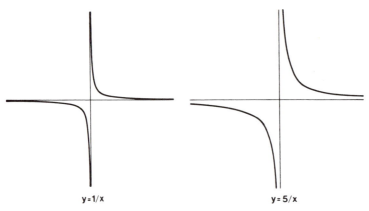

$$y=1/x \qquad\qquad y=5/x$$

Figure 5.9. Two hyperbolas representing the equation $y = A/x$. Notice that the greater A is, the more open the hyperbola is. (Emiliani, 1988, p. 116, Fig. 8.8.)

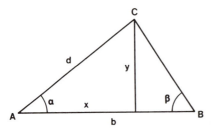 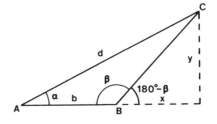

Figure 5.10. Triangulation. The triangles are solved as follows. Left: $y/x = \tan \alpha$; $y/(AB - x) = \tan \beta$; $d = (x^2 + y^2)^{1/2}$. Right: $y/x = \tan(180° - \beta)$; $y/(AB + x) = \tan \alpha$; $d = [y^2 + (AB + x)^2]^{1/2}$. The first two equations for each triangle yield both x and y; the third equation yields the distance d of C from A.

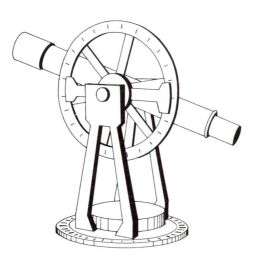

Figure 5.11. A theodolite consists of a small telescope mounted on a vertical graduated circle, which in turn is mounted on a horizontal graduated circle. Horizontal as well as vertical angles can thus be measured.

Vertical and horizontal angles can thus be measured. Standard theodolites can measure angles as small as one minute (1′) of arc and therefore can range objects up to a distance of 340 km with a baseline of only 100 m. Good theodolites can read one second (1″) of arc and therefore can range 60 times farther away using the same baseline. Precision, of course, increases with increasing length of the baseline.

The largest geocentric baseline is the diameter of the Earth; the largest baseline available to us is the diameter of the Earth's orbit. Modern celestial instruments can routinely measure angles as small as $0.004″ = 1.11 \cdot 10^{-6}$ degree of arc, which means that objects as far away as $7.7 \cdot 10^{15}$ km ≈ 800 light-years can in theory be ranged. The error, however, renders unreliable measurement of distances greater than about 300 l.y.

Astronomers determine the parallax (Fig. 5.11) of nearby objects (planets or the

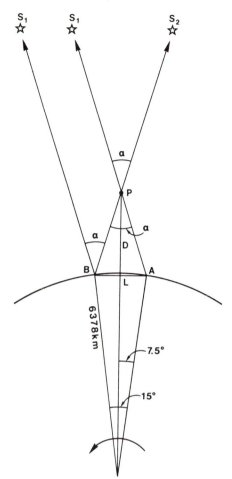

Figure 5.12. Determination of distance using parallax. Parallax is the angle α subtending the baseline L. Point P, stationary above the equator, is sighted from A with reference to star S_1. In 1 hour the Earth rotates 15°, moving the observer from A to B. The length L of the chord which forms the baseline, is equal to $2 \sin 7.5° \times 6,378 = 1,665$ km, half of which (832.5 km) is the parallax of P. The observer, now at B, sights point P against star S_2 and measures the angle α between stars S_1 and S_2. This angle is found to be equal to 40°. Because of the great distances to the stars, A-S_1 is parallel to B-S_1. The angle $S_1BS_2 = 40°$ is equal to the angle BPA. The distance D from P to the center of the chord ($= \cos\frac{1}{2}\alpha = \cos 20°$) is equal to 832.5 km divided by $\tan 20°$, or 2,287 km. The distance to the surface of the Earth above the center of the chord is equal to 2,287 km less 108 km [$= 832.5 \times$ tangent of the angle that the distance from the Earth's surface to the center of the chord subtands from A or B ($= 7.5°$ in this case)].

nearer stars) by measuring the angle between the apparent positions of the object when viewed from two different locations against the backdrop of stars so distant as to be considered at infinite distance. As Fig. 5.12 shows, this angle is the same as that subtending the baseline from the object. (Parallax is the apparent change, when the observer moves, in the position of a nearby object with respect to a distant frame of reference.)

The measured angle, referred to the equatorial diameter of the Earth, is called *horizontal parallax*, because the object is shown as lying on the horizon of point A (Fig. 5.13). If it is referred to the radius of the Earth's orbit, it is called *heliocentric parallax*. The distance from which the average radius of the Earth's orbit ($= 1$ astronomical unit, AU) is subtended

by 1″ of arc is called a parsec (parallax-second, pc) and is equal to 3.26 light-years. The heliocentric parallax in seconds of arc gives directly the distance in parsecs.

THINK

* The circle, ellipse, parabola, and hyperbola are called conic sections because they can be obtained by cutting a cone with a plane. Which other geometric figures can one obtain by cutting a cone with a plane?

* In a triangle shaped like that on the left in Fig. 5.10, $\alpha = 2.86°$, $\beta = 3.14°$, and b = 1 km. Calculate d. In a triangle shaped like that on the right in Fig. 5.10, $\alpha = 2.14°$, $\beta = 85.20°$, and b = 1 km. Calculate d.

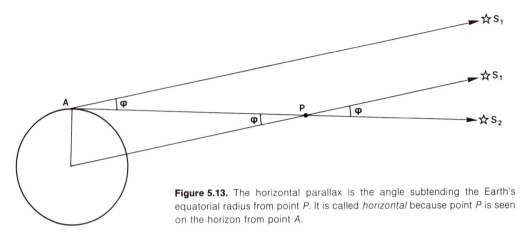

Figure 5.13. The horizontal parallax is the angle subtending the Earth's equatorial radius from point *P*. It is called *horizontal* because point *P* is seen on the horizon from point *A*.

5.2. ANCIENT ASTRONOMY

The ancient Greeks, who did so much to get Western civilization going, tried their best to determine the size of the Earth, the Moon, and the Sun. They succeeded handsomely with the Earth, barely made it with the Moon, and completely failed with the Sun. The first to measure the Earth was Eratosthenes (ca. 276–196 B.C.E.), a Greek mathematician, astronomer, and geographer who was educated in Athens but spent most of his adult life in Alexandria as chief librarian of the great library in that city. Eratosthenes measured with great accuracy the angle between the Earth's axis and the normal to the Earth's orbital plane and found it to be 23.5°, which is the latitude of the tropics.

Having then heard that at noon of the summer-solstice day the rays of the Sun reached the bottom of water wells at Syene (modern Aswân), he concluded that the Sun stood vertical over Syene and that Syene was therefore on the tropic. Eratosthenes thought that the Sun was so large and so distant that its rays would reach the Earth parallel to each other, and so they would reach Alexandria at a slant, because Alexandria was further north (Fig. 5.14).

When the solstice day came, Eratosthenes measured at high noon the angle that the Sun's rays made with the vertical. He probably climbed on top of a temple or something, dropped a plumb line from the edge of the roof, and had an assistant measure the length of the shadow of the line when it became

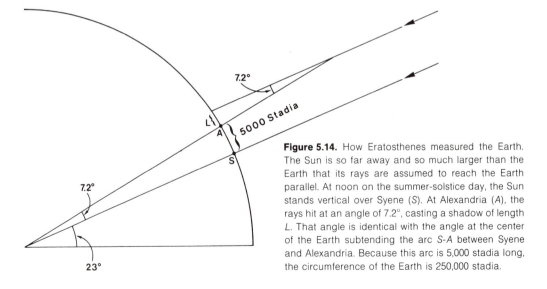

Figure 5.14. How Eratosthenes measured the Earth. The Sun is so far away and so much larger than the Earth that its rays are assumed to reach the Earth parallel. At noon on the summer-solstice day, the Sun stands vertical over Syene (*S*). At Alexandria (*A*), the rays hit at an angle of 7.2°, casting a shadow of length *L*. That angle is identical with the angle at the center of the Earth subtending the arc *S-A* between Syene and Alexandria. Because this arc is 5,000 stadia long, the circumference of the Earth is 250,000 stadia.

shortest (indicating local noon). Eratosthenes then divided the length of the shadow into the height of the temple and obtained the tangent of the angle. That turned out to be 0.126, which corresponds to an angle of 7.2° (Fig. 5.14). The distance between Alexandria and Syene was 5,000 stadia. Eratosthenes reasoned that if the angle of 7.2° subtended 5,000 stadia, the angle of 360° (i.e., the polar circumference of the Earth) would correspond to

$$(5,000/7.2) \times 360 = 250,000 \text{ stadia}$$

The stadion was an ancient measure of length. The Olympian stadion was 600 Greek feet. The Greek foot was 31.6 cm, and so the Olympian stadion was 189.6 m. In Egypt, distances were measured in σχοῖνοι, each σχοῖνος measuring 12,000 royal cubits. A royal cubit was 0.524 m, which made one σχοῖνος equal to 6.288 km. There was also a surveyor stadion, which was 1/40 of the σχοῖνος and therefore 157.2 m long. This was the stadion used by Eratosthenes, which made the circumference of the Earth equal to 250,000 × 157.2 m = 39,300 km, and the radius equal to 250,000/2π = 39,789 stadia or 6,255 km. The modern values are very close: 40,008 km and 6,371 km, respectively.

Poseidonius (135–50 B.C.E.), a Greek philosopher and scientist, observed that the bright star Canopus, when highest in the sky, was barely visible on the horizon at Rhodes, while at the same time it was 7.5° above the horizon at Alexandria. He took the distance between Rhodes and Alexandria to be 3,000 stadia (probably Olympian stadia, making the distance equal to 569 km) and concluded that the polar circumference was 144,000 stadia, or 27,300 km. Although Poseidonius' value of 7.5° was 2.5° larger than it should have been, he had corrected his observations for the refraction of light through the denser layers of the air near the horizon (Fig. 5.15). That so impressed the great Greek astronomer Ptolemy (ca. 100–170 C.E.) that he endorsed Poseidonius' value rather than Eratosthenes'. Based on that endorsement, Poseidonius' value remained unchallenged for 1,500 years. One degree on a great circle (i.e., a circle of maximum size around the Earth) was thought to be 27,300/360 = 75.8 km or 40.9 nautical miles. Martin Behaim's globe of 1492 (Fig. 5.16), made just before Christopher Columbus discovered the Americas, showed the distance between Tenerife in the Canary Islands

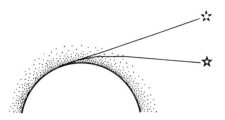

Figure 5.15. The lower atmosphere is denser than the atmosphere higher up, and as a result, a light ray is bent, making the source appear higher on the horizon than it actually is.

(the westernmost Spanish outpost) and Japan to be 90° of longitude, equal to about 3,300 nautical miles at the latitude of 26°N along which Columbus sailed. It would be possible, claimed Columbus, to reach Japan by sailing west. He was lucky to be intercepted by the Americas, for he would never have made it across the vast Pacific.

Hipparchos (ca. 190–120 B.C.E.) was the first to determine the distance to the Moon. He measured its parallax by observing its position with respect to the distant stars at two different times, perhaps an hour or so apart. The parallax, corrected for the Moon's own motion across the sky (0.5°/hour) was 0.2°. In one hour the Earth turns 15°. According to Eratosthenes, 15° along the surface of the Earth are equal to 10,416 stadia. However, Hipparchos was living in Rhodes, which is at a latitude of 36.4°N. The circumference of the parallel at the latitude is equal to cos 53.6° (the colatitude of 36.4°) multiplied by the Earth's circumference (Fig. 5.17):

$$\sin 53.6° \times 250,000 = 0.805 \times 250,000$$
$$= 201,250 \text{ stadia} = 360°$$

Therefore, 15° = 8,385 stadia, and $\frac{1}{2}(15°) = 7.5° = 4,193$ stadia. Referring to Fig. 5.12, where P is the Moon, we have

$$
\begin{aligned}
PA &= 4,193/\sin\tfrac{1}{2}\alpha \\
&= 4,193/0.00174 \\
&= 2,402,413 \text{ stadia} \\
D &= \cos\tfrac{1}{2}\alpha/2,402,413 \\
&= 2,402,409 \text{ stadia} \\
&= 377,659 \text{ km}
\end{aligned}
$$

This compares favorably to the modern value of 384,401 km. (Notice that because of the great distance to the Moon compared with the width of the baseline, PA and D are essentially identical.)

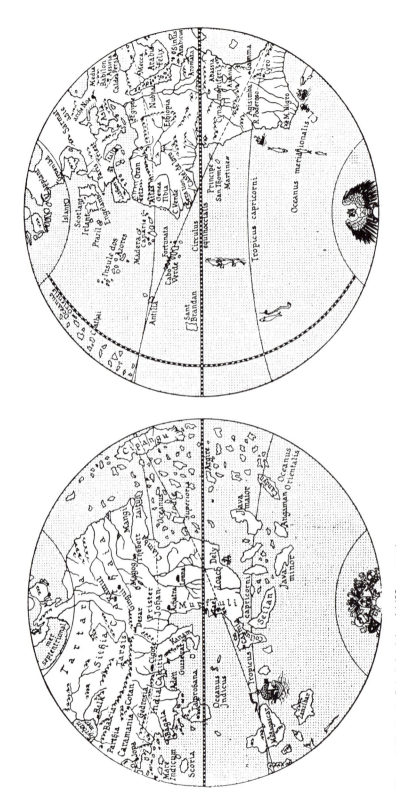

Figure 5.16. Martin Behaim's globe of 1492, summarizing the knowledge current just before Columbus sailed. Japan (Cipangu) and the outer islands of China (Cathaia) look temptingly close. (*Encyclopaedia Britannica*, 1952, 14th ed., vol. 14, p. 842, Fig. 9.)

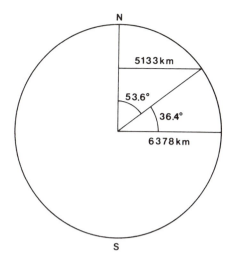

Figure 5.17. The radius of the parallel at the latitude of Rhodes (36.4°N) is equal to the cosine of 36.4° (or the sine of 53.6°, the angle that is complementary to 36.4°). (The sum of two *complementary* angles is equal to 90°; that of two *supplementary* angles is 180°.)

Knowing the distance to the Moon and observing that the subtended diameter of the Moon is 0.5°, Hipparchos could also determine the size of the Moon:

$$
\begin{aligned}
\text{radius of Moon} &= \sin 0.25° \times 2,402,409 \\
&= 0.00436 \times 2,402,409 \\
&= 10,482 \text{ stadia} \\
&= 1,648 \text{ km}
\end{aligned}
$$

This compares well to the modern value of 1738.2 km.

The Greeks failed to determine the distance of the Sun and its size because they could not measure its parallax—when the Sun is up, there are no stars against which to gauge the Sun's position. Hipparchos assumed a maximum solar parallax of 7′ (the minimum observable at the time). The sine of 7′ is 0.002 = 1/491, which placed the Sun at a minimum distance of 491 terrestrial radii, or 491 × 39,789 = 19,536,399 stadia or 3,071,122 km (the modern value of the parallax is 8.8″, and the mean distance is 149,597,870.7 km). Ptolemy used a method devised by Aristarchos two hundred years earlier and obtained a distance of 1,210 terrestrial radii (7,568,550 km). That figure remained in the books for more than 1,500 years.

From the knowledge that a solar eclipse was limited to a small region of the Earth's surface, Aristarchos (310–230 B.C.E.) concluded that the Sun was much larger than the Earth. He therefore proposed that the Sun, not the Earth, was at the center and that around it orbited Mercury, Venus, Earth, Mars, Jupiter, and Saturn (Uranus, Neptune, and Pluto were not known in antiquity; they were discovered in 1781, 1846, and 1930, respectively). The planets were ordered on the basis of the time it took for a planet to return to the same position in the sky. The shortest period was that of the Moon (27.3d), followed by Mercury (88 d), Venus (225 d), Earth (1 y), Mars (1.88 y), Jupiter (11.88 y), and Saturn (29.46 y). These periods had been worked out by Babylonian and Egyptian astronomers and were well known.

Ptolemy rejected the arguments of Aristarchos and put the Earth at the center, with the Moon, Mercury, Venus, the Sun (which has a period of 1 year as seen from the Earth), Mars, Jupiter, and Saturn orbiting around it. The problem was that the outer planets, Mars, Jupiter, and Saturn, moved across the backdrop of the fixed stars in a very strange way. They moved along for a while, but then they stopped, turned back, stopped again, and finally resumed their path. The ancient astronomers had charted this strange dance very accurately, because they believed that the planets had a profound influence on human affairs. Ptolemy explained this celestial dance by stating that the outer planets were actually making a secondary revolution (called *epicycle*) while following their orbit (called *deferent*) around the Earth (Fig. 5.18). What really happens is that the view

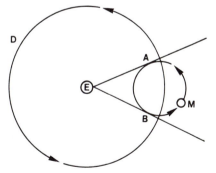

Figure 5.18. According to Ptolemy, the planets revolved around the Sun, following a *deferent D* around which each planet performed a secondary revolution in the same sense, called *epicycle. E* = Earth; *M* = Mars. The motion of Mars appears reversed between *A* and *B*, while the center of the epicycle continues moving along the deferent.

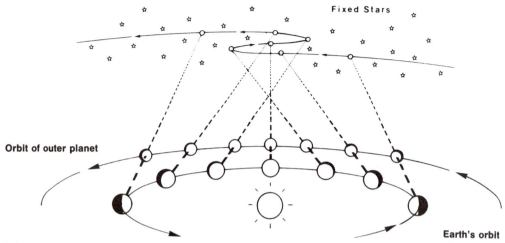

Fixed Stars

Orbit of outer planet

Earth's orbit

Figure 5.19. The apparent retrograde motion of Mars results from the perspective from which Mars is seen as the Earth moves around its orbit. (Littman, 1988, p. 77.)

of an outer planet against the backdrop of the fixed stars changes because the Earth itself is orbiting the Sun (Fig. 5.19).

Key to astronomical measurement is the measurement of angles, but angles are difficult to measure precisely. One degree of arc on a graduated circle 1 m in radius is equal to $6,280/360 = 17.4$ mm, and 1 minute of arc is equal to $17.4/60 = 0.29$ mm. Angles were measured by means of *quadrants*, which were segments of circles graduated in degrees and fractions thereof. A quadrant with a radius of 1 m could barely resolve $1'$ of arc. Unfortunately, astronomical distances are so great that even the largest quadrants of antiquity could not resolve the parallax of any of the planets. Left alone, the Greeks of Alexandria would have undoubtedly developed better astronomical instruments, and probably an entire industrial revolution as well. Unfortunately, they were not left alone. In 389 C.E. Alexandria was sacked by Christian zealots who chased the Greek scholars out and destroyed part of the university library and museum. What was left was burned down by Islamic fanatics in 642 C.E. The spirit of Greece was extinguished forever in the name of God.

5.3. MEDIEVAL ASTRONOMY

During the next 650 years no significant advances were made in astronomy. Ptolemy was hashed and rehashed, especially by Arab experts, until his views became dogma. But then, in 1302, something happened in Padua. There lived a man by the name of Reginaldo degli Scrovegni who had made a bundle in loan-sharking. As he was getting old, he was worried that when he died he would land in Hell. So in order to curry favor with God, he built himself a chapel and commissioned Giotto (ca. 1266–1337), Italy's foremost painter of the time, to illustrate the back wall of the chapel. Giotto painted a scene representing the Last Judgment (Fig. 5.20), with Jesus Christ in the middle, the good people on his right, and the bad people on his left. To his left and toward the bottom is Lucifer gnawing the body of Judas, the world's foremost traitor.

Dante Alighieri (1265–1321), a Florentine politician and poet who had been thrown out of Florence and was meandering through northern Italy, saw the painting. It reminded him of the mosaics on the vault of Florence's Baptistery, a work on the same theme that he had much admired while growing up in Florence. That probably inspired him to write his famous poem *La Commedia* (Italian for *The Comedy* —the title was later upgraded to *La Divina Commedia*). In that poem, Dante describes how he became lost in a dark forest and was rescued by the Latin poet Vergil (70–19 B.C.E.), who suggested that they should visit Hell, Purgatory, and Paradise. The conversation might have gone as follows:

Figure 5.20. The wall of the Cappella degli Scrovegni in Padua, painted by Giotto in 1306. The scene represents the Last Judgment. This painting may have inspired Dante Alighieri to write *The Comedy*. (Courtesy Alinari/Art Resource, New York.)

DANTE (*stumbling on a log*): Porcaccia miseria!!

VERGIL: Ahem!

DANTE (*looking up*): Who are you?

VERGIL: I am Vergil, the famous Latin poet.

DANTE: Sure, Sure.

VERGIL: Really, I am! Look, see? I cast no shadow—I am a pure soul.

DANTE (impressed): I am impressed.

VERGIL: I came to rescue you and guide you back.

DANTE: Very nice of you.

VERGIL: But first you have to go through Hell, Purgatory, and Paradise.

DANTE: Aw c'mon, Verg! Do I have to? It's getting late!

VERGIL: God himself has so decreed.

DANTE: Well!! In that case, let's press on!

And so they went. A lot of nasty things Dante saw in Hell, enough to scare anybody witless. But the most astonishing sight was three popes—the reigning pope, Bonifacio VIII (1294–1303), his predecessor, Niccolò III (1277–1281), and his successor, Clemente V (1305–1314)—stuck head down and feet up in three narrow wells full of flames. As it turned out, the three popes had been guilty of various sins, and so they had been condemned to spend eternity in that most uncomfortable position.

In fourteenth-century Italy, the quality of life was nothing to scream about: For a start, there was no television, no magazines, no newspapers, no electricity, and not much sanitation either. Very few people knew how to read and write, and in any case there was hardly anything to read except the usual holy texts (a crashing bore). So when Dante came up with the *Comedy*, minstrels everywhere would recite the poem on street corners. The people listened in awe, and most came to believe that Dante had *really* been to Hell and had *really* seen three popes there. The mere notion of popes in Hell sent far more shocks up and down the spine of Italy than Khrushchev's destalinization speech in 1956 ever sent across the length and breath of the Soviet Union.

The pope, of course, was furious. The church proscribed all of Dante's books, but could not touch the *Comedy*, because it had become an instant success. To make things worse for the church, Boccaccio (1313–1375), a name that in

Italian means "bad mouth", not only became an ardent propagandist for Dante but also wrote his famous *Decameron*, in which priests and nuns are exhibited in less than saintly pursuits. The end result was that the church lost much of its prestige in Italy (it never really regained it), and that opened the gate for dissent. People rerouted their thinking from the next world to this one, and the Renaissance was under way.

5.4. RENAISSANCE ASTRONOMY

The Renaissance stumbled along for a while, but then it took off like a rocket when Johannes Gutenberg (ca. 1397–1468) invented the printing press. Gutenberg's Bible, a book 1,282 pages long, was published in 300 copies in the year 1454. Copying the Bible by hand had required an expert copyist about nine months. Even at the minimum wage of $4.25 per hour and a 40-hour week, a hand-copied Bible today would cost $8,500. Typesetting the Bible by hand may have taken perhaps twice as much time, but once it was composed, a large number of copies could be printed.

The composing and printing costs of Gutenberg's Bible probably did not exceed $50 per copy. If we double that to allow for profit (though Gutenberg was fleeced by his partners and never made any money), a customer still would have paid less than one-fiftieth of the cost of a manuscript Bible. People suddenly developed a thirst for books. There were many ancient manuscripts to be printed, and soon the presses were working feverishly. One of the most productive was the Aldine press in Venice. Fifty years after Gutenberg's invention, 30,000 works had been printed, averaging 300 copies each.

A Latin translation of Ptolemy's great work, the *Almagest*, was printed in 1462, and numerous editions followed. Ptolemy was a great mathematician, and the *Almagest* is not what one would call "light reading". Only the best minds could cope with it. A twenty-seven-year-old Polish student at the University of Bologna, Nicolaus Copernicus (1473–1543), had just such a mind. He began studying the *Almagest* in the year 1500 at the instigation of Francesco Maria

Novara (1454–1504), the resident astronomy teacher. Novara did not like Ptolemy
very much. He thought that Aristarchos was
probably right in placing the Sun at the center.
Copernicus studied carefully all 13 books of the
Almagest, especially the last five, which deal
with planetary motions, and concluded that
indeed Aristarchos had been right—the
motions of the planets could more readily be
explained (and more closely predicted) by
assuming that the planets moved around the
Sun in circular orbits.

Copernicus spent ten years in Italy (1496–
1505), studying law and medicine. He then returned to Poland to take over essentially eccle-
siastical duties. During the next 30 years, he put
together, working only in his spare time, his
great treatise *De Revolutionibus Orbium Coelestium.* He was not keen to publish it, because he
was afraid that the pope would lean on him.
Eventually the book was published in 1543, a
few weeks before Copernicus died.

The publication of a new book was a much
bigger deal then than now. It was not often that
something new came out, and the appearance
of a new book was eagerly anticipated, often
for many years. Copernicus' treatise was the
first new thing since Ptolemy. It was a great
success, and it greatly spurred interest in
astronomy.

The Danish astronomer Tycho Brahe
(1546–1601), sponsored by King Frederick II
of Denmark, built a gigantic observatory, called
Uraniborg, on the island of Ven, off Copenhagen. The pièce de résistance of his observatory was a quadrant 2 m in radius. Using his
instruments, Tycho was able to reduce the
error of angular measurement from 2′ to 30″.
Tycho was the last astronomer to use exclusively the naked eye (he died eight years before
the invention of the telescope) and, with no
telescope, he could not determine the parallax
of even the closest planets.

Tycho was arrogant and had a bad disposition. He enjoyed making enemies and running
them into the ground. When he was 19 he lost
his nose in a duel and wore a metal nose
for the rest of his life. In 1588 Frederick II died,
and after a few years the new king kicked
Tycho out of Denmark. He was invited to settle
in Prague by Rudolph II, king of Austria, king
of Germany, king of Hungary, king of Bohemia,

and emperor of the Holy Roman Empire.
Rudolph was a bit crazy (it went with the
family), but a lover of the arts and the sciences.
Tycho accepted the invitation, settled in Prague,
and died there in 1601, when, having drunk too
much good beer, his bladder burst (or so they
say).

In Prague, Tycho had managed to acquire
Johannes Kepler (1571–1630), a young German
mathematician, as his assistant. Tycho be-
stowed upon Kepler his tables of planetary
positions, gathered through more than 30 years
of careful observations. Kepler immediately
began trying to make sense of Tycho's data.
After years of trying various geometric combinations, in 1609 he finally announced that the
planetary orbits were not circular, as Copernicus had maintained, but elliptical, with the
Sun occupying one of the two foci (this is called
Kepler's first law). Furthermore, he found that
the speed of a planet along its orbit was not
constant, but was highest at perihelion and
lowest at aphelion

[*Perihelion* is the point along an elliptical orbit
around the Sun that is closest to the Sun.
Aphelion is the opposite point, farthest from
the Sun. *Periastron* and *apastron* are the equivalent points of an orbit around a star.]

Kepler found, specifically, that "the line connecting a planet to the Sun sweeps equal areas
in equal times" (Kepler's second law) (Fig. 5.21).

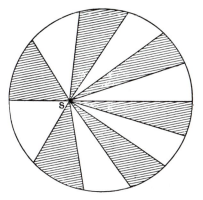

Figure 5.21. Kepler's second law: The planets orbit the
Sun in such a way that the line connecting a planet
with the Sun (*S*) sweeps equal areas in equal times.
This means that the planets move slower at aphelion
and faster at perihelion. (Berry, 1961, p. 186, Fig. 60.)

Table 5.2. Relative astronomical distances (AU)

Planet	According to Kepler			Modern Value (mean)
	Leasts	Greatest	Mean	
Mercury	0.307	0.470	0.389	0.387
Venus	0.719	0.729	0.724	0.723
Earth	0.982	1.018	1.000	1.000
Mars	1.382	1.665	1.523	1.524
Jupiter	4.949	5.451	5.200	5.202
Saturn	8.968	10.052	9.510	9.555

Using those two laws, Kepler was able to establish an accurate scale of relative planetary distances (Table 5.2).

After poring over Tycho's records for another ten years, Kepler discovered that "the square of the sidereal period of a planet [the time a planet takes to return to the same position with respect to the fixed stars] increases with distance from the Sun in proportion to the cube of the semimajor axis of its orbit" (Kepler's third law). The absolute distances of the planets

from the Sun, however, remained a mystery—planetary parallaxes were simply too small to be measured with the instruments available at the time.

Fortunately, the Dutch lens-maker Hans Lippershey (1570–1619) invented the telescope in 1608, just when it had become abundantly clear that nobody could possibly improve on Tycho Brahe's naked-eye measurements. The telescope is essentially a tube with a lens at each end. The telescope is such a simple instrument that it should have been invented two or three thousand years earlier. The Phoenicians were the first to manufacture glass, about 3500 B.C.E. Lenses and their magnifying properties were known in antiquity. "Reading glasses", consisting of a pair of biconvex or planoconvex lenses, are said to have been invented by the Florentine Salvino degli Armati (d. 1317) toward the end of the thirteenth century and were in common use during the Renaissance. But until 1608 it had not occurred to anyone to put two lenses one in front of the other and to look through.

[Light refracts (i.e., it changes direction) when

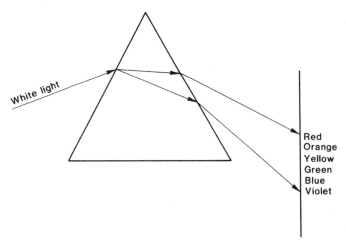

Figure 5.22. The speed of light in vacuo is the same for all wavelengths, but that in transparent media is less, depending on the medium and the wavelength (shorter wavelengths are slowed down more than longer wavelengths). If monochromatic light hits the surface of a prism at an angle other than 90°, light is refracted by an angle whose sine is equal to the sine of the angle of incidence multiplied by the ratio of the speed of that light in the glass to its speed in air (the speed of light in air is close to that in vacuo, 299,705 km/s instead of 299,792 km/s). In glass, the speed of light is about 200,000 km/s (depending on the type of glass). If white light hits the surface of a prism at a slant, the red component is slowed down less and bent less while the blue component is slowed down more and bent more. This results in color dispersion and the formation of a spectrum. The ratio of the speed of light in vacuo to that in a medium is called the *index of refraction* of that medium. The index of refraction of glass is about 1.5. (Emiliani, 1988, p. 22, Fig. 2.5.)

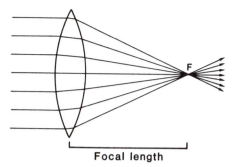

Focal length

Figure 5.23. A biconvex lens focuses parallel rays of monochromatic light on a point called the *focus* (*F*). The distance between the center of the lens and the focus is called the *focal length*. The inverse of the focal length in meters gives the "power" of the lens in *diopters* [power, in this case, has nothing to do with energy/time; the power of a lens is its ability to make parallel light rays converge (positive lens) or diverge (negative lens)—the stronger the convergence or the divergence, the greater the power]. The crystalline lens of the human eye focuses on the retina, 2.5 cm away. Its focal length is 0.025 m, and its power is 40 diopters. Most reading glasses have power between 0.25 and 5 diopters. The power of a negative lens is equal (with a minus sign in front) to that of a positive lens that would compensate it.

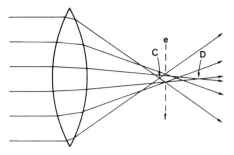

Figure 5.24. Spherical aberration. A lens whose surface is a section of a sphere will focus even monochromatic light at different points along the focal axis. Light passing through the lens near the center focuses farther away (*D*) than light passing through the lens farther away from the center (*C*). The "circle of least confusion" is on the *e-f* plane. (King, 1955, p. 43, Fig. 18.)

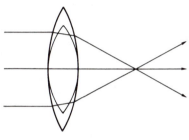

Figure 5.25. In a paraboloidal lens (outer section), the rays away from the center are bent less than in a spherical lens of the same thickness (inner section) and therefore focus farther away from the lens, matching the focus of the rays crossing the lens nearer the center.

passing from one medium to another if the speed of light is different in the two media and if it crosses the interface at a slant (see Fig. 2.11). Fig. 5.22 illustrates the refraction of light passing through a prism. A convex lens is shaped in such a way that parallel light rays passing through it are refracted and converge on a single point, called the *focus* (Fig. 5.23). Conversely, rays diverging from the focus and passing through the lens become parallel. This happens because light rays crossing the lens at different distances from the center encounter different thicknesses of glass and therefore are slowed down differentially. The ray crossing the center of the lens is slowed down the most, while rays crossing the lens at increasing distances from the center are slowed down less and less. The slowing down, decreasing from center to periphery, closely compensates for the increasing length of the paths away from the center of the lens, so that all rays converge approximately on the same point, the focus. Actually, the rays passing close to the periph-

ery converge to a point closer to the lens than rays passing through the center—this is called *spherical aberration* (Fig. 5.24), and it is inherent to any lens whose surface is a spherical section. Spherical aberration can be corrected if the surface is shaped as a paraboloidal section (Fig. 5.25). A paraboloidal lens is difficult to manufacture and could not be made at the time when the telescope was invented.

To make a good lens, a piece of glass is required that is free of internal defects—no bubbles, no cracks, no nothing. First, a slice of glass is cut with a metal saw coated with abrasives. Then the glass slice is shaped by pressing

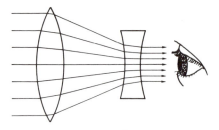

Figure 5.26. The Galilean telescope consists of a positive objective and a negative ocular placed *before* the focus of the objective. The negative lens renders parallel the rays of the objective. The crystalline lens of the observer's eye makes these parallel rays converge on the retina.

it against a cast-iron bowl of suitable curvature mounted on a rotating table, with a slush of water and abrasives. Polishing is done by using finer and finer abrasives, until the surface is smoothed to a precision within one-fourth of the wavelength of visible light (i.e., within about 0.1 μm). It takes a lot of patience and perseverance.]

Lippershey's telescope consisted of two lenses, a positive (i.e., convex) objective and a negative (i.e., concave) ocular. This arrangement (Fig 5.26) collects light impinging on the large surface of the objective and concentrates it on the focus.

VENUS – ANGULAR DIAMETER VERSUS PHASE ANGLE

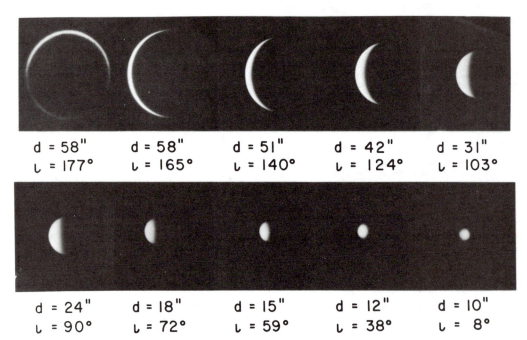

| d = 58" | d = 58" | d = 51" | d = 42" | d = 31" |
| ι = 177° | ι = 165° | ι = 140° | ι = 124° | ι = 103° |

| d = 24" | d = 18" | d = 15" | d = 12" | d = 10" |
| ι = 90° | ι = 72° | ι = 59° | ι = 38° | ι = 8° |

Figure 5.27. The phases of Venus, with new Venus 7 times larger than full Venus. In October 1610, Venus became visible in the evening sky in Florence, and Galileo began observing it. By the beginning of December, Venus' disc had become a half-disc. Without waiting for further shrinking, Galileo wrote a letter (on December 11, 1610) to Giuliano de' Medici, the Tuscan ambassador to Prague, reporting that he had made a discovery that strongly supported the Copernican theory. He added *Haec immatura a me iam frustra leguntur o y*, which translates *These immature things have already been gathered in vain*

by me o y. The phrase not only is meaningless but also includes two letters that are left stranded, *o* and *y*. When, by the end of December, Venus had assumed a crescent shape, Galileo gained confidence in his discovery and revealed that, by rearranging the letters, his Latin phrase would read *Cynthiae figuras aemulatur mater amorum*, which means *The mother of loves* (Venus) *emulates the shapes of Cynthia* (the Moon). The fact that Venus exhibited phases like the Moon proved that Venus, and by inference the other planets, orbited the Sun, not the Earth. (Courtesy Department of Astronomy, New Mexico State University.)

The ocular, placed before the objective's focus, makes the rays converging on the focus parallel again, and the eye's crystalline lens makes them converge on the retina. Magnification is simply the ratio of the focal length of the objective to that of the ocular. Greater magnification can be achieved by using weaker lenses as objectives and stronger lenses as oculars. An objective with a focal length of 10 m and an ocular with a focal length of 10 cm will produce a magnification of $1,000/10 = 100$.

Galileo heard of Lippershey's invention in May 1609 and immediately built himself a telescope. Galileo's telescope consisted of a lead tube with a planoconvex objective and a planoconcave ocular. Its magnification was threefold (i.e. an object appeared 3 times closer and therefore appeared to occupy a surface 9 times larger). Within six months Galileo built additional telescopes with magnifications of 8, 20, and 32. Within one year, Galileo discovered mountains on the Moon, the four major satellites of Jupiter, and the rings of Saturn (although he did not recognize them as rings). He also discovered that the Milky Way was not a band of luminous gas, but a multitude of stars, and that Venus not only had phases like those of the moon, but crescent Venus was almost 7 times larger than full Venus (Fig. 5.27). This discovery proved beyond any doubt that Venus (and, by inference, all other planets as well) circled the Sun, not the Earth. It was the kiss of death for the Ptolemaic system.

Galileo put together his views in a book entitled *DIALOGO DI GALILEO GALILEI LINCEO, MATEMATICO SOPRAORDINARIO DELLO STUDIO DI PISA E Filosofo, e Matematico primario del SERENISSIMO GR. DUCA DI TOSCANA. Dove nei congressi di quattro giornate si discorre sopra i due MASSIMI SISTEMI DEL MONDO, TOLEMAICO E COPERNICANO, Proponendo indeterminatamente le ragioni Filosofiche, e Naturali tanto per l'una, quanto per l'altra parte.* This is an example of the lengthy titles that Renaissance authors usually gave their books. No need to translate it because most of the words are similar to the corresponding English words and are thus readily understood. The title is usually abbreviated to *Dialogo Sopra i Due Massimi Sistemi del Mondo.* This book was published by Landini in Florence in 1632. In the book, a simpleton by the name of Simplicio (which means "simpleton") defends the Ptolemaic system, while an expert by the name of Salviati (Galileo himself) expounds the Copernican system and, of course, wins all the arguments. The pope was not amused. Galileo was summoned to Rome and forced to recant under threat of torture. That was a big mistake for the church, but a lesson was learned. Since then the Roman church has maintained a healthy "hands-off" attitude in matters of science. Even Darwin did not succeed in raising the pope's eyebrows.

Back to the measure of the universe. Kepler, who had received one of Galileo's telescopes in

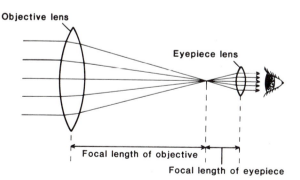

Objective lens

Eyepiece lens

Focal length of objective

Focal length of eyepiece

Figure 5.28. The Keplerian telescope consists of a positive objective and a positive occular of shorter focal length, placed at a distance from the focus of the objective equal to its own focal length. The ocular makes the rays parallel, and the crystalline lens focuses them on the retina. The image is inverted. (Emiliani, 1988, p. 83, Fig. 6.9.)

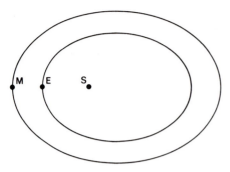

Figure 5.29. Position of Mars when at opposition (i.e., opposite the Sun as viewed from the Earth) and at perihelion. M = Mars; E = Earth; S = Sun.

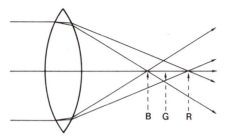

Figure 5.30. Chromatic aberration. Red light (R) focuses farther away than blue light (B), with green light (G) in between. (King, 1955, p. 44, Fig. 19.)

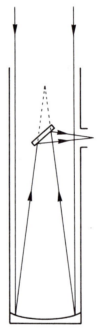

Figure 5.31. The reflecting telescope consists of a reflecting, concave mirror that focuses light without suffering from chromatic aberration. If the reflecting surface is a spherical section, the mirror will suffer from spherical aberration. Paraboloidal mirrors are free of both chromatic and spherical aberration. (Emiliani, 1988, p. 84, Fig. 6.5.)

1610, suggested in a work published in 1611 that if the ocular were replaced with a positive lens (Fig. 5.28), a larger field of view would result. The disadvantage was that the image would be upside down, but then who cares about up and down when viewing the Moon, the planets, and the stars? By the middle of the seventeenth century, Keplerian telescopes (positive objective and positive ocular) had overtaken the Galilean telescopes. Today the Galilean telescope survives only in opera glasses.

5.5. MODERN ASTRONOMY

The first big discovery in what may be considered modern astronomy was made in 1639 by the English astronomer William Gascoigne (1612–1644). A spider, "when it pleased the All Disposer", as Gascoigne wrote, drew its web across his telescope tube between ocular and objective, at a distance such that the web appeared sharply outlined in the field of the telescope. The spider had fortuitously drawn its web precisely across the focal plane between ocular and objective. That brought about the development of micrometers, which allowed much more precise angle measurements than had been possible with the largest quadrants of earlier times.

In the fall of 1672, Mars was both at opposition and near perihelion, which placed it closest to the Earth (Fig. 5.29). Taking advantage of that situation, and using telescopes equipped with micrometers, the Italian astronomer

Giovanni Domenico Cassini (1625–1712) finally succeeded in determining the parallax of Mars. It turned out to be 25″. If φ is taken as the horizontal parallax (Fig. 5.13), the distance to Mars is

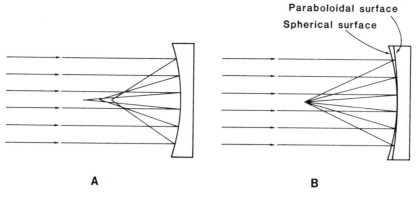

Figure 5.32. (A) Rays reflected from around the center of a spherical mirror are focused farther from the mirror than are rays reflected from farther out. (B) If the mirror is a paraboloidal section (i.e., all cross sections of the mirror passing through the axis are parabolic), all rays focus on the same point. (King, 1955, p. 44, Fig. 19.)

$$\cos \varphi = \sin \varphi / \tan \varphi$$

or

$$\cos 25'' = 6,378 / \tan 25''$$
$$= 52,622,277 \text{ km}$$

Because that distance was 0.382 AU (Table 5.2, least distance from Sun − 1), the astronomical unit (AU) turned out to be 52,622,277/0.382 = 137,754,652 km, which is rather close to the modern value of 149,597,870.7 km. Because Saturn was 9.5 AU from the Sun, the radius of the entire solar system, as known at that time, turned out to be $1.3 \cdot 10^9$ km.

Once the size of the solar system was known, astronomers longed to find the distance to the stars. For that, they needed better telescopes. The Galilean and Keplerian telescopes suffered from the problem of spherical aberration; in addition, there was chromatic aberration. Different colors are refracted by different angles when crossing a lens, with red light focusing farther away than blue light (Fig. 5.30). Spherical and chromatic aberrations combined to make life miserable for the astronomers interested in stars.

Newton, who was keenly interested in optics, among other things, came to the rescue. He constructed the first *reflecting* telescope, that is, a telescope consisting of a concave mirror and an ocular (Fig. 5.31). The reflecting surface was highly polished bronze, consisting of 75% copper and 25% tin. Because the light never crossed a medium with refractive index different from that of air, there was no chromatic aberration. Even the most highly polished metal surface, however, absorbs some light, and so the images of the early reflecting telescopes were dimmer than those of the refracting telescopes. That problem was solved only in the middle of the nineteenth century, when a technique to silver a glass surface was developed.

The mirrors easiest to construct were those shaped as spherical sections. Such mirrors, like spherical lenses, were not free of spherical aberration: Rays hitting the center of the mirror were focused farther away from the mirror than rays hitting the mirror farther away from the center (Fig. 5.32A).

To eliminate spherical aberration, parabolic mirrors were needed (Fig. 5.32B). The first parabolic reflector was constructed in 1721 by the British astronomer John Hadley (1682–1743). Reflectors became increasingly popular because they were easier to manufacture and use. Indeed, it is virtually impossible to make a good lens larger than 1 m in diameter. The largest refractor in use today is at the Yerkes Observatory in Green Bay, Wisconsin, managed by the University of Chicago. Its objective is 40 inches (1.016 m) across, and its focal length is 19.36 m. It has been in service since 1897. Reflectors can be much larger. The largest one is on Mt. Semirodriki, near Zelenchukskaya, on the northern flank of the Caucasus, in Russia. It has a diameter of 6 m and weighs 78 tons. The next largest reflector is the Palomar reflector

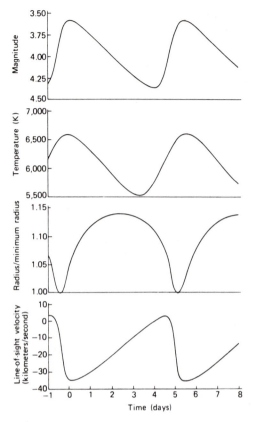

Figure 5.33. Periodic variations in δ Cephei. (Emiliani, 1987, p. 37.)

on Mt. Palomar, California, with a diameter of 200 inches (5.08 m).

For 150 years astronomers explored the sky diligently trying to find a star close enough to have a measurable heliocentric parallax. Finally, in 1838, the German astronomer Friedrich Bessel (1784–1846) succeeded in determining the parallax of the star 61 Cygni, a rather dim star, but one that has the fastest proper motion (5″/year). (*Proper motion* is the motion of a star with respect to neighboring stars, due to its own *peculiar motion* and the motion of the Sun with respect to it.)

Bessel determined that the heliocentric parallax of 61 Cygni was 0.5″, corresponding to a distance of $6.17 \cdot 10^{13}$ km, or 6.5 light-years. By comparison, Kepler thought that the stars were perhaps 0.1 l.y. away, and Newton thought they might be 2 l.y. away. Some stars are even

closer than 61 Cygni. The closest one is Proxima Centauri, 4.26 l.y. away. Most stars, however, are much more distant.

The second half of the nineteenth century was a rather tranquil period for astronomy. The universe seemed to consist of just two principal types of bodies—stars (some of which exhibited a variable brightness) and gaseous masses called *nebulae* (most of which were ellipsoidal or spindle-shaped). The shape of the universe was known to be like a flattened disc, called the Galaxy, with the Milky Way representing the accumulation of stars that one would observe from the inside even if the stars were distributed uniformly. That notion had first been proposed in 1750 by the English astronomer Thomas Wright (1711–1786) and had become generally accepted. The solar system was believed to be in the center of the Galaxy, because the density of the Milky Way seemed to be about the same all around.

There were, of course, a couple of flies in the ointment. One was Newton's contention that the universe must be uniform *and* infinite; otherwise gravity would make it collapse on its center. The second was a paradox enunciated in 1826 by the German physician and astronomer Heinrich Olbers (1758–1840). Said Olbers: Newton must be right—the universe must be infinite and uniform, or else it would collapse. But if it is infinite and uniform, why is the night sky dark?

Good question. If we were inside an infinite universe in which the stars were uniformly distributed, the night sky should be totally packed with stars: No matter in which direction we looked, our line of sight would eventually be intercepted by a star. Beyond that limit there will be many more stars (in fact, an infinite number), but they will be hidden from our view by the nearer stars. Similarly, a person in the middle of a square full of people does not see the people beyond a certain radius, because the people in between obstruct the view.

If Newton and Olbers are right, how come the night sky is not packed with stars, but contains great voids that make it dark? The astronomers of the nineteenth century had no answers to this question. But then, they were used to not having answers for big questions. For instance, they did not know why the Sun and other stars shone, nor what exactly gravity

Figure 5.34. The two Magellanic Clouds form a pair of small, irregular galaxies that are satellite to our Galaxy and visible only from the Southern Hemisphere. The Large Magellanic Cloud (left) is 160,000 l.y. away; the Small Magellanic Cloud is 190,000 l.y. away. The Small Magellanic Cloud contains the Cepheids observed by Henrietta Leavitt in 1912. (Courtesy Harvard College Observatory, Cambridge, Mass.)

was. The answer to Olbers' question came only after the dimension, structure, and composition of the universe became more firmly established.

A method that could be used to determine the distance of stars was discovered in 1912 from a study of certain variable stars called *Cepheids*. All stars are in a sense variable—even the Sun changes its luminosity by 0.1%, with a period of 11 years (see Chapter 7). Some types of stars, however, change their luminosity much more. One such type are the Cepheids, named after the variable star δ Cephei in the Cepheus constellation (stars in a constellation are identified with the letters of the Greek alphabet, beginning with α, in order of decreasing brightness). δ Cephei changes its luminosity by 178%, with a period of 5.2 days (Fig. 5.33); we shall see why in Chapter 6. Other Cepheids have periods as short as 1 day or as long as 100 days.

Cepheids had been known since antiquity, because some are visible with the naked eye. At the beginning of the twentieth century, Henrietta Leavitt (1868–1921), on the staff of the Harvard Observatory, became interested in the Cepheids and began trying to discover what made them pulsate. In 1912 she was in Arequipa, Peru, where Harvard had established an observatory to study the sky of the Southern Hemisphere. She found a number of Cepheids in a nebula called the Small Magellanic Cloud (because it had first been described by the explorer Ferdinand Magellan in 1519) (Fig. 5.34) and noticed that the longer the period was, the brighter the brightness peak was. It was as though the longer a star took to "charge" itself, the larger its "explosion".

[*Brightness* is the power received from a star. *Luminosity* is the power emitted by a star. A luminous but distant star may appear less bright

Table 5.3. Standards of stellar magnitudes

Name	Magnitude
α Canis Majoris (Sirius)	−1.45
α Aurigae (Capella)	+0.06
α Virginis (Spica)	+0.91
α Andromedae (Alpheratz)	+2.06
β Trianguli	+3.00
θ Bootis	+4.07
η Ursae Minoris	+5.05
Faintest visible stars	> +6.00

Table 5.4. Magnitudes of bodies forming the solar system (maximum magnitudes for the Moon and the planets)

Name	Magnitude
Sun	−26.74
Mercury	0.0
Venus	−4.4
Moon	−12.7
Mars	−3.0
Jupiter	−2.6
Saturn	+0.7
Uranus	+5.5
Neptune	+7.8
Pluto	+14.9

than a dim but close star. *Magnitude* is the brightness of a star expressed on a scale devised by Hipparchos. Magnitude 1 represents the brightest stars in the sky; stars of magnitude 2 are 2.512 times less bright than those of magnitude 1; magnitude 3 stars are 2.512 times less bright than the stars of magnitude 2; and so on (Table 5.3). The number 2.512..., equal to $100^{1/5}$, was proposed by the English astronomer Norman Pogson (1829–1891) to define the ratio between one magnitude and the next; it is called the *Pogson ratio*. Stars brighter than magnitude 0 are given negative values. The magnitudes of the bodies forming the solar system are shown in Table 5.4.

Absolute magnitude is brightness reduced to the distance of 10 parsecs (pc). The absolute magnitude of the Sun is +4.84.]

Henrietta Leavitt reasoned that the relationship she found between brightness and period was real, not a result of differences in distance, because all the stars she observed were part of the same system, and therefore all were at about the same distance. It was immediately apparent, therefore, that if one knew the actual distances to at least some Cepheids, one could determine the relationship between luminosity and period and use that relationship to determine the distance to any Cepheid-containing stellar system. Unfortunately, there are no Cepheids close enough to be ranged by the parallax method.

Beyond the limit of the parallax method (about 300 l.y.), the distance to stars that form a cluster (the Big Dipper or the Hyades, for instance) can be determined from the motion of the cluster itself. The motion relative to distant stars, corrected for the motion of the Sun through space, can be determined by photo-

graphing the cluster at intervals of years or decades. A cluster moving away from us seems to progressively shrink, with the stars getting closer to each other, as the cluster moves toward a distant point called *convergent point*. It is simply a matter of perspective, like a pair of railroad tracks that seems to converge at the horizon. Conversely, the stars in a cluster approaching us seem to diverge from a distant point. For any given star in the cluster, the angle α between our line of sight and the convergent point P is also the angle between our line of sight and the actual velocity vector of the star (Fig. 5.35). The radial velocity of the star is determined from the shift in the spectral lines of the light emitted by the star. Knowing the radial velocity v_r and the angles α (measured) and β (equal to 90°), the actual velocity v and the tangential velocity v_t can be calculated. A comparison of v_t with the actual displacement of the star across the sky (measured in seconds of arc) gives the distance of the star.

The Hyades cluster (Fig. 5.36) turns out to be 142.5 ± 6.5 l.y. away. It contains about 100 stars. The distance to other galactic clusters, too far to be ranged by the method used for the Hyades, can be estimated by assuming that the brightest stars in a cluster have luminosities similar to those of the brightest stars in the Hyades cluster and by applying the inverse square law.

Using this method, Harlow Shapley (1885–1972), Ms. Leavitt's boss at Harvard, determined in 1918 the distance to some distant galactic clusters that contained Cepheids (unfortunately, the Hyades contain no Cepheid). Once the

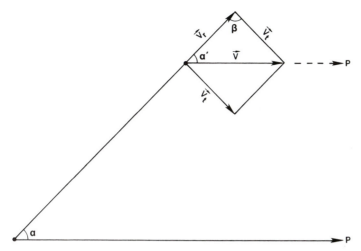

Figure 5.35. How to measure the distance to the stars of an open cluster. The cluster is photographed at intervals of years or decades, and the displacement of the stars across the sky (corrected for the motion of the Sun) is noted. The radial velocity v_r of a star in the cluster is determined from the displacement of its spectral lines. Because the convergent point is at infinity, the angles α and α' are identical. The measurements of v_t and α and the observation that β is 90° yield the velocity v of the star and hence the tangential velocity v_t. The distance to the star is obtained by comparing v_t with the displacement of the star across the sky. For any given v_t, the distance is inversely proportional to the displacement.

distance of the Cepheid-containing clusters was determined, the luminosity of the Cepheids could be calculated using the inverse square law. This made it possible to relate directly period to luminosity (the actual power emitted) (Fig. 5.37). Because the inverse square law says that the power received is equal to the power emitted divided by $4\pi r^2$, the distance r to any Cepheid can be determined from its period (giving the luminosity) and from its brightness (the power received). Using the Cepheid method, Harlow Shapley was able to determine that the radius of our own galaxy was about 150,000 l.y. and that the Sun was about 50,000 l.y. from the center (these two figures are now known to be about 30% too large).

Copernicus had demoted the Earth from the center of action, and now Shapley was demoting the Sun itself. This time, however, the pope not only did not object but received Shapley most cordially (Fig. 5.38).

Enter Edwin Hubble (1889–1953), a young

Figure 5.36. The Hyades. [Lick Observatory]

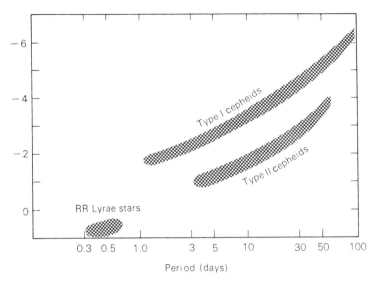

Figure 5.37. Period–luminosity relationship for Cepheids and RR Lyrae stars. Type I Cepheids are younger, metal-rich stars, while type II Cepheids are older, metal-poor stars. The two types exhibit different relationships between luminosity and period. The Cepheids studied by Henrietta Leavitt belonged to Type I. The RR Lyrae are short-period variables. (Abell, Morrison, and Wolff, 1987, p. 417, Fig. 24.3.)

Figure 5.38. Harlow Shapley and Mrs. Shapley in conversation with Pope Pius XII in Rome in 1956. (Gamow, 1958, p. 525, Fig. 20.18.)

Figure 5.39. The Andromeda galaxy, a large, spiral galaxy seen at a slant from the Earth, is 125,000 l.y. across and 2,140,000 l.y. distant. It contains about 10^{11} stars. It is similar to our Galaxy in size, shape, and mass. (Courtesy National Optical Astronomy Observatories, Tucson, Ariz.)

lawyer who at the age of 25 became an astronomer. In 1919 he joined the staff of the Mount Wilson Observatory near Los Angeles. In 1904, J. T. Hooker, a Los Angeles businessman, had given Mount Wilson a pile of money to build a 100-inch reflector. Since the middle of the nineteenth century, reflectors had been made by silvering a concave, paraboloidal surface shaped from a piece of glass. This time, a huge piece of glass was needed. Only the Saint Gobain company in France agreed to undertake the job. Four years later, a glass disk 101 inches across, 13 inches thick, and weighing 4.5 tons arrived in Los Angeles. It took 11 years to polish the surface, silver it, and construct a supporting structure and a housing.

The Hooker telescope was placed in service in 1919, just when Hubble arrived. Hubble was interested in nebulae. In 1924 he announced that he had discovered some Cepheids in the Andromeda nebula (Fig. 5.39) and, using the Leavitt–Shapley yardstick, had determined that it was not a nebula inside our Galaxy, but a

separate galaxy 800,000 light-years away (the figure is now 2,140,000 l.y.). It soon appeared that all ellipsoidal and spindle-shaped nebulae were extragalactic. It was a brand new world, a world much larger than anyone had ever imagined. But that was not all. A few years later, in 1929, Hubble announced that the universe was not static, as everybody, including Einstein, had assumed, but was expanding at a fast clip. He had discovered the *redshift*.

The redshift is the lengthening of the wavelength of light received by an observer when the light source is moving away. If the light source is moving closer, the wavelength is shortened, and the light appears shifted toward the blue end of the spectrum. We are not ordinarily aware of this phenomenon because the speed of even a jet plane is negligible compared with the speed of light. A similar phenomenon, however, occurs with sound waves.

A car horn with a sound frequency of 300 Hz produces 300 waves per second. If we are traveling along a highway at 65 mph, and another car is approaching, also at 65 mph while blowing its horn, the relative speed of the two cars will be 130 mph, or 209 km/h. This is equal to 58 m/s. The speed of sound (at 20°C) is 343 m/s. The relative speed of the two cars is therefore $58/343 = 0.169 = 16.9\%$ of the speed of sound. That is, we will receive 16.9% more waves per second, producing a frequency of 350.7 Hz instead of the original 300 Hz. As soon as the car goes by, we will receive 16.9% fewer waves per second. The frequency will be lower, 249.3 Hz instead of 300 Hz, and we will be hearing a lower pitch (*pitch* is the physiological response to sound frequency). The change in the pitch of sound when source and receiver are in motion relative to each other is called the *Doppler effect*, after the Austrian physicist Christian Doppler (1803–1853), who explained it in 1842.

The Doppler effect applies also to light, except that then it is called the *Doppler shift*. The strongest frequency emitted by excited sodium vapor atoms has a wavelength of 588.9950 nm. This length is equal to $588.9950 \cdot 10^{-9}$ m and corresponds to the frequency of $299,792,458/588.9950 \cdot 10^{-9} = 5.09 \cdot 10^{14}$ Hz. Light of this wavelength and frequency has an energy of 2.10 eV. It is bright yellow in color and forms what is called the D line in the spectrum of sodium. It is emitted by the valence electron in falling back to the lowest energy level of the M shell after having been excited to the next level within the same shell.

Imagine now that a sodium lamp, mounted on a fast vehicle, is approaching an observer at 10% of the speed of light (i.e., at $3 \cdot 10^7$ m/s). The electromagnetic waves that form light are compressed between the lamp and the observer, so that instead of $5.09 \cdot 10^{14}$ waves hitting the observer's eye each second and producing the sensation of yellow, there will be 10% more, or $5.60 \cdot 10^{14}$ waves per second. The wavelength will be 10% shorter, or 530.0955 nm. The color of the light will appear greenish. If the lamp is moving away from the observer at the same speed, 10% fewer waves will hit the observer's eye each second—$4.58 \cdot 10^{14}$ instead of the original $5.09 \cdot 10^{14}$. The wavelength will be 10% longer, or 647.8945 nm. The color of the light will be orange–red. The color of light is analogous to the pitch of sound.

Hubble discovered that the color of distant galaxies was shifted toward the red and that the dimmer a cluster of galaxies was (indi-

cating greater distance), the greater its shift was. This redshift demonstrates that the distances between clusters of galaxies are increasing. In other words, the universe is expanding.

Consider a sphere containing n particles at a distance $r = 1$ m from each other. If the sphere expands, the distance between the particles will increase. If the distance between adjacent particles increases from 1 m to 2 m, the distance between particles once removed will increase from 2 m to 4 m, that between particles twice removed will increase from 3 m to 6 m, and so forth. The relative velocity between adjacent particles will be 1 m/s, that between particles once removed will be 2 m/s, that between particles twice removed will be 3 m/s, and so forth. For a constant rate of expansion, the relative distances between particles increase proportionally to their distances.

The expansion rate of the universe has not yet been exactly established. The reason is that the Cepheid yardstick applies only to the Local Group (a small cluster of about 40 galaxies to which our Galaxy belongs), which has a radius of about 2.5 million light-years. A cluster of galaxies is gravitationally bound and does not expand—what expands is the space between one cluster and the next one. The next one, in our case, is the Virgo cluster, a giant cluster of 2,500 galaxies 49 million light-years away. At that distance, individual Cepheids cannot be resolved. So astronomers study star clusters in a distant galaxy and estimate the distance to that cluster (and hence to its galaxy) by *assuming* that the brightest stars in the cluster have the same luminosity as the brightest stars in clusters within our own Galaxy, followed by application of the inverse square law (Chapter 3). The distances to galaxies that are too far for even the brightest stars to be resolved are estimated by assuming that the brightest galaxies in a cluster of galaxies are as bright as the brightest galaxies in our own cluster—for instance, Andromeda (Fig. 5.39). These methods of estimating distance are essentially the same as gauging the distance of a motorcycle at night from the brightness of its headlight. At distances greater than 300 million light-years, the redshift is measurable with sufficient accuracy to determine distance from the rate of expansion of the universe. Because no cluster of galaxies is close enough to determine the period of Cepheids in it and, at the same time, far enough to measure its redshift, the rate of expansion remains uncertain. At the present time, the probable rate of expansion, called the *Hubble constant* (symbol H_0), appears to center around 18 km/s per million light years of distance. This value means that galactic clusters 300 million light-years apart move away from each other at a speed of $18 \times 300 = 5,400$ km/s.

The redshift parameter z is equal to $\Delta\lambda/\lambda$, where λ is wavelength. A redshift parameter of 0.10, for instance, means that the wavelength is increased by 10%. For nonrelativistic speed, we have

$$z = \Delta\lambda/\lambda = (dr/dt)/c$$

where z is the redshift parameter, λ is wave-

length, r is distance, t is time, and c is the speed of light. For relativistic speeds, we have

$$z = \Delta\lambda/\lambda = [(c + dr/dt)/(c - dr/dt)]^{1/2} - 1 \quad (3)$$
$$v_r = c[(z + 1)^2 - 1]/[(z + 1)^2 + 1] \quad (4)$$
$$r = v_r/H_0 \quad (5)$$

where z is the redshift parameter, λ is wavelength, c is the speed of light, r is distance, v_r is the recessional speed, and H_0 is the Hubble constant. Fig. 5.40 shows the relationship between redshift parameter and recessional velocity.

If $z = \Delta\lambda/\lambda > 0$ (redshift), r is increasing.
If $z = \Delta\lambda/\lambda = 0$ (no shift), r is constant.
If $z = \Delta\lambda/\lambda < 0$ (blueshift), r is decreasing.

The Hubble constant, being equal to (length/time)/length, has the dimension of 1/time. For a constant rate of expansion, therefore, the age of the universe is simply the inverse of the Hubble constant. This quantity is called *Hubble time* (symbol H_t):

$$\begin{aligned} H_t &= 1/H_0 \\ &= 1/[18,000\,\mathrm{m\,s^{-1}}/(10^6 \times 9.46 \cdot 10^{15})\,\mathrm{m}] \\ &= 1/1.90 \cdot 10^{-18}\,\mathrm{s} \\ &= 5.25 \cdot 10^{17}\,\mathrm{s} \\ &= 1.66 \cdot 10^{10}\,\mathrm{y} \\ &= 16.6\ \text{billion years} \end{aligned}$$

The uncertainty about the value of the Hubble constant is such, however, that the age could be as low as 10 billion years or as high as 20 billion years. The Hubble time, therefore, should be rounded to $15 \cdot 10^9$ years.

The *Hubble distance* (symbol H_r) is the distance covered by light during the Hubble time (i.e., since the present universe was created). It is the radius of the visible universe:

$$\begin{aligned} H_r &= c(1/H_0) \\ &= 299,792,458 \times (5.25 \cdot 10^{17}) \\ &= 1.57 \cdot 10^{26}\,\mathrm{m} \\ &= 1.66 \cdot 10^{10}\ \text{l.y.} \\ &= 16.6 \cdot 10^9\ \text{l.y.} \end{aligned}$$

However, if Hubble time were 10^{10} y, the Hubble distance would be 10^{10} l.y.

The discovery that the universe is expanding takes care of Newton's contention that the universe must be uniform and infinite or it would collapse upon itself: It does not collapse because it is not stationary, but is expanding and has been expanding since birth. The

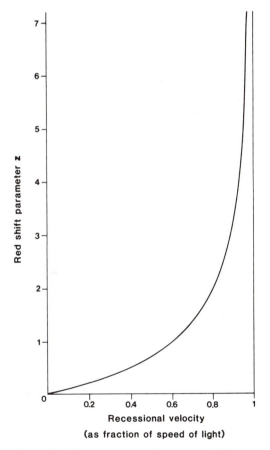

Figure 5.40. Recessional velocity versus redshift parameter z. As the recessional velocity approaches the speed of light, the redshift parameter approaches infinity. (Emiliani, 1988, p. 79, Fig. 6.2.)

discovery also takes care of Olbers' paradox: The sky is dark at night because the most distant reaches of the universe within the *Hubble horizon* (the sphere whose radius is equal to the Hubble distance) are expanding at a rate such that visible light is lengthened to the infrared and beyond, losing energy in the process. More important, the age of the universe is finite ($15 \cdot 10^9$ y), and therefore light emitted by objects beyond a distance greater than the Hubble distance has not yet reached us. If anyone wonders how objects more than $15 \cdot 10^9$ l.y. from each other can exist if the universe is only $15 \cdot 10^9$ y old, one should remember the inflation period between 10^{-33} and 10^{-32} s from time zero (Chapter 4).

As shown by equations (3), (4), and (5), the redshift of a source gives its recessional velocity and hence its distance. The redshift can therefore be used to determine distances. The Dutch astronomer Maarten Schmidt (b. 1929) discovered in 1963 that a star known for its strong radio emission had a redshift parameter of 0.16, corresponding to a recessional speed of 44,000 km/s and a distance of $2.5 \cdot 10^9$ l.y. Obviously, it was not a star within our Galaxy, as everyone had assumed, but a very luminous object at much greater distance. Other similar objects were then found at both lesser and greater distances. These strange objects were named *quasars* (for "quasi-stellar" objects). The most distant quasar yet observed is PC 1247 + 3406, with $z = 4.897$, which, for a Hubble constant of 18 km/s/10^6 y, corresponds to a recessional speed of 283,000 km/s (94.4% of the speed of light) and a distance of $15.7 \cdot 10^9$ l.y. The distribution of quasar distances is shown in Fig. 5.41. The sharp drop at $z = 2.5$ probably is due to the scarcity of quasars being formed during the first billion years of existence of the visible universe.

At the other end of the cosmic scale of distances, very precise measurements of the distances to nearby solid celestial objects, such as the Moon and the inner planets, can be made using radar. Radar (Radio Detecting And Ranging) transmits high-power (10 kW to 10 MW) pulses of radio waves with wavelengths of 1 cm to 1 m (frequency $3 \cdot 10^{10}$ to $3 \cdot 10^8$ Hz) and determines the time of arrival of the echo returning from the target. The duration of a pulse is about 1 μs, and the interval between pulses is 1,000 to 10,000 μs. Radar has

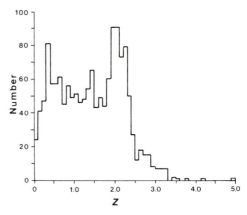

Figure 5.41. Abundance of quasars as a function of redshift parameter *z*. The sharp drop at $z = 2.5$ may indicate that comparatively small numbers of quasars were formed during the first billion years after the Big Bang.

also been used to map the surface features of Venus, which are obscured by the planet's thick, dusty atmosphere (Figs. 7.7 and 7.9).

Even more precise measurements can be obtained with lasers. A laser (Light Amplification by Stimulated Emission of Radiation) is a device consisting of a solid, liquid, or gaseous body whose atoms or molecules are excited to higher energy levels and then stimulated to radiate in phase at the same frequency as the stimulating wave. Lasers require a reflecting surface. In fact, placing a mirror on the Moon was one of the first activities of the first lunar-landing mission (*Apollo 11*, July 20, 1969).

In conclusion:

Eratosthenes discovered that the circumference of the Earth is 40,000 km.

Hipparchos discovered that the Moon is some 380,000 km from the Earth.

Cassini discovered that the Sun is some 140,000,000 km distant.

Bessel discovered that the nearer stars are some 10 light-years away.

Hubble discovered that the nearer galaxies are 2–3 million light-years away.

Schmidt discovered that the most distant objects are more than 10 billion light-years away.

And that's it. This is the measure of the visible universe as we know it today.

THINK

* Why is Venus never seen high in the sky?

* Given that blue light has more energy than red light, why do heat lamps use infrared light?

* Newton was 30 years old when Cassini determined the parallax of Mars and thus obtained a value for the astronomical unit. Knowing the distance of the Sun and measuring its angular diameter, it was easy for him to determine its size. From that, Newton concluded that the distance to the fixed stars was 2 light-years. Can you guess how he may have reached that conclusion? No? I'll give you a hint: Consider Mars.

* The Sun is a star of magnitude -26.74 at a distance of $1.5 \cdot 10^8$ km $= 1.58 \cdot 10^{-5}$ light-years. We receive from the Sun 1,000 watts/m^2 (full sunlight at sea level). Considering that the pupil of the human eye is about 1.5 mm in diameter, calculate how many watts (or joules/ second) the human retina would receive if one were to look directly at the Sun.

[ACHTUNG, ACHTUNG, ACHTUNG!!! Never never never look at the Sun directly, either with the naked eye or, much worse, through a lens or through binoculars. Never do that, even if the Sun looks weak because of clouds or because it's close to setting. The Sun emits 46% of its energy as infrared radiation, which the human eye cannot detect. Infrared radiation can go through clouds and smog, and can literally cook your retina to a nice crisp. To convince you of this, calculate how much more power you would receive from a luminous source by looking at it through standard binoculars with objective lenses 3 cm in diameter, compared with the light received directly by the pupil of your eye (which has a diameter of about 1.5 mm).]

Using the inverse square law and the Pogson ratio, calculate at what distance the Sun would appear to us as a star of magnitude 6 (the weakest magnitude visible with the naked eye), and calculate how many watts you would receive from such a star. The peak power of solar light is at a wavelength of 0.5 μm. Calculate the frequency of light having this wavelength, and, using $E = h\nu$, calculate the energy of a single photon. Finally, calculate how many photons you would receive from a star of magnitude 6.

* A neon lamp 1 liter in volume has a gas pressure of 5 mmHg (760/5 = 1/152 of an atmosphere). One mole of neon gas (20.179 g of Ne) occupies a volume of 22.414 liters at 1 atm. Therefore, the lamp contains $6.022 \cdot 10^{23}/(152 \times 22.414) =$ $0.00177 \cdot 10^{23}$ atoms $= 1.77 \cdot 10^{20}$ atoms. The wavelength of the strongest Ne line is 865.4 nm.

Calculate the frequency of the Ne line and, from that, its energy. If the lamp were a laser, all atoms would be discharging at the same time. Assuming that the discharge takes 10^{-8} s, calculate the power of the laser.

A standard neon lamp has a power of 40 W and a transition time of 10^{-8} s. Calculate the number of atoms discharging at any one given time and the concentration of the discharging atoms.

* Few quasars have been discovered beyond a boundary marked by redshift parameter $z = 2.5$. Using formulas (3), (4), and (5), calculate the distance to that boundary, and assuming a Hubble constant of 18 km/s/10^6 light-years, calculate the *cosmological age* (i.e., the time since the Big Bang) of these quasars.

* Where would you go if you wanted to see the Moon upside down without using a Keplerian telescope and without standing on your head?

* How precisely can one measure distances using radar?

6.1. THE MICROWAVE BACKGROUND RADIATION

The Arabs have a saying: Nothing is more beautiful than a night in the desert. There is truth in that adage. The desert is clean—no traffic, no fumes, no rotting garbage, no smog, no city lights, no nothing—only a crystalline sky studded with stars and soft sand on which to lie in quiet contemplation. But contemplating the starry sky makes us feel like a microscopic piece of nothing lost in space. And indeed we are. The universe is an enormous, expanding sphere that, neglecting virtual particles, is mostly empty: The average density of visible matter and energy in the visible universe is only about 10^{-30} g/cm³. Many scientists believe, however, that the visible universe has a hundred times more invisible matter than visible matter. This *dark matter*, as it is called, could consist of neutrinos with masses of $1.8 \cdot 10^{-5}$ u or of unknown particles that interact with other particles only gravitationally. Even so, the density of the universe still would be extremely low, only 10^{-28} g/cm³, equivalent to 60 hydrogen atoms per cubic meter of space.

There are plenty of photons, however. The density of photons is about a billion times greater than that of hydrogen atoms, or about 60 billion per cubic meter. These photons include radio waves, microwave and infrared radiation, visible light, and ultraviolet, x-ray, and gamma (γ) radiation, which are emitted by stars, interstellar clouds, and galaxies. In addition, there is the *microwave background radiation*, the radiation remaining from the Big Bang that has redshifted all the way down to the microwave band. The spectrum of the microwave background radiation (Fig. 6.1) is that of blackbody radiation (see Section 6.2), at a temperature of 2.735 K. Peak power (per unit wavelength) is emitted at the frequency of 5.25 waves per cen-

timeter (Fig. 6.1A), corresponding to a frequency of $1.57 \cdot 10^{11}$ Hz and a wavelength of 1.90 mm.

[The peak power wavelength λ_t at any time t can be calculated from the equation

$$\lambda_t = ct\lambda_R/R \qquad (1)$$

where c is the speed of light, t is time, λ_R is the peak power wavelength at radius R, the radius of the visible universe.]

If the power received is plotted against wavelength, peak power shifts to a wavelength of 1.26 mm (Fig. 6.1B).

Matter in the universe is concentrated in gas clouds, in stars with their retinues of planets and satellites, in neutron stars, and in black holes. These bodies are assembled into galaxies. The galaxies are grouped in clusters, and the clusters in superclusters that are distributed along lacy filaments throughout the universe (Fig. 6.2). In truth, the structure of the universe is simpler than that of the simplest microbe. Let us explore the structure of each component.

A star is an object that has enough mass for its center to reach a temperature sufficient for nuclear reactions ($> 10^6$ K). The most fundamental characteristic of a star, the property that fixes its life-style, is its mass. If its mass is small, a star may last forever, even though its light may eventually dim. If a star is big, it will burn out rapidly (50 to 500 million years) and then blow itself up.

6.2. THE SUN

The Sun and other common stars produce their energy by two processes: the proton-proton chain and the carbon cycle. The net result of both processes is the fusion of four hydrogen nuclei into a single ^{4}He nucleus. Table 6.1 shows these two systems of nuclear reactions. The

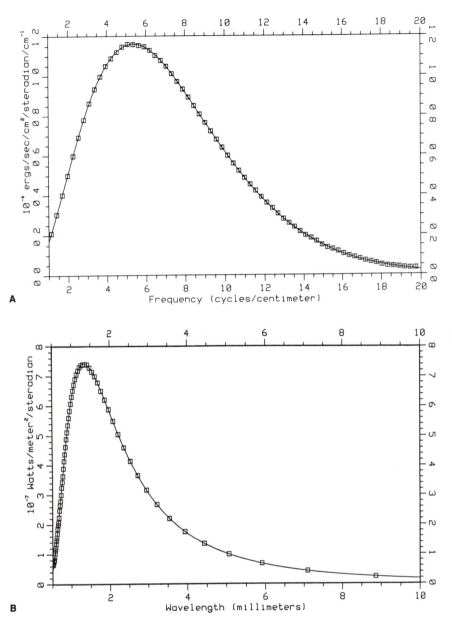

Figure 6.1. About 300,000 y after the Big Bang the temperature of the expanding universe was low enough for the free electrons to be captured by the nuclei of H and He to form neutral atoms. Electron capture was largely completed 800,000 y after the Big Bang. Removal of the free electrons made the universe transparent. The photons emitted by the process of electron capture plus the photons that existed from before formed the background radiation. Over the 15 billion years during which the universe expanded to its present size, the background radiation lost much of its energy because of the redshift. (A) Radiation flux per unit wavelength versus *wave number*, the number of waves (or cycles) per centimeter. Wavelength increases linearly to the left. Peak power occurs at wave number 5.25/cm, corresponding to wavelength 1/5.25 = 0.190 cm = 1.90 mm (which is in the microwave region of the electromagnetic spectrum). The average photon energy is only 0.00062 eV, and the corresponding temperature is 2.735 K. This is the temperature of intergalactic space. (B) Radiation flux versus wavelength. Peak power shifts to 1.26 mm, because the shorter wavelengths have more energy. In both (A) and (B) the little squares are measured values with their errors. The continuous line is the best-fit blackbody spectrum. (Courtesy John Mather, Goddard Space Flight Center, NASA.)

Figure 6.2. The large-scale structure of the universe. Each point represents a cluster of galaxies. The clusters are distributed along filaments that give the universe a lacy structure. (Seldner et al., 1977, Plate 1.)

proton-proton chain requires a temperature of about $15 \cdot 10^6$ K to take place. The carbon cycle requires a higher temperature, because it takes more energy to ram a proton into a carbon nucleus (which contains 6 protons) than attach it to another proton. In the Sun, 91% of the energy is produced by the proton-proton chain, and the rest by the carbon cycle. Stars more massive than the Sun have higher core temperatures and produce energy mainly by the carbon cycle.

Nuclear reactions in stellar cores produce high-energy photons (gamma rays) that interact with free electrons and lose energy while zigzagging toward the star's surface. It takes 1 million years for a photon to reach the surface of the Sun because of continuous absorption and reemission. Fig. 6.3 shows a cross section of the Sun. At the center is the core, with radius of 170,000 km (or 24% of the solar radius), a temperature of $15 \cdot 10^6$ K, and a density of $160 \, \mathrm{g/cm^3}$ (8.5 times greater than that of pure gold, one of the densest substances on Earth). So much hydrogen has been converted to

Table 6.1. **Energy production in stars**

Reaction	Energy produced (MeV)
Proton–proton chain	
$^1H + {}^1H \rightarrow {}^2H + e^+ + \nu_e$	1.442
$^2H + {}^1H \rightarrow {}^3He + \gamma$	5.493
$^3He + {}^3He \rightarrow {}^4He + 2{}^1H$	12.859 or
$^3He + {}^4He \rightarrow {}^7Be + \gamma$	1.586
$^7Be + e^- \rightarrow {}^7Li + \nu_e$	0.861
$^7Li + {}^1H \rightarrow 2{}^4He$	17.374 or
$^3He + {}^4He \rightarrow {}^7Be + \gamma$	1.586
$^7Be + {}^1H \rightarrow {}^8B + \gamma$	0.135
$^8B \rightarrow {}^8Be + e^+ + \nu_e$	15.04
$^8Be \rightarrow 2{}^4He$	3.03
Carbon cycle	
$^{12}C + {}^1H \rightarrow {}^{13}N + \gamma$	1.943
$^{13}N \rightarrow {}^{13}C + e^+ + \nu_e$	2.221
$^{13}C + {}^1H \rightarrow {}^{14}N + \gamma$	7.551
$^{14}N + {}^1H \rightarrow {}^{15}O + \gamma$	7.297
$^{15}O \rightarrow {}^{15}N + e^+ + \nu_e$	2.753
$^{15}N + {}^1H \rightarrow {}^{12}C + {}^4He$	4.966

helium in the core of the Sun since the Sun was formed about 4.7 billion years ago that the core consists of 62% He and 38% H (by mass). Above the core is the *radiative layer* (170,000 to 590,000 km, or 60% of the solar radius). As the name indicates, energy transfer across this layer is by means of photons. Above the radiative layer is the *convective layer* (590,000 to 695,500 km, or 15% of the solar radius), which, as its name indicates, convects and in so doing transfers energy outward. The radiative and convective layers consist of about 72% H, 26% He, and 2% heavier elements (by mass). The average density of the Sun is 1.4 g/cm³, about that of water logged wood. The Sun is kept "fluffy" by the pressure of the photons created in the core and attempting to reach the surface.

The surface layer of the Sun, which emits the light we see, is called the *photosphere*. It is 450 km thick and has a temperature ranging from 8,000 K at the bottom to 4,000 K at the surface (average = 5,770 K). Most of the hydrogen is in the atomic state. Density is similar to that in neon lights, about 10^{-6} g/cm³. At a temperature of 6,000 K, most atoms are neutral—there is simply not enough energy to ionize hydrogen.

[It requires 13.6 eV of energy to ionize hydrogen *from its ground state.* From

$$E = 1.5\,kT$$
$$= h\nu = hc/\lambda$$

where E is energy, k is the Boltzmann constant $(1.38 \cdot 10^{-23}$ J/K), T is temperature, h is Plank's constant $(6.62 \cdot 10^{-34}$ J/Hz), and c is the speed of light (299,792,458 m/s), we have

$$T = E/1.5\,k$$
$$\lambda = hc/E$$

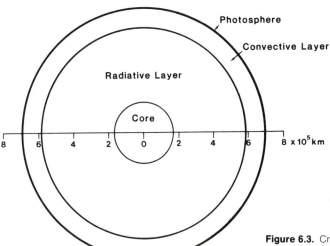

Figure 6.3. Cross section of the Sun. (Emiliani, 1988, p. 48, Fig. 4.8.)

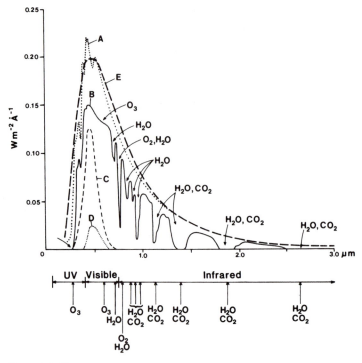

Figure 6.4. The solar spectrum (A) outside the Earth's atmosphere, (B) at sea level, (C) 10 m below the sea surface, and (D) 100 m below the sea surface; E is the spectrum of a blackbody radiating at the same tem- perature as the solar photosphere. The valleys in curve B reflect absorption by molecules in the terrestrial atmosphere. (Adapted from Dietrich et al., 1975.)

For $E = 1 \, eV = 1.6 \cdot 10^{-19} \, J$, we have

$$T = 1.6 \cdot 10^{-19}/(1.5 \times 1.38 \cdot 10^{-23})$$
$$= 0.78 \cdot 10^4 \, K$$
$$= 7{,}800 \, K$$
$$\lambda = 1{,}240 \, nm \quad (= infrared)]$$

In order to ionize hydrogen, a temperature of $7{,}800 \times 13.6 = 106{,}080 \, K$, or ultraviolet (UV) radiation with a wavelength of 91.2 nm, would be needed *if* the hydrogen were in the ground state. That would be the case if the temperature were 0 K. At higher temperatures, however, the hydrogen atoms are in thermodynamic equilib- rium with the photons, continuously absorbing and emitting them. For any given temperature, not all photons have the same energy, but exhibit a spread of energies with the peak some- where near the middle. This distribution is analogous to the Maxwell distribution of molecular velocities (Chapter 3). In the Sun, peak power is emitted at wavelength of 480 nm, which is in the region of blue color (Fig. 6.4).

Less power is emitted at shorter or longer wave- lengths. The sum total of the power emitted at the different wavelengths produces white light. The power distribution of a perfect emitter (called *blackbody emitter*) (Fig. 6.5) was worked out by Max Planck in 1900. It is given by the *Planck radiancy law*:

$$I_\lambda = (2\pi hc^2/\lambda^5)/(e^{hc/\lambda kT} - 1)$$

where I_λ is the radiancy at wavelength λ, c is the speed of light, h is Planck's constant, k is the Boltzmann constant, and T is absolute tem- perature. (It was in deriving this equation that Planck discovered his famous constant.)

A perfect emitter is called a blackbody emitter because it is also a perfect absorber, and when it absorbs, it looks perfectly black. It looks luminous, of course, when it emits. Fig. 6.4 shows that the solar spectrum outside the terrestrial atmosphere (curve A) closely ap- proaches that of a blackbody emitter (curve E) at the same temperature.

Ultraviolet photons in stellar photospheres with temperatures above 10,000 K are sufficiently abundant to almost completely ionize the hydrogen gas. The solar photosphere has a lower temperature (5,770 K) making these energetic photons too scarce for efficient ionization. However, in addition to hydrogen, the photosphere contains a small percentage of metallic atoms (1 in 100,000). Metallic atoms require considerably less energy to ionize than does hydrogen (6–8 eV for Mg, Si, Ca, and Fe in the ground state), and so they are readily ionized by light with wavelengths between 150 and 200 nm, releasing free electrons.

The K shell in an atom is complete when there are two electrons in it. Hydrogen normally has only one electron in it, to balance its single proton. But, if there are free electrons around, hydrogen tends to capture one to complete its K shell, thus forming temporary *negative* ions (H$^-$). The free electrons, which have a wide range of energies, lose energy while being captured, and that energy is what forms the solar spectrum. Negative hydrogen ions last only about 10^{-8} s. At any given instant

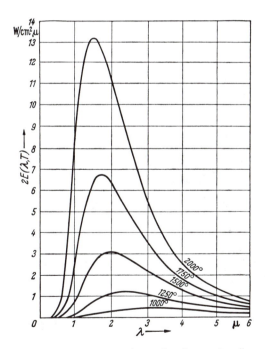

Figure 6.5. Power per unit wavelength per steradian emitted by a blackbody emitter at different temperatures. (Finkelnburg, 1964, p. 45, Fig. 20.)

there is only 1 negative hydrogen per 100 million neutral atoms in the solar photosphere. That is enough, however, to produce all the light emitted by the Sun.

The solar spectrum (Fig. 6.4) approaches that of a blackbody at an effective temperature of 5,770 K. (The *effective temperature* of a body is the surface temperature of a blackbody having the same surface and radiating the same power.) The shape of the blackbody radiancy curve is characteristically asymmetric, with fewer high-energy photons and more low-energy photons (Fig. 6.5). Stars cooler than the Sun have their power peaks shifted toward the longer-wavelength region and appear reddish. Conversely, stars hotter than the Sun have their power peaks shifted toward the shorter-wavelength region and appear bluish. Indeed, the color of a star is a measure of its temperature.

Solar emission amounts to a total of 2.32·10^8 watts/m^2 of solar surface (cf. ~ 1.1·10^6 W/m^2 for tungsten at 2,773 K). The distribution of solar radiation in terms of wavelengths is shown in Table 6.2. Specifically,

UV radiation (λ < 380 nm) = 5.779%,
Visible light (λ 380–765 nm) = 48.572%,
Infrared radiation (λ > 765 nm) = 45.649%.

The visible range is subdivided into colors. Color boundaries (in nanometers) are as follows:

ultraviolet/380/violet/455/blue/492/green/577/yellow/597/orange/622/red/765/infrared

The colors that compose the visible spectrum of the Sun, in their respective power proportions, combine to give white light.

The surface of the solar photosphere is highly turbulent, with granules that form and disappear every few minutes (Fig. 6.6). It is a real inferno up there. The granules are the tops of columns of hot gases, 500–5,000 km across, that rise at speeds of 2–3 km/s, with cooler borders between them where the gases sink back down. The photosphere is broken by sunspots (Fig. 6.7), which are holes up to 50,000 km across that appear dark because their temperature is 1,500°C lower than their surroundings. Sunspots are maintained by strong magnetic fields, 1,000 to 4,000 gauss.

Above the photosphere lies the *chromosphere*, a 2,500-km-thick layer of rarefied gases

Table 6.2.   Intensity of solar radiation in different wavelength bands

Type of radiation	Wavelength (µm)	Power (W/m²)
ultraviolet	0.2–0.3	11
violet	0.3–0.4	91
blue	0.4–0.5	198
green–yellow	0.5–0.6	193
orange–red	0.6–0.7	162
red–infrared	0.7–0.8	127
infrared	0.8–0.9	100
infrared	0.9–1.0	80
infrared	1.0–1.1	66
infrared	1.1–1.2	55
infrared	1.2–1.3	45
infrared	1.3–1.4	36
infrared	1.4–1.5	30
infrared	1.5–1.6	24
infrared	1.6–1.7	20
infrared	1.7–1.8	17
infrared	1.8–1.9	14
infrared	1.9–2.0	12
infrared	>2.0	72

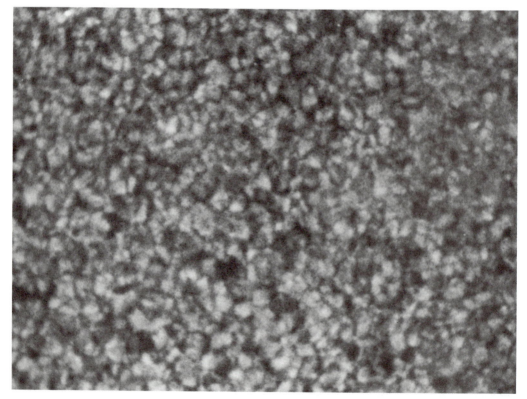

Figure 6.6. Surface of the solar photosphere exhibiting the characteristic granules. The width shown in this illustration is about 9 times the diameter of the Earth. (Courtesy Hale Observatories.)

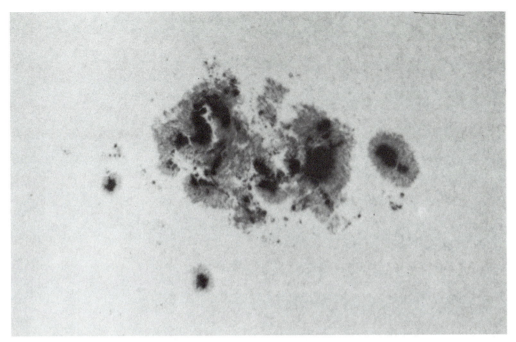

Figure 6.7. Sunspots originate at middle solar latitudes and move toward the solar equator; forming groups. The larger sunspots are 4–5 terrestrial diameters across. (Courtesy Hale Observatories.)

through which density drops from 10^{-8} to 10^{-14} g/cm^3. The chromosphere derives its name from the fact that it appears reddish, because of the strong emission of the hydrogen atom as its electron drops from the M to the L shell. Energy amounting to 1.89 eV ($= 3.02 \cdot 10^{-19}$ J) is emitted, which, using $E = h\nu = h(c/\lambda)$, is seen to correspond to radiation with wavelength $\lambda = hc/E = (6.626 \cdot 10^{-34} \times 3 \cdot 10^8)/3.02 \cdot 10^{-19} = 6.58 \cdot 10^7 = 658$ nm. This wavelength is within the red portion of the spectrum. The surface of the chromosphere exhibits *spicules*, which are jets of hot gases, 1,000 km across, shooting up to a height of 10,000 km at speeds of 20–30 km/s.

The *corona* is the outermost layer of the sun, grading into outer space. The corona displays *prominences*, sheets of ionized gases that move like drapes across the corona (Fig. 6.8). They follow the direction of the solar magnetic field and may reach an altitude of 1 million kilometers above the photosphere. *Solar flares* are ephemeral brightenings in the upper chromosphere–lower corona caused by sudden bursts of energy. These bursts throw particles into space to form the *solar wind*. The particles, which are mainly protons and electrons, travel outward at an average speed of 500 km/s and, in the vicinity of the Earth, are deflected by the magnetic field of the Earth toward the polar areas, where they ionize atmospheric gases at altitudes of 100–200 km and from the *auroras* (*aurora borealis* at high northern latitudes, *aurora australis* at high southern latitudes).

The photosphere, chromosphere, and corona have about the same composition: 78.4% H, 19.8% He, 1.8% havier elements (mainly C, O, N, Ne, Mg, Si, Fe, and S). These figures are mass percentages. In terms of numbers of atoms, hydrogen is 93.9%, helium 5.9%, and heavier elements 0.2%. This composition is similar to the composition of the radiative and convective layers.

6.3. THE STARS

Ancient astronomers grouped the stars into constellations that were perceived as representing animals, mythological personalities, or objects. Within each constellation, the stars are identified by Greek letters, beginning with α for the brightest star. The band of constellations

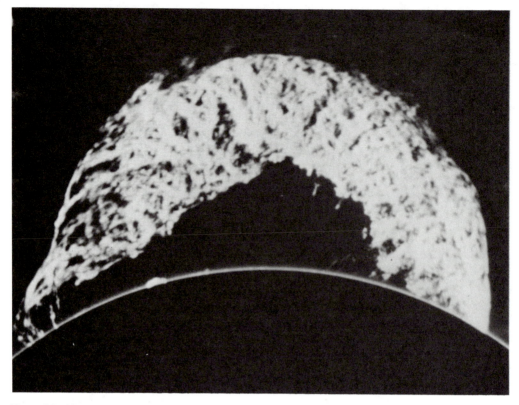

Figure 6.8. A large solar prominence extending to an altitude of 350,000 km (half a solar radius). (Courtesy National Center for Atmospheric Research/National Science Foundation, High Altitude Observatory, Boulder, Colo.)

encircling the sky along the path through which the Sun appears to be traveling during the year is called the *zodiac*. This band contains the 12 zodiac constellations of antiquity (Aries, Taurus, Gemini, Cancer, Leo, Virgo, Libra, Scorpio, Sagittarius, Capricornus, Aquarius, Pisces). These constellations represent mainly animals; hence the name *zodiac* (from the Greek ζῷον, meaning *animal*).

Astronomers have classified stars into spectral types identified by the letters O, B, A, F, G, K, and M. Originally the letters were in alphabetical order, but they were rearranged in the order of temperatures. Table 6.3 shows the colors, temperature ranges, and percentages of the different spectral types. Each spectral type is divided into 10 subdivisions, ranging from 0 to 9. The G type, for instance, ranges from G0 (hotter) to G9 (cooler). The Sun belongs to the G5 type.

Even the closest stars are so far away that the angular dimensions of their discs cannot be resolved. However, the radius of a star can be calculated from its luminosity and temperature, using the Stefan-Boltzmann law. This law states that *radiancy* (power emitted by a luminous body per unit of surface) is proportional to the fourth power of the temperature:

$$R = \sigma T^4$$

where R is radiancy, and σ is the Stefan-Boltzmann constant ($5.67 \cdot 10^{-8}\ \mathrm{Wm^{-2}\,K^{-4}}$).

The luminosity L of a star is the total amount of energy emitted, which is equal to the radiancy times the surface of the star:

$$L = 4\pi r^2 \times \sigma \times T^4$$

L is known from the power received and the

Table 6.3. Stars: spectral types

Spectral type	Percentage	Color	Temperature (K)
O	<1	blue	>30,000
B	3	bluish	10,000–30,000
A	27	white–bluish	7,500–10,000
F	10	white	6,000–7,500
G	16	yellow	5,000–6,000
K	37	red–yellow	3,500–5,000
M	7	dull red	<3,500

distance, σ is known, and T is known from the spectral type of the star. We can thus solve for r.

In 1905 the Danish astronomer Ejnar Hertzsprung (1873–1967) noticed that there was a relationship between the color and the brightness of stars belonging to the same group (Hertzsprung did not know the luminosity of his stars, i.e., the power they emitted, because the Cepheid yardstick had not yet been developed). In 1913 the American astronomer Henry Russell plotted all stars whose distances were known on a graph showing spectral type (and hence color and temperature) versus luminosity. This graph, now known as the Hertzsprung-Russell (H-R) diagram (Fig. 6.9), shows that most stars are crowded along a diagonal line called the *Main Sequence*. The Main Sequence runs from the lower right corner (low-temperature, low-luminosity stars) to the upper left corner (high-temperature, high-luminosity stars). In addition, a concentration of stars is seen above and to the right of the Main Sequence (low-temperature, high-luminosity stars), and scattered stars are seen below the Main Sequence (high-temperature, low-luminosity stars). The H-R diagram is of fundamental importance in understanding the evolution of stars.

Interstellar space contains gas clouds, 0.1 to 100 light-years across, that have temperatures of 20–80 K and densities ranging from 1 to 10^6 particles per cubic centimeter. The particles are mainly molecular hydrogen (H_2) and He atoms in their cosmic proportions (74% H, 26% He by mass; 92% H, 8% He by number of atoms—cf. Table 4.1). Stars are born when a diffuse, cold gas cloud is shocked into collapsing by a passing shock wave (usually generated by a super-

nova explosion, as discussed later). Gas clouds are turbulent, like clouds in the sky, with different portions having different angular momenta, and a small, net angular momentum. When a cloud collapses, heat is generated as matter falls inward, and the shrinking cloud begins to rotate faster and faster in order to conserve angular momentum. The rapid rotation forces some of the material of the original cloud to form a disc rotating around the central body. When the temperature in the core of the central body reaches 10^7 K, the proton-proton chain starts, and the star is born. Depending on the type, size, and dynamics of the original gas cloud, a single star with a planetary system, or a double star, or a multiple star may form. About half of the visible stars are part of double or multiple systems.

When a star begins shining, it is still quite expanded and quite cool at the surface. It is located in the lower right portion of the H-R diagram. As contraction continues, temperature and luminosity rise rapidly, and the star moves up and to the left in the H-R diagram. About 50% of the energy of condensation and much gas is dissipated into space during this stage, which is called the T Tauri stage.

[T Tauri is a variable star of magnitude 9.5 to 13.0 in the Taurus constellation that was the third to be identified as variable in that constellation; within a given constellation, astronomers identify variable stars in order of discovery, beginning with the letter R and, if needed, continuing after Z with double letters.]

Eventually the star comes to equilibrium with the rate of nuclear energy production in its core and settles on the Main Sequence. Where on the Main Sequence a star settles depends on its mass—mass increases from the bottom right to the top left of the H-R diagram. Once settled, a star spends most of its life at its place on the Main Sequence, fusing the hydrogen in its core into helium. Fusion is accompanied by a loss of mass (the Sun loses $4.257 \cdot 10^6$ tons of mass per second), but the loss is so slow that a star hardly moves down and to the right along the Main Sequence as one would expect it to do if the mass loss were more pronounced.

When hydrogen in the core of a star begins to be exhausted, energy production decreases, the

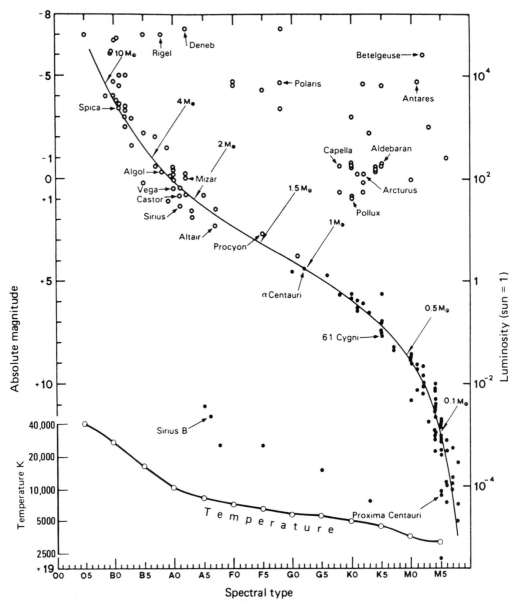

Figure 6.9. The Hertzsprung–Russell diagram. (Emiliani, 1987, p. 99, modified from Rigutti, 1984, p. 212, Fig. 84.)

core contracts, and temperature rises. In small stars, the temperature rise is moderate, and the consequent energy release is slow enough for the energy to be radiated away. In middle-size stars, like the Sun, the temperature rises from 15 to 100 million kelvin. The energy release is too fast for the energy to be radiated away, and the star expands 100-fold, which, in the case

of the Sun, means that its surface will reach halfway between Mercury and Venus. Luminosity increases (because of the increased release of energy), but surface temperature decreases to perhaps 3,000 K (because of the expansion). The star becomes a *red giant*, moving up and to the right in the H-R diagram. When the Sun will reach the red-giant stage, an estimated 5 billion years in the future, the temperature on Earth will rise, all water will be vaporized, the atmosphere will be blown away, and all life (if there is any left by then) will be extinguished.

For stars in the middle region of the Main Sequence, the process of expansion to the red-giant stage is not smooth. As the star expands, its surface layer cools off, and neutral hydrogen and helium atoms are formed by capturing free electrons. This capture radiates energy away, and the surface layer cools further and begins shrinking. As it shrinks, it not only warms up but also gets closer to energetic photons from the interior. The surface layer is ionized, free electrons are produced, and the layer expands. The process repeats itself. The Cepheids are stars of this type, oscillating with periods of 1 to 100 days (Fig. 5.33). Stars that oscillate with periods of less than 1 day are called *RR Lyrae* variables (Fig. 5.37).

The collapse of the core raises the temperature to 10^8 K, which is high enough for two ^{4}He nuclei in the core to overcome their mutual repulsion and combine to form beryllium-8 (^{8}Be). ^{8}Be has a half-life of only 10^{-16} s, but in the dense state of matter in stellar cores, a third alpha particle may hit and be absorbed even within that short time. The result is a ^{12}C nucleus. This process is called the *triple-alpha process*, because it requires the practically simultaneous collision of three ^{4}He nuclei (which are α particles). The formation of one ^{12}C nucleus from three ^{4}He nuclei destroys 0.00507 u of mass and produces 4.7 MeV of energy. ^{12}C is stable. In stars that are smaller than 2 solar masses (i.e., less than 2 times the mass of the Sun), the process of helium burning begins abruptly and produces a *helium flash*.

In the Sun, the fusion of helium to form carbon will last about a billion years. Many other nuclear reactions will occur during that time, some of which produce free neutrons:

$$^{13}C + {}^4He \rightarrow {}^{16}O + \text{neutron}$$
$$^{17}O + {}^4He \rightarrow {}^{20}Ne + \text{neutron}$$
$$^{21}Ne + {}^4He \rightarrow {}^{24}Mg + \text{neutron}$$

These neutrons are absorbed by the nuclei of the various elements, creating heavier isotopes. If more than one or two neutrons are absorbed by a nucleus, the nucleus becomes unstable. A β^- particle is emitted and the next element up is formed. For instance,

$$^{56}Fe + n \rightarrow {}^{57}Fe$$
$$^{57}Fe + n \rightarrow {}^{58}Fe$$
$$^{58}Fe + n \rightarrow {}^{59}Fe$$
$$^{59}Fe \rightarrow {}^{59}Co + \beta^-$$
$$^{59}Co + n \rightarrow {}^{60}Co$$
$$^{60}Co \rightarrow {}^{60}Ni + \beta^-$$

Once β^- decay has created the next element, the process can continue. In this way, all elements and isotopes heavier than boron, up to and including bismuth-209 (^{209}Bi, the heaviest stable isotope), are formed inside stars. This process is called the *s* process (*s* because it proceeds slowly). The process cannot continue beyond ^{209}Bi, however, because polonium, the next element beyond bismuth, is unstable and decays back into lead.

Eventually, all helium in the star's core will have fused into carbon. The star again stops producing energy, and the core again collapses. This time, a much greater amount of energy is released, causing the star to expand into a *red supergiant*, and its luminosity to increase 1,000 to 10,000 times. The Sun will expand beyond the orbit of Mars, and its mean density will decrease to $4 \cdot 10^{-8}$ g/cm^3. The hot solar plasma will engulf the Earth, reducing the Earth's surface from reddish to gray. The orbital velocities of the inner planets will decrease because of drag against the solar plasma, and these planets will spiral into the expanded Sun. The frozen atmosphere of the outer planets will be vaporized and their rocky cores will be exposed. Many stars in our Galaxy are in the red-supergiant stage. One of the largest is Betelgeuse, which can be found in the Star Chart (Fig. 6.10) at Right Ascension (RA) 5 h 53 m and Declination $+7°24'$.

A red supergiant is so extended that it loses about half of its mass to outer space. The material is ejected as an expanding shell of gas. As long as the shell remains rather close to the mother star, UV radiation from the star excites the gas, and the gas appears luminous. From a distance, the gas is seen thickest along the periphery of the shell, forming a ring (Fig. 6.11). What is left of the star is the carbon core, which is intensely hot, but not hot enough to fuse carbon. Nuclear reactions can no longer occur. As a result, the core shrinks, and its density rises to 10^6 g/cm^3 forming a *white dwarf*. The Sun, after its red-supergiant stage, will form a white dwarf similar in size to the Earth. If the

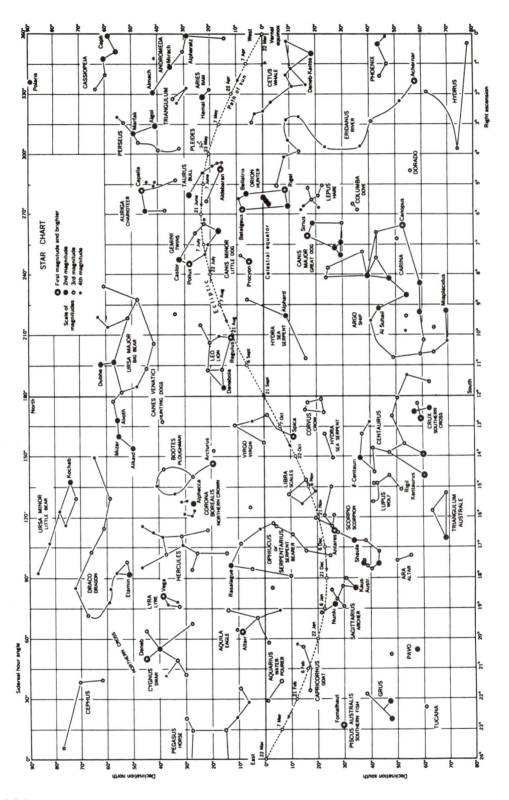

STAR CHART

114

Figure 6.11. The Ring nebula in the Lyra constellation, 4,000 l.y. distant. The ring is an expanding shell of gas ejected 5,000 y ago by the star in the center (the age is known by extrapolating backward the rate of expansion). The shell appears as a ring because that is where the gas is seen thickest from our point of view. The shell is expanding at a speed of 20 km/s and has reached a radius of 0.3 l.y. The ejection marked the beginning of the evolution of the star from a red supergiant to a white dwarf. The gas is made luminous by ultraviolet radiation from the central star. Gaseous structures of this type are called *planetary nebulae*, although they have nothing to do with planetary systems. There are some 50,000 planetary nebulae in our Galaxy at any given time. They remain visible for only about 50,000 y, after which their gases become dispersed in interstellar space. As old planetary nebulae fade away, new ones are born from other evolving stars. (Lick Observatory photography, by permission.)

Figure 6.10. (Facing page) Star Chart: Stars and constellations are positioned in this chart according to the *equatorial coordinate system*. In this system, the reference parameters are the *celestial equator* (which is the projection of the terrestrial equator on the celestial sphere) and the *celestial poles* (which are the projections of the Earth's axis, also on the celestial sphere). The celestial sphere is the imaginary sphere encompassing the entire universe, with its center at the center of the Earth. The celestial latitude is called *declination* (symbol dec) and is measured in degrees north ($+$) or south ($-$) of the celestial equator. The celestial longitude is called *right ascension* (symbol RA) and is measured eastward in hours, minutes, and seconds of time (1 h = 15° of longitude) from the *vernal equinox*. The vernal equinox is the point on the celestial equator where the ecliptic intersects the celestial equator. The vernal equinox has a longitude close to that of the star Alpheratz in the Andromeda con- stellation. In this coordinate system, the red giant Aldebaran is at RA 4h 34m, and dec $+6°55'$; the red supergiant Betelgeuse is at RA 5h 53m, and dec $+7°24'$; and the blue giant Rigel is at RA 5h 13m, and dec $-8°14'$. The right ascension can also be expressed in degrees (1 h = 15°, 1 m = 15', 1 s = 15"), starting at the vernal equinox, in which case the symbol used is α. Another way of expressing the celestial longitude is by means of the *sidereal hour angle* (top of the chart), which is equal to $360° - \alpha$. The dashed line across the chart traces the position of the Sun against the backdrop of the stars during the year, or, with an offset of six months, the highest position of the stars in the sky at midnight. The dot just above the letter *R* in the word ANDROMEDA marks the position of the spiral Andromeda galaxy (RA 0h 41.1m; dec $+41°7'$), the only extragalactic galaxy visible with the naked eye or with a modest binocular. (Strahler, 1971, p. 17, Fig. 1.18.)

core is less than 1.4 solar masses, the collapse will be stopped by electron pressure. This value, 1.4 solar masses for a carbon core, is called the *Chandrasekhar limit*, after the Indian astrophysicist Subrahmanian Chandrasekhar (b. 1910), who has studied the physics of stellar interiors. There are perhaps 10 billion white dwarfs in our Galaxy, with luminosities ranging from 1/100 to less than 1/30,000 that of the Sun. Most of them are not visible, even with powerful telescopes. Only the closer ones are. White dwarfs keep radiating away their residual energy for many billions of years. They keep cooling off, becoming *brown dwarfs*, and eventually *dark dwarfs*.

Stars with masses greater than 3 times the mass of the Sun end up differently. Some have masses up to 100 times that of the Sun. Their cores are hotter, their nuclear reactions proceed a lot faster, and their surface temperatures may be as high as 25,000 K or even more. The most massive ones shine a bluish light and are called *blue giants*. An example is the star Rigel, which can be found in the Star Chart (Fig. 6.10) at RA 5h 13.1m and Declination $-8°14'$.

When the triple-alpha process is completed in the core of a star with mass greater than 3 solar masses, the core contracts and the temperature rises sufficiently for ^{12}C to fuse and form ^{16}O, ^{20}Ne, and ^{24}Mg:

$$^{12}C + {}^{12}C \rightarrow {}^{16}O + 2{}^4He$$
$$^{12}C + {}^{12}C \rightarrow {}^{20}Ne + {}^4He$$
$$^{12}C + {}^{12}C \rightarrow {}^{24}Mg + \gamma$$

If the mass of the star is greater than 8 solar masses, the temperature in the carbon core rises to 10^9 K, which is sufficient to fuse oxygen:

$$^{16}O + {}^{16}O \rightarrow {}^{24}Mg + 2{}^4He$$
$$^{16}O + {}^{16}O \rightarrow {}^{28}Si + {}^4He$$
$$^{16}O + {}^{16}O \rightarrow {}^{32}S + \gamma$$

At temperatures above 10^9 K, ^{20}Ne becomes unstable. The more energetic ^{20}Ne nuclei break up and release alpha particles, which are absorbed by other ^{20}Ne nuclei to form, in succession, magnesium-24 (^{24}Mg), silicon-28 (^{28}Si), sulfur-32 (^{32}S), argon-36 (^{36}Ar), and calcium-40 (^{40}Ca). The temperature in the core rises sufficiently to cause a number of nuclear reactions that produce a flood of alpha particles and free neutrons. These particles react with ^{40}Ca to form the iron group as the temperature in the core reaches about $4 \cdot 10^9$ K. The process stops

with iron, because iron is at the peak of the binding-energy curve (see Fig. 3.4) and, therefore, further fusion consumes energy rather than producing it. A star with a mass 25 times that of the Sun will exhaust the hydrogen in its core in a few million years, will fuse helium for half a million years, and—as the core continues contracting and the temperature continues rising—will fuse carbon for 600 years, oxygen for 6 months, and silicon for 1 day.

At the end of the process, a massive star will have a layered structure, with an iron core surrounded by concentric shells of progressively lighter elements. The iron core, like the carbon core in smaller stars, does not produce any more fusion. As a result, there is a catastrophic collapse, which takes less than one second. The amount of power released is so great that the star literally blows up in a *supernova explosion*. Perhaps 90% of the star's original mass flies away in space, while the core is compressed by the accompanying implosion.

6.4. NEUTRON STARS AND BLACK HOLES

In massive stars with cores between 1.1 (the Chandrasekhar limit for an iron core) and 3 solar masses, the implosion will force the electrons in the core to combine with protons and form neutrons. The core becomes a compact *neutron star*, 5 to 15 km in radius, and with a density similar to that of nuclear matter (10^{14} g/cm^3).

The implosion conserves angular momentum, and so a young neutron star rotates very rapidly. Massive stars rotate at comparatively high speeds to begin with, averaging one turn every six hours. The core of a massive star is smaller than that of the Sun because of the much greater density. A core with a radius of 100,000 km that ends up as a neutron star with a radius of 15 km experiences a radial reduction by a factor of 6,666. The angular momentum of a solid sphere is equal to $(\frac{2}{5})m\omega r^2$, where $m = $ mass, $\omega = $ angular velocity (in radians per second), and $r = $ radius. The angular momentum is therefore proportional to the square of the radius. The square of the radius decreases from $100,000^2 = 10^{10}$ to 225 km^2. The reduction is by a factor of $10^{10}/225 = 4.4 \cdot 10^7$. In order to conserve angular momentum, the neutron star would

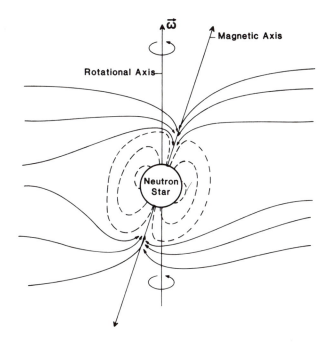

Figure 6.12. A pulsar. Electromagnetic radiation can escape from a neutron star only along its magnetic axis. If that axis is inclined to the star's axis of rotation and if the Earth is in the cone traced by the escaping radiation, a beam of light will hit the Earth each time the neutron star turns. These neutron stars are called *pulsars*. (Emiliani, 1988, p. 74, Fig. 5.13; adapted from Bath, 1980, p. 103, Fig. 5.8.)

have to spin $4.4 \cdot 10^7$ times faster, making 1 turn in 0.5 milliseconds (ms) or 2,000 turns per second.

As a matter of fact, the fastest-rotating neutron star known makes 642 turns per second, or 1 turn in 1.5 ms. Most other neutron stars rotate more slowly, from 10 turns per second to 1 turn in 4.3 s. The reason is interaction between the magnetic field of the neutron star and the ejected matter, which decelerates the rotation of the neutron star while accelerating the ionized particles in the ejected matter.

Neutrons at the surface of a neutron star decay into protons and electrons by β^-. These charged particles are drawn to the magnetic poles of the neutron star and spiral in opposite directions outward along the magnetic axis. Because of the spiraling motion, the particles are accelerated and, being accelerated, emit radiation. This type of radiation is called *synchrotron radiation* because it is observed in particle accelerators on Earth. The power radiated is inversely proportional to the fourth power of the mass. The radiation emitted by the electrons, therefore, is $1,836^4 = 1.14 \cdot 10^{13}$ times stronger than that emitted by the protons. Because the electrons have a whole spectrum of energies, electromagnetic radiation ranging from radio waves through visible light and

x-rays to gamma radiation is produced. This radiation can escape a neutron star only along its magnetic axis. If that axis does not coincide with the axis of rotation, the beam of light traces a cone (Fig. 6.12). If we are in the line of sight, the beam of light will hit us, like a beacon from a lighthouse, each time the neutron star makes one turn. For this reason, neutron stars are also called *pulsars*. If we are not in the line of sight, we see nothing. It is believed, therefore, that there are many more neutron stars than we can see.

The first pulsar was discovered in 1967 by Jocelyn Bell (b. 1943), then a graduate student at Cambridge University in England, using a large array of radio-wave detectors constructed by Anthony Hewish (b. 1924), an English astronomer also working at Cambridge. Since then, more than 300 pulsars have been discovered. Although the amplitude of the pulse of a pulsar may vary considerably, the period remains constant within 1 part in 10^8 over an extended time interval (Fig. 6.13). The pulsar discovered by the Cambridge group had a period of 1.33730113 s.

The magnetic field of a rotating neutron star keeps interacting with charged particles in space long after the neutron star is formed, continuously slowing down its rotation while

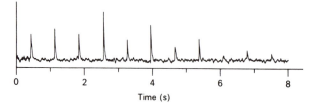

Time (s)

Figure 6.13. Pulse record of pulsar PSR-329 + 54. The amplitudes of these pulses vary widely in all pulsars, but the period is very regular. The period of this pulsar is 714 ms. (Manchester and Taylor, 1977, p. 7, Fig. 1.3.)

accelerating the particles (Fig. 6.14). The rate of slowing can be measured, and the age of a neutron star can thus be determined. It turns out that 80% of the neutron stars are less than 20 million years old. Clearly, a neutron star that rotates too slowly is difficult or impossible to detect. For all practical purposes it is dead.

If the mass of the core of a massive star is greater than 3 solar masses, the implosion forces matter to collapse into a single point at the center, called a *singularity*. There, density is infinite, and space and time are infinitely distorted. The gravitational field in the vicinity of a singularity is so strong that the escape velocity is greater than the speed of light. Not even light can escape. For this reason, the object is called a *black hole*. Only beyond a certain distance from the singularity can light begin to escape. The distance is called the *Schwarzschild radius*, after the German astronomer Karl Schwarzschild (1873–1916) who made the first calculations. A black hole consists of a singularity at the center, where mass is concentrated, and a spherical region of empty space around it from which light cannot escape. The surface of this sphere, whose radius is the Schwarzschild radius, is called the *event horizon*.

The magnitude of the Schwarzschild radius depends on the mass of the black hole. For a nonrotating black hole, the Schwarzschild radius is related to mass very simply:

$$r = 2GM/c^2 \qquad (2)$$

where r is the Schwarzschild radius, G is the gravitational constant ($6.67 \cdot 10^{-11}$ N m^2/kg^2), M is mass, and c is the speed of light. G and c are constants, and so they can be reduced to a single constant:

$$r = 1.48 \cdot 10^{-27} M \qquad (3)$$

A mass of $0.67 \cdot 10^{27}$ kg (the inverse of $1.48 \cdot 10^{-27}$), which is similar to the mass of Saturn, would yield a Schwarzschild radius of 1 m. In other words, if we were to compress Saturn to a single point, we would form a black hole with a Schwarzschild radius of 1 m.

The density of a black hole is defined as the mass of the singularity divided by the volume of the sphere with radius equal to the Schwarzschild radius. The density of the black hole resulting from compressing Saturn to a singularity would be

$$\rho = 0.67 \cdot 10^{27}/\tfrac{4}{3}\pi r^3$$
$$= 0.16 \cdot 10^{27} \text{ kg/m}^3$$
$$= 0.16 \cdot 10^{30} \text{ g/cm}^3$$

This density would be much greater than that of nuclear matter.

Notice that whereas the Schwarzschild radius increases in proportion to the mass of the singularity, as shown by equation (2) or (3), the volume increases in proportion to the *cube* of the radius, because the volume of a sphere is equal to $\tfrac{4}{3}\pi r^3$. Therefore, doubling the mass of a black hole increases the Schwarzschild radius by a factor of 2 and at the same time increases the volume by a factor of 8. As a result, as mass increases, density decreases. The relationship between density and mass is given by the following equation:

$$\rho = 1.36 \cdot 10^{-80}/M^2$$

The supernova explosion of a star 30 times more massive than the Sun might compress a core with a mass equal to 3 times that of the Sun to a singularity within a Schwarzschild radius of 8.8 km. The density would be $1.9 \cdot 10^{19}$ kg/m^3, or $1.9 \cdot 10^{16}$ g/cm^3, which is still greater than the density of nuclear matter. A giant black hole with a mass equal to $4.3 \cdot 10^9$ solar masses would have a radius of 11.6 light-hours (a little more than twice the radius of the solar system) and a density of only 1 g/cm^3 (the density of water).

Because light cannot escape from within a black hole, nothing can be known about it. There are ways by which black holes can be detected, however. Matter, in fact, can enter a black hole, simply by falling into it. The black hole is revealed by the radiation emitted by such matter as it falls in. If a black hole has a

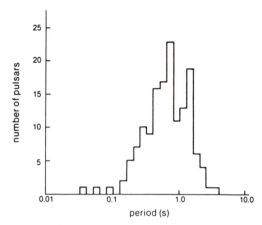

Figure 6.14. Distribution of pulsar periods for 149 pulsars. (Bowers and Deeming, 1984, vol. 1, p. 303, Fig. 16.11.)

companion star, the presence of the black hole is revealed by the gravitational perturbation of the companion star. So far, only three objects in the sky have been tentatively identified as black holes.

A black hole may not last forever, though. The uncertainty principle allows not only virtual force-field particles (photon, graviton, gluon, weak-force particles) but also virtual hadrons and leptons. If a virtual particle-antiparticle pair of hadrons or leptons forms immediately outside the event horizon of a black hole, and if one of the two particles of the pair falls into the black hole before it can recombine with its partner and fulfill the uncertainty principle, the partner may be ejected as a real particle. In that case, the mass-energy of the real particle would come from the black hole, which then would lose that precise amount of mass-energy. According to Stephen Hawking (b. 1942), the English physicist who developed this theory, black holes tend to "evaporate".

6.5. SUPERNOVAE

A supernova explosion is so violent that many nuclei of the heavier elements are broken up. A lot of neutrons are freed, because the heavier elements are comparatively richer in neutrons than the lighter elements. The free neutrons are rapidly captured by lead and bismuth nuclei to form all heavier elements, up to thorium, ura-

nium, and beyond. The process is so rapid (it is called the r process, where r stands for *rapid*) that the unstable elements between bismuth and thorium (polonium, astatine, radium, francium, radon, and actinium) have no time to decay. Other unstable elements, including technetium, promethium, and the transuranic elements, are also made during supernova explosions. At the same time, spallation of heavier nuclei makes lithium, beryllium, and boron, which are uncommon among the lighter elements because they are made by neither the s nor the r process.

It is estimated that in spiral galaxies there are two or three supernova explosions per century. During the past 1,000 years, however, only four supernova explosions have been observed in our Galaxy, instead of the expected 20 or 30. The reason is that half of the explosions, by chance alone, would have been visible only from the Southern Hemisphere (where recorded astronomical observations date only from recent times); and many of those in the other half are expected to have taken place too far from Earth for their light to have survived absorption by the rather dense interstellar medium that exists along the galactic plane.

The brightest supernova explosion on record occurred in the year 1054 and created the Crab nubula (Fig. 6.15). When we say "occurred in the year 1054", we mean that it was seen from Earth in 1054; the Crab nebula is 6,500 light-years away, which means that the explosion actually took place about 5450 B.C.E. Two other supernova explosions were seen in 1572 and 1604. The fourth was detected on February 2, 1987, at 2h 35m 41s A.M. (eastern standard time). It had occurred in the Large Magellanic Cloud (Fig. 5.34), which is 160,000 light-years away. The explosion, therefore, took place 160,000 years ago. It was detected when the light from the explosion and the accompanying burst of neutrinos reached the Earth.

Massive stars have relatively short lives because they burn their nuclear fuel very rapidly. They may last only 10^7–10^8 years, compared with 10^{10} years for a "regular" star like the Sun. Because the universe was created about 15 billion years ago, many generations of massive stars have come and gone. A supernova explosion has two important consequences. First, the new elements and isotopes formed in the star's interior and during the explosion are scattered in an expanding cloud of gases and dust and

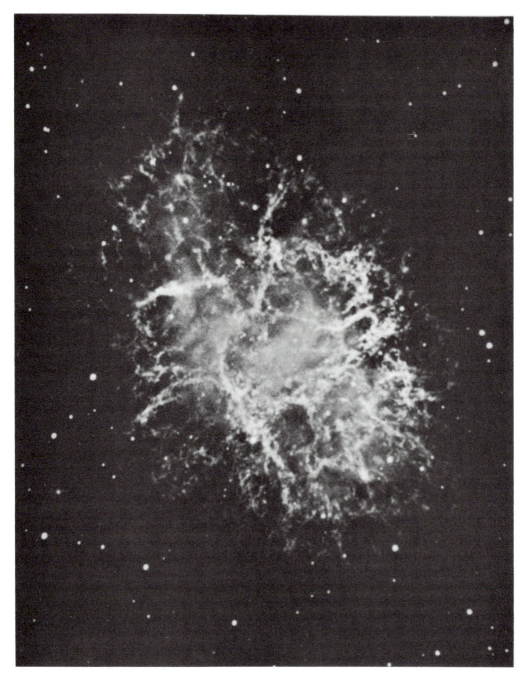

Figure 6.15. The Crab nebula is a turbulent, expanding cloud of gas 6,500 l.y. distant in the Taurus constellation. It originated from a supernova explosion seen in the year 1054. The remnant of that explosion is a pulsar with a period of 33 ms (30 rotations per second). The Crab pulsar is the lower one of the two stars of similar magnitude seen close to each other at 11 o'clock near the center of the nebula. (Lick Observatory photogarph, by permission.)

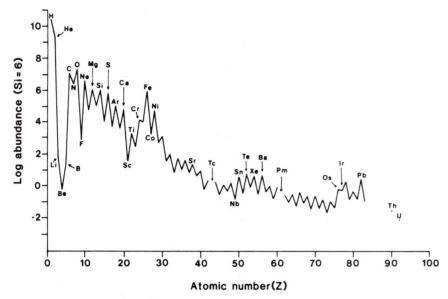

Figure 6.16. The abundances of the elements in the solar system and neighboring regions referred to silicon $= 10^6$. (Emiliani, 1988, p. 94, Fig. 7.1.)

are mixed with the interstellar medium (which is mainly hydrogen and helium atoms). New stars, formed from this mixture, start with a complement of heavy elements, including thorium and uranium. As the process of formation of massive stars, synthesis of heavier elements in their interiors, and supernova explosions continues, the universe as a whole becomes progressively richer in heavy elements. Stars that were formed recently have 100 to 1,000 times more iron and other heavy elements than older stars. The Earth is rich in these elements because the solar system was born "only" 4.7 billion years ago, when the universe was already 10 billion years old. A second, important effect of a supernova explosion is that it can trigger the formation of new stars by sending shock waves through neighboring gas clouds.

Fig. 6.16 shows the abundances of the elements in the solar system, referred to silicon $= 10^6$. The most abundant elements are hydrogen and helium, which account for 98% of the total. There is an overall decrease from hydrogen to uranium. The elements with even numbers of protons are more abundant than those with odd numbers because their nuclei are more tightly bound. The deep trough corresponding

to lithium, beryllium, and boron is due to the fact that these elements are formed mostly by spallation by cosmic rays or during supernova explosions. Carbon and oxygen are relatively abundant because they were formed both by α-particle fusion and by neutron capture followed by β^- decay. Iron is relatively abundant because its nucleus is particularly strongly bound. Nuclear structure is important in explaining other peaks and valleys in the element abundance curve (Fig. 6.16). Nuclei with the number of either protons or neutrons (or both) equal to 2 (He), 8 (O), 20 (Ca), 28 (Fe, Ni), 50 (Sr, Sn), 82 (Pb), or 126 (Pb, Bi) are particularly well bound, and their abundances are enhanced. These numbers are called *magic numbers*. They refer to a shell model for the structure of the nucleus analogous to the shell model for the distribution of electrons in the electron clouds of atoms. The broad rises in the strontium, barium, and lead regions are a magic number effect, increased by the fission and alpha-decay products of uranium and thorium. There is an intimate relationship between the birth of the universe and the dynamics of stars, on one hand, and the abundances of the elements, on the other.

6.6. GALAXIES

The stars are grouped into galaxies. There are three basic types of galaxies:

Elliptical galaxies (60% of all galaxies, Fig. 6.17), containing from 10^6 stars (dwarf ellipticals) to 10^{12} (giant ellipticals)
Spiral galaxies (30%, Fig. 6.18), containing 10^{10} to 10^{11} stars
Irregular galaxies, like the Magellanic Clouds (10%, Fig. 5.34)

In elliptical galaxies, the stars revolve around the galactic center in more or less randomly oriented planes (subspherical galaxies) or along a preferential plane, resulting in an elliptical shape. Elliptical galaxies contain little interstellar gas, and so star formation is not taking place—all their stars are "old" stars.

Spiral and irregular galaxies contain much interstellar gas. They are the sites of ongoing star formation. The spiral arms in a spiral galaxy are not due to the rotation of the galaxy over itself. They result from gravitational interference among the individual elliptical orbits of the stars around the galactic center, which produces a series of *gravity waves*. These waves move through space at a speed of about 30 km/s, which is greater than the speed of sound (~ 10 km/s) in the thin interstellar gas. A gravity wave moves through the gas as a shock wave, causing temporary compressions in the gas and leading to the formation of stars. These shock waves add to the shock waves produced by the supernova explosions discussed earlier.

Stars in spiral galaxies revolve around the galactic center in the same sense, but beyond a distance of 10,000 l.y. from the center their orbital velocities do not decrease as they should. This can happen only if there is a considerable amount of invisible, dark matter *increasing* in concentration with distance from the centers of the galaxies (Fig. 6.19). What this dark matter could be is the subject of intense speculation. Neutrinos with masses of $1.8 \cdot 10^{-5}$ u are a possibility.

Our Galaxy is a spiral galaxy, 50,000 l.y. in radius and 1,000 l.y. thick (at the center). It rotates clockwise, as seen from the north.

[*North* is defined from the Earth, but it applies to space as well. It is the direction toward which the Earth's rotational angular-momentum

Figure 6.17. M84, an elliptical galaxy in Virgo with a diameter of 25,000 l.y. and at a distance of $42 \cdot 10^6$ l.y. (Courtesy National Optical Astronomy Observatories, Tucson, Ariz.)

vector points. The opposite direction is *south*. The sky is divided into two hemispheres, northern and southern. They are separated by the celestial equator, which is the projection of the terrestrial equator from the center of the Earth on the celestial sphere, that is, on the sky.]

Like the other spiral galaxies, our Galaxy does not rotate as a wheel. Its rotational velocity increases radially from 150 km/s near the center to 260 km/s 15,000 l.y. away, and then flattens out to 200–225 km/s farther out (Fig. 6.20). The Sun, about 25,000 l.y. from the center (i.e., half of the way toward the periphery), travels around the center at a speed of 225 km/s. It completes one turn in 200 million years.

In 1932 the American radio engineer Karl Jansky (1905–1950) detected radio waves coming from the center of our Galaxy. Grote Reber (b. 1911), another American radio engineer, built a radio telescope in his backyard in Illinois in 1937 and proceeded to systematically study the sky. Basically, a radio telescope consists of a paraboloidal metal surface that reflects radio waves and focuses them on an antenna. Because radio waves have much longer wave-

Figure 6.18. M74, a spiral galaxy 80,000 l.y. in diameter and $3 \cdot 10^7$ l.y. distant. (Courtesy National Optical Astronomy Observatories, Tucson, Ariz.)

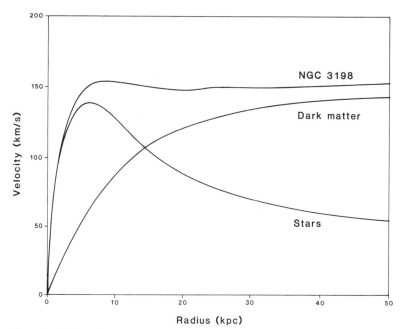

Figure 6.19. The rotation curve of galaxy NGC 3198 shows that the orbital velocity of stars around the galactic nucleus increases from near the center to about 8 kiloparsecs (8,000 parsecs = 26,000 l.y.) and then remains more or less constant as the distance from ther center further increases. The shape of the curve could be explained if dark matter existed, if its concentration increased from the center outward, and if it rotated like a solid sphere. Closer to the center, the stars would follow Keplerian orbits; farther away, they would be dragged along by the dark matter. (Adapted from Gribbin, 1988, p. 139, Fig. 6.2.)

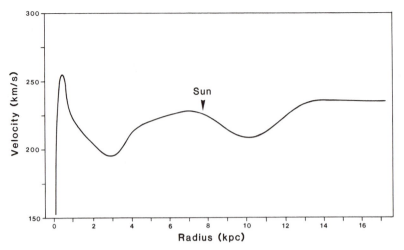

Figure 6.20. The rotation curve of our Galaxy. This curve suggests a more complicated distribution of dark matter (if it exists) in our Galaxy than in NGC 3198 (Fig. 6.19). (Adapted from Clemens, 1985, pp. 428–9, Fig. 3.)

lengths than visible light, the metal surface need not be polished to great accuracy.

Grote Reber discovered that there were other regions in space that emitted powerful radio waves. One was in the Cygnus constellation. That turned out to be an elliptical galaxy with a redshift of 0.057, which corresponds to a recessional speed of $16,600$ km/s and a distance of $9.22 \cdot 10^8$ l.y. The radio power of that galaxy, called Cygnus A, is 10^{38} W, which is 10 times the power of our entire Galaxy. Radio emission does not come directly from the core of Cygnus A. It comes from two extended lobes on opposite sides of the galaxy, at a distance of $160,000$ l.y. from the center. At that distance, the electrons are slowed down sufficiently to emit radio waves by interaction with the magnetic field. About 75% of radio galaxies that have been studied emit radio waves similarly, from extended lobes.

Before World War II it was not possible to pinpoint the direction of a radio source with sufficient accuracy to allow its optical identification. Radio astronomy made tremendous progress after World War II (Fig. 6.21). It soon became possible to pinpoint the positions of radio sources with sufficient accuracy for astronomers to train their optical telescopes on them. Many radio sources turned out to be so-called active galaxies (i.e., galaxies emitting signifi-

cantly more energy than our Galaxy). About 10% of the spiral galaxies are *Seyfert galaxies*, active galaxies that emit mainly in the infrared. The most active Seyferts grade into quasars, which are even more powerful and more distant. Many hundreds of quasars have been discovered since 1963, when the first quasar was identified as an extragalactic source of great power (Section 5.5). The power of quasars ranges from 10^{38} W to more than 10^{41} W (by comparison, the power of our Galaxy is 10^{37} W). Most of the power emitted is in the infrared. Only about 1% of quasars are also radio sources. BL Lac (short for BL Lacertae) objects are similar to quasars in terms of distance and infrared emission, but they are also strong radio sources.

Not only are the nuclei of active galaxies, quasars, and BL Lac objects much more powerful than our entire Galaxy, but the energy they emit exhibits strong variations over time intervals of weeks to months. This indicates that the power source cannot be more than a few light-weeks or light-months across. Astronomers now believe that BL Lac objects are the cores of distant elliptical galaxies, and quasars are the cores of equally distant spiral galaxies. Astronomers also believe that such cores consist of supermassive ($\sim 10^9$ solar masses) black holes that produce energy by gobbling

Figure 6.21. A radio telescope consists of a paraboloidal dish with a metallic surface that reflects the radio waves toward a single point, where a receiving antenna is located. The signal is amplified and recorded. (A) The directional radio telescope at Green Banks, West Virginia, one of the larger, fully steerable radio telescopes. The dish is 42.67 m (140 ft) across and can be rotated to explore different regions of the sky. The rotating weight is 2,359 metric tons. The concrete base houses control instruments. (Courtesy National Radio Astronomy Observatory.) (B) The world's largest radio telescope, 305 m across, built in a natural depression near Arecibo, Puerto Rico. The reflecting surface is a spherical wire mesh laid down with sufficient accuracy to enable imaging with waves as short as 7 cm. This radio telescope obviously cannot be rotated, but the rotation of the Earth brings different parts of the sky into view. In addition, a movable antenna can receive from regions as far as 20° from the zenith. (Courtesy Tony Acevedo, Arecibo, Puerto Rico.) (*Continued*)

up matter in their vicinity. In fact, as matter is accelerated and approaches the Schwarzschild radius of a black hole, a broad spectrum of radiation is released. We can "see" mainly radio waves from the center of a galaxy because shorter wavelengths are scattered by the concentration of gases that form the galactic cores.

C

Figure 6.21 (*Cont.*). (C) The Very Large Array, near Socorro, New Mexico, consists of 27 radio telescopes arranged in a Y-shaped pattern, with a 19-km-long stem and two 21-km-long branches. Radio waves from a given source arrive at slightly different times at different points along the array (unless the source is directly above). For a source close to the horizon and facing the stem directly, for instance, the difference in times of arrival at the closest and the farthest radio telescopes is $19/299{,}792.458 = 63\,\mu s$. The exact location of the source is determined by analyzing the differences in arrival times at different points along the array. (Courtesy National Radio Astronomy Observatory.)

A black hole with a mass equal to 10^9 solar masses has a Schwarzschild radius of about $3 \cdot 10^9$ km or 2.7 light-hours [see equation (2)]. In order to produce 10^{40} W, a black hole needs to transform into energy 1.76 solar masses per year. In fact, from

$$P = 10^{40}\,\text{W}$$
$$= 10^{40}\,\text{J/s}$$

and using $e = mc^2$, we have

$$10^{40} = mc^2/s$$

from which

$$m = 10^{40}/c^2\,\text{kg/s}$$
$$= 1.11 \cdot 10^{23}\,\text{kg/s}$$
$$= 3.51 \cdot 10^{30}\,\text{kg/year}$$
$$= 1.76 \text{ solar masses per year}$$

where P is power, m is mass, c is the speed of light.

The BL Lacs and quasars that we see are not only distant in space but also far back in time. We see an object 10^{10} l.y. away as it was 10^{10} y ago. It is possible that what we see are very young galaxies in the process of becoming stabilized.

Galaxies are grouped into clusters of 10–1,000 galaxies. Elliptical galaxies predominate in regular clusters (which are tightly bound gravitationally), and spiral and irregular galaxies predominate in irregular clusters (which are loosely bound gravitationally). Collisions between galaxies are more frequent in regular clusters than in irregular clusters. When two galaxies collide, only their gas clouds collide—the stars are too small and too far apart to

Figure 6.22. Medieval astronomer taking a peek beyond the curtain of the fixed stars.

collide. Colliding gas clouds form stars, and so the elliptical galaxies formed their stars early and are now depleted of gas. Spiral and irregular galaxies, on the other hand, suffered fewer intergalactic collisions and have kept enough gas to form stars even today.

Galactic clusters are grouped into superclusters (10,000–100,000 galaxies each). The superclusters are then arranged in long, intersecting filaments, giving the universe a lacy appearance (Fig. 6.2).

THINK

* If you feel humbled when comtemplating the starry sky over the desert, consider how truly exclusive is the neighborhood in which you live. To prove the point, calculate the *volume* occupied by humans on Earth (the surface of the Earth, multiplied by a modest thickness of your choice, including soil and trees). Next, calculate the volume of the space halfway to the nearest star (i.e., within a radius of 2.13 l.y.), under the assumption that, for sure, there is no other life within that range. Finally, divide the former into the latter, and compare the result with the area occupied by exclusive neighborhoods in your hometown relative to the total surface of the town.

* The radius of the Sun is 695,990 km, and its surface temperature is 5,770 K. Calculate the luminosity of the Sun using the Stefan-Boltzmann law. The absolute magnitude of the Sun is +4.84; that of the red giant Betelgeuse is −5.6, and its absolute temperature is 3,000 K. Calculate, using the Pogson ratio, the luminosity of Betelgeuse and, using the Stefan-Boltzmann law, its radius. How much larger than the Sun is Betelgeuse?

* Why do you think Hertzsprung and Russell chose to show temperature increasing from right to left in their famous diagram rather than from left to right as is customary?

* What would be the Schwarzschild radius of a black hole formed by the Earth? What would be its density? How much would a mass of 1 kg weigh if it were located on the event horizon?

THE SOLAR SYSTEM

7.1. SUN AND PLANETS

The ruler of the solar system is the Sun itself. To many ancient people the Sun was a god. The Egyptians called him Ra and built him great temples. The Greeks called him Helios, the Persians Mithras, and the Romans Apollo. As early as the fifth century B.C.E., however, the Greek philosopher and natural scientist Anaxagoras (ca. 500–428 B.C.E.) voiced the opinion that the Sun was no god at all, but merely a ball of fire (they ran him out of town for that). Today, we know a lot more about the Sun and the solar system than Anaxagoras did. The major components are:

Sun (mass = $1.9891 \cdot 10^{30}$ kg)
9 planets (combined mass = $2.670 \cdot 10^{27}$ kg)
61 + satellites (combined mass = $7.20 \cdot 10^{23}$ kg)
a large number of minor planets and meteoroids (combined mass = $1.8 \cdot 10^{21}$ kg)
10^{11} (??) comets (combined mass $\approx 10^{23}$ kg).

The total mass of the solar system is $1.9918 \cdot 10^{30}$ kg, of which 99.86% is contributed by the Sun. The Sun contributes little to the angular momentum of the system, in spite of its great mass, because it occupies the center. The total angular momentum is $3.1643 \cdot 10^{43}$ kg m^2 s^{-1}, distributed as shown in Table 7.1. The solar system revolves around a barycenter that can be as far as $1.5 \cdot 10^6$ km (2.15 solar radii) from the center of the Sun at times of planetary alignment. The Sun rotates counterclockwise as seen from the north, and all planets and asteroids revolve around the Sun in the same direction. This direction is called *prograde* and the opposite direction—clockwise as seen from the north—is called *retrograde*. In addition to having prograde orbits, Mercury, Earth, Mars, Jupiter, Saturn, and Neptune have prograde rotations. In contrast, the rotations of Venus, Uranus, and Pluto are retrograde. The rotational period of Venus (the Venusian day) is 243.0 terrestrial days, and its orbital period is shorter, 224.7 days. This means that one day on Venus is longer than one year and that Venus is slowly rotating clockwise, as seen from the north. Uranus and Pluto have their rotational vectors pointing along directions 7.86° and 32.46° (respectively) south of the plane of the ecliptic, which means that from the north, we see Uranus' and Pluto's southern hemispheres rotating clockwise. Their rotation is therefore retrograde.

The solar system is located about 25,000 light-years from the center of the Galaxy, near the inner edge of one of the spiral arms; it revolves around the galactic center at a speed of 225 km/s, making one turn in $\sim 200 \cdot 10^6$ y; and it oscillates across the galactic plane with a period of $33 \pm 3 \cdot 10^6$ y. Important data on the Sun are shown in Table 7.2.

The Sun formed $4.7 \cdot 10^9$ y ago from a cloud of gas and dust whose collapse was triggered by a supernova explosion. The condensing gas and dust that went to form the Sun contained all its original elements plus the elements formed during the supernova explosion. Those

Table 7.1. Angular momentum

Body	Angular momentum (% of total)
Sun	3.100
Mercury	0.028
Venus	0.057
Earth	0.082
Mars	0.011
Jupiter	59.622
Saturn	24.155
Uranus	5.234
Neptune	7.752
Pluto	0.0001

Table 7.2. The Sun

Parameter	Value
radius	= 695,990 ± 70 km
	= 2.32 light-seconds
mass	= $1.9891 \cdot 10^{30}$ kg
mean density	= 1.409 g/cm^3
power	= $3.826 \cdot 10^{26}$ watts
	= $4.257 \cdot 10^6$ tons/second
apparent visual magnitude	= -26.74
absolute visual magnitude	= $+4.84$
effective surface temperature	= 5,770 K
core temperature	= $15 \cdot 10^6$ K
composition	
core	= 38% H, 62% He
other layers	= 72% H, 26% He,
	2% heavier elements
magnetic field	
average	= 1–2 gauss
in sunspots	= 1,000–4,000 gauss
rotational period	
at equator	= 25.53 days
at poles	= 36.61 days
inclination of equator to ecliptic	= $7°15'$
mean distance from earth (= 1 AU)	= 149,597,870.7 km
	= 8.31675 light-mintues
	= 499 light-seconds
velocity relative to nearer stars	= 19.7 km/s toward the *solar apex*, a point at RA = 18h, $\delta = +30°$, near the star Vega
velocity relative to microwave background radiation	= 380 km/s toward a point at RA = 18.5h, $\delta = +30°$, in the Leo constellation
rotational energy	= $3.1 \cdot 10^{35}$ J
translational energy	= $1.4 \cdot 10^{41}$ J

heavier than uranium, from plutonium to californium and beyond, rapidly fissioned into lighter elements or decayed into uranium or thorium. Uranium and thorium are also unstable, but their half-lives are so long that they have survived in considerable amounts to the present day. Many unstable isotopes of the lighter elements were also formed by the supernova explosion (^{3}H, ^{26}Al, ^{36}Cl, ^{60}Fe, etc.). They also decayed rapidly into stable isotopes (^{3}H to ^{3}He by β^-; ^{26}Al to ^{26}Mg by β^+; ^{36}Cl to ^{36}Ar by β^-; ^{60}Fe to ^{60}Co by β^-; ^{60}Co to ^{60}Ni by β^-; etc.). These early radioactive-decay processes contributed to raise temperatures within the Sun and the planetary bodies that were forming.

7.2. FORMATION OF THE PLANETS

The abundances of the different elements in the collapsing cloud of gas and dust that formed the solar system are shown in Fig. 6.16. This cloud also must have contained various types of molecules like those found today in interstellar space (Table 7.3).

The cloud whose collapse formed the solar system probably had a very small net angular momentum, the vectorial sum of the angular momenta of the many different regions of the cloud. As the cloud began collapsing, conservation of angular momentum required it to increase its angular velocity. As a result, the collapse produced a central body (the early Sun) and a planetary ring. In addition to the usual large amounts of H and He, the planetary ring contained all the other elements, lots of molecules, and fine grains of solid matter.

Solid matter in interstellar space today— and most probably also in the gas cloud that formed the solar system—consists of microscopic crystals (0.1 μm across) of frozen gases (including methane, ammonia, water, etc.), iron-magnesium silicates [i.e., minerals rich in silica (SiO_2) as well as in iron and magnesium], and metals (mainly iron and nickel).

[Solid matter can exist in two forms: glasses and crystals. A *glass* is a disordered assemblage of atoms or molecules that are bound strongly enough to be solid at room temperature or below. In common glass, the atoms of the silica group (SiO_4) are arranged in the shape of a tetrahedron (a pyramid with four faces and four corners) (Fig. 7.1). The four oxygen atoms are at the four corners, with the silicon atom in the center. The tetrahedra, however, are linked at random and do not form the kinds of orderly structures seen in crystals. A glass is essentially a supercooled liquid that has solidified without properly arranging its atoms or molecules. A *crystal*, on the other hand, is an orderly assemblage of atoms or molecules according to a pattern that repeats itself over and over in space. As bonds

Table 7.3. Interstellar molecules: abundances relative to hydrogen (H = 1)

10^{-2}–10^{-5}	10^{-6}	10^{-7}	10^{-8}	10^{-9}	10^{-10}
H_2	HCN	OH	CH	HNCO	H_2CS
CO	NH_3	CS	CN	NH_2CN	HCOOH
N_2		SO	OCS	CH_3CH_2	CH_2NH
H_2O		C_2H	H_2S		$HCONH_2$
		SO_2	HCO		CH_3NH_2
		CH_3OH	H_2CO		C_2H_3CN
			HC_3N		CH_3CHO
					HC_5N
					CH_3COOH
					CH_3C_3N
					CH_3OCH_3
					C_2H_5OH
					CH_3CH_2

Note: Numerous ionic and isotopic species have also been identified.

are formed, bonding energy is released. A crystal is better bonded and therefore has less internal energy than a glass. As a result, glasses tend to crystallize as time goes by. The iridescence often seen in ancient glasses is the result of this crystallization.]

Knowing the approximate chemical composition of the planetary ring and assuming reasonable temperatures and pressures, one can deduce the types of minerals that could form. As it turns out, most of the material appears to have crystallized into only a few major types of minerals:

Olivine: a green mineral consisting of iron-magnesium silicate. Its chemical formula is $(Mg, Fe)_2SiO_4$. The expression in parentheses signifies that olivine may range from pure magnesium silicate (Mg_2SiO_4) to pure iron silicate (Fe_2SiO_4). Most olivines are somewhere in between.

Pyroxenes: dark silicate minerals containing varying amounts of magnesium, iron, and calcium. The most common pyroxenes are hypersthene $[(Mg, Fe)SiO_3]$, diopside $(CaMgSi_2O_6)$, and augite $[(Ca, Na) (Mg, Fe, Al)SiAl_2O_6]$.

Kamacite: a body-centered cubic mineral of iron (*body-centered* means that the iron atoms are distributed at the corners of a hypothetical cubic array, with an iron atom at the center of each cube; the iron atoms at the corners of the cubic cells are shared among 8 adjacent cubic cells, whereas that at the center of each cell is unshared).

Taenite: a face-centered cubic mineral of iron and nickel (*face-centered* means that in addition to the iron atoms in the cubic array, there is an iron atom at the center of each face of the cubic cell; these are shared between two adjacent cells).

Collapse of the cloud to form the Sun transformed a lot of gravitational energy into heat. Although most of this heat of accumulation was radiated away during the process of accumulation, the young Sun had a surface temperature considerably higher than today. The strong solar radiation ionized the gases in the planetary ring. The sweep of the Sun's magnetic field speeded up the revolution of the ionized gases around the Sun, while slowing down the rotation of the Sun itself (in order to conserve angular momentum). The contribution of the Sun to the total angular momentum of the solar system was reduced to a mere 3.1% of the total (Table 7.1).

During that period, the solar ring broke down into 10 concentric rings at specific radial distances from the Sun. The spacing of the rings apparently is a result of gravitational effects and resonances that are summarized in a numerical formula called the *Titius–Bode law.*

[The Titius–Bode law is not really a law, but a numerical relationship that gives the distances of the planets from the Sun in astronomical units. It was devised by the German astron-

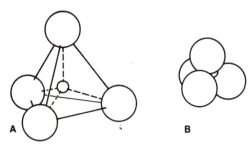

Figure 7.1. Silicon and oxygen bind to each other, forming a tetrahedron, with silicon ion at the center surrounded by four oxygen ions (A). (A *tetrahedron* is a pyramid with four sides.) Each oxygen ion is shared with an adjacent silicon ion, so that the number of oxygen ions is double the number of silicon ions, giving the chemical formula SiO_2 for silica. The oxygen ion is $1.32 \cdot 10^{-10}$ m across, and the silicon ion is $0.42 \cdot 10^{-10}$ m across. The oxygen ion is therefore more than three times as large as the silicon ion. As a result, the silicon ion is almost entirely hidden (B). (Berry, Mason, and Dietrich, 1983, p. 383, Fig. 15.1.)

omer Johann Titius (1729–1796) and popularized by another German astronomer, Johann Bode (1747–1826). According to this law, the distance from the Sun r (in astronomical units) is given by the formula

$$r = 0.4 + 0.3 \times 2^n$$

where $n = -\infty$ for Mercury, 0 for Venus, and 1 to 8 for the planets from Earth to Pluto, including the asteroids. The distance predicted for the planets from Mercury to Uranus (and the asteroidal belt) are quite accurate, but those predicted for Neptune and Pluto are, respectively, 29% and 95% too large. It is possible that these two, being the outermost planets, have been perturbed by a passing star. Neptune and Pluto were not known at the time when Titius and Bode lived, having been discovered, respectively, in 1846 and 1930.]

The mere existence of the Titius–Bode law, a numerical relationship for which we have no explanation, indicates that planetary distances are not random.

In addition to ionizing the planetary gases, the strong solar radiation raised the temperature within the planetary ring system. According to a model first proposed by Harold Urey in 1952, water, methane, and ammonia liquefied and wetted the solid particles, which became stuck together and formed mudballs one kilometer or more across, called *planetesimals*. Meanwhile, the young Sun was going through the T Tauri stage, the first stage in the life of a star of solar mass. During that stage, the Sun ejected large masses of gas, and its strong radiation blew all the gases that had not been trapped inside the planetesimals away from the inner rings toward the outer ones.

In a few million years, the swarm of planetesimals within each ring agglomerated into a single body, with the largest planetesimal within each ring exerting the strongest gravitational attraction. The docking of the first few planetesimals did not develop much heat, because the gravitational field of the masses involved was small. As agglomerations proceeded, however, the gravitational field increased proportionally, causing agglomeration to become increasingly more rapid and also more violent. This is an example of *positive feedback:* The cause (increasing mass) produces an effect (greater gravitational pull) that increases the cause (greater mass). Agglomeration soon became catastrophic, and in a short time, perhaps a few million years, the four inner rings formed the four inner planets (Mercury, Venus, Earth, Mars), and the five outer rings formed the five outer planets (Jupiter, Saturn, Uranus, Neptune, and Pluto). The fifth ring from the center failed to form a planet, for the reasons to be explained later.

Formation of the planets was so rapid that the relatively short-lived radioactive isotopes formed during the supernova explosion—especially ^{26}Al ($t_{1/2} = 720,000$ y), ^{36}Cl ($t_{1/2} = 301,000$ y), and ^{60}Fe ($t_{1/2} = 1,490,000$ y)—had no time to decay before being trapped inside the newly formed planets. The same is true for ^{107}Pd ($t_{1/2} = 6.5 \cdot 10^6$ y) and ^{129}I ($t_{1/2} = 15.6 \cdot 10^6$ y), which left an excess of their daughter isotopes ^{107}Ag and ^{129}Xe (respectively) as found in meteorites.

The young planets must have consisted of a chaotic mixture of silicates, metals, and trapped gases. Although the center of each planet probably was cold, the bulk was hot because of the decay of the short-lived radioactive isotopes and because much of the heat of accumulation was trapped inside. Silicates, in fact, are poor conductors of heat.

At that time, the planets were certainly

plastic, and possibly liquid. The metals, which were heavier than the silicates, began to sink to the center, while the silicates floated above. The sinking of the metals raised the temperature even higher, which made the sinking even faster—another example of positive feedback. Formation of metallic cores in the planets overturned the interior of each planet and forced out the gases that had been trapped inside. These gases formed the early planetary atmospheres.

The gases that came out from the interiors of the young planets had been in chemical high-temperature equilibrium with the metallic iron and the silicate minerals. They were, in order of abundance, hydrogen (H_2), helium (He), methane (CH_4), water (H_2O), nitrogen (N_2), ammonia (NH_3), and hydrogen sulfide (H_2S). The smaller planetary bodies, Mercury and the Moon, quickly lost all their gases because of their relatively small gravitational fields. As a result, their surfaces were not subsequently modified by atmospheric phenomena and still exhibit the scars of the fierce planetesimal bombardment that formed them (Figs. 7.6 and 7.10). Venus and Earth, on the other hand, were much larger and had a stronger gravitational field that could retain most of the gases that emerged from their interiors. In addition, their greater amounts of internal heat resulted in strong volcanic and tectonic activity. As a result, any trace of the primordial bombardment on their surfaces has been lost forever. Mars retained an atmosphere for a period much longer than did Mercury or the Moon, but eventually lost most of it.

The giant planets (Jupiter, Saturn, Uranus, and Neptune) retained their light gases and have thick atmospheres that consist predominantly of hydrogen ($\sim 85\%$) and helium ($\sim 15\%$). Pluto, the outermost planet, is smaller than the Moon and has polar caps of solid CH_4 that extend to 45° of latitude. It may also have a very thin atmosphere of CH_4.

7.3. COMETS

Planetesimals formed not only within the solar ring system but also farther out in space, far beyond the outermost ring. Those planetesimals could not form planets because they were too far apart from one another. During the $4.6 \cdot 10^9$ of the solar system's history, the orbits of those outlying planetesimals became perturbed by nearby stars and are now randomly oriented with respect to the main plane of the solar system. The planetesimals are randomly directed along their orbits, with 50% of them orbiting in the same direction as the planets (prograde motion), and the other half in the opposite direction (retrograde motion). There may be billions of them out there, forming a sparse cloud called the *Oort cloud*, after the Dutch astronomer Jan Oort (b. 1900) who proposed its existence in 1950.

If one of these bodies happens to have a prograde orbit close to the main plane of the solar system, it may be captured by the gravitational pull of an outer planet and thrown into an orbit that will bring it close to the Earth. At that distance from the Sun, its frozen gases are vaporized and ionized by solar radiation and shaped into an expanded head (called *coma*) and a long tail pointing away from the Sun. These bodies are the familiar comets (Fig. 7.2). About 750 comets are known. Of these, 150 have prograde motions and orbits that lie largely or totally within the orbit of Pluto. They are called *short-period comets* because their periods are less than 200 years. The most famous short-period comet is Halley's comet, which was first sighted in 239 B.C.E. Other appearances were recorded in 1066, 1145, 1222, 1301, 1378, 1456, 1531, 1607, and 1682. Originally those sightings were believed to relate to different comets, but in 1705 the English astronomer Edmund Halley (1656–1742) concluded that they had been appearances of the same comet, returning every 75–76 years. He predicted that the comet would return in 1759, and it did. More recent appearances have occurred in 1875, 1910, and 1986. The next appearance is due in 2061.

In addition to the 150 short-period comets, there are 600 comets that have almost parabolic orbits, with periods greater than 200 years. Their orbits have random orientations with respect to the main plane of the solar system, and their motions are either prograde or retrograde. Many of these long-period comets have been seen only once. No comets have yet been reported with hyperbolic orbits, which would indicate that they are freely roaming through the stars.

Comets consist of the primordial matter

Figure 7.2. Comet Kohoutek, January 1974. (Courtesy Hale Observatories, Pasadena, Calif.)

that formed the solar ring. The gases forming the coma and tail are very tenuous—one can still see stars through comets. The light they emit results from excitation of their electrons by solar radiation and from the reflection of solar light by solid dust grains. Spectroscopic analysis of cometary light has shown the presence of the atoms and molecules listed in Table 7.4. We can assume that these atoms and molecules are primordial, representing the earliest composition of the solar system. The metals in the comas are parts of the mineral grains that shield the gases in the coma from solar radiation and prevent their ionization. In contrast, the gases in the tail are ionized.

When comets come within the orbit of Mars, the gases start vaporizing. The comets begin losing matter and leave behind a trail of dust particles, which are, on the average, less than 1 μm across. A trail is around $10-50 \cdot 10^6$ km

in diameter and remains in the same orbit as the comet that generated it. The particles are slowed down by the *Poynting-Robertson effect*, caused by the orbiting particles impacting solar radiation at an angle as the radiation

Table 7.4. Chemical composition of comets

Comas	Tails
Nonmetals	
H, OH, O, S, S_2, H_2O, H_2CO, $(H_2CO)_n$, C, C_2, C_3, CH, CN, CO, CS, N_2, NH, NH_2, NH_3, HCN, CH_3CN	CO^+, CO_2^+, H_2O^+, OH^+, H_3O, CH^+, CN^+, N_2^+
Metals	
Na, K, Al, Mg, Si, Ca, Ti, V, Cr, Mn, Fe, Co, Ni, Cu	C^+, Ca^+

Figure 7.3. A small comet exploded 8.5 km above ground over the Tunguska River (eastern Siberia) on June 30, 1908. The trees in the forest below were flattened radially to a distance of 15 km. (Negative no. 126035, courtesy Department of Library Services, American Museum of Natural History, New York.)

passes by on its way out. This slowing down causes the particles to spiral toward the Sun and eventually fall into it. Particles up to millimeters or even centimeters in size are subject to this effect. The larger the particle, the smaller the effect, and the longer it takes for the particle to fall into the Sun. A 1-mm particle may take 10 million years to fall into the Sun. As old particles disappear into the Sun, new comets bring in new particles.

When the Earth, in its orbit around the Sun, crosses one of the cometary trails, particles are captured, and as they enter our atmosphere at 20–70 km/s, they burn. The *shooting stars* or *meteors* that streak across the night sky at specific times of the year are the luminous trails left by these particles. The richest meteor shower is the Perseid shower, which lasts for three days around August 12. The Perseids originate from the direction of $\alpha = 46°, \delta = +58°$ in the Perseus constellation (see the Star Chart, Fig. 6.10). The direction from which a meteor shower appears to originate is called its *radiant*. Other showers, in order of decreasing display, and their radiants, are as follows: Quandrantids (January 3; $\alpha = 231°, \delta = +49°$), Geminids (December 13; $\alpha = 112°, \delta = +32°$), δ Aquarids (July 30; $\alpha = 339°, \delta = -10°$), Orionids (October 21; $\alpha = 95°, \delta = +15°$); η Aquarids (May 4; $\alpha = 336°, \delta = 0°$).

Comets and asteroids collide with the Earth from time to time. On June 30, 1908, the nucleus of a small comet entered the Earth's atmosphere at high speed (30–40 km/s) and blew up 8.5 km above ground over the Tunguska River basin in northeastern Siberia. The trees on the ground below were blown down radially from ground zero (the point directly below impact) for a distance of 15 km (Fig. 7.3). At the exact center under the explosion, the trees were found still standing, although shorn of their limbs. It has been calculated that the cometary body was perhaps 50 meters across and had a mass of 60,000 tons. Its kinetic energy ($\frac{1}{2}mv^2$) was therefore $\frac{1}{2} \times 6 \cdot 10^7 \times (35 \cdot 10^3)^2 = 3.7 \cdot 10^{16}$ joules $= 8.8$

megatons (Mt), 704 times the energy of the 12.5-kiloton (kt) atomic bomb dropped on Hiroshima.

[Atomic bombs are gauged in terms of the equivalent energy of a conventional explosive rated at 1,000 calories (cal) per gram $= 4.1868 \cdot 10^3$ J/g. Thus, 1 kt $= 1,000$ t of explosive $= 10^9$ g $= 4.1868 \cdot 10^{12}$ J; 1 Mt $= 10^6$ t of explosive $= 4.1868 \cdot 10^{15}$ J. In the nitro compounds, such as TNT [short for trinitrotoluene, $(NO_2)_3C_6H_2CH_3$] and nitroglycerin $(CH_2NO_3CHNO_3CH_2NO_3)$, and in gunpowder, which is a mixture of charcoal (10%), sulfur (15%), and potassium nitrate (KNO_3, 75%), the oxygen is transferred from the nitrogen to the carbon, releasing $\sim 1,000$ cal/g. Because the oxygen is in the same molecule as the carbon, the reaction is much faster than in standard fuels, where the oxygen is outside. Paraffin wax, a mixture of hydrocarbons ranging from $C_{25}H_{52}$ to $C_{30}H_{62}$, delivers 10,000 cal/g when burning. A candle weighing 60 g and taking 6 hours to burn delivers 600,000 calories $(= 2,500,000$ J) in $6 \times 3,600 = 216,000$ seconds, or $2,500,000/216,000 = 12$ watts. By comparison, 60 g of TNT will deliver only 60,000 calories $(= 250,000$ J), but it will deliver them in 10 microseconds. The power delivered is therefore $60,000/10 \cdot 10^{-6} = 6 \cdot 10^9$ watts. The energy is much less, but the power is much greater because the time is much shorter.]

A cometary impact much larger than the one over Tunguska occurred 700,000 years ago, when a comet hit somewhere south of Australia. Molten rocks were splashed northward all the way to the Philippines and Vietnam. The molten fragments, an inch or so across, solidified in flight. These objects are called *tektites* (Fig. 7.4). Tektites may contain inclusions of meteoritic materials. Their average chemical composition (mass %) is $SiO_2 = 75.6$, $Al_2O_3 = 13.0$, $FeO + Fe_2O_3 = 4.1$, $Na_2O + K_2O = 3.5$, $MgO = 1.7$, $CaO = 1.4$, $TiO_2 = 0.8$, $H_2O = 0.005$. The very small amount of water, unusual for surface terrestrial rocks, may be explained by the high temperature at which these bodies solidified. No crater has been found associated with the Australasian tektites (it may be under water between Antarctica and Australia). Other tektite falls have occurred:

Figure 7.4. A typical tektite from Australia. The tektite flew through the air in the direction shown by arrow. Molten material from the front was blown toward the back and solidified in flight, forming a raised rim. (Courtesy Virgil E. Barnes, University of Texas, Austin.)

around the Aral Sea $1.1 \cdot 10^6$ y ago, associated with the Zhamanshin crater (10–15 km across); in Ghana $1.3 \cdot 10^6$ y ago, associated with the Bosumtwi crater (10.5 km across); in central Europe $14.8 \cdot 10^6$ y ago, associated with the Ries crater (24 km across); and in Texas $35 \cdot 10^6$ y ago (no associated crater identified yet). Some of these events also produced swarms of *microtektites*, spherical bodies less than 1 mm across that are found in deep-sea sediments. The Australasian microtektite field covers the western Pacific and the entire Indian Ocean; the Ghana microtektites cover an area of the eastern equatorial Atlantic off Ghana; and the Texas microtektites range from the Caribbean westward across the equatorial Pacific to Indonesia. A microtektite field about $2.3 \cdot 10^6$ y old has been discovered in the sea west of the Drake Passage.

Tektites are fused rock and are therefore glasses. Glasses have no crystalline structure. They tend to crystallize, causing cracks in their structure that again facilitate weathering. They do not last forever, which is the reason that no tektites older than the ones from Texas have been found.

7.4. PLANETS AND SATELLITES

There are 9 planets and at least 61 satellites in the solar system. The planets orbit the Sun,

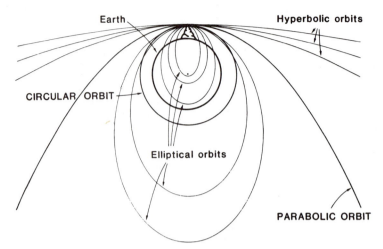

Figure 7.5. A cannonball shot horizontally from a mountaintop will reach farther the greater its initial speed, following segments of elliptical orbits as it falls back to Earth. If the cannonball is shot at a speed of 7.91 km/s, it will circle the Earth and return to the cannon. At speeds between 7.91 and 11.18 km/s, the ball will follow elliptical orbits of increasing ellipticity. At a speed of 11.18 km/s the ball will follow a parabolic orbit and escape. At speeds greater than 11.18 km/s the ball will follow hyperbolic orbits increasingly more open. (Emiliani, 1988, p. 119, Fig. 8.10.)

and the satellites orbit the planets. All orbiting celestial bodies move along elliptical orbits. All planetary orbits are counterclockwise, as seen from the north. Table A12 lists the planets and pertinent physical data.

The reason that the orbits of planets and satellites (including artificial satellites) are not circular is that these bodies perturb each other. But even under ideal conditions, that is, in the absence of perturbations, as would be the case of a very small body orbiting a much larger one, the orbit of the small body is very likely to be elliptical. Consider, for instance, a cannonball shot horizontally from a cannon placed on top of a mountain on an Earth free in space (Fig. 7.5). The greater the initial speed of the cannonball, the farther it will go. Eventually the cannonball will circle the Earth in the lowest possible orbit, which is circular. To achieve that, the velocity of the cannonball will have to be

$$v_c = (gr)^{1/2}$$

where v_c is velocity, g = gravitational field of the Earth (9.81 m/s²), and r is the radius of the Earth (6,371 km = $6.371 \cdot 10^6$ m). Using these values, we obtain

$$v_c = 7.9 \text{ km/s}$$

If the velocity is greater, the orbit will become more and more elliptical. At the escape velocity, the orbit becomes a parabola, which is an open curve, and the cannonball escapes the Earth's gravitational field. If the velocity is greater than the minimum for escaping, the orbit will be a hyperbola, and the ball will escape along an even more open curve. The minimum escape velocity v_e is given by the equation

$$v_e = 2^{1/2} v_c = 11.172 \text{ km/s}$$

Fig. 5.6 shows that when a plane intersects a cone at various angles (except through the apex), the circle and parabola are boundary cases, whereas the ellipse and hyperbola are not. Even if a planet, miraculously, were formed in a circular orbit, interactions with the other planets would soon perturb it into an elliptical orbit or would cause it to spiral into the Sun.

The laws governing the motions of the planets were worked out by the German astronomer Johannes Kepler (1571–1630), as mentioned in Section 5.4. Because of their importance, we present them here in a more systematic form:

First law: "The orbit of a planet is an ellipse, with the Sun at one of the two foci." This law establishes the shape of planetary orbits.

Second law: "A planet revolves around the Sun, with the connecting line sweeping equal areas in equal times." This law establishes that the planets move faster when they are closer to the Sun along their elliptical paths and more slowly when they are farther away.

Third law: "The square of the sidereal period of a planet is proportional to the cube of the semimajor axis of its orbit." This law establishes that the sidereal periods of the planets (the time it takes to go around the Sun once) increase as the distance from the Sun increases.

[The *sidereal period*, also called *sidereal year*, is the time a planet takes to return to the same position along its orbit with respect to the Sun and a distant star. See Section 8.1.]

If all the angular momenta of the Sun and the planets related to both rotation and revolution are added vectorially, the angular momentum of the entire solar system is obtained. This vector has a magnitude of $3.16 \cdot 10^{43}\,\mathrm{kg\,m^2\,s^{-1}}$ and a direction close to the rotational axis of the Sun (but not exactly coinciding with it) and pointing north. The plane normal to the total angular-momentum vector is called the *invariable plane.*

The orbital planes of the planets are within a few degrees of the invariable plane, with the exception of Pluto, whose orbital plane forms an angle of 15°33' (the orbital plane of the Earth forms an angle of 1°39'). This high inclination may have resulted from gravitational interaction with some passing celestial body, alien to the solar system. The rotational axes of the planets are within 30° of the normal to the invariable plane, with the exception of Uranus, whose rotational axis lies practically on the invariable plane.

Mercury is the innermost planet. It was given the Roman name for the Greek god Hermes, an illegitimate child of Zeus who grew up to become Zeus' messenger and personal secretary, as well as the protector of travelers. The planet Mercury is so small (only 5.5% of the mass of the Earth) that its internal heat is dissipated by conduction. Mercury has virtually no atmosphere. Its surface, therefore, has preserved the original features, the craters produced by the impacts of the last planetesimals (Fig. 7.6). The most conspicuous feature on the surface of Mercury is the enormous Caloris basin, 1,300 km across. This basin apparently resulted from a giant impact. Antipodal to the basin is a region of disturbed terrain apparently produced by seismic waves related to the giant impact. Other than that, Mercury's surface resembles that of the Moon. Mercury has no satellites.

Helium, nitrogen, oxygen, and hydrogen have been detected on Mercury, but in extremely small quantities (Table 7.5). These gases are evaporating from the interior of the planet and are on their way to being lost in space.

Venus is our sister planet, with a mass equal to 81.5% of that of the Earth. It was given the Roman name for the Greek goddess of love, Aphrodite. Venus is too large to dissipate its internal heat by conduction, and therefore its mantle convects. Convection in the solid state, as occurs in the mantle of both Venus and Earth, is very slow (a few centimeters per year),

Figure 7.6. Mercury has no atmosphere and no water. Its surface still preserves the features produced by the final bombardment of planetesimals falling in. (Courtesy National Space Science Data Center, Greenbelt, Md. Principal Investigator for the *Mariner* data is Dr. Bruce C. Murray.)

Table 7.5. Planets: mean surface temperatures and atmospheres

Planet	Surface temperature (K)	Gas Pressure (bars)	Gas	Volume percent
Mercury	440	$2 \cdot 10^{-15}$	He	42
			N	42
			O	15
			H	1
Venus	730	90	CO_2	96.4
			N_2	3.4
			SO_2	0.015
			H_2O	0.010
			Ar	0.007
Earth	288	1.013	N_2	78.804
			O_2	20.946
			Ar	0.934
			H_2O	variable
Moon	257	$2 \cdot 10^{-14}$	Ne	40
			Ar	40
			He	20
Mars	218	0.007	CO_2	95.7
			N_2	2.7
			Ar	1.6
Jupiter	165	$\gg 100$	H_2	82
			He	17
			Other	1
Saturn	140	$\gg 100$	H_2	82
			He	17
			Other	1
Uranus	57	$\gg 100$	H_2	85
			He	15
Neptune	57	$\gg 100$	H_2	85
			He	15
Pluto	42	10^{-5}	CH_4	100

but it nevertheless transfers heat from the interior to the surface. Early in the history of these two planets, internal heat was much greater and convection a lot faster. Gases that had been trapped in the interior during the process of accumulation were brought to the surface. As mentioned earlier, these gases were, in order of decreasing abundance, H, He, CH_4, H_2O, N_2, NH_3, and H_2S. Venus and Earth lost H and He, but retained the other gases. Ultraviolet radiation has enough energy to break up molecules in a process called *photolysis*. Photolysis of the H_2O, NH_3, and H_2S molecules produced free oxygen, free nitrogen, and free sulfur (the hydrogen, again, escaped). The free oxygen oxidized CH_4 and H_2S to CO_2 and SO_2, respectively. Venus rapidly acquired an atmosphere consisting mostly of CO_2, with

some nitrogen and smaller amounts of SO_2 and water vapor (Table 7.5).

The total pressure of Venus' atmosphere is now 92 bars. The partial pressure of CO_2 is 96.4% of 92 bars or 88.7 bars, which is equal to 90.4 kg/cm².

[The bar (b) is a unit of pressure; $1 \, b = 10^5$ pascals (exactly) = 0.9869 atmospheres = 1.0197 kg/cm².]

The surface of Venus is $4.6 \cdot 10^{18}$ cm², and so the total amount of CO_2 on Venus is $90.4 \times 4.6 \cdot 10^{18} = 4.16 \cdot 10^{20}$ kg. This vast amount of CO_2 creates a giant greenhouse effect, raising the average surface temperature to 460°C. This temperature is much higher than that of Mercury (160°C), although Venus is twice as far from the Sun and therefore receives only

Figure 7.7. Venus, photographed from a distance of 58,000 km, shows the cloud layer and its horizontal circulation. North is up, and east is to the right. Air rises along the equator and travels poleward at a slant toward the northwest (northern hemisphere) or the southwest (southern hemisphere). Rising air is replaced by air from the dark side traveling along the equator from east to west. (Courtesy NASA.)

one-fourth of the solar radiation that Mercury receives (per unit of surface). Venus' sky is red because CO_2 scatters red light. The Sun appears bluish because of loss of the red component. There is no liquid water on Venus and therefore no life.

The atmosphere includes a layer of clouds (Fig. 7.7) between 45 km of altitude (where temperature is about 25°C) and 70 km of altitude (where temperature is about −50°C). The clouds consist of droplets, 1 to 10 μm across, of concentrated (80%) sulfuric acid. Above and below the clouds there is haze, which consists mainly of sulfur microcrystals. The bottom 30 km of the atmosphere is clear of particles.

Temperature in the atmosphere of Venus drops from 460°C at the surface to about −90°C at an altitude of 100 km. Higher up it rises again, reaching about 25°C above 150 km of altitude. Absorption of solar radiation occurs in the hemisphere facing the Sun. Heat absorption within the atmosphere, from the Sun as well as from the ground as infrared radiation, occurs within the cloud layer.

The rotation of Venus is slower than its revolution (243 d versus 225 d; see Table A12), making the rotation of Venus retrograde. As a result, the atmospheric circulation on Venus is totally different from that on Earth. The circulation is driven by a cell within the cloud layer, where, as just mentioned, most of the heat absorptiom takes place. Heated air within the cloud layer rises at the equator to an altitude of about 70 km, sucking in air from the east and traveling poleward in an east-to-west direction (Fig. 7.8). Upon reaching the high latitudes, the air sinks down to an altitude of 45 km and returns to the equator as a counterflow. This cell drags along air both above and below, forming two adjacent cells that have opposite motions (poleward below, with return flow toward the equator above). The overall direction of air flow is the same as that of the planet, east-to-west. Wind speeds at the bottom of the cloud layer reach 350 km/h but decrease steadily to 3.5 km/h near the ground. The rapid circulation maintains the shady side of Venus hot, with a temperature only 4°C lower than on the sunny side. Furthermore, because descending air compresses, the temperature at high latitudes can be higher than at the equator.

Together with the haze below and above, the clouds conceal the solid surface of Venus. Radar beamed from spacecraft, both Russian

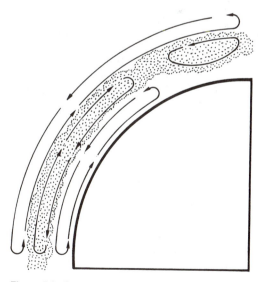

Figure 7.8. Cross section showing the vertical circulation in the atmosphere of Venus.

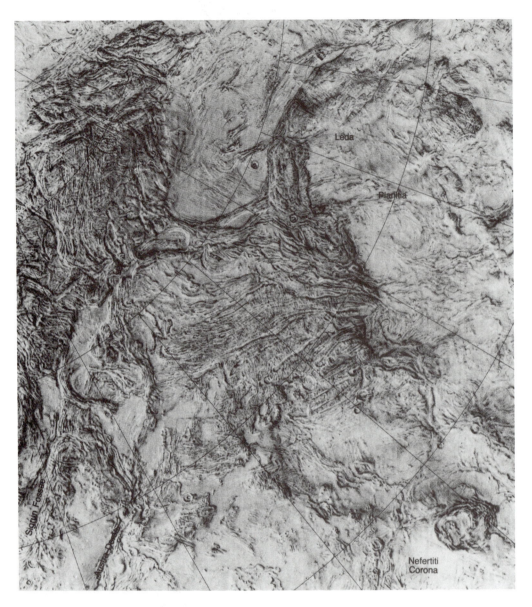

Figure 7.9. Folded mountain systems on Venus (scale: 1 cm = 200 km). (Courtesy NASA.)

and American, has been used to study the topography of Venus. There are lowlands covering about 30% of the surface of the planet, with scattered depressions reaching depths of 2 km below the mean radius of the planet. There are rolling uplands, averaging about 2.5 km of altitude and covering 60% of the surface, with folded mountain ranges (Fig. 7.9). The folding resulted from compression due to convective motions in the mantle and from the upwelling of magma. Wrinkling of the crust is facilitated by the high surface temperature, which impedes heat flow from the interior. There are also high plateaus at an average altitude of 4 km, covering about 10% of the surface. Above the plateaus rise some high mountains, the highest of which reach to 12 km above the mean radius of the planet. Detailed radar mapping of the surface of Venus by the *Magellan* spacecraft has revealed a host of volcanic and tectonic features, ranging from circular extrusions to ridges and swarms of faults. The surface of Venus exhibits relatively few impact craters, indicating that the crust is renewed by volcanic and tectonic activity. Venus has no satellites.

The real name of our planet is not Earth, but Gea (poetically Gaia), the cantankerous wife of Uranus and mother of Zeus (Gea is the root of the word *geology*, of course). Earth receives half as much radiation from the Sun as does Venus, and therefore the process of photolysis is less efficient than on Venus. Nevertheless, CO_2 and SO_2 also accumulated in the Earth's atmosphere. As more CO_2 accumulated, temperature rose. The sky turned red and the Sun turned bluish.

The Earth has 1.23 times the mass of Venus and should have produced 1.23 times more gases than Venus, in approximately the same proportions. If so, the amount of CO_2 in the earliest terrestrial atmosphere was $4.16 \cdot 10^{20} \times 1.23 = 5.1 \cdot 10^{20}$ kg. The surface of the Earth is $5.1 \cdot 10^{18}$ cm². The partial pressure of CO_2 on the surface of the young Earth was therefore $5.1 \cdot 10^{20}/5.1 \cdot 10^{18} = 100.0$ kg/cm² $= 97$ atmospheres. With that much CO_2 in the atmosphere, the surface temperature was about 90°C.

Unlike Venus, Earth had a surface temperature sufficiently low for liquid water to exist and accumulate in natural depressions to form the primitive ocean. That made it possible for CO_2 to dissolve in the water, react with silicates, and precipitate as carbonate. With a still high partial pressure of CO_2 in the atmosphere and hot water on the ground, the hydrologic cycle of evaporation, precipitation, and weathering, together with the lithologic cycle of sediment formation, transport, deposition, and recycling, must have been much more rapid than today. In a few hundred million years, most of the CO_2 was removed from the atmosphere, leaving nitrogen as the dominant gas. Nitrogen molecules scatter blue light, and therefore the sky turned blue. The carbonate that was precipitated at that time has been recycled many times over. It is now locked in the living and fossil coral reefs of the world,

in carbonate banks, in deep-sea muds, in limestone, and in marble.

The oldest rocks thus far discovered on Earth date from $3.96 \cdot 10^9$ y ago, "only" 700 million years after the formation of the solar system. It is likely that life originated during the early, relatively short time when the Earth's atmosphere still consisted of CH_4, H_2O, N_2, NH_3, and H_2S (see Chapter 19). Evidence of abundant bacterial life has been found in rocks as old as 3.5 billion years.

Bacteria ruled the Earth for the first three billion years and evolved most of the basic biochemical reactions upon which life rests. A major innovation was photosynthesis, the complex process by which CO_2 and H_2O are reduced to carbohydrates, utilizing solar light as a source of energy (see Chapter 19). Photosynthesis by the more advanced bacteria, the cyanobacteria, releases free oxygen to the atmosphere. The free oxygen went first to oxidize the rocks on the surface of the Earth, changing their color from grayish, like the rocks on the Moon, to reddish, the color of rust. After that, oxygen began accumulating in the atmosphere. Over a period of 4 billion years, the atmosphere of the Earth reached its modern composition, 78% N_2, 21% O_2, and 1% Ar.

The surface of the Earth is very different from that of Venus because the Earth has liquid water in its oceans, crust, and upper mantle. Water vapor (and CO_2 gas) in the atmosphere traps back-radiation from the ground and maintains the surface temperature at a balmy average of 15°C. Without water vapor and CO_2 gas in the atmosphere, the Earth's surface temperature would average -15°C. Water lowers the melting points of minerals and rocks and lubricates the internal motions of the Earth. The Earth's mantle convects more smoothly than that of Venus, moving continents around, opening new oceans, and closing old ones (see Chapter 12).

There are at least 61 satellites orbiting the planets of the solar system. Mercury and Venus have none. The Earth has the Moon (Fig. 7.10). (The Moon, *Luna* in Latin and Italian, *Selene* in Greek, was the sister of the Sun-god, Helios; the Germans got it all wrong, for they have the Moon masculine—*der Mond*—and the Sun feminine—*die Sonne*.) The mass of the Moon is 1.23% that of the Earth, which makes

Figure 7.10. The Moon, like Mercury, has no atmosphere and no water. Its surface, too, preserves the features produced by the final bombardment of planetesimals falling in. The side facing the Earth (A) contains vast areas flooded with basalt from the interior. These areas are called *maria*. The ages of the lunar basalts range from 3.9 to $3.1 \cdot 10^9$ y ago. (Lick Observatory). The opposite side of the Moon (B) contains no maria and resembles the surface of Mercury. (Courtesy National Space Science Data Center, Greenbelt, Md. Principal Investigator for the *Appollo* data is Fredrick J. Doyle.)

the Moon the second largest satellite with respect to its planet (the largest is Charon, as discussed later). The mean distance of the Moon from the Earth is 384,401 km, or 1.282 light-seconds. The Moon is at the same mean distance from the Sun as Earth, but because it does not have the thermal protection of an atmosphere its surface temperature ranges from a searing 125°C at the lunar noon to a chilling −160°C during the lunar night.

The gases released from the interior of the Moon during or shortly after its formation have escaped into space, because the Moon's gravitational field (16.5% of that of the Earth) is too weak to hold an atmosphere. As a result, the surface of the Moon is still pockmarked with the craters left by the last planetesimal impacts. Most of the craters are older than 1 billion years, but some are considerably younger: Copernicus, near the lunar equator, is about 800 million years old, and Tycho, in the southern hemisphere, is 270 million years old. Each is about 100 km in diameter. Both are perfectly visible with a modest pair of binoculars. They were caused by asteroids similar in size to the one that created havoc on Earth 65 million years ago.

The *Apollo* missions (1969–1972) placed a number of instruments on the Moon and brought back 382 kg of lunar rocks. The Russian *Luna* program also returned samples. We now know that the Moon consists of a small (200–300 km in radius) iron–nickel core, a 1,400-km-thick mantle made of peridotite (a rock rich in iron and magnesium silicates), and a crust 60 km thick consisting of anorthosite and basalt. Anorthosite, a rock uncommon on Earth, consists largely of the mineral anorthite ($CaAl_2Si_2O_8$) and forms the lunar highlands. Basalt consists of pyroxenes (calcium-iron-magnesium silicate minerals) and anorthite; it forms the floor of the oceans on Earth and some large expanses on the continents. Basalt also forms the floor of the lunar *maria* (singular: *mare*, Latin for *sea*), the vast, dark basins that are visible from Earth with the naked eye. The density of the Moon, 3.3 g/cm³, is much less than that of the Earth (5.52 g/cm³), indicating that the Moon has comparatively less iron and nickel than the Earth. The ages of the lunar anorthosites range from $4.6 \cdot 10^9$ to $3.1 \cdot 10^9$ y, whereas those of the basalts range from $3.9 \cdot 10^9$ to $3.1 \cdot 10^9$ y. Apparently, the formation of rocks from rock melts ended about 3 billion years ago. Since then the Moon has been essentially dead. The only activities are *moonquakes*, which are internal rock re-adjustments in response to terrestrial tides. Moonquakes were detected by instruments left on the Moon by the *Apollo* astronauts.

The orbit of the Moon is inclined 18.3° to 28.6° to the equatorial plane of the Earth. If it had formed from a ring of material orbiting the young Earth, its orbit would lie on the equatorial plane of the Earth or close to it. The high inclination of the lunar orbit speaks against that possibility. Although the lunar rocks bear many similarities to rocks common on Earth, they differ on one basic point—they contain no water, no hydrated minerals, and no minerals with the OH group in their crystal structure. In contrast, minerals that are hydrated or contain the OH group are plentiful on Earth. So where did the Moon come from? There is no clear answer to this question, in spite of all explorations and analyses. Perhaps the Moon is a piece of Mercury's mantle that somehow escaped Mercury and was later captured by the Earth. This would explain the high residual density of Mercury, the low density of the Moon, and the absence of hydrated minerals in lunar rocks—the young Sun certainly blew water vapor away from the innermost solar ring that produced Mercury.

Mars was given the Roman name of the Greek god of war, probably because it looks red. Mars has a mass that is only 11% of that of the Earth. Originally, its mantle probably convected, but convection soon ceased because the small size of the planet allowed much heat to escape directly into space. Mars has a tenuous atmosphere (pressure = 0.007 bar) consisting mainly of CO_2, with some nitrogen and argon (Table 7.5). It also has two ice caps over its poles. Each ice cap consists of a permanent, central portion made up of water ice and a seasonal, thin (~ 30 cm) layer of frozen CO_2 that extends down to latitude 60° during the winter. During the summer, the CO_2 vaporizes and condenses over the opposite pole, which is experiencing winter. Its atmosphere is so thin that the surface of Mars is clearly visible from Earth and from orbiting spacecraft (Fig. 7.11).

In 1877 Mars was at opposition, which means it was at its closest approach to the Earth (Fig. 5.29). The Italian astronomer Giovanni Schiaparelli (1835–1910) made a detailed study of the surface of Mars, as seen

Figure 7.11. Mars photographed from Earth, showing the north polar cap and the division of Mars into two hemispheres with sharply different crustal characteristics. (Lick Observatory photograph, by permission.)

through his telescope, and concluded that Mars was criss-crossed with straight lines, which he called *canali*. The popular press immediately came up with the notion of canal-digging Martians. The American astronomer Percival Lowell (1855–1916), a Bostonian of considerable wealth, established his own observatory near Flagstaff, Arizona, in 1894, at an altitude of 2,200 m, where the air is thin and dry, and settled down to study Mars. He, too, saw canals, more than 500 of them, intersecting in "oases" and sometimes changing in number and direction; indeed, he thought, they must be the works of Martians. Astronomers were a bit skeptical, but Lowell offered thousands of drawings. The matter was settled quite conclusively on July 14, 1965, when *Mariner 4* transmitted the first close-up picture of Mars from a distance of 10,000 km: There were no canals, only impact craters, just as on the Moon. The "canals" were explained as an illusion created by the tendency of the human eye, at the limit of visibility, to align scattered points.

Exploration of the Martian surface by subsequent spacecraft revealed that the surface of Mars is far more complicated than that of the Moon. To begin with, there are marked differences between the northern and southern hemispheres. The northern hemisphere is largely

Figure 7.12. Martian river system in the sourthern highlands flowing into a depression now filled with fine sediment. The two large craters are about 35 km across. (Courtesy NASA.)

a lowland, sparsely cratered; the southern hemisphere is a heavily cratered highland at an average altitude of 2 km above the mean radius of the planet. The northern hemisphere, however, contains a broad plateau 8,000 km across and 12 km high, above which rise huge volcanic structures. One of these, Olympus Mons, is the tallest mountain in the solar system. It rises to 25 km above the mean radius of the planet. Space explorations have in a way vindicated Schiaparelli and Lowell: There are channels on Mars (Fig. 7.12). These channels are not the huge canals envisioned before, but are somewhat similar to the dry river beds seen on Earth in desert areas—the arroyos of the North American southwest and the *widian* (singular *wadi*) of the Libyan and Arabian deserts. Connected to some of the larger channels are areas of "chaotic terrain" that contain giant ripples. They may have been created by sudden floods resulting from the collapse of ice dams. Similar features are seen in the American northwest (see Section 24.3). There are also volcanic islands surrounded by streamlined sedimentary deposits. All these features clearly show that once there was abundant water on Mars, and that water was running.

Mars has broad dune fields, especially around the ice caps. In spite of its thin atmosphere, Mars has windstorms capable of raising dust and moving it across long distances (Fig. 7.13). In conclusion, Mars is not a dead world like the Moon, but it is not very much

Figure 7.13. A Martian desert landscape in the northern hemisphere photographed by the *Viking 1* lander. Notice the dunes shaped by wing blowing from the right. (Courtesy NASA.)

alive like the Earth either. Is there life on Mars? Experiments carried out during the *Viking 1* and *Viking 2* landings yielded inconclusive results. However, primitive life in the form of bacteria may well have evolved there. Oxygen produced by the photosynthetic activities of these bacteria may have been the principal cause of the highly oxidized state of the Martian surface, which has gained Mars its name of "red planet".

Mars has two small satellites (Fig. 7.14), called Phobos ("Fear") and Deimos ("Terror"), after the two children of Mars who always accompanied him on his bloody forays. The two satellites are chunks of rock only about 20 km across. They probably are captured asteroids.

Jupiter (Fig. 7.15) is the Latin name for Zeus, the ruler among the twelve Olympian gods. It is fitting that the largest planet was named for him. Jupiter is believed to have a core of silicates and metals at its center, perhaps 10,000 km in radius, with an average density of some 40 g/cm³ because of the great pressure. Surrounding the core is a 60,000-km-thick hydrogen mantle, and above it a 1,000-km-thick atmosphere consisting of hydrogen and helium (Table 7.5). The hydrogen in the lower mantle, up to 25,000 km from the surface, is in the metallic state because of the high pressure—the electrons are loose within an

ocean of protons. The upper mantle, from 25,000 km of depth to the surface, consists of liquid hydrogen. The density of the mantle decreases from 15 g/cm³ at the base to 1 g/cm³ at the top. Jupiter has a dipole magnetic field that, like that of Earth, is inclined about 10.8° to the rotational axis. Its direction, however, is northward, which is opposite the direction of the magnetic field of the Earth. Jupiter's field appears to result from convective motions in the hydrogen mantle. The dipole moment is almost 20,000 times greater than that of the Earth, but the surface field at the equator is only 10 times greater (because of the greater distance from the source).

Jupiter's atmosphere consists of 82% molecular hydrogen (H_2), 17% helium, and 1% other gases (by volume) (Table 7.5). It is divided into latitudinal bands of low pressure (rising air) alternating with bands of high pressure (sinking air), with strong winds (500 km/h) at the boundaries between the belts. In a broad, high-pressure band along the equator, the circulation is eastward, with strong eastward winds along both boundaries with the adjacent low-pressure bands on either side of the equator. The winds between bands farther away from the equator flow alternately eastward and westward. At 22° south of the equator is Jupiter's Great Red Spot, a large (15,000 × 30,000 km) oval vortex consisting of a mass of

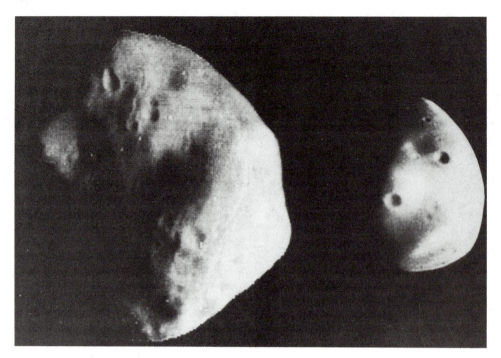

Figure 7.14. Phobos (left) and Deimos (right) reproduced in their relative dimensions. (Beatty and Chaiken, 1990, p. 225, Fig. 6.)

warm gases rising 8 km above the surrounding clouds. The red color may be due to organic molecules, including amino acids, synthesized by electrical discharges. The Great Red Spot has been there for more than 200 years. The circulation pattern in Jupiter's atmosphere is ruled by the Coriolis effect (see Section 13.4), which is particularly strong because Jupiter rotates quite rapidly (one turn in 9.8 hours). The energy derives mainly from the original heat of accumulation, which is still being released, and, in addition, from the sinking of helium through the layer of molecular hydrogen (remember that anything that falls toward a center of gravity releases energy). In fact, Jupiter radiates away more energy (as infrared radiation) than it receives from the Sun as visible light.

Jupiter has at least 16 satellites, whose distances from the equator range from 56,500 km to $24 \cdot 10^6$ km, as well as a system of three tenuous rings that extend from 30,000 km to 140,000 km, also from the equator. The innermost ring is about 20,000 km thick, and the outermost about 4000 km thick, but the middle ring is very thin. The particles that form the rings are exceedingly small, less than 10 μm across. They may be cometary particles similar to those responsible for the "shooting stars" in our sky. The four innermost satellites are within the ring system. They are small (20 to 200 km across) and are called "shepherd satellites" because they shepherd the particles that form the rings by guiding them along narrow paths. The next four satellites, called Galilean satellites, because they were first observed by Galileo in 1610, are the largest. In increasing distance from Jupiter, they are Io (son of Apollo and the daughter of an Athenian king), Europa (daughter of the king of Sidon or Tyre and mistress of Zeus), Ganymede (cupbearer of the gods and lover of Zeus), and Callisto (girl friend of the hunting goddess Diana). Their diameters range from 3,100 to 5,300 km (by comparison, the diameter of the Moon is 3,500 km). These satellites are very different from one another. Io and Europa have densities of 3.55 and 3.04 g/cm³, respec-

Figure 7.15. Jupiter photographed from *Voyager 1.* Notice the Great Red Spot south of the equator. (Courtesy NASA.)

tively, and appear to consist mainly of silicates. The surface of Io is covered with sulfur and sulfur compounds that erupt from the interior. The energy for the eruptions derives from the strong tidal forces induced by Jupiter and Europa that stress the interior of Io and keep it hot. The surface of Europa is an ice crust about 100 km thick, covered with dust particles. The ice must have surfaced from the interior of Europa after Jupiter had dissipated its heat of accumulation. The surface has as few impact craters as the lunar maria. This suggests a similar age, about 3 billion years.

Ganymede and Callisto have much lower densities (1.94 and 1.86 g/cm³, respectively) than Io and Europa. Each consists of a silicate core extending to 70% of the radial distance from the center, a mantle of convecting ice, and a 100-km-thick crust of solid ice. The icy crust of Ganymede is corrugated by the convective motions of the mantle below, cratered by the impacts of asteroids and meteorites, and covered with dust. The outer Galilean satellites have lower densities than the inner ones for the same reason that the outer planets have lower densities than the inner planets: The heat of accumulation of the central body (Jupiter, in this case) blew the gases outward.

The innermost four satellites of Jupiter may be fragments of a larger predecessor, possibly broken up by Jupiter's tides. That predecessor and the next four satellites, the Galilean satellites, probably formed around Jupiter in a manner similar to that by which the planets formed around the Sun. In fact, these satellites have orbits that are within 1° of the equatorial plane of Jupiter. The next four satellites, with diameters ranging from less than 20 km to 190 km, have orbits that are inclined 27° to 29°. The outermost four satellites, with diameters ranging from 20 to 60 km, have orbits that are inclined 147° to 163° (making their orbits retrograde). So satellites 9 through 12 have prograde orbits that are inclined an average of 28°, while satellites 13 through 16 have retrograde orbits that are inclined an average of 152.5°. The supplement of 152.5° is 27.5°, which, within the limits of error, is identical to the average inclination of the preceding four satellites. This must mean something, but what exactly remains unknown. It would seem, in any case, that the outer 8 satellites of Jupiter are captured asteroids or cometary cores.

Saturn, named after the Roman god of fertility and agriculture, has a mass 95 times that of the Earth (Fig. 7.16). It is similar to Jupiter in internal structure, with a core of silicates and metals surrounded by a mantle of metallic hydrogen below and liquid hydrogen above. It has a dipole magnetic that is 20

Figure 7.16. Saturn photographed from *Voyager 1.* The wide dark band in the ring system is the Cassini division, which separates the inner rings from the outer rings. The thinner black band farther out is the Encke gap. The three white dots are, from left to right, Tethys, Dione, and Rhea. The black dot on the surface of Saturn is the shadow of Tethys. (Courtesy NASA.)

times weaker than that of Jupiter. The field is aligned with the rotational axis and points northward (like that of Jupiter). Above the mantle lies the 1,000-km-thick atmosphere, which consists of 82% molecular hydrogen (H_2), 17% helium, and 1% other gases (Table 7.5). The atmosphere has a broad equatorial band ranging from 30°N to 30°S, where pressure is high, wind flow is eastward, and top wind speed is 1,800 km/h (more than triple that of a very strong tornado). Like Jupiter, Saturn radiates more energy than it receives from the Sun, and as in the case of Jupiter, that energy derives from residual heat of accumulation and the sinking of helium through the liquid outer mantle.

Saturn has at least 18 satellites, the largest of which is Titan (5,150 km in diameter, almost as large as Ganymede). Titan (named after the collective name of the children of the primordial couple, Uranus and Gaia) has a density of 1.88 g/cm^3, similar to that of Ganymede and Callisto. Like Ganymede and Callisto, Titan appears to consist of a rocky core extending from the center to 70% of the radius, surrounded by an 800-km-thick mantle of ice. In this, it is similar to Ganymede and Callisto. Above the ice, however, Titan has a 1-km-thick "ocean" of liquid ethane, methane, and nitrogen, and above that a 100-km-thick atmosphere of nitrogen, methane, and argon. Atmospheric pressure at "sea" level is 1.5 atmospheres.

Six other satellites of Saturn have diameters between 200 and 1,500 km. They have densities between 1.2 and 1.4 g/cm^3 and cratered surfaces with high reflectivity. They appear to consist of a mixture of water ice and rocks. Tethys, 540 km across, has a density of 1.2 g/cm^3 and therefore is mainly ice (Fig. 7.17). Its heavily cratered surface indicates that it has not changed much since it formed early in the history of the solar system. All but the two outermost satellites have orbits with inclinations within 1.5° of the equatorial plane of Saturn. They obviously were formed at the same time as the planet itself. The 17th satellite (Iapetus, with a diameter of 1,440 km) has a prograde orbit inclined 14.7°, and the 18th satellite (Phoebe, the outermost one) has a retrograde orbit inclined 175°.

Saturn is famous for its ring system (Fig. 7.18), first noticed by Galileo (but not

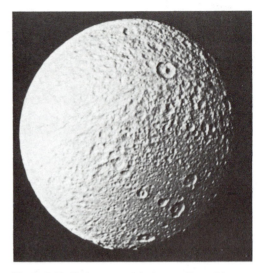

Figure 7.17. Tethys, one of the icy satellites of Saturn. (Courtesy National Space Science Data Center.)

recognized as a ring). It consists of seven concentric rings separated by gaps, extending from 7,000 km to 420,000 km from the equator. The inner five rings contain nearly all the mass of the ring system (about 10^{19} kg) and extend to 80,000 km from the equator, while the two outer rings are extremely tenuous and extend from 105,000 to 420,000 km. The first 10 satellites, ranging in diameter from 50 to 1,120 km, are within the ring system. The rings are made of ice particles, variously coated with dust, that range in size from less than one centimeter to several meters. The smaller ones are, of course, the more abundant. Compared with the area they cover (about $7 \cdot 10^{17} \text{ m}^2$), the rings are extremely thin—a few tens of meters, like a sheet of paper covering Mexico. The rings remain thin because any particle straying above or below the ring will have a sinusoidal orbit that will cut across the ring twice each orbit. The vertical excursion will be dampened by collision with other particles in the ring, so that the straying particle will rejoin the ring (this does not happen in Jupiter's rings because the particles are much smaller, and the rings much more tenuous).

The rings may derive from comets that passed too close to Saturn and were disrupted and captured. That would explain why Saturn's rings are much more conspicuous than those of Jupiter: Saturn, being farther out

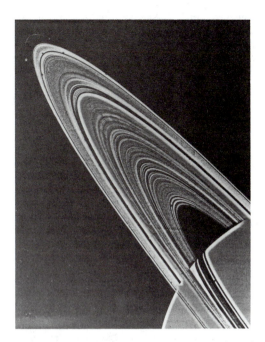

Figure 7.18. The ring system of Saturn consists of many more rings than can be seen from Earth. The outermost ring is the F ring, about 80,000 km from the planet's equator. The two outermost rings are not shown. The dark dots in the ring system are satellites. (Courtesy NASA.)

Poseidon, the Greek god of the sea) are also large planets, with masses that are, respectively, 14 and 17 times the mass of the Earth. They appear blue because of absorption of red light by CH_4 molecules scattered through their atmospheres. Their rocky cores are probably similar to the solid Earth in size and structure. Their mantles are believed to be poorly differentiated mixtures of ices and rocks below, followed above by a layer of liquid hydrogen with helium in it. They differ from the other planets in having their magnetic axes at considerable angles ($55°$ and $47°$, respectively) to the rotational angular-momentum vector.

Uranus is quite anomalous in having its rotational angular momentum vector lying across the orbital plane and pointing $8°$ *south* of it. As seen from the north, Uranus' rotation is therefore retrograde. The tilt of Uranus' axis probably is the result of a giant impact. The atmosphere of Uranus consists of about 85% hydrogen and 15% helium. Because of the inclination of its rotational axis, the summer Sun is alternately over the opposite poles. Ideally, warmer air should rise above the pole during the summer. Because of the Coriolis effect (see Chapter 13), it should spiral toward the opposite pole, changing direction across the equator, and underneath there should be a spiraling return flow that again should change direction across the equator. The polar summer (or winter) lasts some 20 years. During the equinoxes, hot air should rise at the equator and move poleward, again under the influence of the Coriolis effect. Undoubtedly, the circulation is more complicated than that, but the details are not yet known. Uranus emits about the same amount of energy as it receives from the Sun.

Uranus has 15 satellites extending to a distance of 557,000 km from the surface of the planet. The first 10 have diameters of 100 km or less, and the remaining 5 have diameters ranging from 500 to 1,600 km. The densities of the larger satellites range from 1.5 to 1.7 g/cm³, indicating that they consist of a mixture of ice and rocks in about equal proportions. Of the larger satellites, Miranda is the closest to Uranus. It is unique in the solar system in that it shows evidence of having seen shattered more than once and having reassembled itself each time (Fig. 7.19).

Uranus has a system of 11 rings; they extend

than Jupiter, is in a better position to capture comets. The total mass of Saturn's rings is estimated at 10^{19} kg, and the mass of a comet at 10^{15} kg. Saturn's rings, therefore, would have required the capture of some 10,000 comets. Alternatively, the rings of Saturn may be original material that failed to aggregate into a satellite because it was inside the *Roche limit*, the distance at which the tidal force (see Chapter 14) is stronger than the self-gravitation of the satellite. For a satellite with no tensile strength and with a density equal to that of its planet, the Roche limit is $2.4R$, where R is the equatorial radius of the planet. The equatorial radius of Saturn is 60,268 km, and so its Roche limit is 145,000 km. The inner five rings, which contain nearly all the mass of the ring system, extend to 140,000 km from the center of the planet and thus are within the Roche limit.

Uranus (named after the Greek god of the sky) and Neptune (the Roman name for

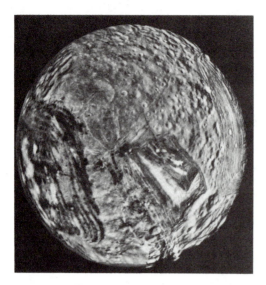

Figure 7.19. Miranda and its chaotic surface. Miranda probably was shattered and reassembled more than once. (Courtesy NASA.)

from 12,500 km to 25,000 km from the equator of the planet and appear to consist of particles 10 cm to 1 m in size. The outermost ring is shepherded by two small satellites, one just inside it, and the other just outside. All other satellites are outside the ring system. All satellites, excluding Miranda, have orbital inclinations within half a degree of the equatorial plane of Uranus. Miranda has an inclination of 3.4°.

Neptune's axis has an inclination of 29.6° with respect to the normal to the orbital plane, which is not greatly different from that of the Earth (23.5°). As in the cases of Jupiter and Saturn, the circulation in Neptune's atmosphere is *zonal* (along parallels). At 22°S there is a Great Dark Spot, an oval vortex with counterclockwise circulation and wind speed of 2,200 km/h.

Neptune has eight satellites. The six inner satellites are less than 400 km across and have orbits of very low inclination. The next two, Triton (son of Poseidon and Amphitrite, and trumpeter of the sea) and Nereid (collective name the 50 daughters of Nereus—the Old Man of the Sea—and his wife Doris) are different. Triton is 2,700 km across, and its orbit is inclined 157° with respect to Neptune's equator (which makes the orbit retrograde). Triton has a density of 2.07 g/cm³, suggesting that it consists of more rock than ice. Its surface,

however, consists of ice sculpted by tidal stresses. Nereid is small (perhaps 350 km across) and very far from Neptune ($5.5 \cdot 10^6$ km), with an orbital inclination of 29°. There is speculation that both Triton and Nereid were formed elsewhere and then captured by Neptune.

Neptune has a tenuous ring system with four rings that extend from 17,000 km to 38,000 km from the equator of the planet. The four inner satellites are within the ring system. The remaining four are outside.

Finally we come to Pluto, named for the Roman god of the underworld (the Greek Hades), who was also the god of wealth, because gold is found underground. Pluto was discovered on February 18, 1930, by the American amateur astronomer Clyde W. Tombaugh (b. 1906), who studied 90 million star images on photographic plates at the Lowell Observatory in Arizona. Pluto has not yet been visited by a spacecraft, and therefore we do not know much about it. Pluto has an orbit with a high inclination on the plane of the ecliptic (122.46°) and a high eccentricity (0.25).

[The *eccentricity* of an ellipse is defined as $(a^2 - b^2)^{1/2} = c/a \, (<1)$, and *ellipticity* is defined as $(a - b)/a$, where a is the semimajor axis, b is the semiminor axis, and c is the semidistance between foci. It is obvious that c must be smaller than a, or else we do not have an ellipse.]

The orbit of Pluto is so eccentric that the distance of the planet from the Sun at perihelion is 60% of that at aphelion. In fact, a portion of Pluto's orbit around perihelion is inside Neptune's orbit, making Neptune temporarily the outermost planet: Between 1979 and 1999, Pluto, as it crosses about its perihelion, will be closer to the Sun than Neptune.

Pluto is $2,290 \pm 40$ km in diameter, smaller than our Moon, and has a density of 1.84 g/cm³. It probably consists of a mixture of rocks and ice. The rotational axis of Pluto, like that of Uranus, lies close to the plane of the ecliptic. Pluto has a satellite, Charon (named after the boatman who ferries the souls across the River Styx to Hades). Charon is $1,190 \pm 40$ km across. Relative to its planet, Charon is the largest satellite in the solar system. Charon is in synchronous orbit around Pluto on Pluto's equatorial plane. (A secondary body orbiting a primary body is in *synchronous orbit* when its revolution has the same period as the rotation of the primary.) Charon has the same density as Pluto and probably the same composition.

Pluto is 39.8 times more distant from the Sun than is Earth, and so it receives only 0.06% of the solar radiation that we receive. It appears to have a very tenuous atmosphere of methane (Table 7.5).

7.5. ASTEROIDS

Asteroids are minor planetary bodies that orbit the Sun between the orbits of Mars and Jupiter. They were not able to assemble into a single body because of Jupiter's perturbations. These bodies are still there, circling the Sun between the orbits of Mars and Jupiter, hitting each other, and continually fragmenting into smaller and smaller pieces. The larger bodies are called *asteroids*. The smaller pieces, when captured by the Earth, are called *meteorites*.

The asteroids orbit the Sun in the same sense as the major planets, and they rotate over their own axes with periods on the order of hours (~ 8 h for the larger asteroids). The largest asteroid, Ceres (named after the Roman goddess of agriculture), is 974 km across. Altogether, there are 7 asteroids with diameters greater than 300 km, 200 with diameters greater than 100 km, some 2,000 with diameters greater than 10 km, and probably 500,000 with diameters greater than 1 km. The larger asteroids are spherical or spheroidal in shape, but the smaller ones are shaped like irregular potatoes (cf. Fig. 7.14). About 75% of the asteroids consist of iron-magnesium silicates. These asteroids are gray in color, are more common in the outer portion of the asteroidal belt, and may be similar to the meteorites called chondrites (see Section 7.6). Another 15% consist of iron-magnesium silicates plus iron-nickel metal; they are reddish in color, are more common in the inner portion of the asteroidal belt, and may be similar to the stony-iron meteorites. About 5% of the asteroids appear to be completely metallic and may be similar to iron meteorites. The remaining 5% are different and may represent other types of meteorites. The total mass of the asteroids is about 2% of the mass of the Moon.

The asteroids are distributed in a belt ranging in width from 2 to 4 astronomical units (mean distance 2.4 AU). Within this belt, the asteroids are distributed in concentric rings separated by gaps (Kirkwood gaps). These gaps result from the absence of asteroids orbiting with periods represented by simple fractions ($\frac{1}{4}$, $\frac{1}{3}$, $\frac{1}{2}$, $\frac{2}{5}$, etc.) of the orbital period of Jupiter. Asteroids with orbits of such periods would come too close to Jupiter each 4th, 3rd, 2nd, etc., passage at aphelion. Two groups of asteroids (called *Trojans*) orbit the Sun along Jupiter's own orbit, preceding and following Jupiter by 60° of arc. One asteroid, Chiron, discovered in 1977, orbits the Sun beyond the orbit of Saturn. Chiron (named after a centaur) has a radius of about 200 km and may be but the largest asteroid in a second asteroidal belt between Saturn and Neptune. Alternatively, it may be a giant comet that does not look like a comet because it stays too far from the Sun to have its gases ionized. Its surface is dark in color. Another cometlike asteroid is Hidalgo, 40 km across, which has a highly eccentric orbit, with perihelion close to the orbit of Mars and aphelion near the orbit of Saturn. Hidalgo, too, could be a very large comet that does not look like a comet.

Some asteroids have orbits that bring them close to the Earth. The Amor group contains 1,000–2,000 asteroids whose orbits have a perihelion between the orbits of Mars and Earth. Their rate of collision with the Earth is estimated at 0.5 per million years. The Aten group includes perhaps 100 asteroids with orbits close to the Earth's orbit; the estimated rate of collision is 1 per million years. The Apollo group includes perhaps 1,000 asteroids, with perihelion *inside* the Earth's orbit; the estimated rate of collision is 2 per million years. All these asteroids have prograde orbits, with inclinations that usually are less than 20° with respect to the plane of the ecliptic. One of the Amor asteroids, Icarus, 1.9 km across, came within 470 terrestrial diameters in 1968. Another, much larger Amor asteroid, Eros, with dimensions of $7 \times 16 \times 35$ km, came within 1,800 terrestrial diameters in 1975. (Icarus was the son of Daedalus, an Athenian craftsman. Eros, the god of love, existed right from the beginning.) Regardless of their classy Greek names, if either one of them were to hit the Earth, we would have a major, major problem.

An asteroid that is estimated to have been 10 km across struck the Earth 65 million years ago. The energy delivered was about one billion megatons, and a crater 180 km across and 100 km deep was excavated. The asteroid and an equivalent amount of rock material were instantly vaporized, and a giant cloud of dust and rock fragments shot through the atmosphere into the stratosphere and beyond. Some fragments were thrown into space with sufficient energy to escape (some may have landed on the Moon). Most of the fragments, however, had insufficient velocities to escape; they ignited upon reentering the atmosphere and kindled a global firestorm. A dark cloud of dust and smoke enveloped the Earth, blocking solar radiation and causing a night that lasted for weeks. Many groups of animals and plants died, including the dinosaurs. A thin layer of microtektites found in some of the deposits dating from that time mark the event. It was one of the greatest mass extinctions that our planet had suffered in its long history (see Chapter 23).

Asteroids have hit the Earth before, and they will hit it again. The difference is that now we have a large human population spread across the globe. A hit almost anywhere will have catastrophic consequences. An "asteroid watch" is in effect to sound the alert should an asteroid be discovered to be on a collision course with the Earth. Modern space technology makes it possible to board such an asteroid and redirect it away from the Earth.

The smaller asteroids are so numerous, and their collisions so frequent, that smaller fragments are produced in increasing numbers. Some of these fragments are perturbed (probably by the passage of comets) and thrown into orbits that bring them close to the Earth. Some of them are captured. The smaller fragments are the familiar meteorites.

7.6. METEORITES

Meteorites are classified into three major groups: *stony meteorites* (92.8%), *stony-iron meteorites* (1.5%), and *iron meteorites* (5.7%). The stony meteorites, as the name indicates, consist of silicate minerals. The stony-iron meteorites consist of mixtures of silicates and metals, with some iron sulfide (troilite, FeS) and iron-nickel phosphide [schreibersite, $(Fe, Ni)_3P$]. The iron meteorites consist of

Table 7.6. Meteorites: Classification and mineralogical compositions

Name	Mineral composition
Stony (92.8%)	
chondrites (85.7%)	
carbonaceous (5.7%)	hydrated Fe–Mg silicates, organic compounds, 1–9% H_2O
ordinary (67.6%)	pyroxenes, olivine, Fe–Ni
other (12.5%)	pyroxenes, olivine, Fe–Ni
achondrites (7.1%)	
Ca-rich (4.7%)	pyroxenes, olivine, plagioclase
Ca-poor (2.4%)	pyroxenes, olivine
Stony-iron (1.5%)	
mesosiderites (0.9%)	pyroxene, plagioclase, Fe–Ni
pallasites (0.5%)	olivine, Fe–Ni
others (0.1%)	pyroxenes, olivine, Fe-Ni
Iron (5.7%)	
octahedrites (4.3%)	kamacite, taenite
hexahedrites (0.6%)	kamacite
ataxites (0.8%)	taenite, kamacite

Fe–Ni alloy. The stony meteorites are subdivided into chondrites and achondrites. The chondrites contain *chondrules*, which are minute (0.5–1-mm) spherical bodies consisting of olivine, pyroxene, or plagioclase minerals. They were formed in a gravity-free environment, such as a turbulent fluid or a heavy vapor.

The carbonaceous chondrites contain low-temperature hydrated minerals (e.g., serpentine), organic compounds, and 1–9% water. They are believed to represent the earliest, most primitive, least differentiated condensate of planetary matter. They date from $4.6 \cdot 10^9$ y ago, the same age as the oldest lunar rocks. The achondrites, as the name indicates, contain no chondrules. Table 7.6 shows the classification and mineralogical compositions of meteorites. On average, 550 meteorites fall on Earth each year. About 1,700 meteorites have been recovered. The masses of meteorites vary tremendously. An iron meteorite 30 m across and weighing 150,000 tons fell in the Arizona desert 50,000 years ago. The impact excavated Meteor Crater, near Winslow, Arizona, which is 1,200 m across and 180 m deep (Fig. 7.20),

Figure 7.20. Meteor Crater (Arizona), 180 m deep and 1,200 m wide, was excavated 50,000 y ago by an iron meteorite that was about 30 m across and weighed about 110,000 tons. (Courtesy U.S. Air Force.)

and the meteorite was shattered into innumerable fragments. The largest meteorite thus far discovered is an iron meteorite weighing 60 t, found near Grootfontein in Namibia. The smallest fragments are the sizes of pebbles. Iron meteorites tend to be larger than stony meteorites, probably because they are less subject to fragmentation in both the asteroidal belt and upon impact on Earth. The smallest meteorites capable of surviving entry through the atmosphere are of centimeter size. In passing through the atmosphere, meteorites shed molten droplets that then solidify and accumulate on Earth as metallic or silicate spherules. Their sizes range from 0.01 to 1 mm across, and their total flux is estimated at about $2 \cdot 10^4$ tons per year. These *cosmic spherules*, as they are called, are found in ice and in deep-sea sediments, where contaminants from the terrestrial sources are minimized. Particles from outer space, smaller than 20 μm, can survive entry through the terrestrial atmos-

phere because they are sufficiently decelerated by rarefied gases in the outer atmosphere to radiate away the heat generated. Heterogeneous mixtures of such particles, many irregular in shape, have been found in Antarctic ice.

Some 40 known meteorites are basaltic in composition, indicating derivation from the surface of a planetary body large enough to sustain basaltic vulcanism. Of these, 2 came from the Moon, 8 from Mars (age $1.3 \cdot 10^9$ y), and 30, with an age of $4.5 \cdot 10^9$ y, from an ancient planetoid or asteroid. Their ejections from their parent bodies apparently were by asteroidal or cometary impacts.

The iron, stony-iron, achondritic, and basaltic meteorites (other than those from the Moon and Mars) came from the fragmentation (by mutual collision) of two or more early planetoids that formed in the asteroidal belt. They represent, respectively, the metallic core, the transition zone, the silicate mantle, and the crust of these bodies. In order to form a metallic core, a planetoid must be heated. The two possible sources of heat are the heat of accumulation and the heat produced by decay of relatively short-lived radioactive isotopes of common elements formed during the supernova explosion that produced the solar system (see Section 7.2). In a few half-lives those radioactive isotopes would have been practically gone, which means that the accumulation of planetesimals in the solar system must have occurred almost immediately after the supernova explosion.

THINK

* The mass of the Earth is $5.97 \cdot 10^{24}$ kg, and its mean radius is 6,371 km. The angular momentum of a rotating solid sphere of uniform density is $\frac{2}{5} m \omega r^2$, where m is mass, ω is angular velocity (in radians per second), and r is radius. Assuming that the Earth is such a sphere, calculate its rotational angular momentum $m \omega r^2 = m v^2/r$, where r is the radius of Earth. The angular momentum of a mass revolving around a center of revolution is $m \omega R^2$, where m is mass, ω is angular velocity (in radians per second), and R is the distance between the mass and the center of revolution. The mean distance between Earth and Sun is 149,597,871 km ($= 1$ AU). Calculate the angular momentum of the revolving Earth, and compare it with the rotational angular momentum. How long would one day have to be for the rotational angular momentum to be equal to the orbital angular momentum?

* Newton's law of gravitation says that the gravitational force between two masses is equal to $G(mM)/R^2$, where G is the gravitational constant ($6.67 \cdot 10^{-11}$ N m^2 kg^{-2}), m and M are the masses of the two bodies, and R is their distance.

Newton also said that the gravitational attraction between two bodies works as though the masses of the two bodies are concentrated at their barycenter (center of mass). For a uniform, solid sphere, the barycenter is the center of the sphere. Assume the Earth to be such a sphere. Calculate the gravitational attraction exerted by the Earth on a mass of 1 kg at its surface on the equator (mass of Earth $= 5.97 \cdot 10^{24}$ kg; equatorial radius of Earth $= 6,378$ km). The centrifugal force is not really a force—it is the reaction felt by a revolving body (a mass at the end of a string, for instance) to the *centripetal force* ($= m \omega R^2 = m v^2/R$), that is, the force directed from the mass toward the center of attraction that keeps the mass in orbit. If this force disappears, the mass flies off *by the tangent*, not outward. The centrifugal force has the same magnitude as the centripetal force. Calculate this magnitude for an Earth spinning at its normal rate, as well as for an Earth with a rotational period of 5 m and 15 s (as previously calculated). If the Earth were to spin that fast, what would fly into space? What would not?

* We said in the text that, relative to the Earth, the mass of the Moon is 1.23%, and its surface gravitational acceleration is 16.5%. How come the latter is comparatively so high?

* Neptune's distance from the Sun is 30.1 AU, compared with the 38.8 AU predicted by the Titius–Bode law. Suppose that Neptune had originally been at a distance of 38.8 AU. Using Kepler's third law, calculate the sidereal period and the orbital velocity of Neptune at both distances (38.8 AU and 30.1 AU). Next calculate the orbital angular momentum for each distance. Is the angular momentum conserved? If not, what kind of gravitational interaction could have reduced the distance from 38.8 to 30.1? Assume the orbits to be circular.

8 EARTH IN SPACE

8.1. MOTIONS OF THE EARTH

The orientation of the Earth's axis and the motions of the Earth in space determine the length of the day, the length of the year, and the season. These parameters are of fundamental importance for the welfare of life on Earth, and therefore it is important that their significance be clearly understood.

Before Copernicus, most astronomers believed that the Earth was stationary and that the Sun, the Moon, the planets, and the stars were rotating around it from east to west. The Greek astronomer Heracleides (ca. 388–315 B.C.E.) was first to suggest that the apparent east-to-west motion of the celestial bodies could be explained by a west-to-east rotation of the Earth. Heracleides' observation, although reported by the Roman politician and orator Cicero (106–43 B.C.E.), remained unnoticed until Copernicus brought it back to light.

In order to describe the positions of the celestial bodies, the ancient astronomers developed systems of celestial coordinates. Today, four systems of celestial coordinates are used. In the *equatorial system* of coordinates (see Fig. 6.10), the celestial latitude (symbol δ), called declination, is measured in degrees of arc north $(+)$ or south$(-)$ of the celestial equator, and the celestial longitude (symbol α), called right ascension (RA), is measured in degrees or in hours (1 h = 15°) eastward starting at the longitude of the vernal (spring) equinox. In the *ecliptic system* of coordinates, celestial latitude is referred to the ecliptic (the intersection of the plane of the Earth's orbit with the celestial sphere), and celestial longitude is measured eastward starting at the longitude of the vernal equinox. In the *galactic system* of coordinates, galactic latitude is referred to the galactic equator, and galactic longitude is measured eastward starting from the direction of the galactic center ($\alpha = 265°36'$,

$\delta = -28°55'$). In the *horizon system* of coordinates, latitude is referred to the celestial horizon (the projection of the observer's horizon on the celestial sphere), and longitude is the *azimuth*, the clockwise angle measured from 0° to 360° starting from the northern direction. Fig. 8.1 shows the four coordinate systems.

The Earth *rotates* about its axis at a speed of $7.27 \cdot 10^{-5}$ rad/s and *revolves* around the Sun at a speed of 29.78 km/s. The Sun, in turn, moves with respect to neighboring stars at a speed of 19.5 km/s toward the star Vega and, together with the neighboring stars, around the galactic center at a speed of 225 km/s. The net motion of the Sun (and hence of the Earth) with respect to the microwave background radiation (obtained by determining the redshift–blueshift of the radiation and its direction in space) is 380 km/s toward the constellation Leo. This is our absolute velocity in space.

The rotation of the Earth about its axis and its revolution around the Sun are both prograde. Astronomical and geophysical data pertinent to the Earth are shown in Table A3 in Appendix A.

The *year* or *period* of a planet is the time it takes to orbit the Sun. There are, however, different types of years. Fig. 8.2 shows the Earth aligned between the Sun and a distant star (A). The Earth revolves in the direction shown by the arrows. The time the Earth takes to make one complete revolution and return to A is called the *sidereal year* (*sidereal* derives from the Latin word *sidus*, which means *constellation*). The sidereal year is equal to 365.25636566 d or 31,558,149.993 s. Suppose summer solstice in the Northern Hemisphere (June 21) occurs at A. The next summer solstice will not occur at A, but at B, becuase the Earth's axis precesses (Fig. 8.3). The time the Earth takes to move from A to B is called the *tropical year* (*tropical*

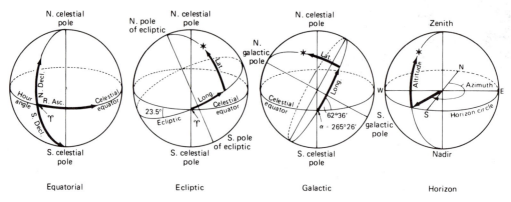

Figure 8.1. Celestial coordinate systems (see text). (Emiliani, 1987, p. 49.)

derives from the Greek verb τρέπειν, which means *to turn*). The tropical year is equal to 365.24219342 d or 31,556,925.511 s. The tropical year is what counts for us humans, because it measures the return to the same point in the seasonal cycle (officially, vernal equinox to vernal equinox, but any other time of the year will do). When people, including scientists, talk about years, they mean tropical years.

The tropical year is shorter than the sidereal year because of the precession of the Earth's axis. *Precession* is the motion of the axis of a spinning top. The Earth, like a top, has an equatorial bulge—the equatorial radius is

21.360 km longer than the polar radius. This bulge experiences a torque applied by the Moon, the Sun, and the planets. The direction of the torque changes with time, depending on the position of the the acting bodies. The result is a precessional motion of the Earth's axis that is *clockwise* as seen from the north (Fig. 8.3). This motion was discovered by Hipparchos about 150 B.C.E.

The precessional angle is 46°52′56″, twice the angle between the Earth's axis and the normal to the orbital plane. The lunisolar precession amounts to 50.40″ per year. The torques applied by the planets, together with relativistic

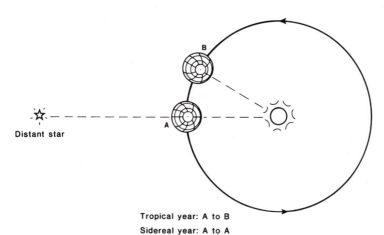

Distant star

Tropical year: A to B
Sidereal year: A to A

Figure 8.2. Here the Earth is located between the Sun and a distant star at the northern summer solstice (Northern Hemisphere inclined toward the Sun). The time the Earth takes to return to position *A* (between the Sun and the distant star) is called the *sidereal year*. The next summer solstice, however, does not occur at position *A*, but at position *B*, before the Earth has completed one turn around the Sun, because meanwhile the Earth's axis has precessed (by 50.8″). The time taken by the Earth in going from *A* to *B* in the sense shown by arrows is called the *tropical year*. (Emiliani, 1988, p. 121, Fig. 8.11b.)

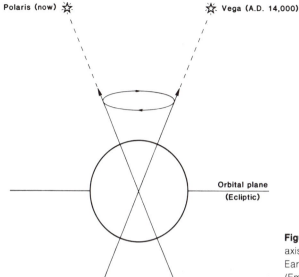

Polaris (now) ☆   ☆ Vega (A.D. 14,000)

Orbital plane
(Ecliptic)

Figure 8.3. Precessional motion of the Earth's axis. In 12,000 y, the northern direction of the Earth's axis will point toward the star Vega. (Emiliani, 1988, p. 120, Fig. 8.11a.)

effects, cause the Earth's orbit as a whole to precess counterclockwise (as seen from the north) by 0.12″ per year. The net precession, called *general precession*, is clockwise (as seen from the north) by 50.28″, and its period is $360°/50.28″ = 25,800$ y.

Consider now Fig. 8.2, showing the Earth between the Sun and a distant star. If the Earth's axis did not precess, the Earth would return to that position in exactly one year. Because of precession, however, the Earth returns to the same position (summer solstice) with respect to the Sun, completing one tropical year, before reaching the same position with respect to the distant star (which defines the sidereal year). The tropical year, therefore, is shorter (by 1,225 seconds = 20 m 25 s) than the sidereal year.

Because of the opposite senses of the axial and orbital precessions, the seasonal precessional period (e.g., summer solstice at perihelion to the next summer solstice at perihelion) is equal to sidereal year/(anomalistic year − tropical year) = 21,000 years. (The *anomalistic year* is the time interval between successive passages of the Earth at perihelion.)

8.2. THE CALENDAR

The astronomical motions of the Earth and the Moon provide the basis for the calendar—the

subdivisions of time into years, months, weeks, and days. The problem with these subdivisions is that they are based on phenomena that have different periods—the revolution of the Earth around the Sun, the revolution of the Moon about the Earth, and the rotation of the Earth. The periods are not commensurate: there are 29.53 days in a lunar month, 4.22 weeks in a lunar month, and 365.24 days in a tropical year. The development of an accurate calendar has been a vexing problem since antiquity (the name *calendar* derives from the Latin name for the day of the new Moon, marking the beginning of the lunar month).

The Egyptians started their year in the latter part of July, when the Nile began flooding the delta. It was a big event, because the flooding fertilized the delta. The Egyptians divided the year into 12 months of 30 days each, giving a total of 360 days, to which they added an extra 5 days at the end. In the years around 4241 B.C.E., the Egyptians noticed that the rising of the Nile coincided with the day when the star Sirius (known as Sothis) became again visible in the southern sky as seen from the capital city of Memphis. The Roman author Censorinus reported in his work *De die natali* (written in 238 C.E. and dedicated to his patron on his birthday) that in 139 C.E. the rising of Sothis had occurred on July 20. Now, the Egyptian year of 365 days was shorter by 0.242 day than the tropical year. As a result, the Egyptian calen-

Table 8.1.   Names of the months

Month	Name origin	Number of days
January	Janus, the two-faced god of gates	31
February	Februa, day of purification (February 15)	28
March	Mars, god of war	31
April	Apru, Etruscan goddess of love	30
May	Maia, eldest daughter of Atlas	31
June	Juno, wife of Jupiter	30
July	Julius Caesar	31
August	Augustus, adopted son of Julius Caesar	31
September	the seventh month of the earlier Roman calendar	30
October	the eighth month of the earlier Roman calendar	31
November	the ninth month of the earlier Roman calendar	30
December	the tenth month of the earlier Roman calendar	31

dar fell behind by 1 day in 4 years, making one full circle (called *Sothic cycle*) in $4 \times 365 = 1,460$ years. Because the rising of Sirius at Memphis and the beginning of flooding of the delta coincide only once in 1,460 years, the Egyptian calendar must have been developed at one of three times: $1460 - 139 = 1321$ B.C.E., or $1321 + 1460 = 2781$ B.C.E., or $2781 + 1460 = 4241$ B.C.E. This last date is the most probable, because the calendar was already in use about 3000 B.C.E.

The early Greeks adopted the Egyptian calendar, but in the fifth century B.C.E. the ὀκταετηρίς was introduced, a period of 8 years in which the year consisted of 6 lunar months of 30 days and 6 lunar months of 29 days, plus 1 lunar month of 30 days in 3 out of the 8 years. The total number of days was thus 2,922, giving an average of 365.25 per year (exactly). The Greeks set the beginning of the year at either the summer or winter solstice (depending on which Greek city).

According to legend, Rome was founded by Romulus, its first king, on April 21, 753 B.C.E. At that time, the Romans had a calendar of 10 months (*Martius, Aprilis, Maius, Junius, Quintilis, Sextilis, September, October, November, December*), of which four (*Martius, Maius, Quintilis, October*) had 31 days and the other six had 30 days. In addition, there were two unnamed months of 30 days each in the winter. The second king of Rome, Numa Pompilius (ca. 715–672 B.C.E.), is said to have named the two winter months *Ianuarius* and *Februarius* and to have moved the beginning of the year to January 1. In the reform recommended by the Alexandrine astronomer Sosigenes and promulgated by Julius Caesar (100–44 B.C.E.), the year remained equal to 365.25 days, by having 365 days per year but adding to the month of February one day every fourth year (leap year). In addition, the 12 months were given unequal durations, and the names *Quintilis* and *Sextilis* were changed to July and August (Table 8.1). The first Julian year began on January 1, 709 AUC (*Ab Urbe Condita*, i.e., *since the founding of the City*), which translates into January 1, 45 B.C.E.

In 526 C.E., the Byzantine emperor Justinian asked a monk named Dionysius Exiguus (he must have been very short and very thin) to change the calendar, reckoning time no longer from the founding of Rome, but from the birth of Jesus Christ. Dionysius figured out that Jesus had been born 753 years after the founding of Rome, apparently not knowing that Herod, under whom Jesus was born, had died 749 years after the founding of Rome. Accordingly, Jesus had to be born at least four years earlier than the good monk thought. This error was discovered long after the Justinian calendar had been adopted by all the Christian nations of Europe. To complicate things, some recent research indicates that Jesus Christ was crucified on April 3, 33 C.E. Because he was 33 years old when he died, his birth could be fixed at the end of the year 1 B.C.E. or at the beginning of the year 1 C.E. (there is no year zero), four years after Herod had died. At this point we notice that there are discrepancies among the four gospels. Most important, the earliest gospel, Mark, says nothing about the birth of

Christ or about Herod, and neither does John. It is still possible, therefore, that Dionysius Exiguus made no mistake and that the birth of Chirst took place at the beginning of year 1, after the death of Herod. In any case, we still reckon the years of the Common Era (C.E.) and the years before the Common Era (B.C.E.) from the fix provided by Dionysius Exiguus. May his soul rest in peace.

The number of days in a tropical year, as fixed by Julius Ceasar's reform, is not quite exact—it is too large by 0.0078 days = 11.23 minutes. By the middle of the sixteenth century, the calendar was off by 11.7 days with respect to the seasons. Pope Gregory XIII ordered that the day after Thursday, October 4, 1582, should be Friday, October 15, 1582; in addition, from that year on, centennial years should not be leap years, even though divisible by 4, unless divisible by 400. In a 400-y cycle, therefore, there would be $(365 \times 400) + 97 = 146,097$ days, equal to 365.242 days per year. The Gregorian calendar was not the work of Gregory XIII (just as the Julian calendar was not the work of Julius Caesar), but of the Jesuit priest Christopher Clavius and the Neapolitan astronomer and physician Luigi Lilio Ghiraldi. The new calendar was soon adopted in all Catholic countries, but only in 1752 in England and the British Empire (by which time those countries were 12 days behind), and only in 1918 in Russia, when that country was 13 days behind.

In the same year, 1582, that Pope Gregory XIII promulgated his reform of the calendar, the scholar Joseph Scaliger (1540–1609) devised the Julian period, which he named after his father, Julius Caesar Scaliger. The Julian period is a period of 7,980 years that is obtained by multiplying three cycles, the *solar* (28 y), *Metonic* (19 y), and *indictio* (15): $28 \times 19 \times 15 = 7,980$. The solar cycle is a period of 28 years that derives from the fact that in 365 days there are 52 weeks $+ \frac{1}{7}$ of a day. In successive years, therefore, the same day of the year will fall on the following day of the week. Were it not for the leap years, in which the day of the week advances by two, the cycle would be completed in 7 years. It is instead completed in $7 \times 4 = 28$ years. The Metonic cycle (discovered by the Greek astronomer Meton in the fifth century B.C.E.) is a period of time that is divisible into both a whole number of years and a whole number of lunar months; it is equal to 19 years

and includes 235 lunar months. The *indictio* cycle is an ancient Egyptian cycle that, beginning with January 1, 313 C.E., was adopted by the emperor Constantine as the Roman taxation cycle.

Scaliger adopted the three cycles not for astronomical reasons but because they were used in Hellenistic, Roman, and Byzantine calendars. Thoroughly familiar with the Egyptian calendar and its probable origin around 4241 B.C.E., Scaliger set the beginning of the current Julian period at 12:00 noon Greenwich time, 4713 B.C.E. Days are counted in continuity from that date. The Julian period is important because astronomers use it to record the dates of celestial phenomena. Thus, January 1, 2000, at 00:00 h, will be Julian date (JD) 2,451,544.50.

The *sidereal day* is the time it takes for a distant star to return to the same celestial meridian (or to facing the same terrestrial meridian). It is, therefore, the time that it takes for the Earth to make one full turn and is equal to 23h, 56m, 4.1s ($= 86,164.1$ s). The *solar day* is the time it takes for the Sun to return to facing the same terrestrial meridian and is equal, by definition, to 24 hours. Because, again by definition, there are 60 minutes in an hour and 60 seconds in a minute, the number of seconds in one day is $24 \times 60 \times 60 = 86,400$.

The solar day is longer than the sidereal day because the Earth revolves around the Sun in the same direction it rotates—as the Earth makes one turn, it advances along its orbit by $0.9856° = 59'8.2''$, which is equal to 360° divided by the number of tropical days in one tropical year. This advance makes it necessary for the Earth to turn an additional $59'8.2''$ in order for the sun to return to the same meridian. The additional turning requires 3m 56.5 s of time, which is equal to one solar day (86,400 s) divided by the number of sidereal days in one sidereal year (365.256). This is so because one synchronous revolution (a revolution during which the same hemisphere faces the Sun all along) is geometrically equivalent to one rotation.

As indicated by Kepler's second law, the Earth does not move along its orbit at a constant rate: It moves fastest at perihelion, slowest at aphelion, and at intermediate speeds in between. Because the rotational velocity remains essentially constant throughout the year, the length of the day is greater at perihelion than at aphelion. Rather than correcting for this effect, we use the *mean solar day*, defined as the time interval between successive passages of the *mean Sun* at meridian. The mean Sun is the hypothetical position of the Sun if the Earth were moving along its orbit at a steady pace. This convention averages the length of the day throughout the year. The second, defined as 1/86,400 of the mean solar day, is called *ephemeris second* (symbol s_E; $\dot{\varepsilon}\varphi\dot{\eta}\mu\varepsilon\rho o\varsigma$ means *daily* in Greek). The ephemeris second was replaced in 1967 by the *atomic second* (s_A), defined as the duration of 9,192,631,770 cycles of the radiation corresponding to the transition between the two hyperfine levels of the ground state of ^{133}Cs (see Section 1.2). The number of cycles was chosen, of course, to make the atomic second identical to the ephemeris second.

Greenwich Mean Time (GMT) is the mean solar time counted from midnight at the Greenwich meridian. This time is also called universal time (UT) or Zulu (Z) time. Greenwich Mean Astronomical Time (GMAT) is Greenwich Mean Time counted from noon at the Greenwich meridian. Reckoning from noon (as is done also for the Julian Date) is advantageous in astronomy because most astronomical observations are made at night. Because most observatories are within 90° east or west of Greenwich, astronomical observations made at night in these observatories are not divided into separate days.

The rotation of the Earth established the day and night. In antiquity, the day was divided into 12 day hours and 12 night hours. Depending upon the season, the day hour was longer or shorter than the night hour. The Greeks simplified things by dividing the day into 24 equal hours. The rotation of the Earth dictates that different longitudes have different hours of the day. Thus, the meridian directly facing the Sun is at noon, the opposite meridian is at midnight, and intermediate meridians have intermediate times. In 1883, standard time was adopted, a scheme by which the Earth's surface was divided into zones 15° of longitude wide. Each zone represents one hour of time, because there are 24 hours in one day ($15° \times 24 = 360°$). Starting with time zone Z, bisected by the Greenwich meridian, the time zones are lettered A to M eastward and N to Y westward. The boundary between zones M and Y is the international date line, across which one passes from one day to the following day (going westward) or to the preceding day (going eastward). The zoning is far from precise because of the desirability of following political boundaries.

The lunar tides slow down the rotation of the Earth by $1.5 \cdot 10^{-5}$ s/y, requiring that the distance Earth–Moon increase by 4 cm/y in order to conserve angular momentum (see Section 14.5). Studies of ancient laminated fossils and sedimentary formations indicate that, 400 million years ago, there were about 400 days in one year.

The seven-day week dates from antiquity— it appears in Genesis, for instance. It may have originated from the observation that the lunar cycle, 29.5 days long, exhibits four phases— new Moon, half Moon, full Moon, half Moon. Each phase, therefore, lasts about 7 days. In fact, in many languages the name *week* derives from the word for *seven* ($\dot{\varepsilon}\beta\delta o\mu\dot{\alpha}\varsigma$ in Greek, *septimana* in Latin, for instance). Noticing, in addition, that there were seven known planets, the ancients related each planet to a day of the week: Sun for Sunday, Moon for Monday, Mars for Tuesday (*dies Martis* in Latin), Mercury for Wednesday (*dies Mercurii*), Jupiter for Thursday (*dies Jovis*), Venus for Friday (*dies Veneris*), and Saturn for Saturday (*dies Saturni*).

8.3. THE SEASONS

The axis of the Earth is inclined 23° 26' 28" from the normal to the orbital plane. This inclination is the cause of the seasons: During the northern summer, the Northern Hemisphere is inclined toward the Sun, and the Southern Hemisphere is inclined away, thereby experiencing winter (Fig. 8.4). During the northern summer solstice (June 21), the Sun's rays are perpendicular to the Tropic of Cancer, the northern polar cap (the area north of the Arctic Circle) receives maximum illumination, and the southern polar cap receives no sunlight at all. During the southern summer solstice (December 21), the rays of the Sun are perpendicular to the Tropic of Capricorn, the southern polar cap receives maximum illumination, and the northern polar cap receives no sunlight at all. During the equinoxes (March 21, September

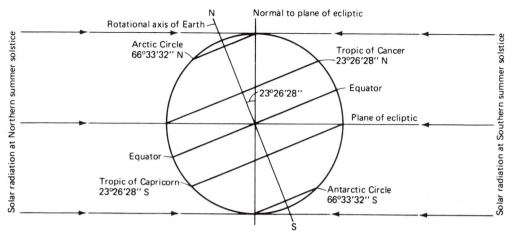

Figure 8.4. The latitudes of the tropics and the seasons are determined by the inclination of the Earth's axis from the normal to the orbital plane. As a result of this inclination, the Sun at local noon appears to move up and down during the year, with an excursion of 46°52′56″. At the vernal (spring) and fall equinoxes, the Sun stands vertical over the equator, the two hemispheres are equally lighted, and nights and days have the same duration at all latitudes except at the poles, which are grazed by the Sun's rays for 24 hours. As the Sun moves toward a tropic, the days become longer than the nights in that hemisphere, reaching a maximum of 6 months at the pole. Meanwhile, the nights become longer than the days in the opposite hemisphere, reaching a maximum of 6 months at the pole. (Emiliani, 1987, p. 216.)

23), the Sun's rays are perpendicular to the equator, and both hemispheres receive the same amount of sunlight. The hydroatmosphere has considerable thermal inertia, and as a result, yearly temperature extremes are experienced about one month after the solstices.

Because of the general precession, the seasons move along the orbital plane. Northern summer, for instance, is now at aphelion, but in $21,000/2 = 10,500$ y it will be at perihelion. The aphelion distance is 1.0167 AU, and the perihelion distance is 0.9832 AU. The difference in insolation is therefore $(1.0167)^2 - (0.9832)^2 = 0.0693 = \sim 7\%$. Needless to say, the northern summers will be warmer when the Earth will be at perihelion, in 12,500 C.E.

The precessional angle, which is now 46° 52′ 56″, changes with time between a minimum of 43° 18′ and a maximum of 49° 12′. The corresponding inclination of the Earth's axis to the normal to the orbital plane (which is the same as the inclination of the equatorial plane to the plane of the ecliptic) is half as much, ranging from 21° 39′ to 24° 36′. This change in inclination has a periodicity of 40,600 y. The orbit of the Earth is elliptical, with the semimajor axis

a equal to 149,597,870.7 km ($=$ mean Sun–Earth distance $= 1$ AU by definition), and the semiminor axis b equal to 149,576,880.8 km $=$ 0.999860 AU. The eccentricity (e) is therefore

$$e = (a^2 - b^2)^{1/2}/a = 0.01675104$$

The eccentricity of the Earth's orbit changes from a minimum of 0.01 to a maximum of 0.07, with periodicities of 95,000 and 413,000 years. It is now close to a minimum, with a difference of only 0.014% between the major and minor axes of the orbit. The Earth's orbit, therefore, is now almost circular. When eccentricity reaches a maximum, the difference will be 0.25%, and the perihelion will be 5% closer to the Sun. Altogether, seasonality (hot summers, cold winters) will be extreme when eccentricity is high, summer is at perihelion, and the inclination of the Earth's axis is large. Conversely, seasonality will be minimal when eccentricity is at a minimum, northern summer is at aphelion, and the inclination of the axis is minimum. As we shall see in Chapter 24, these changes in seasonality have an important bearing on the timing of the ice ages.

THINK

• Consider the simplest planetary system possible, consisting of a star like the Sun and a planet like the Earth. If the planet were an ellipsoid of revolution, it would be free to rotate at any rate it wished. But if the equatorial bulge were larger in one direction, the star's attraction on that protrusion would slow down the rotation of the planet until the protrusion is permanently directed toward the star. From then on, a planetary day would be the same as a planetary year. (That is, in fact, what happened to the Moon, which has a marked protrusion, about 780 m high, directed toward the Earth.) If the irregularity is less marked, the planet will be slowed down and made to rotate a simple fraction of its orbital period. Compare the rotational and orbital periods of Mercury (Table 7.6). Is there a simple fractional relationship between the two? Can you find a similar relationship for Venus? How about between the rotational period of Venus and the orbital period of the Earth?

• If the Earth's axis were exactly normal to the Earth's orbital plane, how would the Sun appear from the north pole? from the south pole?

• If the Earth's orbital eccentricity were higher, would the northern summer be hotter or cooler?

• If the Earth were perfectly spherical, would it precess?

IV Geology

H AVING learned, in Parts I–III, about the structure and dynamics of matter and energy, and about the origin and evolution of the universe and the celestial bodies that are in it, we are now ready to consider what is undoubtedly one of the most complex systems of all—Planet Earth. Part IV deals with the Earth as a physical body, and Part V deals with the evolution of life and environment from the birth of our planet to the present.

Part IV presents the principles of physical geology in a way that is succinct in form but not in content. The subject matter is arranged in 10 chapters:

Chapter 9 discusses the chemical composition of the Earth in terms of percentages of atoms, rather than by weight, because minerals contain atoms in characteristic numerical proportions. Similarly, the solar atmosphere and the major solid phases of the solar system are compared in terms of their relative percentages of atoms, not, as is customary, in terms of the weights of elements, molecular species, or oxides. This section is followed by a rather detailed study of the crystallographic and physical properties of minerals—density, hardness, luster, refractivity, x-ray optics, streak, luminescence, heat capacity, electrical resistivity, strain electricity, and magnetic susceptibility. These properties can be covered in some detail here because the necessary background was set out in Parts I and II. The chapter concludes with a discussion of gems. Several tables in the chapter and two tables in Appendix A enable the reader to gain a feeling for the magnitudes and ranges of the quantities involved.

There are hundreds of different types of igneous rocks. Fortunately, those that compose most of the Earth's crust and mantle can be grouped into a relatively few families. These rocks and their modes of emplacement are treated in Chapter 10, together with volcanoes, volcanic structure, and volcanic activities.

Gravitational and seismic phenomena involve concepts that were explained in Parts I–III, as well as some that are new. Geophysics, therefore, receives a treatment that is reasonably quantitative. Chapter 11 discusses the geophysics of the solid Earth—gravity, seismology, geothermics, geomagnetism, and paleomagnetism. On the basis of what was discussed in Part II, the physical principles of magnetism and the magnetic properties of bulk matter are reviewed, the magnetic polarity scale is introduced, and the workings of the proton magnetometer are explained. These topics lead into Chapter 12, which deals with plate tectonics—the great scientific revolution that emerged from work at sea during the 1950s and 1960s and has provided us with a global theory of the Earth's dynamics. Plate motions and plate interactions are shown to explain the distributions of mineralization, fossil fuels, ice, and biotas.

The atmosphere (Chapter 13) and the hydrosphere (Chapter 14) are introduced next, as the agents responsible for the formation and deposition of sediments. Because of the background provided in Part II, it is possible to treat quantitatively the radiation balance of the Earth, atmospheric and oceanic chemistry, the motions of air masses and ocean currents, the Coriolis acceleration and the Coriolis effect, and tides.

Sediments, sedimentary rocks, and the spectrum of sedimentary environments are discussed in Chapters 15 and 16. Chapter 17 covers diagenesis and metamorphism, the processes by which loose sediments and organic matter are transformed into rocks and fossil fuels. Also included is a discussion of how rocks are stressed and deformed by forces arising from mantle motions. Chapter 18 deals with the principles of stratigraphy, the science that studies the layers of sediments and sedimentary rocks on land and on the ocean floor in order to reconstruct the ancient environments and to date and correlate the sedimentary formations. This chapter effectively connects Part IV with Part V, which is dedicated to the an analysis of the parallel evolution of life and environment on Earth.

MINERALS AND GEMS

9.1. CHEMICAL COMPOSITION OF THE EARTH

We have seen (Chapter 6) that the most abundant elements in the cosmos and in the Sun are hydrogen and helium. As shown in Fig. 6.16, heavier elements are much rarer. Not so for the inner planets. The fierce solar wind during the T Tauri stage blew away the gases from the inner region of the solar ring, leaving behind solid matter and whatever gases were trapped in it. Element abundances on the Earth are very different from those for the Sun and the cosmos, with oxygen, iron, silicon, and magnesium forming 94% of the total in terms of percentage of atoms (Table 9.1). The relative abundances of the nonvolatile elements in the Sun, in chondritic meteorites, and on Earth are shown in Table 9.2. The least differentiated materials are shown in the first four columns. There are marked similarities in nonvolatile composition among the Sun, the meteorites, and the Earth as a whole, indicating a similar origin. Chemical differentiation sets apart the materials in the remaining five columns. Pyrolite is a rock that would exist if the Earth's mantle and crust were mixed together—in other words, it is average Earth rock. What has been removed from it is the core, and therefore pyrolite is depleted of iron and comparatively enriched in silica, magnesium, and calcium. Basalt, which forms the oceanic crust, and granite, which forms the bulk of the continental crust, are enriched in sodium, aluminum, and silica, and greatly depleted of magnesium, in comparison with pyrolite. Shale, which is the product of weathering of igneous rocks, has a composition intermediate between those of basalt and granite. Basalt, granite, and shale have highly variable compositions. Those shown in Table 9.2 are averages.

Table 9.1. The most common elements on Earth

Element	Percentage of atoms
O^a	48.86
Fe	18.84
Si	13.96
Mg	12.42
S^a	1.39
Ni	1.39
Al	1.31
Na	0.64
Ca	0.46
P	0.14
H^a	0.12
Cr	0.11
C^a	0.10
K	0.05
Mn	0.05
Co	0.05
Cl^a	0.04
Ti	0.03
all other	0.04
total	100.00

^{a}Elements that also form volatile compounds (H_2O, H_2S, HCl, SO_2, CO_2, CO, OH, etc.).

THINK

* Table 9.1 shows that the ratio Fe/O in the Earth is $48.86/18.84 = 2.59$, in terms of number of atoms. Calculate what this ratio would be in terms of mass.

9.2. MINERAL STRUCTURES

A mineral is any naturally occurring, chemically definable substance; it can be solid crys-

Table 9.2. Nonvolatile elements: relative abundance (percentage of atoms)

Element	Solar atmosphere	CI chondrites	Ordinary chondrites	Whole Earth	Pyrolite	Basalt (av.)	Crust (Continental)	Granite (av.)	Shale (av.)
Na	1.52	1.97	1.67	1.36	0.74	3.44	5.94	5.02	5.36
Mg	31.40	32.58	34.04	28.48	41.28	7.54	3.22	1.07	4.12
Al	2.61	2.57	2.58	2.21	3.29	13.26	16.57	15.00	15.28
Si	35.28	30.54	36.26	29.48	37.07	40.88	56.32	63.78	52.00
P	—	0.37	0.26	0.18	—	0.26	0.16	0.16	0.25
K	0.11	0.10	0.13	0.10	0.19	1.03	3.40	6.64	4.89
Ca	1.76	2.08	1.85	1.53	4.14	12.87	6.40	2.96	7.32
Ti	0.09	0.07	0.08	0.06	0.75	2.35	0.69	0.45	1.18
Cr	0.39	0.36	0.33	0.26	0.52	—	—	—	—
Mn	0.21	0.25	0.27	0.22	0.19	0.23	0.12	0.18	—
Fe	25.05	27.57	21.35	33.79	11.56	18.13	7.26	4.83	9.61
Co	0.06	0.07	0.05	0.12	—	—	—	—	—
Ni	1.52	1.45	1.02	2.21	0.28	—	—	—	—
Total	100.00	99.98	99.99	100.00	100.01	99.99	100.09	100.09	100.01

talline (e.g., quartz) or noncrystalline (e.g., opal, coal), liquid (petrolium, water), or gaseous (methane, CO_2, air).

A crystal is a solid substance in which the atoms of one or more elements form a regularly repeating three-dimensional pattern called *unit cell*. For instance, the unit cell of common kitchen salt (NaCl), which is one of the simplest, consists of one sodium ion (Na^+) at the center of an imaginary cube and 12 additional Na^+, each one in the middle of an edge and therefore shared by four adjacent cells (Fig. 9.1). The total number of Na^+ per cell is therefore $12/4 + 1 = 4$. In addition, there is one chloride ion (Cl^-) at each corner (shared by eight adjacent cells) and one Cl^- at the center of each face (shared by two adjacent cells). The total number of Cl^- per cell is therefore $8/8 + 6/2 = 4$. In crystals, the charges of cations equal the charges of anions.

The salt crystal is formed by translating the unit cell along three mutually perpendicular axes (*a* axis, front to back; *b* axis, left to right; and *c* axis, top-to-bottom), forming a cubic array of alternating Na^+ and Cl^- (Fig. 9.1). In fact, if you look at the salt grains from your salt shaker with a magnifying glass, you will see a prevailing cubic structure. Atoms in crystals are linked by ionic bonds (e.g., Na^+Cl^-) or by covalent bonds (e.g., diamond), or by bonds of intermediate type (e.g., sphalerite, ZnS; iodargyrite, AgI). In most inorganic crystals it is not possible to identify to which cation a given anion belongs (or vice versa). If a salt crystal is dissolved, the solution does not contain NaCl molecules, but separate Na^+ and Cl^- ions. In fact, there is no such thing as a molecule of sodium chloride.

There are also molecular crystals, that is, crystals in which the identity of the original molecule is preserved. Ice and iodine are examples among inorganic substances. Most organic substances can form molecular crystals, including such comparatively large "particles"

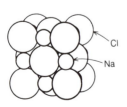

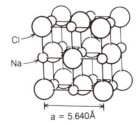

$a = 5.640Å$

Figure 9.1. Structure of halite. Left: Ions drawn proportional to their sizes. Right: Expanded view to show the interior of the unit cell. (Berry, Mason, and Dietrich, 1983, p. 326, Fig. 11.3.)

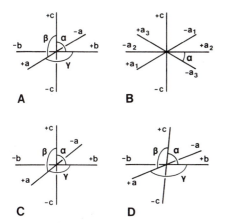

Figure 9.2. The crystallographic axes (A) for the cubic, tetragonal, and orthorhombic systems, (B) for the hexagonal system, (C) for the monoclinic system, and (D) for the triclinic system. See Table 9.3.

as virions, which are the viral units (see Chapter 19). Molecular crystals are kept together by hydrogen bonds or van der Waals forces and therefore have relatively low ($< 200°C$) melting points. When a molecular crystal melts or is dissolved, the original molecules reappear intact.

The orientation of the faces of a crystal is described with respect to the three mutually perpendicular axes *a*, *b*, and *c* (Fig. 9.2A,C,D) or, in cases of threefold or sixfold symmetry, three horizontal axes intersecting at 120° angles (the a_1, a_2, and a_3 axes) and one vertical axis (the *c* axis) (Fig. 9.2B). Crystals grown without interference exhibit planar faces that form characteristic angles with each other. The faces intersect the crystallographic axes in such a way that the lengths of the intercepts are represented by simple integers. For instance (Fig. 9.3):

A face intersecting	But parallel to	Is represented as
a	the *b*–*c* plane	(100)
b	the *a*–*c* plane	(010)
c	the *a*–*b* plane	(001)

A face intersecting *a*, *b*, and *c* at equal distances is represented by the symbol (111). A face intersecting *a* at a distance of $\frac{1}{2}$ or $\frac{1}{3}$ that at which it intersects *b* and *c* is represented by (211) or

(311); that is, the index is the reciprocal of the relative intercept distance.

The orientation of crystal faces in crystals with threefold or sixfold symmetry is expressed using four indices, three for the three horizontal axes and one for the vertical axis (the *c* axis) (Fig. 9.2B). A face intersecting a_1, a_2, and *c* at equal distances will necessarily intersect $-a_3$ at half that distance. It is represented by the symbol ($11\bar{2}1$).

The arrangement of atoms (considered as geometric points, regardless of type) in space forms a *space lattice*. There are 14 space lattices (*Bravais lattices*) partitioned among six *crystallographic systems* (Table 9.3; Fig. 9.4).

9.3. SYMMETRY

A space lattice can be created by replicating in space a point or a system of points by means of *symmetry operations*. In addition to translation, there are 10 basic symmetry elements: 5 rotation axes (1, 2, 3, 4, 6) and 5 rotoinversion

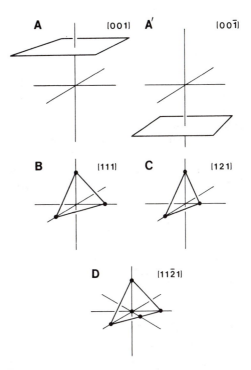

Figure 9.3. Crystallographic axes and examples of crystal faces and their symbols.

Table 9.3.   Crystal systems and lattices

Name of system and/or lattice	Axes $(a \neq b \neq c)$	Angles of axes
1. Triclinic	a, b, c	$\alpha, \beta, \gamma \neq 90°$
2. Monoclinic	a, b, c	$\alpha = \gamma = 90°; \ \beta \neq 90°$
end-face-centered monoclinic		
3. Orthorhombic	a, b, c	$\alpha = \beta = \gamma = 90°$
end-face-centered orthorhombic		
body-centered orthorhombic		
face-centered orthorhombic		
4. Hexagonal	a, a, c	$\alpha = 60°; \ \gamma = 90°$
rhombohedral	a, a, c	$\alpha = 60°; \ \gamma = 90°$
5. Tetragonal	a, a, c	$\alpha = \beta = \gamma = 90°$
body-centered tetragonal		
6. Cubic (isometric)	$a, a, a,$	$\alpha = \beta = \gamma = 90°$
body-centered cubic		
face-centered cubic		

axes ($\bar{1}, \bar{2}, \bar{3}, \bar{4}, \bar{6}$). Here the numbers $1, 2, 3, 4,$ and 6 refer to 1-fold rotation (rotation by 360°), 2-fold rotation (2 successive rotations by 180° each), 3-fold rotation (3 successive rotations by 120° each), 4-fold rotation (4 successive rotations by 90° each), or 6-fold rotation (6 successive rotations by 60° each). The symbols with the bars above them (to be read *bar one*, etc.) refer to 1-fold, 2-fold, 3-fold, 4-fold, or 6-fold rotation with inversions through a center of symmetry. Notice that $\bar{1}$ has the same result as inverting a point (or system of points) through a center of symmetry (Fig. 9.5A), and $\bar{2}$ (a rotation by 180°, followed by inversion through the center of symmetry) has the same result as reflection through a mirror plane normal to the axis of rotation and passing through the center of symmetry (Fig. 9.5B). The symbol for the mirror plane is *m*; that for the center of symmetry is *i*.

Combination of rotation and translation generates 11 helication elements. If *t* is the distance along the rotation axis between a point (or group of points) and the next, a 2 axis can combine with a translation of $\frac{1}{2}t$ (Fig. 9.5C); a 3 axis can combine with a translation of $\frac{1}{3}t$ or $\frac{2}{3}t$; a 4 axis can combine with a translation of $\frac{1}{4}t$, $\frac{2}{4}t$, or $\frac{3}{4}t$; a 6 axis can combine with a translation of $\frac{1}{6}t$, $\frac{2}{6}t$, $\frac{3}{6}t$, $\frac{4}{6}t$, or $\frac{5}{6}t$. The total number of helication elements is thus 11.

Combination of reflection through a mirror plane with translation adds one more symmetry element, called *glide plane* (Fig. 9.5D).

Only 22 combinations are possible. These, together with the 5 rotation axes and the 5 rotoinversion axes, form a total of 32 *point groups* (Table 9.4). For instance:

Point group $\bar{1}$ (read *1 bar*): a single 1-fold rotoinversion axis, equal to a center of symmetry (Fig. 9.5A). This is the symmetry of the plagioclases, a family of common igneous minerals (minerals formed from rock melts).

Point group 32 (read *three, two*): one 3-fold axis and, normal to it, one 2-fold axis that is made into a set of three 2-fold axes by the rotation of the 3-fold axis (Figure 9.5E, E′). This is the symmetry of the class to which quartz belongs.

Point group 2/m (read *2 over m*): a 2-fold rotation axis with mirror (Fig. 9.5B). This is the symmetry of clinopyroxenes, clinoamphiboles, orthoclase, muscovite, biotite, chlorite, and kaolinite, which are common igneous minerals or minerals derived from alteration of igneous minerals.

Point group $\bar{4}3m$ (read *4 bar, 3 m*): a 4-fold rotoinversion axis and a 3-fold axis with parallel mirror. This is the symmetry of diamond.

Point group $\bar{3}2/m$ (read *3 bar, 2 over m*): a 3-fold rotoinversion axis and a 2-fold axis with perpendicular mirror (Fig. 9.5F, F′). This is the symmetry of corundum, hematite, siderite, and calcite, which are common igneous

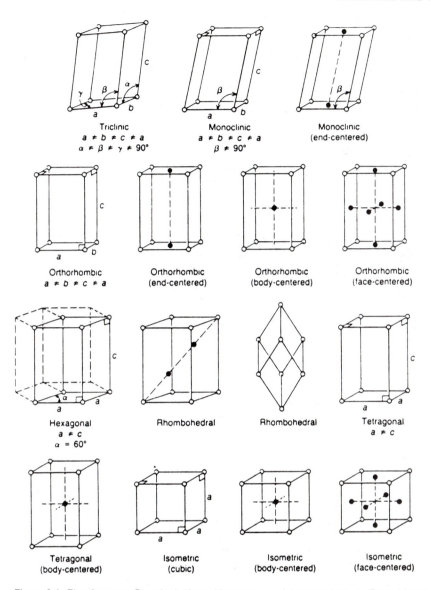

Figure 9.4. The fourteen Bravais lattices (the two rhombohedral illustrations are different expressions of the same lattice). (Berry, Mason, and Dietrich, 1983, p. 18, Fig. 1.7.)

minerals or minerals precipitated from solutions.

Point group $4/m\bar{3}2/m$ (read *4 over m, 3 bar, 2 over m*): one 4-fold axis with perpendicular mirror, one 3-fold rotoinversion axis, and one 2-fold axis with perpendicular mirror. This is the symmetry of numerous common minerals, including copper, gold, halite (the common kitchen salt), fluorite, galena, spinel, magnetite, and garnet.

Point group $2/m2/m2/m$ (read *2 over m, 2 over m, 2 over m*): three 2-fold axes mutually perpendicular, each one with a perpendicular mirror (Fig. 9.5G). This is the symmetry of sulfur, olivine, orthopyroxenes, aragonite, andalusite, and sillimanite.

The 32 point groups characterize the 32 *crystal classes*. Translations of Bravais lattices, together with symmetry operations, produce

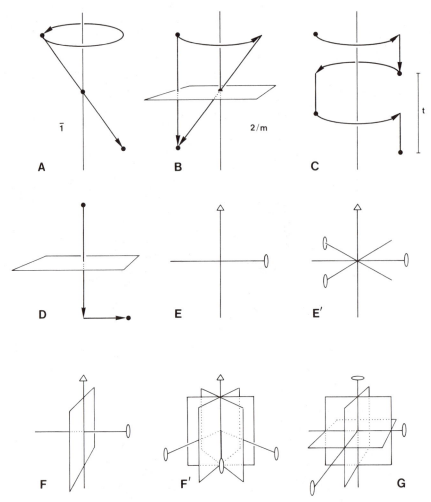

Figure 9.5. Symmetry operations. (A) Rotation by 360° plus inversion through center of symmetry (= inversion through center of symmetry, symbol *i*). (B) Rotation by 180° and inversion through center of symmetry (= reflection through mirror, symbol *m*). (C) Rotation by 180° and translation by $\frac{1}{2}t$. (D) Reflection through mirror, followed by translation parallel to mirror (glide plane). (E) Ternary *c* axis with binary axis perpendicular to it, resulting in a set of three binary axes due to the rotation of the ternary axis (E'). (F) Ternary axis with parallel mirror and perpendicular binary axis, resulting in a set of three binary axes and three mirrors (F'). (G) Three mutually perpendicular binary axes with mirrors.

a total of 230 combinations called *space groups*.

9.4. CRYSTAL STRUCTURES

Ions forming crystals approximate spheres. In order to form crystals, cations and anions must be in physical contact. Cations generally are smaller than anions (e.g., $Si^{4+} = 0.42 \text{Å}$ vs. $Si^{4-} = 2.71 \text{ Å}$) because they have lost one or more electrons, and generally fit into the voids between anions (which, having gained one or more electrons, are larger). The anions, considered as spheres, can be closely packed in two different ways, hexagonal close packing (HCP) and cubic close packing (CCP). The first layer is the same for both: each sphere is surrounded by 6 spheres on a plane. The second layer, above the first, is also the same: the spheres fit

Table 9.4. **Symmetry elements**

Type	symbol	Number
rotation	$1, 2, 3, 4, 6$	5
rotoinversion	$\bar{1}(=i), \bar{2}(=m), \bar{3}, \bar{4}, \bar{6}$	5
combination of rotation axes	$222, 23, 32, 422, 432, 622$	6
rotation axis with parallel mirror(s)	$2mm, 3m, 4mm, 6mm$	4
rotation axis with perpendicular mirror	$2/m, 3/m\,(=\bar{6}), 4/m, 6/m$	3
rotation axes with perpendicular mirrors	$2/m2/m2/m, 4/m2/m2/m, 6m/2m/2m$	3
rotoinversion with rotation and mirror	$\bar{3}2/m, \bar{4}2m, \bar{6}m2$	3
rotation, rotoinversion, and mirror(s)	$2/m\bar{3}, \bar{4}3m, 4m\bar{3}2/m$	3
Total		32

on the voids of the first layer. The third layer is different. In HCP packing, the spheres fit on the voids of the second layer, but lie directly above the spheres of the first layer (Fig. 9.6, location 1); in CCP packing, the spheres also fit on the voids of the second layer, but lie directly above the voids between the spheres of the first layer (Fig. 9.6, location 2). In either case, the cations occupy the voids between the anions. The number of anions in contact with a cation is called the *coordination number* (Table 9.5). Cations can substitute for each other if their ionic radii are similar and they have the same coordination number (e.g., Mg^{2+} for Fe^{2+}; Al^{3+} for Fe^{2+}, Fe^{3+}, Si^{4+}; Na^{+} for Ca^{2+}).

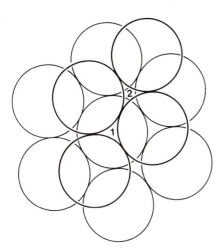

Figure 9.6. Packing of spheres. In HCP, the third-layer spheres lie directly above the first-layer spheres (location 1); in CCP, the third-layer spheres lie directly above the holes between the first-layer spheres (location 2). In either case, the amount of empty space is 25.95%.

In single-element metals, the ions are all equal in size and approximate spheres in shape, commonly forming HCP or CCP structures. These structures have the closest packing possible for spheres, but still contain $\pi/18^{1/2} = 25.95\%$ empty space.

Crystals can also be grouped according to chemical structure. The major types of chemical structures are as follows:

1. **a.** γ-iron structure (γ-Fe, Cu, Ag, Au) (Fig. 9.7A): FCC (face-centered cubic).
   b. Diamond structure (Fig. 9.7B): FCC, with four additional carbon atoms inside the cubic unit cell, giving a total of 18 carbon atoms per cell. Each carbon atom is bound to four adjacent carbon atoms, forming a tetrahedron. Although diamond is strongly bonded, it is not close-packed—the C atoms occupy only 34% of the available space. At pressure less than 20,000 atmospheres, the normal mode of crystallization for carbon is graphite, which has a hexagonal structure (Fig. 9.7C).
2. AX structure (A = cation; X = anion)
   a. CsCl structure: BCC (body-centered cubic) (Fig. 9.7D). Unit cell: cube with one Cl^- at each corner and a Cs^+ at center; the 8 corners are shared by 8 adjacent cells; thus, the unit cell holds 1 Cl^- and 1 Cs^+.
   b. NaCl (halite) structure: FCC (Fig. 9.1). Cubic array of alternating Na^+ and Cl^- ions. Unit cell:
   Cl^- at corners and at center of imaginary cube:
   • the 8 corners are shared by adjacent cells = 1 Cl^-/cell

Table 9.5.   Coordination numbers

Cation/anion radius ratio	Arrangement of anions around cation	Coordination number of cation
$0.15–0.22\,(C^{4+})$	corners of equilateral triangle	3
$0.22–0.41\,(Be^{2+})$	corners of tetrahedron	4
$0.42–0.73\,(Si^{4+}, Mg^{2+}, Fe^{2+}, Fe^{3+}, Al^{3+})$	corners of octahedron	6
$0.74–1\,(Na^{+}, Ca^{2+})$	corners of cube	8
$>1\,(K^{+})$	midpoints of cube edges	12

- the 6 faces are shared by 2 adjacent cells = 3 Cl^-/cell

The total number of Cl^- per cell is 4.

Na$^+$ at center of imaginary cube and at center of each edge:

- 1 at center (not shared) = 1 Na^+/cell
- 12 on edges, each shared by 4 adjacent cells = 3 Cl^-/cell

The total number of Na^+ per cell is 4.

Common minerals with the halite structure are most alkali halides and most alkali-earth (Mg, Ca, Sr, Ba) halides. Under pressures greater than 18 kilobars ($\sim 50\,km$ of depth inside the Earth), the CsCl structure converts to the halite structure.

3. AX_2 structure
   a. Fluorite (CaF_2) structure (Fig. 9.7E). Unit cell: FCC array of Ca^{2+} cations with two F^- anions inside the cell along diagonals between opposite corners of the cube, each accompanied by one Ca^{2+} cation.
   b. Rutile (TiO_2) structure (Fig. 9.7F). Unit cell: BCT (body-centered tetragonal) array of Ti^{4+} cations with six O^{2-} anions inside. Examples: rutile (TiO_2), cassiterite (SnO_2), stishovite (SiO_2).
   c. Quartz structure (Fig. 9.7G): framework of $Si^{4+}O_4^{2-}$ tetrahedra sharing all four O^{2-} anions with adjacent tetrahedra (hexagonal system).
4. A_2X_3 structure. Corundum (Al_2O_3) structure: HCP packing of O^{2-} anions with Al^{3+} cations occupying two-thirds of the voids. Another important mineral with the corundum structure is hematite (Fe_2O_3).
5. ABX_3 structure
   a. Calcite ($CaCO_3$) structure (Fig. 9.7H). The Ca^{2+} cation has the octahedral coordination number 6, and the C^{4+} cation has the triangular coordination number 3. All carbonates with cations smaller than Ca^{2+} crystallize in the calcite structure. Examples are magnesite ($MgCO_3$) and siderite ($FeCO_3$).
   b. Aragonite ($CaCO_3$) structure (Fig. 9.7I): similar to calcite, but the CO_3^{2-} anion is oriented in such a way that the Ca^{2+} cation is surrounded by nine O^{2-} anions instead of six as in calcite. Aragonite is denser than calcite (2.9 vs. 2.7). All carbonates with ions larger than Ca^{2+} crystallize in the aragonite structure. Examples are witherite ($BaCO_3$), strontianite ($SrCO_3$), and cerussite ($PbCO_3$).

The *silicates* are characterized by independent or linked SiO_4^{4-} tetrahedra balanced by cations. The linked tetrahedra form complex anions characteristic of the different classes of silicates:

1. Nesosilicates (Fig. 9.8A): independent SiO_4^{4-} tetrahedra held together by cations (most commonly Mg, Al, Ca, Fe). Examples:

   Olivine: $(Mg, Fe)_2SiO_4$: SiO^{4-} tetrahedral anions held together by Mg^{2+} and Fe^{2+} cations; the chemical composition of olivine is intermediate between that of two end members, forsterite (Mg_2SiO_4) and fayalite (Fe_2SiO_4); commonly, olivine contains 70–90% Mg_2SiO_4.

   Garnets: Mg^{2+}, Ca^{2+}, Fe^{2+}, and Mn^{2+} cations surrounded by eight O^{2-} anions, and Al^{3+}, Fe^{3+}, and Cr^{3+} cations surrounded by six O^{2-} anions.

2. Sorosilicates (Fig. 9.8B): pairs of tetrahedra held together by one corner and balanced by cations. Example: ackermanite, $Ca_2Mg(Si_2O_7)$.

3. Alternating nesosilicate and sorosilicate

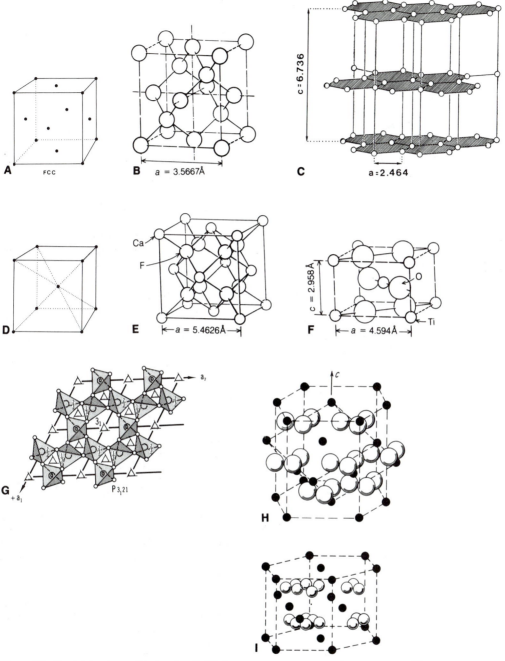

Figure 9.7. Crystal structures: (A) γ-iron (FCC); (B) diamond; (C) graphite; (D) CsCl (BCC); (E) fluorite; (F) rutile; (G) quartz, showing the array of SiO$_4$ tetrahedra projected on the 0001 plane (tetrahedra marked $\frac{1}{3}$ and $\frac{2}{3}$ are, respectively, $\frac{1}{3}$ and $\frac{2}{3}$ of the distance along a helix, with axis perpendicular to the page, positive helicity, and a pitch of about 1 nm that connects successive layers of unmarked tetrahedra); (H) calcite; (I) aragonite (black dots represent Ca^{2+} ions, the carbon ion being hidden within the three O^{2-} ions). (B, E, and F from Berry, Mason, and Dietrich, 1983, p. 244, Fig. 8.12, p. 326, Fig.11.3, p. 305, Fig. 10.11; C from Andreatta, 1943, p. 386, Fig. 378; G from Bloss, 1971, p. 319, Fig. 10.18A; H and I from Frye, 1974, p. 63, Fig. 2.9.)

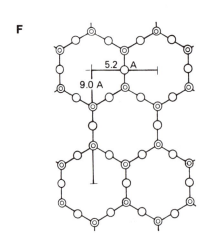

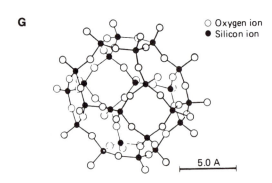

structure. Example: epidote (Ca, Fe, Al silicate with the OH group).

4. Cyclosilicates (Fig. 9.8C): closed rings of $3(Si_3O_9^{6-})$, $4(Si_4O_{12}^{6-})$, or $6(Si_6O_{18}^{6-})$ tetrahedra, each sharing two corners. Example: beryl, $Be_3Al_2(Si_6O_{18})$.

5. Inosilicates:

   a. Single chains of SiO_4^{4-} tetrahedra, each sharing two corners and therefore forming SiO^{2-} anions (Fig. 9.8D): Example: hypersthene, $(Mg,Fe)SiO_3$, a pyroxene.

   b. Double chains of SiO_4^{4-} tetrahedra, alternately sharing two and three corners, forming the complex anion $Si_4O_{11}^{6-}$ (Fig. 9.8E). Example: hornblende (Ca, Na, Mg, Fe, Al silicate with the OH group).

6. Phyllosilicates (Fig. 9.8F): sheets of SiO_4^{4-} tetrahedra, each sharing three corners and forming the complex $Si_2O_5^{-}$ anion. Example: muscovite, $KAl_2(Si_2O_5)_2(OH)_2$; kaolinite, $Al_4(Si_2O_5)_2(OH)_8$.

7. Tektosilicates (Fig. 9.8G): three-dimensional framework of SiO_4^{4-} tetrahedra, each sharing four corners. Example: quartz, SiO_4^{4-}.

9.5. CRYSTAL DEFECTS

In a perfect crystal, all loci prescribed by the symmetry elements are occupied. This state is approached only by *whiskers*, which are extremely thin (1–2 μm in diameter) spontaneous growths, up to 1 mm long, on metal surfaces, produced by the metal itself to reduce internal strain. Larger whiskers (a few micrometers across, up to several centimeters long) can be grown artificially from supersaturated solutions of metals, metal oxides, and metal chlorides. Whiskers exhibit up to 500 times the strength of natural crystals.

All natural crystals exhibit crystal defects to various extents. There are six classes of crystal defects:

Figure 9.8. Structures of the silicates: (a) nesosilicate; (b) sorosilicate; (c) cyclosilicate; (d) inosilicate, showing the single-chain structure of pyroxenes; (e) inosilicate, showing the double-chain structure of amphiboles; (f) phyllosilicate, showing the sheet structure; (g) tectosilicate, showing the three-dimensional framework of SiO_4 tetrahedra. (Bragg, 1955, pp. 134–5, Fig. 82.)

1. Interstitial defect: the occurrence of a foreign ion (e.g., C in steel, amounting to $\approx 1\%$) within the regular lattice.
2. Schottky defect: an ion is missing.
3. Frenkel defect: an ion is relocated to an interstitial position, leaving a hole.
4. Edge defect: a plane of ions extends only partly into the crystal structure.
5. Screw dislocation: one or more layers of ions grow by rotating around an axis normal to the growing face.
6. Disorder defect: ions are disordered at a locus within a crystal.

9.6. MINERAL PROPERTIES

Minerals have specific properties that derive from their chemical composition, structure, and chemical bonding. The principal physical properties are density, hardness, color, luster, streak, refractivity, luminesence, heat capacity, and thermal conductivity. Their electric properties include resistivity and strain electricity (piezoelectricity and pyroelectricity). Their magnetic properties include susceptibility and the Curie point. Many substances have the capacity to crystallize in more than one form, depending upon the ambient physical and chemical conditions. This property is called *polymorphism*. Carbon, for instance, can crystallize in the hexagonal system as graphite, or in the cubic system as diamond. The physical properties of these two polymorphs are vastly different. In graphite, the common form of carbon at low temperatures and pressures, the carbon atoms are bonded in sheets, with weaker bonds between one sheet and the next (Fig. 9.7C). At high pressures and temperatures, all carbon atoms are bonded to each other in the diamond structure (Fig. 9.7B). Density is 2.3 g/cm³ for graphite and 3.5 g/cm³ for diamond.

Density (ρ) is mass per unit volume (Table 9.6). It is more conveniently expressed in grams per cubic centimeter (which makes the density of water = 1) than in the SI units of kilograms per cubic meter (which makes the density of water = 1,000). Minerals tend to crystallize in denser phases as pressure increases. For example, quartz ($\rho = 2.65\,g/cm^3$) transforms into coesite ($\rho = 2.91$) at 20 kilobars (kb); coesite in turn, transforms into stishovite ($\rho = 4.28$) at 100 kb; olivine changes

Table 9.6. Densities of common minerals

Mineral	Density (g/cm³)
ice (air free)	0.91
halite	2.16
orthoclase	2.55
albite	2.62
quartz	2.65
calcite	2.71
anorthite	2.76
muscovite	2.8–2.9
biotite	2.8–3.2
aragonite	2.93
augite	3.2–3.5
hypersthene	3.4–3.5
olivine	3.3–4.3
kamacite	7.3–7.9
taenite	7.8–8.2
iron	7.87
nickel	8.9

its structure from orthorhombic ($\rho = 3.3$–4.3) to cubic, with $\sim 10\%$ increase in density; and graphite changes its structure from hexagonal ($\rho = 2.27$) to cubic and becomes diamond ($\rho = 3.51$).

[1 kilobar (1 kb) = 989 atm, a pressure that exists under approximately 3 km of rock.]

Hardness measures the resistance of a mineral to bond rupture. Ionic and covalent bond strengths are on the order of 3 to 8 eV. Covalent bonds are directional, and therefore bonding strengths commonly are different in different directions. In general, bonding strength is inversely proportional to ion size and directly proportional to bond density (number of bonds per unit of volume). Most crystals are bonded by a combination of ionic and covalent bonds. Metals are bonded by the metallic bond, which is weaker (1–3 eV), making metals susceptible to plastic deformation rather than rupture. Alloys are bonded by a combination of metallic and covalent bonds.

A hardness scale (Table 9.7) was devised in 1822 by the German mineralogist Friedrich Mohs (1773–1839). A mineral of a given hardness can scratch all minerals of lower hardness and, in turn, is scratched by minerals of greater hardness. The hardest mineral is

Table 9.7. The Mohs hardness scale

Hardness	Mineral
1	talc
2	gypsum
3	calcite
4	fluorite
5	apatite
6	orthoclase
7	quartz
8	topaz
9	corundum
10	diamond

diamond, in which all carbon atoms share all their valence electrons. In contrast, talc, the softest mineral, consists of two layers of SiO_4^{4-} tetrahedra, with their vertices pointing toward each other and bonded by an intervening layer of Mg^{2+} and OH^- ions. This system forms a sheet that has no net charge. Sheets are stacked above each other and held together by weak van der Waals forces. As a result, talc flakes easily.

The hardness of metals it assessed by pressing a standardized stylus on the metal surface. The hardnesses of gold, silver, and copper are $2\frac{1}{2}$–3, and that of iron is $4\frac{1}{2}$.

Cleavage is the property of many minerals to cleave preferentially in one or more directions. Cleavage planes are planes where chemical bonding is weakest.

Color and *luster* are properties that result from the interaction of photons with the electrons in a mineral. The free electrons in metals have available an enormous number of energy states. When white light strikes a clean metal surface, nearly all photons are absorbed and reemitted (within the usual 10^{-8}s). Clean metal surfaces are therefore specular, although many metals absorb some wavelengths, giving them their color.

A pure, perfect, nonmetallic mineral with its electrons in the ground state is unable to absorb white light and is therefore perfectly transparent. But natural minerals are never perfect, and never pure either. There may be some free electrons in holes, or alien atoms ("impurities") stuck in the crystal, with one or more extra electrons that cannot find a bond. The result is that when white light strikes, some wavelengths may have the right energy to be absorbed by electrons here and there within the crystal. The electrons thus energized reemit, of course, but reemission is on a 4π geometry. Much less energy is transmitted in any given direction. If, say, a photon of red light is absorbed from the incident white light and almost immediately reemitted with the usual 4π geometry, the white light striking your eye will be depleted of red and will no longer appear white, but bluish. If the mineral structure is anisotropic, reemission may occur on a distorted 4π geometry, with different colors being preferentially scattered in different directions. It usually takes only minute amounts of impurities to produce color in crystals (or glasses), because our eyes are extremely sensitive to color differences.

Luster is a property resulting from the interaction of photons with the electrons in the surface layer of a substance or with the small-scale structure of the surface itself. Crystal surfaces, like crystal interiors, are never perfect. Increasing roughness increasingly scatters light and reduces luster. There is, therefore, a

Table 9.8. Indices of refraction and speed of light

Substance	Index of refraction (Na light; $\lambda = 0.5893\ \mu m$)	Speed of light in substance (m/s)
vacuum	1	299,792,458
air (dry, 0°C, 1 atm)	1.0002926	299,704,764
pure water (20°C)	1.33335	224,841,533
seawater (35‰ salinity, 20°C)	1.339	223,892,799
fused quartz	1.4584	205,562,574
diamond	2.42	123,881,181
rutile	2.61	114,863,011
iodine	3.34	89,758,221

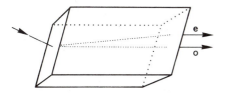

Figure 9.9. Birefringence, showing the ordinary (o) and extraordinary (e) rays produced when a light ray strikes normal to the face of a calcite crystal.

complete range of lusters, from metallic to mat.

Refractivity of a medium relates to the speed of light in that medium. The lower the speed, the higher the refractivity. The *index of refraction* is the ratio of the speed of light in vacuo to that in the medium (Table 9.8). When a light ray enters a transparent medium with parallel surfaces (a window pane, for instance) normal to one of the faces, the ray slows down, and when it emerges from the other side, it continues in the same direction at the original speed. (Here and in the following, "light" is meant to mean monochromatic light.) If, on the other hand, light

strikes the surface at an angle other than $90°$, it is refracted (see Fig. 2.11). This is true for substances that are optically isotropic (i.e., substances through which light travels at the same speed in all directions). These include amorphous substances and crystals belonging to the cubic system. In crystals belonging to the other systems, a ray of light is split into two rays that usuallly propagate at different velocities within the crystal. The difference can be as high as 17.2% (calcite) or 26% (orpiment) (Fig. 9.9). One ray (the ordinary or o ray) travels at the same speed in all directions; the other ray (the extraordinary or e ray) has a speed that can be lower than that of the o ray ("positive" crystals) or higher ("negative" crystals). The two rays are polarized normally to each other.

[A beam of light consists of photons; photons consist of pulses of electromagnetic radiation; and a pulse consists of an electric field **E** oscillating on a plane and a magnetic field **B** oscillating on another plane normal to the first, both propagating at the speed of light along the intersection of the two planes (Fig. 9.10A). Because the directions of the

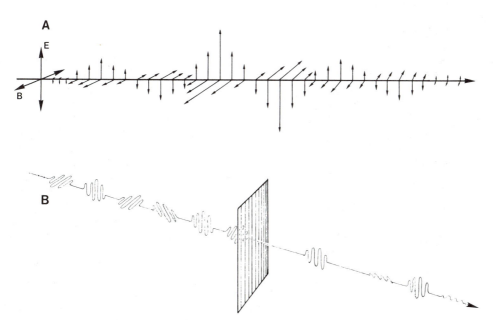

Figure 9.10. Light as a stream of photons. (A) A photon can be visualized as an electromagnetic wave disturbance, consisting of an oscillating electric field **E** and a magnetic field **B** oscillating normal to it. (B) As the photons, variously oriented with respect to each

other, reach a polarizer, only the photons whose vector **E** is in the same direction as the atoms or molecules of the polarizer can get through with full amplitude. Those at an angle will emerge with their amplitude reduced proportionally to the cosine of the angle.

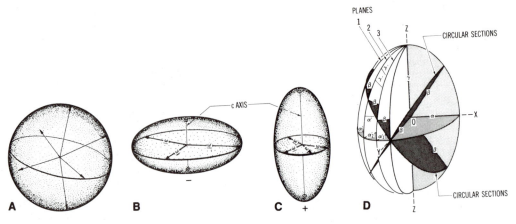

Figure 9.11. The indicatrix for (A) cubic crystals, (B) negative or (C) positive uniaxial crystals, and (D) biaxial crystals. (Bloss, 1971, p. 420, Fig. 12.3; p. 423, Fig. 12.5; p. 425, Fig. 12.7A.)

electric and the magnetic oscillations are at 90° from the direction of propagation, electromagnetic waves are *transversal waves.* White light consists of photons of different wavelengths, vibrating along planes with different orientations in space. Monochromatic light consists of photons of the same wavelength, still vibrating along planes with different orientations. If either white light or monochromatic light is passed through a crystal or other atomically or molecularly ordered substance (such as a Polaroid sheet), only light whose vector **E** is vibrating on a plane that can pass through the layers of atoms or molecules of the substance will emerge (Fig. 9.10B). This phenomenon is called *polarization.* Because of their complex atomic structures, crystals can

Let polarized light through,
Let some polarized light through (if the polarization planes of the light ray and the substance are at an angle between 0° and 90°),
Stop polarized light (if the polarization planes of the light and the substance are at 90° to each other), or
Rotate the polarization plane with either positive or negative helicity.]

Minerals can be identified by means of these optical properties.

In crystals belonging to the tetragonal and hexagonal systems there is one preferred direc-

tion, called the *optic axis,* along which light propagates without splitting into two rays. These crystals are called *uniaxial.* In the orthorhombic, monoclinic, and triclinic systems there are two such directions. These crystals are called *biaxial.* The index of refraction associated with the *o* ray is identified by the symbol ω; that associated with the *e* ray is identified by the symbol ε. The *o* and *e* rays propagate at the same speed along the optic axes.

Because in amorphous minerals or minerals belonging to the cubic systems the speed of light is the same in all directions, the index of refraction is also the same in all directions. The *indicatrix* is an imaginary surface centered inside a mineral whose distance from the center represents the index of refraction for light vibrating along that direction (and therefore propagating normal to it). In cubic minerals, the indicatrix is a spherical surface—light travels at the same speed in all directions (Fig. 9.11A). In minerals belonging to the tetragonal and hexagonal systems the indicatrix is an ellipsoid of revolution, with its axis of revolution coincident with the *c* axis of the crystal (Fig. 9.11B, C). If the mineral is "positive", the *o* ray's velocity is greater than that of the *e* ray, the index of refraction ω is smaller than ε, and the ellipsoid is prolate (elongated along its axis of rotation) (Fig. 9.11B); if the crystal is "negative", ω is greater than ε, and the ellipsoid is oblate (Fig. 9.11C). In either case, the equatorial circumference is a circle.

In minerals belonging to the orthorhombic, monoclinic, and triclinic systems the indicatrix is a triaxial ellipsoid (i.e., an ellipsoid whose equatorial circumference is elliptical rather than circular). In such an ellipsoid there are two circular sections (Fig. 9.11D), perpendicular to which, and passing through the center of the indicatrix, are 2 optic axes. The highest index of refraction is identified by the letter γ, and the lowest by the letter α. The intermediate one, normal to the other two, is represented by the letter β (Fig. 9.11D). It forms the radius of the two circular sections.

Interference is the phenomenon by which two light rays may reinforce, cancel, or variously affect each other (see Fig. 2.9). Interference is produced when light from a given source is split into rays that follow different paths and then converge on the same target (a screen, for instance). For visible light, interference is caused when pinholes, slits, edges, and similar impediments are in the path of the light ray. The dimensions of the impediment must be of the same order as the wavelength of the radiation; that means a few micrometers at most for visible light.

Atoms in crystals are spaced about 1 to 4 angstroms (Å) apart (i.e., a few tenths of a nanometer apart). Light does not exhibit interference when passing through a crystal because its wavelength ($\sim 0.5 \mu m$) is some 10,000 times longer than the separation of the atoms; on the other hand, x-rays, which have wavelengths similar to the radii of atoms, do exhibit interference. The interference pattern of x-rays passing through a crystal is the fundamental method by which the structures of crystals are determined.

[X-rays are produced by bombarding a metal surface with electrons accelerated through potential differences of 20,000–100,000 V. An x-ray tube is a diode with a tungsten filament as cathode and a metal (usually Cr, Fe, Co, Ni, Cu, or Mo) as anode. When energetic electrons impinge on the metal surface, the free electrons, as well as the bound electrons, are raised to higher energy levels, and in returning to ground they emit a continuous spectrum ranging from x-rays to the infrared. Superimposed on the continuous spectrum are sharp peaks produced when one of the more energetic impinging electrons manages to

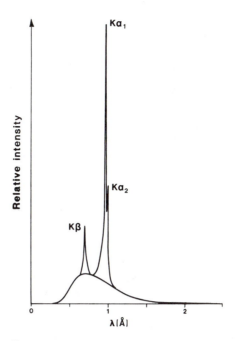

Figure 9.12. X-ray diffraction by crystals. The asymmetric curve at the base represents the continuous x-ray spectrum diffracted by a crystal. The $K\alpha_1$ peak and the adjacent $K\alpha_2$ peak represents the filling of holes in the K shell by electrons from the two energy levels of the L shell. The $K\beta$ peak represents filling by electrons from the lowest energy level of the M shell. Peak height represents number of events. The $K\beta$ transition is less frequent than the $K\alpha$ transitions, but it has greater energy (shorter wavelength).

penetrate deep into the electron cloud of an atom and dislodge an electron from the K shell. The hole is immediately filled from the L shell, giving the so-called $K\alpha_1$ and $K\alpha_2$ peaks (Fig. 9.12), one from the lower energy level of the L shell and the other from the higher one. The hole may also be filled from the M shell, giving the $K\beta$ peak that has higher energy (shorter wavelength), but it is less frequent, and therefore it has lower intensity or peak height. Or the hole may be filled from higher shells or energy levels within shells. The available energy levels are 1 (K shell), 3 (L shell), 5 (M shell), and 7 each for the higher shells (many unfilled, see Table A6). Each transition is represented by a specific line (on a photographic plate) or by a specific peak (on a power spectrum). Lines or peaks are characteristic of specific elements, and so x-rays can be used for elemental analysis (this is the principle of the *electron microprobe*, an instrument that can focus electrons on a surface as small as $1\,\mu m^2$ and reveal which elements are there).]

Crystal structure can be determined by focusing a beam of x-rays on a crystal that is rotated about an axis parallel to the edge between two crystal faces, the axis of rotation being normal to the x-ray beam. A film is wrapped around the crystal, forming a coaxial cylinder, for the purpose of recording the diffraction pattern. A series of regularly spaced spots will appear when the film is developed. Alternatively, the intensity of the diffracted rays can be determined using a scintillation counter (see Section 3.4) instead of a film. The intensity is recorded on a strip chart as the counter is revolved about the crystal. Spots and peaks represent the points where x-rays emitted by electrons that have been excited by the primary beam are in phase with each other. Coherence occurs only along specific directions, which thus reveal the orientation and spacing of the various atomic planes within the crystal.

Streak is the color of a powdered mineral. Pulverization disperses the "impurities" that may give a mineral its color, and therefore the color of the powdered mineral is often more diagnostic than that of the mineral when whole. The streak of a colored mineral is often white. Sometimes the streak has a color unexpectedly different from that of the mineral. Hematite, which has a metallic black luster, yields a power that is reddish brown.

Luminescence is the emission of visible light by a crystal that has been excited by ultraviolet light. In the type of luminescence called *fluorescence*, the energy absorbed is released in two or more steps, one of which is in the visible range. In the type called *phosphorescence*, the excited electron falls into an electron trap (such as a hole), where it may remain for up to several hours, until the thermal energy of the system causes its recombination with a positive ion. Light is emitted during the recombination.

Heat capacity is the quantity of heat (calories or joules) needed to raise the temperature of a body by 1°C, which will depend on the type of body and its size. *Specific heat* is heat capacity per gram of substance.

Table 9.9. Specific heats

Substance	Specific heat (cal/g) (at 25° C unless specified)
pure water	
3.98°C	1.0043
14.5°C	1.0000
100.0°C	1.0070
Ice	
−4.5°C	0.4984
−24.5°C	0.4605
albite	0.185
andalusite	0.271
calcite	0.169
copper	0.082
diamond	0.180
granite	0.192
graphite	0.223
hypersthene (pyroxene)	0.233
iron	0.106
marble	0.21
nickel	0.106
orthoclase	0.222
quartz	0.182
sandstone	0.26
shale	0.17
silver	0.056

The specific heat of water (one of the highest) is $1\,cal/g = 4.1868 \cdot 10^3$ J/kg. Specific heat is also defined as the ratio of the heat capacity of a substance to that of water at 14.5°C, both per unit of mass. In that case, the specific heat is a dimensionless number. Specific heats of representative substances are shown in Table 9.9. Notice the following:

The specific heat of water is around 1
The specific heat of ice is around 0.5
The specific heats of silicate minerals and of rocks are on the order of 0.2–0.3
The specific heats of metals are on the order of 0.01

(all referred to the specific heat of water).

Thermal conductivity is an important property regarding the thermal regime of the Earth. Representative thermal conductivities are shown in Table 9.10. Notice the following:

Silver, copper, and aluminum have the highest heat conductivities
Sulfur has the lowest heat conductivity among the elements in the solid state

Table 9.10. Thermal conductivities

Substance	Thermal conductivity at 25°C (W cm^{-1} K^{-1})
aluminum	2.37
calcite	0.04–0.05
copper	4.02
diamond (max.)	1.21–1.63
dolomite	0.05
gabbro	
0°C	0.023
200°C	0.021
Globigerina ooze	0.010
granite	
0°C	0.028
200°C	0.023
ice (0°C)	0.02
igneous rocks	0.02–0.03
iron	0.80
limestone	0.02–0.03
marble	0.05
nickel	0.91
orthoclase	0.04
quartz	0.07–0.11
quartzite	0.04–0.07
red clay	0.006
sandstone	0.02–0.03
seawater (17.5°C, 35‰ salinity)	0.00562
shale	0.02–0.03
silver	4.29
snow	0.004
sulfur	0.003
water (pure)	
0°C	0.00536
25°C	0.00607
50°C	0.00636

Table 9.11. Electrical resistivities

Substance	Electrical resistivity (room temperature) (Ω·m)
aluminum	$2.65 \cdot 10^{-8}$
ceramics	10^{13}–10^{17}
copper	$1.678 \cdot 10^{-8}$
diamond	10^{11}
germanium	0.46
ice	$4.7 \cdot 10^{5}$
igneous rocks	10^{4}–10^{7}
iron	$9.71 \cdot 10^{-8}$
marble	10^{6}–10^{8}
muscovite (c axis)	$4 \cdot 10^{15}$
nickel	$6.84 \cdot 10^{-8}$
porcelain (unglazed)	$3 \cdot 10^{12}$
quartz (fused)	10^{17}
sedimentary rocks	10^{2}–10^{4}
sediments (dry)	10^{3}–10^{4}
silicon	$3 \cdot 10^{-2}$
silver	$1.586 \cdot 10^{-8}$
sulfur	$2 \cdot 10^{15}$
water	
fresh	1–80
sea	0.15–0.20

Minerals and rocks have low heat conductivities, mostly between 0.02 and 0.05 W cm^{-1} K^{-1}

Loose, wet deep-sea sediments (Globigerina ooze, red clay) have heat conductivities on the order of that of seawater

In anisotropic minerals, heat conductivity may vary in different directions by up to 50%. In muscovite, conductivity along the c axis (across foliation planes) is very low (0.008 W cm^{-1} K^{-1}), but the conductivities along the a and b axes are, respectively, 5.8 and 6.3 times greater. The thermal conductivity in rocks decreases with increasing temperature by 5–15% per 100°C.

Electrical resistivity is measured in ohms· meter. It ranges from 10^{-8} to 10^{-6} Ω·m for conductors; from 10^{-6} to 10^{6} Ω·m for semiconductors, and from 10^{6} to 10^{16} Ω·m for insulators (Table 9.11).

Silver has the lowest resistivity (= highest conductivity) of all elements, with copper a close second. This is the reason, of course, that copper is extensively used in electric circuitry. At the other extreme we have ceramics, muscovite (in the direction of the c axis), and fused quartz, which are used as insulators.

Strain electricity is electric polarization caused in a crystal by mechanical strain.

[The *electric dipole moment* **p** of two opposite, equal charges is the product of either charge times the distance **l** between the two charges: $\mathbf{p} = q \cdot \mathbf{l}$. It is a vector directed from the negative to the positive charge. *Electric polarization P* is the electric dipole moment per unit volume: $P = q \cdot l/l^3 = q/l^2$. It is measured in coulombs per square meter.]

Strain electricity works this way: Electric charges in crystals are contributed by ions. The distribution of ions in three-dimensional space depends on the structure of the unit cell. In crystals that have no center of symmetry (21 of the 32 classes) it is possible that positive and negative charges will not be uniformly distributed—there may be more positive charges in one direction and more negative charges in the other. In 11 of these classes, other symmetry elements exist whose effect is to produce mirror images of any polar vector, thereby precluding the occurrence of strain electricity. That leaves 10 classes. In these classes, the vectorial sum of the polar vectors, produced by the symmetry elements, is not zero, and a residual vector, called the *unique polar vector*, remains, and so the charge distribution exhibits polarity. Quartz belongs to class 32, and so it has a 3-fold axis of symmetry and, normal to it, a 2-fold axis that is made into a set of three uniterminal 2-fold axes by the rotation of the 3-fold axis (Fig. 9.5F, F'). The three 2-fold axes emerge from the edges of the hexagonal prism perpendicular to the 3-fold symmetry axis. The edges of the prism become alternately positively and negatively charged when the crystal is strained.

The inherent electric polarization would remain permanently in a perfect crystal in vacuum at $0°$ K. In the natural environment, the inherent electric polarization is destroyed by charge migration within the crystal via crystal defects and by either absorption of charges from the environment or release of charges to the environment. Normally, a crystal does not exhibit electric polarization.

If, however, a crystal is stressed (by pressure or tension, or by a temperature change), the neutralizing charges are no longer in the appropriate positions, and electric polarization reappears. A stress of 1 N/m along the polar axis produces a polarization of 1.8 in quartz, 1.2 in tourmaline, 36 in barium titanate ($BaTiO_3$), and 435 in Rochelle salt ($KNaC_4H_4O_6 \cdot 4H_2O$), all in units of 10^{-12} C/m^2. A change in temperature of $1°C$ in tourmaline stresses the crystal sufficiently to produce a polarization of 10^{-5} C/m^2. Electric polarization produced by mechanically induced strain is called *piezoelectricity*. That produced by thermally induced strain is called *pyroelectricity*.

[*Piezoelectricity* derives from πιέζειν, which means *to squeeze*, and from ἤλεκτρον, which means *amber* (see Section 1.2); *pyroelectricity* derives from πῦρ, which means *fire*.]

The process is reversible: Application of an external electrical field produces strain or a temperature change. As a result, a crystal may function as a generator (transforming mechanical energy or heat into electricity) or as an electric motor or heater (transforming electricity into mechanical energy or heat).

Induced strain electricity lasts only a short time before thermal energy restores charge equilibrium. If, however, an external, periodic electrical field is applied, piezoelectricity or mechanical deformation of the same period will result.

Quartz laminae or rods that are cut from a quartz crystal at specific angles to the *c* axis react to an ac current of a frequency comparable to the resonant period of the lamina or rod by flexing, thickening and thinning, or twisting. Each lamina or rod geometry and deformation mode has its own resonant frequency—the thinner the lamina or rod, the higher the frequency. Quartz laminae or rods, therefore, can be used as frequency regulators. In quartz watches, for instance, a quartz lamina inserted into an oscillating circuit will stabilize the oscillating frequency of the circuit and therefore "keep time".

Underwater sound detectors consist of quartz laminae and/or rods that generate an electrial signal when distorted by the arrival of a sound wave in the water. Conversely, an echo sounder consists of a quartz plate that is made to deform in resonance with an ac frequency. The deformation is transmitted to the water as a sound wave of the same frequency. The sound wave travels to the target (the bottom, for instance) and the returning wave sets into motion a quartz receiver that transforms the sound frequency back into an electric frequency. The maximum power that can be projected in water by an oscillating quartz lamina is 0.5 W/m^2, the limitation being the mechanical strength of the lamina.

Some minerals, most notably galena (PbS), exhibit semimetallic properties, that is, properties intermediate between those of metals and those of semiconductors.

[Metals have free electrons, and therefore the valence band (the outermost electron shell) is continuous with the conduction band. In insulators, the valence band is separated from the conduction band by a wide energy gap (2–10 eV). Examples are carborundum, 2.86 eV; corundum, 8.3 eV; fluorite, ~ 10 eV; perovskite, 3.7 eV; and rutile, 3.05 eV. In semiconductors, the valence band is separated from the conduction band by a smaller energy gap (silicon, 1.09 eV; germanium, 0,72 eV; galena, 0.37 eV; pyrite, 1.2 eV). When electrons in the valence band are sufficiently energized, they will cross the energy gap and enter the conduction band, leaving behind vacancies (called *holes*) that act as positive particles. Under the influence of an applied voltage, electrons and holes move in opposite directions.]

A radio frequency cutting across a galena crystal moves electrons in one direction, but hardly in the opposite direction. A galena crystal can therefore act as a radio detector. An amplitude-modulated (AM) radio transmitter produces a radio wave, called the *carrier wave*, with a frequency normally between 500 kHz and 1.6 MHz. The amplitude of the carrier wave is modified by an electric signal produced by a microphone. On the receiving end, the galena crystal rectifies the radio frequency and lets through the amplitude variations, which are transformed back into sound via a loudspeaker.

[A capacitor is an electric device consisting of two parallel plates separated by a gap. If one plate has a positive charge and the other one a negative charge of the same magnitude, a voltage difference will exist between the two plates. *Capacitance* (C), defined as charge q divided by voltage V ($C = q/V$), is measured in farads (F), the SI unit of capacitance. One farad is equal to 1 C/V. [The farad is far too large a unit of capacitance for common usage; in its place, the microfarad ($\mu F = 10^{-6}$ F), the nanofarad (nF = 10^{-9} F), and the picofarad (pF = 10^{-12} F) are used.] For the same charge, the potential difference between the two plates of a capacitor is proportional to their distance. A variable capacitor is a capacitor in which the distance between the two plates, and therefore the voltage, can be changed. Microphones and loudspeakers are essentially variable capacitors, with one plate flexible and the other rigid. In microphones, sound waves hitting the flexible plate change the capacitance, which is transformed into an electric signal that is superimposed on the carrier wave and changes its amplitude. In a loudspeaker, the amplitude changes cause the flexible plate to vibrate, thus reproducing the sound. A galena crystal charges the fixed plate of a speaker with changing charge, which causes the flexible plate to vibrate and thus to reproduce the original sound.]

Magnetic susceptibility (χ) is the ratio of magnetization of a material to the strength of the applied magnetic field:

$$\chi = \mathbf{M}/\mathbf{H}$$
$$= (\mu_r - \mu_0)/\mu_0$$

where χ is magnetic susceptibility, **M** is magnetization, **H** is the intensity of the applied magnetic field, and μ_r is the relative permeability, that is, the ratio of the permeability μ of a substance to the permeability of vacuum μ_0 (which is also called the permeability constant). Susceptibility is expressed in gauss per oersted and is therefore a dimensionless number.

Substances are classified as diamagnetic (negative susceptibility, reaching a maximum of $-3 \cdot 10^{-4}$ for Bi), paramagnetic (positive susceptibility, reaching a maximum of 0.1 for Tb), or ferromagnetic (positive susceptibility ranging from 0.1 to 10^4) (see Chapter 2). Diamagnetism and paramagnetism arise from the structure of the electronic clouds in the atoms of the pertinent substances. Ferromagnetism is a bulk property of some solid metals (Fe, Co, Ni) and alloys.

The magnetic susceptibility of an element depends on its electronic structure. In diamagnetic substances, the electrons are paired, and susceptibility is small or even zero because paired electrons have opposite spins. An external magnetic field will accelerate one electron in a pair and decelerate the other, creating a magnetization in the substance opposite that of the field. Paramagnetic substances have a significant net magnetic moment, due to the spin and orbital motions of unpaired electrons. Among the elements, Li, O, Na through Al, K through Mn, Rb through Pd, Sn, and Cs through Pt are paramagnetic. When a paramagnetic substance is placed within an external magnetic field, the net magnetic moment of each atom becomes aligned with the external field. Thermal energy at room temperature is

Table 9.12.   **Magnetic susceptibilities**

Mineral	Magnetic susceptibility ($\times 10^{-6}\,emu_{cgs}/g$)
amphiboles	13–75
biotite	53.78
bismuth	−1.4
calcite	−0.38
copper	−0.09
diamond	−0.49
fayalite	100
galena	−0.34
garnets	31–159
halite	−0.52
hypersthene	73
ice	−0.70
ilmenite	0.87
manganese	11
pyrolusite	40
quartz	−0.44
rutile	0.07
siderite	98
silver	−0.18
sulfur	−0.48
water	−0.72

10^2 to 10^3 times greater than the induced magnetic energy, and it quickly randomizes the orientations of the atoms in the substance if the field is removed.

In ferromagnetic substances (Fe, Co, Ni, Gd, and Dy among the elements, and several special alloys), adjacent atoms form *domains* ($\sim 10^{15}$–10^{18} atoms each), within which the net magnetic moments of the atoms are aligned in the same direction. Adjacent domains have different orientations. The application of an external field results in all domains acquiring the same orientation. Removal of the external field leaves the substance magnetized, because thermal energy at room temperature is insufficient to randomize the domains. Randomization, however, can be achieved by raising the temperature to a sufficiently high level (the Curie point, equal to 770°C for Fe).

There are also substances that are *antiferromagnetic* or *ferrimagnetic*. In these substances, the atomic spins are aligned, but alternating between parallel and antiparallel directions. Antiferromagnetic substances (e.g., ilmenite below 57 K) have equal numbers of parallel and antiparallel spins. In ferrimagnetic substances, the numbers are unequal (e.g., magnetite). The magnetic susceptibilities of typical minerals are shown in Table 9.12. Negative numbers indicate diamagnetic substances. All others are paramagnetic. The susceptibilities of ferromagnetic elements and alloys and of ferrimagnetic minerals are much higher.

9.7. PHASE DIAGRAMS

Minerals may have different phases at different temperatures and pressures. Phase diagrams relate the phases to temperature and pressure. Consider, for instance, water. Water has three phases—solid, liquid, and gaseous (Fig. 9.13A). If water is placed in a closed vessel, the air removed, and the temperature adjusted to 0.01°C, the three phases will coexist. There will be liquid water, some ice floating on the

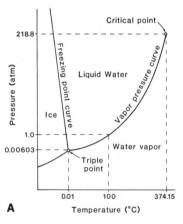

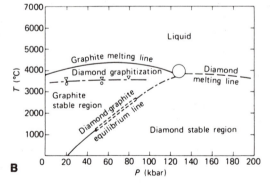

Figure 9.13. See caption on page 185.

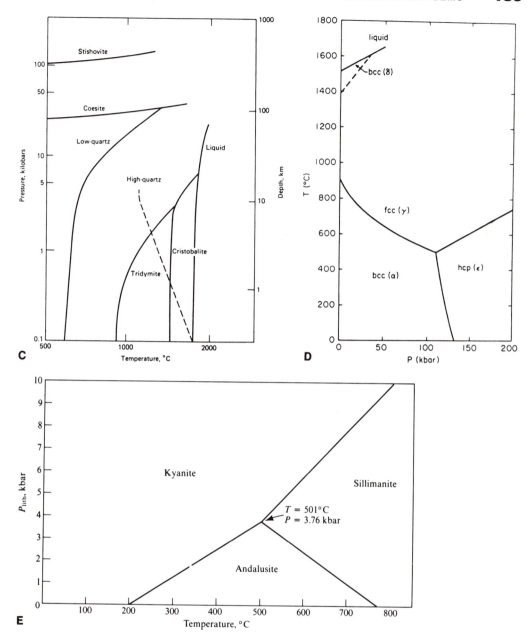

Figure 9.13. Phase diagrams for (A) water, (B) carbon, (C) silica, (D) iron,(E) the andalusite-kyanite-sillimanite system. Diagrams B–E are for dry conditions—even a small amount of water will significantly change the boundary lines separating the different phases. (A from Krauskopf, 1979, p. 289, Fig. 13.1; B and D from Jayaraman and Cohen, 1970, p. 268, Fig. 9, p. 261, Fig. 4. E from Ehlers and Blatt, 1982, p. 561, Fig. 19.1)

surface, and water vapor above. As shown in Fig. 9.13A, the pressure of the water vapor will be only 0.00603 atm, or 4.57 mmHg. The temperature and pressure at which the three phases coexist is called the *triple point*. Three lines radiate from the triple point, one separating ice from water vapor, another separating ice from liquid water, and a third separating liquid water from water vapor. At the temperature and pressure values falling along these lines, two of the three phases coexist; only one phase exists in the space between the lines. The area labeled "ice" in Fig. 9.13A shows that ice can exist under a broad range of temperature and pressure values. As pressure increases, however, the water molecules rearrange themselves in more compact crystalline patterns. Each pattern represents a different solid phase (not shown in the illustration). A total of 11 different solid phases of ice have been identified between 1 and 20,000 kb and between 0°C and −150°C.

Fig. 9.13B shows the phase diagram of carbon. Graphite is the stable, low-pressure form of carbon, and diamond is the stable form at high pressure. Similar phase diagrams are shown for silica (Fig. 9.13C), iron (Fig. 9.13D), and the andalusite-kyanite-sillimanite system (Fig. 9.13E).

All these diagrams are for dry systems. In the Earth's interior there is always some water, which can change the phase boundaries considerably. The dashed line in Fig. 9.13C shows the solid–liquid boundary in the presence of a small amount of water.

9.8. GEMS

Diamond is by far the most popular gem. Its name derives from the Greek ἀδάμας for diamond. Diamond is simply carbon, and therefore it should retail for no more than $1 per kilogram. However, it is a rather special type of carbon. In diamond, all carbon atoms are tightly bound to each other, leaving no free valences. This give diamond its extreme hardness. Being carbon, diamond can burn. That was proved in 1694–1695 by two members of the Accademia del Cimento, a scientific academy in Florence, Italy. They persuaded the grand duke of Tuscany, Cosimo III (1642–1723), to give them some small diamonds. Some were placed in a furnace, and some were placed at the focus of a large lens that concentrated sunlight. The diamonds did not melt, but decreased in size and finally disappeared. One might think that the substance simply vaporized (this process, the direct transition from solid to vapor, or vice versa, is called *sublimation*). But the carbon of the diamonds did not sublimate, a process that would require a temperature of 1500°C in vacuo; instead, it burned by combining with oxygen to form CO_2 at a temperature of 850°C. This was conclusively demonstrated in 1772 by the French chemist Antoine Lavoisier (1743–1794), who showed that diamonds disappeared only when heated in air, not when heated in vacuo.

Diamond is the stable form of carbon at pressures of more than 20 kb at room temperature, rising to 40 kb at 1,000°C. The stable form of carbon at low pressure is graphite. Increasing temperature favors graphite, whereas increasing pressure favors diamond (Fig. 9.13B). Only at depths greater than about 150 km in the Earth's interior is diamond the stable form of carbon. The diamonds that are found at the surface of the Earth have been shot up from that depth by rising magmas moving at high speed and forming *diamond pipes* some 50–200 m across (Fig. 9.14). Diamond crystals usually are euhedral, that is, they have properly formed crystal faces on all sides. This indicates that the crystals were formed within a high-density, turbulent melt. Diamonds at the temperature and pressure prevailing at the Earth's surface are under mechanical stress because surface conditions are very different from those under which they formed. The stress is insufficient to break the diamond apart in perfect or almost perfect crystals, but diamonds with significant crystal defects or inclusions have been known to shatter spontaneously.

Diamond has a very high index of refraction (Table 9.13). The difference between the index of refraction for the A line and that for the H line is called the *color dispersion coefficient*. As can be calculated from the table, the dispersion coefficient for diamond is 0.05741, whereas that for the glass is 0.02037. This means that a prism made out of diamond will produce a spectrum 2.8 times as wide as an identical prism made out of glass, accounting for the optical property that jewelers call "fire".

Diamonds are found in the matrix of the original rock (called *kimberlite*) in diamond pipes, or in sedimentary deposits derived from the original rock. Diamonds, having a hardness of 10 and a high density (most commonly 3.51 g/cm^3), are very resistant to abrasion and tend to accumulate in pockets in the floor of river beds. Diamonds were discovered in antiquity in sedimentary deposits in central and eastern Deccan (peninsular India) and, during the Renaissance, in Borneo. Until diamonds were discovered in Brazil in 1728, India and, to a much lesser extent,

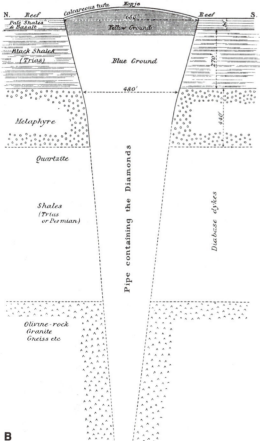

Figure 9.14. (A) The Big Hole, the Kimberley diamond pipe (South Africa), which was excavated to a depth of 495 m. It is now partly filled with water. (B) Cross section of the Kimberley diamond pipe. (A: Cotterill 1985, p. 118; B: Bauer 1968, p. 189, Fig. 30.)

Table 9.13.   Indices of refraction of diamond and a common glass for light of different colors

				Index of refraction	
Color	Line No.[a]	Atom	Wavelength (nm)	Diamond	Glass
red	A	Li	670.8	2.40735	1.52431
yellow	D	Na	589.3	2.41734	1.52798
violet	H	Ca	396.8	2.46476	1.54468

[a]The line numbers are based on the nomenclature devised by German physicist Joseph von Fraunhofer (1787–1826).

Borneo were the only sources of diamonds. The Indian diamonds are in thin conglomerate beds within the Vindhyan formation, which grades from marine below to continental above, and is of Late Proterozoic to Early Cambrian age (600 to 550 million years ago). Diamonds are also found in river deposits derived from these beds.

The world's larger diamonds are famous not only because of their size but also because of their often colorful histories. The Great Mogul is a famous Indian diamond that was described in detail in 1665 by the French jewel dealer and adventurer Jean Baptiste Tavernier (1605–1689) as weighing about 190 carats [1 carat (ct) = 0.2 g]. The diamond was then in possession of the Great Mogul Aurangzeb of India (1618–1707). It subsequently disappeared. The Orloff, another famous Indian diamond that weighs 194.75 ct, is said to have been stolen by a French soldier from the eye of an idol in a Brahmin temple in southern India, and stolen again from him by an English sea captain, who took it to Europe. The Orloff was eventually bought in Amsterdam by the Russian prince Orloff for the empress Catherine II (1729–1796) and is still part of the Russian crown jewels. Other famous Indian diamonds are the Moon of the Mountains, which weighs 120 ct and is also part of the Russian crown jewels, and the Koh-i-noor, which weighs 106.06 ct and is part of the British crown treasury. These two diamonds once came into the possession of Nadir Shah (ca. 1700–1747), leader of a Persian robber band who later became the shah of Persia. Nadir Shah conquered Delhi in 1738, massacred the population when they attempted to revolt, stole the diamonds from the reigning Great Mogul, had his only son blinded when he suspected him of sedition (nice daddy), and was himself killed by his own bodyguards. Following his assassination, an Afghan soldier grabbed the Moon of the Mountains and sold it to an Armenian, who sold it to the empress Catherine II.

Perhaps the most nearly perfect of the Indian diamonds is the Regent, found in a mine in southern India in 1701. It was bought in 1717 by the duke of Orléans, then regent of France. It was stolen during the French Revolution, but it was recovered and is still part of the French crown jewels.

Another diamond that was stolen from the French crown collection during the revolution was a blue diamond weighing 67.125 ct, which comes with a legend of misfortune. The diamond had been cut from a rough stone weighing 112.18 ct and sold by Tavernier to Louis XIV in 1668 (it might be said that the acquisition of that diamond marked the beginning of the decline of France as the major power in Europe). In 1812, a cut diamond of exactly the same color as the one that had been stolen surfaced in London in possession of a prominent diamond dealer by the name of Daniel Eliason. Eliason sold it in 1830 to the English banker Henry Philip Hope (whose ancestral firm, Hope & Co., had helped finance the Louisiana Purchase). Hope died in 1839 and left the blue diamond (by then known as the Hope diamond) to his nephew, Henry Thomas Hope. H. T. Hope died at the age of 54 and left the diamond to his daughter, who left it to her son, Lord Francis Hope. Lord Francis promptly went bankrupt, married an American vaudeville singer by the name of May Yohe, accidentally shot himself in the foot, and was dumped by May Yohe in favor of the son of a former mayor of New York. In 1901 Lord Francis sold the diamond to Simon Frankel, an American diamond dealer. After passing through some other hands, the diamond wound up with Pierre Cartier, who sold it in 1912 to the American heiress Evalyn Walsh McLean. Evalyn took the diamond right away to a priest and had it blessed. It didn't work. Her eldest son, Vinson, died in a car accident at age 9; her husband became an alcoholic and ended up insane; and her daughter died of an overdose of sleeping pills. Evalyn died in 1947, intending that her jewelry go to her grandchildren. A court, however, ordered the jewelry sold to pay for debts. The whole lot was purchased by the New York jeweler Harry Winston, who donated the diamond to the Smithsonian Institution in Washington.

The Smithsonian formally took possession of the Hope diamond on November 10, 1958. Within six month a U-2 spy plane was shot down over the Soviet Union, soon followed by the Bay of Pigs disaster, the assassination of President Kennedy, the riots in Watts, a giant blackout in the Northeast, the Pueblo affair, the Tet offensive, the assassinations of Martin Luther King and Robert Kennedy, the Chappaquiddick incident, the Watergate affair, the rout out of Saigon, the hostage crisis in Teheran and the disastrous rescue attempt, the carnage of marines and sailors in Beirut, the emergence of the AIDS epidemic, the explosion of the space shuttle *Challenger*, and the expanding drug epidemic. I say that it is time for the U.S. government to get rid of the Hope diamond.

The largest diamonds have been found in South Africa, with the record belonging to the Cullinan diamond, found in 1905 and weighing 3,106 ct (621.2 g). It was bought by the Transvaal government and presented to King Edward VII, who had it cut into nine large diamonds and several small ones. The two largest cuts weigh 516.5 and 309.18 ct. Diamonds were discovered in South Africa only in 1867, and so there are few colorful stories about the African diamonds.

The South African diamond pipes that brought the diamonds up from the interior of the Earth date from 65 million years ago. Several hundred kimberlite pipes have been found in Africa, but only 29 (probably coming from the deepest sources) contain diamonds. Diamonds are also found in alluvial deposits, especially in Zaire, but also in Angola and in offshore sand deposits along the southwest-

ern African coast. The Brazilian diamonds also occur in alluvial deposits, derived from the sedimentary Roraima formation, which is about 2 billion years old.

Diamonds, either associated with kimberlites in pipes or scattered in sedimentary deposits, have been found in Botswana, Namibia, Angola, Zaire, Ghana, Ivory Coast, Sierra Leone, Guinea, Liberia, the Soviet Union, Spain, Guyana, Venezuela, Australia, and Arkansas. Of these, only Zaire, Ghana, and the Soviet Union are major producers.

Artificial diamonds, about 0.1 mm across, are now produced in great quantity (several tons per year) for industrial purposes. One process involves the direct transformation of graphite into diamond at pressures above 130 kb and temperatures above 3,300 K. Another process uses shock waves generating transient pressures of 300 kb at a temperature of about 1,300 K. Still another process involves dissolving graphite in molten nickel and the crystallization of diamonds at pressures above 50 kb and temperatures above 1,500 K. Carat-size, gem-quality diamonds have also been produced, but thus far their production costs have far exceeded their market value.

Other highly prized gems are ruby, sapphire, the oriental emerald, and the oriental topaz. All these are simply *corundum*, aluminum sesquioxide (Al_2O_3), which is the white dust that forms on aluminum surfaces exposed to weathering. Next to diamond, corundum is the hardest mineral known, measuring 9 on the Mohs scale. It is commonly used as an abrasive. Corundum has a density of 3.99 g/cm^3, which is 12% greater than that of diamond. In its pure form, corundum is transparent and colorless. Iron and titanium impurities give corundum the colors that characterize the sapphire, the oriental emerald, and the oriental topaz (see Table A7). The red color of ruby is due to chromium impurities. This occurrence is rare, and so rubies of a good red color are more expensive than diamonds of equivalent size. The corundum gems are found in compact limestones that have been recrystallized into marble by contact metamorphism (see Chapter 17). In the process, the "impurities" are left out and recrystallized as corundum gems. The most important deposits of corundum gems are in Myanmar (formerly Burma), Thailand, and Sri Lanka, where the gems are mined from the mother rock or from secondary deposits. A mine active during the Middle Ages was on the eastern side of the Oxus River in southern Tajikistan. Gems from that mine furnished the treasury of many Mongol rulers.

All corundum gems can be made artificially by fusing pure Al_2O_3 powder in an oxygen-hydrogen flame and adding appropriate metallic oxides as colorants (5–6% Cr_2O_3 for rubies). Spinel ($MgAl_2O_4$) can be made by adding 20 mole percent MgO to the Al_2O_3. These artificial gems are mineralogically identical with their natural counterparts and, in addition, have fewer crystal defects. However, the natural ones, crystal defects and all, are still vastly more valued (it's all in the eye of the beholder).

Emerald, as contrasted with "oriental emerald," is beryl, an aluminosilicate of beryllium colored green by chromium impurities; topaz, as contrasted with "oriental topaz", is an aluminosilicate with the (OH,F) group colored yellow or, less commonly, red or blue by impurities. Emerald and topaz, therefore, are totally different, mineralogically speaking, from their oriental homonyms. Emerald occurs almost exclusively in the primary rock, most commonly crystalline schist, of which it is a component. Emeralds were mined in southern Egypt in antiquity. Today, emeralds come mainly from two areas, the Takovaya River valley in the Urals and the Muzo deposit in Colombia. In the latter, emeralds are found embedded in a calcitic matrix (an exceptional occurrence). Perfect emeralds of good green color are very rare. Topaz was mined in antiquity in the same mine in southern Egypt where emeralds were mined. Today, the major producer of topaz is Brazil, where topaz is common in the Ouro Prêto region.

Alexandrite, an aluminum oxide of beryllium, is green in daylight and purple in artificial light. It is found in the Takovaya Valley in the Urals, where emeralds are mined. Spinel is colored red, yellow, green, or blue by cobalt or other impurities. Spinel occurs in northern Myanmar together with ruby in marble or in secondary deposits. It also occurred in the Tajik mines that produced rubies (now depleted). Marco Polo (ca. 1254–1324), the famous Venetian traveler, picked up some spinels and rubies there.

The minerals mentioned so far have hardnesses above $7\frac{1}{2}$. Many other minerals are used as gems, but they are softer and, therefore, less prized. Among these are the garnets, aluminosilicates of iron, calcium, magnesium, or manganese, that are most commonly colored red by iron impurities; chrysolite or peridot, gem olivine with a greenish color due to iron impurities; tourmaline, a complex silicate variously colored; zircon, a silicate of zirconium that is usually colorless; and quartz, which bears different names (amethyst, smoky quartz, etc.) according to its color (see Table A7).

Lazurite, a complex silicate of sodium and calcium, is an opaque mineral with an intense blue color. Lazurite, produced in limestones by contact metamorphism, is usually microcrystalline and usually embedded in a matrix of calcite and other minerals. This complex is called lapis-lazuli. It was highly prized during the Renaissance for intaglio work. Major deposits are in Badakhshan (northeastern Afghanistan), on the western end of Lake Baikal in Siberia, and high in the Chilean Andes northeast of Santiago.

Jade, microcrystalline nephrite (an amphibole) or microcrystalline jadeite (a pyroxene), is highly prized for carving objects. Nephrite is found in crystalline schists and as pebbles in secondary deposits. Before metal smelting was invented, jade was commonly used to make axes and other cutting tools. The largest jade deposits are found in China. Another important deposit is on the South Island of New Zealand. Jadeite is found in western China and in northern Myanmar. Microcrystalline quartz, variously named according to its color (see Table A7), is widely used, like jade, to make a variety of objects.

Turquoise, a microcrystalline hydrated sulfate of aluminum and copper, is a secondary mineral formed as an incrustation on rock surfaces by the action of surface water in arid regions. The most important deposits are in Iran, the Sinai Peninsula, and New Mexico. Turquoise, being a hydrated mineral, is quite delicate and should not be exposed to excessive heat.

Also delicate is opal, which consists of discontinuous layers of tightly packed microscopic spheres of hydrated silica ($SiO_2 \cdot nH_2O$) about 0.3 μm across (Fig. 9.15). The voids are filled with air or water, with dimensions ranging from 0.3 to more than 30 μm (which encompasses the wavelength range of visible

Figure 9.15. Scanning electron micrograph of opal, showing tightly packed opal spherules arranged in discontinuous sheets. Diameter of the spherules is about 0.3 μm. (Darragh, Gaskin, and Sanders, 1976, vol. 234, no. 4, p. 86.)

light, 0.3–0.7 μm). Light striking opal is diffracted in the voids and color separation occurs, giving opal its appearance. The water content ranges from 5 to 10%. Opal is delicate because excessive heat draws the water out and the opal becomes white dust. Gem-quality opal is formed from percolating water in rock cavities. It is found in andesites in northern Hungary, in trachytes in Central America, and in sandstones in Australia. Large quantities of common opal (not gem-quality) are deposited by silica-rich water around hot springs. Common opal forms sponge spicules in many sponges, as well as the shells of radiolaria and diatoms (see Sections 21.5 and 23.1)

Lastly we should consider two gems that are products of organic activity: amber and pearls. Amber is a fossil tree resin common around the world in continental sediments ranging in age from the Carboniferous to the Pleistocene. It often contains as inclusions insects or plant parts. Along the Baltic and the North Sea coasts, amber is washed ashore by the sea from deposits of Eocene to Miocene age that outcrop below sea level. Baltic amber was traded in antiquity to Greece and Rome.

The pollen content of amber helps to identify the plant that produced it.

Pearls are also products of organic activity. They are spherical, spheroidal, or botryoidal concretions consisting of aragonitic lamellae alternating with an organic film (*botryoidal* means shaped like a bunch of grapes). The thicknesses of the lamellae and the organic film are not uniform, thus creating interference and color dispersion of incident light. Pearls are formed by marine mollusks belonging to the genus *Pinctada* and by several genera of freshwater mollusks as coatings around a foreign particle (a sand grain, for instance) intruding into the body cavity of the mollusk. The coatings are similar to the mother-of-pearl coating (called *nacre*) on the inner surface of many shells. The formation of pearls can be induced artificially by inserting spherical beads up to 6 mm across into the body cavity of the pearl-bearing mollusk *Pinctada martensii*. The tissue of the mollusk secretes a protective coating around the bead that grows at a rate of about 0.15 mm/year. After 3 years the mollusk will have produced a pearl 6.5 mm across (of which only the outer 0.5 mm is natural).

Natural pearls are most commonly found in shells that are malformed, indicating that the mollusks had suffered some stress. The largest pearl, found in the Philippines inside a specimen of *Tridacna* (the giant clam), weighs 6.378 kg and measures 24 cm in length and 14 cm across.

THINK

* Can a pigeon dropping be called a mineral when it (a) is still inside the pigeon, (b) is on its way to the ground, (c) is splattered on the ground, (d) has been dried by the Sun, (e) has been buried with sediments, (f) has been fossilized?

* Symmetry axes are 1-fold, 2-fold, 3-fold, 4-fold, and 6-fold, but not 5-fold. Why?

* If space were two-dimensional, which of the 11 basic symmetry elements would apply?

* Olivine commonly consists of 80% Mg_2SiO_4 and 20% Fe_2SiO_4 (by mass). Calculate the percentages of Mg and Fe atoms.

* In diamond, the distance between nuclei of adjacent carbon atoms is $1.54452 \cdot 10^{-8}$ cm. The radius of a carbon atom is therefore $1.54452 \cdot 10^{-8}/2 = 0.772 \cdot 10^{-8}$ cm. Calculate the volume of a carbon atom, and knowing that its mass is 12 u (by definition), calculate its density. Calculate the density of diamond, assuming that the carbon atoms in diamond are spheres and are as closely packed as possible. Does the density thus derived match the density of diamond? If not, why not?

* Tables 9.9 and 9.10 show that metals have high heat conductivities but low heat capacities. Why?

* Why does glass tend to crystallize?

* If you wanted to get rid of the Hope diamond, how would you do it? Throwing it into a canal, a river, a lake, or an ocean might create misfortune for fishermen or sailors. Casting it deep into a desert might endanger caravans. Shooting it into space might invite a catastrophic asteroidal hit. Burying it deep in the ground might cause a disastrous earthquake. How about burning it? Knowing the mass of the Hope diamond, calculate how many molecules of CO_2 would be produced by burning it. Measure (or estimate) the volume of air that you inhale with each breath. Knowing that the density of air is $1.225 \cdot 10^{-3}$ g/cm^3, that the concentration of CO_2 in air is $0.32 \cdot 10^{-3}$ by volume, and that the mass of the atmosphere is $5.136 \cdot 10^{21}$ g, calculate how many molecules of Hope CO_2 you would inhale with each breath, assuming that the atmosphere is well mixed. Does your result show that it is safe to burn the Hope diamond? If not, can you find another way to get rid of this accursed gem?

10 IGNEOUS ROCKS AND VOLCANISM

10.1. ROCK-FORMING MINERALS

Several thousand different minerals have been described; the more common are listed in Table A10. Rock-forming minerals can be classified in three broad categories: igneous, sedimentary, and metamorphic. Igneous minerals are formed from melts and solutions at temperatures higher than 200–300°C. Sedimentary minerals are mainly precipitates from water solutions at low temperatures, forming bulk rocks (e.g., limestone) or cementing mineral particles directly or secondarily derived from igneous rocks. Metamorphic minerals are produced by structural and chemical reorganization of preexisting igneous or sedimentary minerals. The minerals that form the bulk of the igneous, sedimentary, and metamorphic rocks are comparatively few. They are listed in Table 10.1.

Quartz is the low-temperature, low-pressure polymorph of silica. At higher temperatures and pressures, coesite and then stishovite are the stable forms (Fig. 9.13C). Andalusite, kyanite, and sillimanite are pure aluminosilicates. Andalusite is the low-pressure polymorph, kyanite the high-pressure polymorph, and sillimanite the high-temperature polymorph (Fig. 9.13E). The feldspars are aluminosilicates with K, Na, and/or Ca as cations. Orthoclase is the K feldspar. The Na–Ca feldspars (plagioclases) range from 100% Na plagioclase (albite) to 100% Ca plagioclase (anorthite) (Table 10.2).

The pyroxenes are silicates of Mg, Ca, and Fe. The Mg-rich enstatite and hypersthene are *orthopyroxenes*, crystallizing in the orthorhombic system. The Ca-rich diopside, hedenbergite, and augite are *clinopyroxenes*, crystallizing in the monoclinic system. The most common pyroxene is augite, which contains all three cations plus some Na and Al. Olivine is a silicate of Mg and Fe, and is a major component of the Earth's mantle. Kamacite

and taenite are Ni–Fe alloys that form iron meteorites. Their composition presumably reflects that of the Earth's core. Troilite also is found in meteorites, indicating that the Earth's core may contain some sulfur.

Serpentine, amphiboles, epidote, and micas, all containing the hydroxyl group (OH), are products of alteration of igneous or sedimentary minerals under different conditions of temperature and pressure. Zeolites result from the alteration of feldspars at low temperatures (<200°C) and pressures (<2 kb), and serpentine derives from low-temperature, low-pressure alteration of olivine and enstatite. Glaucophane results from low temperatures (100–300°C) and intermediate pressures (3–8 kb), epidote from higher temperatures (300–500°C) and pressures of 1–10 kb, and hornblende from even higher temperatures (400–600°C) and pressures of 1 kb to more than 12 kb.

Garnet and graphite are formed by contact metamorphism, that is, in sedimentary rocks that have been heated by contact with hot igneous rocks. Garnet is also common in rocks formed at high temperatures and pressures. Muscovite and biotite occur commonly in both igneous and metamorphic rocks. Chlorite is a common product of alteration of biotite and other Mg–Fe minerals; it is also common in clays.

In a rock melt containing the ingredients for the major rock-forming minerals, the first minerals to crystallize will be those with the highest melting points (olivine and anorthite). Their crystals will sink to the bottom of the melt, and other minerals will follow as temperature decreases. The last to crystallize will be quartz. The minerals form a two-branched series called the *Bowen reaction series* (Table 10.3). One branch (called the *discontinuous series*) ranges from olivine (which is a nesosilicate and therefore consists of single

Table 10.1.   Common rock-forming minerals

Name	Chemical composition	Name	Chemical composition
Igneous			
Quartz	SiO_2 ($\rho = 2.65$)	Epidote	$Ca_2(Al, Fe)Al_2O(SiO_4) \cdot (Si_2O_7) \cdot$
Coesite	SiO_2 ($\rho = 2.91$)		(OH)
Stishovite	SiO_2 ($\rho = 4.29$)	Garnet[a]	$Q_3R_2(SiO_4)$ ($Q = Ca, Mg, Fe^{2+};$
Andalusite	Al_2SiO_5 ($\rho = 3.15$)		$Mn^{2+}; R = Al, Fe^{3+}, Cr)$
Sillimanite	Al_2SiO_5 ($\rho = 3.24$)	Graphite	C
Kyanite	Al_2SiO_5 ($\rho = 3.63$)	Micas	
Feldspars		Muscovite[a]	$KAl_2(AlSi_3O_{10})(OH)_2$
Orthoclase	$KAlSi_3O_8$	Biotite[a]	$K(Mg, Fe)_3(AlSi_3O_{10})(OH)_2$
Albite	$NaAlSi_3O_8$	Chlorite[b]	$(Mg, Fe)_3(Si, Al)_4O_{10} \cdot$
Anorthite	$CaAl_2Si_2O_8$		$(OH)_2 \cdot (Mg, Fe)_3(OH)_6$
Pyroxenes			
Enstatite	$MgSiO_3$	**Sedimentary**	
Hypersthene	$(Mg, Fe)SiO_3$	Clays	
Diopside	$CaMgSi_2O_6$	Illite	$(K, H_3O)(Al, Mg, Fe)_2(Si, Al)_4O_{10} \cdot$
Hedenbergite	$CaFeSi_2O_6$		$[(OH)_2H_2O]$
Augite	$(Ca, Na)(Mg, Fe, Al)(Si, Al)_2O_6$	Montmorillonite	$Q_{0.33}Al_2Si_4O_{10}(OH)_2 \cdot nH_2O$
Olivine	$(Mg, Fe)_2SiO_4$		$(Q = Na^+, K^+, Mg^{2+},$ or $Ca^{2+})$
Kamacite	$Ni_{0.6-0.7}Fe_{0.94-0.93}$	Kaolinite	$Al_4Si_4O_{10}(OH)_8$
Taenite	$Ni_{0.27-0.65}Fe_{0.73-0.35}$	Carbonates	
Troilite	FeS	Calcite	$CaCO_3$ ($\rho = 2.712$)
		Aragonite	$CaCO_3$ ($\rho = 2.930$)
Metamorphic		Dolomite	$CaMg(CO_3)_2$ ($\rho = 2.85$)
Zeolites		Sulfates	
Analcime	$NaAlSi_2O_6 \cdot H_2O$	Anhydrite	$CaSO_4$
Laumontite	$CaAl_2Si_4O_{12} \cdot 4H_2O$	Gypsum	$CaSO_4 \cdot 2H_2O$
Serpentine	$(Mg, Fe)_3Si_2O_5(OH)_4$	Halites	
Amphiboles		Halite	$NaCl$
Glaucophane	$Na_2(Mg, Fe)_3Al_2Si_8O_{22}(OH)_2$	Sylvite	KCl
Hornblende[a]	$(Ca, Na)_{2-3}(Mg, Fe, Al)_5 \cdot$		
	$(Si, Al)_8O_{22}(OH)_2$		

[a] Common also in igneous rocks.   [b] Common also in clays.

SiO_4 tetrahedra) through pyroxene (single chains of tetrahedra), amphibole (double chains of tetrahedra), and biotite (sheets of tetrahedra), ending up with orthoclase and quartz (which are tektosilicates, consisting of a three-dimensional networks of tetrahedra). It is apparent that as temperature decreases, more complex structures of silica tetrahedra can form. The other branch (called the *continuous series*) starts with anorthite, which grades into albite and ends up also with orthoclase and quartz.

The sedimentary minerals include clays, carbonates, sulfates, and halites. The clays are the products of weathering of igneous minerals. The most highly leached is kaolin, which consists only of Al and silica.

Calcite and aragonite are precipitated in

marine or fresh waters, either inorganically or through the activity of organisms (skeleta). Although aragonite is the high-pressure polymorph of calcium carbonate, it is commonly precipitated at low pressure (1 atm to a few atmospheres) by organic activity (pearls, mother of pearl, oolites, coral skeleta, the skeleta of many molluskan species). Dolomite is formed by secondary replacement of 50% of the Ca^{2+} with Mg^{2+}. Sulfates and halites are precipitated from water bodies that have reached saturation.

10.2. IGNEOUS ROCKS

Igneous rocks are rocks solidified from rock melts. Those solidified inside the Earth are

Table 10.2. The plagioclases

Name	Composition (%)	
	Albite	Anorthite
albite	90–100	0–10
oligoclase	70–90	10–30
andesine	50–70	30–50
labradorite	30–50	50–70
bytownite	10–30	70–90
anorthite	0–10	90–100

called *intrusive*. Those pouring out on the surface of the solid Earth, either subaerially or subaqueously, are called *extrusive*. The Earth's outer core is a liquid mixture of Fe (90%) Ni (9%), and probably S (1%). The solid inner core consists of Fe–Ni alloy of composition similar to that of the outer core. It is solid because it is under high pressure. No material from the Earth's core has ever been found in the Earth's crust. The deepest rocks that have reached the Earth's surface are kimberlites in diamond pipes. Those that contain diamonds come from depths of about 150 km, which is only 1/40 of the distance to the core. Fig. 10.1 shows a schematic cross section of the Earth and its chemical (left) and mechanical (right) subdivisions.

There is a great variety of igneous rocks. Only a few types, however, form most of the solid Earth. They are listed in Table 10.4 in order of decreasing amount of silica, together with the Fe–Ni–S minerals found in meteorites. Notice the double nomenclature for a given composition, depending on whether the rock is intrusive or extrusive. The difference is that intrusive rocks cool slowly and form crystals clearly visible to the naked eye (1–2 mm across),

whereas extrusive rocks cool very rapidly and form only microcrystals (1–100 μm across). *Porphyritic rocks* are rocks consisting of conspicuous crystal and crystals aggregated in a matrix of fine microcrystals or glasses. These rocks began solidifying at depth, but were then forced out and completed their solidification rapidly at low pressures and low temperatures.

When the Earth formed, $4.5 \cdot 10^9$ y ago, its average composition was similar to that of average meteoritic matter (see Table 9.2). The metals, being heavier, soon trickled downward to form the core, leaving behind a mantle of average rock composition (*pyrolite*, enriched in Mg and Si, but depleted of Fe). The minerals forming the mantle have different melting points (Table 10.5), and so do the rocks that are formed from these minerals (Table 10.6).

The SiO_4^{4-} tetrahedra are not destroyed by melting. They still form polymers (chains, sheets, rings) that increase in complexity with increasing silica content and decreasing temperature. As a result, viscosity increases from basaltic to rhyolitic magmas.

[*Viscosity*, the resistance of a substance to flow, is due to the chemical bond—molecular bonds in liquids and glasses, ionic and covalent bonds in crystalline substances—or to momentum transfer among atoms and molecules moving at different speeds in gases. Consider two parallel surfaces of area A bounding a fluid and separated by a distance Z (Fig. 10.2). One surface is fixed, and the other is made to slide with velocity v by the force F. We have

$$F = \eta A (dv/dZ) \qquad (1)$$

where η is the coefficient of viscosity. From (1) we have

$$\eta = F/A(dv/dZ) \qquad (2)$$

Table 10.3. The Bowen reaction series

Temperature (°C)	Discontinuous branch		Continuous branch
1,880	olivine		
1,525–1,550	enstatite (orthopyroxene)		anorthite
1,390	diopside (clinopyroxene)		
1,200–1,250	amphiboles		$\downarrow$
1,100–1,150	micas		albite
1,000		orthoclase	
687		quartz	

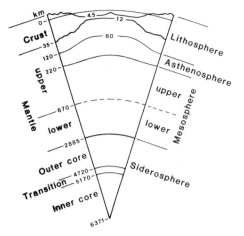

Figure 10.1. Cross section of the Earth: petrographic subdivisions to the left; mechanical subdivisions to the right; depths in kilometers. The continental crust averages 35 km in thickness, but the oceanic crust is only 7.5 km thick. The asthenosphere is about 1% liquid. The overlying lithosphere and the underlying upper mesosphere are more rigid. The lithosphere is broken up into plates (see Chapter 12). No earthquakes deeper than 670 km have been observed, suggesting that the lower mesosphere is more plastic than the upper mesosphere or the lithosphere.

The coefficient of viscosity has the dimensions of force/area = pressure divided by velocity/length = length/time/length = 1/time, or pressure·time. It is expressed in poises (P) in the cgs system $[P = (dyn/cm^2)\cdot s]$ and in poiseuilles (Pl) in the SI system $[Pl = (N/m^2)\cdot s = Pa\cdot s]$. Because $1\,N = 10^5$ dyn and $1\,m^2 = 10^4\,cm^2$, $P = 0.1\,Pl$. Both the poise and poiseuille are named after the French physician Jean Poiseuille (1797–1869), who, while studying the flow of blood in blood vessels, developed the math of fluid flow in pipes. Table 10.7 gives examples of viscosities in nature.

Viscosity as defined above is called *dynamic viscosity*. Dynamic viscosity divided by the fluid's density is called *kinematic viscosity*. Kinematic viscosity is measured in stokes (St) in the cgs system $[St = P/(g/cm^3)]$. A 30-weight lubricating oil has a kinematic viscosity of 1.8–2.8 St at 25°C, a density of 0.9 g/cm^3, and therefore a dynamic viscosity of 1.6–2.5 P.]

Viscosity depends also on the amount of water in a magma. Wet magmas have lower viscosities than dry magmas because water depolymerizes SiO_4 structures by replacing a binding O atom with a nonbinding OH group.

Between 4.5 and 3.8 billion years ago, the heat produced by the accumulation process, the formation of the core, and the decay of radioactive elements that were more abundant then than now probably caused the mantle to melt. Fractional crystallization produced an early basaltic crust enriched in Ca and Al with respect to pyrolite. The depleted pyrolite below formed the peridotites that now underlie the crust. It is not known when the mantle solidified. It certainly was already solid at the beginning of the Phanerozoic (570 million years ago). Today, there is an Earth-encircling shell 100 km thick, with its upper surface at a depth of 60 km beneath the oceans and 120 km beneath the continents, where pressure and temperature combine to keep the rocks close

Table 10.4. Igneous rocks and metal alloys

Intrusive	Extrusive	SiO$_2$ (%)	Mineral composition	Density (intrusive rocks) (g cm^{-3})
granite	rhyolite	70	quartz, feldspar, biotite	2.67
granodiorite	dacite	65	Na feldspar, quartz, biotite	2.72
syenite	trachyte	60	K feldspar, clinopyroxene, biotite	2.72
diorite	andesite	55	Na/K feldspar, hornblende, biotite	2.84
gabbro	basalt	50	Ca feldspar, clinopyroxene	2.98
peridotite	komatiite	40	olivine, orthopyroxene	3.23
kamacite	—	0	$Ni_{0.5-0.7}Fe_{0.95-0.93}$	7.3–7.9
taenite	—	0	$Ni_{0.27-0.65}Fe_{0.73-0.35}$	7.8–8.2
troilite	—	0	FeS	4.83

Table 10.5. Melting points of common rock-forming minerals at a pressure of 1 atm

Mineral	Melting point (°C)
periclase (MgO)	2,642
forsterite (Mg$_2$SiO$_4$)	1,910
olivine ($\sim$70–90% Mg$_2$SiO$_4$)	1,880
cristobalite (SiO$_2$)	1,713
anorthite (Ca plagioclase)	1,553
enstatite (Mg pyroxene)	1,524
diopside (Ca–Mg pyroxene)	1,390
fayalite (Fe$_2$SiO$_4$)	1,205
amphiboles (Fe–Mg–Ca Al-silicates)	1,200 (?)
muscovite	1,140
albite (Na plagioclase)	1,118
quartz (SiO$_2$)	867

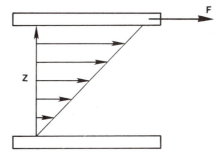

Figure 10.2. Viscosity is the resistance of a fluid to flow. Force F is moving the upper plate in the direction shown, while the bottom plate is stationary. Velocity within the fluid changes as shown. The fluid has a viscosity of 1 poise when a force of 1 dyne applied to a surface of 1 cm^2 maintains a velocity difference of 1 cm/s between the upper and lower plates when the distance between the two is 1 cm. Olive oil at room temperature has a viscosity of about 1 poise.

Table 10.6. Melting-temperature ranges of igneous rocks (dry)

Rock	Percentage of silica (SiO$_2$)	Melting-temperature range (°C)
komatiite	40	1,600–1,800
basalt	50	1,000–1,200
andesite	60	900–1,000
rhyolite	70	800–900

Table 10.7. Viscosities of natural substances

Material	Temperature (°C)	Coefficient of viscosity (P)
air	20	$1.86 \cdot 10^{-6}$
water	20	0.010
olive oil	20	0.84
pitch	15	10^{10}
ice	0	10^{13}
halite	20	10^{16}
glass	20	10^{20}
asthenosphere	1,000	10^{21}
marble	20	10^{22}
lower mantle	2,400–3,800	10^{22}–10^{23}

to melting. This layer is called the *asthenosphere*. The overlying layer, containing the uppermost 50 km of the mantle under the oceans and the uppermost 90 km under the continents, plus the crust, is called the *lithosphere* (Fig. 10.1).

[*Lithosphere and asthenosphere* are not very good names. *Lithosphere* derives from the Greek λίθος (stone) and means "sphere of stone". As such, it used to denote mantle + crust, both of which consist of stone. It is now taken to mean the mechanically hard, cool, brittle, outermost layer of the Earth, including crust and uppermost mantle. *Asthenosphere* derives from ἀσθενής, meaning *without strength*, which, of course, is not true. Because these subdivisions are mechanical, better names would be *sclerosphere*, from σκληρός (hard), for the lithosphere, and *hygrosphere*, from ὑγρός (weak), for the asthenosphere.]

Pockets of molten rocks exist here and there in the mantle, especially in the asthenosphere, feeding existing volcanoes. The deepest and largest ones (up to an estimated 100 km across) are close to the mantle–core boundary and may be a leftover from the time when the entire mantle was molten. They are called *hot spots*.

The reason that pockets of molten rock form in the upper mantle is that the mantle is not perfectly homogeneous. Subduction of surface materials, heat trapped by continental covers, irregularities in the distribution of minerals and rocks with slightly different thermodynamic properties, and irregularities in the distribution of radioactive elements can result in appreciable temperature differences for areas at the same depth. The heat conductivity of silicates decreases with increasing temperature—the higher the temperature, the smaller the amount of heat that can get through. Heat accumulates along the bottom of any area hotter than its neighboring regions, and melting starts. The first to melt will be the minerals with lower melting points, whereas the minerals with higher melting points (olivine, Ca–Fe and Ca–Mg pyroxenes, Ca plagioclase) will remain solid and, being denser, will tend to accumulate toward the bottom of the developing magma chamber. A density stratification develops that impedes convection and therefore efficient heat transfer. A magma chamber, therefore, acts as a heat block. This means that the heat flux per unit of cross section around its periphery will be higher than normal, leading to enlargement of the chamber. It would seem that a magma chamber, once started, is self-enlarging.

The volume of the liquid phase, however, is about 5–10% greater than that of the solid phase. An increase in melting means increasing pressure, which slows down melting because increasing pressure favors the solid phase. With continued heat accumulating at the bottom of the magma chamber, melting wins. Eventually, the lithostatic pressure is overcome, and the magma breaks through the overlying rocks, releasing the pressure.

Most basaltic magmas form from dry pyrolite at depths of 100–200 km and at 1,500–2,000°C. The presence of water in a rock melt lowers the melting point. Rhyolitic magmas are formed from the recycling of wet sediments at depths of 30–40 km and at 600–800°C. Andesitic magmas form along the edges of continents from mixtures of basaltic and rhyolitic magmas (the name andesitic derives from the Andes Mountains, where andesitic rocks are common).

10.3. EMPLACEMENT OF IGNEOUS ROCKS

Igneous rocks (Fig. 10.3) can be emplaced inside the Earth (intrusive igneous rocks) or at the surface (extrusive igneous rocks). Intrusive

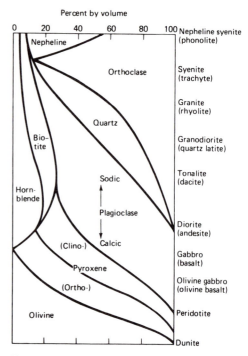

Figure 10.3. Mineral compositions of igneous rocks. (Emiliani, 1987, p. 106; adapted from Mason and Moore, 1982, p. 96, Fig. 5.3.)

igneous rocks form bodies of various sizes and shapes when emplaced. The more common types (Fig. 10.4) are as follows:

Dikes are sheets of magma originating from a magma chamber and cutting across overlying rock layers (Fig. 10.4A). Dikes range in width from centimeters to several meters and may extend to several kilometers in the other two directions. Along the axes of the mid-ocean-ridge systems, dikes are continuously pushing up. A new dike will split a preceding dike precisely in half, because the rock in the center of the dike will be hotter and mechanically weaker. In volcanic structures, dikes often are seen to radiate away from the duct.

Sills are sheets of magma intruded between the beds of a preexisting, layered rock (Fig. 10.4B). Sills range in thickness from centimeters to hundreds of meters and in area from a few square meters to several thousand square kilometers. The Palisades sill on the Hudson River (Fig. 10.5) is 300 m thick.

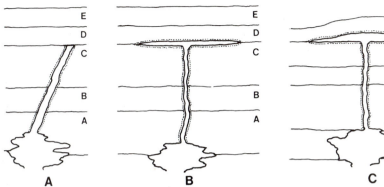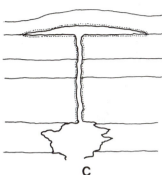

Figure 10.4. Intrusive structures; the dots represent alteration of the intruded rock by the heat introduced by the dike. (A) Dike; layers *A*, *B*, and *C* are older than the dike; layers *D* and *E* are younger. (B) Sill fed by a dike; layers *A*, *B*, *C*, and *D* are all older than the dike and sill. (C) Laccolith; the intruded rock has been updomed.

Laccoliths are magmatic injections, produced by dikes from below, that have a flat lower surface and a domal upper surface (Fig. 10.4C).

Batholiths are granitic-to-dioritic *plutons*— large intrusions most common along continental margins (Fig. 10.6). Often they are thousands or even tens of thousands of square kilometers in areal extent and up to 15–20 km thick. Batholiths probably are mixtures of recycled sediments and magmatic differentiates. Compared with gabbro-basalt, they contain 5 times more uranium, as much thorium, and 10 times more

Figure 10.5. The Palisades sill on the Hudson River, New York and New Jersey. This extensive layer of columnar basalt, covering an area of about 2,500 km² and averaging 330 m in thickness, was formed when North America separated from Europe and Africa, about 200 million years ago. The columns are 30–50 cm across. (Photograph by Horace Gilmore, courtesy of Interstate Park Commission, Bear Mountain, New York.)

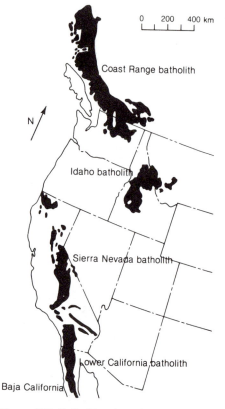

Figure 10.6. Batholiths along the western margin of North America. (Press and Siever, 1986, p. 387, Fig. 15.9; adapted from King, 1965.)

potassium-40; they produce 6.6 times more heat per unit mass. Being hotter and less dense, batholiths creep upward and emplace themselves close to the surface.

Extrusive igneous rocks flow out as lavas along the continental or submarine surfaces of the Earth or blast off into the atmosphere as pyroclastics:

Lavas are microcrystalline, vesicular volcanic rocks that have been rapidly chilled on contact with air or water. The surface of a lava extruded subaerially can be jagged if the lava is viscous (*aa*), or "ropey" if the lava is less viscous (*pahoehoe*). When basalt pours out on the ocean floor, as along the mid-ocean-ridge system, it comes in contact with cold seawater and forms structures known as *pillow basalts*. The surface does indeed look like a spread of pillows (Fig. 10.7).

Pyroclastics are igneous rock fragments ejected by volcanoes in the course of volcanic explosions. They are classified according to the sizes of the fragments (Table 10.8).

Volcanic ash consists of glassy fragments—there is not time to form crystals. An accumulation of loose pyroclastics is called *tephra* (from the Greek τέφρα, singular feminine noun meaning *ashes*). *Tuff* is cemented tephra. *Ignimbrite*

Figure 10.7. Pillow basalt. (Courtesy Robert Embley, Oregon State University.)

Table 10.8. Size ranges of pyroclastics

Name	Diameter (mm)
	<0.0625
volcanic ash	
	0.063
cinder	
	4
lapilli	
	64
bombs	
	>64

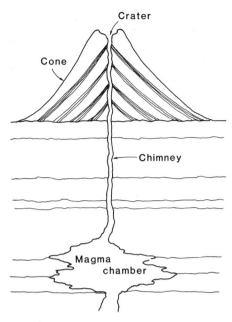

Figure 10.8. Cross section of a stratovolcano.

(from *ignis*, Latin for *fire*, and *imber*, Latin for *rain*; also called *welded tuff*) is tephra deposited while the glassy fragments are still hot; thus the fragments are cemented together as the tephra cools.

10.4. VOLCANISM

A volcano consists of magma chamber, chimney, cone, and crater (Fig. 10.8). A volcano may have secondary feeders, with secondary craters. Volcanoes have different shapes, sizes, and activities, depending on the viscosity of the lava feeding them.

Shield volcanoes are formed by low-silica (50% SiO_2), low-viscosity, hot basaltic lavas that pour out of fissures and flow down the sides of the volcano. Classical examples of shield volcanoes are the giant volcanoes that built Hawaii–Mauna Kea (4,205 m high) and Mauna Loa (4,170 m). The Hawaiian lavas rise from a primary magma chamber about 60 km below sea level to a secondary magma chamber just below the volcanoes. A standard Hawaiian eruption may produce 0.1 km^3 of lava and empty the secondary magma chamber. The radius of the secondary chamber, when full, is on the order of a few hundred meters. Eruption occurs when pressure from the inside overcomes lithostatic pressure. The rising lava follows the path of least resistance, which usually is the path of a preceding eruption, along which the rocks are still warm and therefore more yielding. Rising lava must break rocks, nevertheless, and so eruptions usually are preceded and accompanied by swarms of small earthquakes.

As the eruption continues, lithostatic pres-sure collapses the chamber. At the end, what remains of the chamber is a pancake of hot rocks that are mechanically weaker than the surrounding cooler rocks. Because of this weakness, the emptied chamber remains a weak locus for new lava from the primary chamber to accumulate and grow into a new magma chamber. The sizes of both the primary and secondary magma chambers are limited by the strength of the overlying rocks. When an eruption occurs and the magma chamber empties, the overlying structure subsides. The Hawaiian lava is so fluid that the average slope of Hawaii is only 1.3° from the submarine base to the top.

Stratovolcanoes are characteristic of ande-sitic magmas (55% SiO_2) and consist of alternat-ing layers of lavas and pyroclastics (Fig. 10.8). Examples are Mount Vesuvius, Mount Fuji, and the volcanoes of the West Indies, Indonesia, and the Pacific rim. Slopes average more than 10°.

Pyroclastics are produced by explosive volcanism. High-silica melts can remain fluid at lower temperatures (800–900°C) than can basaltic melts (1,000–1,200°C) and therefore can contain more gases in solution. Table 10.9 list the gases from Showashinzan, in Japan.

Table 10.9. Volcanic gases from Showashinzan (Japan)

Gas	Percentage of number of molecules
H_2O	99.250
CO_2	0.470
H_2	0.181
HCl	0.039
N_2	0.026
HF	0.020
SO_2	0.012
H_2S	0.001
CH_4	0.0006
NH_3	0.0004

Water is by far the dominant fluid in magmas. It commonly amounts to a small percentage by mass, equivalent to some 15% by number of molecules. As a magma cools, more and more crystals form, the gases are compressed into less and less volume, and vapor pressure builds up, until a duct is blasted open. The immediate decrease in pressure has the effect of exsolving water and the other gases— the magma becomes a froth as it is blasted into the atmosphere. A mass of hot gases and pyroclastics rises above the volcano, reaching into the stratosphere if the explosion is powerful enough. This mass is called a *nuée ardente* (blazing cloud in French). Temperatures inside a nuée ardente are on the order of 250–500°C. Density is greater than that of the surrounding air, and so the nuée ardente rolls downslope, attaining speeds of up to 200 km/h (55 m/s).

Water condensing out of a volcanic cloud and raining through the pyroclastics still in the air, as well as on the pyroclastics already on the ground, may form a flow of hot mud that rolls downslope. This flow, called a *lahar*, does not move as fast as a nuée ardente, but it is deadly to anyone caught in it. Some of the most famous volcanic eruptions were due to explosive volcanism accompanied by nuées ardentes and lahars.

Mount Vesuvius, after having been dormant since prehistory, erupted in 79 C.E. and destroyed Pompeii, 9 km to the southeast, and Herculaneum, 7 km to the southwest. Pompeii, the playground of the Roman Empire, was buried under 6 m of volcanic ash. Compacted volcanic ash has high porosity but low permeability. Rain cannot seep through it, because the voids are not in communication with each other. As a result, much of Pompeii was preserved under that layer of ash, including buildings, paintings, artifacts, and the bodies of humans and animals. Herculaneum, a city built on the seashore at the foot of Mount Vesuvius, 15 km northwest of Pompeii, was covered by a lahar some 13 to 19 m thick.

Krakatau, a volcanic island 800 m high in the Sunda Strait between Sumatra and Java, erupted at 1 P.M. (local time) on August 26, 1883. A plume of volcanic ash rose to a height of 30 km, well into the stratosphere. Next morning, at 10:02 A.M., a side of the volcano collapsed, the sea rushed in, and the island disappeared in a giant explosion. It did not disappear into thin air, however; it disappeared by sinking into the emptied magma chamber and forming a vast submarine *caldera*, a depression reaching 300 m below sea level. The force of the explosion has been estimated at 100 megatons. That explosion produced a plume of ash that rose to a height of 80 km, passing right through the ozone layer. Twenty cubic kilometers of volcanic ash were produced

Table 10.10. Major basaltic floods

Location	Volume (km³)	Area covered (km²)	Thickness (m)	Age 10⁶ (× 10⁶ y)
Western Siberia	1,500,000	400,000	3,700	248
Karroo (South Africa)	1,400,000	2,000,000	700	160–150
Paraná (South America)	770,000	1,200,000	300	160–150
Deccan (India)	500,000	500,000	1,000	66
South Australia	400,000	400,000	1,000	40
Columbia (United States)	190,000	220,000	500	14–17

from the magma chamber. Fine volcanic ash remained in the atmosphere for more than three years. It scattered the red portion of the solar spectrum and produced vivid sunsets. The volume of the ash indicated a magma chamber some 2.5 km across. The collapse of the magma chamber launched a sea wave (called *tsunami*, see Chapter 14) that reached a height of 40 m along the western coast of Java. That wave killed 36,000 people. What survived of Krakatau were three small islets along the rim of the caldera.

Mount Pelée in Martinique erupted on May 8, 1902, and destroyed the harbor-city of Saint-Pierre. Thirty thousand people died, and one survived (he was in jail, in a dungeon without windows).

Mount St. Helens, in the state of Washington, blew up on May 18, 1980, with a force of 25 megatons. Volcanic ash shot up to a height of 25 km, a huge nuée ardente with a temperature of 250°C rolled downslope northward, and lahars formed all around the volcano and flowed down the valley floors. Only 60 people lost their lives, however, because of the many advance signals, including local earthquakes and changes in the shape of the mountain, as well as the close watch and the many warnings by the U.S. Geological Survey.

Pyroclastic cones derive from viscous rhyolitic magmas. They usually are small and steep-sided (slope $> 20°$). An example is Paricutín, in Mexico, a cone 412 m high built from February 24, 1943, to March 4, 1952.

Calderas are depressions left when volcanoes blow up. Among the largest are Crater Lake, Oregon, which is 8 km across. It was created by the Mount Mazama explosion (about 6400 B.C.E.), which threw out 75 km^3 of material. Yellowstone Park is a very large caldera, 45×70 km, produced by a prehistoric eruption that ejected 1,000 km^3 of rhyolitic lavas and ashes.

Fissure volcanism occurs when basalt pours out along elongated fissures produced by updoming of continental crust during incipient rifting. Major flows are shown in Table 10.10. Individual flows measure about 15 m in thickness. Each of the major outpourings listed in Table 10.10 included tens or hundreds of individual flows. Basalt flows tend to break up into columnar structures when they solidify (Fig. 10.5).

Mid-ocean-ridge volcanism evolves from fissure volcanism as the continental crust is rifted and the opposite blocks move apart. Basalt, fed from elongated secondary magma chambers below the center of the ridge, pours out along the ridge axis. Seafloor basalts form the surface of the oceanic crust, which is about 7 km thick. It consists of

0–1 km of sediments (top),
1 km of pillow basalts,
5 km of gabbro sills fed by dikes (bottom).

Posteruptive phenomena that can result from interaction of phreatic waters with buried hot rocks include the following:

Hot springs, which are formed when phreatic water is heated and mineralized in contact with hot rocks

Geysers, which are periodic eruptions of water due to energy stored in water superheated at depth (Fig. 10.9)

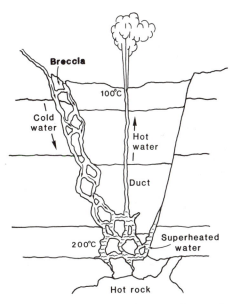

Figure 10.9. Cross section of a geyser. A geyser forms when hot rock is sufficiently close to the surface (<200 m) and when the overlying rock is fractured. Water percolating down from the surface fills the fractures and becomes superheated. Its rising temperature causes the water to expand, and the duct overflows, releasing pressure. That release transforms the entire water column into steam, and a blowout takes place.

Fumaroles, which are gaseous exhalations of, in order of abundance, H_2O, CO_2, CO, SO_2, H_2S, HCl, and HF

Solfataras, which are fumaroles rich in sulfur compounds.

THINK

* A lava flow pours out of a coastal volcano in Iceland, races down the slope, and ends up in the sea. Which part of this lava flow will form the largest crystals? Which the smallest?

* Olivine commonly consists of 80% forsterite and 20% fayalite. Forsterite and fayalite belong to the same crystallographic system (the orthorhombic system, like olivine) and have similar formulas. Yet the melting point of forsterite is 1,880°C and that of fayalite is 1,205°C. Keeping in mind the electron structures of magnesium and iron (Table A6), can you speculate why this is so?

* It is possible that very early in the history of the Earth its mantle was liquid. If so, and neglecting any other consideration, was the Earth rotating faster or slower than today?

* When a batholith comes up, an equal volume of presumably denser rock must go down, because there cannot be empty spaces inside the Earth. What do these motions do to the angular velocity of the rotating Earth?

* You are rushing to move your possessions out of your house because it is threatened by a lava flow. By which type of lava would you prefer to be threatened?

* Which types of lavas make the better soils? The faster developing soils?

* If you think that the presence of water in molten rock seems strange, consider the strength of the H–HO bond (5.16 eV) and that of the H–O bond (4.43 eV), and calculate (using $E = kT$, where k = Boltzmann constant) the temperature needed to break up the water molecule.

THE EARTH'S INTERIOR

11.1. GRAVITY AND ISOSTASY

Except for the outermost 10–12 km, the interior of the Earth is inaccessible to direct observation. Geologists and geophysicists therefore probe the interior of the Earth by studying the shape of the Earth's gravitational field, the patterns of propagation of natural and artificial seismic waves, the dynamics of the magnetic field and its record in rocks and sediments, and the amount and distribution of internal heat reaching the surface.

If the Earth were a uniform or uniformly layered spherical body, its gravitational field would be spherical and radially decreasing according to the inverse square law. The Earth, however, is neither spherical nor uniform. Its gravitational field therefore varies from place to place in complex ways.

The gravitational force F between two masses m_1 and m_2 is proportional to the product of the two masses divided by the square of the distance between the two centers of mass. In order to express the force F in terms of Newton's second law $F = ma$ (where m is mass and a is acceleration), we need to include the gravitational constant G. This constant is equal to $6.67 \cdot 10^{-11}\,\mathrm{N\,m^2\,kg^{-2}}$, because the gravitational force between two masses of 1 kg each at a distance $r = 1$ m is $1/6.67 \cdot 10^{-11}$ or $1.5 \cdot 10^{10}$ times *smaller* than the force needed to accelerate a mass of 1 kg by 1 m/s². We thus obtain Newton's law of gravitation:

$$F = Gm_1m_2/r^2 \qquad (1)$$

The constant G was first measured in 1798 by Henry Cavendish (1731–1810) using the apparatus shown in Fig. 11.1. Knowing r and measuring F, Cavendish was able to find the value of G. His result was within 1.2% of the modern value of $6.67206(\pm 8) \cdot 10^{-11}\,\mathrm{N\,m^2\,kg^{-2}}$.

Knowing that a mass m_2 of 1 kg is attracted by the Earth at sea level with a force of 9.81 N, knowing the radius of the Earth, and knowing G, Cavendish found the mass m_1 of the Earth:

$$
\begin{aligned}
m_1 &= Fr^2/Gm_2 \\
&= 9.81 \times (6.371 \cdot 10^6)^2/6.67 \cdot 10^{-11} \times 1 \\
&= 5.97 \cdot 10^{24}\,\mathrm{kg}
\end{aligned}
$$

Having found the mass of the Earth, Lord Cavendish could also calculate its average density. He found that to be $5.52\,\mathrm{g/cm^3}$, twice that of common rocks. Lord Cavendish concluded that inside the Earth there was denser materal, perhaps an iron core.

The symbol for the gravitational acceleration at the surface of the Earth is g. The reference value, called *standard* g (symbol g_0), is 980.665 cm/s² (exactly). A pilot easing an airplane out of a steep dive may experience a force equal to 2–3 g's.

If the Earth were a uniform or concentrically uniform sphere, g would be normal to the surface, and the equipotential surfaces would be spherical. The Earth is instead a triaxial spheroid, approximating in shape an oblate ellipsoid of revolution.

The gravity equipotential surface most closely approximating mean sea level is called the *geoid*. It is the reference surface for all gravity measurements. The surface of the geoid differs from the theoretical three-axial ellipsoid by as much as $+67$ m (North Atlantic) and -100 m (just south of India) because of lateral density differences within the mantle (Fig. 11.2). A ship sailing from India to the North Atlantic would have to climb 167 m, but because it would be sailing along an equipotential gravitational surface, it would consume no energy. Geoid highs and lows bear no clear relationships to the distribution of the continents and oceans, although some of the strongest gradients are related to plate margins.

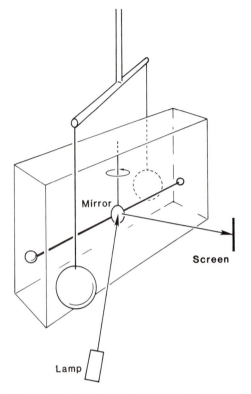

Figure 11.1. How Cavendish measured the universal gravitational constant and derived the mass and density of the Earth. Two lead balls, 2 inches in diameter, were attached to a 6-ft-long bar, and the bar was suspended inside a case from a torsion fiber. Two lead balls, 12 inches across, were attached to a second 6-ft-long bar, outside the case, that was suspended from a rigid mounting. When the rigid mounting was lowered, bringing the two larger balls close to the smaller balls, the bar inside the case rotated until the torsion of the fiber brought the rotation to a stop when the smaller balls were at a distance r from the larger balls. The rotation was measured from the motion of a light beam reflected by a mirror mounted on the bar inside the case. Knowing, from independent experiments, the force needed to twist the fiber by that amount, Cavendish was able to determine G and, from that, the mass of the Earth and its mean density.

The following units are commonly used to express gravitational acceleration:

m/s^2 (SI)
cm/s^2 = galileo (gal) (cgs)
milligalileo (mgal) = 10^{-3} gal
gravity unit (g.u.) = 10^{-6} m/s^2 = 0.1 mgal

[The cgs unit of acceleration is the galileo, usually shortened to gal, whose symbol is *gal.*]

The value of g at sea level changes with latitude (Table 11.1).

The Earth's gravitational field obeys the inverse square law. Because the equatorial radius is 1.00336 times the polar radius, the gravitational acceleration at the equator should be $1/(1.00336)^2 = 1/1.00673$ that at the pole, or $983.2/1.00673 = 976.6$ gal. It is, instead, 978.0 gal, because the equatorial bulge is rock, not empty space, and therefore it adds to the gravitational acceleration.

The Earth is rotating. Therefore, at all latitudes other than 90°, the gravitational acceleration, which pulls a body downward, is reduced by the centrifugal force, which tends to make the body fly off by the tangent. The greatest reduction is at the equator. A mass at the equator moves eastward at $40,075,000/86,400 = 463.8$ m/s. The centripetal acceleration, which prevents it from flying into space by the tangent, is

$$a_r = \omega^2 r = (6.28/86,400)^2 \times 6,378,100$$
$$= v^2/r = (463.8)^2/6,378,100$$
$$= 0.0337 \text{ m/s}^2$$
$$= 3.37 \text{ cm/s}^2 \text{ (gal)}$$

If the Earth were not rotating, the gravitational acceleration at the equator would be $978.0 + 3.37 = 981.4$ gal.

The Earth's gravitational field increases with depth below the surface of the Earth, reaching a maximum of 1,070 gal at the mantle–core boundary. It then begins decreasing (because of the gravitational attraction by the overlying rocks), until it reaches zero at the center. That was well understood by Dante Alighieri, who described how, having reached the center of the Earth, he had to turn upside down to climb up on the other side (*Divine Comedy, Inferno*, xxxiv, 76–90).

If the Earth's interior were a uniform or uniformly stratified sphere, the geoid would show no anomalies. But anomalies exist because the interior is not uniform nor uniformly stratified. Gravity measurements provide important information on the internal structure and dynamics of the planet.

Gravitational acceleration is measured by means of gravimeters. A gravimeter, stripped to its essentials, is a mass suspended from a spring (Fig. 11.3): The stronger the gravitational acceleration, the greater the force stretching the spring. Modern gravimeters have masses and springs made of fused quartz, and

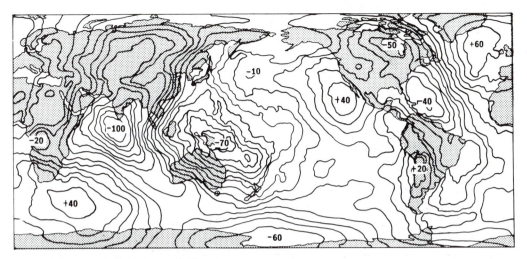

Figure 11.2. The surface of the geoid. Elevations are in meters. (Spencer, 1988, p. 40, Fig. 3.8.)

temperature is controlled to within 0.002°C by heating coils with negative-feedback circuits to keep it stable. With these instruments, gravitational acceleration can be measured with a precision of 0.01 mgal (10^{-5} gal). The change in gravitational acceleration with altitude amounts to 0.3086 mgal/m. A good gravimeter, therefore, is sensitive to altitude differences as small as 3 cm.

Gravity measurements can also be made at sea and in the air. The first gravity measurements at sea were made in 1923 on board a Dutch submarine. The engines were cut off, the sailors were ordered not to move, and measurements were made. A submarine at rest at a depth of more than 50 m is a very stable

platform. Submarines were used extensively by the Dutch geophysicist Felix Vening Meinesz (1887–1966) in the 1930s in the seas around Indonesia. Today, the gravimeter is mounted on a stabilized platform, the motion of the inertial mass is dampened by magnets or other devices, and the short-term changes in acceleration due to wave motions and ship or aircraft vibrations are filtered out electronically. Although accelerations due to wave motion can be as large as 100 gal in rough seas (about 10% of the standard gravitational acceleration), measurement at sea can be as accurate as ± 1 mgal. Gravity measurements from airplanes or helicopters are less accurate (± 2–3 mgal). Gravity

Table 11.1. Gravitational acceleration *g* at sea level

Latitude	*g* (gal)
0°	978.04
10°	978.19
20°	978.64
30°	979.32
40°	980.17
45°	980.62
50°	981.06
60°	981.91
70°	982.60
80°	983.05
90°	983.21

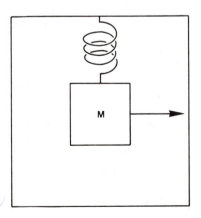

Figure 11.3. A gravimeter is, in essence, an inertial mass *M* suspended from a spring. The stronger the gravitational field, the farther the spring is stretched.

measurements from moving platforms (ships, airplanes, helicopters) must be corrected for the centrifugal acceleration resulting from the motion of the platform along the curved surface of the Earth and for the centrifugal acceleration resulting from the motion of the ship around the Earth's axis at a speed different from that of the surface. These two corrections combined are called the Eötvös correction, after the Hungarian physicist Roland von Eötvös (1848–1919).

Gravimetric measurements are important in assessing subsurface structures and dynamics. For this purpose, gravimetric measurements are referred to the geoid by the application of corrections:

Latitude correction is a correction to account for the oblateness of the Earth. Gravitational acceleration increases from equator to pole simply because the distance to the center of the Earth decreases (Table 11.1).

Elevation correction is a correction that reduces the reading to sea level. A reading on top of a mountain will give a smaller g than a reading at sea level simply because of the greater distance from the center of the Earth. This correction is generally called *free-air correction* because it considers altitude only, not the material between the gravimeter and the geoid surface. Elevation can be negative if the station is below the geoid surface (as on the shore of the Dead Sea, which is 392 m below sea level), in which case a higher g will be recorded. In order to fully utilize the precision of a modern gravimeter (0.01 mgal), the elevation must be known within 3 cm. This can be obtained with a precision topographic survey.

Bouguer correction is a correction applied after the elevation correction, to account for the fact that at any station above sea level there is rock between gravimeter and geoid. The gravimeter's reading is decreased by the amount of attraction provided by the rock. If the station is below sea level, the gravimeter's reading is increased by the amount of the gravitational attraction that would be produced by a rock layer between the gravimeter and the geoid surface. The Bouguer correction, named after the French mathematician and geophysicist Pierre Bouguer (1698–1758), amounts to 0.04185 ρh mgal/m

(ρ = density of the rock; h = vertical distance in meters between station and geoid surface).

Topographic correction is a correction to account for the local topography. A hill or valley close to a station may affect the reading significantly.

Tidal correction is a correction for the tidal motions of the solid Earth surface, amounting to a maximum of 0.2 mgal (depending on latitude and lunar time). The vertical displacement of the Earth's surface ranges up to 0.6 m.

Isostatic correction is a correction accounting for the presence of less dense rocks under mountains in isostatic equilibrium.

Because the Bouguer correction accounts for the mass excess above the geoid and the isostatic correction accounts for the mass deficiency below the geoid, the two cancel out for a system in isostatic equilibrium.

Isostasy is the condition of buoyant equilibrium. It is based on Archimedes' principle:

In a gravitational field, a body floating on or totally immersed in a fluid experiences an upward force equal to the weight of the fluid displaced.

In floating bodies at isostatic equilibrium, the mass percentage above the fluid's surface is equal to the density difference between body and fluid. For ice floating in water;

$$\rho_{ice} = 0.9$$
$$\rho_{water} = 1.0$$

The density difference is 10%; therefore, 10% of the mass of the ice is above water.

The Earth is generally in isostatic equilibrium. Therefore, a gravimeter on an airplane flying level will generally register only small gravity anomalies ($\pm$ 20 mgal). These are due to small variations in mass distributions in the Earth's crust and uppermost mantle. There are highly significant exceptions, however, although even the strongest isostatic anomalies (250 mgal) amount to only $250/980,000 = 0.00025 = 0.025\%$ of the total field. The most notable isostatic exceptions are as follows:

Strong negative isostatic anomalies occur along deep-sea trenches, amounting to -50 to -150 mgal (but reaching as high as -250 mgal in some spots). These anomalies are caused by mass deficiencies that are maintained because the lithosphere is being actively "pulled down" by convection currents in the mantle.

A strong positive isostatic anomaly (+200 mgal) occurs on the island of Cyprus, apparently resulting from deep-seated rocks (peridotites-gabbros-basalts) being actively pushed up.

Bouguer anomalies, in contrast to isostatic anomalies, result from observation, not theory (in this case, isostatic theory). Bouguer-anomaly maps reveal mass deficiencies or mass excesses after the altitude, topography, and Bouguer corrections have been applied. Bouguer-anomaly maps, therefore, reveal much about the structures and compositions of sub-surface rocks. For instance:

Strong negative Bouguer anomalies (−150 mgal) occur under the Alps and other mountain ranges, indicating mass deficiency and revealing the "roots" that support the ranges themselves.

Strong negative Bouguer anomalies (−150 mgal) occur along deep troughs filled with sediments (e.g., the Po Valley).

Strong negative Bouguer anomalies (−150 mgal) occur along the crest of the mid-ocean-ridge system, implying mass deficiency below. This deficiency is best explained by the presence under the ridge axis of rocks that are hot and therefore less dense.

Broad negative Bouguer anomalies of −50 to −60 mgal are found around the Gulf of Bothnia and in eastern Canada. These two regions supported approximately 3 km of ice until about 15,000 years ago. The weight of the ice displaced deep-seated rocks along the asthenosphere. When the ice melted, depressions were left that are now being uplifted by the returning deep rocks. The magnitudes of these anomalies suggest that an additional 200 m of uplift is forthcoming.

Negative Bouguer anomalies amounting to −50 mgal exist along rift valleys, such as the rift valley of East Africa. The anomaly is due to the downfaulting of crustal blocks, followed by sediment infilling (Fig. 11.4). In fact, the bottom portion of the downfaulted wedge is less dense than the material surrounding it. Similar anomalies exist over granitic batholiths and salt domes, both of which usually are about 2–5% less dense than surroundings rocks.

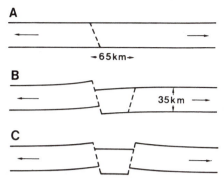

Figure 11.4. Formation of a graben. When the litho-sphere is stretched by mantle motions, it breaks because it is rigid. As shown in this illustration, produced by Vening Meinesz in 1958, a slanted break (A) causes the lithosphere to the left of the break to rise (B) because of the mass deficiency created above the lower lithosphere adjacent to the break (there is air instead of rock). At the same time, the lithosphere to the right of the break sinks because of mass excess created by the upper lithosphere adjacent to the break. A second break occurs at the point of inflection (B). The detached block has the shape of a wedge pointing downward. The wedge is top-heavy, and so it sinks, with its bottom part penetrating the asthenosphere (C). Sediment infilling over the top surface of the wedge follows.

Local Bouguer anomalies may be indicative of mineral deposits, including metals (plus a few milligals) or salt domes (minus a few milligals).

THINK

* Fig. 11.5 shows a stamp issued by the Republic of Paraguay to celebrate the centenary of the

Figure 11.5. Paraguayan stamp commemorating the 100th anniversary of the Epopeya Nacional, the great war of 1864–1870 that saw Paraguay pitted against Brazil, Argentina, and Uruguay, and ultimately defeated.

Epopeya Nacional, the great national war of 1864–1870 that saw Paraguay, under the leadership of Francisco Solano López, simultaneously fighting Brazil, Argentina, and Uruguay. In the end, Paraguay was defeated and López was killed. The Paraguayans fought with legendary bravery—death in battle reduced the population from 1.3 million in 1864 to less than 300,000 in 1870, of which only 28,746 were men (the Catholic church allowed polygamy for a spell, in order to repopulate the devastated country). Anyway, to properly celebrate the hundredth anniversary of the event, the Paraguayans chose Newton. Anything wrong with this stamp?

* According to the *Guinness Book of World Records*, the fastest dragster accelerated from 0 to 268.01 mph in 5.391 s. Assuming a uniform acceleration, calculate how many g's the driver experienced.

* You are traveling at a conservative 80 mph when your car hits a wall. Your car is equipped with an air bag that makes your body decelerate to zero speed over a distance of 25 cm. Calculate how many g's your body will experience.

* Mount Everest, 8,848 m high, is located at 27.6°N. Estimate, from Table 11.1, the value of g at sea level at that latitude, and use the inverse square law to calculate what it would be at the altitude of Mount Everest. This is your free-air result, because it takes into consideration only altitude above sea level, not any intervening mass. Assuming that Mount Everest is a cone with 30° slope and an average density of 2.6 g/cm³, calculate its mass (the volume of a cone is given by the area of the base times $\frac{1}{3}$ of the height). Assume now that the entire mass of Mount Everest is concentrated at its center of mass, which obviously is along the cone's axis at a point where a plane cutting across it normal to the axis would divide the mass into two equal masses. Find out where this point is, calculate the gravitational attraction produced by Mount Everest on a gravimeter on its top, and correct your free-air result.

11.2. STRESS AND STRAIN

Rocks are compressible and deformable to a certain extent. The heat flowing outward from the Earth's interior maintains the rocks of the mantle and crust under stress. If stress exceeds

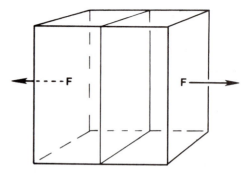

Figure 11.6. A cube of homogeneous material being stressed by equal forces normal to a pair of opposite faces. The longitudinal strain s_l is the relative elongation in the direction of the two forces. The transverse strain s_t is the relative reduction in cross section. For most materials, s_t/s_l is a constant that lies between 0.20 and 0.45. It is called, with reversed sign, the Poisson's ratio σ (i.e., $\sigma = -s_t/s_l$).

a certain limit, which depends on the type of rock and its physical conditions, fracture will occur. Seismic waves are mechanical waves produced in the Earth's interior by rock fracture.

Stress is the force per unit area (F/A) applied to a solid. It has the same dimensions as pressure and is expressed in pascals (1 Pa = 1 N/m² = 10^{-5} bar = $9.87 \cdot 10^{-6}$ atm). *Strain* is the response of a solid body to stress. Strain is the change in volume and/or shape of a body due to applied stress(es). Strain is a ratio of equivalent dimensions and therefore is dimensionless. A total of nine possible strains can affect an anisotropic body: three tensile, three shear, and three volumetric.

Fig. 11.6 shows a cube stressed by two equal tensile forces F (considered positive if directed outward) applied normal to two opposite faces. The tensile stress on an internal surface normal to the applied stress is F/A.

If the internal surface is inclined (Fig. 11.7), the tensile force F, and hence the tensile strain s, can be decomposed into a normal component s_n and a shear component s_t. The shear component s_t is equal to

$$s_t = \tfrac{1}{2}s_n\sin 2\theta$$

where s_t is the shear component, s_n is the normal component, θ is the angle of the plane to the normal to the external tensile force. Notice that

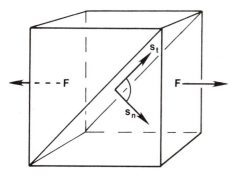

Figure 11.7. The strain on an internal surface at an angle to the two external forces can be decomposed into two components: s_n normal to the surface, and s_t tangential to it.

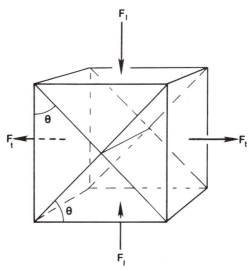

Figure 11.9. A cube subjected to two sets of equal forces normal to each other and to the cube's faces, one compressional and the other tensional. The planes of maximum shear strain are at an angle $\theta = 45°$ from the directions of the forces.

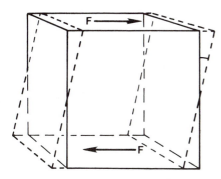

Figure 11.8. A cube subjected to shear stress.

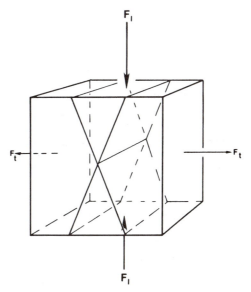

Figure 11.10. A cube subjected to two sets of unequal forces normal to each other and to the cube's faces, one compressional and one tensional. The planes of maximum shear stress are at an angle less than 45° from the direction of the set of the two larger forces.

s_n is maximum when $s_t = 0$ (i.e., when s_n coincides with the external tensile force F)

s_t is maximum when $\theta = 45°$ (i.e., when $\sin 2\theta = \sin 90° = 1$).

Fig. 11.8 shows a cube stressed by two opposite, tangential forces. The cube is deformed by a rotational strain; that is, every particle in the cube undergoes a small clockwise rotation. The angle θ is called the *shear angle*.

Fig. 11.9 shows a cube of rock inside the Earth stressed by the lithostatic load (F_l) and by tensile forces (F_t) of similar magnitude, all normal to the faces of the cube. The shear stress is maximum at 45° from the direction of the external forces. If the forces of one set are stronger than those of the other set, the angle between shear stress and the set of stronger forces will be less than 45° (Fig. 11.10).

Within limits, strain is proportional to stress. This means that, within limits, solids behave elastically—if the stress is removed, the solid returns to its original shape. However, if the stress rises above the *limit of elasticity*, the solid deforms plastically; that is, when the stress is removed, the solid exhibits permanent

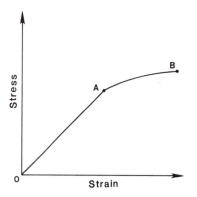

Figure 11.11. Stress–strain curve for a typical rock. $0A$ = elastic deformation; A = elastic limit; AB = plastic deformation. Rupture occurs at point B.

deformation. Beyond the point of maximum stress tolerable by the medium, rupture occurs, sharply releasing the accumulated strain energy. The limits of elastic and plastic deformations depend on the rate of stress application. Slow application of stress leads to rock deformation that would not be possible with a faster rate.

Fig. 11.11 shows the relationship between stress and strain for a typical rock. The proportional region (OA) reflects *Hooke's law*, which says that within the limit of elasticity, strain is proportional to stress. This law was enunciated in 1678 by the English physicist Robert Hooke (1635–1703). The proportionality factor, which depends on the material, is called *Young's modulus* after the English physicist Thomas Young (1773–1829). Because strain is dimensionless, Young's modulus has the dimension of force/area = pressure, and is expressed in pascals.

[*Modulus* means *measure*.]

In Fig. 11.11, AB represents plastic deformation, and B is the point of rupture.

[Young's modulus of structural steel is $20 \cdot 10^{10}$ Pa. If a tensile force of 9,800 N (the weight of 1 ton) is applied to a steel rod 1 cm in diameter and 1 m long, the stress S will be

$S = F/A$
$= F/\pi r^2$
$= 9,800/7.8 \cdot 10^{-5}$
$= 1.25 \cdot 10^8$ Pa

but

$F/A = 1.25 \cdot 10^8$ Pa $= E(\Delta l/l)$

from which

$\Delta l = 1.25 \cdot 10^8/20 \cdot 10^{10}$
$= 6.25 \cdot 10^{-4}$ m
$= 0.625$ mm

In the above S is stress, F is force, A is area, r is the radius of the rod, l is the length of the rod, and E is Young's modulus.]

Rocks are more susceptible to shear stress than to compression stress. The *shear modulus* (symbol μ) is the proportionality factor relating shear stress to the shear angle θ (Fig. 11.8); it is also called the *rigidity modulus*. The shear angle is dimensionless, and so the shear modulus is expressed in pascals.

The *bulk modulus of elasticity* (or simply *bulk modulus*) (symbol β) is the reciprocal of the compressibility (symbol λ), which is defined as the fractional change in volume per unit of pressure. Because the fractional change in volume is a ratio of two quantities with the same dimensions and therefore is dimensionless, the bulk modulus, too, is expressed in pascals.

$\lambda = -(\Delta V/V)/P$
$\beta = 1/\lambda$
$\quad = -pV/\Delta V$

where λ is compressibility, V is volume, p is pressure, and β is the bulk modulus of elasticity (the negative sign indicates that the volume has been reduced). A compressed body decreases in volume. Therefore, $\Delta V/V < 0$. Table 11.2 shows the elastic moduli for various materials.

The bulk modulus of structural steel is $16 \cdot 10^{10}$ Pa. The pressure at the bottom of the Marianas Trench (10,915 m) is $1.10 \cdot 10^8$ Pa. If a structural steel rod 1 cm in diameter and 1 m long (volume = $7.85 \cdot 10^{-5}$ m³) is dropped to the bottom, its change in volume due to hydrostatic compression will be

$\Delta V = -pV/\beta$
$\quad = -(1.10 \cdot 10^8 \times 7.85 \cdot 10^{-5})/16 \cdot 10^{10}$
$\quad = -8,635/16 \cdot 10^{10}$
$\quad = -5.40 \cdot 10^{-8}$
$\quad = -0.07\%$

In contrast, a liter of water will be changed by

$\Delta V = -pV/\beta$
$\quad = -(1.10 \cdot 10^8 \times 1)/2.18 \cdot 10^9$
$\quad = -0.05$
$\quad = -5\%$

Table 11.2. Elastic moduli

Material	Young's modulus (E) (10^{10} Pa)	Shear modulus (μ) (10^{10} Pa)	Bulk modulus (β) (10^{10} Pa)
basalt	6–8	—	4–5
copper	12.4	4.5	13.1
diabase	9	—	—
dolomite	7–9	—	—
Earth			
0–20 km of depth		3	4–5
45–120 km of depth	6	—	14
gabbro	9–10	—	6
glass	5.5	2.3	3.1
gneiss	1–2	—	—
granite	3–5	—	4
granodiorite	6	—	—
ice	1	—	—
iron	19	7.6	12.7
limestone	5–7	—	—
marble	3–5	—	—
quartz	8.2	3.0	10.5
quartzite	6–10	—	4
sandstone	0.6–1	—	—
serpentinite	0.2–0.4	—	—
schist	0.4–0.7	—	—
shale	0.1–0.4	—	—
structural steel (ASTM-A36)	20	—	16
water	—	—	0.218

As expected, liquids (and glasses, see Table 11.2) are more compressible than crystalline substances.

11.3. MECHANICAL WAVES

Pyrite (FeS_2) crystallizes in the cubic system and commonly forms well-shaped cubes. Consider one such cube resting on a rigid surface. If one hits this cube from above with a mallet (not hard enough to break it, though), the cube reacts like a spring that is suddenly compressed. The atomic bonds are temporarily compressed in the vertical direction and react by rebounding and overshooting the initial position. A series of compressions and decompressions in the vertical direction takes place until the energy is dissipated into heat and the crystal returns to its original rest. In effect, the crystal "rings". Consider now the same crystal and imagine that its opposite faces are fixed to two parallel, rigid surfaces. If one surface is moved horizontally with respect to

the other surface, the cubic crystal deforms a little bit by allowing its atomic bonds to be stretched. If the motion continues, however, bonds break along a plane passing through the crystal, and the two pieces react with a series of compressions and expansions until they come to rest. Compression reduces the standard bond length, whereas expansion or shearing increases it. These expansions, compressions, and shearings, amounting to roughly 0.1% of an atomic diameter, are mechanical and periodical and therefore are mechanical waves. Rocks behave in a more complicated way than do single crystals, because rocks consist of many different crystals variously bonded to each other; their compositions change from place to place, and they are broken up by joints.

Mechanical waves are cyclic phenomena, a model of which is the harmonic oscillator. A harmonic oscillator is a physical system bound to a position of equilibrium by a restoring force that is proportional to the displacement. Wave motion may be expressed by the equation

$$y = y_m \sin(\omega t + \phi)$$

where y_m is the absolute value of the maximum displacement along the y axis ($=$ amplitude), ($\omega t + \phi$) is phase, ω is angular frequency ($= 2\pi v = 2\pi/T$), ϕ is the phase constant, v is frequency ($= 1/T$), and T is period ($= 1/v = 2\pi/\omega$). Wave characteristics are as follows:

Amplitude: maximum displacement from zero
Wavelength: shortest distance in space between successive points that have the same phase angle
Period: the time it takes for one full wave to pass by a fixed reference point
Frequency: number of periods per second
Phase constant: initial phase of the wave
Phase velocity: velocity at which the phase of a wave is traveling

[It is important to understand that waves are forms, not material objects, although they can transport energy. The lunar tide is a wave about 0.5 m high in the open ocean. It travels from Panama to the Philippines, a distance of 16,500 km, in 14 hours. Its speed is therefore almost 1,200 km/h. It is the *shape* of the water surface (i.e., the *phase* of the wave) that moves, not the water. The water only moves up and down 50 cm every 24h 50m, corresponding to an average up-and-down speed of only 2 cm/h.]

11.4. SEISMIC WAVES

There are two families of seismic waves: body waves and surface waves. Each family has two members:

Body waves

P (for *primary* or *pressure*) waves are waves produced by a push (or pull) on a solid, liquid, or gaseous body (Fig. 11.12). Inter-atomic or intermolecular distances are disturbed, and the disturbance propagates through the body with a speed that is characteristic of the structure and composition of the body and its physical conditions. The displacement of atoms in solids and liquids produced by the passage of a P wave amounts to 10^{-13} m (1/1,000 of an atomic diameter).

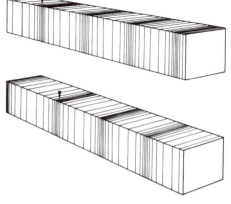

Figure 11.12. P waves are successive compressions and rarefactions of the medium. The restoring force is the mutual repulsion of the electron clouds of the atoms and molecules of the medium. (Emiliani, 1988, p. 170, Fig. 10.4.)

S (for *secondary* or *shear*) waves are waves produced by shear deformation. Propagation is at 90° from the direction of deformation (Fig. 11.13). The speed of S waves in the solid Earth is about 60% of that of P waves.

Surface waves

Rayleigh waves are P waves propagating along the free surface of an elastic medium. The motion of a point on the free surface is elliptical on a vertical plane, and retrograde (Fig. 11.14). The motion decreases exponentially with increasing distance below the surface. Rayleigh waves travel at 0.9 the speed of S waves, or about 55% of the speed of P waves.

Love waves are S waves propagating along a surface layer by multiple reflection between the surface and the top of an underlying layer. A Love wave striking the interface at a low angle will travel a shorter path and will arrive earlier at a recording station than Love waves striking the interface at a higher angle (Fig. 11.15).

Seismic waves are generated by the abrupt release of strain energy. P waves are transmitted through solids, liquids, and gases. S waves are transmitted only through solids. The velocities of P and S waves depend on the elastic moduli of the medium. Specifically,

$$V_P = [(\lambda + 2\mu)]/\rho]^{1/2}$$

and

$$V_S = (\mu/\rho)^{1/2}$$

where V_P is the velocity of the P waves, V_S is the velocity of the S waves, λ is compressibility, μ is the shear modulus, and ρ is density. Examples of P-wave velocities are shown in Table 11.3. Characteristic sound frequencies and wavelengths are shown in Table 11.4.

The human ear has evolved to be extremely sensitive to the cyclic pressure changes accompanying sound waves. These pressure

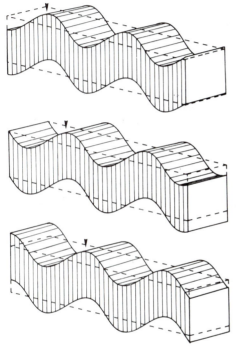

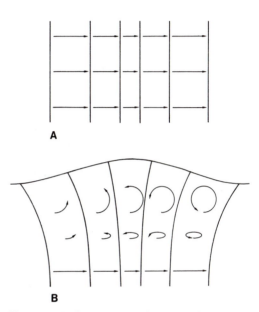

Figure 11.13. S waves are generated by shear stress and propagate normally to the direction of stress. The restoring force is the resistance to deformation by the atomic and molecular bonds of the medium. (Emiliani, 1988, p. 171, Fig. 10.5.)

Figure 11.14. Rayleigh waves form at the free surface of a solid medium. (A) Passage of a P wave across a medium unbounded above and below. (B) Passage of a P wave through a medium bound by a free surface. The medium is less compressed near the surface, and so it expands upward with the particle moving in elliptical, retrograde paths.

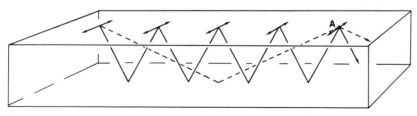

Figure 11.15. Love waves are S waves produced by multiple reflections between the surface and an underlying layer. The waves of longer wavelength (dashed line) undergo fewer reflections than do those of shorter wavelength, and so they arrive first at the recording station A, producing the phenomenon known as phase dispersion.

changes cause a cyclic displacement of the eardrum, which is picked up by ciliary sensors and transmitted to the brain. At the frequency of 1,000 Hz, the loudest sound tolerable by the

Table 11.3.   Speeds of P waves

Medium	Speed (km/s)
air (dry, 0°C)	0.331
fresh water (25°C)	1.509
seawater (7°C, 34.5‰ salinity, 50 atm)	1.479
shale	~3
sandstone	~3.5
limestone	~4
granite	~5.5
basalt	~5.5
peridotite	~8
iron	5.9
nickel	6.0
diamond	18.1

human ear causes a cyclic pressure change of 280 μbar (1 μbar = 1 dyn/cm^2) above and below ambient atmospheric pressure. The amplitude of the displacement of the eardrum is 11 μm. At the other extreme, the faintest audible sound causes a pressure change of $2 \cdot 10^{-4}$ μbar and a displacement of only $8 \cdot 10^{-6}$ μm ($= 8 \cdot 10^{-4}$ Å $= \sim 1/1,000$ of the diameter of an atom).

Like electromagnetic waves (see Section 2.6), seismic waves reflect, refract, and diffract. Seismic reflection and seismic refraction are two methods of geophysical exploration by which the layering of the Earth's crust and upper mantle can be studied. Seismic reflection shows the layering (Figs. 11.16 and 11.17), but in order to determine the depth of the layers, the velocity of seismic waves in the different layers must be known or at least closely estimated. In contrast, seismic refraction yields direct information on the thicknesses of the various layers. Fig. 11.18 shows how to deter-

Table 11.4.   Sound frequencies and wavelengths

Type of sound	Frequency (Hz)	Wavelength
lowest audible frequency	~16	20.7 m
highest audible frequency	~20,000	1.65 cm
A$_4$ (piano)	440	0.75 m
male speech range	100–9,000	3.31 m to 3.48 cm
female speech range	150–10,000	2.20 m to 3.31 cm
earthquake P wave	~1	4–8 km
free oscillations of the Earth	0.2–0.02	10^4 km
highest sound frequency produced	$6 \cdot 10^8$	$5.5 \cdot 10^{-7}$ cm

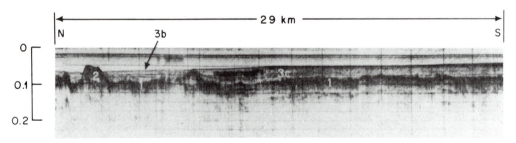

Figure 11.16. A seismic reflection profile in the western Gulf of Maine. The vertical scale gives half the time (in seconds) taken by the sound-wave pulse to return to the ship, that is, the time taken by the sound pulse to reach bottom and the subbottom reflectors. Depth can be calculated if the sound speed in seawater (corrected for temperature and pressure) and that in the sediment layers are known. Here, sound velocity in seawater is 1,480 m/s. The bottom is seen to slope from a depth of about 50 m on the right to a depth of about 100 m on the left. The layer labeled 1 identifies basement rocks (of Paleozoic age) at a subbottom depth of about 50 m. The hill on the left (labeled 2) is a mound left by the glaciers that covered this area during the last ice age. Layers 3b and 3c are postglacial marine deposits, 50 m thick. (Emery and Uchupi, 1972, p. 185, Fig. 148.)

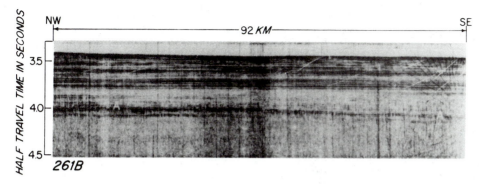

Figure 11.17. Seismic reflection profile across the Sohm Abyssal Plain off the southeastern United States. Here the surface drops from a depth of 5,030 m on the left to a depth of 5,100 m on the right, equal to 0.76 m/km, or a slope of $0.00076 = 0.076\% = 0.04°$. The prominent reflector labeled A is the top of a chert layer deposited in middle Eocene time (about $45 \cdot 10^6$ y ago). About 1,100 m of sediments have accumulated on top of this layer at this site during the past $45 \cdot 10^6$ y, giving an average rate of accumulation of 2.4 cm/1,000 years. (Emery and Uchupi, 1972, p. 200, Fig. 163.)

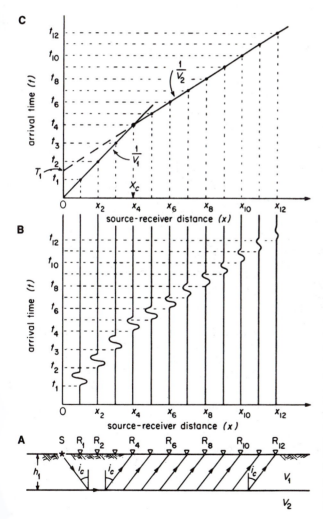

Figure 11.18. Wave paths (A), records from an array of 12 geophones (B), and travel-time curve (C) subsequent to an artificial explosion at source S. First arrival times at receiving geophones R_1, R_2, and R_3 are due to waves traveling along the top layer of thickness h_1. First arrivals at receiving geophone R_5 and beyond are from waves that have been refracted at the critical angle along the interface between the top layer and the layer below, and back to the surface. Waves propagating along the top layer travel at a slower speed than the refracted waves, as indicated by the different slopes of the lines joining the first arrival times (C). Direct waves and refracted waves arrive at the same time at receivig geophone R_4. The distance $0 - x_4$ is called the *critical distance* (x_c). (Robinson and Coruh, 1988, p. 44, Fig. 3.3.)

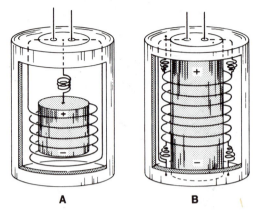

Figure 11.19. A geophone (schematic). (A) The magnet, suspended from a spring, acts as an inertial mass. (B) The magnet is attached to the case, and the coil is suspended by springs. If the case moves up and down, the coil moves with respect to the magnet, or vice versa, and a current is generated through the coil. The current is amplified and registered by a galvanometer. The system geophone-amplifier-galvanometer froms a single *channel*. Commonly, multichannel systems are used. (Robinson and Coruh, 1988, p. 34, Fig. 2.24.)

mine the thickness *h* of surface layer 1, through which P waves travel at the speed V_1, overlying layer 2, through which the speed is V_2. At a certain distance, called the *critical distance*, the waves traveling along the surface layer and those refracted at the critical angle and traveling at V_2 will arrive at the recording geophone R_4 at the same time.

[A *geophone* is essentially a magnet suspended from a spring with a coil around it (Fig. 11.19). Ground motion causes the case to move up and down with respect to the magnet. This develops a current in the coil, which activates a galvanometer. A *galvanometer* consists of a coil with a pointer. The coil is placed between the poles of a magnet. When a current flows through the coil, the coil rotates in the magnetic field, and the pointer leads to a reading. *Hydrophones* are recording devices that detect pressure pulses through water by means of a piezoelectric sensor (see Chapter 9).]

Geophones at points between *S* and R_4 (excluded) will receive first the surface wave and second the refracted wave. Geophones

beyond R_4 will receive first the refracted wave and second the surface wave. Only geophone R_4 will register a single arrival. The distance SR_4 is the critical distance (x_c). The thickness h_1 of layer 1 is given by the equation

$$h_1 = \tfrac{1}{2}x_c[(V_2 - V_1)/(V_2 + V_1)]^{1/2}$$

If the interface between the surface and the underlying layer is not parallel to the surface, or if there are faults or folds, matters become more complicated.

Diffraction of seismic waves occurs when lithology changes abruptly over a short distance.

Reflection and refraction of seismic waves produced either naturally (by earthquakes) or artificially (by chemical or nuclear explosives) provide important information on the internal structure of the Earth.

11.5. EARTHQUAKES

Earthquakes are tremors produced by rock failure (usually at depth) and the sudden release of accumulated strain. The earthquake site is called the *focus* or *hypocenter*. The point on the surface directly above is called the *epicenter*. Seismic waves produced by earthquakes are detected by means of seismometers (Fig. 11.20). The distance of the epicenter from a seismographic observatory is obtained from the difference in arrival times of P and S waves. In order to determine the location of the epicenter, the distances from the epicenter to three stations are needed (Fig. 11.21).

The Richter scale is commonly used to express earthquake intensity (called *magnitude*). On this scale, magnitude is given by

$$M = \log(A/T)$$

where *M* is magnitude, *A* is the maximum amplitude of ground motion in micrometers, registered by a standard Wood–Anderson short-period seismometer and reduced to the standard distance of 100 km from the epicenter (special tables are used to reduce observations to this standard distance); *T* is the dominant wave period in seconds. The smallest earthquakes have Richter magnitudes of −3 or −2, and the largest ones have magnitudes of about 9. Earthquakes cannot have magnitudes

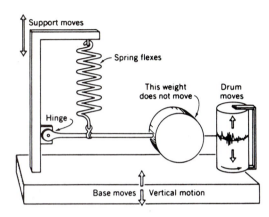

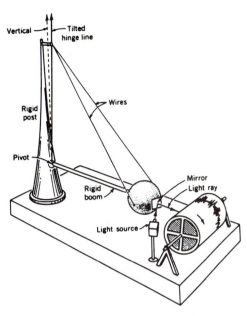

Figure 11.20. Seismometers, like geophones, are instruments used to measure the motions of the ground. The seismometer on the left registers vertical motions. The one on the right registers horizontal motions. (Strahler, 1971, p. 394, Figs. 23.14B,C.)

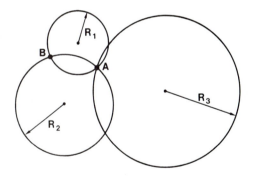

Figure 11.21. Locating the epicenter of an earthquake requires at least three stations. The distance of each station from the epicenter is calculated from the difference between first arrival of the P waves and that of the slower-moving S waves. A single station will not be able to identify the location of the epicenter, only its distance R_1—the epicenter could be anywhere along the circle of radius R_1. For a second station, the epicenter could be anywhere on a circle of radius R_2. Combining these two records shows that the epicenter could be only at either point A or point B, where the two circles intersect. A third station at a distance R_3 from the earthquake will pinpoint the location of the epicenter, because at only one point, in this case A, will the three circles intersect.

much higher than 9 because the rocks are not sufficiently strong to store the necessary energy without breaking. Magnitude is related to the energy dissipated by the equation

$$\log_{10} E = 5.24 + 1.44 M \tag{1}$$

where E is energy and M is magnitude.

The amplitude of ground motion decreases away from the epicenter because of the spreading of the energy along an essentially cylindrical wave front (Fig. 11.22). The decrease is proportional to the radius of the cylinder. Therefore,

$$A = A_0/r \tag{2}$$

where A is the amplitude at a distance r from

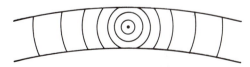

Figure 11.22. Propagation of a seismic wave along a confined layer.

the epicenter, and A_0 is the amplitude at the epicenter.

In addition to the decrease because of spreading, there is a decrease due to energy absorption by the medium (i.e., conversion of seismic energy into heat):

$$A = A_0 e^{-\alpha r} \qquad (3)$$

where α is the attenuation coefficient (see Section 3.3). Equations (2) and (3) can be combined into equation (4):

$$A = A_0 e^{-\alpha r} / r \qquad (4)$$

Attenuation coefficients α for some rocks are shown in Table 11.5. The energy dissipated by a large earthquake is on the order of 10^{18} joules. At or near the epicenter, accelerations can approach $10\,\text{m/s}^2$, and ground motion can range up to 1 m vertically and several meters horizontally. Earthquakes in coastal areas, along continental slopes, or on the deep-sea floor generate sea waves, call *tsunamis*, that can reach heights of 10 m or more when approaching land if properly funneled by the local morphology (see Chapter 14).

In addition to seismic motions, the entire Earth can be made to oscillate by a large earthquake or by an asteroidal or cometary impact. The periods of such free oscillations range from 3 to 54 minutes, with wavelengths ranging from 1,500 to 25,000 km, and amplitudes on the order of 0.1 mm.

Earthquake magnitudes follow a Poisson distribution.

[The Poisson distribution (after Siméon Denis Poisson, French mathematician, 1781–1840) is a frequency distribution for random events of increasing unlikelihood per unit of measurement. In the case of earthquakes, the increasing unlikelihood is magnitude, and the unit of measurement is the year. The probability $P(M)$ for a magnitude-M earthquake to occur is given by the Poisson distribution equation:

$$P(M) = e^{-\bar{M}} \cdot \bar{M}^M / M!$$

where P is probability, M is magnitude, and $\bar{M}$ is mean magnitude.]

Taking 5.5 as the mean magnitude and 9 as the maximum, the probability for a 9-magnitude earthquake to occur anywhere on Earth is 0.05 per year, or once in 20 years, which is about right.

Table 11.5. Attenuation coefficients for 50-Hz seismic waves

Rock	Velocity of P waves (km/s)	Attenuation coefficient $\alpha\,(\text{km}^{-1})$
shale	2.15	2.32
sandstone	4.0	1.77
limestone	6.0	0.4
granite	5.0	0.3
diorite	5.8	0.21
basalt	5.5	0.4

An earthquake of magnitude 9 releases about $1.6 \cdot 10^{18}\,\text{J}/20\,\text{y} = 8 \cdot 10^{16}\,\text{J/y}$. All lesser earthquakes combined, numbering more than 500,000 per year, release a similar amount of energy per year, giving a total of $1.6 \cdot 10^{17}\,\text{J/y}$. This is only 0.01% of the energy released as heat flow from the Earth's interior, which amounts to $0.0699\,\text{W/m}^2 = 1.125 \cdot 10^{21}\,\text{J/y}$.

Fig. 11.23 shows the distribution of earthquakes around the world. The highest concentration is around the Pacific, with a band extending northward and westward from Indonesia to Myanmar (Burma), the Himalayas, Iran, Turkey, the Balkans, and Italy. The crest of the mid-ocean-ridge system is also a site of earthquake activity.

Fig. 11.24 is a map of earthquake hazard for the United States (see also Chapter 12). It shows the expected intensity of ground shaking as a percentage of the standard gravitational acceleration g.

The internal structure of the Earth is revealed mainly by the patterns of seismic-wave propagations. Both P and S waves are reflected and refracted in the Earth's interior. First-arrival seismic waves minimize their travel time by following curved paths through deeper rock layers, where they can travel faster (Fig. 11.25). Refraction of P waves in crossing the mantle–core interface creates a *shadow zone* between 103° and 143° from the focus of an earthquake. In addition, no S waves are received beyond 143°, indicating that the intervening body, the outer core, is liquid.

Fig. 11.26 shows the velocities of P and S waves as functions of depth inside the Earth. The velocity maximum near the top (lower lithosphere) is followed by a sharp decrease

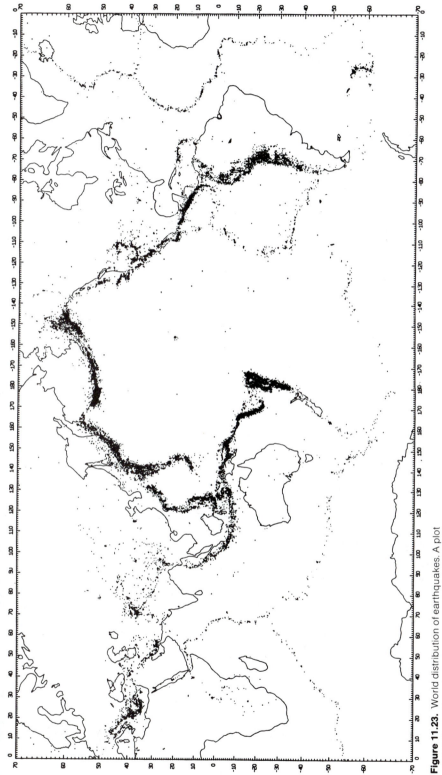

Figure 11.23. World distribution of earthquakes. A plot of 20,000 earthquakes, ranging in depth from 0 to 700 km, that occurred between 1961 and 1967. (Barazangi and Dorman, 1969.)

221

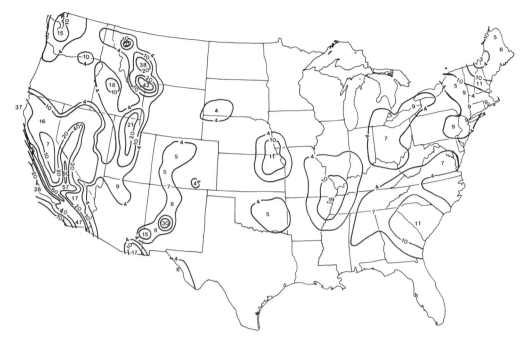

Figure 11.24. Map of earthquake hazard in the United States, in terms of ground shaking, expressed as expected percentage of g. (Algermissien and Perkins, 1976.)

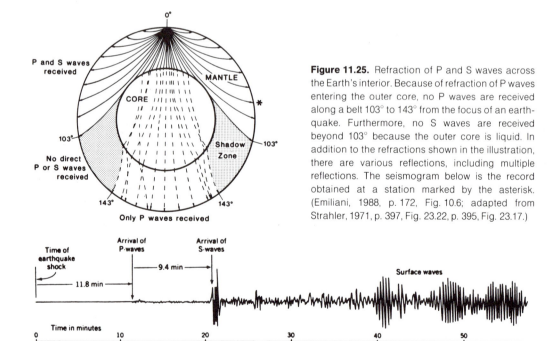

Figure 11.25. Refraction of P and S waves across the Earth's interior. Because of refraction of P waves entering the outer core, no P waves are received along a belt 103° to 143° from the focus of an earthquake. Furthermore, no S waves are received beyond 103° because the outer core is liquid. In addition to the refractions shown in the illustration, there are various reflections, including multiple reflections. The seismogram below is the record obtained at a station marked by the asterisk. (Emiliani, 1988, p. 172, Fig. 10.6; adapted from Strahler, 1971, p. 397, Fig. 23.22, p. 395, Fig. 23.17.)

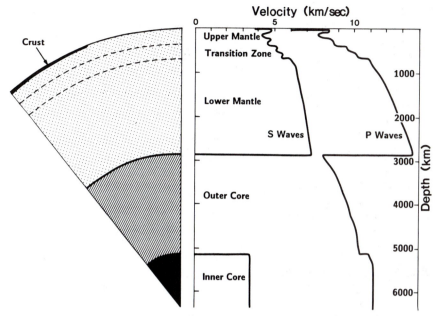

Figure 11.26. Velocities of P and S waves as functions of depth. The maximum near the surface represents the lower lithosphere, and the minimum just below it the asthenosphere. Velocities increase irregularly to a depth of 700 km because of phase changes. Below that depth, velocities increase smoothly because there are no more phase changes. S waves disappear at the mantle–core boundary, but reappear in the inner core, where they are generated by impinging P waves. The speed of the P waves decreases sharply at the mantle–core boundary, but it increases farther down, with a gradient through the transition zone into the inner core. (Spencer, 1988, p. 12, Fig. 2.1.)

representing the asthenosphere. Velocities increase with a few sharp jumps to 700 km, and then smoothly to the base of the mantle (2,885 km). S waves disappear at 2,885 km, and the P waves show a large decrease in velocity, because the outer core is liquid. S waves reappear in the inner core (where they are called J waves) because they are generated by impinging P waves.

Within each seismic layer there are lateral irregularities. Some places are hotter or cooler than normal, and there seismic waves will travel slower or faster, respectively. Travel-time anomalies reveal the thermal structure of the mantle. The study of these anomalies is called *seismic tomography*, which means seismic *sectioning* (from the Greek τόμος, which means *cut* or *slice*). Seismic tomography has revealed, for instance, that the asthenosphere is anomalously warm, and the lower mantle is anomalously cool under the Atlantic and anomalously warm under the Pacific. These studies are of great importance in elucidating the structure and dynamics of the Earth's interior.

THINK

* Why is Young's modulus for sedimentary rocks (Table 11.2) much lower than that for igneous rocks? What does that mean?

* The intensity I of sound is given by

$$I = 2\pi^2 \rho v A^2 v^2$$

where ρ is the density of the medium, v is the velocity of sound in that medium, A is amplitude, and v is frequency. This equation shows that intensity is proportional to the squares of both amplitude and frequency. Assuming that men and women can generate screams of similar amplitudes, a woman's scream will be heard much farther away because of its higher frequency. What are the advantages? Any disadvantage?

* Considering Fig. 11.18, do you expect the critical distance to be directly or inversely proportional to the difference in seismic-wave velocities between the top layer and the underlying layer?

* There is a tremor, and your seismograph shows that P and S waves arrive at the same time. Where is the earthquake?

* Why is equation (3) not of the form $A = A_0/r^2$?

* You discover a planetary system that is 8 billion years old, with a planet similar to the Earth. Would you expect the seismic shadow zone of that planet to be larger or smaller than that of the Earth? Explain.

11.6. INTERNAL STRUCTURE OF THE EARTH

The interior of the Earth is layered both chemically and mechanically (see Fig. 10.1). The chemical and mechanical subdivisions are shown in Table 11.6. The oceanic crust is only 7 km thick in the deep oceanic basins, and it consists of pillow basalts above and gabbro below. The basaltic layer is crossed by dikes from below and is capped by a layer of sediments 0–2 km thick.

The continental crust averages 35 km in thickness, but it ranges up to 60 km under mountain ranges and up to 80 km in regions of crustal duplication (Tibet). The sediment cover ranges in thickness from zero to more than 15 km. Below the sediments, the crust consists of volcanic, plutonic, and metamorphic rocks, presumably grading downward into gabbro. The deepest drilling into the continental crust is in progress on the Kola Peninsula, in northwest Russia. It has reached a depth of 12 km. The top 7 km consist of alternating layers of sediments and volcanic rocks. The underlying 5 km consist mainly of gneisses. Presumably, gabbro lies deeper.

The discontinuity in the speed of the P waves that marks the boundary between gabbro above and peridotite below is called the *Mohorovičić discontinuity*, Moho for short, named after the Croatian geophysicist Andrija Mohorovičić (1857–1936), who discovered it. The velocity of the P waves jumps from 7–7.5 km/s to 8.0–8.2 km/s. Under the mid-ocean ridges, the rocks are hotter and no such jump is evident. The dominant rock below the Mohorovičić discontinuity is peridotite. The peridotitic layer extends to about 30 km under the ocean and to about 120 km under the continents. The peridotitic layer approximates the residue of the fractional crystallization process that formed the oceanic and continental crusts. Mixing the crust and the peridotitic layer would form an "average" mantle rock, to which the name *pyrolite* has been given, consisting of pyroxene, olivine, and garnet. This composition corresponds to a mixture of 3 parts of alpine peridotite (79% olivine, 20% orthopyroxene, 1% spinel) and 1 part of tholeiitic basalt. The entire mantle below the base of the peridotitic layer is believed to consist of pyrolite.

Two prominent seismic discontinuities occur in the upper mantle, at 400 km and at 670 km. The 400-km discontinuity is presumed to be due to a phase change by which pyroxene changes into olivine + stishovite, and olivine changes its structure from the orthorhombic $2/m2/m2/m$ to the cubic $4/m\bar{3}2/m$ (which is the structure of spinel), with a 5% increase in density. At 670 km of depth, olivine breaks down to the perovskite structure (monoclinic, $2/m$) and periclase (MgO, cubic, $4/m\bar{3}2/m$).

The lower mantle, from 670 to 2,885 km, does not exhibit either phase changes or earthquakes. Density and the velocities of the P and S waves continue increasing downward because of the increasing pressure—the atoms are simply squeezed closer together, but no further changes in crystal structure are possible.

There is a sharp but not very smooth boundary between mantle and core. The core is believed to have a chemical composition similar to that of iron meteorites (Table 11.7). Because no S waves are transmitted through the outer core, the outer core must be liquid. Halfway down toward the center, the metals begin solidifying, forming a 450-km-thick mushy layer. The inner core is solid because of the dominant effect of pressure over temperature: Increasing pressure tends to solidify the liquid, whereas increasing temperature tends to liquefy the solid—and in this case pressure wins. Physical data pertinent to the Earth's interior are shown in Table A3.

Table 11.6. **The chemical and mechanical subdivisions of the Earth's interior**

Chemical subdivisions[a]
Crust
  oceanic: 7 km thick
    0–0.5 km: sediments
    0.5–1 km: basalt
    1–7 km: gabbro
  continental: 35 km thick (average); maximum thickness 80 km under Tibet
    0–1 km: sediments
    1–30?: granite-granodiorite
    30?–35: gabbro
Mantle: 7 or 35 km to 2,885 km of depth
  upper (to 670 km of depth): peridotite → pyrolite
  lower (670–2,885 km): pyrolite
Core: 2,885 km to center; Fe (90%), Ni (9%), S + P + C (1%)

Mechanical subdivisions[a]
Lithosphere: rigid outer shell fragmented into *plates*
  oceanic: 0–65 km
  continental: 0–120 km
Asthenosphere: 65 or 120 km to ∼200 km; less rigid (1% melt?)
Mesosphere: 200–2,885 km; rigid
  400 km: pyroxene → garnet structure + stishovite; olivine → spinel structure
  670 km: garnet → perovskite structure; olivine (spinel structure) → perovskite + periclase; no further phase
    changes below 670 km
Outer core: 2,885–4,720 km: liquid
Transition: 4,720–5,170 km: mushy
Inner core: 5,170–6,371 km: solid

[a]The depths are average values.

11.7. THE EARTH'S INTERNAL HEAT

The internal temperature of the Earth increases from 15°C at the surface (average) to 6,000°C at the center. In the Kola Peninsula ultradeep borehole the average temperature increase is 16.67°C/km down to 12 km. By extrapolation, the temperature at the base of the crust (35 km) should be 583°C. The sources of internal heat are as follows:

1. Residual heat of accumulation.

2. Continued trickling of heavy metals through the mantle into the core, and

3. Radioactive decay of long-lived radioactive isotopes (^{238}U, $t_{1/2} = 4.468 \cdot 10^9$ y; ^{235}U, $t_{1/2} = 704 \cdot 10^6$ y; ^{232}Th, $t_{1/2} = 14.05 \cdot 10^9$ y; ^{40}K, $t_{1/2} = 1.277 \cdot 10^9$ y).

The Earth formed by the infalling of a large number of planetesimals toward a common center of gravitation. The kinetic energy of the infalling planetesimals was transformed into heat upon impact.

If the Earth had accumulated from matter dispersed to infinity, its energy of accumulation should be $2.49 \cdot 10^{32}$ J $= 40,160$ J/g $= 9,600$ cal g.

Table 11.7. **Chemical composition of iron meteorites (average)**

Element	Mass percentage
Fe	90.6
Ni	7.9
Co	0.5
S	0.7
P	0.2
C	0.04
other	0.06
total	100.00

Table 11.8.   **Heat production by radioactive elements in rocks**

Rock type	Element concentration (10^{-6})			Heat production ($10^{-6}\,cal\,g^{-1}\,y^{-1}$)			
	U	Th	K	U	Th	K	Total
granite	4.7	18.0	40,000	3.4	4	1.08	8.48
diorite	2.6	9.0	25,000	1.9	1.8	0.67	4.37
gabbro	0.9	2.7	4,600	0.66	0.5	0.12	1.28
peridotite	0.015	0.01	300	0.011	0.002	0.008	0.021
chondrites	0.012	0.04	845	0.009	0.008	0.022	0.039
siderites	0.0001	0.0001	—	$9 \cdot 10^{-4}$	$8 \cdot 10^{-4}$	—	$1.7 \cdot 10^{-3}$

The specific heat of pyrolite is $0.2\,cal\,g^{-1}\,°C^{-1}$ (at $0°C$). Its temperature, therefore, should have risen to $9,600/0.2 = 48,000°C$. That did not happen, because

- the Earth did not accumulate from matter dispersed to infinity,
- early infalling planetesimals found only a small gravitational field and thus accumulated cold, and
- a lot of heat was dispersed during accumulation.

Nevertheless, the infalling of planetesimals, after the Earth had reached lunar size, must have created a lot of heat, most of which was radiated away, but some of which must have been trapped inside. At the end of the process of accumulation, the Earth may have had a temperature distribution inverse to the present one—coldest at the center, and hottest near the surface. It may have had an undifferentiated "core" of lunar size, an entirely molten mantle above, and a thin and probably broken "crust" consisting of refractory floats. That would have been a stable stratification. The Earth, however, contained a lot of metals (about one-third of its mass), and the metals, being heavy, trickled down, transferring heat downward and producing more heat by their fall. Today the internal heat of the Earth is believed to come from three sources, as mentioned above: the residual heat of accumulation, continuing accretion of the core, and the decay of long-lived radioactive isotopes. Table 11.8 shows the heat produced by radioactive decay. By analogy with the siderites, very little radiogenic heat is produced in the core. If pyrolite averages the chemical composition of chondrites, radiogenic heat production is presently $0.039 \cdot 10^{-6} \times 4.076 \cdot 10^{27}$ (mass of mantle + crust in grams) $= 1.590 \cdot 10^{20}$ cal/y. This would correspond to a heat flux through the Earth's surface of $1.590 \cdot 10^{20}/5.100 \cdot 10^{18} = 31.176\,cal\,cm^{-2}\,y^{-1} = 0.99 \cdot 10^{-6}\,cal\,cm^{-2}\,s^{-1}$. The measured geothermal flux is about 50% higher (Table 11.9). The measured value is higher because (1) there was more uranium, thorium, and potassium-40 in the past, (2) some of the original heat of accumulation is still coming out, and (3) the core may still be accreting.

The heat flux q is given by

$$q = \kappa(\Delta T/z) \qquad (5)$$

where q is heat flux per unit time, κ is thermal conductivity, z is vertical distance, and ΔT is temperature difference (K or $°C$). Measuring heat flux, therefore, involves measuring the temperature difference between two points along a vertical line within the solid Earth and determining the conductivity of the rock or sediment between the two points. (Clearly, a large temperature difference between two points separated by a thickness z of material with high thermal conductivity indicates a large flux; conversely, a small temperature difference coupled with low thermal conductivity of the intervening material indicates a

Table 11.9.   **Terrestrial heat flux**

oceanic $= 1.469 \cdot 10^{-6}\,cal\,cm^{-2}\,s^{-1}$
continental $= 1.460 \cdot 10^{-6}\,cal\,cm^{-2}\,s^{-1}$
world $= 1.467 \cdot 10^{-6}\,cal\,cm^{-2}\,s^{-1} = 6.142 \cdot 10^{-2}\,W\,m^{-2}$

low flux.) In the field, a probe is inserted into the ground, and after the heat produced by the insertion has dissipated, the temperature difference between two points along the probe is measured. The temperatures are measured by means of *thermistors*, which are ceramic semiconductors whose resistivity decreases with increasing temperature. The thermal conductivity of the rock is determined by heating a rock core at one end and by measuring the temperature rise along the core caused by that heating. The speed at which temperature rises is proportional to the conductivity of the rock. Equation (5) can then be solved and the flux determined. Because rocks may be inhomogeneous, probes with more than two thermistors usually are employed.

Determining the heat flux on the continents, other than at equatorial latitudes and low elevation, requires measurements in mines or boreholes deeper than 100 m, to eliminate the effect of seasonal temperature changes at the surface. Heat flux measurements at sea are made by dropping a probe on the ocean floor and allowing it to sink into the sediments (areas floored by hard rocks are avoided). The sediments on the ocean floor are soaked with, and covered by, seawater that has a low, constant temperature. The frictional heat produced by the insertion of the probe into the sediment is dissipated in a few minutes. Probes for work at sea are about 5 m long. Over vast areas of the deep-sea floor, the temperature difference between top and bottom of the probe is about 0.25°C or 0.05°C per meter.

The heat generated inside the Earth is dissipated by convection. Hot rocks are less dense and tend to rise, while rocks that have been cooled near the surface are denser and tend to sink. A convection cell may be 5,000 km across at the top of the mantle and 2,500 km across at the bottom, for a total perimeter of about 13,000 km (including the rising and sinking limbs). Mantle motions are on the order of 3 cm/y. At an average speed of 3 cm/y, a round-trip will take about 450 million years. The average rate of radiogenic heat production during the past 450 million years was 15% greater than the present rate, or about $1.83 \cdot 10^{20}$ cal/y, corresponding to a heat flux of $1.14 \cdot 10^{-6}$ cal cm^{-2} s^{-1}. That flux still is lower than the amount measured, indicating that about 0.3 cal cm^{-2} s^{-1} is due to continued dissipation of the primordial heat of accumulation and core accretion.

It should be noted that the present average flux of $6.14 \cdot 10^{-2}$ W/m^2 from the Earth's interior compares with a flux of 159 W/m^2 from the Sun (24-hour average at ground level). The flux from the Sun is therefore 2,589 times that from the interior of the Earth. Solar energy is used in part to energize the hydroatmosphere and the biosphere, with the balance returned to space. Internal energy is used to energize the solid Earth, including core and mantle convection, plate motions, mountain building, earthquakes, and volcanism.

THINK

* Consider an iron ball, 1 kg in mass, free-falling from a height of 100 m. The vertical distance z covered by the falling ball is given by

$$z = \tfrac{1}{2} g t^2$$

where g is the gravitational acceleration at the Earth's surface (9.81 m/s²), and t is time in seconds. Calculate how long it will take for the ball to hit the ground. The velocity of the ball at any time is given by

$$v = gt$$

Calculate the speed with which the ball hits the ground. Kinetic energy is given by

$$E = \tfrac{1}{2} m v^2$$

Calculate the energy that the ball has when hitting ground, and reduce the joules to calories (use Table A2). The specific heat of iron is 106 cal kg^{-1} °C^{-1}. Calculate the temperature increase of the iron ball that results from hitting the ground.

* Repeat the preceding calculation using an iron ball whose mass is 1 g. If your result is the same as before, how come? If not, how come?

* From the information that the temperature difference between top and bottom of a 5-m-long probe for work at sea is 0.25°C, and that the heat flux is $1.469 \cdot 10^{-6}$ cal cm^{-2} s^{-1} (Table 11.9), use equation (5) to calculate the heat conductivity of the sediment. Does your value match the value given in Table 9.10 for Globigerina ooze, a common deep-sea sediment? If not, why not?

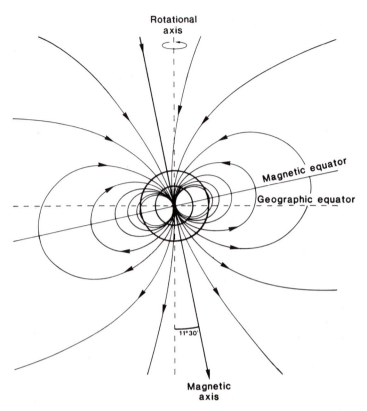

Figure 11.27. The Earth's geomagnetic field. The geomagnetic field is derived from the observed magnetic field by removing irregularities. The geomagnetic field is a dipole field, with the geomagnetic north pole in Antarctica (78°30′ S, 110°10′ E) and the geomagnetic south pole on the northwestern coast of Greenland (78°30′ N, 68°50′ W). The geomagnetic axis forms an angle of 11°30′ with the rotational axis, but it coincides with it if averaged across 10,000 y. (Emiliani, 1988, p. 183, Fig. 10.16.)

11.8. THE EARTH'S MAGNETIC FIELD

As we discussed in Sections 1.2 and 2.1, moving electric charges that are either free or bound to atoms in matter create magnetic fields. The SI unit of magnetic-field intensity is the tesla (T), equal to 1 newton/ampere·meter. Other units in common use in geophysics are the gauss (G), equal to 10^{-4} T, and the gamma (γ), equal to 10^{-5} G or 1 nanotesla (nT). The magnetic field through a point in space is represented by a vector pointing *from magnetic south to magnetic north*. The magnetic north and south poles are defined on the basis of the compass needle—the end of the needle that points toward geographic north is called the *north magnetic pole*, and the opposite pole of the needle is called the *south magnetic pole* (see Section 2.1). The north pole of a compass needle points north because it is attracted by the Earth's north magnetic pole, located in eastern Canada (77°18′N, 101°48′W). However, a north magnetic pole attracts a south magnetic pole, and vice versa. The pole in Canada that attracts the north magnetic pole of the needle (which we have used to define north magnetic pole) must, by definition, be the Earth's south magnetic pole. This is indeed the case (Fig. 11.27), even though everyone, including scientists and the *National Geographic*, calls it the "north magnetic pole." What follows, of course, is that the "south" magnetic pole of the Earth, located in eastern Antarctica (65.8°S, 139.0°E), is in fact the Earth's north magnetic pole.

In a bar magnet (Fig. 11.28), the north pole is sometimes called "positive" and the south pole "negative" because the magnetic field

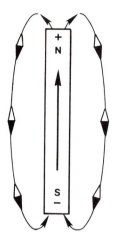

Figure 11.28. Directions of the magnetic field inside and in the vicinity of a bar magnet. Notice that the direction of the magnetic field is S to N inside the bar magnet (same direction as in a spinning proton), but N to S outside.

revealed by, for instance, a small compass needle placed in the vicinity of the bar magnet was supposed to result from a "flux" of magnetic "current" that would, of course, flow downhill, that is, from positive to negative (but notice that inside the magnet the flow would be uphill, i.e., from negative to positive). Fig. 2.2 shows the magnetic field created by the positive charge of a spinning proton. The direction of the magnetic vector (magnetic south to magnetic north, as defined above) is the same as that of the spin vector (m_s). For the electron, the magnetic vector is opposite that of the spin vector, because the electron has a negative charge. The magnetic field of the bar magnet is created by a flow of electrons in orbits that would appear counterclockwise as seen along the direction of the magnetic field within the bar.

Electromagnetic interaction is mediated by photons. Four basic points need to be remembered in order to understand magnetism. The first point is that a charged particle moving with *uniform* motions creates around itself a magnetic field due to continuous emission and reabsorption of virtual photons; the second point is that a magnetic field forces a change of direction in a charged particle moving at an angle through it; the third point is that a charged particle moving with *accelerated* motion

(except electrons within orbitals), produces an electromagnetic wave; the fourth point is that an electromagnetic wave accelerates a charged particle.

If electrons flow with uniform motion through a straight wire, a circular magnetic field is created around the wire that has a counterclockwise direction when viewed along the direction of motion of the electrons.

[Remember that the direction of an electrical current is by convention opposite the direction of the electron flow that generates it (see Section 1.2).]

Conversely, if the electrons move with uniform motion around a loop in a counterclockwise direction, a magnetic field is created along the axis of the loop that is directed along the line of sight when viewing the electron flow in the loop as counterclockwise (Fig. 11.29).

The magnetic properties of atoms, molecules, and bulk matter derive from the motions of the constituent particles. Protons spin at a constant rate equal to $\frac{1}{2}h/2\pi$ (where h is Planck's constant). Because the proton is charged with one unit of positive electricity, the spinning proton has a magnetic moment, which is equal to 2.7926 nuclear magnetons (see Section 2.1).

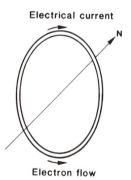

Electrical current

N

Electron flow

Figure 11.29. The electron-flow arrow shows the actual direction of motion of the electrons in a circular loop. The arrow labeled "electrical current" shows the direction of the current, which is, by convention, opposite that of the electron flow. The axis of the magnetic field created by the electron flow is normal to the loop and passes through the loop's center; its S-to-N direction is that in which the electron flow appears counterclockwise (or the current appears clockwise). (Emiliani, 1988, p. 186, Fig. 10.18.)

Table 11.10. **Average magnetic susceptibilities of common rocks**

Material	Susceptibility (gauss/oersted)
limestone	$1.4 \cdot 10^{-5}$
sandstone	$1.7 \cdot 10^{-5}$
granite	$2.7 \cdot 10^{-3}$
basalt	$1.4 \cdot 10^{-2}$
gabbro	$7.2 \cdot 10^{-3}$
peridotite	$1.4 \cdot 10^{-2}$

Electrons also spin at a constant rate of $\frac{1}{2}h/2\pi$. Because they, too, are charged particles, they have a magnetic moment, which is equal to 1.0011 Bohr magnetons (see Sections 1.2 and 2.1). For the same spin direction, the magnetic moments of the electron and the proton are opposite, because the charges are opposite (see Fig. 2.2). Electrons also orbit. The orbital motion of an electron creates a magnetic field through the center of the orbit, with direction along the line of sight when viewing the orbital motion as counterclock-

wise (Fig. 11.29). The magnetic moment of the revolving electron is equal to the magnetic moment of the spinning electron.

The magnetic susceptibility of a substance depends on the magnetic properties of its constituent atoms (or molecules) and their interactions. The magnetic susceptibilities of common minerals are shown in Table 9.12. The magnetic susceptibilities of common rocks vary considerably within a given rock type. Representative values are shown in Table 11.10.

The intensity of the magnetic field of the Earth at sea level ranges from 0.25 G at the magnetic equator to 0.60 G at the magnetic poles. By comparison, that of a small pocket magnet may range around 100 G.

The Earth's magnetic field is a distorted dipole field (Fig. 11.30), with the north magnetic pole (the actual south magnetic pole) presently at 77°18′N, 101°48′W (Bathurst Island, Canadian Arctic), and the south magnetic pole (the actual north magnetic pole) at 65°48′S, 139°6′E (on the Antarctic coast, south of Adelaide, Australia). The field presumably is created by electrical currents in the Earth's core, with a dominant net counterclockwise

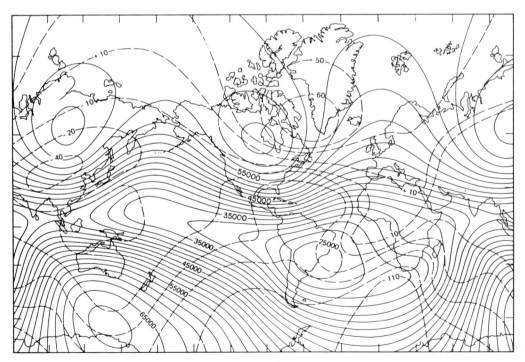

Figure 11.30. Intensity of the earth's magnetic field (solid lines) and annual rate of change (dashed lines). Values are in gammas. (Robinson and Coruh, 1988, p. 359, Fig. 10–22.)

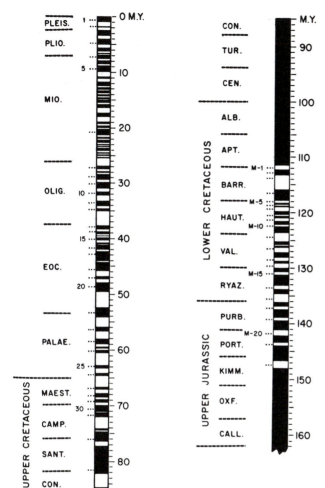

Figure 11.31. Geomagnetic polarity scale for the past 162·10⁶ y. (Larson and Pitman, 1972, p. 3,651.)

component (as seen from the geographic north). Convection in the outer core distorts this motion, superimposing on the dipole field a nondipole component and causing the relatively rapid motions of the magnetic poles ($\sim 5\,$km/y since 1831).

If short-term variations are filtered out, the dipole *geomagnetic* field is obtained, with the "north" geomagnetic pole at 78°30′N, 68°50′W, and the antipodal "south" geomagnetic pole at 78°30′S, 110°10′E. If averaged across 10,000 years of time, the geomagnetic axis coincides with the rotational axis, and the geomagnetic poles therefore coincide with the geographic poles.

The intensity of the Earth's magnetic field may rise as high as 1 G (at the magnetic poles). Occasionally the intensity decreases to zero and the polarity reverses itself. Reversals ap-

parently are caused by the electrical currents in the core decreasing to zero in speed and then starting again in the opposite direction. Reversals occur at intervals ranging from 10,000 years to 25 million years. During a reversal, the magnetic field decreases over a period of 10,000 years, reverses over a period of 1,000–2,000 years, and then returns to its previous intensity over another 10,000 years. The cause of these reversals is not known. The present polarity epoch (called *normal*) was established 730,000 years ago. It has been given the name Brunhes after the French geographer Jean Brunhes (1869–1930), who was the first to observe ancient magnetization in rocks (1906).

Fig. 11.31 shows the geomagnetic polarity scale for the past $162·10^6$ y. Two major polarity epochs, one in the Jurassic and the

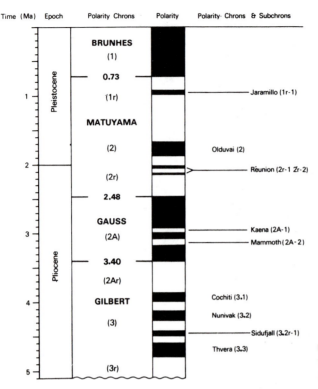

Time (Ma)   Epoch   Polarity Chrons   Polarity   Polarity· Chrons & Subchrons

Figure 11.32. Geomagnetic polarity scale for the past $5 \cdot 10^6$ y. (Harland et al., 1982, p. 66, Fig. 4.2.)

other in the Cretaceous, lasted more than $15 \cdot 10^6$ y each. Both happened to be normal. In the rest of the time, polarity epochs lasted less than a few million years. Fig. 11.32 shows in detail the behavior of the geomagnetic field during the past $4.5 \cdot 10^6$ y. As may be seen, some polarity events lasted less than 10,000 years. One such event (in this case, reverse polarity) occurred between 30,000 and 20,000 years ago. Polarity events are shorter than polarity epochs, as the illustration indicates. The last four epochs, ranging back to $6.5 \cdot 10^6$ y ago, are named after Jean Brunhes, Motonori Matuyama (1884–1956), Karl Gauss (1777–1855), and William Gilbert (1544–1603), all of whom studied terrestrial magnetism.

The direction of the Earth's magnetic field is measured with a free-swinging magnetic needle. The intensity of the field is commonly measured with a proton magnetometer.

[The *proton magnetometer* consists of a container filled with water (or other hydrogen-containing substance) and surrounded by a coil. The magnetic axes of the spinning protons, which normally are oriented at random, become aligned when a dc current flows through the coil. When the current is interrupted, the protons precess around the direction of the ambient magnetic field to resume random orientation (unless the induced field happens to be parallel to the ambient field). The frequency of precession, called *Larmor frequency*, is given by

$$v = HM/2\pi L \qquad (6)$$

where v is frequency, H is the intensity of the ambient magnetic field, M is the magnetic moment of the proton, and L is the angular momentum of the proton. $G = M/L$ is the gyromagnetic ratio of the proton $= 26.75221 \cdot 10^7$ Hz/tesla $= 0.2675221$ Hz/gamma. From (6) we have

$$H = 2\pi v/G = 23.48660\, v \qquad (7)$$

Equation (7) says, in other words, that a precessional frequency of 1 Hz is produced by an ambient field of 23.48660 gammas. For an ambient field of 0.5 G ($= 50,000\,\gamma$), a value common at midlatitudes, we have $v = 2,129$ Hz.

Precession lasts about 1 s. Precision is better than 0.5 γ, limited by the accuracy to which the proton gyromagnetic ratio is known.]

The origin and dynamics of the Earth's magnetic field still are not clear. The electron flow needed to support the magnetic field at its present intensity is about $4.4 \cdot 10^9$ amperes if it is in the form of a toroidal current in the outer core. In order to maintain this flow, an energy source is needed. Heat is the most obvious choice, but mechanical torques related to the precessional motions of the Earth's axis have also been suggested. The trick is to transform heat and/or mechanical energy into an electron flow. It is obvious that the Earth's rotation has something to do with the orientation of the magnetic field, because, as we have seen, the geomagnetic axis comes to coincide with the axis of rotation if averaged across 10,000 years. So, again averaged across 10,000 years, the electron flow is circling the rotational axis along the equatorial plane. Perhaps the flow that generates the Earth's magnetic field is only a *net* flow. There are a lot of free electrons in both the inner core and the outer core; there is a temperature gradient; and there is convection, at least in the outer core. The system is highly complicated in terms of electrical properties. It is likely that there are electrons flowing in all directions, except that there may be a minute excess flowing in a direction related to the Earth's rotation, that is either overtaking the Earth (and producing a normal magnetic epoch) or lagging behind (and producing a reversed magnetic epoch).

A popular model is that of the self-exciting dynamo: A conductor rotating in a magnetic field creates an electron flow. If the flow passes through a coil, the coil itself creates a magnetic field that keeps the flow moving as long as the conductor keeps rotating. Well, the Earth rotates. The core is conducting, but probably more here and less there because of temperature and/or compositional inhomogeneities. The more conducting portions could act as coils by funneling the electron flow. Although we may have here the ingredients for the creation of a magnetic field, we still do not know exactly how it works.

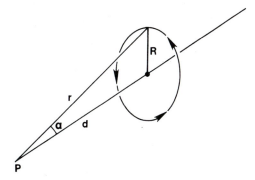

Figure 11.33. The intensity B of a magnetic field at point P located at distance d from the center of a circular loop of radius R along which a steady electron flow i occurs is given by $B = (\mu_0/2)(iR \sin \alpha/r^2)$, where $\mu_0 =$ permeability constant $(= 4\pi \cdot 10^{-7} = 12.566 \cdot 10^{-7}$ henry/meter), i = electron flow, $\alpha =$ angle at P between d and r $(= \arctan R/d)$, and r = distance between P and any point on the loop.

THINK

* Table 11.10 shows an increase in magnetic susceptibility from limestone to magnetite. Which component of the various rocks or minerals increases in the same direction?

* The intensity B of the magnetic field at a point P located at a distance d from the center of a circular loop of radius R along which an electron flow i occurs (Fig. 11.33) is given by

$$B = (\mu_0/2)(iR \sin \alpha/r^2)$$

where μ_0 is the permeability constant $(= 4\pi \cdot 10^{-7} = 12.566 \cdot 10^{-7}$ henry/meter), i is electron flow, α is the angle at P between d and r = arctan R/d, and r is the distance between P and any point on the loop. The intensity of the magnetic field at P along the axis of the loop is equal to B sin α. At the Earth's magnetic poles, B is presently equal to 0.6 gauss $(= 6 \cdot 10^{-5}$ tesla). For an electron flow along the equator of the outer core, we have R = $3.486 \cdot 10^6$ m; and r = $(R^2 + d^2)^{1/2} = 7.262 \cdot 10^6$. Calculate the magnitude of the electron flow (amperes), and compare it with the value mentioned in the text. Repeat the same calculation for a point on the Earth's magnetic equator, where B is 0.3 gauss. Any comment?

12 PLATE TECTONICS

12.1. PLATES

Plate tectonics is the unifying theory that shows how the Earth works. Many scientists have contributed to the development of this theory. A paper published in 1879 by Sir George Darwin (1845–1912), third son of Charles Darwin, suggested that the Moon may have split from the Earth in early times. That suggestion provoked others, most notably the Reverend Osmond Fisher, to propose (in 1882 and again in 1889) that the Pacific was the fission scar left by the departing Moon, that the continents on the opposite sides of the Pacific had moved toward each other to fill the scar, and that the westward motion of the Americas had split them from the Old World and opened the Atlantic. When those ideas were proposed, the Earth's mantle was presumed to be liquid. Later, in 1912 and 1915, the German meteorologist Alfred Wegener (1880–1930), using the geological evidence that had accumulated during the preceding thirty years, made a strong case for continental motion. Geologists balked, however, because seismic evidence had by then demonstrated that the mantle was solid. It would be impossible, it was maintained, to slide a continent sideways without breaking up the adjacent crust and mantle: just try to slide around a brick stuck in the middle of a brick floor! But then, in the early 1920s, Beno Gutenberg (1889–1960), a German geophysicist, demostrated that the viscosity of the solid mantle, although high, was insufficient to prevent convective motions *in the solid state*. Those motions could move continents around. Still, geologists were incredulous, at least those in the Northern Hemisphere (South African and Australian geologists, impressed with the evidence of simultaneous glaciation in their distant lands, were more favorable to the idea of continental motion). After World War II, geologists went to sea and began probing the morphology and structure of the deep-sea floor. It was soon found that the ocean floor, far from being a structureless plain, was highly structured (Fig. 12.1). Many great discoveries followed in rapid succession:

Flat-topped seamounts in the Pacific (Hess 1946)
A layer of soft rocks in the upper mantle (Gutenberg 1948)
The astonishing thinness of sediments on the deep Pacific floor (Pettersson 1949)
The absence of a sialic crust under the Atlantic (Ewing 1950)
The comparatively high heat flow in the Pacific (Revelle and Maxwell 1952)
Quasi-sinusoidal magnetic anomalies across the equatorial Atlantic (Heezen 1953)
Earthquake zones all around the Pacific, plunging from the ocean to under the continents (Benioff 1955)
A continuous band of shallow earthquakes under the crests of the mid-ocean ridges (Ewing and Heezen 1956)
Magnetic stripes on the Pacific floor (Mason 1958)
The global extent of the soft rock layer discovered by Gutenberg in 1948 (Press 1959)

In 1961, those discoveries led Robert Schmalz (b. 1929) at Pennsylvania State University and Harry Hess (1906–1969) at Princeton independently to suggest that the Earth's mantle was indeed convecting, that hot rocks surfaced along the axis of the mid-ocean-ridge system, and that the ocean floor spread away from the crests of the ridges and plunged under the continents. That model was named *sea-floor spreading* by Robert Dietz (1961). More discoveries followed:

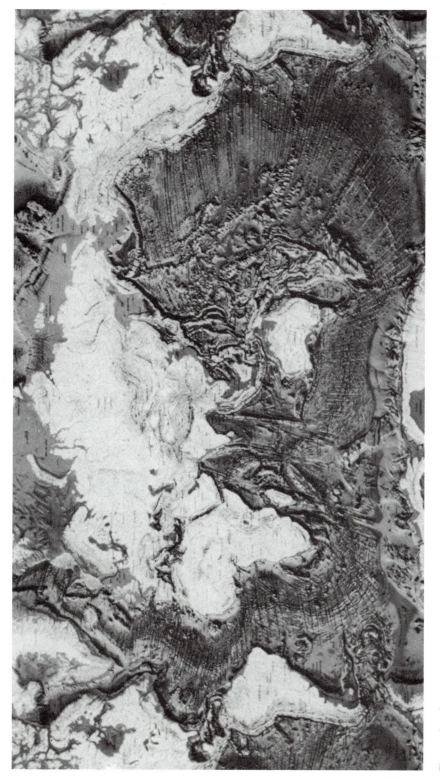

Figure 12.1. Physiography of the surface of the lithosphere. (Courtesy Marie Tharp, 1 Washington Ave., South Nyack, N.Y. 10960.)

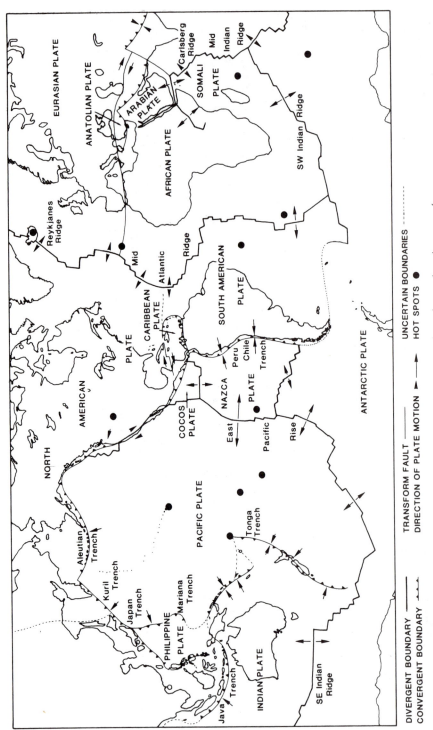

Figure 12.2. The plates of the Earth's lithosphere. The length of an arrow in millimeters gives the motion in centimeters per year. The black dots indicate the positions of the major hot spots active today. The litho-

spheric plates move on the asthenosphere because of its relatively lower viscosity (10^{21} poise vs. 10^{22} poise for the lithosphere or the mesosphere). (Emiliani, 1988, p. 176, Fig. 10.10.)

DIVERGENT BOUNDARY ——————
CONVERGENT BOUNDARY ⤙⤙⤙⤙

TRANSFORM FAULT ——————
DIRECTION OF PLATE MOTION ——→

UNCERTAIN BOUNDARIES ----------
HOT SPOTS ●

The paleomagnetic demonstration that North America had moved westward with respect to Europe by 20° (Runcorn 1962)

The demonstration that lavas from hot spots fixed in the mantle pierced the overlying crust as it moved across (Wilson 1963)

The explanation of the magnetic stripes of the oceanic crust as the result of crust being formed while the magnetic field kept reversing itself (Vine and Matthews 1963)

The astonishing symmetry of the magnetic stripes on the opposite sides of the Reykjanes Ridge crest (Heirtzler, Le Pichon, and Baron 1966).

Finally, in 1967, Dan McKenzie (b. 1942) of Cambridge University put it all together in the theory of plate tectonics: The Earth's lithosphere, the rigid layer above the soft rocks discovered by Beno Gutenberg in 1948, was broken into plates that moved with respect to one another, by sliding, by collision, by subduction, or simply by moving apart along the mid-ocean-ridge axes. In 1970 that model was proved correct by Arthur Maxwell and colleagues in the course of a deep-sea drilling expedition across the South Atlantic. They found that the ages of the deepest sediments coincided with the radiometric ages of the basalt immediately below, and those ages increased regularly away from the Mid–Atlantic ridge toward the South American continent. A mystery that had defied solution for more than 2,000 years—how the Earth works—was solved.

The plates into which the lithosphere is broken are shown in Fig. 12.2. The major plates are the African, Eurasian, Pacific, Indian, North American, South American, Antarctic, and Nazca. Lesser plates are the Cocos, Caribbean, Somali, Arabian, Philippine, and Scotia. In addition, there are a number of minor plates, microplates, plate fragments, and plate remains. Examples of the latter are the Juan de Fuca and Gorda plates off northwest North America, the remnants of the Kula plate that once floored most of the North Pacific.

Plates are about 120 km thick under the continents and 65 km thick under the oceans. The major plates range in size from 12,000 km across (Pacific plate) to 3,000 km across (Nazca plate). Compared with their surface areas, they are rather thin. If a plate had a surface equal to that of this page, its thickness would be 15

pages under the oceans and 60 pages under the continents. Aside from occasional magma chambers, the plates are not only solid but also rather rigid—they break if stressed too rapidly. Of the major plates, the Pacific and Nazca are entirely oceanic, while the Eurasian plate is mainly continental. The other plates are partly continental and partly oceanic.

Beneath the lithosphere lies the asthenosphere, which is about 1% molten and therefore acts as a lubricant layer over which the plates can slide. The underlying mantle is solid, but not perfectly rigid. Its viscosity is high, but like all solids, except perfect crystals at 0 K, it can flow if slowly stressed and given enough time.

Because of their high viscosity (see Section 10.2), rocks normally do not appear to be readily deformable, at least not on a human time scale. They seem to remain as they are forever or, if stressed too much, to break. However, closely folded rock layers are clearly visible in many mountain ranges, indicating that rocks are readily deformable (see Section 17.3). Bodies that respond rigidly to stresses that have a short time range will respond plastically if the same stresses are applied over a long time range. A glass rod will respond to an attempt to bend it by breaking, but the same glass rod will bend (and remain bent) if the stress is applied through a period of months. Given enough patience and time, one could twist a glass rod into a pretzel, even at room temperature (but one would have to exert the stress continuously for years).

Because the mantle has a high but not infinite viscosity, hot rocks in the hot spots deep in the mantle (Fig. 12.2), being less dense, tend to rise. Today, a complete convection cycle requires about 450 million years (Chapter 11). During the Precambrian, when the Earth's interior was hotter (because of the greater residual heat of accumulation and the greater amounts of radioactive isotopes, especially ^{235}U and ^{40}K), convection was faster, and all phenomena associated with it were also faster.

12.2. PLATE MOTIONS

Convection in the mantle forces the overlying lithospheric plates to move with respect to one another. Five hundred million years ago the continents were dispersed across the surface of the Earth (Fig. 12.3), but not in the positions

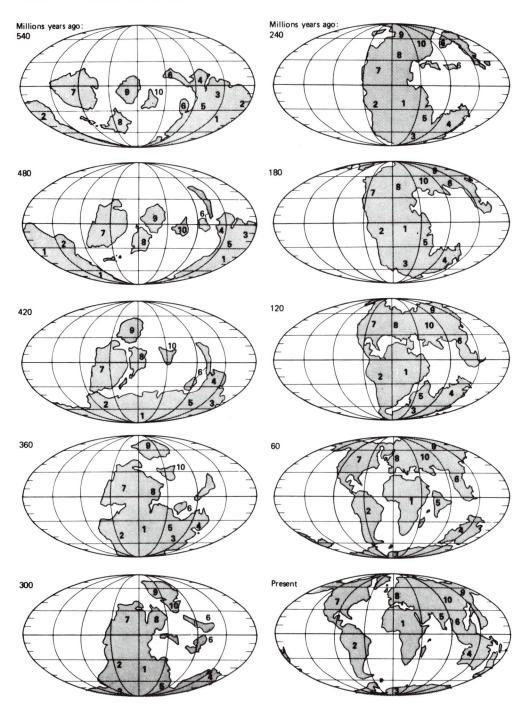

Figure 12.3. The motions of the continents during the past 540 million years. (Emiliani, 1987, p. 48; adapted from Siever 1983.)

they occupy today. The system of convection currents then operating in the mantle ended up by pushing all the continents together 200 million years ago. A single supercontinent, called Pangea, came into existence, together with a single superocean called Panthalassa. Panthalassa extended westward partway into Pangea, forming a broad gulf called Tethys. Tethys divided Pangea into two portions, a northern portion called Laurasia, and a southern portion called Gondwana. Such was the geography of the world 200 million years ago.

[About these names: Pangea derives from πᾶν (all) and γή (earth); thalassa derives from θάλασσα (sea); Tethys (Τήθυς) was a Greek Titaness, daughter of Uranus (Οὐρανός, Heaven) and Gaia (Γαῖα, poetic form of Γή, Earth), sister as well as wife of the Titan Oceanus (Ὠκεανός) and aunt of Zeus (Ζεύς); Laurasia is a combination of Laurentia (the Canadian shield) and Eurasia; Gondwana, meaning "the land of the Gonds", is a region in central India inhabited by the Gonds, a native tribe that ruled that area between the twelfth and eighteenth centuries.]

The huge landmass of Pangea, with its high content of uranium, thorium, and potassium, acted as a hot lid, impeding heat flow from inside the Earth. The trapped heat produced a welt right in the middle of Pangea, roughly where the Bahamas are now. The welt cracked, and the crack propagated along the axis of the welt both northward (separating North America from North Africa about 180 million years ago and from Europe about 150 million years ago) and southward (separating South America from Africa about 110 million years ago). The last cut was the separation of Greenland from Norway 65 million years ago, when Iceland began forming. When the breakup began, Tethys extended further westward, separating Laurasia from Gondwana. The original welt is now the Mid-Atlantic Ridge, whose axis is still an active site of basaltic outpourings. It is part of a globe-encircling system of active mid-ocean ridges (Fig. 12.1) where new lithosphere is created.

Four types of plate interactions are recognized:

Spreading: two plates move apart, leading to the surfacing of hot rocks and the formation of new lithosphere between the two

Subduction: oceanic lithosphere subducts (dives under) when colliding with less dense lithosphere (usually continental)

Collision: two plates with lithosphere of equal density collide

Transform: two plates slide by each other

The westward motion of the American plates is pushing them against the Pacific and Nazca plates. These two plates are entirely oceanic, and so they consist of thinner and denser lithosphere than the continental lithosphere they confront along the western margin of the American plates. The Pacific and Nazca plates are forced to subduct under the American plates, at an angle of 45–60° (Fig. 12.4). The Pacific plate also subducts under the Aleutians and all along the western border of the Pacific. There, the Pacific plate subducts under the less dense oceanic lithosphere of the marginal seas bordering the western Pacific. The Indian plate subducts under Indonesia, and portions of the Atlantic lithosphere subduct under the Caribbean and the Scotia plates.

Subduction under oceanic lithosphere produces lines of volcanic islands that are more or less arcuate, depending upon the angle of subduction. A high angle of subduction produces an almost straight line of volcanic islands (e.g., the Tonga–Kermadec line), whereas a shallower angle produces an arcuate structure (e.g., the West Indies island arc). You can see this for yourself: If you cut an apple at an angle of 90° (i.e., through the center), the trace of the cut on the surface appears as a straight line when seen from above (as on a map projection), but if you cut the apple at a shallow angle, the trace will be arcuate.

Subduction does not proceed smoothly and continuously; it proceeds by jerks, and each jerk produces an earthquake. Thus, the Pacific is ringed not only by subduction zones and accompanying volcanoes but also by earthquake zones (see Fig. 11.23). All subduction belts are marked by deep-sea trenches with strong (-200 mgal) negative gravity anomalies. The straightest subduction zones are along the Kuril–Kamchatka and the Tonga–Kermadec island lines. There, the oceanic plate descends at a steep angle of 60° to at least 700 km of depth before losing its thermal and seismic identity.

A cross section of a subduction zone involving two oceanic plates (Fig. 12.5) shows what

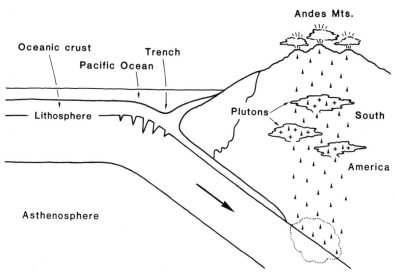

Figure 12.4. Subduction of the Nazca plate under the South American plate. (Emiliani, 1988, p. 180, Fig. 10.13.)

is going on. The subducting oceanic plate bends and cracks along the subduction margin; the cracks fill up with sediments, hydrated minerals, and water, and a deep-sea trench forms at the bend. The subducting plate plasters against the front of the opposite plate a mixture of sediments called *mélange* (which in French means *mixture*); in addition, the subducting plate carries down on its back sediments, hydrated minerals, and water. The water vapor is released and produces andesitic volcanism. The descending plate drags along some of the adjacent plate, creating tension behind the volcanic arc; tension breaks the lithosphere, and a back-

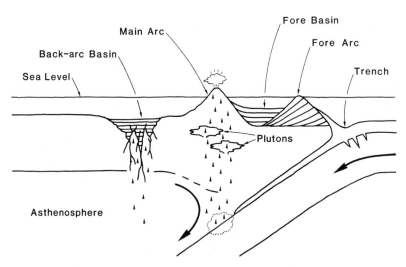

Figure 12.5. Subduction of the Pacific plate under the Philippine plate.

arc basin is formed; lavas pour out through cracks in the bottom of the basin. Granodioritic plutons separate by magmatic differentiation and are emplaced in the crust of the volcanic arc. The volcanoes keep producing lavas while weathering produces sediments that are deposited in both the trench and the back-arc basin. Continued pressure from the subducting plate may wrinkle the sediment blanket in the fore-arc basin and form a fore-arc sedimentary ridge that may rise above sea level. Continued sedimentation may fill up the back-arc basin. In the high-pressure, low-temperature belt beneath the fore-arc ridge, the sediments are metamorphosed into blueschists. Blueschists are laminated rocks (schists) characterized by the presence of the the metamorphic mineral glacophane, a hydrous Na–Mg aluminosilicate that gives them a bluish color. Toward the main arc, where temperatures are higher, the blueschists grade into greenschists. The greenschists are characterized by the mineral chlorite, a hydrous Mg–Fe aluminosilicate that gives them a greenish color. (See Chapter 17.)

When an oceanic plate subducts under a continental plate, as the Nazca plate does under South America, the same development takes place, except that tension caused by the subducting plate behind the line of volcanoes does not produce a back-arc basin (the continental lithosphere is too thick for that), but rather breaks up the continental lithosphere into a series of grabens and horsts.

[Plates are curved because the surface of the Earth is curved. The radius of curvature of a plate is normally the radius of the Earth. If a welt forms because a plume of hot rock rises beneath a plate, the welt will have a smaller radius of curvature (normally several hundred kilometers). This will create tension, and the plate will break into blocks with slanted sides (Fig. 12.6; see also Fig. 11.4). Blocks bound by surfaces slanted inward are top-heavy and sink a bit into the asthenosphere in order to maintain isostatic equilibrium. The depressions are called *grabens*. Blocks that are bound by surfaces slanted outward are bottom-heavy and tend to rise. They are called *horsts*.]

When two plates meet and each is carrying a continent at its leading margin, neither plate can subduct. Plate collision follows, and the lithosphere is wrinkled and thickened (Fig. 12.7). Examples of collision margins include the margin between Africa and Europe and that between India and Tibet. The push of Africa has created the Alps, and the push of India has created the Himalayas and has thickened the crust under Tibet to a record 80 km.

When two plates collide, a section of oceanic crust and upper mantle, called an *ophiolite complex*, may be caught between the two. The structure of a typical ophiolite complex is shown in Table 12.1. Below lherzolite is pyrolite (olivine, pyroxene, Ca plagioclase), which is undifferentiated mantle rock. Serpentinite, the pro-

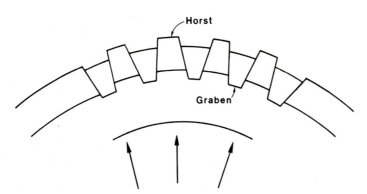

Figure 12.6. The formation of grabens and horsts. Swelling breaks the lithosphere into elongated blocks. In order to obey Archimedes' principle and to maintain isostatic equilibrium, blocks tapering upward (*horsts*) rise, and those tapering downward (*grabens*) sink (cf. Fig. 11.4).

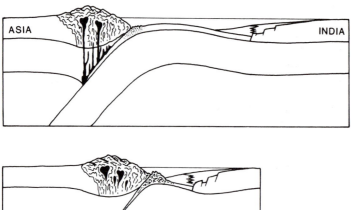

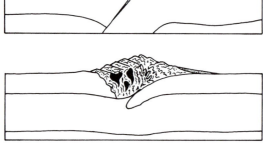

Figure 12.7. The collision of India with Asia began 40 million years ago (top) and formed the Himalayas (bottom). (Emiliani, 1988, p. 179, Fig. 10.12.)

duct of hydrothermal alteration of submarine basalts and ultramafics, is common in ophiolite complexes. Serpentinite consists mainly of the mineral serpentine $[(Mg, Fe)_3Si_2O_5(OH)_4]$, with accessory chlorite, talc, and carbonates. The name *ophiolite* derives from the Greek ὄφις, meaning *serpent*; *serpentine* derives from the Latin *serpens*, also meaning *serpent*. Polished serpentinite, in fact, has the appearance of a shiny snake-skin surface. Examples of ophiolite complexes are the Troodos Massif of Cyprus and the Bay of Islands, Newfoundland.

Hot spots deep in the mantle produce plumes that can reach the surface in two ways. If the plume is under a plate, the rising rocks will pierce the lithosphere and form a midplate volcano. As the plate moves, the rising rocks will keep piercing the plate, forming a series of volcanoes. A classic example is the Hawaiian–Emperor chain, which began forming at least 65 million years ago. The hot spot is now directly under Hawaii, but 65 million years ago the area that is now abutting Kamchatka was above the hot spot. At that time, the Pacific plate was moving almost directly northward. Between 65 and 40 million years ago, a series of volcanoes

was formed, the remnants of which form the Emperor chain of dead, sunken volcanoes (Fig. 12.8). Forty million years ago the direction of the Pacific plate changed from north to northwest, and the Hawaiian chain formed. The volcanism has not been continuous: a new volcano has formed every two million years or so.

If a hot spot lies under a spreading axis, the plume changes shape, from cylindrical to highly elliptical, because the line of least resistance for the plume to the surface is along the elongated spreading axis. Presumably, a single hot spot may feed a ridge segment hundreds or even thousands of kilometers long on either side of the spot. The basaltic lavas are emplaced as dikes along the axis of the mid-ocean-ridge system.

When a dike solidifies, the iron particles align themselves along the magnetic field of the Earth at the time of emplacement. A new dike will cut right through the middle of an earlier dike, because that is where the older lava will still be warm and therefore more easily intruded.

As spreading continues and new dikes are

Table 12.1. Structure of a typical ophiolite complex

Depth below sea floor (km)	Rock	Mineral composition
0–0.5	basalt	Ca plagioclase, pyroxene
0.5–2	diabase dikes	Ca plagioclase, pyroxene
2–7	gabbro	Ca plagioclase, pyroxene
7 to ~20–25	harzburgite	clinopyroxene, orthopyroxene, olivine
20–25 to ~40	lherzolite	olivine, orthopyroxene, clinopyroxene

emplaced, the direction of the magnetic field of the Earth is recorded in the rocks. If the magnetic field reverses itself, the rocks will record reverse magnetization. Fig. 12.9 shows the magnetic record of the sea floor across the axis of the Mid-Atlantic Ridge south of Iceland. The black stripes indicate rocks emplaced during times of normal magnetization, and the white stripes indicate rocks emplaced during times of reversed magnetization. The record is symmetrical about the axis of the ridge. The central stripe of normal magnetization is 14 km wide. The last magnetic reversal occurred 730,000 years ago. The rate of spreading south of Iceland is therefore $14/730,000 = 2 \cdot 10^{-5}$ km/y $= 2$ cm/y.

The magnetic stripes of the ocean floor were discovered by towing magnetometers on a course normal to a ridge axis. It was found that the ambient magnetic field (say 0.5 gauss) had superimposed on it oscillations amounting to $\pm 250 \ \gamma$ (1 $\gamma = 10^{-5}$ gauss), or about $\pm 0.5\%$. The ambient field decreases by ~0.5% each time the magnetometer passes over a strip of ocean floor with lavas magnetized in the sense opposite that of the present field; it increases by the same amount each time the magnetometer passes over a strip of ocean floor with lavas magnetized in the same direction as the present field. Knowing the ages of the magnetic reversals, it is possible to determine the age of

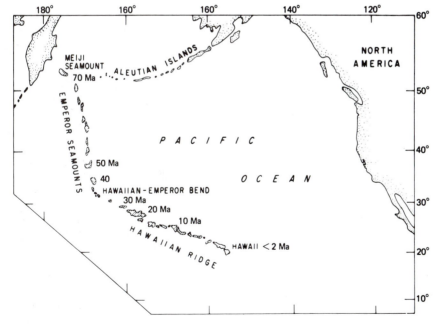

Figure 12.8. The Hawaiian–Emperor chain. (Kennett, 1982, p. 164, Fig. 5.19.)

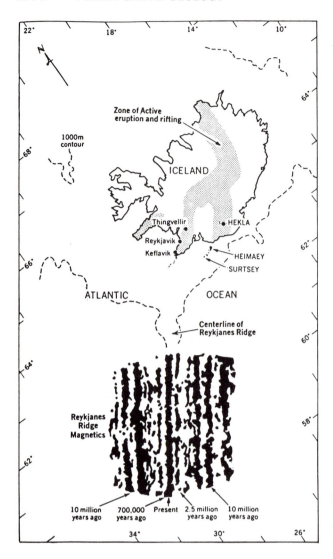

Figure 12.9. Sea-floor spreading on the Reykjanes Ridge, a portion of the Mid-Atlantic Ridge south of Iceland. (Anderson, 1986, p. 55, Fig. 3.24.)

any area of the ocean floor simply from its magnetic pattern (Fig. 12.10). The youngest floors, less than one million years old, are along the crests of the mid-ocean ridges. The oldest sea floors, dating from 160 million years ago, are found off the Bahamas and off Japan. The sea floor is continuously being renewed, with new crust forming along the mid-ocean ridges, and old crust subducting. By comparison, the continental lithosphere, which cannot subduct because it is too light, includes rocks that are more than three billion years old.

Spreading rates range from 2 cm/y (northern Mid-Atlantic Ridge) to 18 cm/y (East Paci-

fic Rise). Plate motions are half as much, ranging from 1 to 4 cm/y. Plate motions are so sluggish, and the inertia so large, that geologists can predict where the continents will be 50 million years in the future; the map (Fig. 12.11) is certainly more accurate than the map predicting tomorrow's weather.

The spreading along a mid-ocean-ridge crest may not be uniform along the length of the ridge. If there is a change in the spreading rate, a break parallel to the direction of spreading will occur, because surface rocks cannot deform plastically (Fig. 12.12A) at the speed of sea-floor spreading. Moreover, if the spreading

Figure 12.10. The age of the sea floor. (Larson and Pitman, 1988.)

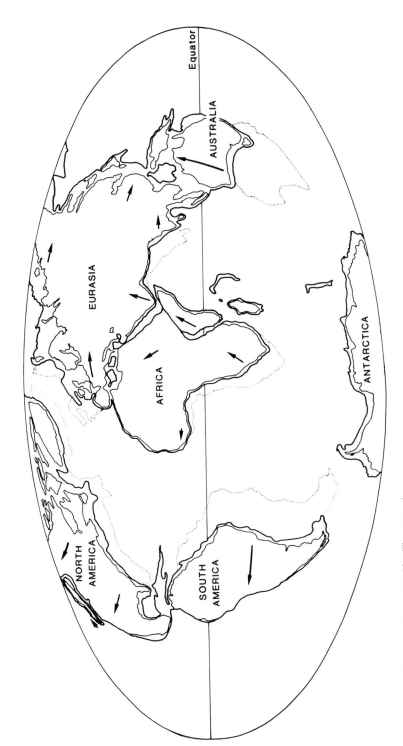

Figure 12.11. The map of the world 50 million years in the future. (Emiliani, 1988, p. 192, Fig. 10.22.)

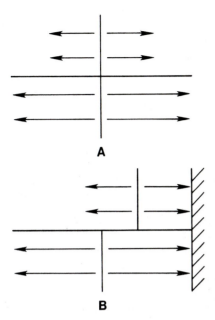

A

B

Figure 12.12. Transform faults form between spreading blocks with different spreading rates. If there is no confinement (A), there is no ridge offset. Confinement on one side (B) leads to ridge offset. The spreading motion of the two blocks are in opposite directions within the offset, but in the same direction beyond the offset.

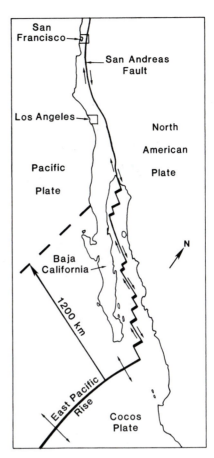

Figure 12.13. The Gulf of California was formed starting $15 \cdot 10^6$ y ago by a series of transform faults that have chopped up the crest of the East Pacific Rise. The spreading rate has been and is 8 cm/y. Spreading segments and intervening transform faults separate the Pacific plate to the west from the North American plate to the east. (Adapted from Elders et al., 1972.)

lithosphere cannot subduct (e.g., the lithosphere of the eastern Atlantic) a change in the spreading rate will result in a ridge offset (Fig. 12.12B). The resulting fault is called a *transform fault*, because motions that are in opposite directions within the offset transform into parallel motions beyond.

In three places around the world a mid-ocean ridge abuts against a continent: in Baja California, in the Gulf of Aden, and off the Lena River valley in Siberia. In all three cases the continental lithosphere is being rifted. Fig. 12.13 shows what happens in the Gulf of California. The crest of the East Pacific Rise approaches the coast of Mexico at an angle. The right-limb push against the Mexican lithosphere displaces the crest of the Rise northwestward by a series of transform faults, the longest of which is the San Andreas fault (Fig. 12.14). At a rate of spreading of 8 cm/y, the segmented crest of the East Pacific Rise has created the 1,200-km-long Gulf of California in 15 million years.

The Mid-Atlantic Ridge continues northward beyond Iceland, crosses the Arctic, and hits Siberia at the mouth of the Lena River, suggesting that the Lena River valley may be an incipient rift valley.

The Carlsberg Ridge (the northern continuation of the Mid-Indian Ocean Ridge) enters the Gulf of Aden and divides into two branches (Fig. 12.15). One branch enters Ethiopia and continues southward to form the rift valley of East Africa. The other branch continues northward through the Red Sea, the Gulf of Aqaba, the Dead Sea, and the Jordan River valley.

Figure 12.14. The San Andreas fault looking north-west. The fault separates the Pacific plate to the left from the North American plate to the right. (Courtesy Robert E. Wallace, U.S. Geological Survey, Menlo Park, Calif.)

Until 15 million years ago, Arabia was part of Africa. The Red Sea began forming at about that time and has since grown to its present size.

12.3. CONTINENTAL MARGINS

Continents are carried around by plates. A continental margin that is along a plate margin where subduction, collision, or transform-fault motion is occurring is termed *active*. A continental margin that is moving away from a spreading axis is termed *passive*. An example of a passive continental margin is the east coast of North America. A passive margin keeps accumulating sediments as the continent moves away from the spreading axis. The lithosphere becomes increasingly cooler and denser, and so the ocean floor off the passive margin becomes increasingly deeper. When the Atlantic was still narrow, it was also shallow. Coral reefs began growing along the eastern coast of North America because of the favorable oceanic circulation (warm water currents from the south). The reefs kept growing as the continent moved away from the ridge axis and the basement

deepened. South Florida and the Bahamas are parts of an enormous carbonate platform 5 km thick and still growing. The underlying lithosphere is cold and therefore quite stable—no earthquakes have ever been registered in south Florida or the Bahamas.

Clastic sediments brought in by rivers also accumulate along passive margins. Sediments have an average density of $2.5 \, \text{g/cm}^3$ compared with a density of $3.3 \, \text{g/cm}^3$ in the underlying mantle. If 5 km of seawater are replaced with 5 km of sediments off a passive margin, the sea floor receives an extra load. Because the density of the upper mantle is 25% greater than that of the sediments, a sediment column 5 km thick will cause subsidence by $5 \times 0.25 = 1.25$ km. If sediments continue to be added, a column of sediment may be deposited with a terminal thickness that will be $5 + (5 \times 0.25) + [(5 \times 0.25) \times 0.25] + \ldots = 6.65$ km. This thick belt of sediments along the border of a passive margin is called a *geosyncline*. The name implies that there was a trough before the sediments started pouring in (*syncline* derives from the Greek σύν, which means *together*, and κλίνειν, which means *to slant*). Actually there was no trough at all; it was the load of sediments that created the trough. The name geosyncline has now been replaced by the name *geocline*.

The landward portion of a geocline contains predominantly sands that have lost much of the less resistant minerals (feldspars, pyroxenes, amphiboles) and thus are enriched in quartz together with shales and, where appropriate, reefs. This portion is called a *miogeocline*. The seaward portion of the geocline contains rapidly deposited sands that have conserved the less resistant minerals, together with deep-water muds. This portion, called a *eugeocline*, often is intruded with oceanic magmas from below.

More than 6.6 km of rocks can accumulate, however. Geoclinal sediment accumulations up to 10–15 km in thickness are known, both fossil (e.g., Swaziland) and modern (e.g., the eastern coast of North America, the Gulf Coast). In order to accommodate this great thickness, the bottom of the geocline must sink for reasons other than load. A possible mechanism may result from the motion of the plate itself away from the spreading axis. There is a sharp transition in plate thickness between the continental portion, which is 120 km thick, and the oceanic portion, which is 65 km thick. As a

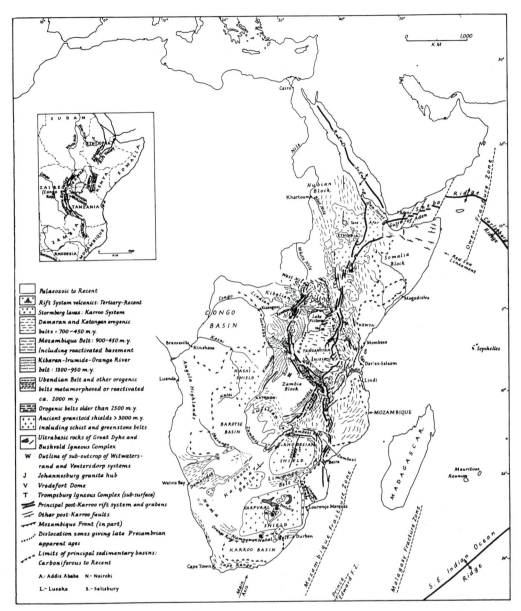

Figure 12.15. A westward extension of the Carlsberg Ridge began opening the Gulf of Aden and the Red Sea 15·10^6 y ago. A branch of the spreading axis extends southward through Ethiopia and forms the rift valley of East Africa. (McConnell, 1972, pp. 2549–72, Fig. 1.)

plate moves away from a spreading axis, the sloping underside of the transition zone creates a mass deficiency that is compensated by accommodating more sediment (Fig. 12.16). Compensation is not total, however, as is indicated by the persistence of negative isostatic anomalies.

The transitional lithosphere is stressed, especially where it thins out and becomes oceanic, allowing volcanic intrusions in the eugeocline. If the regime of mantle convection changes and the spreading ceases, a new convection system may transform a passive continental margin into an active margin, either by

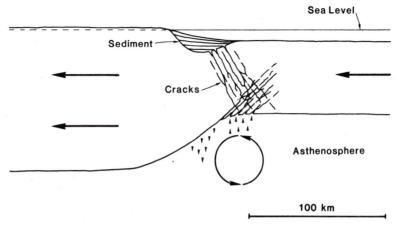

Figure 12.16. The North American plate is continental to the west, with a thickness of 125 km, and oceanic to the east, with a thickness of 65 km. As the plate moves westward, the sloping underside of the transition zone generates a mass deficiency in the adjacent astheno-sphere. The geocline is pulled down, leading to thick sediment accumulation. Stressing cracks the transitional lithosphere, allowing intrusions to rise and penetrate the seaward portion of the geocline (eugeocline).

developing subduction against it or by ramming it against another continental margin. In either case, the pile of sediments will be wrinkled up and transformed into a mountain range. The Alps, for instance, have been shortened by about 150 km with respect to the width of the original geocline into which the sediments accumulated.

12.4. MIDPLATE SEISMICITY

Because of the intense activity occurring at their margins, plates are under continuous stress. The continental lithosphere, being thick and relatively cold, can accumulate a large amount of strain energy. Eventually a break will occur, resulting in a midplate earthquake that can be very strong. The New Madrid earthquakes, which struck the middle Mississippi Valley in the winter of 1811–1812, are a classic example of this phenomenon. The town of New Madrid is located in Missouri, on the west bank of the Mississippi River, about 250 km north of Memphis. It had fewer than a thousand inhabitants at that time; it has 3,000 inhabitants today. The area was densely forested and sparsely settled. The first earthquake struck at 2 a.m. on December 6, 1811. Log cabins splintered and collapsed, and trees crashed

to the ground. Fissures up to 10 m across and generally trending N–S opened in the ground. Between fissures the ground was compressed into ridges, the tops of which exploded, throwing loose sediment into the air. The intense shaking turned the Mississippi River into a stormy sea, and flashes like lightning were seen in the sky. A second earthquake of similar magnitude struck on January 23, 1812. Aftershocks of decreasing magnitude lasted a few days after each of those two earthquakes. Then, on February 7, 1812, a third, more powerful earthquake occurred. The ground was in practically constant motion for several days, and aftershocks continued for more than a year. Those earthquakes produced profound changes in the landscape, including vertical displacements of up to several meters. Some of these changes are still visible today. One of the fissures is now Lake St. Francis, 60 km long and less than 1 km wide; another is Reelfoot Lake in Tennessee, 30 km long, 10 km wide, and 8 m deep. In spite of all that havoc, only a few people were killed.

The fact that a continental plate is relatively cold and rigid not only makes it a likely site for strong earthquakes but also enables the earthquake waves to propagate to great distances. The New Madrid earthquakes rattled windows and shook chandeliers as far as New Orleans,

Montreal, and Washington. Church bells started ringing in Richmond, Virginia, and clock pendulums were stopped in Charleston, South Carolina. The New Madrid earthquakes were the strongest earthquakes to hit North America in historical time. There were no seismographs at the time, but it has been estimated that the earthquakes ranged around magnitude 9 on the Richter scale.

The great Lisbon earthquakes on November 1, 1755, are another example of midplate seismicity. The first earthquake struck at 9:40 A.M., destroying much of the city. A second earthquake struck at 10 A.M., and a third at noon. The epicenter was at sea, some 50 km west of Lisbon. The first shock generated a tsunami that was 15 m high when it crashed on the waterfront during the second shock, demolishing a pier on which hundreds of people had sought refuge from the destruction. The great sea wave raced across the center of the city, bounced against the hills surrounding it, and withdrew, carrying along a full load of debris and corpses. The Lisbon earthquakes were felt to a distance of more than 2,000 km across both Europe and North Africa. Standing waves (called *seiches*) were formed in European lakes as far away as Finland. The destruction of Lisbon was completed by fire, which raged through the city for six days. Some 60,000 people died.

After a major break occurs within a plate, strain energy begins accumulating again. The next break is likely to occur in the same place, because there the lithosphere has been weakened. Indeed, Lisbon had seen major earthquakes before: in 1009, in 1344, and in 1531. Intervals of 335, 187, and 224 years separate these earthquakes and that of 1755. Another 235 years have passed since 1755, suggesting that another catastrophic earthquake may soon hit Lisbon. The same may be said of the middle Mississippi Valley. Midplate earthquakes are less frequent, but stronger; plate-margin earthquakes are weaker, but more frequent.

12.5. WHAT PLATE TECTONICS EXPLAINS

Plate tectonics explains not only the distribution of continents and oceans, the distribution of mountain ranges, volcanoes, earthquakes, and so forth, but also the distribution of economic resources, the distribution of life on Earth, and the distribution of sediments and geomorphic features relating to climate. Here are four examples.

El Dorado

El Dorado was a legendary kingdom or city-state in Latin America, said to be rich in gold and other precious materials. It was the goal of the conquistadores. Although there was no El Dorado, there were great mineral resources. Plate tectonics tells why. Fig. 12.3 shows that the western coast of the Americas has been abutting oceanic lithosphere for the past 300 million years, and the geological evidence indicates that the western coast of the Americas has been an active continental margin for that long. The subducting oceanic lithosphere originates on the crest of a mid-ocean ridge, in this case the East Pacific Rise. Welting and consequent cracking along the axis of the ridge allow seawater to penetrate through the cracks to a depth of at least 5 km. Water is heated in contact with hot rocks and returns to the surface through vents along the ridge crest (Fig. 12.17). These vents eject hot water plus a wealth of mineral particles (especially sulfides and chlorides) that have been leached by the hot water from the rocks below. Some of the particles accumulate in situ; others are scattered around by deep ocean currents and accumulate on the adjacent sea floor, together with clay particles and the remains of planktic organisms. The oceanic lithosphere reaching the western coast of the Americas subducts, carrying down a layer of sediments and metal particles several hundreds meters thick. As temperature rises, this mixture melts, magma is made, volcanoes erupt, and mineralizations are formed. It is a sort of giant still. The ascending fluids also leach the overlying continental crust, adding to the overall mineralization. That is why the western borders of the Americas are rich in many minerals. Ancient mineral districts originated in a similar way—along ancient subduction margins.

[*Mineralization* is simply the concentration in one spot of elements and/or compounds that are economically valuable. The concentration needed for a mineral deposit to be commercially exploitable depends on the mineral. Iron ore

Figure 12.17. A vent along the East Pacific Rise. Hot, mineralized water rises through the vent, carrying along H_2S and other metal sulfides. Sulfur bacteria thrive in this environment, drawing their energy from the oxidation of H_2S. These bacteria are at the bottom of a food pyramid that includes worms, giant clams, and fishes. This picture was taken by Dudley Foster, Woods Hole Oceanographic Institution, from the submersible *Alvin*. (Courtesy Dudley Foster, Woods Hole Oceanographic Institution.)

normally contains 30% or more of iron; zinc and lead ores contain 2–3% of these metals; copper ore contains 0.5% copper. If a rock is to be exploitable for gold, it must contain at least 0.0003% gold. The concentration process in nature that leads to the formation of ores, is carried out by percolating fluids. Weathering and transport by running water can concentrate mineral particles, forming *placer* deposits. A concentration of only 0.00003% gold may be sufficient to make a placer deposit exploitable.]

Black Gold

Black gold (i.e., petroleum) is making some nations quite wealthy. Plate tectonics explains why there is so much petroleum in the Middle East. The map of the world as it was 180 million years ago (Fig. 12.3) shows that a wide gulf (Tethys) was separating southwestern Asia from North Africa. Planktic life was thriving in the warm water of Tethys, and the re-

mains of those organisms accumulated on the sea floor, together with the sediments that were pouring in from southwestern Asia and North Africa. In due time the organic matter deep in the sediments was cooked into petroleum and natural gas (see Section 17.2). As the African landmass (which included Arabia) approached Eurasia, the sediment layers were gently folded and fractured. Petroleum seeped into the fractures from the sediments below, where it had formed. Layers of salt, anhydrite, and gypsum, deposited during times when Tethys had become temporarily isolated from the world ocean, could not be fractured, because these rocks deform plastically. They formed a tight cap, trapping the oil below. The gentle folds of the Middle East now contain the world's largest petroleum reserves.

Marsupials

Marsupials evolved about 90 million years ago in North America, and from there they expanded to South America, western Antarctica, Australia, New Guinea, and Celebes, all of which were adjacent to one another (Fig. 12.3). Placentals evolved 20 million years later in Eurasia. From there, they spread to North America, Africa, Madagascar, and India. They could not reach South America or Australia, however, because those two continents had become separated from the rest. There the marsupials flourished.

[*Marsupials* are primitive mammals in which the baby is born immature and is kept in a pouch on the mother's abdomen. *Placentals* are advanced mammals, including humans, in which the fetus draws nutrients from the mother and returns to her its waste products via a specialized exchange organ called the *placenta*; this allows the fetus to reach a more advanced stage before birth.]

About 4 million years ago North America and South America became joined by the Central American isthmus, which arose from the sea floor. Through that land bridge the North American placentals invaded South America. There they replaced all marsupials except opossums. There are still 65 species of opossums in South America, one of which has crossed into North America. Australia has remained isolated, and today it has 183 species of marsupials. Placentals have arrived, though, imported by immigrants. Eventually the Australian marsupials may suffer the same fate as their South American cousins.

Ice

Landforms clearly produced by the actions of ice have been found in the western Sahara, dating from about 400 million years ago, and in eastern South America, South Africa, Madagascar, India, and southern Australia,

dating from 300 to 250 million years ago. These landforms are similar to those produced in northern North America and northern Europe during the recent ice ages.

These findings are unexplainable in terms of the present geography of the world. However, if you look at Fig. 12.3, you will see that 400 million years ago North Africa lay right at the South Pole. Later, around 300 million years ago, South America, South Africa, Madagascar, India, and Australia were all close to the South Pole. When Pangea broke up and all continents (except Antarctica) began moving northward, ice largely (if not totally) disappeared from Earth. Today, North America and Eurasia are jammed against each other around the North Pole, and another major glaciation is in progress (see Chapter 24).

THINK

* Which planets, in addition to the Earth, should have plates and plate tectonics?

* The African plate has a surface of $6\cdot10^7$ km^2. Assuming an average thickness of 100 km, calculate its volume (in cm^3), and assuming a density of 3 g/cm^3, calculate its mass (in kg). Assuming an average northward velocity of 2 cm/y, and knowing that kinetic energy $= \frac{1}{2}mv^2$, calculate how many 100-watt light bulbs you could operate, and for how long, with the power obtained by stopping the African plate.

* An earthquake of magnitude 9 on the Richter scale may dissipate about 10^{19} joules of energy. Knowing that common asteroidal velocities are 20 km/s, that the density of a common-type asteroid is 3 g/cm^3, and that $E = \frac{1}{2}mv^2$, calculate the diameter of an asteroid that would dissipate a similar amount of energy if it hit the Earth.

13 THE ATMOSPHERE

13.1. RADIATION BALANCE

The atmosphere is the gaseous envelope of the Earth. It is energized by the Sun. The amount of solar power received by a unit surface perpendicular to the Sun's rays at a mean distance of 1 AU is called the *solar constant*. The actual power received by the Earth's surface is called *insolation*.

The solar constant is not really constant. The power emitted by the Sun was much greater in the early stage, when the Sun was dissipating its own heat of accumulation. Since the Sun stabilized, more than 4 billion years ago, the solar constant has increased by about 25% because of core contraction as more and more helium formed. The increase since the beginning of the Cambrian, 600 million years ago, has been about 3%.

In addition to this long-range increase, the solar constant is quite variable in ultraviolet, radio, and corpuscular emission on a short time scale (days to weeks). Because this emission accounts for only a minute percentage of the total solar power, the total power is practically constant. The solar flux at the distance of the Earth from the Sun amounts to $1,360 \, \text{W/m}^2 = 1.950 \, \text{cal/cm}^{-2} \, \text{m}^{-1} = 1.950$ langleys per minute (the *langley* is a unit of insolation equal to $1 \, \text{cal cm}^{-2} \, \text{m}^{-1}$). The solar flux is therefore much larger than the geothermal flux, which averages $0.614 \cdot 10^{-2} \, \text{W/m}^{-2} = 1.467 \cdot 10^{-6} \, \text{cal cm}^{-2} \, \text{s}^{-1}$). The average insolation of the total Earth surface if there were no atmosphere would be $1,360/4 = 340 \, \text{W/m}^2$ (because the cross section of a sphere is equal to πr^2, whereas the surface of a sphere is equal to $4\pi r^2$).

Atoms and molecules of atmospheric gases absorb and reemit, and/or scatter, solar radia-tion of specific wavelengths. A small charged particle with negligible inertia (e.g., an electron) in the path of an electromagnetic wave will oscillate with the frequency of the wave and emit radiation of the same frequency. Reemission is on a 4π geometry (meaning in all directions, 4π being the angle at a center of a sphere that subtends the sphere; compare with 2π geometry, which means in all directions on a plane, 2π being the angle at the center of the circle that subtends the circle).

Light is scattered when it encounters parti-cles or transient, random molecular groupings of a size similar to its wavelength (such group-ings form because of statistical density fluctua-tions of the medium). Scattering is maximum along the direction of the beam or in the op-posite direction; it is minimal at 90°. If the diameter of the scattering particle is 0.1λ, that is, less than 1/10 of the wavelength λ of the incident radiation ($< 50 \, \text{nm}$ for solar light), scattering is inversely proportional to the fourth power of the wavelength (*Rayleigh law*). The predominant scattering of the shorter-wavelength component of solar light makes the sky blue and the Sun yellowish when high on the horizon. The Sun is red when low on the horizon because solar light has to cross more atmosphere and because there are more aero-sols near the ground. Aerosols are small particles, including dust from the world's deserts and from volcanic eruptions, smoke from forest fires, pollutants of various types, and a lot of microscopic salt crystals derived from the evaporation of sea spray. If the diameter of the scattering particles is more than 30 times the incident wavelengths, the scattering will be independent of wavelengths (which makes clouds white). Scattering by particles with diameters between 1/10 and 30

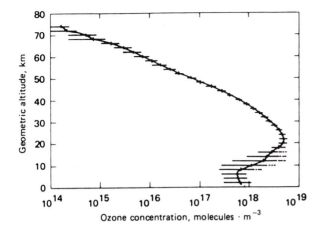

Figure 13.1. The concentration of ozone in the mid-latitude atmosphere as a function of altitude. (*U.S. Standard Atmosphere 1976*, p. 29, Fig. 31.)

times the wavelength of the incident radiation is partly dependent on the wavelength.

Radiation with $\lambda < 300$ nm is absorbed by ozone (O_3) and oxygen (O_2) molecules, which fly apart. Nitrogen (N_2) has its strongest absorption at 399.5 nm (blue light). N_2 does not break up; it reradiates on a 4π geometry, adding to the blue of the sky. Radiation with $\lambda > 700$ nm is largely absorbed (and reradiated) by H_2O and CO_2 molecules.

The ozone molecule (O_3) consists of 3 oxygen atoms. The central oxygen atom is bonded to two oxygen atoms by two resonant hybrid bonds (i.e., bonds that switch between single and double) that have an average length of 1.28 angstroms and form an angle of 117° [1 angstrom (Å) = 10^{-10} m]. The molecule vibrates and rotates. Absorption band peak is at 230–280 nm (UV) and absorption/emission at 560–620 nm (green–orange). Ozone forms during the night by the reaction

$$O_2 + O + M \rightarrow O_3 + M \qquad (1)$$

where M is any other molecule, needed to conserve momentum. Atomic oxygen is produced during the day by photodissociation of O_2:

$$O_2 + h\nu \rightarrow 2O$$

where $h\nu$ is energy. The O_2 bond has an energy of 5.16 eV; that is, 5.16 eV of energy are needed to break the bond and dissociate O_2 into 2O. This energy is supplied by UV radiation from the Sun ($\nu = 1.249 \cdot 10^{15}$ Hz, corresponding to a wavelength λ of 240 nm). Formation of ozone requires a three-molecule collision, which

becomes more probable the denser the atmosphere. At low altitudes, however, where the atmosphere is denser, there is not enough UV radiation. The maximum concentration of ozone (10^{-5}–10^{-6}) is therefore at 30 km of altitude at the equator, sloping down to 18 km of altitude at the poles (Fig. 13.1). By comparison, the concentration in the troposphere is only 10^{-8}. The amount of ozone in the atmosphere is equal to a layer 3 mm thick if reduced to the pressure of one atmosphere.

Ozone is destroyed by three processes:

1. photodissociation during the day by UV radiation:

$$O_3 + h\nu \rightarrow O_2 + O$$

   but O is immediately reconverted to O_3 by reaction (1);

2. reaction with NO:

$$O_3 + NO \rightarrow O_2 + NO_2$$

   releasing 1.7 eV of bond energy (= radiation with $\lambda = 712$ nm, i.e., infrared radiation);

3. reaction with chlorofluorocarbons:

$$Cl + O_3 \rightarrow ClO + O_2$$
$$ClO + O \rightarrow Cl + O_2$$

   (notice that Cl is reformed, and the reaction can be repeated n times).

The Earth's back-radiation (Fig. 13.2) ranges mainly from $\lambda = 1.6\,\mu m$ to $\lambda = 50\,\mu m$, with peak power at $\lambda = 10\,\mu m$, corresponding to a mean surface temperature of 288 K (= 15°C). The back-radiation spectrum (Fig. 13.2) diverges

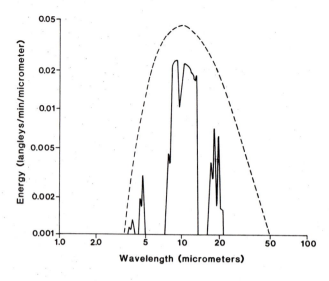

Figure 13.2. The Earth's radiation spectrum (solid line) compared with the spectrum of a blackbody radiating at 300 K (dotted line). (Adapted from Sellers, 1965, p. 21, Fig. 6.)

considerably from that of a blackbody because of the strong absorption and back-radiation by the "greenhouse" gases, especially H_2O, CO_2, and CH_4. The concentration of H_2O in the atmosphere is controlled by equilibrium with the ocean; CO_2 is introduced into the atmosphere by volcanism, and, under equilibrium conditions, an equivalent amount is removed by precipitation as $CaCO_3$ in the ocean (then recycled by means of subduction); CH_4, twenty times more effective than CO_2 as a greenhouse gas, but comparatively very scarce, is produced by anaerobic, cellulose-metabolizing bacteria in the guts of grass- or wood-eating animals (cattle, roaches, termites, etc.). The residence time of CH_4 in the atmosphere is only 2.6–4 years, because methane is rapidly oxidized to CO_2. Human activities have added CH_4 and CO_2 to the atmosphere for centuries, by the burning of fossil fuels and by various agricultural practices. During the past 100 years, the concentration of CH_4 has more than doubled (from $7.0 \cdot 10^{-7}$ to $15.5 \cdot 10^{-7}$), and that of CO_2 has increased by 19% (from $2.90 \cdot 10^{-4}$ to $3.46 \cdot 10^{-4}$). The added CO_2 slowly dissolves in the ocean, which contains 58 times more CO_2 (in solution) than the atmosphere. The time delay involved, however, causes the observed, continuing increase in concentration in the atmosphere. Today there is much concern that the continuing addition of greenhouse gases to the atmosphere will produce a runaway global increase in temperature, causing ice to melt in Greenland and Antarctica and causing sea level to rise. The principal absorption/emission bands by H_2O and CO_2 are at $\lambda = 5.0$–7.5, 14.5–16.4, and $> 22\,\mu m$ (Fig. 13.2). The Earth's radiation balance is shown in Table 13.1 and Fig. 13.3. Table 13.1 shows that 47% of the incoming solar energy is absorbed by the ground (or the surface of the ocean). Of that amount, 9% is reradiated to outer space, and 42% is retained by the atmosphere (13% as infrared radiation to atmospheric gases and clouds, 24% as latent heat, 5% as sensible heat).

[*Latent heat* is the energy content of a body because of its molecular or atomic bonding state (see Section 13.4); *sensible heat* is the energy content of a body related to its atomic or molecular motions, recorded as temperature.]

Table 13.1 and Fig. 13.3 show, as expected, that outgoing radiation balances incoming radiation (otherwise there would be a net, secular temperature increase or decrease). This balance will change if the solar output or the Earth's albedo changes. That has indeed happened in the past.

[*Albedo* is the "whiteness" or reflectivity of surfaces, expressed as the ratio of energy reflected in all directions to total incident energy.

Table 13.1. The Earth's radiation balance

13.2. COMPOSITION

Surface + atmosphere energy balance

Incoming solar radiation (100%)	
Absorption by O_3	3
Absorption by H_2O and aerosols	13
Absorption by clouds	2
Backscatter by clouds	24
Backscatter by air and aerosols	7
Backscatter by surface	4
Total	53
Absorption by ground	
from Sun	25
from atmosphere and clouds	22
Total	47
Total incoming radiation	100
Outgoing terrestrial radiation	
Short wave	
Back-radiation from H_2O, CO_2, and clouds to outer space	31
Back-radiation from ground to outer space	4
Total	35
Long wave	
Back-radiation from ground	5
Back-radiation from H_2O, CO_2, and clouds	60
Total	65
Total outgoing radiation	100

Surface energy balance (as percentage of incoming solar radiation outside the atmosphere)

Incoming from	
Sun	25
Atmosphere	26
Total	51
Outgoing to	
Atmosphere	
Long wave to H_2O, CO_2, and clouds	13
Latent heat	24
Sensible heat	5
Outer space	
Backscattered from ground	4
Long wave	5
Total	51

Examples of albedo values: open, smooth ocean with Sun at zenith, 0.02–0.04; forest, 0.1; grassland, 0.2; desert, 0.2; ice, 0.7; fresh snow, 0.8; smooth cloud tops, 0.8; Earth, 0.39; Moon, 0.068.]

13.2. COMPOSITION

The earliest atmosphere of the Earth consisted of gases in high-temperature equilibrium with iron-nickel metal and iron-magnesium silicates. The major gases were, in order of decreasing abundances, CH_4, H_2O, N_2, NH_3, and H_2S. This composition did not last long, because photolysis split H_2O into H_2, which escaped (it still does), and O, which went to oxidize CH_4 to CO_2. As long as the surface temperature remained above the critical temperature of water (374.15°C), no liquid water could exist. Atmospheric pressure on Earth was as high as 100 atm (see Section 7.4). As temperature dropped and liquid water came into existence, CO_2 reacted with silicates to form carbonates, and CO_2 was gradually removed from the atmosphere. With the evolution of cyanobacteria, about $3.5 \cdot 10^9$ y ago, widespread photosynthesis produced abundant free oxygen. That free oxygen went to oxidize the reduced surface of the Earth. When that was largely accomplished, oxygen began accumulating in the atmosphere. It has been estimated that at the beginning of the Cambrian, the concentration of oxygen in the atmosphere was about 2%, or about one-tenth of its present concentration. The composition of the modern atmosphere is shown in Table 13.2. Up to an altitude of 100 km, the composition of the atmosphere is uniform in terms of the three major gases (N_2, O_2, and Ar). Above 100 km, the composition begins to change, with progressive relative increases of the lighter gases (Fig. 13.4).

13.3. STRUCTURE

The atmosphere is subdivided into four thermal layers, separated by boundaries identified by terms ending with -*pause* (Table 13.3 and Fig. 13.5). Of the incoming solar radiation, 53% is absorbed and reradiated or backscattered by the atmosphere, 4% is backscattered from the ground, and 47% is absorbed by the ground (Table 13.1). The ground warms up and reradiates infrared radiation that is absorbed by atmospheric water vapor and CO_2 and warms up the atmosphere. Temperature in the troposphere decreases with altitude because

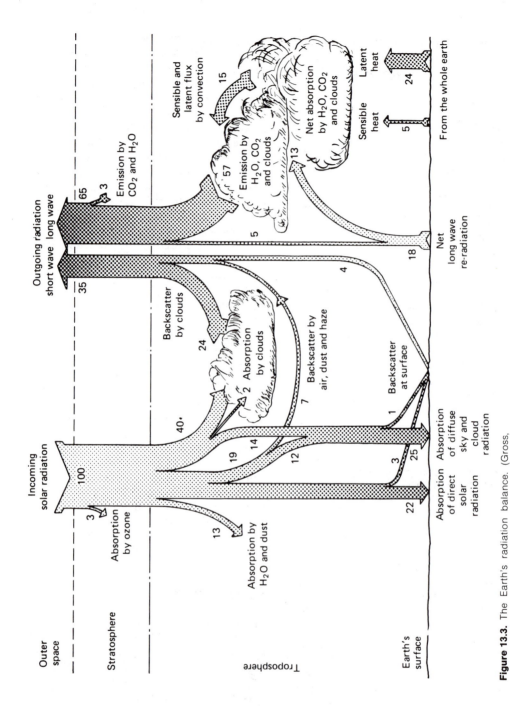

Figure 13.3. The Earth's radiation balance. (Gross, 1977, p. 147, Fig. 6.4.)

Outer space

Stratosphere

Troposphere

Earth's surface

Incoming solar radiation

100

Absorption by ozone

3

Absorption by H_2O and dust

13

40

Absorption by clouds

2

19

14

Backscatter by air, dust and haze

7

12

Backscatter at surface

1

3

25

22

Absorption of direct solar radiation

Absorption of diffuse sky and cloud radiation

Outgoing radiation short wave long wave

65

3

Emission by CO_2 and H_2O

35

Backscatter by clouds

24

4

57

Emission by H_2O, CO_2 and clouds

13

Net absorption by H_2O, CO_2 and clouds

Sensible and latent flux by convection

15

5

18

Net long wave re-radiation

Sensible heat

5

Latent heat

24

From the whole earth

258

Table 13.2. Composition of the atmosphere

Component	Concentration	Residence time
N_2	0.78084	$4 \cdot 10^8$ y for cycling through sediments
O_2	0.20946	6,000 y for cycling through biosphere
H_2O	$(4-0.004) \cdot 10^{-2}$	—
Ar	$9.34 \cdot 10^{-3}$	largely accumulating
CO_2	$0.346 \cdot 10^{-3}$	10 y for cycling through biosphere
Ne	$1.818 \cdot 10^{-5}$	largely accumulating
He	$5.24 \cdot 10^{-6}$	$2 \cdot 10^6$ y for escape
CH_4 (methane)	$1.55 \cdot 10^{-6}$	2.6–8 y
Kr	$1.14 \cdot 10^{-6}$	largely accumulating
H	$5.5 \cdot 10^{-7}$	4–7 y
N_2O	$3.3 \cdot 10^{-7}$	5–50 y
CO	$(2-0.6) \cdot 10^{-7}$	0.5 y
Xe	$8.7 \cdot 10^{-8}$	largely accumulating
O_3 (ozone)	$(3-1) \cdot 10^{-8}$	—
CH_2O (formaldehyde)	$<1 \cdot 10^{-8}$	—
NH_3 (ammonia)	$(20-6) \cdot 10^{-9}$	about 1 d
SO_2	$(4-1) \cdot 10^{-9}$	hours to weeks
$NO + NO_2$	10^{-9}	<1 mo
CH_3Cl (methyl chloride)	$5 \cdot 10^{-10}$	—
CCl_4 (carbon tetrachloride)	$(2.5-1) \cdot 10^{-10}$	—
CF_2Cl_2 (Freon 12)	$2.3 \cdot 10^{-10}$	45–68 y
H_2S	$\leqslant 2 \cdot 10^{-10}$	$\leqslant$ l d
$CFCl_3$ (Freon 11)	$1.3 \cdot 10^{-10}$	45–68 y

Concentrations, by volume, of components at ground level (excluding local pollutants) and their residence times.

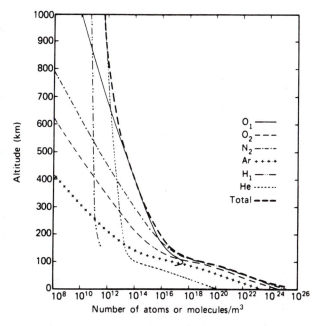

Figure 13.4. Concentrations of atmospheric gases as functions of altitude. (*U.S. Standard Atmosphere 1976*, p. 13, Fig. 5.)

Table 13.3.   **Thermal structure of the atmosphere**

Name of layer	Altitude (km)
troposphere	0–10 (at the poles); 0–16 (at the equator)
tropopause	10 (poles) to 16 (equator)
stratosphere	10–16 to 50
stratopause	50
mesosphere	50–85
mesopause	85
thermosphere	85+

the distance from the heat source (the ground) increases. A minimum of about −60°C is reached at the tropopause. Warmer air lying under colder air is unstable and so it rises while the colder air sinks—the atmosphere convects.

In fact, *troposphere* derives form the Greek τρέπειν, which means *to turn*.

The troposphere is 16 km thick at the equator and 10 km thick at the poles. The tropopause does not continuously slope down from equator to poles, but exhibits a sharp incline at middle latitudes. Air moving poleward near the top of the troposphere is forced to drop in altitude along this incline. To conserve angular momentum, velocity increases, while the Coriolis effect (discussed later) forces the air to move eastward. This high-speed eastward flow of high-altitude air is called the *jet stream*.

Above the troposphere lies the stratosphere. Temperature rises through the stratosphere, from −60°C at its base to 0°C at the top (the stratopause). This temperature increase is due to the absorption and conversion of UV solar radiation by ozone molecules. Ozone is more abundant below, reaching a maximum concen-

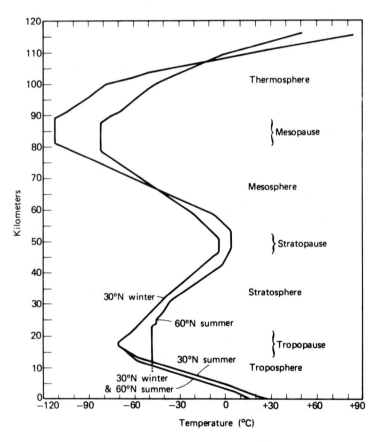

Figure 13.5. Thermal structure of the atmosphere, 0–120 km. (Emiliani, 1987, p. 16.)

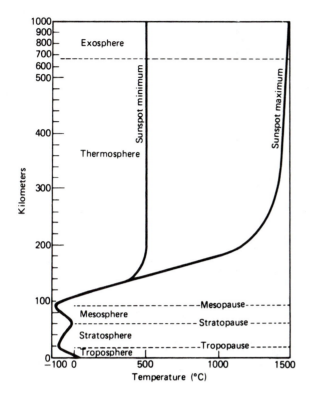

Figure 13.6. Thermal structure of the atmosphere, 0–1,000 km. (Emiliani, 1987, p. 17.)

tration at a height of 22 km (Fig. 13.1), but UV radiation, necessary to produce the atomic oxygen that forms ozone, is more plentiful above. The height of 50 km for the temperature maximum is a compromise.

Temperature decreases again through the mesosphere, reaching a minimum of −80°C to −120°C at an altitude of 80–90 km (the mesopause). From there, temperature rises to 500°C at an altitude of 120 km (lower thermosphere). The temperature of the thermosphere remains at 500°C from 150 to 1,000+ km of altitude during sunspot minima; during sunspot maxima it increases to 1,500°C (Fig. 13.6). The mean particle speed follows the thermal structure below the thermosphere, but in the thermosphere it rapidly increases upward (Fig. 13.7).

The mean free path continuously increases upward as density decreases (Fig. 13.8; see also Section 2.2). At an altitude of 650 km the mean free path along the horizontal direction is 650 km. This altitude, where height is equal to mean free path, is called the *critical level*.

The *ionosphere* is a layer in the thermo-sphere where ionization of atmospheric gases is produced by UV, x-ray, and Hα radiation from the Sun (the Hα line, with $\lambda = 121.6$ nm = 10.15 eV, is caused by a drop from $n = 2$ to $n = 1$ of the electron bound to a hydrogen atom). The ionosphere ranges from 100 to 400 km of altitude, with peak concentration of charged particles at 250 km of altitude. There, the concentration of charged particles is $10^{12}/m^3$, which, when compared with a total density of $1.9 \cdot 10^{15}$ particles per cubic meter, indicates that only 1 in about 2,000 particles is ionized.

[In 1894, Guglielmo Marconi (1874–1937), the Italian physicist who invented radio, sent a radio signal across the unheard-of distance of 10 m. In 1897 he succeeded in sending a signal from land to an Italian warship 20 km away, well beyond the horizon. That convinced Marconi that, somehow, radio waves could follow the curvature of the Earth. In 1901 he proved that by sending a radio signal clear across the Atlantic. In 1909 he received a Nobel Prize. The British-American engineer

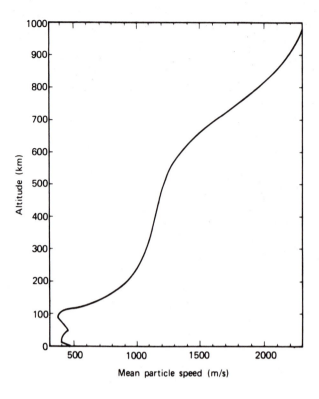

Figure 13.7. Mean particle speed as a function of altitude. (*U.S. Standard Atmosphere, 1976*, p. 17, Fig. 10.)

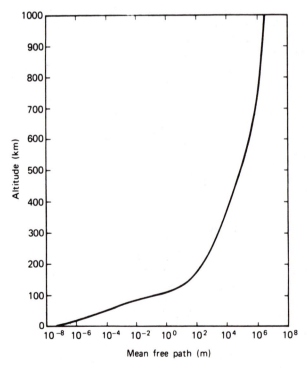

Figure 13.8. Mean free path of atmospheric particles with mean molecular mass (28.964 u) as a function of altitude. (*U.S. Standard Atmosphere 1976*, p. 18, Fig. 11.)

Figure 13.9. Aurora borealis as seen from the Fairbanks campus of the University of Alaska. (Courtesy L. Snyder.)

Arthur Kennelly (1861–1939) and the British physicist Oliver Heaviside (1850–1925) suggested that there might be a layer of ionized particles high in the atmosphere capable of reflecting Marconi's waves back to Earth. That layer, now called the ionosphere, was discovered in 1924 by the British physicist Edward Appleton (1892–1965), who received a Nobel Prize for his discovery.]

The *auroras* occur within the ionosphere when free electrons are recaptured by the ionized gases (Fig. 13.9). Here, ionization is produced mainly by energetic electrons of the solar wind interacting with the magnetic field of the Earth at high latitudes and altitudes. With energy commonly in the range of a few thousand electron volts, a single electron is capable of ionizing several hundred atoms and molecules of atmospheric gases before coming to rest at an altitude of about 100 km (and being itself captured).

The *exosphere* is that portion of the thermosphere above the ionosphere. It grades into outer space.

The *Van Allen belts* (Fig. 13.10) are two toroidal belts surrounding the Earth around the equator and containing protons and electrons trapped there by the magnetic field of the Earth. The inner belt, centered at 3,200 km of altitude, consists of high-energy protons ($>15\,\mathrm{MeV}$) and electrons ($>1.5\,\mathrm{MeV}$) of probable galactic or extragalactic origin. The outer belt, centered at 25,000 km of altitude, contains low-energy protons ($\sim 200\,\mathrm{eV}$) and electrons ($\sim 400\,\mathrm{eV}$) of probable solar origin.

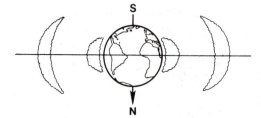

Figure 13.10. The Van Allen belts. The arrow represents the geomagnetic axis.

13.4. MOTIONS

The motions of the atmosphere are energized by the Sun and governed by density differences. They range from local winds and breezes to the great planetary wind system. Local winds include *coastal breezes* and *catabatic winds*. Coastal breezes result from alternating daytime warming and nighttime cooling of coastal areas. This happens because water has a higher heat capacity than do rocks and soil (1 cal/°C vs. 0.2 cal/°C) and therefore it maintains a more uniform temperature. During the day, the land warms up more than the water, air over the land rises, and air flows inland from over the water to replace the air that had been rising. This flow is called a *sea breeze*. At the same time, a reverse flow occurs at a higher altitude. During the night, the land cools more rapidly than the water, and a reverse flow, called *land breeze*, occurs. Catabatic winds are winds that become warmer and drier while descending the side of a mountain range. They become warmer because atmospheric pressure increases as they descend, and they become drier because the increasing temperature lowers the relative humidity. The rate of warming is about 1°C/100 m of drop. Examples are the Foehn, down the southern slope of the Alps; the Santa Ana, down the southwestern slope of the coastal ranges of southern California; and the chinook, down the eastern slope of the Rocky Mountains. These winds can cause sudden melting of the snow and extensive flooding in the lowlands below.

Regional winds are conditioned by seasonal contrast between continents and seas. This is particularly evident across the Mediterranean,

which is bordered on the south by a major desert and on the north by a land that extends to the Arctic. In Homer's time, four winds were recognized: Boreas, a cold, winter wind from the north; Notos, a hot wind from the south; Euros, from the east; and Zephyros, from the west. By the seventeenth century, the recognized directions from which the wind could blow had grown to 32. Today, the direction of the wind is simply expressed in degrees measured clockwise from the north (Fig. 13.11). Regional winds have different names in different regions. Standardized symbols (Fig. 13.11) are used to show local wind speed and direction, as well as the local cloud cover. Table 13.4 shows the Beaufort wind scale.

The large-scale motions of the atmosphere fall under the influence of the *Coriolis effect*. The Coriolis effect derives from insufficient application of the *Coriolis acceleration* (Fig. 13.12).

[Coriolis acceleration

The Earth is rotating with respect to a fixed frame of reference *oxyz* (o = origin). A point moving along the surface of the Earth is rotating with respect to the rotating frame of

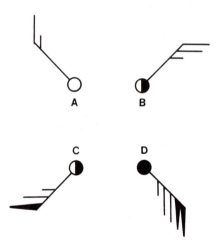

Figure 13.11. Wind and cloud-cover symbols used in weather maps. Empty circle = clear sky; half-full circle = partly cloudy; black circle = cloudy. Wind speeds: flag = 50 mph; full barb = 10 mph; half barb = 5 mph. Stem indicates direction *from which* the wind is coming. (A) Clear, with wind from the NW at 15 mph. (B) Partly cloudy, with wind from the NE at 25 mph. (C) Partly cloudy, with wind from the SW at 65 mph. (D) Cloudy, with wind from the SE at 125 mph (hurricane conditions).

Table 13.4. The Beaufort wind scale

Beaufort scale	Velocity (knots)	Marine term	Description	
			Sea	Land
0	<1	calm	sea like mirror	smoke rises vertically
1	1–4	light air	gentle ripples (<30 cm high)	smoke drifts slowly
2	4–7	light breeze	small waves (<1 m)	gentle leaf rustling
3	7–11	gentle breeze	1–1.5-m waves	leaves and twigs in motion
4	11–17	moderate breeze	1.5-m waves	small branches moving
5	17–22	fresh breeze	2-m waves	small trees waving
6	22–28	strong breeze	2.5-m waves, whitecaps everywhere	large branches in motion
7	28–34	moderate gale	3-m waves	whole trees swaying
8	34–41	fresh gale	3.5-m waves; foam in streaks	twigs broken off trees
9	41–48	strong gale	5–6-m waves; strong foam streaks	branches broken off trees
10	48–56	whole gale	6–12-m waves; spray	smaller trees uprooted
11	56–64	storm	12–15-m waves; strong spray	large trees uprooted
12–17	>64	hurricane	waves > 15 m; very strong spray	heavy structural damage

reference of the Earth $\omega\xi\upsilon\zeta$ (ω = origin). The velocity **v** of the point with respect to the fixed frame is

$$\mathbf{v} = (\boldsymbol{\omega}_\omega \times \mathbf{r}) + (\boldsymbol{\omega}_o \times \mathbf{r})$$

where $\boldsymbol{\omega}_\omega \times \mathbf{r}$ is the angular velocity of the point with respect to the Earth frame $\omega\xi\upsilon\zeta$, and $\boldsymbol{\omega}_o \times \mathbf{r}$ is the angular velocity of the Earth frame $\omega\xi\upsilon\zeta$ with respect to the fixed frame $oxyz$. The centripetal acceleration is

$$\begin{aligned}
\mathbf{a} &= \omega^2 \mathbf{r} \\
&= (\boldsymbol{\omega}_\omega + \boldsymbol{\omega}_o)^2 \mathbf{r} \\
&= \omega_\omega^2 \mathbf{r} + \omega_o^2 \mathbf{r} + 2\omega_\omega\omega_o \mathbf{r} \\
&= \mathbf{a}_\omega + \mathbf{a}_o + 2\omega_\omega\omega_o \mathbf{r}
\end{aligned}$$

but $\mathbf{v}_\omega = \boldsymbol{\omega}_\omega \times \mathbf{r}$; therefore,

$$\mathbf{a} = \mathbf{a}_\omega + \mathbf{a}_o + 2\boldsymbol{\omega}_o \times \mathbf{v}_\omega$$

The term $2\boldsymbol{\omega}_o \times \mathbf{v}_\omega$ (or simply $2\boldsymbol{\omega} \times \mathbf{v}$, the vector product of the angular-velocity vector $\boldsymbol{\omega}$ of the Earth times the velocity vector **v** of a body moving on the surface of the Earth) is called *Coriolis acceleration*. In summary:

The Coriolis acceleration is the acceleration needed for an object to stay on track when moving with respect to a rotating frame of reference.

The Coriolis acceleration is

Eastward for a body moving along a meridian toward the equator

Westward for a body moving along a meridian toward either pole

Downward for a body moving eastward along the equator

Upward for a body moving westward along the equator

Downward or upward at a slant toward the pole for a body moving along a parallel, with the angle of slant from the horizontal equal to the colatitude (*colatitude* is the complementary angle of the latitude, i.e., $90° - \phi$, where ϕ is the latitude)

Zero for a body moving north or south across the equator.

Coriolis effect

The *Coriolis effect* results from inadequate application of the Coriolis acceleration. Winds and ocean currents are insufficiently coupled with the solid surface of the Earth, and they strongly feel the Coriolis effect. The result is

A clockwise motion in the Northern Hemisphere,

A counterclockwise motion in the Southern Hemisphere.]

The motions of the atmosphere include the general circulation, frontogenesis, storms, and local winds.

Warm air rises at the equator from sea level to the upper troposphere and travels poleward on either side of the equator (Fig. 13.13). The Coriolis effect forces air to deflect eastward,

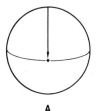 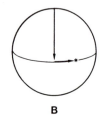  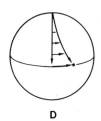

A   B   C   D

Figure 13.12. Coriolis acceleration and Coriolis effect. A bowling ball thrown from the pole toward a target (indicated by the asterisk) on the equator of a perfectly smooth, nonrotating Earth (A) will reach the intended target. On a perfectly smooth but rotating Earth (B), the bowling ball will reach the equator *west of* the target, because while the ball was traveling, the target moved *eastward*. From the point of view of the Earth's rotating frame of reference, the ball *appears* to have veered to the right (C). In order to stay on track and hit the target, the ball will have to be accelerated eastward all along, as it moves toward the equator (D). (Remember that acceleration can be a change in either the magnitude or the direction of a velocity vector, or changes in both. In the present case the magnitude of the velocity vector is assumed to remain constant.) The acceleration needed to keep the ball on track is the *Coriolis acceleration*. Inadequate application of the Coriolis acceleration results in veering, which is called *Coriolis effect*. A ball thrown from equator to pole will start with a strong velocity component toward the east (1,670 km/h), which will cause the ball to overtake the Earth as the ball moves poleward. In order to reach the pole, the ball will have to be *decelerated* (by application of a westward acceleration). Veering is to the right in the Northern Hemisphere and to the left in the Southern Hemisphere. The acceleration needed to stay on track is eastward in both hemispheres for a body traveling toward the equator; it is westward in both hemispheres for a body traveling poleward. The exact direction of the Coriolis acceleration vector is obtained by multiplying the vector **ω**, representing the angular velocity of the Earth, by the vector **v**, representing the velocity of the moving body. For a body moving along the equator, the two vectors are perpendicular to each other and lie on a plane tangential to the equator. Therefore, if the body is moving eastward, the product of the two vectors is directed straight down; if the body is moving westward, the product vector is directed straight up.

until at 30° of latitude (N or S) air is moving eastward. There the air descends, and a major portion returns to the equator following a 225° course (north of the equator, forming the *northeasterly trades*) or a 135° course (south of the equator, forming the *southeasterly trades*). These cells on either side of the equator between 0° and 30° of latitude are called *Hadley cells*.

A portion of the descending limb of a Hadley cell continues poleward, forming the *prevailing westerlies* in both hemispheres. Cold air sinks at the poles and streams out of the polar caps, following a 225° course (Northern Hemisphere) or a 315° course (Southern Hemisphere). These winds form the *polar easterlies* in both hemispheres. The prevailing westerlies override them along the 50° parallel (polar front), rise, and return to 30° of latitude as a high-altitude flow. There they descend to ground level opposite the descending limbs of the Hadley cells, forming the *Ferrel cells*.

The descending limbs of the Hadley and Ferrel cells form a high-pressure belt around 30° of latitude on both hemispheres. Air is compressed and therefore heated by the descent. Relative humidity is low, and deserts form on the land below (Saharan, Arabian, and Iranian deserts in the Northern Hemisphere; Atacama, Kalahari, and Australian deserts in the Southern Hemisphere).

The flows of the Hadley and Ferrel cells are greatly distorted by the continents, high mountain ranges, high plateaus, the albedo of the ground, and so forth. As a result, large permanent gyres (clockwise in the Northern Hemisphere, counterclockwise in the Southern Hemisphere) form over the oceans on either side of the equator (Fig. 13.14). These gyres are responsible for the wind-driven circulation of the oceans.

The *monsoon* is a wind system that reverses its course in opposite seasons. The largest monsoon is the Indian monsoon (Fig. 13.14). During the winter, central Asia is very cold, and a strong high-pressure center develops there. Winds stream out, blowing southwestward across the Himalayas, India, the Arabian Sea, and across the equator. During the summer, southwestern Asia becomes very hot, a

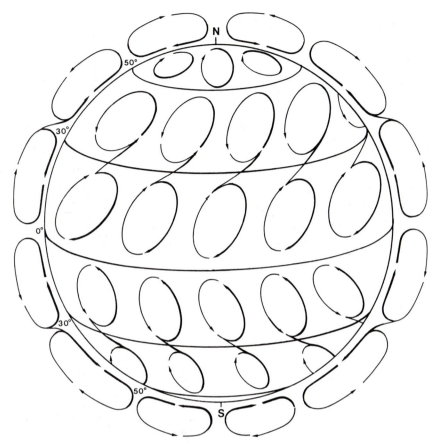

Figure 13.13. Global circulation (vertical). Thicker lines indicate ground flow. Thinner lines indicate upper flow. Warm, humid air rises at the equator, creating a low-pressure belt (*doldrums*) and dropping rain. The warm air travels to 30° of latitude, where it descends. This descent warms the air and decreases its relative humidity, creating a dry belt where the major deserts of the world lie (*horse latitudes*). A branch of the descending air travels as ground flow to 50° of latitude. There it meets and overrides the cold polar air (*polar front*), dropping rain and snow. It returns to 30° of latitude as upper airflow. The direction of airflow is governed by the Coriolis effect; the speed is governed by channel cross section and conservation of momentum. The convection cells between 0° and 30° of latitude are called *Hadley cells*; those between 30° and 60° are called *Ferrel cells*.

strong low-pressure center develops, and air is sucked in. The summer monsoon moves hot, humid air from south of the equator, across the Arabian Sea, India and the Himalayas. The southern slopes of the Himalayas receive abundant rainfall and, higher up, snowfall. A weaker monsoon develops over the North American Southwest.

The *jet stream* (Fig. 13.15) is a narrow, high-velocity wind flowing eastward in the upper troposphere at middle latitudes in both hemispheres. It results from the conservation of angular momentum as air masses drop from the higher tropical tropopause (16 km of altitude) to the lower polar tropopause (10 km of altitude) at approximately 50° of latitude. The contrast is greater in the winter, and so the jet stream is faster. The average speed is 60 km/h during the summer, and 150 km/h during the winter. The highest speed is 400 km/h.

The jet stream does not run around the Earth on a circle. It develops waves, called *Rossby waves*, that follow a sinuous course

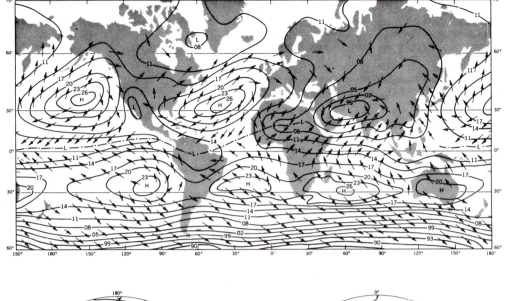

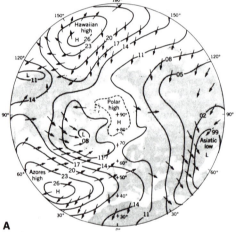

A

Figure 13.14. Global circulation at ground level during the northern summer (A) and during the northern winter (facing page, B). Ground-level circulation is strongly influenced by the distribution of continents and oceans and by landforms. As a result, great gyres form over the oceans, and a strong monsoon circulation develops over India and adjacent regions. (Strahler, 1971, p. 252, Fig. 15.32, p. 253, Fig. 15.33.) (*Continued*)

with loops extending southward to the tropics (Fig. 13.15). A loop may become detached, isolating a cold or warm air mass.

An *air mass* is a large body of air with fairly uniform properties of temperature and humidity. An air mass acquires its properties by remaining stationary over a given area for a period long enough to come to equilibrium with the surface below. The properties of an air mass (temperature, humidity, and abun-

dance of minor components) are determined by the nature of the underlying surface.

Air masses are classified as polar (P), tropical (T), or equatorial (E) in terms of temperature. They are classified as continental (c) or maritime (m) in terms of humidity. A continental air mass is relatively dry, and a maritime air mass is relatively humid.

Different air masses have different densities. The surface along which they meet is called a

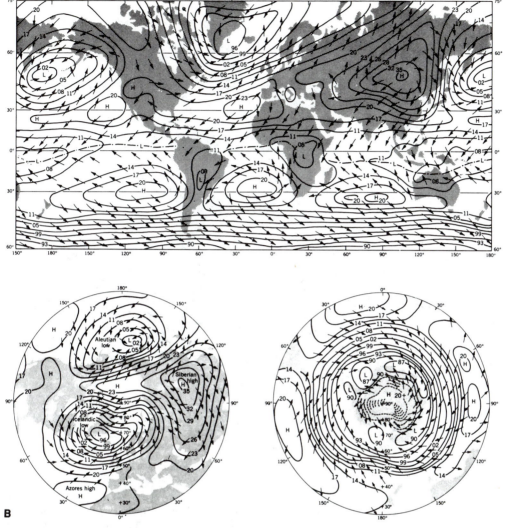

Figure 13.14. (*Cont.*). (B) see caption on preceding page.

frontal surface. The intersection of the frontal surface with the ground is called a *front.* There are five different types of fronts: *cold fronts, warm fronts, stationary fronts,* and two types of *occluded fronts* (Fig. 13.16). A frontal surface usually is 1 km thick. It is a highly turbulent zone where cold air and warm air mix. The temperature drop across the frontal zone may be as great as 10°C or more. The difference between a cold front and a warm front is the direction in which the front moves. A cold front moves toward the warm area; a warm front

moves toward the cold area. The slope of a warm front usually is much less than the slope of a cold front.

A *stationary front* is a front that does not move. Stationary fronts do not last more than a day or so. An *occluded front* develops when a cold front overtakes a warm front. If the air under the warm front is colder than the air of the advancing cold front, an *occluded warm front* will form. If the air of the advancing cold front is colder than the air under the warm front, an *occluded cold front* will result. In either

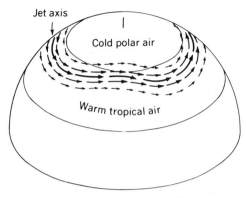

A. Jet stream begins to undulate

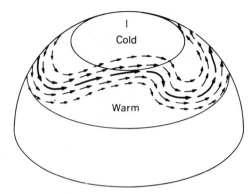

B. Rossby waves begin to form

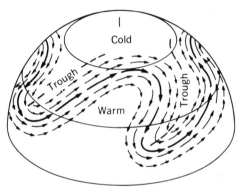

C. Waves strongly developed

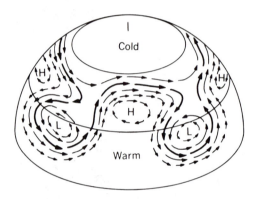

D. Cells of cold and warm air bodies are formed

Figure 13.15. Evolution of the jet stream and formation of cyclones (L, low pressure) and anticyclones (H, high pressure). (Strahler, 1971, p. 247, Fig. 15.26.)

case, the warm air is squeezed aloft, and precipitation occurs.

A major front is the *polar front*. The polar front separates polar air from tropical air. The polar front often is broken up into a number of segments, each behaving more or less independently. Each segment advances and retreats according to the season and the weather pattern. Polar fronts reach farther toward low latitudes during the winter, as one would expect. All fronts move eastward, following the general circulation of the jet stream.

13.5. ATMOSPHERIC PRESSURE

A column of air that is 1 cm² in cross section and reaches from sea level to outer space

weighs 1.033 kg. This is the same weight as a column of water 10.33 m high or a column of mercury 0.760 m high (both 1 cm² in cross section). We are not squashed by this weight because we are immersed in the atmosphere. In the same way, a sponge at the bottom of a bathtub full of water is not squashed by the weight of the water above.

Pressure is force (or weight) per unit area. The SI unit of pressure is the pascal (Pa), which is equal to 1 N/m². It is a very small pressure, about 10 mg/cm². Other, more practical units of pressure for studying the atmosphere are the bar (b) = 10^5 Pa (exactly), the millibar (mb) = 10^2 Pa, the millimeter of mercury (mmHg) = 1.3329 mb, and the atmosphere (atm) = 1.013 b = 1,013 mb = 760 mmHg (with ρ_{Hg} set at 13.5951 g/cm³). Normally, atmospheric pressure at sea

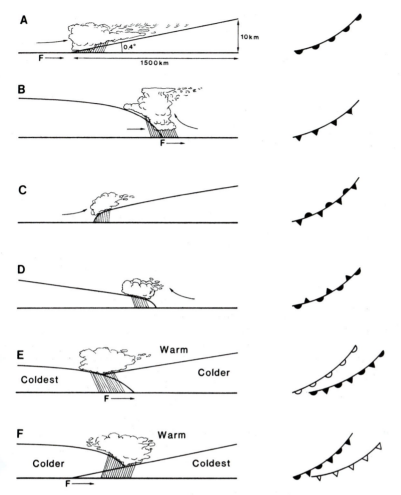

Figure 13.16. Fronts (F = direction of motion of front): (A) warm front, (B) cold front, (C) and (D) stationary fronts, (E) occluded cold front, (F) occluded warm front. Front symbols to the right.

level remains within 1,000–1,030 mb. Newspaper weather maps in the United States give the pressure in inches of mercury (1 atm = 29.92 inHg).

Atmospheric pressure was first measured by the Italian physicist Evangelista Torricelli (1608–1647). In 1644 Torricelli took a glass tube about 1 m long, sealed one end, and filled it with mercury. He then held the other end closed with a finger and turned the tube upside down into a bowl full of mercury. When the finger was removed, the mercury in the tube began emptying into the bowl, but then it stopped. Torricelli measured the height of the mercury column left in the tube and found

it to reach 760 mm above the surface of the mercury in the bowl (Fig. 13.17). He reasoned that what kept the column of mercury up inside the tube was the pressure of the atmosphere on the surface of the mercury in the bowl. There was no corresponding pressure on the surface of the mercury inside the tube because the tube was sealed at the top.

Torricelli was one of a small group of scientists gathered by Grand Duke Ferdinand II of Tuscany (1610–1670) at his court in Florence. Ferdinand himself invented the thermometer, a modification of Torricelli's barometer. He did not use mercury in his bowl, but spirit of wine, and, most importantly, he sealed the

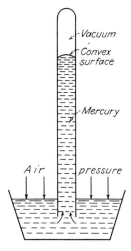

Figure 13.17. Torricelli's barometer. (Donn, 1965, p. 146, Fig. 8.1.)

Figure 13.18. The Grand Duke's thermometer.

bowl, making it into a bulb. The liquid sealed in the bulb would not feel pressure differences, but would respond to temperature differences by expanding or contracting and therefore changing its level in the thin tube extending above the bulb (Fig. 13.18).

Pressure decreases with altitude (Table 13.5). If you were to repeat Torricelli's experiment on top of Mount Everest (8,848 m high), you would find that the surface of the mercury inside the tube would be only 236 mm above the surface of the mercury in the bowl. But very likely the mercury in the bowl, the mercury in the tube, and you, too, would be frozen solid.

Atmospheric pressure is measured by means of barometers (from the Greek βαρύς, which means *heavy*). Mercury barometers are cumbersome to handle. More commonly used are aneroid barometers [*aneroid* means "without fluid", from the Greek ἀ-, meaning *without*, and νηρός (more commonly ναρός), meaning *fluid*]. In fact, an aneroid barometer consists of a

Table 13.5. Physical properties of the atmosphere[a]

Height (km)	Pressure (mb)	Temperature (K)	Density (kg/m^3)	Mean molecular mass (u)
0	$1.01 \cdot 10^3$	288	$1.23 \cdot 10^0$	28.96
5	$5.40 \cdot 10^2$	256	$7.36 \cdot 10^{-1}$	28.96
10	$2.65 \cdot 10^2$	223	$4.14 \cdot 10^{-1}$	28.96
20	$5.53 \cdot 10^1$	217	$8.89 \cdot 10^{-2}$	28.96
40	$2.87 \cdot 10^0$	250	$4.00 \cdot 10^{-3}$	28.96
60	$2.20 \cdot 10^{-1}$	247	$3.10 \cdot 10^{-4}$	28.96
80	$1.05 \cdot 10^{-2}$	199	$1.85 \cdot 10^{-5}$	28.96
100	$3.20 \cdot 10^{-4}$	195	$5.60 \cdot 10^{-7}$	28.40
150	$4.54 \cdot 10^{-6}$	634	$2.08 \cdot 10^{-9}$	24.10
200	$8.47 \cdot 10^{-7}$	855	$2.54 \cdot 10^{-10}$	21.30
300	$8.77 \cdot 10^{-8}$	976	$1.92 \cdot 10^{-11}$	17.73
400	$1.45 \cdot 10^{-8}$	996	$2.80 \cdot 10^{-12}$	15.98
500	$3.02 \cdot 10^{-9}$	999	$5.21 \cdot 10^{-13}$	14.33
600	$8.21 \cdot 10^{-10}$	1,000	$1.14 \cdot 10^{-13}$	11.51

[a] Above 120 km of altitude, the atmospheric parameters given here depend on the phase of the sunspot cycle. The values given refer to the year 1976 (sunspot minimum).

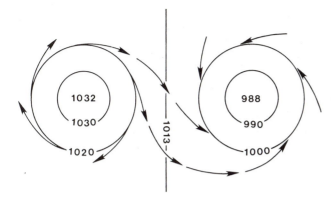

Figure 13.19. Dynamics of an anticyclone (left) and a cyclone (right). In the Northern Hemisphere, wind attempting to remove the pressure difference between a pressure high and a pressure low is forced to turn to the right. Air getting sufficiently close to the low is sucked in, in a counterclockwise fashion. In the Southern Hemisphere the circulation is reversed.

thin-walled metal chamber from which a portion of the air has been removed. As atmospheric pressure changes, the chamber becomes more or less squashed. The change in shape activates a pointer that indicates the pressure on a gauge.

13.6. CYCLONES AND ANTICYCLONES

Pressure differences arise when adjacent air masses have different densities. The denser air mass is the site of a pressure high, while the lighter air mass is the site of a pressure low. High-pressure centers are called *anticyclones*, and low-pressure centers are called *cyclones.*

Wind moves from a high to a low to smooth out the pressure difference. Because of the Coriolis effect, the wind in the Northern Hemisphere is deflected to the right as it moves out of the high. As it gets closer to the low, the wind is pulled in, and a counterclockwise circulation is established around the low (Fig. 13.19). In the Southern Hemisphere, circulation is in the opposite sense—counterclockwise around a high, and clockwise around a low. Winds flowing from highs to lows under the influence of the Coriolis effect are called *geostrophic winds.* The pressure differences between highs and lows normally are only 1–2%.

Cold fronts, warm fronts, and occluded fronts are formed when two or three air masses having different temperatures collide. Two air masses flowing in opposite directions and colliding at an angle are deflected away from each other by the Coriolis effect. A divergence forms. The divergence is characterized by a low-pressure center around which, in the Northern

Hemisphere, a counterclockwise circulation develops. Precipitation occurs along the advancing warm front. More intense precipitation, but on a narrower band, occurs along the steeper, overtaking cold front. Eventually the cold front overtakes the warm front, the warm air is squeezed out, and an occluded front forms. This development forms a cyclone (Fig. 13.20).

Cyclone development is particularly frequent along the polar front during the winter, when the polar front is particularly well developed. Cyclones form directly under the jet stream, and like the jet stream, they move eastward in both the Northern Hemisphere

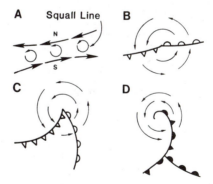

Figure 13.20. Formation of a cyclone (Northern Hemisphere). Polar air flowing south is deflected to the right, and so is warm air flowing north. (A) A belt of divergence, and hence low pressure, develops between the two, with counterclockwise eddies. Counterclockwise circulation (B) causes the cold front to overtake the warm front (C), resulting in an occluded cold front with strong cyclonic (i.e., counterclockwise) circulation (D). Circulation in the Southern Hemisphere is in the opposite sense.

and the Southern Hemisphere. The average speed is about 40 km/h. A cyclone may last several days and drop several centimeters of water. As discussed later, cyclones derive their energy from the latent heat released when water vapor condenses. An average cyclone develops 10 to 100 megatons of energy.

13.7. EVAPORATION AND PRECIPITATION

Most of the atmospheric water vapor that forms clouds comes from the tropical belt of the oceans, where temperatures average 25°C. Water molecules above the water surface move in all directions. Those close to the water surface and moving downward may return to the liquid. When the number of molecules that evaporate equals the number of molecules that return to the liquid, the vapor is said to be *saturated*. The pressure of saturated vapor depends upon the temperature. The pressure of water vapor is 4.58 mmHg at 0°C, 92.56 mmHg at 50°C, and 760 mmHg at 100°C. As you see, the vapor pressure at 100°C is equal to one atmosphere. This is the reason that water at sea level boils at 100°C. At 100°C the average energy of the molecules is sufficient to transform all the liquid water into vapor. On the top of Mount Everest, pressure is 236 mmHg, and water boils at 70°C.

Evaporation takes place mainly in the tropics because of the high temperatures and because of the steady trade winds that blow water vapor away. The amount of water vapor in the atmosphere ranges from almost zero above the deserts of the world and above the ice caps of Greenland and Antarctica to 2.5 weight percent in the humid equatorial areas. Warm air holds more water vapor than cold air. Air at 30°C, for instance, can hold up to 30 g/m^3, while air at 0°C can hold only 4.8 g/m^3.

The amount of water vapor per cubic meter of air is called the *absolute* humidity. Air humidity, however, usually is expressed as *relative humidity*, the percentage of the maximum amount of water vapor that air can hold at a given temperature. Air at 30°C with 30 g of water vapor per cubic meter contains the maximum amount of water vapor it can hold at that

temperature. Its relative humidity, therefore, is 100%. If it had only 15 g of water vapor, its relative humidity would be 50%.

Warm, humid air is less dense than the cooler air above. As it rises, it expands and temperature decreases. If relative humidity reaches 100%, water begins to condense. The temperature over the ocean in the tropics during the summer is around 28°C, and relative humidity is around 90%. As hot, humid air rises, temperature drops, and pressure is reduced. Saturation over the tropical ocean during the summer is achieved when temperature drops to 24°C or below, that is, at an altitude of 600 m and above. When water cools, more molecules bond, releasing energy (latent heat). Ice at 0°C is fully hydrogen-bonded, but water at 0°C is only 52% hydrogen-bonded. To reduce hydrogen bonding from 100% to 52% requires 80 cal/g. Adding 100 cal/g to water at 0°C raises the temperature to 100°C and reduces the bonding to 33%. Another 539 cal/g need to be added to break the remaining bonds and vaporize the water. Conversely, 539 cal/g are released when vapor becomes liquid at 100°C; 100 cal/g are released as water cools from 100°C to 0°C; and an additional 80 cal/g are released when water solidifies to ice at 0°C. The total amount of energy needed for water to change from the fully free state to the fully bonded state is thus 719 cal/g. This energy, absorbed to break bonds or released when bonds are formed, is the latent heat.

Water evaporates at any temperature, not at 100°C only, because, on account of the Maxwell-Boltzmann distribution of molecular velocities (Fig. 2.6), there are always molecules with enough energy to leave the surface of liquid water or even ice (Table 13.6). Of course, the lower the temperature, the greater the energy needed. The latent heat of vaporization of water at 0°C, for instance, is 597 cal/g, compared with 539 cal/g at 100°C.

Condensation of atmospheric water is facilitated by the presence of aerosols in the atmosphere. Water molecules are polar and form weak bonds with the aerosol particles, which usually have a nonuniform charge distribution or are altogether charged (by having lost or gained some electrons). More water molecules are added by hydrogen bonding. Aerosols, usually less than 1 μm in size, act as

Table 13.6. Pressure of water vapor over ice and liquid water

Temperature (°C)	Vapor pressure (mmHg)
−50	0.029
−40	0.097
−30	0.286
−20	0.776
−10	1.950
0	4.579
10	9.209
20	17.535
30	31.824
40	55.324
50	92.51
60	149.38
70	233.79
80	355.28
90	525.76
100	760.00

condensation nuclei, and water droplets grow around them, reaching a size of about $20\,\mu m$ in about 100 seconds.

A cloud consists of a concentration of water droplets in amounts of about 1,000 droplets per cubic centimeter, or one droplet per cubic millimeter. Fog is similar to a cloud, except that it hovers at ground level when there is a density inversion. The droplets that form a cloud are so small that air turbulence within the cloud keeps them in suspension. Droplets tend to fuse with each other to form larger droplets and eventually raindrops. Raindrops are drops that are too big to remain in the air—their average size is about 2 mm. It takes 10^6 droplets to make one raindrop.

Meteorologists recognize several types of clouds (Fig. 13.21). When hot, humid air rises, it forms cumulus clouds. Large cumulus clouds rising to the top of the troposphere expand horizontally and form cumulonimbus clouds. These are the familiar "thunderheads" of the summer. Cumulus and cumulonimbus clouds are characterized by high turbulence, with updraft velocities ranging from 2 m/s in the lower part of the cloud to 20 m/s in the upper part. There is also strong electrical activity, with much lightning and thunder within the cloud, between adjacent clouds, and between clouds

and the ground. The average diameter of a cumulonimbus cloud is 15 km.

Stratus clouds form an extended cloud layer with a rather even base. They range from 2 to 12 km of altitude. They may cover the entire sky. If the cloud is several kilometers thick, the sky is dark, and the cloud is called nimbostratus. There is no significant electrical activity within stratus or nimbostratus clouds.

Clouds at altitudes of 6–8 km are called altocumulus or altostratus, depending on their shapes. Similar clouds at altitudes of 8–12 km are called cirrocumulus or cirrostratus. Altocumulus and cirrocumulus clouds often are aligned in rows, separated from each other by bands of clear sky. Convection cells several hundred meters across are responsible for this pattern.

Cirrus clouds are the highest clouds in the troposphere. They consist of ice crystals and appear as fine, white filaments.

Two types of clouds occur in the stratosphere, both formed by tiny ice crystals. Nacreous clouds are translucent sheets of ice crystals at 20–30 km of altitude. Noctilucent clouds are even less dense than nacreous clouds and occur at altitudes of 80–90 km. No clouds have been recorded at greater altitudes because water molecules are too scarce.

Precipitation is solid or liquid water that falls back on the earth's surface. A part of it evaporates, a part is absorbed by the ground, and a part returns to the ocean as river discharge. River discharge is called *runoff*. The evaporation–precipitation cycle is called the *hydrologic cycle*. The hydrologic cycle is illustrated in Fig. 13.22.

Precipitation is highly variable. In the Atacama desert (northern Chile) it almost never rains. At the other extreme is Mount Waialeale, Kauai, Hawaii, with an average yearly precipitation of 11.45 m. Temperature and precipitation are the most important climatic factors.

The type of rainfall depends on the type of cloud. Nimbostratus clouds produce quiet rain or gentle drizzle that may last several hours. Cumulonimbus clouds, on the other hand, tend to produce torrential rainstorms that last less than one or two hours. Ice crystals form in the upper portions of cumulonimbus clouds. If turbulence is high and the temperature in the lower part of the cloud is higher than 0°C, the

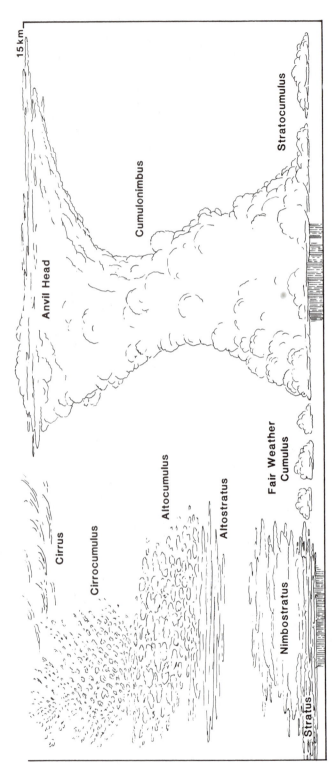

Figure 13.21. Types of clouds.

15 km

Anvil Head

Cumulonimbus

Stratocumulus

Cirrus

Cirrocumulus

Altocumulus

Altostratus

Fair Weather Cumulus

Nimbostratus

Stratus

276

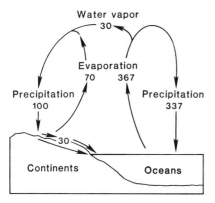

Figure 13.22. The hydrologic cycle. The mean annual precipitation is 85.7 cm. The heat of vaporization of water at 15°C is 582 cal/g. The figures are in 10^{15} kg y^{-1}. (Data from Barry and Chorley, 1987, p. 60.)

ice crystals may grow into hailstones. Ice crystals formed in the upper part of the cloud gather a layer of water as they fall down. Strong updrafts then carry the growing hailstones upward, where the temperature is sufficiently low for freezing. Several roundtrips between the upper and lower parts of a cloud can produce hazelnut-size hailstones. Larger hailstones are not uncommon. The largest on record weighed 750 g. It fell in Kansas in 1970.

Ice crystals formed by condensation of water vapor at temperatures below 0°C may grow to form snowflakes. If the ground temperature is close to 0°C or below 0°C, snow will fall and accumulate on the ground. The amount of snow falling is measured in centimeters of water. The density of snow is 0.10–0.15 g/cm³. One meter of snow, therefore, equals 10–15 cm of water. The greatest snowfall on record dropped almost 2 m of snow in a single day on Silver Lake, Colorado, in 1921. The snow that fell was equal to 20 cm of rain. The heaviest rainfall on record, on the other hand, dropped 1.87 m of water in 24 hours on Cilaos, La Réunion, in the Indian Ocean, in 1952. A heavy rainfall drops much more water than does a heavy snowfall.

Sleet is frozen rain. Sleet falls when there is a temperature inversion and the ground temperature is lower than the temperature above. This happens, for instance, when cold wind blows along the ground below a layer of warmer air.

Precipitation is measured by means of rain

gauges. A rain gauge is simply a cylindrical chamber with a funnel at the top. Water falling on the funnel is collected in the chamber below, and its height is measured. The opening at the bottom of the funnel is very small so as to maintain saturation humidity in the chamber below and to prevent evaporation of the water collected. Automatic gauges collect the precipitation in a container that is placed on a spring balance. The weight is continuously registered on a strip chart.

13.8. STORMS

Thunderstorms derive from cumulonimbus clouds formed by strong updrafts. Upward velocities within the cloud range from 2 m/s in the lower part to 20 m/s and more in the upper part. The top of the cloud is drawn into an anvil shape by high-altitude winds. Meanwhile, water droplets grown to raindrop size begin falling. Air is drawn downward by the falling drops, and a downdraft develops. Downdraft velocities are about 10 m/s. Because of the downdraft, air moves with the raindrops. Large drops can reach the ground without breaking. Within half an hour, updraft slows down, and the downdraft is replaced by a diffused downdraft at slower speed. The initial rainstorm is replaced by gentle rain that may last one or two hours.

Thunderstorms derive their energy from the heat of condensation of water. An average thunderstorm may deliver 500 megawatts of power, which is similar, while it lasts, to the output of an average nuclear power plant. There are about 45,000 thunderstorms per day around the world.

Turbulence within cumulonimbus clouds is high. Commonly, the bottom of the cloud is rather flat and several kilometers across. The top may reach the stratopause, in which case is spreads out in an anvil shape. The temperature in the midlayer of the cloud, at an altitude of 6 km, is about -15°C (Fig. 13.5). Ice grains, formed from the agglomeration of ice microcrystals, tend to fall. In the upper portion of the cloud, these falling grains rub against very cold microcrystals and strip them of some of their electrons. As a result, the upper part of the cloud becomes electron-deficient or positively charged. When the grains reach the $-15°C$ isotherm within the cloud, the grains are stripped of their extra electrons, and some of their own, by the warmer ice crystals at that altitude. As a result, the midlayer of the cloud becomes strongly negatively charged. The ice grains, now somewhat positive, keep falling, causing the bottom of the cloud to become weakly positive. The ground, a few hundred meters to a kilometer or two below the bottom of the cloud, feels mainly the effect of the negatively charged midlayer of the cloud. Acting like a plate in a capacitor, the ground becomes positively charged. If the ground is not perfectly flat and uniform, elevations, such as trees, buildings, chimneys, steeples, and ship masts, will be more electron-deficient than the surrounding areas, thereby exerting an extra attraction on the electrons in the cloud.

In fair weather, the ground is negatively charged to balance the positively charged ionosphere. The ionosphere is centered at about 250 km, and the voltage drop is 250,000 V, giving an electrostatic gradient of about 100 V/m. This gradient is insufficient to produce a spark, because of the high resistivity of air. During a thunderstorm, however, the potential difference between cloud and ground may rise to $500 \cdot 10^6$ V or so, and that is sufficient to trigger a spark.

Lightning is nothing more than a huge electrical spark. The length of a lightning discharge ranges from 500 m to more than 5 km. Typically, some 10^{20} electrons (about 10 coulombs) dart from cloud to ground in a series of rapidly succeeding strokes through the same channel, designed to pierce a hole through the atmosphere. Each successive stroke advances some 50 m in 1 μs and is separated from the next stroke by an interval of about 50 μs. The speed of a single 50 m stroke is thus 50,000 km/s, and the electron flow averages $10 \, C/10^{-6} \, s = 10^7$ A. For a height of 3 km, the number of strokes would be 60, lasting a total of 60 μs = 0.06 ms (not counting the intervals). The duration of the event (including the intervals) is about 3 ms, giving an average speed of 1,000 km/s and an average electron flow of 3,300 A. A channel a few centimeters across is formed, surrounded by an envelope of ionized gases. The descending electrons leave behind a large electron deficiency. In fact, a descending negative charge has the same effect as an ascending positive charge, like electrons and holes moving through a semiconductor (see Section 9.6). The ascending positive front travels at much higher speed (10,000 km/s) through the open channel, raising the temperature in the channel to 30,000°C and pressure to 10–100 atmospheres. This sudden pressure increase causes a shock wave that propagates through the atmosphere and is heard as thunder. The charges keep bouncing up and down between cloud and ground as many as 50 times, at intervals of 10 ms, for a total duration of about 0.5 second for the entire event. Electrons and holes are lost during each trip, however. When not enough are left to keep the channel open, the flow stops, and the channel collapses in one last blast of thunder.

There are, of course, many variations on this theme, depending on the length of the path, the type of cloud, the amount of charge, the relative humidity of the air, and so forth. The figures that were given above illustrate an average event.

The light of lightning is due to the neutralization of ionized gases in the envelope around the lightning stroke. Lightning emits not only visible light but also x-rays, ultraviolet light, and radio waves. This is the reason that a radio "crackles" during a thunderstorm. The sound is due to a shock wave that develops when the stroke channel forms and subsequently collapses. Whereas light travels practically instantaneously, sound travels in air at a speed of 331 m/s (at 0°C). If you count the number of seconds between the time you see lightning and the time you hear the thunder, and divide by 3 (it takes about 3 seconds for sound to travel 1 km), you can tell the approximate distance of the lightning in kilometers. Because the electron deficiency under a charged cloud is particularly high at elevated points, lightning rods are placed on buildings and connected to the ground by means of a thick wire, a rod, or a pipe to lead the charge into the ground. The return stroke is led through the lightning rod out of the ground to discharge into the cloud.

Tornadoes are violent weather disturbances along cold fronts that lead into cyclonic lows. Warm air rising along the cold front condenses into a line of thunderclouds called a *squall line*. The energy liberated by the condensation of water vapor energizes turbulent motion along

Figure 13.23. A typical tornado. (Donn, 1965, p. 229.)

the squall line. In the Northern Hemisphere, warm air flowing into a cyclonic low is deflected to the right. This deflection creates secondary pressure lows along the squall line (Fig. 13.20). The adjacent warm air moves into these secondary depressions, following a counterclockwise path. As the air approaches the pressure minimum, it rotates faster and faster. The reason is, of course, conservation of angular momentum.

A tornado develops when the vortex forms a funnel extending downward from the bottom of the thundercloud (Fig. 13.23). This extension downward is not caused by the wind moving downward, but by the air below being drawn into the vortex. The motion of air within the vortex is *upward*. The pressure at the center of the vortex is estimated to be 10–15% less than that outside. This sudden pressure decrease causes the air drawn into the vortex to become

immediately saturated. The water vapor condenses, and that makes the funnel visible.

Horizontal wind speeds within a tornado can reach 200–500 km/h. Upward wind speeds may be as high as 300 km/h. There are no precise measurements of wind speeds within tornadoes, because the instruments themselves are destroyed. Wind speeds are estimated from the physical damage and other physical evidence. Updraft speeds are estimated on the basis of objects such as roofs, cattle, farm machinery, and even people that have been uplifted and carried for hundreds of meters.

The vortex is counterclockwise in the Northern Hemisphere because it is around a pressure low. The widths of vortices at ground level average about 300 m, with a range from less than 20 m to more than 1 km. Within the main vortex are smaller, temporary vortices that can rotate either clockwise or counterclockwise. They are created by turbulence within the main vortex and are responsible for the hissing noise of tornadoes.

The energy developed by an average tornado is about 0.1 kiloton (10^{11} joules). Although this is only one thousandth of the energy developed by an average thunderstorm, the energy is highly concentrated in both space and time. The power of a tornado is therefore very intense. Tornadoes move parallel to the cold front at speeds of 40–60 km/h. The path of a tornado along the ground averages 5 km in length and often is discontinuous. A tornado lasts, on the average, only 5–10 minutes.

In the United States, tornadoes are most frequent in an area called *tornado alley* that extends from northern Texas to Illinois and Indiana. In this area, 1,200 km long and 600 km wide, as many as 300 tornadoes touch ground each year. No structure can survive the fury of a tornado. What makes tornadoes so destructive is the same effect that holds airplanes in the air. Air blowing on an airplane wing creates a low-pressure zone above the wing that forces the airplane up. Tornado winds blowing against structures create pressure differences that blow them apart. Fig. 13.24 shows an example of tornado damage.

Waterspouts are tornadoes that develop at sea when seawater is sucked up from the ocean surface. Waterspouts are much less powerful than tornadoes because the temperature con-

Figure 13.24. Damage at Hubbard, Ohio, caused by a tornado that touched ground on the evening of May 31, 1985. The diameter of a tornado funnel at ground level may be only a few tens to a few hundreds of meters. The path of destruction is therefore very narrow. (*Akron Beacon Journal.*)

trast between adjacent air masses at sea usually is much less than that over land. Wind speeds within waterspouts rarely exceed 80 km/h. Waterspouts are common in the waters off Florida (Fig. 13.25).

Hurricanes are similar to tornadoes in their mode of formation, except that the scale is much larger because the planetary wind system is involved rather than local air masses.

Fig. 13.14 shows that there are some regions of the world where adjacent planetary winds move in opposite directions. One of these regions is the area around the Cape Verde Islands off West Africa during the northern summer. There the southern trade winds cross the equator and veer to the right to join the counterclockwise circulation around the pressure minimum over North Africa. As they veer to the right, the southern trades are bordered to the west by the northern trades, which flow in the opposite direction. The two diverge, and the zone in between is not only a center of low

pressure but also a site of counterclockwise motion. Many such depressions develop each year, but on the average only six of them develop into hurricanes.

Other areas where planetary wind conditions favor the generation of hurricanes are off the west coast of Central America and off the Philippines during the northern summer. During the southern summer, conditions are favorable to hurricane development in the western Pacific and in the Indian Ocean south of the equator. There they are called *typhoons*.

A fully developed hurricane is an almost perfectly circular vortex, 600 km in diameter (Fig. 13.26). The pressure difference between center and periphery is 5–10%. At the center of the vortex is the eye, a core of warm, descending air about 10–20 km across. The highest winds circle the core in a band 10–50 km wide while rising from sea level to the tropopause. The highest wind speeds are found at 1–3 km of altitude on the right quadrant, reckoned

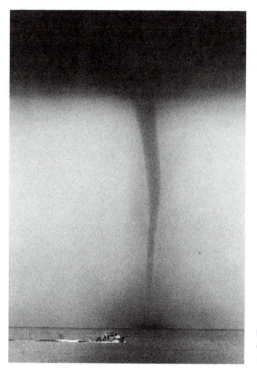

Figure 13.25. A waterspout off the Florida east coast. (Courtesy Bob Eighmie, *Miami Herald.*)

with respect to the direction of motion of the hurricane. The wind there may attain speeds of more than 350 km/h. At sea level, wind speed decreases rapidly beyond a distance of 50 km from the eye.

Condensation of water vapor is the source from which hurricanes derive their energy. Hurricanes develop only within the tropics, where absolute humidity is high. An average hurricane develops about a million megatons of energy. In most hurricanes, the highest sustained winds are 100–120 km/h, but some may reach 150 km/h or more.

Outside the 10–50 km-wide central vortex of high winds lie spiral bands of clouds. They are produced by convection cells that extend out to the periphery of the hurricane. The strength of the wind decreases from center to periphery. Fig. 13.27 shows the structure of a typical hurricane. Turbulence within a hurri-cane generates tornadoes, which make a hurricane even more dangerous.

Once developed, hurricanes move following the general circulation at speeds of 5–25 km/h. Atlantic hurricanes move from the Cape Verde region westward and then curve northward. The predominant hurricane tracks are shown in Fig. 13.28. Christopher Columbus was the first European to witness western Atlantic hurricanes. His ships encountered two, in 1494 and in 1495.

Hurricanes produce waves 10–12 m high in the open ocean. When a hurricane crosses a coastline in the Northern Hemisphere and moves inshore, the winds in the right quadrant push seawater inland. The sea level may rise as much as 5 m, causing extensive flooding. Waves on top of the advancing sea can cause almost total destruction of coastal areas (Fig. 13.29). Fortunately, the arrival of a hur-

Figure 13.26. Atlantic hurricane Gladys photographed on October 18, 1968, by the crew of *Apollo 7*. (Courtesy NASA.)

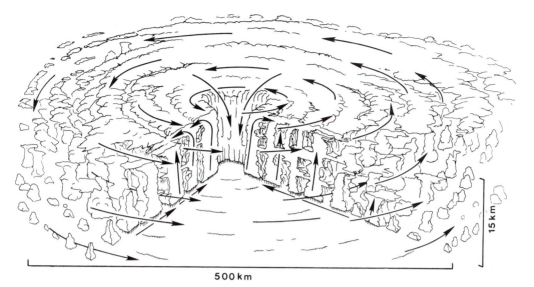

Figure 13.27. Structure and circulation of a Northern Hemisphere hurricane. The horizontal circulation is in the opposite sense in the Southern Hemisphere.

ricane, unlike that of an earthquake, can be fairly accurately predicted today.

As you can see by examining Fig. 13.19, when you face the wind in the Northern Hemisphere, the center of the hurricane is to your right. When you face the wind in the Southern Hemisphere, the center of the hurricane is to your left. This simple rule has helped many people to flee from the path of a hurricane, both on land and at sea. It did not help Admiral Halsey one bit in December 1944, because he did not follow it. Admiral Halsey

commanded Task Force 38 in the western Pacific. Task Force 38, with 90 ships, was part of a force assembled to invade the Philippines, which were then occupied by Japanese forces. On December 15 strong winds were blowing from the northeast. Halsey should have known that the center of the storm had to be to the southeast, but his weather expert told him that it was to the northwest. The weather expert may have reached that conclusion thinking that winds turn to the right in the Northern Hemisphere because of the Coriolis effect. This

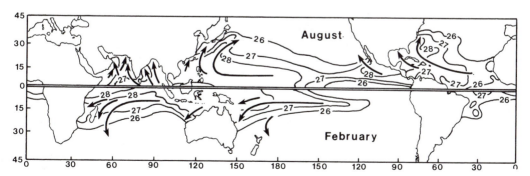

Figure 13.28. Paths and frequencies of hurricanes and typhoons. Because these storms occur mainly during the summer, the illustration shows the two hemispheres in their respective summers.

Figure 13.29. Hurricane damage in Galveston, Texas (September 8, 1900). Galveston, built on a sandbar off the Texas coast, had a population of 40,000 people when the hurricane struck. The hurricane, packing winds of 200 km/h, pushed ashore waves 5 m high. No structure could resist such pounding. The city was leveled, and 6,000 people died. The city was rebuilt behind the protection of a powerful seawall. Low-lying, unprotected coastal areas from the Gulf of Mexico to New England remain prone to hurricane destruction by pounding waves. (Courtesy Rosenberg Library, Galveston, Tex.)

is true for winds streaming out of a pressure high, but it is not true for winds streaming into a pressure low (Fig. 13.19). Following the advice of his expert, Admiral Halsey turned toward the southeast. On December 17, at noon, the wind was still blowing from the northeast and was strengthening. At that point, Admiral Halsey must have recognized the error and realized that the eye of the typhoon was to the southeast, not to the northwest. Knowing that the paths of typhoons and hurricanes tend to turn to the right in the Northern Hemisphere, Admiral Halsey ordered his fleet to turn west. But the typhoon was not turning right fast enough, and so the next day Admiral Halsey found himself crossing the path of the typhoon. He lost three destroyers and 790 men. But that was not all. Admiral Halsey found himself threatened by another typhoon six months later, while preparing to invade Okinawa. In an attempt to escape from that storm, the admiral followed a course that again took him directly into the eye of the typhoon. This time he lost 76 airplanes and six men. Fig. 13.30 shows Admiral Halsey's course in December 1944 and the path of the typhoon.

Today, storms are monitored by weather satellites. Daily reports from the weather bureaus keep people across the world informed so that appropriate precautions can be taken, both on land and at sea.

THINK

* The solar constant for the Earth is 1,360 W/m². Calculate the solar constants for Mars and Pluto.

* For the continuing addition of CO_2 to the atmosphere to lead to a catastrophic temperature increase, a positive-feedback mechanism must be operating. Can you figure out one? How about two?

* List the inner planets in order of increasing divergence of their back-radiation spectra from a blackbody spectrum at their respective temperatures.

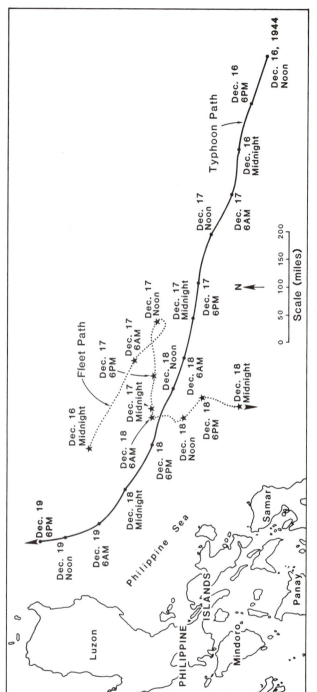

Figure 13.30. The path of typhoon Cobra and the path of Admiral Halsey's fleet in December 1944. (Courtesy U.S. Navy.)

* Redo Fig. 13.5 for an atmosphere containing no ozone.

* What happens to the Van Allen belts during a geomagnetic reversal?

* When, according to the Bible, did the Coriolis effect disappear?

* You are traveling by train from Chicago to New Orleans, and you tell the conductor that you are just *dying* to experience the Coriolis acceleration. What could he do to accommodate you?

* Would you expect the monsoon circulation at the time when Pangea was in existence to be more or less widespread than today?

* Compare the upper illustrations on pages 268 and 269 with the round illustrations just below them. What does each of the round illustrations represent? How come the round illustrations on the left side of each page are so different while the ones on the right side are so similar?

* The energy released by condensation of water in the tropical atmosphere, where most of the evaporation occurs, is 539 cal/g, plus about 100 cal/g because the water droplets are cooled down to 0°C (and below) in rising through the troposphere. The total energy released is thus 639 cal/g $= 2.7 \cdot 10^3$ J/g. Calculate the energy (in megatons) released by a tropical rainstorm dropping 5 cm of water over an area of 1,000 km².

* The energy needed for water to pass from the fully free state to the fully bonded state is 719 cal/g. Calculate the strength (in electron volts) of the bond between H_2O molecules.

* How do the droplets in the water vapor issuing from a boiling teakettle form?

* A lightning discharge may last 0.55 s and carry an average current of 1,500 A across a voltage drop of 10^9 V. Calculate the energy dissipated, and compare it with that of a B-52H Stratobomber flying at full speed (see Section 2.6).

* Calculate the radius of a spherical drop of water that would contain as many electrons as are involved in a giant lightning discharge of 100 coulombs.

14

THE HYDROSPHERE

Planet Earth has variously been called Planet Ocean, Planet Water, or the Blue Planet, because of the abundant water on its surface. Liquid water, which we take for granted, is a rare commodity in the cosmos. There is no liquid water on the Sun nor anywhere else in the solar system. Not a drop of water can be found in the immensity of interstellar space. Only on a planet of the right mass and chemical composition, and at the right distance from a neighboring star, can liquid water be found. There life may evolve and possibly flourish. It certainly has on Earth (see Chapter 19). Here we have essentially pure water (water consisting of only H_2O) in droplets condensing from a clean atmosphere, fresh water (water with a fraction of 1% dissolved salts) from springs, seawater (water with a few percent dissolved salts), and brines (solutions saturated with salts). These different waters are found in different geological environments. The largest reservoir is the ocean.

The Oceans

14.1. COMPOSITION

The oceans (and their marginal seas) cover 71% of the Earth's surface ($362.03 \cdot 10^6 \, km^2$ out of a total of $510.05 \cdot 10^6 \, km^2$). The volume of the oceans is $1,350 \cdot 10^6 \, km^3$. The mean depth of the oceans is 3,729 m, and that of the oceanic basins is 4,500 m. The areas and mean depths of the different oceans are shown in Table 14.1.

The oceans have an average salinity of 34.72‰. By comparison, *fresh water* has a salinity ranging from 0.2‰ to 4‰, and *pure water* consists only of H_2O molecules (with or without dissolved atmospheric gases in equilibrium with ambient temperature and pressure).

In the water molecule (Fig. 14.1A), the two hydrogen bonds form an angle of 104° 31', with the oxygen at the apex. In pure ice, the angle is 109° 28'. As a result, pure-ice density is 0.92, compared with 1.000 for water (at 3.98°C). The density of pure water increases as temperature is lowered. At 3.98°C, pure water reaches its maximum density (1.000 g/cm³). If the temperature is lowered, density decreases, until, at 0°C, ice forms. The decrease in density between 3.98°C and 0°C is due to the formation of domains with ice structure within the liquid water. These domains increase in number and size until ice forms. The electron density around the water molecule and the modes of vibration of the molecule are shown in Figs. 14.1B and 14.1C.

Ions dissolved in seawater prevent the formation of domains with the ice structure. As a result, the density of seawater increases as temperature is lowered until freezing occurs. Seawater of average salinity (34.72‰) has a density of 1.025 (at 15°C) and freezes at −1.775°C. The concentrations and residence times of major ions in seawater are shown in Table 14.2. Those ions that do not enter biologically precipitated minerals, such as Cl^- and Na^+, are subducted with wet sediments and resurface through submarine or sub-aerial volcanic vents. This cycle is very long. At the other extreme, the bicarbonate ion, which enters biologically precipitated carbonates, undergoes, in part, the precipitation–solution cycle within the ocean and has a much shorter residence time.

Surface temperatures in the oceans range from 25–28°C in the tropics to 0°C in the polar seas (Fig. 14.2), and salinities range from 35–36‰ to 32–33‰ (Fig. 14.3). Bottom temperatures and salinities are about 1–2°C and 34.7–34.9‰, respectively (Fig. 14.4). Temperature and salinity ranges in the marginal seas of the Atlantic Ocean are shown in Table 14.3.

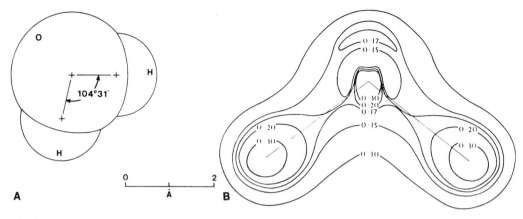

Figure 14.1. (A) The water molecule consists of two hydrogen atoms bound to an oxygen atom. The illustration shows the relative sizes of the hydrogen and oxygen components. (Emiliani, 1988, p. 32, Fig. 2.4.) Bond length is 0.95717 Å. Bond strength is 9.69 eV (4.43 eV to remove one hydrogen, and 5.19 eV to remove the other). The two hydrogen atoms form an angle of 104°31′ with the oxygen atom in liquid water or water vapor. In ice the angle is 109°28′, making ice a more open structure than liquid water and giving it a lower density (0.91 g/cm^3 vs. 1.00 g/cm^3). (B) Electron density in the water molecule. (Eisenberg and Kauzman, 1969, p. 26, Fig. 1.6.) (*Continued*)

14.2. CIRCULATION

The ocean is subject to four types of mass motions—surface currents, deep circulation, tides, and tsunamis. The energy source for the surface currents and deep circulation is solar radiation. The energy source for tidal motions is the differential gravitational attraction by the Moon and the Sun on different parts of the Earth's surface. The energy source for tsunamis is the Earth's internal energy.

The mean amount of solar radiation per unit surface normal to the Sun's rays at the mean distance of the Earth from the Sun, that is, the solar constant, is equal to 1.950 cal cm^{-2} min^{-1} = 1,360 W/m^2 (Section 13.1).

This amounts to 340 W/m^2 of the Earth's surface. Incoming solar radiation heats the ocean from above. This creates a stable stratification, with warmer, lighter water above, and colder, denser water below. There would not be any vertical circulation were it not for the fact that as water is warmed, evaporation increases, salinity increases, and the surface water becomes denser than the water below. During the Mesozoic era, when there was little or no ice on earth,

Table 14.1. Areas and mean depths of the oceans

Name	Area (10⁶ km²)	Mean depth (m)
Pacific	166.241	4,188
Atlantic	86.577	3,736
Indian	73.427	3,872
Arctic	12.257	1,117
All oceans and seas	362.033	3,729

Table 14.2. Concentrations of major ions in average seawater (salinity = 34.72‰)

Ion	Average Concentration (g/kg)	Residence Time (years)
Cl^-	19.107	10^8
Na^+	10.624	$4.8 \cdot 10^7$
SO_4^{2-}	2.677	$7.9 \cdot 10^6$
Mg^{2+}	1.278	10^7
Ca^{2+}	0.407	$8.5 \cdot 10^5$
K^+	0.394	$5.9 \cdot 10^6$
HCO_3^-	0.143	$8 \cdot 10^4$
Br^-	0.066	10^8
Sr^{2+}	0.008	$4 \cdot 10^6$
B^{2+}	0.0045	10^7
F^-	0.0015	$5 \cdot 10^5$

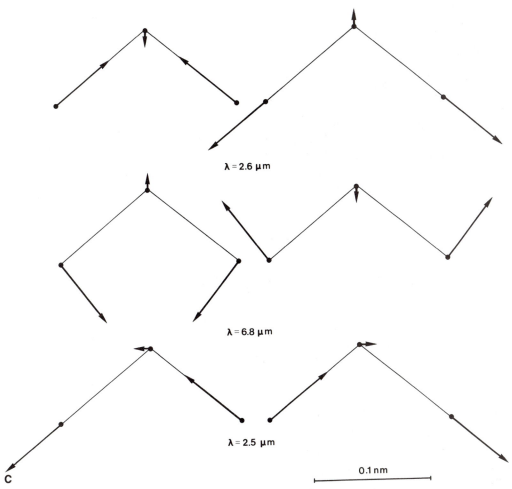

$\lambda = 2.6 \; \mu m$

$\lambda = 6.8 \; \mu m$

$\lambda = 2.5 \; \mu m$

C

0.1 nm

Figure 14.1 (cont.). (C) The water molecule, like all molecules, vibrates, with bond length changing by about 7%. The illustration shows the modes of vibration of the water molecule and their wavelengths. The length of an arrow indicates the amount of displacement (magnified 10 times with respect to the mean bond length). Notice that the vector sum of momentum vectors for the two hydrogen atoms and the oxygen atom, which is eight times more massive than the two hydrogen atoms combined, is zero.

bottom water probably formed by the sinking of warm, saline surface water. At that time, the Atlantic Ocean was narrower than today, and planktic life was plentiful. The sinking warm, saline water was rich in organic matter, and the bottom of the Atlantic became anaerobic from time to time. During the Cenozoic era, ice returned at the high northern and southern latitudes, and global temperature decreased. Sea ice began forming at high latitudes during the winter.

When sea ice forms, the ions dissolved in seawater cannot enter the ice structure—these ions are left behind, increasing the salinity of the seawater just below the sea ice. First, pure ice crystals separate, increasing the concentration of salts in the remaining brine. If the temperature is lowered to $-8.2°C$, the solubility of Na_2SO_4 is exceeded, and Na_2SO_4 begins to precipitate. At $-23°C$, NaCl begins to crystallize, together with a trace of $CaCO_3$. At $-30°C$, the composition of the system is as shown in Table 14.4. At $-30°C$, the brine is a saturated solution (31‰) of Cl^-, Na^+, Mg^{2+}, Ca^{2+}, K^+, Br^-, and SO_4^{2-}. If freezing occurs very rapidly, seawater is trapped between the ice crys-

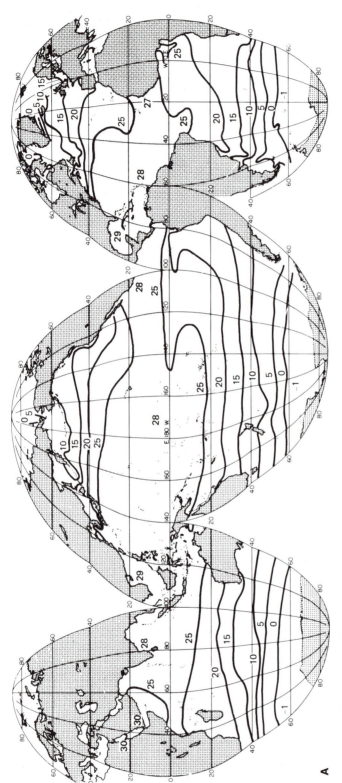

Figure 14.2. (A) Surface temperatures of the oceans in August. (Gross, 1987, pp. 156–7, Fig. 7.2.)

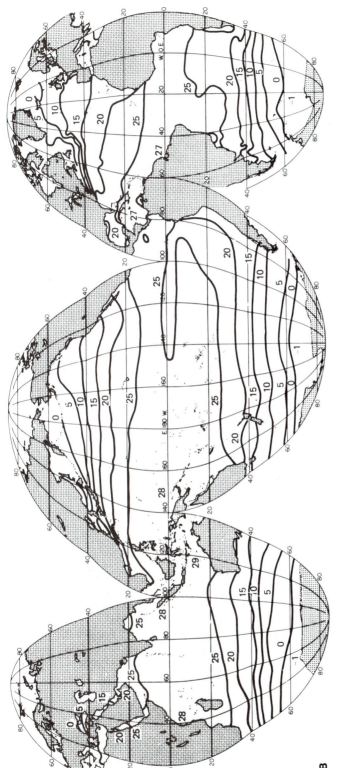

Figure 14.2. (B) Surface temperatures of the oceans in February. (Gross, 1987, pp. 156–7, Figure 7.2.)

291

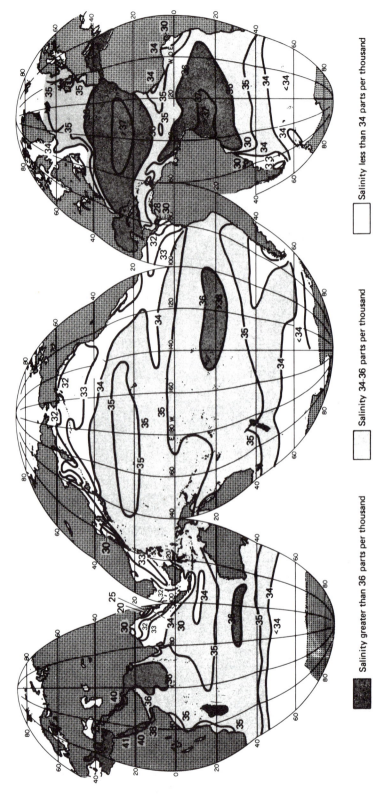

Figure 14.3. Surface salinity during the northern summer. (Gross, 1987, p. 159, Fig. 7.4.)

Salinity less than 34 parts per thousand

Salinity 34-36 parts per thousand

Salinity greater than 36 parts per thousand

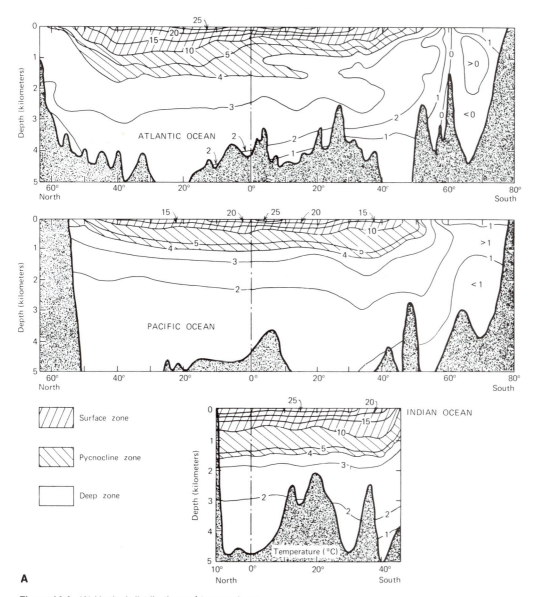

Figure 14.4. (A) Vertical distributions of temperatures in the oceans. (Gross, 1987, p. 166, Fig. 7.12.) (*Continued on p. 294.*)

tals, and sea ice with a salinity approaching that of seawater may form. Normally, however, the salinity of sea ice ranges from 3‰ to 10‰.

The brine below sea ice is dense and may sink all the way to the bottom. Bottom-water formation brings dissolved atmospheric gases to the ocean floor, renewing the oxygen supply and making life on the ocean floor possible.

The solubilities of dissolved gases are shown in Table 14.5. Today, bottom-water formation occurs during the winter in the Labrador Sea and in the Weddell Sea. From the Labrador Sea the North Atlantic Deep Water (NADW) flows south, clinging to the slope of the North American continent because of the Coriolis effect. The Weddell bottom water forms the

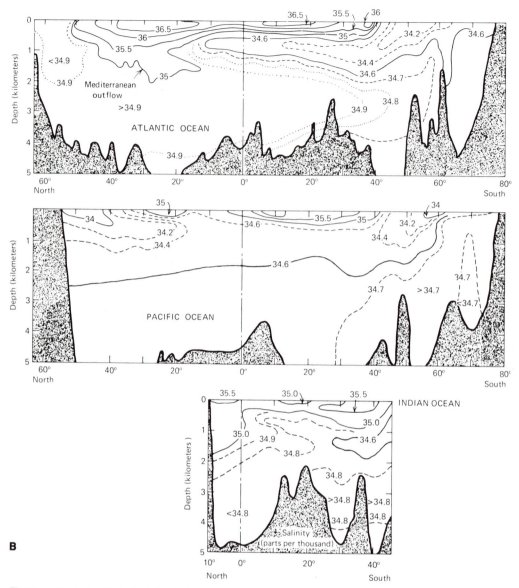

Figure 14.4 (*cont.*). (B) Vertical distributions of salinities in the oceans. (Gross, 1987, p. 166, Fig. 7.12.)

Table 14.3.   Atlantic marginal seas

Sea	Surface water		Deep water	
	Salinity (‰)	Temperature (°C)	Salinity (‰)	Temperature (°C)
Caribbean–Gulf of Mexico	36.5	20–29	35.0	3.8
Mediterranean	37.5–38.5	13–25	38.5–38.7	13
Black Sea	17	4–24	22	9
Baltic Sea	7	0–16	12–15	5

Table 14.4. Composition of frozen seawater (at $-30°C$)

Component	Concentration (g/kg)
ice crystals	931.9
NaCl crystals	20.23
Na_2SO_4 crystals	3.95
$CaCO_3$ crystals	trace
brine	43.95

Table 14.5. Gases dissolved in seawater at 1 atm

Gas	Amount (mL/L)		
	0°C	10°C	20°C
N_2	14.3	11.4	9.5
O_2	8.4	6.3	5.2
CO_2	1.76	1.63	1.50
Ar	0.4	0.3	0.25

Antarctic bottom water (AABW), which flows eastward along the floor of the Weddell abyssal plain. There it divides into two branches. One branch enters the eastern Atlantic through fracture zones in the Atlantic–Indian Ridge and flows northward, clinging to the eastern slope of the Mid-Atlantic Ridge south of the equator, and to the continental slope of Africa and Europe north of the equator (again because of the Coriolis effect). The other branch continues eastward and then flows northward along the deep basins of the Indian and Pacific oceans. It eventually reaches the Gulf of Alaska. During this long voyage, benthic animals consume oxygen. The concentration of oxygen in the bottom water drops from about 6 mL/L in the North Atlantic to 2 mL/L in the North Pacific. Bottom water has an age of 250 y in the Atlantic and 1,500 y in the North Pacific. The vertical circulation of the ocean is due to two factors— temperature and salinity—and therefore is called *thermohaline circulation* to distinguish it from the surface circulation, which is wind-driven.

Fig. 14.5 and Table 14.6 show the major ocean currents of the world and the amounts of water they transport. All of them are driven by the prevailing winds. The speed of a current is highest at the surface, amounting to about 2% of the speed of the wind that generates it. By comparison, the maximum flood discharge of the Amazon, the largest river in the world, amounts to only 200,000 m^3/s. It would be difficult to visualize a volume of 1 million m^3 were it not for the fact that that is just about the volume of the Empire State Building in New York (which is 1,057,723 m^3). So we now introduce a new unit of volume, the gigaliter (GL), defined as being equal to $10^6 m^3 = 10^9$ liters. The gigaliter per second already exists as a unit of oceanic current transport. It is called the sverdrup (sv), after the Norwegian oceanographer Harald U. Sverdrup (1888–1957).

Surface currents are influenced by the Coriolis effect. Surface flow is offset by 45° (to the right in the Northern Hemisphere, to the left in the Southern Hemisphere) with respect to the wind direction because of this effect. The velocity vector is further offset as depth increases, until, at a depth called the *friction depth*, it reaches a direction opposite that of the wind. As the magnitude of the velocity vector decreases with depth, the net transport, called *Ekman transport*, is at 90° (to the right in the Northern Hemisphere, to the left in the Southern Hemisphere) to the wind direction. The velocity vector forms a spiral called the *Ekman spiral* (Fig. 14.6), which has positive helicity in the Northern Hemisphere and negative helicity in the Southern Hemisphere. The spiral is named after the Swedish oceanographer V. Walfrid Ekman (1874–1954), who explained the effect in 1902.

Atmospheric pressure lows and converging ocean current tend to elevate (by 1 m or so) the ocean surface. Conversely, pressure highs and diverging currents tend to depress it (also by 1 m or so). Water flowing from highs to lows does not take the direct path of maximum declivity, but follows contour lines, again because of the Coriolis effect. These currents are called *geostrophic currents*.

Air and water masses moving along the Earth's surface have both momentum and angular momentum (because the Earth's surface is a sphere). We can imagine an instantaneous vector representing the net sum of all angular momenta of both atmosphere and ocean. When a current, atmospheric or oceanic, slows down, accelerates, or changes size or course, the net

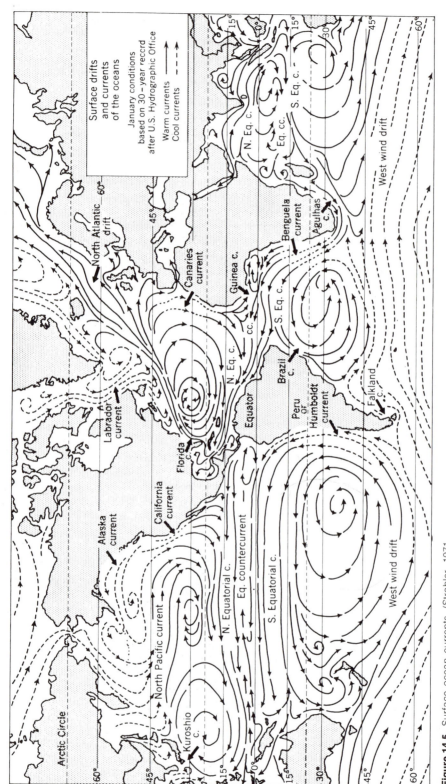

Figure 14.5. Surface ocean currents. (Strahler, 1971, p. 260, Fig. 16.6.)

Table 14.6. Surface ocean currents

Name	Transport ($\times 10^6$ m³/s)
Antarctic circumpolar	125
Antilles	12
Benguela	16
Brazil	10
California	10
Canaries	16
Caribbean	26
East Greenland	7
Equatorial Countercurrent	25
Florida	26
Gulf Stream	55
Kuroshio	65
Labrador	6
North Atlantic	10
North equatorial (Pacific)	30
South equatorial (Pacific)	10
West Spitsbergen	7

angular momentum will be affected. The result is a contribution to the *Chandler wobble* (Fig. 14.7), a wobble of the Earth's axis with an amplitude of 8 m around the mean north pole and periods of 12 and 14 months. Interference between these two periods (84 months) causes the radius of the wobble to go to zero every 7 years. The 12-month period is obviously seasonal, and the 14-month period may originate from motions in the fluid outer core.

14.3. MARGINAL SEAS

The major marginal seas are shown in Table 14.7. The Caribbean and Gulf of Mexico are flushed with $30 \cdot 10^6$ m³/s of surface Atlantic seawater that enters through the passages between the islands of the West Indies island arc, continues through the Yucatan passage, and exits through the Florida Straits as the Florida current (which becomes the Gulf Stream after being reinforced by the Antilles current). The bottom water is at about 3.8°C, the temperature of the North Atlantic Deep Water at the level of the Jungfern Passage in the Virgin Islands ($-1,815$ m), through which the North Atlantic Deep Water enters the Caribbean.

The Mediterranean communicates with the ocean only through the 320-m-deep Strait of Gibraltar. The temperature at the top of the sill (-320 m) is 13°C, and so is the temperature of the bottom water that forms during the winter off the Franco–Italian Riviera. As a result, the Mediterranean maintains this temperature to the bottom, which reaches a depth of 3,719 m in the Tyrrhenian Sea and 5,530 m in the Ionian Sea. In spite of contributions of 6,300 m³/s from the Black Sea and another 3,000 m³/s from the Nile, the Mediterranean loses a net of 76,500 m³/s because of excessive evaporation, followed by precipitation over areas that drain directly into the Atlantic or the Indian Ocean, not back into the Mediterranean. This loss, amounting to 4.5 times the average Mississippi discharge of 16,800 m³/s, is made up by a net influx from the Atlantic. The excessive evaporation raises the salinity of the surface water of the Mediterranean to 37.5–38.5‰ (Table 14.3). The Mediterranean water at the level of the Gibraltar sill is denser than the adjacent Atlantic water, and so it flows out under an inflowing current of Atlantic water. The inflowing current is 1,680,000 m³/s, and the underlying outgoing current is 1,603,500 m³/s. The net inflow is thus 76,500 m³/s, which balances the evaporation excess. Because of its distinctive characteristics (high salinity and temperature) the Mediterranean water can be traced across the Atlantic, both northward (to Greenland)

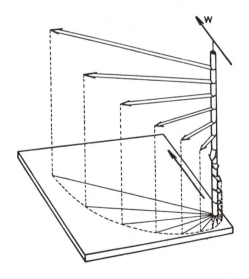

Figure 14.6. The Ekman spiral (W = wind). (Sverdrup, Johnson, and Fleming, 1942, p. 493, Fig. 121.)

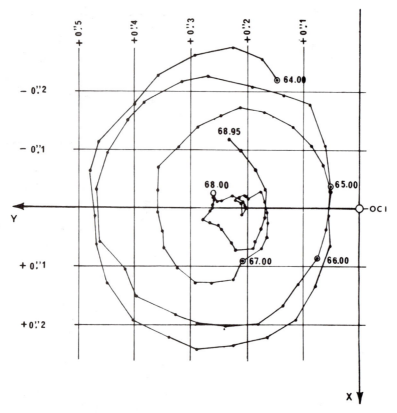

Figure 14.7. The Chandler wobble. The Earth's axis of rotation precesses with a period of 25,800 y. Super-imposed on this motion is short-period wobble of the Earth with respect to its own axis of rotation. The result of this wobble is a circular motion of the pole around a center that represents the axis of the greatest moment of inertia. The wobble has two superimposed periodicities: 12 and 14 months. The two coincide every 84 months (7 years), 84 being the minimum common multiple of 12 and 14, with the result that the wobble vanishes every 7 years. The maximum distance from the center is about 0.2″ of arc or about 6 m. The dots indicate time intervals of 1/20 of a year. (Stacey, 1977, p. 62, Fig. 3.5.)

and southward (around South Africa and even into the Indian Ocean).

The Black Sea has an excess of precipitation and runoff over evaporation, amounting to $6,300 \, m^3/s$, and so it exports water to the Mediterranean. However, a situation analogous to that at Gibraltar develops across the Bosporus, but on a much smaller scale because the Bosporus is shallow (40–90 m). A thin tongue of Mediterranean water, being denser than the adjacent Black Sea water, leaks along the bottom of the Bosporus into the Black Sea and spreads along the bottom. The amount is negligible by oceanic standards, only $6,100 \, m^3/s$. Enough organic matter accumulates on the floor of the Black Sea to immediately remove whatever oxygen comes in with the Mediterranean water, and so the bottom of the Black Sea is anaerobic. Salinity increases from 17.5‰ at the surface to 22.5‰ at the bottom, producing a stable stratification. Because no bottom water forms in the Black Sea, the entire water column up to 150 m from the surface is anaerobic. There is no dissolved oxygen. In its place there is H_2S in the amount of 6–8 mL/L. Temperature is constant at 8.7°C from −150 m to the bottom. Because of the anaerobic conditions, the organic matter that accumulates on the floor of the Black Sea is not oxidized. In due time it will be transformed into petroleum.

The Caspian Sea ($370,800 \, km^3$, with a mean depth of 1,025 m) and the Aral Sea ($64,500 \, km^2$,

Table 14.7.   **Marginal seas**

Name	Surface ($\times 10^6$ km²)	Mean Depth (m)
South China Sea	2.795	1,437
Caribbean Sea	2.515	2,575
Mediterranean Sea	2.510	1,502
Bering Sea	2.261	1,492
Gulf of Mexico	1.507	1,614
Sea of Okhotsk	1.392	973
Sea of Japan	1.013	1,667
Black Sea	0.461	1,197
Baltic Sea	0.154	55

with a mean depth of 67 m) are two large inland bodies of salty water (13‰ and 11‰ salinity, respectively) that were connected with the Black Sea during the last deglaciation (~ 15,000–10,000 y ago).

The Baltic Sea has an excess of precipitation and runoff over evaporation, amounting to 15,000 m³/s. It drains into the North Atlantic through the Denmark strait. Surface salinity decreases from 15.20‰ in the Kattegat to 1‰ at the head of the Gulf of Bothnia, and bottom salinity decreases from 35‰ to 2‰.

14.4. WAVES

Waves are produced by wind stress on the surface of the water. A minimum wind speed of 1 m/s is needed to create a wave. In waves with wave heights greater than 2–3 mm, the restoring force is gravity; in waves with wave heights less than 2–3 mm, the restoring force is surface tension.

In the open ocean, waves commonly have wavelengths of 60–120 m, velocities of 10–15 m/s, periods of 6–9 s, and heights of 1–2 m.

[*Wave height* is the vertical distance between the bottom of a trough and the top of the adjacent crest; *wave amplitude* is the vertical distance between the mean level and the top of the crest or the bottom of the trough. Wave amplitude is thus one-half of wave height.]

The velocity of a wave is a function of wavelength L. If the depth of the bottom is greater than the wavelength L, the velocity v is given by

$$v = (gL/2\pi)^{1/2} = (1.56L)^{1/2}$$

where g is gravitational acceleration (9.81 m/s²). If the bottom depth is less than the wavelength L, the velocity v is given by

$$v = (gh)^{1/2}$$

[Remember that *wave velocity* actually means *phase velocity*. What moves is not the water, but the crests and troughs, that is, the *phase* of the water surface.]

The water particles on the surface of the ocean move in a vertical, almost circular, prograde orbit with a radius equal to the amplitude of the wave. This radius decreases with depth. At a depth equal to $\frac{1}{2}L$, the radius is 4% of that at the surface, and at a depth equal to one wavelength, the radius is 0.2% (Fig. 14.8). The orbits are not exactly closed, with the result that there is a net movement of the water forward. This transport is proportional to wave amplitude and inversely proportional to wavelength. It ranges from a few centimeters per second to more than 1 m/s. This motion is not related to wind friction on the water surface, although wind friction is what creates the wave.

Storms with strong, sustained winds can produce waves 10–15 m high (the maximum on record is 34.14 m, measured on February 6–7, 1933, by the USS *Ramapo* in the Pacific during a typhoon).

The water moves forward significantly only when it approaches land, forming breakers. The forward speed of breakers is on the order of 1 m/s. Strong breakers move a lot of water— 1 m³ of water (about the size of a desk) has a mass of 1 ton. Moving water has a lot of momentum. Wind of hurricane force (120 km/h or more) can pile up water 10–15 m high against a sloping coastline, and create breakers on top of that. Structural damage usually is extremely severe.

A *swell* is a broad wave characterized by a long period (30–300 s), a long wavelength (several hundred meters), and an appreciable height (a few meters) created by major, prolonged storms at sea. It travels at high speed (several hundred kilometers per hour) out of the storm area and across the ocean for as

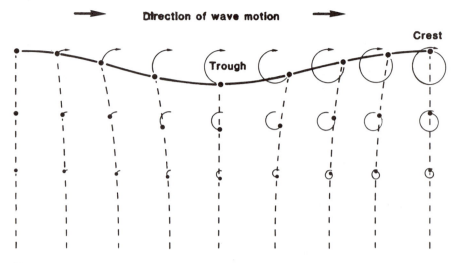

Direction of wave motion

Crest

Trough

Figure 14.8. Circular, prograde motion of particles in an open-ocean wave.

much as 10,000 km without appreciable loss of energy. Upon approaching land, giant breakers form, to the delight of surfers.

Tsunamis are surface waves of very long period (3 minutes to 3 hours), long wavelength (up to 200 km), low amplitude (centimeters to decimeters away from the source and terminus), and high speed (600–800 km/h) produced by submarine or coastal earthquakes. A tsunami 200 km across and 0.25 m high delivers 50,000 tons of seawater per meter of coastline. Damage is particularly severe if the water is funneled up a narrowing bay. At the head of such a bay, where often a harbor is located, seawater may splash to a height of 50 m.

[*Tsunami* is a Japanese word meaning "harbor" (*tsu*) "wave" (*nami*).]

14.5. TIDES

Tides are caused by the *tide-generating force*, which results from the combined actions of the gravitational attractions and centrifugal forces in the Earth–Moon and Earth–Sun systems. For a body rotating on its axis at an angular velocity ω, the centrifugal force $\omega^2 r$ on any point at distance r from the rotational axis must equal the centripetal force $\omega^2 r$ produced by the body's own gravitational acceleration. As the Earth turns on its axis, the centrifugal

force causes the solid surface of the Earth and the surfaces of the ocean and atmosphere to bulge up toward the equator. The centrifugal force at sea level at the equator is equal to $\omega^2 R$, where R is the radius of the Earth. It decreases to zero at the poles.

In a system consisting of two nonrotating bodies of the same mass $m = 1$ and radius $R = 1$ orbiting around a common axis equidistant between the two centers of gravity (Fig. 14.9), the centripetal acceleration at the center of mass of each body is equal to $\omega^2 r$, where r is the distance between the center of mass of each body and the rotational axis. The centrifugal force is also $\omega^2 r$. The centrifugal force on each point mass of the two bodies is therefore linearly proportional to r.

The gravitational acceleration between the gravitational centers of the two bodies is inversely proportional to the square $(4r^2)$ of the distance $2r$ between the two centers. Therefore, the center of mass of each of the two bodies feels a gravitational attraction to $1/(2r)^2$. This attractive force must balance the centrifugal force. Therefore,

$$\omega^2 r = 1/(2r)^2$$

from which

$$\omega^2 r \times (2r)^2 = \omega^2 4r^3 = 1$$

for the center of mass of each body.

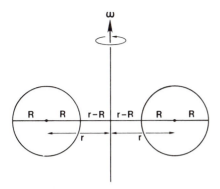

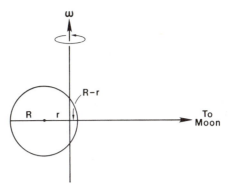

Figure 14.9. A system of two identical, nonrotating bodies revolving about a common axis.

Figure 14.10. The Earth–Moon system. The Earth and the Moon rotate around each other about a common axis that passes through the Earth, down to 1,649 km below the sublunar point.

The tide-generating force results from disequilibrium between gravitational attraction and centrifugal force at all points of the two bodies other than their centers of gravitation (if either body or both are also rotating about their own axes, the circumequatorial bulge produced by the centrifugal force $\omega^2 r$ does not count because it has the same height all around the body).

The point closest to the orbital axis feels a force equal to $\omega^2 \times 4(r - R)^3$, and the opposite point feels a force equal to $\omega^2 \times 4(r + R)^3$. The two forces are identical because r is the same for each of the two points, and the two R's are equal but in *opposite* directions [you will agree that $r - R = r + (-R)$].

In the Earth–Moon system, one of the bodies (Earth) is 81.286 times more massive than the other. The common axis around which the two bodies orbit lies, therefore, 1/81.282 of the distance between the centers of mass of the two bodies, or 4,729 km from the center of the Earth (Fig. 14.10). The sublunar point feels a force equal to $\omega^2 \times 4(R - r)^3$, where R is the radius of the Earth, and r is the distance from the center of the Earth to the common axis. The opposite point on the Earth's surface feels a force $\omega^2 \times 4(R + r)^3$. Again, the two forces are identical, because r is the same for each of the two points, and the R's are equal but in opposite directions.

In the Earth–Moon–Sun system, the tide-generating force of the Sun is 2.17 times smaller than that of the Moon, but its gravitational pull on the center of the Earth is 180 times greater.

In a land-free Earth there would be two tides per lunar day, one under the Moon and the

other on the opposite side of the Earth (semidiurnal tides). The continental shorelines, however, highly distort the progress of the tidal bulge.

[The lunar day is 24h 50m 28s long. It is longer than the solar day because the Moon revolves around the Earth in the same sense as the Earth's rotation. It takes an extra 50m 28s for a sublunar point to catch up with the Moon the next time around.]

Collisions of the tidal bulge with continental shorelines (especially the eastern coast of the Americas), and the fact that friction causes the tidal bulge to lag by 1° behind the Moon (i.e., the tidal crest is 1° ahead of the Moon because its formation is retarded, Fig. 14.11), combine to slow down the rotation of the Earth by $1.5 \cdot 10^{-5}$ s/y. In order to conserve the angular mo-

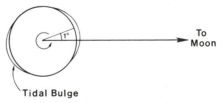

Figure 14.11. The Earth–Moon system seen looking down on the north pole. The Moon exerts its attraction on the sublunar point, but the inertia of the ocean causes the bulge to form 4 minutes later, by which time the Earth has rotated 1°. The attraction by the Moon on the bulge contributes to slowing down the rotation of the Earth.

mentum of the Earth–Moon system, the Earth–Moon distance increases by about 4 cm/y.

THINK

* If all the men and women on Earth decided to jump into the oceans at the same time, how much would sea level rise?

* With reference to Table 14.2, why does the Br$^-$ ion have such a long residence time in the ocean?

* An average nuclear reactor delivers 500 MW of power. The Florida current transports water in the amount of $26 \cdot 10^6$ m³/s at an average speed of 2 knots. If the total energy of the Florida current could he harnessed, how many nuclear reactors could be replaced?

* Each winter about 10^7 km² of the sea around Antarctica become covered with sea ice. Calculate how much heat and how much power are released by the freezing of that ice (assuming that the freezing takes 6 months). Also calculate how many nuclear reactors (of an average power of 500 MW) this source could replace.

* Calculate the volume of the Caribbean Sea (from Table 14.7), and knowing that the Caribbean is flushed with $3 \cdot 10^7$ m³ of Atlantic water each second, calculate how long it takes to replace the Caribbean water entirely.

* As explained in the text, the Moon recedes from the Earth 4 cm/y in order to conserve angular momentum. That recessional velocity probably has changed over time as the continents have moved around. Assuming, nevertheless, that it has averaged 4 cm/y, calculate how much closer to the Earth the Moon was at the beginning of the Cambrian (570 million years ago) and how much higher the lunar tides were at that time.

Groundwater and Surface Water

14.6. GROUNDWATER

Precipitation over the United States averages 76 cm/y, of which 45 cm evaporates, 30 cm flows back into the ocean (*runoff*), and 1 cm goes into groundwater. Groundwater percolates downward and accumulates above an impermeable layer to form a layer (*phreatic*

water) with a free surface (*water table*). The free surface is called the *piezometric* or *potentiometric surface*. *Vadose water* is the water between the surface of the ground and the water table. An *aquifer* is a rock layer of sufficient porosity and permeability to convey water for common usage.

Fresh water contains dissolved ions (~ 0.2–4 g/L). An example of superior fresh water is the San Pellegrino spring water (Italy), whose composition is shown in Table 14.8. Rainwater not only dissolves atmospheric CO_2, but also picks up more CO_2 and humic acids in percolating through soils. If the water is undersaturated with respect to a soluble substance (e.g., $CaCO_3$), it will tend to dissolve that substance. Water can circulate through cracks as narrow as 10 μm and enlarge them to 1 cm when turbulent flow takes over. This increases water flow and dissolution. If the country rock is limestone, solution systems that may include large caverns can form. The largest such cavern is the Mammoth Cave system in Kentucky, with a total of 530 km of interconnected passages. Passages are actually excavated by pressure erosion by the flowing water and exhibit a circular or elliptical cross section. Water flow usually is seasonal—during the dry season, passages may be only partly full of water, with the result that the ceiling and walls may collapse into rubble, which, having greater surface, is more easily eroded and dissolved. Large chambers can thus form. The largest one is the Sarawak Chamber in the Gunung Mulu

Table 14.8. Chemical composition of San Pellegrino spring water

Substance	Amount (mg/L)	Percentage
SO_4^{2-}	538.00	47.49
HCO_3^-	219.60	19.38
Ca^{2+}	201.60	17.80
Cl^-	67.40	5.95
Mg^{2+}	57.40	5.07
Na^+	41.20	3.64
Sr^{2+}	3.80	0.33
K^+	2.90	0.26
F^-	0.70	0.06
Li^+	0.18	0.02
Fe, Al, Cr, Mn, Ni, Cu, Zn	traces	—
Total	1,132.78	100.00

Figure 14.12. Stalactites (hanging from ceiling) and stalagmites (rising from floor) in the Carlsbad Caverns system, New Mexico. (Courtesy National Park Service, Carlsbad Caverns and Guadalupe Mountains National Parks, Carlsbad, N. Mex.)

National Park, Sarawak, which is 700 m long, 300 m wide, and 70 m high. The deepest cave reaches 1,535 m below ground in the French Alps.

If the water table drops, a cave may dry up. Water may keep dripping in, however, forming *speleothems* (i.e., stalactites, stalagmites, and crusts) as the water evaporates and the carbonate precipitates (Fig. 14.12). These precipitates are calcitic or aragonitic or both.

Solution by rain and vadose water in limestone terrain forms surface solution features that range from tiny pits to large sinkholes with vertical walls called *dolinas* or *cenotes* (Fig. 14.13). This type of terrain is called *karstic* (after Karst, a region in western Slovenia where such features are common).

During the ice ages, sea level was 120 m lower than today because of the ice accumulated on land. Water tables were generally deeper, especially where surface rocks were porous limestones (e.g., the Florida–Bahamian platform). Vadose waters dissolved the limestone and formed many caverns, sinkholes, and blue holes (submarine sinkholes) in and around Florida, the Bahamas, and Yucatan. Little Salt Spring in Florida is a shallow, water-filled basin 75 m across, with its floor sloping to a depth of 12 m in the center, where a hole, 25–30 m across, communicates with a 60-m-deep underwater cave (Fig. 14.14). The water is anaerobic. The cave was dry at peak of the last ice age, 20,000 years ago. As ice melted, the sea levels rose, and the cave became partly filled with fresh water. It was a most valuable resource in the otherwise dry Florida plateau, and it attracted populations of early Indians. There are ledges in the cave, now under water, where ancient Indians lowered themselves to catch animals that had been chased into the hole, had fallen into the water below, and were thrashing around. In the early 1970s, anthropologist Carl Clausen found, on a ledge 26 m below the present lake surface, the overturned shell of a giant tortoise with a sharp-pointed wooden stake in a position indicating that it had been driven through the heart. There were carbonized remains around the shell, showing

Figure 14.13. A large sinkhole near Montevallo in central Alabama (100 × 130 m at the surface, 45 m deep), formed in December 1972. (Courtesy U. S. Geological Survey.)

that the tortoise had been killed and cooked right on that ledge. The stake was found to be 12,600 years old, which makes it one of the oldest artifacts in the Western Hemisphere.

An aquifer underlying an impermeable rock layer may contain water under pressure and may form springs and artesian flows (Fig. 14.15).

14.7. STREAMS

Streams are water courses that channel runoff. The midstream line where the two sides of a valley intersect is called a *thalweg* (German for *path of the valley*). Streams have different names according to size and/or behavior: A river is a large stream; a creek is a small stream; a brook is a stream smaller than a creek; a torrent is a stream that is dry during the dry season.

In desert areas, torrents excavate flat-bottomed, steep-sided beds called *arroyos* in Spanish or *widian* (singular *wadi*) in Arabic. *Do not picnic on the bottom of an arroyo*: In the middle of your picnic you may hear a rumble and, looking up, see a 5-m-high wall of red muddy water laden with rocks and debris of all kinds racing toward you at very high speed. What happens is that sometimes it rains even in the desert, and when it does, usually it is a torrential rain on a restricted area. This area may be miles and miles upstream from you, and you may have a clear sky for your picnic. So there is no warning. The sight of an arroyo in full flood is a sight to behold. While the Mississippi in full flood may be termed by a poet "majestic" or even "awesome", an arroyo in full flood is pure fury—you will have no time to gather your victuals and split.

Stream gradients range from 100 m/km at their heads to 0.1 m/km near the mouths of large rivers. Streams erode their own beds where stream gradients are significant, deposit-

Figure 14.14. Little Salt Spring, Florida: cross section. (Clausen et al., 1979.)

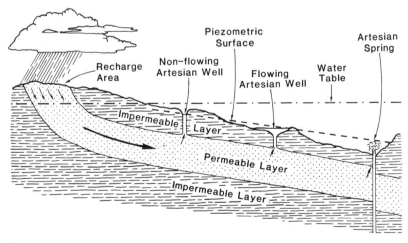

Figure 14.15. If an aquifer is covered by a sloping, impermeable layer, the water will be under pressure away from the recharge area. The horizontal line is the projection of the water-table surface in the recharge area. The sloping line is the piezometric surface to which water will rise if allowed to. Water will not rise naturally to the top line because of friction in rising through the overlying rock layers. If the piezometric surface is higher than the topographic surface, artesian flow will occur either through natural fractures (forming an artesian spring) or through a well (artesian well). The name *artesian* derives from Artois, a region in northwestern France.

Figure 14.16. Alluvial fan in Death Valley, California. (Courtesy J. S. Shelton, La Jolla, Calif.)

ing sediments and forming alluvial plains where the gradient is negligible. *Competence* is the ability of a stream to transport rock fragments (the larger the fragment, the greater the competence), and *capacity* is the ability of a stream to carry a sediment load. Streams are capable of transporting even large boulders if their water is laden with mud, because mud increases the density of the fluid, making the rocks easier to transport (Archimedes' law).

In its upper course, a stream usually has large competence (it can move large boulders downstream) but little capacity (the sediment load is comparatively small). As a stream exits from the mountains onto a plain, it drops sediments, forming an *alluvial fan* (Fig. 14.16). Adjacent alluvial fans may coalesce to form an *alluvial apron*. The *base level* is the level below which the stream connot erode any more. It then begins depositing sediments, forming a floodplain along the valley floor. If the base

level drops, the stream will incise its own bed and leave stream terraces on either side of the valley. Sea level dropped by about 120 m many times during the most recent ice ages, and so stream terraces were formed (Fig. 16.4).

A river system consists of a main river and its tributaries and subtributaries. The floodplains of the tributaries usually merge with the floodplain of the main river. Rivers in floodplains usually form *meanders* within the river channel. A meander is initiated by a minor topographic obstacle that deflects the stream sideways. This deflection forces the water to change its momentum profile—momentum is highest *below* the surface along the downstream side of the bend (causing maximum erosion) and lowest at the surface along the inside of the bend (where deposition occurs) (Fig. 14.17). The deflected water is bounced against the opposite side of the river channel while moving downstream. Because erosion is

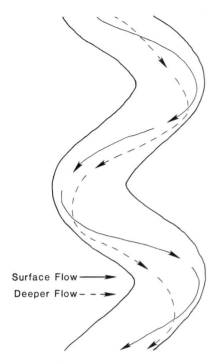

Surface Flow ⟶
Deeper Flow ⟶ (dashed)

Figure 14.17. Formation of meanders.

even minuscule slope gradients that the course of a river across an apparently perfectly flat floodplain can reveal even deeply buried structures. In the late 1940s, the Italian geologist Marco Piero Marchetti used this method to locate deeply buried hydrocarbon reservoirs in the Po Valley in northern Italy.

A plot of flood discharge versus flood frequency follows a Poisson distribution. Therefore, record floods are rare. One possible explanation for the flood legends so common in ancient traditions is that record floods have occurred in different regions at different times, and so each tradition has a memory of at least one record flood. The universal flood would then be simply a coalescing of those independent traditions.

Some floodplains, like the Po Valley in northern Italy, are so flat that the curvature of the Earth can be seen by climbing a belfry and observing with binoculars the belfries of distant towns—only the upper parts are visible. About 3 m of sediment have been deposited on the lower Po Valley since Roman times. As flooding (now constrained by levees) maintains the surface flat, tectonic motions produce increasing folding and faulting at increasing depth. In the Po Valley, the 5-million-year-old Miocene surface, now buried under 2 km of post-Miocene sediments in the center of the valley, exhibits folds with vertical amplitudes of several hundred meters.

Rivers emptying into the ocean form deltas or estuaries. *Deltas* usually are triangular piles of sediments deposited by a river at its mouth. The beds that form deltas have characteristic orientations (Fig. 14.18). A *topset bed* is a horizontal bed deposited on top of a delta; a *foreset bed* is a bed deposited on the slope of the delta front; a *bottomset bed* is a horizon-

maximum below the surface on the downstream side of the bend, the bend is not only extended laterally but also forced to migrate downstream. The downstream side of a bend may overtake the upstream side of the second bend below, causing the meander between the two to be cut off. The name *meander* derives from the river Meander, a meandering river in western Turkey that empties into the Aegean Sea (it used to empty opposite Miletus in what used to be Miletus Bay, which is now filled up with silt). Flowing streams are so sensitive to

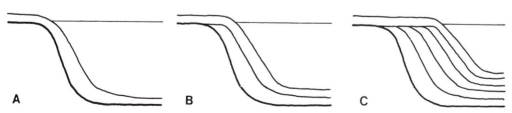

A   **B**   **C**

Figure 14.18. Model structure of a delta. (A) Deposition of sediment bed by flood; the topset bed, 10–20 cm thick, is above sea level. (B) Erosion of the topset bed and deposition of sediment layer by the next flood; because of the preceding erosion, this layer truncates the top of the preceding foreset bed. (C) A series of foreset and bottomset beds has formed, with the topset bed truncating the foreset beds. If subsidence is active, some of the topset beds, besides the most recent one, may be preserved.

Figure 14.19. The Nile River valley and delta from space. (Courtesy NASA.)

tal bed deposited on the deep flat in front of a delta and then buried by the advancing delta—it is a continuation of the foreset bed.

A classic example of a delta is the Nile River delta (Fig. 14.19). It was called *delta* by the ancient Greeks because, seen from the point of view of a Greek arriving by sea, it appeared to have the shape of the capital Greek letter delta (Δ). The Mississippi River delta, on the other hand, is shaped like a bird's foot and consists of seven overlapping deltas formed

from 5300 B.P. to the present as the river changed its course repeatedly (Fig. 14.20).

An *estuary* is a broad valley with a river emptying at its head and the ocean occupying the lower portion. Tidal currents move sediments offshore and keep the estuary open. An example is the Gironde, on the western coast of France, into which the Garonne River empties.

Table 14.9 shows the yearly amounts of sediment carried to the oceans by major rivers.

Streams form networks with characteristic patterns that depend on the geological structure of the terrain. A *dendritic* pattern forms when the rock substrate is homogeneous and flat-lying; a *radial* pattern is generated by conical topographic highs; a *rectangular* pattern is generated when the substrate consists of a jointed rocks; a *parallel* pattern forms on gentle slopes or where parallel synclines exist; an irregular pattern develops in formerly glaciated areas.

A *consequent* stream is a stream that follows the geological trends of an area (it runs along the bottom of a syncline, for instance); an *antecedent* stream is a stream that formerly followed such trends but later cut across structures (anticlines, horsts) as they were slowly raised by tectonic processes.

Waterfalls and *rapids* are nonequilibrium features produced by the occurrence of a ledge of hard rock along the course of a river. The tallest waterfall is the Cherun–Meru (Angel Falls) in Venezuela, with a single drop of 807 m and a total drop of 979 m. The largest waterfall

Table 14.9. Sediment Transport to Oceans by Major Rivers

River	Sediment Transported to Ocean (10^6 tons/y)
Ganges	1,600
Mekong	1,000
Brahmaputra	726
Yangtze	500
Amazon	499
Indus	400
Irrawaddy	300
Mississippi	213
Paraná	121
Zambezi	100

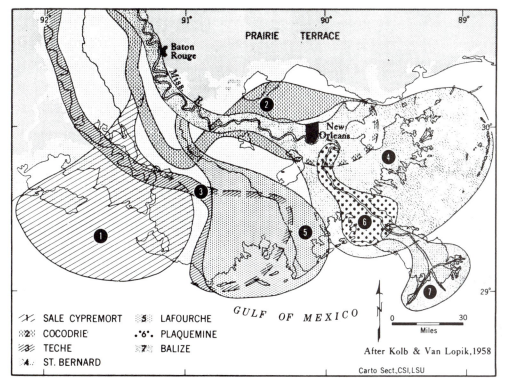

Figure 14.20. The deltas of the Mississippi. Names and ages (in years before the present) are as follows: 1, Sale Cypremort (5300–4400); 2, Cocodrie (4600–3600); 3, Teche (3900–2700); 4, St. Bernard (2800–2200); 5, Lafourche (900–700); 6, Plaquemine (1200–500); 7, Balize (500–0). The Mississippi is now attempting to start a new delta, the Atchafalaya, approximately on top of delta 1, but is being prevented by the U. S. Army Corps of Engineers. (Coleman, 1968, p. 257, Fig. 2.)

volume was the Salto das Sete Cuedas, between Brazil and Paraguay, with a maximum volume of 55,000 m³/s, before the ltaipu dam was constructed (1982).

THINK

* Does Fig. 14.6 refer to the Northern or the Southern Hemisphere?

* It was mentioned earlier that a stake found on a ledge in Little Salt Spring, Florida, gave a ^{14}C age of 12,600 y. Using the basic radiometric dating equation $N = N_0 e^{-\lambda t}$ (Section 3.4), calculate the percentage of ^{14}C atoms remaining in that sample.

* The Bible says that the great flood covered even the tallest mountains with 15 cubits of water. Taking Mount Everest (8,848 m) as the tallest mountain, and knowing that the average height of the land is 840 m and that 1 cubit = 52 cm, calculate the amount of water that rained. Considering that the deluge lasted 40 days and 40 nights (Genesis 7:12), calculate the rainfall rate per 24 hours. If the rate seems too high, read Genesis 7:11 for a way out.

* The Himalayas may be reduced to a prism about 2,000 km long, 200 km wide, and 3 km high. Using Table 14.9, calculate the volumes of sediment transported each year to the sea by the Indus, the Ganges, and the Brahmaputra (assume a density of 2.5 g/cm³). Now calculate how long it would take to erode the Himalayas completely. Is your result reasonable? If not, why?

* Knowing that Niagara Falls drops 7,000 m³ of water each second over a cliff 50 m high, calculate the power that could be extracted from the falls, and compare it with the power of an average nuclear power plant (500 MW).

15 SEDIMENTS AND SEDIMENTARY ROCKS

Sediments are ubiquitous on the surface of the solid Earth, both above and below sea level. They are produced from both igneous and sedimentary rocks by a complex process called weathering.

15.1. WEATHERING

Weathering is the mechanical and/or chemical alteration of minerals and rocks at the surface of the Earth by natural agents. There are two major types of weathering—mechanical and chemical (biochemical, if mediated by organisms).

Mechanical weathering is the breaking down of rocks by mechanical processes. When an igneous rock contracts because of solidification, or when a mud layer dries up, cracks develop that are called *primary joints*. Joints are rock fractures without relative motion of the two opposing faces. Examples are the columnar joints in basaltic lava and mud cracks. *Diastrophic joints* are formed in rock layers subject to tension, bending, or expansion due to removal of overburden by erosion or to tectonic processes. Joints provide ready access of weathering agents to the rock interior.

Water in rock pores and fissures expands by 10% when it freezes. *Frost action* occurs in rocks in wet climates that are exposed to repeated freeze–thaw cycles. Plant roots that penetrate rocks through fissures become swollen by osmotic pressure, enlarging the fissures and breaking the rock. This process is called *root pry*. Both frost action and root pry can rapidly reduce rocks to rubble.

If the wind is laden with sand (especially quartz sand), exposed surfaces are sandblasted and eroded (Fig. 15.1). *Wind action* is a powerful erosion agent in deserts.

Fragmentation of rocks by mechanical weathering greatly increases the exposed surface and therefore facilitates the processes of chemical and biochemical weathering. A cubic block $1\,m^3$ in volume has a surface of $6\,m^2$. If it is divided into fragments that are $1\,mm^3$ each, its surface increases to $6\cdot10^9\,mm^2 = 6,000\,m^2$.

Inorganic chemical weathering was the only type of chemical weathering during the first half billion years of the Earth's history. Because the rates of chemical reactions double with each increase of $10°C$ in temperature, the rates of chemical weathering on the surface of the young Earth were perhaps several times higher than today. With the evolution and spreading of cyanobacteria, biochemical weathering became important.

Pure, neutral water ($pH = 7$) contains 10^{-7} Av of H^+ ions and 10^{-7} Av of OH^- ions per liter, which is equivalent to $6\cdot10^{16}\ H^+$ ions and $6\cdot10^{16}\ OH^-$ ions per liter, or concentrations of $1.8\cdot10^{-9}\ H^+$ ions and $1.8\cdot10^{-9}\ OH^-$ ions .

[pH derives from *puissance d'hydrogène* and is expressed as the *negative* of the logarithm to base 10 of the molar concentration of hydrogen ions in an aqueous solution at $25°C$. Because there are 55.5 avogadros of water in 1 liter, the concentration of a solvent, even in a concentrated solution, is generally represented by the number 10 with a negative exponent, making the pH a positive number (because the exponent is the negative of a negative). In 1 liter of pure water at $25°C$, 10^{-7} Av of water naturally dissociate, releasing 10^{-7} Av of H^+ ions and an equal number of OH^- ions. Neutral water thus has a pH of 7. Acidic solutions have more H^+ ions ($pH < 7$) and fewer OH^- ions; basic solutions have fewer H^+ ions ($PH > 7$) and more OH^- ions.]

Figure 15.1. The step pyramid at Sakkarah, 50 km south of Cairo. Desert sandstorms have weathered the sides and accumulated talus on each step. (Fakhry, 1969, p. 21, Fig. 9.)

The solubility of CO_2 in water as a function of temperature and pressure is shown in Table 15.1. As Table 15.1 indicates, the solubility of CO_2 decreases with temperature and increases with pressure. CO_2 dissolves in rain droplets and forms the ionic species shown in Table 15.2.

The basic weathering reactions, without balancing the equations, are as follows:

A. Mafic minerals: *oxidation.* Example:

$$(Mg, Fe)SiO_3 + O_2 + H_2O$$
$$\rightarrow Mg^+(sol) + Fe^2 + SiO_2(sol)$$
$$Fe^{2+} + O_2 \rightarrow Fe^{3+}$$
$$Fe^{2+} + O_2 \rightarrow Fe_2O_3$$

B. Felsic minerals: *hydrolysis.* Example:

$$KAlSi_3O_8 + H^+ + HCO_3^- + H_2O \rightarrow K^+(sol)$$
$$+ Al_4Si_4O_{10}(OH)_8 \text{ (kaolinite)} + SiO_2(sol)$$

C. Carbonate minerals: *solution.* Example:

$$CaCO_3 + H_2CO_3 \rightleftarrows Ca^{2+}(sol) + 2HCO_3^-$$

D. Oxides (SiO_2, Al_2O_3, TiO_2, ZrO_2, etc.): not readily weathered.

In conclusion:

- The cations (Na, K, Mg, Ca, Fe) and silica go into solution
- The aluminosilicates remain as clays
- The oxides remain as resistates.

Table 15.1. CO_2 **solubility in water at 1 atm**

Temperature (°C)	Solubility (g/L)		
	1 atm	100 atm	400 atm
0°C	3.35	—	—
12°C	1.92	72.7	81.2
25°C	1.45	62.8	75.4

Table 15.2. Ionic species in the H_2O–CO_2 system

Reaction	Quantity
$CO_2(gas) + H_2O(liquid) \rightleftarrows H_2CO_3$ (carbonic acid)	$6 \cdot 10^{-4}$ g of H_2CO_3 per liter
$H_2CO_3 \rightleftarrows H^+ + HCO_3^{3-}$ (bicarbonate ion)	10^{-3} molecules of H_2CO_3 dissociate
$HCO_3^- \rightleftarrows H^+ + CO_3^{2-}$ (carbonate ion)	10^{-5} molecules of HCO_3^- dissociate

Figure 15.2. Exfoliation of a mass of granite into spheroidal boulders. (Courtesy Sheldon Judson, Princeton University.)

The major types of clay minerals are illite, kaolinite, and montmorillonite. Illite [K, H$_3$O)-(Al, Mg, Fe)$_2$(Si, Al)$_4$O$_{10}${(OH)$_2$, H$_2$O}] consists of two layers of SiO$_4$ tetrahedra with vertices pointing toward each other forming Si$_4$O$_{10}$ sheets. The sheets are joined by Al and OH ions that, together with O from the SiO$_4$ vertices, form an intervening Al$_4$O$_4$(OH)$_8$ sheet. About 15% of the Si atoms in the tetrahedra are replaced by Al; the resulting charge deficiency is balanced by K ions bonding the facing bases of adjacent tetrahedral sheets. The total thickness of a packet is 10.0 Å. Illite is an intermediate weathering product between the original K-Mg-Fe-Al silicate minerals and kaolinite. It is the dominant clay mineral at middle latitudes.

Kaolinite [Al$_4$Si$_4$O$_{10}$(OH)$_8$] derives from the deep weathering of feldspars and other aluminosilicates. It is the dominant clay mineral at low latitudes. Kaolinite has a layered structure consisting of a layer of SiO$_4$ tetrahedra bound to a layer of Al and OH ions. The thickness of the two-layer packet is 7.37 Å.

Montmorillonite   [(Na, Ca)(Al, Mg)$_6$-(Si$_4$O$_{10}$)$_3$(OH)$_6$·nH$_2$O] is a hydrated, expandable clay mineral derived from the alteration of Fe–Mg minerals and volcanic ash. Montmorillonite consists of two layers of SiO$_4$ tetrahedra pointing toward each other and joined by Al and OH ions, which, together with O from the SiO$_4$ tetrahedra, form an intervening octahedral layer. The total thickness of a packet is 9.6 Å. Water molecules and exchangeable cations can insert themselves between the facing bases of the SiO$_4$ tetrahedra, expanding the packet to 21.4 Å.

The stability of minerals to weathering is the inverse of the Bowen reaction series (Section 10.1):

Figure 15.3. Rates of weathering are strongly dependent on local climate. An Egyptian obelisk as it was in Alexandria (A) and as it is now in New York (B), where it was taken in 1879; 100 y in New York caused more damage than 3,000 y in Egypt. (Courtesy The Metropolitan Museum of Art, New York City, Department of Parks and Recreation.)

olivine (least stable)
anorthite
pyroxenes
amphiboles
albite
biotite
orthoclase
muscovite
clays
oxides (most stable)

Chemical weathering smoothes the corners and edges of rock fragments because corners are attacked from three sides, and edges from two, whereas a surface is attacked from only one direction. This exfoliates angular blocks into rounded shapes (Fig. 15.2).

Rates of weathering are strongly dependent on the local conditions of temperature, humidity, wind load, and the chemical composition of the precipitation (Fig. 15.3). Rough examples of weathering rates are shown in Table 15.3. Weathering in mountain regions produces mass wasting, the downslope movement of rock fragments and soils. A *rockfall* is what the name says, the falling of loose rocks down steep mountainsides. The rocks accumulate at the foot of the slope and form a deposit called *talus*

(Fig. 15.4). *Landslides* are downslope movements of soil and loose rocks over slippery bedrock, which usually is shale made wet by percolating waters. A *mudflow* is (surprise!) a flow of mud. If the mud is volcanic, the mudflow is called a *lahar*. A lahar is a nasty business—not only very messy but also boiling hot. A lahar buried Herculaneum during the eruption of Mount Vesuvius in 79 C.E. (Section 10.4).

Areas of the world consisting of mélanges are in continuous motion. Mélanges are heterogeneous mixtures of rocks of different ages and facies usually formed by tectonic compressions along active margins. They flow like porridge

Table 15.3. Rates of weathering (μm/1,000 y) of clean rock surfaces

Rock	Climate	
	Cold	Warm, Humid
basalt	10	100
granite	1	10
marble	20	200

Figure 15.4. Talus cones at the base of a peak in the Madison Mountains of Colorado. (Spencer, 1965, p. 281, Fig. 17.4.)

in response to tectonic stresses and to the deepening of drainage systems. They adjust by continuous creep (on the order of 1 cm/year), and by frequent landslides. A famous example is the *argille scagliose* (Italian for *scaly clays*) of the northern Apennines that have a scaly appearance because of continuous internal flow. The force resisting flow is provided by the hydrogen bonding of clay particles to each other and to the ever-present water molecules. Hydrogen bonding gives wet, fine sediment the property of *thixotropy* (from the Greek ϑιγγάνειν, *to touch*, and τρόπος, *a turn*).

Thixotropy can be tricky. Have you ever walked on quicksand? Everything is fine until you stomp your foot: The hydrogen bonds break, the sediment liquefies, and you sink. But do not panic, even if your horse refuses to pull you out. Your body is less dense than the quicksand. So ditch your jacket (which is undoubtedly weighted down with your trusty Colt, some ammunition, and a bunch of silver dollars), slowly but steadily swim out of the quicksand (the backstroke is recommended), and once out, have a word or two with your horse.

15.2. SOILS

Soil is produced on the land surface by the interaction of living organisms and organic matter with weathering products.

[The name *soil* should not be applied to the surface layer of the Moon, which has been weathered purely mechanically by meteoroid impacts over the past 4.5 billion years. The better name is *rhegolith*, usually misspelled *regolith*, which derives from the Greek ῥέγος, meaning *cover*, and λίϑος, meaning *stone*.]

Archaebacteria, photosynthetic bacteria, cyanobacteria, green algae, and all plants need elements that are sequestered in minerals and rocks. Chemical weathering places some of these elements in solution, but bacteria can

Table 15.4. Soil orders and suborders (USDA)

Order	Suborder	Order	Suborder	Order	Suborder
Alfisols	aqualfs	Inceptisols	andepts	Oxisols, *cont.*	torrox
	boralfs		aquedepts		ustox
	udalfs		ochrepts	Spondosols	aquods
	ustalfs		plaggepts		ferrods
	xeralfs		tropepts		humods
Aridisols	argids		umbrepts		orthods
	orthids	Mollisols	albolls	Ultisols	aquults
Entisols	aquents		aquolls		humults
	arents		borolls		udults
	fluvents		rendolls		ustults
	orthents		udolls		xerults
	psamments		ustolls	Vertisols	torrerts
Histosols	fibrists		xerolls		uderts
	folists	Oxisols	aquox		usterts
	hemists		humox		xererts
	saprists		orthox		

extract them directly from the minerals by creating an acidic microenvironment. Lichens, consisting of the association of a fungus with a cyanobacterium or a green alga, can also extract cations directly from minerals by means of acidic solutions. This type of biochemical weathering is extremely slow, but it is the first step toward the formation of soils. Once some soil is formed, higher plants can take over and greatly expedite biochemical weathering. Plant root tips are capped by a protective cell layer, called *mycorrhiza*, that includes, in most higher plants, an association with a fungus. Just above the root tip, the epidermal cells extend into thin, absorbing hairs that tremendously increase the root surface in contact with mineral particles and water. A single plant of winter rye (*Secale cereale*), 50 cm high, was found to a have a root system consisting of 143 main roots, 35,600 secondary roots, $2.3 \cdot 10^6$ tertiary roots, and $11.5 \cdot 10^6$ quaternary roots. The root system was found to have a total length of 600 km and a total surface of about 250 m².

Real soil exists only on Earth, where life is abundant on land. In my opinion, Mars is very likely to have had early forms of life, but only marine or freshwater forms. No life exists or existed on any of the other planets.

There are, of course, a great many different types of soils, depending on the local rocks, physicochemical conditions, and the communities of organisms interacting with the rocks. In 1975, the Soil Survey Staff, U.S. Department of Agriculture, having concluded that there were more than 10,000 different types of soils in the United States alone, produced a 754-page manual to classify the soils. In their classification there are 10 orders and 47 suborders. Below suborders are great groups, subgroups, families, and soil series, with six phases each. The total is 60,000. In the process of producing this classification, a wonderful series of new names were created that are now in the literature. The names of the 10 orders and 47 suborders are shown in Table 15.4. I list them here so that you can use them when playing Scrabble. If you insist on knowing what they mean, you will have to buy the manual (Soil Survey Staff, 1971. *Soil Taxonomy*, U.S. Department of Agriculture, Handbook 436, U.S. Government Printing Office, Washington, D.C.).

Soil is the place where atmosphere, hydrosphere, biosphere, and lithosphere meet and interact. It is an extremely complicated biogeo-lithochemical system. Serious study of soils requires a profound knowledge of all the basic sciences, with the result that serious studies in soil science are uncommon. We have, instead, an orgy of classification. For our purposes, we shall stick to the older, simpler classification shown in Table 15.5.

A well-developed soil consists of three horizons:

Table 15.5. Older, simpler soil classification

Soil Name	Climate	Soil Characteristics
pedalfer	humid mid-latitudes	rich in Al and Fe; no $CaCO_3$
pedocal	warm, dry	rich in $CaCO_3$
laterite	warm, humid	all soluble minerals gone; residual Al_2O_3 and Fe_2O_3

A horizon (topsoil): humus, organic remains, living organisms (especially nematodes), clay minerals, quartz, rutile, ZrO_2

[*Humus*, the Latin name for *soil*, is non-living, finely divided organic matter; *humic acid* is a collective name for a family of organic acids derived from the humus.]

B horizon: little organic matter; Fe oxides + silica (redeposited from above)
C horizon: little-altered bedrock

Soil science is actually a very serious matter— no soil, no food. Once an area has been denuded, it takes centuries to rebuild even a modest soil. The soils in the breadbaskets of the world, the North American midcontinent, central Europe, Ukraine, and Argentina, took 10,000 years to form. So soil conservation is also a very serious matter.

15.3. SEDIMENTS AND SEDIMENT TRANSPORT

Sediments come in two basic types: *clastic* (from the Greek κλαστός, meaning *broken*), and *precipitates* derived from marine or fresh waters. *Clastic sediments* are formed from the breakdown of preexisting source rocks, which can be igneous, metamorphic, or sedimentary. *Precipitates* are formed by biogenic or inorganic precipitations from marine or fresh-water solutions.

The type of clastic sediment that is produced depends on the parent rock type:

Light weathering of igneous rocks of granitic to dioritic composition produces gray ("dirty") sands that include, in addition to quartz and clays, minerals that are poorly resistant to weathering (feldspars, pyroxenes, amphiboles, biotite, muscovite).
Intensive weathering of the same rock types produces white ("clean") sands consisting mainly of quartz and clays.
Weathering of gabbro-basalt produces clays and oxides.
Light weathering of clastic sedimentary rocks produces sediments similar to those forming the source rock, including chips of the source rock.
Intensive weathering of clastic sedimentary rocks produces sediments that have been "cleaned" of the minerals that are less resistant to weathering.
Weathering of limestones and dolomites produces carbonate fragments that range from boulder size to clay size.

Sediments are transported by rivers into open oceans (Hudson, Amazon, Congo, Indus, Ganges), into marginal seas (Mississippi, Orinoco, Mekong, Yangtze, Huang Ho, Lena, Yenisey, Ob, Mackenzie), or into semi-isolated or isolated bodies of water (Danube, Dnieper, Don, Volga). The sediments may form a delta at the river mouth, or they may be moved sideways along the coast by longshore currents. In either case they are eventually moved downslope as *turbidity currents*.

Sediments on a delta front are held by thixotropy. When the angle of repose is exceeded (or a big storm or an earthquake occurs), the water-laden sediment is mobilized and rolls downslope as a mass of turbid water. The density of the muddy water at the bottom of the flow is about $1.2\,g/cm^3$ (the mineral particles therefore weigh half as much as in air).

The coarser particles are deposited first, forming a *graded bed*, with sand below grading upward into silt and mud. The mud surface soon becomes the living floor of benthic fauna until the next turbidity current occurs. The frequency of turbidity currents may average one in 500 years. Turbidity currents do not erode, as they are already fully loaded with sediment. As a result, delicate traces of benthic life on the soft mud are preserved. Turbidity currents also have the important function of transporting nearshore minerals, organic matter, and nutrients to the nutrient-deficient deep-ocean floor.

Graded beds (*turbidites*) of uniform thickness (15–30 cm) form sequences of hundreds or thousands of beds (Apennines, Appalachians, etc.). Graded beds are also common in lake sediments. Ice-age lakes in the western United States filled with turbidites have flat surfaces called *playas*.

Turbidity currents down the eastern continental slope of the Americas and the western continental slopes of Europe and Africa extend across the Atlantic floor to the flanks of the Mid-Atlantic Ridge. The surface of the sediment filling is extraordinarily flat (slopes < 1 m/km) and forms the *abyssal plains*. Abyssal plains also exist in the Gulf of Alaska. Most of the Pacific, however, is protected from turbidity currents by marginal seas and deep-sea trenches that act as sediment traps.

Turbidite deposits may attain great thickness. Along active margins, the thickness is accommodated by subduction. Along passive margins it is accommodated by the mechanism discussed in Section 12.3 (see Fig. 12.16). Thick sediment accumulations occur not only along continental margins but also within cratons.

[A *craton* (from the genitive κρατός of the Greek κράς, meaning *head, height, peak*) is the elevated central portion of a continent.]

Cratonic basins are filled up with shallow-water sediments to a total thickness of 5 km or more. The Michigan Basin, for instance, began subsiding about 500 million years ago and kept subsiding for 200 million years while accumulating a total thickness of 5.5 km of shallow-water sediments. The floor of the basin must have been continuously subsiding (at an average rate of 2.75 cm/1,000 y). Such subsidence may be caused by phase changes in the mantle—a mantle block 50 km across shrinking 10% will produce 5 km of subsidence at the surface. Cratons also exhibit domes. Domes may be produced by the opposite process—a phase change from denser to less dense mineral structures in the mantle.

During transport, the mechanically and chemically weaker minerals are destroyed, and the remaining ones are progressively rounded. *Sphericity* is a measure of how close a particle approaches a sphere. Theoretical sphericity is equal to P/S, where P is the surface of the particle, and S is the surface of a sphere of the same volume. The surface of a particle is not

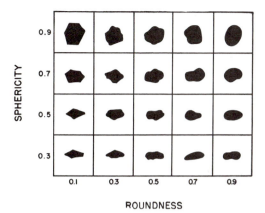

Figure 15.5. Chart for visual estimation of sphericity and roundness of sand grains. (Powers, 1953.)

easy to measure. A more practical measure of sphericity is $(V_P/V_S)^{1/3}$, where V_P is the volume of the particle, and V_S is the volume of circumscribing sphere. *Roundness* is a measure of how smoothed the surface of a particle has become. It is given by r_P/s_C, where r_P is the average radius of the corners and edges of the particle, and r_S is the radius of the maximum inscribed circle.

[*Note:* A cube has high sphericity and low roundness; a highly flattened ellipsoid of rotation has low sphericity and high roundness.]

A chart for visual estimation of sphericity and roundness is shown in Fig. 15.5.

The Wentworth grade scale ranks clastic sediments according to particle size (Table 15.6). A ϕ (phi) grade scale is also used to express particle sizes. On this scale, $\phi = -\log_2(d/d_0)$, where d is the diameter of the particle (in millimeters), and d_0 is a diameter of 1 mm. The size (average diameter) of 1 mm corresponds, therefore, to $\phi = 0$; the size of 4,096 mm corresponds to $\phi = -12$; and the size of 1/256 mm corresponds to $\phi = +8$. Notice that ϕ is a ratio of two lengths and therefore is a dimensionless number.

15.4. SEDIMENTARY ROCKS

Sedimentary rocks are derived from sediments by compaction and cementation. Like sediments, sedimentary rocks include clastics and precipitates. The major types of clastic sedi-

Table 15.6. Wentworth Grade Scale

Name	Size	Diameter (mm)	
boulder	very large	2^{12}	4,096
	large	2^{11}	2,048
	medium	2^{10}	1,024
	small	2^{9}	512
	very small	2^{8}	256
cobble	large	2^{7}	128
	small	2^{6}	64
pebble	very large	2^{5}	32
	large	2^{4}	16
	medium	2^{3}	8
	small	2^{2}	8
	very small	2^{1}	2
sand	very coarse	2^{0}	1
	coarse	2^{-1}	1/2 (500 μm)
	medium	2^{-2}	1/4 (250 μm)
	fine	2^{-3}	1/8 (125 μm)
	very fine	2^{-4}	1/16 (62.5 μm)
silt	coarse	2^{-5}	1/32 (31.25 μm)
	medium	2^{-6}	1/64 (15.625 μm)
	fine	2^{-7}	1/128 (7.8125 μm)
	very fine	2^{-8}	1/256 (3.90625 μm)
	clay	0	0

mentary rocks in terms of component sizes are as follows:

Conglomerate: boulders, cobbles, pebbles
Sandstone: cemented sand
Siltstone: cemented silt
Mudstone: cemented mud

In addition, the following types are recognized:

Quartzarenite: sandstone consisting predominantly ($> 95\%$) of quartz
Graywacke: sandstone containing appreciable amounts of dark minerals
Arkose: sandstone containing little dark minerals, but an appreciable amount ($> 25\%$) of feldspars (indicative of dry-climate weathering)
Litharenite: sandstone with an appreciable amount ($> 15\%$) of fragments of other rocks (usually shale)
Shale: mudstone with oriented clay minerals, exhibiting *fissility* (the property of breaking into laminae).

The cement that binds particles to form sandstones is quartz (if the sandstone consists of $> 90\%$ quartz), calcite, and/or hematite. The cement form scattered ionic bonds with the mineral particles.

Limestones and evaporites are precipitates from salt or fresh water. Marine limestones consist most commonly of the remains of marine organisms. The original material ranges from reef limestones and shell beds to pelagic limestones consisting of the remains of planktic protophyta and protozoa. The aragonitic components are rapidly changed to calcite and

Table 15.7. Evaporitic succession

Mineral	Composition	Percentage of total thickness in Zechstein Sea
bischofite (last)	$MgCl_2 \cdot 6H_2O$	1
sylvite	KCl	1
kieserite	$MgSO_4 \cdot H_2O$	2
halite	$NaCl$	80
gypsum	$CaSO_4 \cdot H_2O$	13
carbonates/dolomites (first)	$CaCO_3/CaMg(CO_3)_2$	3
Total		*100*

cements with calcitic cement. *Calcarenites* are limestones consisting of cemented sand-size particles. In *calcilutites* the particles are of micrometer size (originally algal needles or coccoliths). Freshwater limestones include compact limestones of algal origin, speleothems, and travertines precipitated from hot springs.

Evaporites are sequences of mineral deposits formed when seawater trapped in a basin evaporates. Major ion concentrations in seawater are shown in Table 14.2. Evaporation of seawater to saturation (at about 330°/₀₀ salinity) produces a characteristic succession of minerals that, in order of increasing solubility, range from carbonates and dolomites at the bottom to bischofite (hydrated Mg chloride) at the top. Table 15.7 shows the succession of minerals in the evaporites deposited in the Zechstein Sea (Permian of Germany, see Section 22.8). The Permian salt deposits in the Zechstein Sea of Europe, on the Russian platform, and in the United States (Kansas to West Texas), and the Middle Jurassic salt deposits along the Gulf Coast, range in aggregate thickness from less than 100 m to more than 1,000 m.

Along the Gulf Coast and offshore, sediments several kilometers thick accumulated on top of the salt layers. When sediments reach a thickness of 3 km, the salt, which behaves like a plastic, is squeezed out and forms *salt domes*. A salt dome (Fig. 15.6) is a carrot-shaped structure that may be several kilometers across and as much as 10 km high. In rising, the salt domes upturn and pierce the overlying sediments. Any petroleum contained in the sediment layers migrates upward, following the bedding planes that rise toward the salt domes and coming to a halt against the salt (which is impermeable). Salt domes, therefore, often have petroleum deposits around their sides.

Salt domes are capped by anhydrite (CaSO₄), a diagenetic product of the kieserite that was deposited on top of the main salt layer. The calcium sulfate in turn is capped by a layer of porous limestone. The porous limestone often contains sulfur. This sulfur is derived from the calcium sulfate below in contact with hydrocarbons. Anaerobic bacteria first reduce CaSO₄ to H₂S:

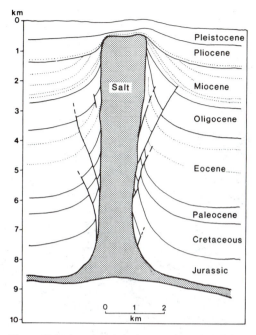

Figure 15.6. A typical salt dome on the Gulf Coast. Some salt domes reach the surface, but most do not. The salt (Louann Salt) was deposited in evaporitic basins in the Gulf area in Middle Jurassic time. The great load of the sediments subsequently deposited squeezed the salt out in the form of domes that pierced the overlying sediment cover wherever the resistance was least. In piercing the sediments, the salt domes upturned the layers' margins around the domes. Salt domes are capped by a layer of calcium sulfate, either gypsum (CaSO₄·2H₂O) or anhydrite (CaSO₄), which in turn is capped by a layer of porous limestone that may contain sulfur. There are about 150 salt domes along the Gulf Coast, and another 100 offshore under the continental shelf. (Emiliani, 1988, p. 206, Fig. 11.9.)

$$CaSO_4 + 2CH_2O \text{(organic matter)}$$
$$\rightarrow CaCO_3 + H_2S + H_2CO_3$$

Aerobic bacteria then oxidize H₂S to S:

$$2H_2S + O_2 \rightarrow 2H_2O + 2S$$

For the aerobic bacteria to operate, the top of the salt dome must be no more than 500 m deep. Subsurface sulfur is recovered by the *Frasch process*: Three concentric pipes are driven into the sulfur. Water heated under pressure to 140°C is forced down between the outer pipe and the middle pipe; the sulfur melts

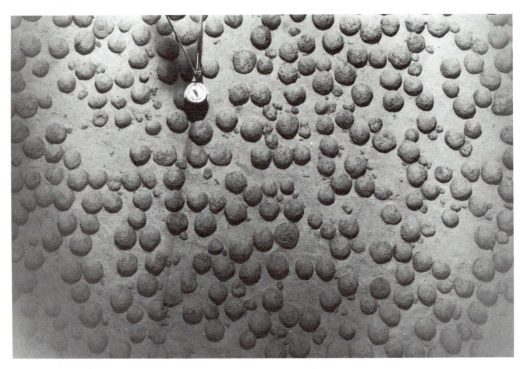

Figure 15.7. A spread of manganese nodules on the floor of the Blake Plateau, east of Jacksonville, Florida. These nodules are 8–10 cm in diameter. (Courtesy Ed Fisher, University of Miami.)

(melting point $= 112.8 \cdot 8°C$); air pressure applied through the inner pipe forces the molten sulfur up between the inner pipe and the middle pipe. There are about 150 salt domes along the Gulf Coast and another 100 offshore.

Thick evaporitic beds can form if a basin is periodically refilled. Deposits 300–500 m thick formed on the floor of the Mediterranean Sea five million years ago when communication with the Atlantic was repeatedly interrupted (a layer only 26 m thick would form if the Mediterranean were entirely evaporated once—see Section 24.2).

Dolomite [$CaMg(CO_3)_2$] is both a carbonate mineral and a rock in which 50% of the calcium has been replaced by magnesium. It has been suggested that the formation of dolomite is due to percolating solutions. The Mg/Ca ratio averages 0.46 in fresh water and 5.2 in seawater. A mixture of 95% fresh water and 5% seawater is undersaturated with respect to calcite, but oversaturated with respect to dolomite. Such a mixture could form in coastal areas where phreatic water lies above seawater. Another way for dolomite to form is by evaporation of seawater: When 50% of the seawater is gone, calcite precipitates, followed by gypsum when 80% of the seawater is gone. The remaining brine has a salinity of $125°/_{oo}$, a density 9% greater than that of standard seawater, and a Mg/Ca ratio of 10. This dense brine sinks through the porous underlying limestone and may replace Ca with Mg.

Chert is microcrystalline quartz. Chert beds are most abundant and widespread in Middle Proterozoic deposits (around two billion years ago). Silica (normally as opaline silica, $SiO_2 \cdot nH_2O$) can precipitate inorganically from marine or fresh water if the water is supersaturated for silica. That seems to have been the case for most of the Precambrian. The association of chert layers with cyanobacteria (see Section 19.7) indicates a shallow-water environment. With the evolution of opaline-silica-depositing organisms (siliceous sponges in the Late Proterozoic, radiolaria in the Cambrian, and diatoms in the Cretaceous), the

Table 15.8. Manganese nodules: average chemical composition (weight percent, the balance being detrital particles and organic components)

Element	Atlantic Ocean	Indian Ocean	Pacific Ocean
Mn	16.18	18.3	19.75
Fe	21.2	16.25	14.29
Ni	0.297	0.510	0.722
Co	0.309	0.279	0.381
Cu	0.109	0.223	0.366
SiO_2	14.0	5.8	9.4
Al_2O_3	5.4	6.8	5.8
Ti	0.4	0.4	0.7
Ca	3.3	1.4	1.9
Sr	0.1	0.09	0.08
Ba	0.5	0.4	0.2
Na	2.3	—	2.6
K	0.4	—	0.8

oceans became undersaturated with respect to silica. The siliceous skeletal parts that drop to the sea floor upon reproduction of the organism that produced them can accumulate there, in spite of the fact that seawater is undersaturated with respect to silica, because dissolution is extremely slow. Opaline silica diagenizes to quartz. Deposits of opaline silica free of other sedimentary components (clastics, etc.) can form chert beds up to one meter in thickness. A deposit consisting of opaline silica elements mixed in with carbonate elements forms a limestone bed, with chert nodules of pebble size.

Phosphorites are limestone deposits in which the $-CO_3^{2-}$ group has been replaced by the $-PO_4^{3-}$ group. The average concentration of phosphate in seawater is very low (0.088 mg/kg), but sufficient to replace the carbonate ion and, together with the fluorine ion, form fluorapatite [$Ca_5(PO_4)_3F$]. The concentration of fluorine is 1.2 mg/kg. For this process to occur, however, the phosphate-bearing water must be circulated within the carbonate sediment. The phosphate ion is an integral part of the structure of both DNA and RNA. The surface seawater, where most of the organic syntheses occur, is thus depleted of phosphorus. Not so the deep water, which contains about 0.1 g/kg of phosphorus leached from the fecal pellets and other organic remains dropped on the ocean floor. Deep water is brought to the surface in areas of upwelling, where prevailing winds flow from land to sea and peel off the surface layer. Phosphorites average 10% phosphorus by mass. Phosphorus is a vital ingredient for fertilizers. Large phosphate resources in the United States are the Phosphoria formation in Idaho, Wyoming, and Utah, and deposits in Florida and North Carolina.

Manganese nodules are potato-shaped bodies widespread on the deep-ocean floor, where sedimentation rates are low (Fig. 15.7). Their chemical composition is shown in Table 15.8. The main sources of the metals probably are marine exhalates (as chlorides), although contributions from the land (as hydrated oxides) are obvious. The metals are leached by hydrothermal circulation along the mid-ocean-ridge system and oxidized in contact with the oxygenated bottom water. Precipitation requires a solid surface, which can be a foraminiferal shell, a fish tooth, or a mineral grain. Manganese nodules often exhibit a concentric, layered structure, clearly indicating radial growth (Fig. 15.8). In addition to the metals in a variety of mineral phases, the nodules contain detrital matter (as indicated by the high concentration of aluminum) and some organic structures (worm tubes, etc.) abandoned by organisms that once lived attached to the surface of the nodule during its accretion. Because the structure of the nodules is rather spongy, their average density is rather low, making them somewhat buoyant

Figure 15.8. Cross section of a manganese nodule. (Gross, 1987, p. 86, Fig. 4.9.)

in seawater. This explains how the nodules can grow radially, while resting (or rather floating) on the sediment. Growth rates are exceedingly low, about $1–4 \, mm/10^6 \, y$.

Oolites are small (diameter $< 1 \, mm$) spherical or ovoidal bodies (Fig. 15.9) that consist of concentric layers of aragonitic microcrystals and organic matter, grown on a detrital nucleus (usually a shell fragment). They are formed by inorganic (?) precipitation of calcium carbonate from supersaturated seawater in a milieu where wave and tidal motions are sufficiently strong to keep the spheroids rolling around. Some oolitic formations have been mineralized by iron solutions that have replaced aragonite with goethite [$FeO(OH)$], limonite (hydrated goethite), or hematite (Fe_2O_3). Examples are the Clinton Formation in the eastern United States (Silurian) and the *minettes* of the Ruhr Valley in Germany (Jurassic).

15.5. POROSITY AND PERMEABILITY

Porosity and permeability are important properties of sedimentary rocks (and of fractured igneous rocks). *Porosity* is the volume per-centage of pores in a solid. The intergranular porosities of common clastic rocks range from 10% to 35%. Spheres of identical sizes with cubic packing have a porosity of 47.64%, and with rhombohedral packing have a porosity of 25.95% (see Section 9.4). Porosity is not enough for petroleum recovery from a rock. The rock must also be permeable; that is, the pores must be interconnected so as to allow the fluid to move. The *permeability* of a rock depends on the number, geometry, size, and interconnections of the pores. Permeability is measured in darcys. One darcy is equal to the passage of $1 \, cm^3$ of a fluid with a viscosity of $1 \, cP$ (centipoise) in $1 \, s$ under a pressure difference of $1 \, atm$ through a porous medium having a cross section of $1 \, cm^2$ and a length of $1 \, cm$. By the way, the centipoise is $1/100$ of the poise, which is the cgs unit of dynamic viscosity. The poise is equal to $1 \, dyn \, s \, cm^{-2} = 0.1 \, Pa \, s$ (because dynes/square centimeter = pressure). Permeability ranges from 5 to 1 darcy in loose sediments, and from 1 to 0.05 darcy in rocks. The lowest value (0.05 darcy) applies to cherts, which are practically impermeable masses of microcrystalline quartz. A rock may have a high porosity but no permeability because the holes are not interconnected (e.g., a vesicular lava) or because they are too small (e.g., clay).

Figure 15.9. Oolitic sand. These aragonitic oolites, each about 0.5 mm across, have been embedded in plastic and thin-sectioned. They are viewed with polarized light. (Courtesy Robert N. Ginsburg.)

THINK

* Would you expect frost action to be more efficient in Chicago or on Victoria Island in the Canadian Arctic?

* If 1 g of CO_2 is dissolved in 1 liter of water, how many CO_2^{2+} ions are formed in the solution?

* It was mentioned in the text that the root system of a single plant of winter rye was found to be 600 km long and to have a surface of 250 m². Calculate the average diameter of a root strand.

* If a cube of rock 1 m in edge length were to be reduced to fine dust (i.e., made into cubic fragments 5 μm across), how large would the exposed surface become?

* Referring to Section 2.1 for the radius of the hydrogen atom in the ground state, calculate the ϕ value of that atom.

* How much net energy is expended in raising a salt dome 10 km high and 3 km across?

* Table 14.2 shows that in 1 kg of seawater there are 0.407 g of Ca^{2+} and 0.143 g of HCO_3^-. Only 10^{-5} molecules of HCO_3^- dissociate to form CO_3^{2-} (Table 15.2). Calculate how many grams of $Ca^{2+}CO_3^{2-}$ there are in solution in the oceans and how thick a layer of $CaCO_3$ would form on the ocean floor if all that $CaCO_3$ were precipitated in one instant.

16 SEDIMENTARY ENVIRONMENTS

Sediments are formed in all kinds of environments, from the slopes of Mount Everest to the bottom of the Marianas Trench. Because different environments produce different types of sediments, the application to ancient sediments of the information obtained from the study of modern sedimentary environments enables us to assess the environments of the past and thus reconstruct the environmental history of our planet. This effort is of fundamental importance for the management of our planet both today and in the future.

Sedimentary environments can be grouped into three major families: continental, glacial-marine, and marine. To these we add the recently discovered vent environment (Section 16.4).

16.1. THE CONTINENTAL ENVIRONMENT

The continental environment includes pyroclastic, desertic, fluvial, lacustrine, and glacial deposits.

Pyroclastic deposits

Volcanic explosions produce volcanic bombs, lapilli, cinder, and ash. The coarsest fragments fall back in the immediate vicinity of the volcano, but significant amounts of the finer particles can be carried farther than 10 km as part of a nuée ardente, and for hundreds of kilometers by the wind. The Pierre shale, a rock formation of Late Cretaceous age, extends from Kansas to South Dakota and westward to Colorado and Wyoming, and contains several layers of bentonite. (Bentonite is a montmorillonitic clay derived from the alteration of andesitic ash.) The Pearlette ash layer of Pleistocene age

extends from South Dakota to Texas. A single ash layer represents an episode well defined in time. Ash layers, therefore, have been used as time markers, giving rise to the field of *tephrachronology*, or "chronology by means of ash layers" (τέφρα means *ash* in Greek).

Deserts

If the supply of sand in a desert is sufficient, dunes form. The wind rolls the sand grains upslope and across the crest and deposits them on the lee side. The major types of dunes are *transversal*, *longitudinal*, and *barchans*. If the sand supply is abundant, transversal or longitudinal dunes form. Transversal dunes (Fig. 16.1A) have crests normal to the wind and form if the wind is blowing steadily from a given direction. Transversal dunes range in height up to 100 m and in length up to 100 km. Longitudinal dunes (Fig. 16.1B) form if the wind is shifting about one dominant direction, shaping the dune crest in that direction. Each wind shift changes the direction of deposition, and so longitudinal dunes are cross-bedded (Fig. 16.2). Longitudinal dunes include the largest dunes in the world, the seif dunes of the western Sahara, which range in height up to 300 m and in length up to 300 km. Barchans (Fig. 16.1C) are crescent-shaped dunes convex windward that form when the wind is steady or shifting about one dominant direction but the sand supply is limited. The heights of barchans range up to 30 m, and the widths up to 400 m. The lee side of barchans is the angle of repose of dry sand grains, about 35°. With a greater sand supply, barchans grade into transversal dunes.

[*Cross-bedding* is produced when the transporting medium changes direction. The trans-

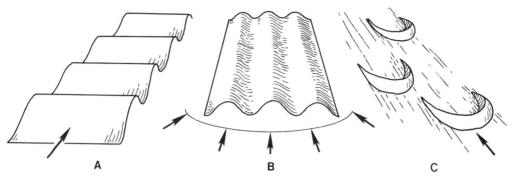

Figure 16.1. The three major types of dunes: (A) transversal, formed by steady wind when the supply of sand is abundant; (B) longitudinal, formed by winds shifting about a dominant direction when the supply of sand is abundant; (C) barchans, formed by steady wind or by winds shifting about a dominant direction when the supply of sand is limited.

porting medium is air in the case of dunes, and water in the case of rivers or coastal marine deposits.]

Irregular winds and an irregular sand supply produce irregular dunes or dunes of types intermediate between the three listed above. Barchans and transversal dunes migrate downwind, at speeds up to 25 m/y for small barchans. Sandy deserts are sites of frequent sandstorms. A *hammada* is a rocky desert with little sand.

Fluvial and lacustrine deposits

Fluvial deposits are deposits laid down by rivers. A river's competence decreases with decreasing water speed, which leads to *sorting*: Coarser elements (boulders, cobbles, pebbles) are left upstream, and finer elements (sands,

Figure 16.2. Cross-bedding in the Navajo Sandstone, a vast dune field of Early Jurassic age that spread across Utah and adjacent states. (Moore, 1958, p. 356, Fig. 14.7.)

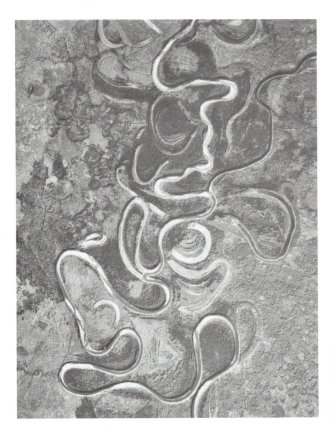

Figure 16.3. The meanders of the Hay River, a tributary of the Great Slave Lake in Canada. (Courtesy National Air Photo Library, Surveys and Mapping Branch, Canada Department of Energy, Mines, and Resources.)

silts, muds) are carried downstream. During transport, the mechanically and chemically weaker minerals are destroyed, and the remaining ones are progressively rounded.

A river issuing from a mountain range onto a plain forms an alluvial fan (Fig. 14.16). On the plain itself, the river will begin meandering, with frequent and repetitive changes of its course (Fig. 16.3). These changes produce deposits (mainly fine sands, silts, and clays) that exhibit cross-bedding. River floods will deposit sediments on either side of the river, often to considerable distances, forming *floodplains*. A classical example of a floodplain is the Po River valley in northern Italy, which is bounded on its northern and western sides by the Alps and on its south side by the Apennines. The longitudinal slope between Turin and the Po River delta in the Adriatic Sea is only 0.56 m/km, similar to the slopes of abyssal plains (see Section 15.3). Floodplain deposits consist largely of silts and clays, with an occasional sandy or even pebbly layer. Sedimentation rates are several millimeters per year (9 mm/y

for the Nile River floodplain; 4.5 mm/y for the Ohio River floodplain; 3 mm/y for the Po River floodplain). If the base level of a river drops, the river will incise its own floodplain and leave river terraces on either side (Fig. 16.4).

Lacustrine deposits are formed on the bottoms of lakes. Sedimentation rates are on the order of millimeters per year (1.2 mm/y in Lake Geneva, Switzerland; 3 mm/y in Lake Michigan). The western part of the United States is dotted with small basins, which are surrounded by mountains with steep slopes and are filled with sediment deposited by turbidite currents during the recent ice ages. Sedimentation rates were very high: 4–5 mm/y. The surfaces of the lake fillings are remarkably flat and are called *playas*. The fillings include pebbly layers, coarse sands, silts and clays, and evaporites. A major lake, called Lake Bonneville, 50,000 km^2 in area and 300 m deep, was formed in Utah during the ice ages. The size of the lake increased and decreased with the waxing and waning of the ice. It dried up around 6,000 years ago, leaving an evaporitic deposit on the

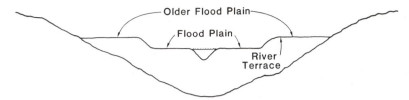

Figure 16.4. River terraces are formed when base level drops and a river incises its own floodplain.

dry surface. The Great Salt Lake, 3,900 km² and averaging 4–5 m of depth, is a recent refilling. Much more extensive lake deposits form when tectonic motions prevent an entire region from draining into the sea. An example is the lacustrine Green River Formation of Eocene age that covers about 125,000 km² with oil-rich shales 600 m thick in southwestern Wyoming, northwestern Colorado, and northeastern Utah (Fig. 16.5).

The Green River deposits and the deposits on the bottoms of many glacial lakes are varved (i.e., they consist of *varves*).

[*Varves* are yearly laminae consisting of an organic-rich layer and an organic-poor layer. Depending on the setting, either layer could form during the summer or during the winter.

In the Green River Formation, summer algal blooms provided the material for the organic-rich layers. In Alpine lakes, the lake surface and the river or rivers feeding a lake freeze during the winter. Enough solar light penetrates through the ice surface, however, to allow the continuation of algal growth in the water beneath the ice. During the spring, the ice melts, and much fine sediment is brought in by the rivers through spring and summer. The sediment dilutes the organic matter produced by the algae, so that although algal production is greater in spring and summer, there is less organic matter in the sediment. The organic-rich layer is the winter layer, and the organic-poor layer is the summer layer, which is opposite to what the Green River Formation shows. In any case, a

Figure 16.5. The Green River Formation in northeastern Utah. These beds were deposited on the bottom of a vast lake that covered southwestern Wyoming, northwestern Colorado, and northeastern Utah. The formation is 600 m thick. The dark layers are rich in organic matter, the remains of algal blooms that were preserved on the anaerobic bottom. (Moore, 1958, p. 444, Fig. 17.8.)

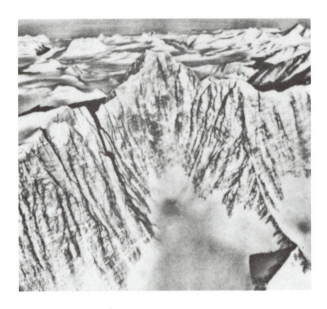

Figure 16.6. Two glacial cirques separated by a sharp ridge called *arête*. The peak in the center is Mount Tyree in the Sentinel Range, Antarctica. (Courtesy U. S. Geological Survey).

varve, consisting of two laminae, one formed during the summer and the other formed during the winter, represents one year. The time that it took to form a varved deposit can be determined simply by counting the number of varves.]

Varve counts show that the Green River Formation took $4 \cdot 10^6$ y to accumulate.

Glaciers

Glaciers begin forming on high mountainsides, carving bedrock to form a *cirque* (Fig. 16.6). The downslope movement of ice in a glacial valley erodes the valley floor and sides by means of rocks embedded in the underside of the ice. The valley acquires a U-shape, contrasting with V-shape normal to valleys cut by rivers. The bedrock under the ice becomes striated and grooved.

Rubble forms underneath a glacier (*ground moraine*) and is piled along both sides (*lateral moraines*) as well as in front of the glacier (*frontal moraine*). When two glaciers coalesce, their lateral moraines form a *medial moraine* (Fig. 16.7). Glacial deposits are unsorted, like landslide deposits.

Increasing pressure tends to melt ice, because water is the higher-density phase. A water layer may form under a glacier, allowing the glacier to surge (with speeds up to 15 m/day).

Coalescing mountain glaciers form an *ice cap*. Coalescing ice caps form an *ice sheet*. Table 16.1 shows the amount of ice now present on Earth. If all ice were to melt, the sea level would rise 77 m (54 m with isostatic adjustment). During the last ice age, the sea level was 120 m lower than it is today. Many of the continental-shelf areas of the world were dry land, crossed by river valleys and accumulating continental sediments.

Ice sheets are floored with a vast ground moraine call *till* or *drift*, less than one meter to a few meters thick. Ice sheets can carry boulders for hundreds of kilometers and drop them far away from their sources (*erratics*). An advancing ice-sheet front may shape till into

Figure 16.7. (*Facing page, bottom*) Coalescing mountain glaciers in the Barnard Glacier, Alaska. (Goudie, 1989, p. 75, Fig. 3.7.)

Table 16.1. Ice on Earth

Ice Body	Surface (km²)	Mean Thickness (km)	Volume (× 10⁶ km³)
Antarctic ice sheet	$13.3 \cdot 10^6$	1.8	24.0
Greenland ice sheet	$1.7 \cdot 10^6$	1.6	2.7
Other glaciers	—	—	1.3
Total ice on Earth	$15.0 \cdot 10^6$	1.8	28.0
Additional amount of ice on Earth during last ice age	25.8	1.9	49.0

ellipsoidal hills (*drumlins*) tens of meters high, 1–2 km long, and 400–600 m wide.

The major ice-age ice sheets are shown in Table 16.2 (ice volumes estimated from the glaciated surfaces and a global sea-level decrease of 120 m). The "other glaciers" shown in Table 16.2 include the ice caps that formed on all high mountain ranges of the world. The additional ice then in existence was about $49 \cdot 10^6$ km³, equivalent to $44.4 \cdot 10^6$ km³ of water.

During the summer, ice meltwater flowing out from under an ice sheet (pressure favors melting) carries out fine rock particles in suspension (*rock flour*) that are deposited in front of the ice. Reworking of such deposits by catabatic winds spreads out the sediment, forming *loess*. Loess deposits range across North America, central Europe, Siberia, China, and

Table 16.2. Ice-age ice

Location	Surface (10⁶ km²)	Mean Thickness (km).	Volume (10⁶ km²)
Canada	14.8	2.0	29,6
Greenland	1.8	2.0	3.6
Scandinavia	6.7	1.8	12.0
Antarctica	13.8	2.0	27.6
Patagonia	1.0	1.2	1.2
Other glaciers	2.7	1.1	3.0
Total			77.0

Figure 16.8. Loess near the Mississippi River in southern Illinois. (Courtesy James E. Patterson, Normal, Ill.)

Patagonia. Their thicknesses range up to 150 m (in China).

Loess does not remain powdery (Fig. 16.8). Loess wetted by rain compacts under its own weight, and the mineral particles form hydrogen bonds. When the loess dries up, the bonds do not break (like a dry clay ball). Wet loess is impermeable, and so the top layer protects the layers below when it rains. The Chinese cut caves in loess and use them for habitation (bad news when an earthquake strikes).

Permafrost is permanently frozen ground in formerly glaciated or periglacial areas. Permafrost reaches a thickness of up to 1–1.5 km in northern Canada and northern Siberia.

During an interglacial age there is no ice (except in Antarctica and Greenland), and the loess belt becomes forested. During an ice age, the forest is killed, and a loess sheet is deposited on the fossil soil. In 1936, the Hungarian geologist Emil Scherf (1889–1967) found 11

fossil soils in a brickyard pit at Paks, Hungary. It is now known that there have been about 30 ice ages during the past three million years, increasing in severity toward the present.

16.2. THE GLACIAL-MARINE ENVIRONMENT

When a glacier reaches the ocean, the ice continues its advance until it begins to float. Debris drops to the sea floor as ice melts and becomes mixed with normal marine sediments. Fluctuations of the ice front cause the deposits to interfinger between mainly glacial sediments landward and mainly marine sediments seaward. The glacial component is called *diamicton* (unsorted glacial debris), and the marine component is ooze of the type characteristic for the region. Glacial-marine sediments are common around Antarctica and were common, during the ice ages, around the northern North Atlantic and in the Gulf of Alaska.

Formerly glaciated valleys ending in the sea are called *fjords*. Debris dropped by the ancient glaciers at the mouths of fjords forms submarine ridges that now restrict the exchange of water with the open ocean. Fjords are thus particularly susceptible to pollution.

Table 16.3 Depth zones in the marine environments

Zone	Depth Range
intertidal	between high tide and low tide
neritic	between low tide and -130 m
bathyal	-130 to $-2,000$ m
abyssal	$-2,000$ to $-6,000$ m
hadal	below $-6,000$ m

16.3. THE MARINE ENVIRONMENT

The marine environment ranges from the intertidal zone to the hadal zone. The different zones are shown in Table 16.3. The environment is so rich in facies that it is better seen in tabular form (Table 16.4).

During marine ingressions, the marine environment left on land a rich record of fossil-

Table 16.4. The marine environments

1. *Intertidal*: narrow zone with specialized fauna
    a. high-energy (meaning strong wave motion), rocky coastlines: limpets, chitons, barnacles, *Mytilus*, sabellid worms, pholads (Fig. 16.9)
    b. low-energy (meaning weak wave motion), sandy-muddy coastlines: crabs, shrimps, worms, mollusks, echinoderms; mainly infaunal detritus or filter-feeders
2. *Neritic*: from low tide to edge of continental shelf [average 130 m of depth; average width of continental shelf 75 km; average slope of shelf $0°7'$ (2 m/km)]
    a. with supply of clastic materials:
    sand bars
    lagoons behind bars with silts and muds
    changes in currents (including tides) produce cross-bedding (Fig. 16.10)
    wave and current motions produce ripple marks (Fig. 16.11)
    rich epidauna of detritus-feeders (echinoderms) and infauna of filter-feeders (mollusks)
    b. without supply of clastic materials:
    coral reefs   (corals, mollusks, echinoderms, foraminifera, calcareous algae, crustaceans, fish)
3. *Bathyal* (-130 to $-2,000$ m): on continental and island slopes; slopes range from precipitous (1 km/km along the eastern margin of the Bahamian platform) to gentle (20 m/km); slopes are cut by submarine canyons (deep gorges that cut across the continental shelf and slope)
    a. along passive continental margins: in continuations of major rivers (Hudson, Congo, etc.) (Fig. 16.12)
    b. along active continental margins (e.g., California, Baja California): cut subaerially and then drowned by subsidence (Fig. 16.13)
    c. Mediterranean type: cut 6 to $5 \cdot 10^6$ y ago when the Mediterranean went dry (Fig. 16.14)
    submarine canyons funnel turbidity currents to the deep-ocean floor
    most common sediments: sands, silts, muds
    fauna: crinoids (filter-feeders) with long stems; abundant mud-feeders (species of mollusks, crustaceans, echinoderms) with long life cycles
4. *Abyssal* ($-2,000$ to $-6,000$ m): lower submarine slopes and deep-sea floor; silts and clays with abyssal fauna (ophiuroids, holothuroids, foraminifera)
    a. Globigerina ooze: ($>30\%$ $CaCO_3$; mainly shells of planktic foraminifera (~ 32 living species) and coccoliths (~ 60 living, coccolithforming species) (Figs. 16.15 and 23.10B)
    b. radiolarian and diatom oozes: deep-sea clays with abundant shells of radiolarians (Section 21.1) and diatoms (Section 23.5)
    c. red clay ($>4,500$ m): clay with Fe oxides and quartz particles (wind-transported)
5. *Hadal* ($>6,000$ m in deep-sea trenches): red clay with turbidites; impoverished abyssal fauna

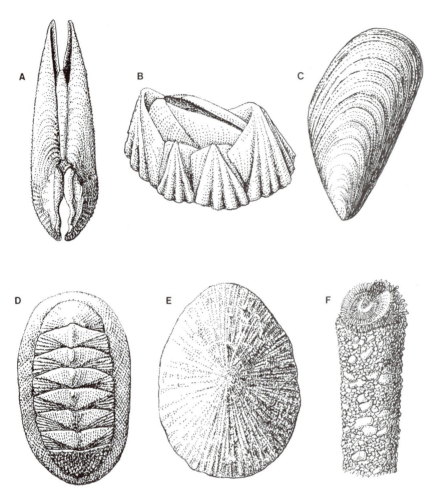

Figure 16.9. Intertidal organisms inhabiting a high-energy, rocky coastline: (A) *Pholas*, a boring bivalve; (B) *Balanus*, a barnacle; (C) *Mytilus edulis*, a bivalve that attaches itself to pilings and other structures by byssus threads (the *byssus* is a tuft of threads, secreted by the *byssus gland* on the foot of the animal, by which some species of bivalve mollusks attach themselves to a solid substrate; the name *byssus* derives from the Greek βύσσος, for *fine flax*); (D) *Chiton*; (E) *Patella*, a limpet; (F) *Sabellaria*, a sedentary worm.

containing rocks that spans the entire Phanerozoic (the past 600 million years of the Earth's history). It is from the study of this record that we derive a fairly clear picture of the evolution of life and environment on Earth (Part V).

The coastal environment ranges from plunging rocky shores beaten by stormy seas to beaches and quiet lagoons. The sediments range from accumulations of gravels to sands that are shifted and often cross-bedded by waves and currents (Fig. 16.10) and to finely laminated tidal deposits rich in organic matter formed in quiet waters. The associated organisms are listed in Table 16.4. If sediment flux from land is negligible or absent, and if the yearly temperature minimum does not drop below 18°C, coral reefs may develop just offshore.

Coral reefs are marine structures of extraordinary interest. Because they have been around for 500 million years, they have evolved into an extremely rich and highly complex ecosystem. In addition, fossil reefs may be repositories for hydrocarbons or may have become mineralized with metals by percolating solutions. Examples of living reefs are the reefs off the Florida Keys (Fig. 16.16A), which consist of the following habitats and inhabitants:

Figure 16.10 Cross-bedded oolitic marine sand deposited in shallow water 120,000 y. ago and now forming a lithified ridge along Biscayne Bay, Miami, Florida. The top of the ridge is 6.5 m above the present sea level. The sediment was formed when sea level was about 6 m higher than today, and was lithified shortly afterward when sea level dropped and exposed the deposit. A subsequent rise in sea level, possibly about 80,000 y ago, carved the notches that are visible in the illustration.

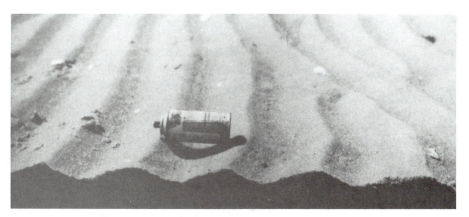

Figure 16.11. Ripple marks on a tidal flat of the North Sea. Water motion is from left to right. (Reineck and Singh, 1973, p. 24, Fig. 19.)

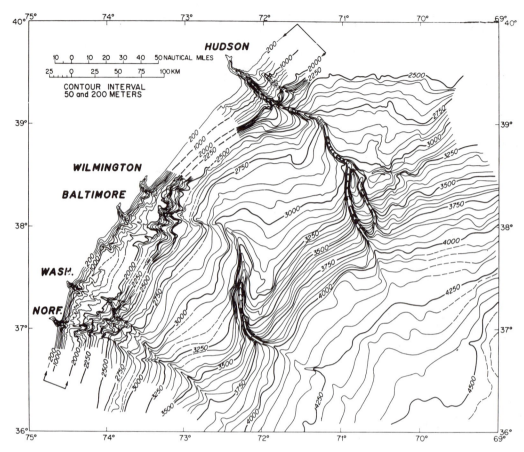

Figure 16.12. Submarine canyons along a passive continental margin (eastern coast of the United States). These canyons are incised in soft sediments. (Emery and Uchupi, 1972, p. 77, Fig. 66.)

1. Fore-reef: *Millepora* (Fig. 16.16B), *Acropora palmata* (elkhorn coral) (Fig. 16.16C)

2. Reef flat: *Acropora cervicornis* (staghorn coral) (Fig. 16.16D)

3. Back-reef: *Montastrea annularis* (star coral) (Fig. 16.16E), *Diploria strigosa* (common brain coral) (Fig. 16.16F), *Meandrina meandrites* (brain coral) (Fig. 16.16G); mollusks, sponges, echinoderms; calcareous sands consisting of foraminifera and molluskan shell fragments; *Penicillus* and *Halimeda* (Fig. 16.17) producing aragonitic needles (10 μm long, 1 μm thick)

4. Back-reef:
    a. oolitic shoals: sediment consists of aragonitic oolites (Fig. 15.9)

b. grapestone shoals: sediment consists of grapestone grains, 0.5–2.5 mm across, made of cemented ooids

c. pellet flats: sediment consists of 50% pellets (fecal excretes, mainly of polychaete worms) (Fig. 16.18), 50–200 μm across

d. mud flats: mud consists mainly of aragonitic algal needles (produced by *Halimeda* and *Penicillus*) (Fig. 16.17)

e. complex shoals: mangrove swamp to open shelf; 140 species of mollusks, 235 species of Foraminifera

A coral reef may be attached to a coastline (*fringing reef*) or may be separated from it by a lagoon (*barrier reef*) (Fig. 16.19 and 16.20). If

A

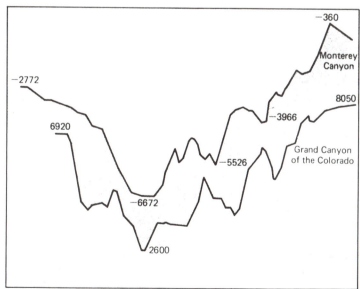

B

Figure 16.13. (A) The Monterey and Carmel submarine canyons off the central California coast, incised in sedimentary rocks. (Courtesy U. S. Geological Survey.) (B) The Monterey Canyon (upper curve) exhibits a profile similar to that of the Grand Canyon (lower curve) when drawn to the same scale. (Stowe, 1987, p. 46, Fig. 2.37.)

Figure 16.14. Submarine canyons off the Franco-Italian Riviera and the west coast of Corsica. These canyons were cut in hard rocks when the Mediterranean became dry between $6.2 \cdot 10^6$ and $5.1 \cdot 10^6$ y ago. (Moullade, 1978, p. 70, Fig. 2.)

sea level rises (or the coast sinks), such reefs can prograde landward. If the land rises (or sea level falls), ribbons of fossil reefs are left to drape the rising coast (as in Barbados or New Guinea). *Atolls* are low-latitude, extinct oceanic volcanoes whose tops have been flattened by weathering and wave action and have subsided as plate motion has moved them farther away from a spreading center. Coral-reef growth forms a carbonate cap (1,500 m thick at Eniwetok; 800 m thick at Bikini), with a circular, ellipsoidal, or polygonal reef rim

(better developed on the windward side) and a lagoon that may range in depth from a few meters to more than 100 m (Fig. 16.21). Atolls range in diameter from 1 km to 130 km. *Guyots* are flat-topped, extinct, sunken volcanoes at latitudes too high for coral growth (Fig. 16.22). As soon as a volcano begins sinking, the flat top becomes covered first with a veneer of shallow-water organisms and then, as subsidence continues, with a thin cover of pelagic muds. In contrast, *seamounts* are extinct submarine volcanoes that never reached the surface.

A

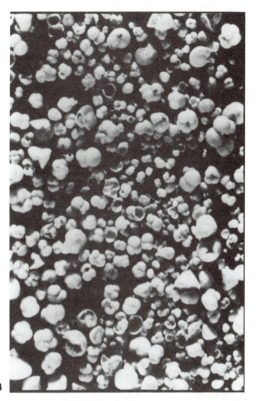

B

Figure 16.15. Globigerina ooze consists mainly of the shells of planktic foraminifera embedded in a fine matrix of coccolith and clay particles. The illustration shows the foraminiferal shells from an equatorial Atlantic deep-sea core after the fine matrix has been washed out. The largest shells are about 1 mm across. (A) postglacial (about 3,000 y B.P.); (B) glacial (about 20,000 y B.P.). (Microphotographs by José Leal.)

Below the edge of the continental shelf (average depth, −130 m), sand- and clay-size sediments accumulate. The type of sediment depends, of course, on the source. Off carbonate platforms (Bahamas, Florida, and Yucatan, for instance) and all around atolls the particles are mainly carbonate. Elsewhere, the particles are mainly silicates and clay minerals. The sediments are transported to the edge of the continental shelf by currents (Section 15.3). Turbidity currents are often channeled along deep-sea canyons (Fig. 16.12 to 16.14), commonly forming broad sedimentary fans or aprons as they issue from the mouths of the canyons.

Between the foot of the continental slope and the adjacent deep-sea floor one often finds a broad, low ridge consisting of fine sediments. This ridge parallels the foot of the slope and is called *continental rise*. The sediments are deposited by deep *boundary currents*, deep-water currents that flow equator-ward or pole-ward and cling to the continental slopes because of the Coriolis effect. The continental rise is particularly well developed along the east coast of North America because of the significant sediment flux injected into the North Atlantic from the Davis Strait and by rivers emptying all along the coast.

On the deep-sea floor of the world ocean, clays and quartz particles accumulate. These mineral particles derive from the weathering of continental rocks transported to the ocean by rivers and winds. In addition, the clay mineral montmorillonite forms from the submarine oxidation and breakdown of the products of submarine volcanism. The clay and quartz particles are small enough to be distributed all over the oceans. Their rate of accumulation is about 1 mm/1,000 y. The bulk sediment

338

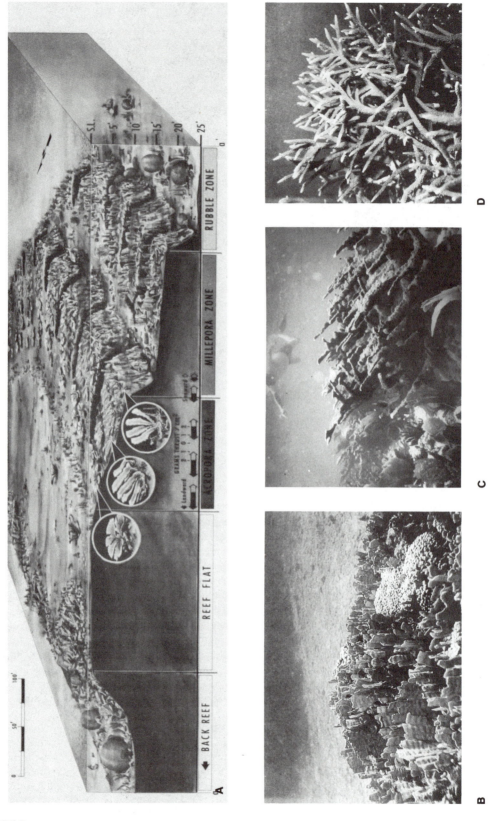

A

BACK REEF | REEF FLAT | ACROPORA ZONE | MILLEPORA ZONE | RUBBLE ZONE

GRAINS THROAT / cm²
← landward   2 1 0 1 2   seaward →

100'   50'   0

S.L.
5'
10'
15'
20'
25'
0

B

C

D

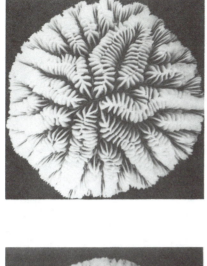

G

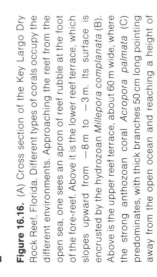

F

Figure 16.16. (A) Cross section of the Key Largo Dry Rock Reef, Florida. Different types of corals occupy the different environments. Approaching the reef from the open sea, one sees an apron of reef rubble at the foot of the fore-reef. Above it is the lower reef terrace, which slopes upward from −8 m to −3 m. Its surface is encrusted by the hydrozoan *Millepora complanata* (B). Above is the upper reef terrace, about 60 m wide, where the strong anthozoan coral *Acropora palmata* (C) predominates, with thick branches 50 cm long pointing away from the open ocean and reaching a height of 2–3 m. In places protected by the growth of *A. palmata* one finds *Acropora cervicornis* (D), with branches 1–2 cm thick and reaching a height of 1–2 m. On the reef flat beyond the upper reef terrace *A. palmata* and *A. cervicornis* are common, but the branches of *A. palmata* are not oriented. The floor is strewn with branches of the weaker *A. cervicornis* torn off by storms. Beyond the reef flat is the back-reef area, where one commonly finds the hemispherical colonies of *Montastrea annularis* (E), up to 2 m across, *Diploria strigosa* (F), up to 2 m across, and *Meandrina meandrites* (G), up to 1 m across. These and other species of both stony corals and soft corals (*Gorgonia* etc.) form patch reefs, which are smaller reefs rising above the sandy or muddy floor. The sand consists of benthic foraminiferal shells and fragments of molluskan shells and other calcareous organisms. The mud consists of aragonitic algal needles secreted by the green algae *Penicillus* and *Halimeda*. (A from Shinn, 1963, pp. 291–303. B–G from von Prahl and Erhardt, 1985, p. 222, Fig. 131; p. 221, Fig. 131; p. 2, p. 150, Fig. 78a, p. 134, Fig. 63a; p. 158, Fig. 86a.)

Figure 16.17. The back-reef environment: white sands with *Halimeda* (fingered) and *Penicillus* (tufts). The knife is for scale. (Courtesy Robert N. Ginsburg.)

Figure 16.18. Fecal pellets are the ejecta of benthic and planktic organisms. Most are microscopic (50–250 μm) ovoidal aggregates of fecal matter covered with a pellicle. (Newell, 1956.)

Figure 16.19. A fringing reef around a small island off Java. (Courtesy Java Air Force.)

Figure 16.20. A barrier reef surrounding Tahaa Island in the Society Islands, about 140 nautical miles northwest of Tahiti. (Courtesy U. S. Navy.)

Figure 16.21. Rongelap Atoll, Marshall Islands. The atoll is about 80 km across. The lagoon is about 40 m deep, but it has many patch reefs rising to the surface. (Courtesy NASA.)

consisting of these particles is called *red clay*, because of the maroon color imparted by iron hydroxides.

The euphotic zone of the ocean is rich in life. Some taxa produce durable skeleta that are dropped onto the ocean floor, where they accumulate together with the red clay matrix. At depths less than 3,700 m in the Pacific and 4,400 in the Atlantic a sediment rich in the shells of planktic foraminifera (Figs. 16.15 and 21.15) and in coccoliths (Fig. 23.10), called *Globigerina ooze*, accumulates at an average rate of 2.5 cm/1,000 y. Planktic foraminiferal shells and coccoliths are calcitic, and calcite begins to dissolve at depths greater than 3,700–4,400 m because of the increase in partial pressure of CO_2 related to the decrease in temperature (cold water holds more CO_2 in solution than does warm water). Aragonite is more soluble than calcite, and the aragonitic shells of pteropods (planktic gastropods) are not preserved below a depth of −3,000 m. Pteropod oozes are found in restricted areas of

the ocean, mainly in tropical seas around atolls at depths less than −3,000 m.

Radiolaria (Section 21.1, Fig. 21.2) are planktic protozoa that deposit lacy shells consisting of hydrated silica ($SiO_2 \cdot nH_2O$). Radiolarian ooze, rich in radiolarian shells, forms, a band that stretches across the Pacific along the equator. Around Antarctica and in the northern North Pacific diatom oozes are common. Diatoms are protophyta that, like radiolaria, form shells (called *frustules*) that consist of hydrated silica (Section 23.5, Fig. 23.16).

Iron and manganese exhalates from submarine volcanism are rapidly oxidized in contact with sea water and, if a solid surface is available (a foraminiferal shell, for instance, or a rocky outcrop), precipitation of the oxides on that surface occurs. If the surface remains free of sediments by being swept clean by steady bottom currents, continued accumulation forms black crusts that grow at an exceedingly slow rate (a few millimeters per million years). Concentric growths around solid nuclei form manganese nodules (see Section 15.4).

16.4. THE VENT ENVIRONMENT

The *vent environment* is a submarine hydrothermal environment capable of supporting luxuriant but specialized life. Hydrothermal circulation is established at many places along the crests of mid-ocean ridges where hot rocks are close to the ridge surface and surface rocks are fissured (Fig. 12.17). Sea water descends from the flanks to depths as great as 5 km and emerges through vents at temperatures up to 300°C. The water is rich in sulfides (H, Cu, Pb, Zn, Ag). An extraordinary community of animals lives around these vents, including archaebacteria. The most abundant macroorganisms are giant worms (up to 3 m long, 2–3 cm across) without mouth or gut, and giant clams. These organisms feed on chemotrophic bacteria that grow in their tissues. The bacteria draw their energy by oxidizing H_2S from the vents:

$$H_2S + O_2 \rightarrow 2H_2O + S$$
$$2S + 3O_2 + 2H_2O \rightarrow 2H_2SO_4$$

Similar communities have been found in other

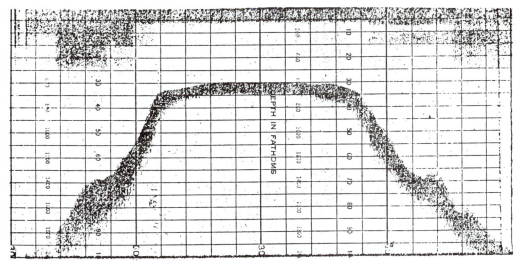

Figure 16.22. Echo-sounding profile of one of the flat-topped seamounts discovered by H. H. Hess during World War II, which he named *guyots*. (Hess, 1946, p. 772, Fig. 2A.)

areas where sulfide-bearing waters discharge in the ocean, such as along the slopes of passive continental margins and in subduction areas.

THINK

* Pumice has a very high porosity, and yet it floats in water. Why?

* During the last ice age the sea level was 120 m lower than it is today. If isostatic equilibrium had been maintained, how many meters of water would have had to be removed from the ocean to produce a sea-level drop of 120 m?

* There are extensive coral reefs in the Bahamas and the West Indies island arc, but none on the opposite African coast. Why?

* Suppose you are a holothurian fooling around the floor of the Sohm abyssal plain and feeding on the finely disseminated organic matter in the mud. Suddenly you find yourself engulfed in a turbidity current. Is that bad or good for you? Explain.

DIAGENESIS, METAMORPHISM, AND ROCK DEFORMATION

If a mineral or organic substance, formed in one physicochemical environment, later finds itself in a different environment it may undergo a physicochemical change (so does your skin when you go the beach). The change may be purely physical, such as a phase change, or it may include a chemical change, such as ionic diffusion. Most sediments and the organic substances they may contain are complex chemical systems formed under a specific range of physicochemical conditions. A different range will lead to diagenesis and metamorphism.

17.1. DIAGENESIS OF SEDIMENTS

Diagenesis, the postdepositional alteration of a sediment, includes compaction and cementation, but not extensive recrystallization. Some diagenetic changes take place subaerially. Drying of mud, for instance, leads to the formation of hydrogen bonds between the mud particles. On tropical and subtropical beaches where the sand contains an appreciable amount of carbonate particles, the sand may become cemented and form *beach rock* (Fig. 17.1). Although the exact process is not clear, cementation may be caused by evaporation of water supersaturated with respect to calcite during the tidal cycle.

Carbonate deposits can be diagenized very rapidly even subaerially. The Key Largo reef limestone, formed 125,000 years ago and exposed subaerially since then, has had nearly all its aragonite replaced by calcite, has had its calcite largely recrystallized, and has become sufficiently cemented to be used for the facing of buildings and for paving.

Most diagenetic changes, however, take place below the surface. The major cementing agents are calcite, silica, hematite, and clay

minerals. Percolating water at 60–70°C may have 200 ppt silica, 600 ppt calcium, and 10 ppt iron (ppt means parts per thousand; unless specified otherwise, it means parts per thousand by mass).

An average sandstone may contain 65% quartz grains, 15% shale fragments, 15% feldspar, and 5% clay minerals. These components commonly are cemented at depth by calcite deposited from percolating water solutions. Individual calcite crystals range in size from 10 μm to several centimeters, in which case the detrital grains are engulfed in the crystal. Hematite crystals are less than 1 μm in size. Hematite usually is precipitated together with the calcite cement, giving the rock a reddish coloration. Clay minerals are ubiquitous in sandstones, forming the rock matrix. Most are detrital, but when buried, they can grow and help cement the surrounding grains. Sandstones that contain more than 90% quartz often are cemented by silica precipitated around the grains, producing overgrowths that can connect one grain to the next.

The cement binds to the sand grains by means of hydrogen bonds and also by means of ionic or covalent bonds when a cementing atom fits an adjacent atom on a crystal surface. In friable rocks the bonding is poor, and many pores ramain unfilled. At the other extreme, a well-cemented quartz sandstone can be very hard and may have a porosity of less than 0.5 millidarcy.

Cement commonly represents 20–30% of the mass of a sandstone. Cementation requires that the original sediment contain materials that can be dissolved and reprecipitated. A pure quartz sand may remain uncemented even at depth. However, if the pressure is sufficiently high, silica for cementation of a pure quartz sand may be derived from *pressure solution* of the grains themselves.

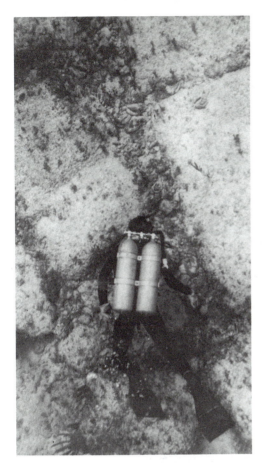

Figure 17.1. Beach rock off Bimini, Bahamas. Beach rock forms in the intertidal and near-shore zone when mixing with fresh water produces supersaturation with respect to calcite. The sediment can be cemented to a depth of up to 1 m. The surface is incised by grooves that commonly are at right angles to each other and normal to the beach. The formation shown in this illustration was claimed by some enthusiasts to be the pavement of an ancient temple. (Shinn, 1978, p. 131.)

[*Pressure solution* is a bad term because it gives the impression that one can dissolve a substance by pressure. That is true for ice, but for most substances the solid phase has a higher density than the liquid phase. As a result, increasing pressure solidifies a liquid, rather than liquefying a solid.]

Consider a quartz sand under pressure so that all quartz grains are squeezed against each other. When a corner of one grain is pressing on a face of an adjacent grain, the pressure of the point of contact will be much higher than when two grains are pressed against each other face to face. At the points of highest pressure, the ionic bonds are strained, and if the pressure is sufficient, the weaker ones will break. SiO_4 tetrahedra will thus be free to migrate and will move toward the areas of lower pressure, following the chemical-potential gradient. Given enough time, an assembly of spherical quartz grains will assume the aspect of an interlocked mosaic.

[*Chemical potential* (symbol μ) is the tendency of a component to react or move along a chemical gradient. It is expressed in joules per avogadro.]

Cementation depends more on the availability of cementing material and on the chemistry of the percolating waters than on pressure and temperature, both of which increase with depth. Increasing temperature facilitates solution, but increasing pressure compacts the particles and impedes the flow of cementing solutions. Temperature and pressure increase downward. The temperature through the continental crust increases by 33°C/km in the top 6 km. That is reduced to 7.4°C/km in the bottom 6 km. The average is 17°C/km. The pressure increases by 0.250 kb/km in the top 2 km, increasing to 3.6 kb/km at the bottom. The average is 0.28 kb/km. There is, therefore, a wide variety of *P-T* conditions available in the crust.

17.2. DIAGENESIS OF ORGANIC MATTER AND FORMATION OF FOSSIL FUELS

Coal and hydrocarbons are the two primary fossil fuels on which much of the economy of the world rests. Coal derives from diagenesis of land plants, and hydrocarbons derive primarily from diagenesis of marine phytoplankton.

Coal

Coal is carbon plus various amounts of vegetable matter (mainly cellulose and lignin).

[Cellulose, the main constituent of plant tissue,

Table 17.1. The ranks of coal

Name	Percentage C (dry)	Heat of Combustion (cal/g)	Density Real	Density Bulk	Characteristics
peat	40	3,000	—	—	plant matter recognizable; free cellulose
lignite	50	4,000	—	0.70	plant matter partly recognizable; no free cellulose
subbituminous	60	6,000	—	0.75	marked compaction; some plant matter recognizable
bituminous	80	8,000	1.3	0.80	no plant matter recognizable
subanthracite	90	8,500	1.4	0.85	no plant matter recognizable
anthracite	95	7,900	1.6	0.90	no plant matter recognizable
graphite	100	7,901	2.27	2.27	crystalline carbon

is a polysaccharide: $(C_6H_{10}O_5)_{2,000-4,000}$, with molecular mass $= (162.142)_{2,000-4,000}$. Lignin is a polymer consisting of 12 phenol rings (C_6H_4OH) with various substitutions and embellishments. It is the main noncarbohydrate component of wood, providing support for the cellulose fibers.]

Coal is formed from vegetable matter buried by sediments. As temperature and pressure rise, H_2O and volatiles (e.g., CH_4) are lost, leading to the formation of coal. Coals are ranked in terms of the amount of change from vegetable matter (Table 17.1). Compare heats of combustion of coals with those of other substances (Table 17.2). Most of the coal in the world was formed in the lowlands of the northern continents that were repeatedly flooded by the ocean and drained as sea level rose and fell in response to the waning and waxing of ice in the Southern Hemisphere

Table 17.2. Heat of combustion of various substances

Substance	Heat of Combustion (Cal/g)
dynamite	1,290
wood	4,700
carbon	7,901
egg yolk	8,100
olive oil	9,400
butter	9,532
hydrogen	29,150

during the Pennsylvanian–Permian glaciation (see Chapter 22). When ice forms on Earth, equilibrium is reached between precipitation and ablation. The precessional motions of the earth, however, change the amount of solar radiation reaching the high northern or southern latitudes. These motions have periodicities of 25,800, 41,000, 95,000, and 412,000 years. The dominant periodicity for the Permo-Carboniferous glaciations appears to have been the 412,000-y periodicity. Each time ice formed in the Southern Hemisphere, the sea level dropped, and the rivers built up floodplains in the Northern Hemisphere. Soil developed and forests grew. Each time ice melted, the sea level rose, the lowlands were flooded, the forests were killed, and the vegetal remains were buried under marine clays and limestones. The sediment sequence representing one ingressive–regressive cycle is called a *cyclothem* (Fig. 17.2). More than 100 cyclothems occur in West Virginia. From the number of cyclothems and the periodicity of 412,000 years, we can conclude that the southern glaciation lasted more than 40 million years. It ended when Pangea split up and South America, Africa, Madagascar, India, and Australia started moving northward.

The formation of coal does not necessarily require cyclic sea-level changes. Important coal deposits formed in China during the Jurassic, in central Europe during the Cretaceous, and in Alaska during the Eocene. The important parameter in coal formation is rapid burial of vegetal remains, as might occur in a

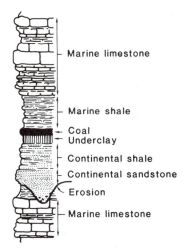

Marine limestone

Marine shale

Coal
Underclay

Continental shale

Continental sandstone

Erosion

Marine limestone

Figure 17.2. A complete cyclothem. The sea level is minimum during the phase of erosion. As the sea level rises, river waters are backed up, depositing sands and clays. As the sea level continues to rise, marshes form, with forests. The rising sea eventually inundates the marshes, kills the forests, and deposits marine clays and limestones. The average thickness of a cyclothem is 6 m, half of which is continental and the other half marine. Underlying is the ancient soil. (Emiliani, 1988, p. 204, Fig. 11.8.)

swamp periodically flooded with sediments from neighboring rivers. The accumulation of coal in significant amount requires, in addition, continued subsidence. Fig. 17.3 shows the world distribution of coal deposits.

Hydrocarbons

Hydrocarbons are compounds consisting exclusively of C and H. They are classified in the following families (see also Appendix B):

1. *Alkanes*: straight or branched chains of carbon atoms linked to each other by single covalent bonds. The remaining valences in alkanes and all other hydrocarbons are taken up by H atoms. General formula: C_nH_{2n+2}. Example: methane (CH_4). See Table 17.3.

2. *Alkenes*: straight or branched chains of carbon atoms in which there is in each chain one double bond between two carbon atoms. General formula: C_2H_{2n}. Example: ethene (C_2H_4).

3. *Alkynes*: straight or branched chains of carbon atoms in which there is in each chain a triple bond between two carbon atoms. General formula: C_2H_{2n-2}. Example: acetylene (C_2H_2).

4. *Alkadyenes*: straight or branched chains of C atoms with two double bonds, each linking two adjacent carbon atoms. Example: propadiene (C_3H_5).

5. *Arenes (aromatic hydrocarbons)*: hydrocarbons with one or more benzene rings.

   [The *benzene ring* is a planar, hexagonal ring of 6 atoms, numbered 1 to 6 clockwise starting from the 12 o'clock position. The bonds are intermediate between single and double, so that all bonds are equivalent in strength. The internal angles between the bonds are 120°, and the length of the bond is 1.39 Å.]

The simplest aromatic hydrocarbon is benzene (C_6H_6), consisting of a single benzene ring, with each carbon bound to the two adjacent carbons by hybrid bonds (i.e., bonds resonating between single and double). This makes a total of three bonds for each carbon. The fourth bond is taken up by a hydrogen atom. Naphthalene consists of a double benzene ring, and anthracene of a triple ring.

Natural hydrocarbons consist of mixtures of all the compounds listed above. The chains run from one carbon up to tens of carbons. In general, the longer the chain, the higher the melting and boiling points (Table 17.3).

   [*Bitumen* is a mixture of hydrocarbons with more than 15 C atoms in each molecule; *asphalt* is heavy bitumen or heavy bitumen plus crushed rock.]

Hydrocarbons are formed by the diagenesis of (mainly) marine phytoplankton matter buried in sediments. Organic matter accumulates in all types of sediments, usually amounting to 1 to 5 percent by mass. If accumulation occurs in a basin in which the bottom water is stagnant or poorly ventilated, the organic matter will not be oxidized and will be retained in the sediment. The floors of such basins usually are environments of quiet sedimentation, where clays tend to be the dominant sediment type.

   The sediment in which organic matter accumulates is called *source rock*. Usually it is

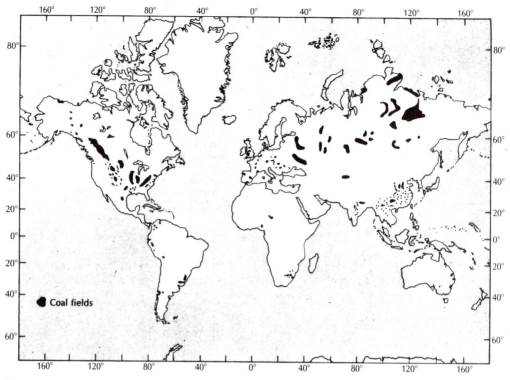

Figure 17.3. World distribution of coal deposits.
(Brookins, 1981, p. 126, Fig. 7.14.)

rich in montmorillonitic clay. Montmorillonite consists of two layers of SiO_4 tetrahedra pointing toward each other and jointed by Al and OH ions, which together with O from the SiO_4 tetrahedra form an intervening octahedral layer. The thickness of the packet is 9.6 Å. In freshly deposited montmorillonite, however, a layer of water molecules and exchangeable cations is commonly present between the SiO_4 tetrahedra, expanding the thickness of a packet to 21.4 Å (Section 15.1). This water, plus the water between the clay flakes, results in the uncompacted sediment having a water content as high as 100%. As more sediment accumulates, temperature rises, and lithostatic pressure forces much of the water out, reducing the content in the sediment to less than 25%. Under these conditions, the organic matter is diagenized to hydrocarbons. Optimum temperature and pressure ranges for the production of hydrocarbons are between 20°C, 0.2 kb (= 600 m of depth) and 90°C, 2 kb (= 6 km of depth). Notice that 1 kb is the pressure exerted by about 3 km of rock. If the temperature is too high, only CH_4 forms, together with a graphitic residue. The process is very slow, and so time is also involved.

The water, carrying along soluble organic matter as well as hydrocarbons of low molecular mass (up to 10 carbons), migrates laterally and upward until it is trapped in a porous and permeable *reservoir rock* covered by an impermeable layer or until it reaches the surface and dissipates.

Once in the reservoir rock, the hydrocarbons float above the water because they are lighter.

[In the liquid-fuel industry, the density of liquid hydrocarbons is expressed in degrees A.P.I. (°A.P.I.), where A.P.I. means American Petroleum Institute. The °A.P.I. is defined as being equal to $141.5/\rho_{w,60°} - 131.5$, where $\rho_{w,60°}$ is the relative density of the oil with

Table 17.3.   The Alkanes

Name	Composition	Melting Point (°C)	Boiling Point (°C)
methane	CH_4	−182	−164
ethane	C_2H_6	−183.3	−88.6
propane	C_3H_8	−189.7	−42.1
butane	C_4H_{10}	−138.4	−0.5
pentane	C_5H_{12}	−130	36.1
hexane	C_6H_{14}	−95	69
heptane	C_7H_{16}	−90.6	98.4
octane	C_8H_{18}	−56.8	125.7
nonane	C_9H_{20}	−51	150.8
decane	$C_{10}H_{22}$	−29.7	174.1
Petroleum fractions			
rigolene	C_4H_{10}, C_5H_{12}	—	18–21
petroleum ether	C_5H_{12}, C_6H_{14}	—	40–60
gasoline	C_6H_{14} to $C_{10}H_{22}$	—	60–200
naphtha	C_6H_{14}, C_7H_{16}	—	60–90
ligroin	C_7H_{16}, C_8H_{18}	—	90–120
benzine	C_8H_{18}, C_9H_{20}	—	120–150
kerosene	C_9H_{20} to $C_{16}H_{34}$	—	150–300
lubricating oils	$C_{12}H_{26}$ to $C_{20}H_{42}$	—	300+
petrolatum	$C_{20}H_{42}$ to $C_{30}H_{62}$	—	300+
asphalt	$C_{30}H_{62}$ to $> C_{90}H_{182}$	—	300+

respect to water at 60°F (15.5°C). The "30-weight oil" you hear about at your gas station refers to SAE 30, where SAE means Society of Automotive Engineers, and 30 means 30°A.P.I.]

17.3. ROCK DEFORMATION

Rocks are deformed in two ways—plastic deformation and rupture. Stresses derive from the internal energy of the Earth. Stress can be tensile, compressive, or shearing, or a combination thereof. Rocks with fissile minerals (e.g., schists) deform more readily than rocks without (e.g., granite). Higher temperatures and water content facilitate deformation.

Syngenetic deformation is the deformation of sediment layers while they are undergoing diagenesis, metamorphism, and tectonization. Syngenetic deformation leads mainly to plastic deformation. An example is provided by the folds buried under the Po Valley in Italy—the deeper the folds, the greater their amplitude.

[An upwardly convex fold is called an *anticline*, and an upwardly concave fold is called a *syncline* (Figs. 17.4A and 17.5). If the crest of the fold is not horizontal, the fold is called *plunging* (Fig. 17.4B,C).]

If the axial plane of a fold is tilted, the fold is called *asymmetric* (Fig. 17.4D). More tilting produces *overturned folds* (i.e., folds with one limb rotated beyond the vertical) (Fig. 17.4E), and then *recumbent folds* (i.e., folds with axial planes 90° or more from the vertical) (Fig. 17.4F).

Postgenetic deformation is the deformation of rocks after they have formed and become emplaced. It can be caused by tension, compression, or shear. Postgenetic deformation usually causes rupture and the formation of faults. The fault plane can be vertical, horizontal, or, more commonly, oblique. Different types of faults are produced by different stresses.

Updoming (circular, ellipsoidal, or elongated, as in incipient rifting) causes tension. Rocks cannot deform plastically under tension:

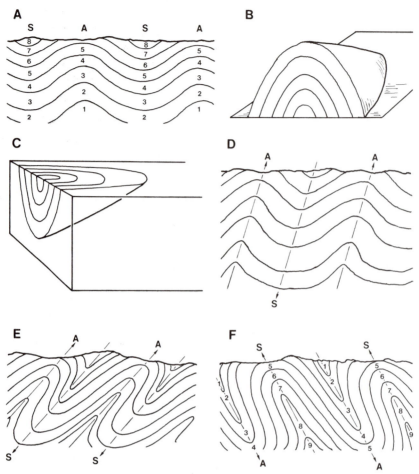

Figure 17.4. (A) Series of adjacent anticlines and synclines. The beds decrease in age from 1 to 8. The erosional surface cutting across the series shows that the ages of the layers decrease horizontally from the core of an anticline (*A*) to that of a syncline (*S*). (B) A plunging anticline. The angle that the line marking the crest of the anticline makes with the horizontal plane is called the *plunge*. (C) A plunging syncline. (D) Asymmetric folds. The arrows show the axes of the fold (*A* = anticline; *S* = syncline). (E) Overturned folds. The right limbs of the anticlines are turned beyond the vertical, so that the top of a layer may actually be below the bottom of the same layer. (*A* = anticline; *S* = syncline). (F) Recumbent folds, apparently making anticlines into synclines, and vice versa. An examination of the sediments or ages of the layers reveals that the apparent bottom of a syncline is in fact the top of an anticline, and vice versa. The numbers indicate the order of deposition of the beds, 1 being the oldest and 8 the youngest (*A* = anticline; *S* = syncline).

They rupture and form grabens and horsts (see Fig. 12.6) bounded by *normal faults* (Fig. 17.6A). The different characteristics of faults have led to the fault nomenclature shown in Table 17.4 and illustrated in Fig. 17.6. In normal faults, older rocks on the footwall side face younger rocks on the hanging-wall side.

Compression produces *reverse faults* and *thrust faults*. In reverse faults (Fig. 17.6B),

younger footwall layers face older hanging-wall layers. Thrust faults are reversed faults with the fault plane dipping less than 45°. Thrust faults are formed along active continental margins facing subduction zones: Wedges of sedimentary rocks from the geocline, plastered against the continental margin, can slide a long distance over a basement that is gently sloping landward, supported by fluid pressure (Fig. 17.7).

Figure 17.5. Robert Bridges inside an anticline in the Appalachians, 3 miles west of Hancock, Maryland.

(Courtesy U. S. Geological Survey, C. D. Walcott, no. 328C.)

Shear causes adjacent blocks to move horizontally in opposite directions (Fig. 17.6C). A shear fault is called *right-lateral* if the block opposite the one an observer is standing on appears to have moved to the right; it is called *left-lateral* if the block appears to have moved to the left. The motion remains in the same direction regardless of the block one chooses as stationary.

There are three types of shear faults: *strike-slip faults*, *transcurrent faults*, and *transform faults*. Strike-slip faults (Fig. 17.6C) range in horizontal displacement from a few meters to hundreds of kilometers. The strike-slip fault illustrated is a left-lateral fault. Transcurrent faults are large strike-slip faults. Transform faults are strike-slip faults offsetting the sea-floor spreading axis—beyond the offset zone (called a *fracture zone*) the opposite blocks move in the same direction (see Fig. 12.12).

Other types of faults that one encounters in the field are *oblique faults*, *hinge faults*, and *scissors faults*. Oblique faults are strike-slip fautls combined with down-dip or up-dip

motion (Fig. 17.6D). Hinge faults are normal or reverse faults that start at a point, with displacement increasing away from the origin (Fig. 17.6E). Scissors faults are double-hinge faults, with the displacement increasing in opposite directions from a central origin (Fig. 17.6F).

Faults range in length from microfractures to the great transcurrent and transform faults that may be thousands of kilometers long. The larger faults are not single planes, but rather fault zones, belts of closely spaced, intersecting, and ramifying fault planes. They constitute a long-lasting zone of weakness in the Earth's crust where strain energy is released periodically by means of earthquakes.

17.4. TRAPPING HYDROCARBONS

Petroleum

Hydrocarbons can be trapped by structural changes or by diagenetic changes. When reservoir rocks are folded, the hydrocarbons migrate

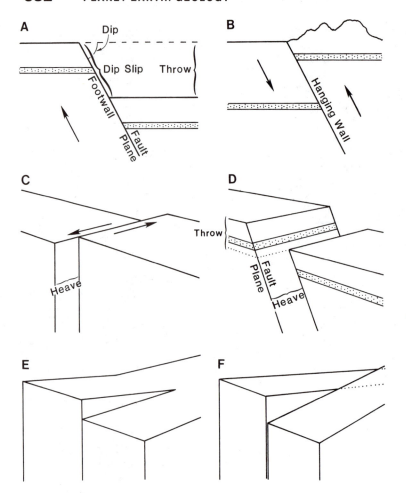

Figure 17.6. (A) Normal fault. (B) Reversed fault. (C) Strike-slip fault. (D) Oblique fault. (E) Hinge fault. (F) Scissors fault.

upward under the anticlinal fold (Fig. 17.8A). If the anticline contains an impermeable rock layer (e.g., shale), the hydrocarbons are trapped. Gaseous hydrocarbons are found impregnating the highest, permeable layers along the crest of the anticline; liquid hydrocarbons may be found below the gaseous ones; and, still below, one usually finds water. *Faults* can form hydrocarbon traps if motion along a fault causes an upward-slanted reservoir layer to abut against an impermeable layer (Fig. 17.8B). *Unconformities* are breaks in sedimentation (Section 18.5). If a permeable rock layer is truncated by erosion, overlain by an impermeable layer, and then mineralized, a trap is produced. Salt domes pierce the overlying

Table 17.4. Fault nomenclature

Name	Explanation
footwall	the underlying side of a fault
hanging wall	the overlying side of a fault
strike	the orientation of the fault-plane intersection with the horizontal; it is best expressed in degrees, 0° to 360°, measured clockwise from the north
dip	the angle of the fault plane with respect to the horizontal
dip-slip	the displacement along the fault plane normal to the strike
throw	the vertical displacement of a fault
heave	the horizontal displacement of a fault

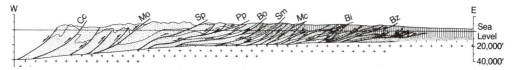

Figure 17.7. Thrust faulting in the Canadian Rockies. The wedges, supported by fluid pressure, moved to the right when the basement was sloping in that direction. (Dahlstrom, 1969.)

layers and in so doing upturn reservoir layers and seal them, because salt is impermeable to fluids.

Stratigraphic traps are produced when there is a lateral facies change from permeable reservoir rock (e.g., sandstone) to impermeable rock (e.g., shale) and when the direction of the facies change is tilted upward. Hydrocarbons attempting to rise become trapped by the facies change. *Cementation* of a portion of a permeable layer under an impermeable layer, followed by tilting upward, also traps the hydrocarbons that are attempting to rise.

Marine sediments contain from less than 2% to more than 10% organic matter. Nearly all geoclinal and basinal sediment accumulations are potential sources of petroleum. The origin of the giant oil fields of the Middle East

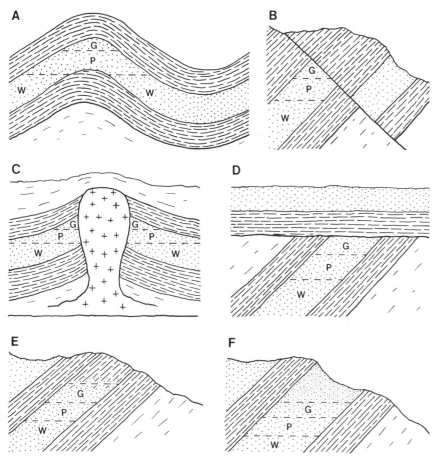

Figure 17.8. Hydrocarbon traps (dashed strata = shale; dotted strata = sandstone): (A) anticline; (B) fault; (C) salt dome; (D) uncomformity; (E) stratigraphic (sandstone grades into shale); (F) cementation (W = water; P = petroleum; G = gas).

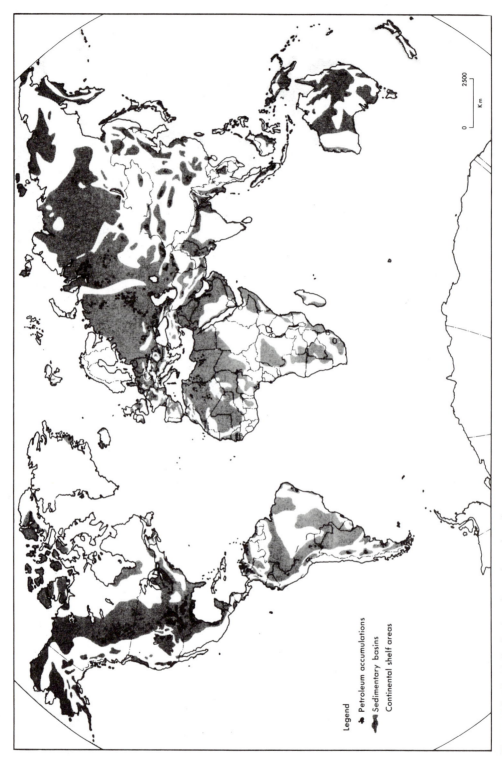

Figure 17.9. World distribution of petroleum deposits. (Skinner, 1969, p. 115, Fig. 7.8.)

Legend

Petroleum accumulations

Sedimentary basins

Continental shelf areas

was discussed in Section 12.5. Sixty percent of the recoverable oil reserves known today are in the Middle East. Fig. 17.9 shows the world distribution of known petroleum deposits. Current production (and concomitant consumption) is about 20 billion barrels (2.9 billion tons) per year.

[A barrel (bbl) is equal to 42 U.S. gallons or 158.987 liters. Because the average density of oil is 0.91, 1 bbl is equal to 145 kg.]

At the current rate of extraction, and barring the discovery of major new oil fields, the world petroleum reserves will last 35 years.

Oil shales are fine-grained, often laminated shales containing more than 10% organic matter deposited under anaerobic conditions in either a marine or lacustrine environment. *Kerogen* is an organic substance with a high molecular mass that is largely insoluble in hydrocarbon solvents but that breaks down to yield hydrocarbons. The yield can be as high as 70% of the total organic carbon in the kerogen. A rich shale containing 40% kerogen (which amounts to 53% of the volume of the rock) will have a density of 1.8 g/cm^3 and will yield about 225 liters (1.4 bbl) of oil per ton of rock.

The world's largest oil-shale deposit is the Green River Formation (see Section 16.1). The oil concentration averages 125–150 liters per ton of shale. At the present time, the cost of retrieving that oil would be greater than the price of the oil from the large oil field of the Middle East. Nevertheless, the Green River shales constitute an oil reserve amounting to 10^{13} t or 10^{11} bbl (about 20% of the proven oil reserves of the world). At some future time we will be glad we have it (personally, I am glad even now).

The first production of oil from oil shale was in France in 1838. Oil from shale reached a production peak in Europe and Japan during World War II. Production was discontinued in the early 1960s when the price of oil dropped. It no doubt will resume if the price of oil rises high enough. Producing oil from oil shales poses some serious problems, including air pollution and a huge amount of waste rock. The world reserves of oil from oil shale are estimated at about $2.9 \cdot 10^{14}$ t or $2 \cdot 10^{15}$ bbl. At the current rate of consumption, those reserves would last 100,000 years.

In addition to oil shales, there are bituminous sands, also called "oil sands" or "tar sands", which are uncemented deltaic sands impregnated with bitumen. The lighter hydrocarbons have been lost, but the bitumen is preserved, because even though the cover is not truly impermeable, the vapor pressure is low. The world reserves of bitumen are estimated at about 10^{12} t. Major deposits are in Alberta (20% of the world total). The largest single deposit is that of the Athabasca Tar Sands of Early Cretaceous age, which contain $1.5 \cdot 10^{11}$ t of heavy oils and bitumen. A similar amount is contained in the Orinoco Tar Belt, which is 60 km wide and stretches for 700 km along the Orinoco River in Venezuela.

Natural gas

Natural gas consists of a mixture of light hydrocarbons that are gaseous at room temperature and pressure (Table 17.5). Natural gas usually is found above petroleum in oil deposits, but because it is a gas, it may migrate away from the petroleum and form its own deposit; permeability need not be as high as for liquid hydrocarbons, because of the much lower viscosity of the gas.

Traps capable of trapping natural gas also trap other gases attempting to escape from the interior of the Earth. These gases include CO_2, water vapor, Ar, He, and trace amounts of the other rare gases. Argon and helium are particularly abundant, argon being the decay product of ^{40}K, and helium being former alpha particles emitted by uranium and thorium inside the Earth. Once in the atmosphere, argon stays. Argon needed by industry is simply extracted from the atmosphere. Helium, on the other hand, escapes, leaving only a transient concentration of $5.24 \cdot 10^{-6}$ (by volume). Helium needed by industry is extracted

Table 17.5. Composition of natural gas

Compound	Volume Percentage
methane (CH_4)	85
ethane (C_2H_6)	10
propane (C_3H_3)	3
butane (C_4H_{10}) and heavier	2

from natural gas at the well, where it may have a much higher concentration. The highest concentrations ($>0.5\%$) are found in gas reservoirs under highly impermeable cover, and where the gas was derived from Paleozoic source rocks (giving them more time to accumulate helium).

17.5. METAMORPHISM

Diagenesis grades into metamorphism at increasing temperature and pressure. Generally, minerals react with each other and with H_2O, forming new minerals characterized by the OH group. Table 17.6 shows the metamorphic facies that result from increasing metamorphism. Low-temperature, high-pressure metamorphism occurs in subduction zones, where the rocks of a subducting plate have no time to warm up. High-temperature, low-pressure metamorphism occurs in the vicinity of hot intrusions. This type of metamorphism is called *contact metamorphism.*

The pressure in metamorphism is unidirectional, not isotropic. The growth of phyllosilicates (chlorite, muscovite) is favored, leading to the development of *schistosity.*

"Clean" sandstones metamorphose to quartzite, and "dirty" sandstones metamorphose to gneiss. The growth of phyllosilicates during the metamorphism of shales leads to the formation of *phyllites* and, with further

metamorphism, to the formation of *schists.* Quartzites, gneisses, and schists are the products of high-grade metamorphism of clastic siliceous rocks.

Limestones generally contain some clay. At 300–400°C, talc (Mg silicate, OH) and tremolite (Ca-Mg silicate, OH) are formed, while the dominant calcite recrystallizes to form *marble.*

The metamorphism of sediments on the bottom of a geocline and during subsequent tectonization into mountain ranges is called *regional metamorphism* because it affects a large region.

Migma is a mixture of solid rocks and magma, and *migmatites* are mixtures of igneous and metamorphic rocks (Fig. 17.10). *Migmatization* is the highest grade of metamorphism; it takes place at the bottom of the continental crust, where granitic magmas form. Mixtures of gneiss, quartzite, and schist form granitic magma, which crystallizes into granite. Granite is less dense than the surrounding rocks in the lower continental crust and tends to rise, forming batholiths (Section 10.3).

Metamorphism in the oceanic crust involves percolating seawater. The basalts of the oceanic crust are fractured, especially along the mid-ocean-ridge crests, allowing seawater to penetrate to depths up to 5 km. Low-temperature alteration of gabbro-basalts produces serpentine (Mg-Fe silicates, OH), brucite [$Mg(OH)_2$], and talc (Mg silicate, OH). Low-grade metamorphism produces mineral suites

Table 17.6. Metamorphic facies

Temperature (°C)	Pressure (kb)	Type of Mineral	Composition	Facies
Continental crust				
Temperature proportional to pressure				
< 200	< 3	zeolites	(Na, K, Ca)(Al-sil.)OH	zeolite
200–400	3–8	chlorite	(Mg, Fe)(Al-sil.)OH	greenschist
400–500	8–10	epidote	(Ca, Fe)(Al-sil.)OH	epidote
500–600	10–12	amphibole	(Mg, Fe, Ca, Na)(Al-sil.)OH	amphibolite
> 600	> 12	sillimanite	Al_2SiO_5	granulite
Low temperature, high pressure				
100–200	3–8	glaucophane	(Na, Mg, Fe)(Al-sil.)OH	blueschist
High temperature, low pressure				
300–800	~ 2	andalusite	Al_2SiO_5	hornfels
Oceanic crust				
< 150	0.5–2	serpentine	$(Mg, Fe)_3Si_2O_5(OH)_4$	ophiolite

Figure 17.10. This was once a pile of sediments lying at the bottom of a 16-km-thick-geocline in eastern Greenland. Granitic magma, injected in thin sheets from below, migmatized zone 3 and, to a lesser extent, zone 2. Zone 1 consists of sedimentary rocks that were metamorphosed but not migmatized. (Haller 1971, photo 46.)

characteristic of the greenschist facies (i.e., chlorite, plagioclase, and calcite). Intermediate-grade metamorphism produces amphibolites. High-grade metamorphism dehydrates the amphiboles and produces pyroxenes and silli-manite (*granulite facies*).

THINK

* Given the same content of cementing fluid, would you expect a quartz sand with grains of low roundness to be more or less readily cemented than a quartz sand with high roundness?

* The lunar rhegolith covering the maria has an average thickness of 4–5 m. It consists of particles ranging around 50–100 μm in size, with interspersed pebble- to boulder-size blocks. The lunar rhegolith was formed by the settling of materials following impacts. Most of it formed more than 3 billion years ago. The bulk density of the finer rhegolith increases from 1.50 g/cm^3 in the top 15 cm to 1.74 g/cm^3 in a layer only 30–60 cm below the surface. Porosity averages 40%. By comparison, loess, with particle size and bulk density in the same ranges as the surface rhegolith, has a porosity of 60%. How do you account for the high compaction gradient and the low porosity of the lunar rhegolith in view of the low gravitational field and the absence of surface agents?

* The Sun has a mass of 1.989·10^{30} kg. Its distance from the Earth is 149.6·10^6 km. The solar constant is 1,360 W. Using these figures and the data in Table 17.2, calculate how long the Sun would last if it were made of butter.

* Calculate the density of the SAE 30 oil.

* Assuming that the Green River oil shale yields an average of 1 bbl of oil per ton of rock, and assuming that the rock has an average density of 2 g/cm^3, calculate how many cubic meters of oil you could obtain from the entire formation and how big a cylindrical tank (with width equal to height) you would need to store it.

* Is the San Andreas fault left-lateral or right-lateral?

Stratigraphy is that branch of the geological sciences dedicated to the study of the succession of sedimentary rocks, their fossil contents, their environmental significance, and their positions in time. A sedimentary rock, derived from the lithification of sediment, provides us with information on the geological environment in which the original sediment was deposited. As we saw in Chapter 16, different geological environments produce different types of sediments. The fossil content of a sedimentary rock yields additional, often critical information on the ancient environment; in addition, because fossils reflect evolutionary changes over time, they also reveal the approximate age of the rock. Fossils, therefore, are of key importance to geological studies.

18.1. DEVELOPMENT OF STRATIGRAPHY

Fossils attracted the attention of humans early on—a neolithic grave in England was found to contain a skeleton surrounded by fossil echinoderms, and a silicified tree trunk was used as a pedestal in an Etruscan tomb. Some ancient Greeks not only had an idea of evolution but also understood what fossils were. Anaximander (ca. 610–540 B.C.E.), for instance, believed that life originated at sea and that land animals derived from marine animals left stranded on the beach; Empedocles (ca. 490–430 B.C.E.) thought that fossils were the remains of extinct species and that only organisms fit for the environment in which they lived survived; and Herodotus (ca. 484–425 B.C.E.) stated that the fossil echinoderms he saw strewn around the Egyptian desert were the remains of marine organisms that had lived there at a time when Egypt was under water (see Part VI).

As far as fossils are concerned, Aristotle (384–322 B.C.E.) represented a setback—he believed that fossils were *lusi naturae* (plays of nature), born and grown in mud, that never made it to the living world. The Roman historian Titus Livius (59 B.C.E. to 19 C.E.), the philosopher Seneca (55 B.C.E. to 39 C.E.), and the scientist Pliny the Elder (23–79 C.E.) believed, like Herodotus, that fossils were the remains of marine organisms that had lived when the areas where they were found had been below sea level.

Because, later, Aristotle was held in such high regard by the church, his views on fossils prevailed until the Renaissance. In Germany, for instance, the famous mineralogist and mining engineer Georgius Agricola (1494–1555) maintained in his book *De natura fossilium* ("On the Nature of Fossils") that fossils were mineral concretions made by "special processes". In Italy, on the other hand, Giovanni Boccaccio (1313–1375), observing the geological formations around Florence, maintained that fossils were the remains of marine organisms; so did Leonardo da Vinci (1452–1519) and Nicolaus Steno (1638–1686), a Danish physician and geologist who migrated to Florence in 1660. There he wrote a treatise that established the *principle of superposition*, namely, the principle that younger rock layers lie above older rock layers.

Giovanni Arduino (1714–1795), a professor at the University of Padua, made numerous observations on the Alps and on the hills and plains below. He subdivided the rock formations as follows:

Primary: the granites and granodiorites that form the core of the Alps;
Secondary: the sedimentary rocks to the south of the high Alps, consisting of minerals derived from the primary rocks;

Tertiary: the loose sands and clays of the Po
   Valley derived from the secondary rocks.

The granites and granodiorites of the Alps are
far from being the oldest rocks on earth. They
probably are, at least in part, ultrametamor-
phosed sediments deposited on the bottom of
the Alpine geocline less than 500 million years
ago. Nevertheless, Arduino's names have survived to this day, but with different meanings.
The name *Primary*, still used in the French and
Italian literature (*Primaire*, *Primario*), is now
taken to represent the geological era called
the *Paleozoic* (570–245·10^6 y ago). The name
Secondary, also still used in the French and
Italian literature (*Secondaire*, *Secondario*), is
now taken to represent the next geological era,
called the *Mesozoic* (245–65·10^6 y ago). The
name *Tertiary* is universally used to represent
that portion of the present geological era, the
Cenozoic, that preceded the beginning of
major glaciations in the Northern Hemisphere
(65–1.6·10^6 y ago).

   The science of stratigraphy was placed on
a solid foundation by the English William
("Strata") Smith (1769–1839). Smith was a
canal digger. He carefully observed the succession of sedimentary rocks and fossils as he
was digging his canals and noticed that the
same successions of fossil faunas could be seen
in different parts of the country. He thus
established the principles of faunal succession
and stratigraphic correlation—the correla-
tions between sedimentary layers in distant
regions characterized by the same fossil content. Smith's work prepared the foundation for
Charles Lyell's *Principles of Geology* (1830–
1833) and for Charles Darwin's *On the Origin
of Species by Natural Selection* (1859).

   Lyell (1797–1875) subdivided geological time
(the Phanerozoic, the time since the appearance
of abundant fossils) into 13 periods, recognized
as being different because of the marked
changes in fossils between one period and the
next. Lyell's classification has undergone
numerous changes and refinements. A current
scheme is shown in Table A8.

18.2. ANCIENT SECTIONS

The oldest crustal sections are geoclinal deposits (Table 18.1). The 14.3-km-thick Swaziland

Table 18.1. The oldest crustal sections

Name	Location	Age (10^9 y)
Isua section	Southern Greenland	3.8
Pilbara block	Northwestern Australia	3.6
Swaziland Sequence	Eastern South Africa	3.4

Table 18.2. The Swaziland Sequence

B. Sedimentary Sections
   4. Banded Iron Formations (BIF) (highest)
   3. shales
   2. graywackes
   1. conglomerates

A. Volcanic Sections
   3. andesites
   2. basalts
   1. komatiites (lowest)

Sequence is an impressive textbook picture of
the evolution of the early crust. Table 18.2
summarizes the stratigraphy of this section. As
is customary in stratigraphic representation,
the table presents the sequence of layers as
they appear in the field, from the bottom up.
Thus, the komatiites are the oldest rocks in
the sequence, and the Banded Iron Formations
(BIF) are the youngest. It is not known how
much time the Swaziland Sequence represents—
perhaps 100 million years or so. The age of the
sequence is about 3.4·10^9 y. There are several
komatiite-basalt cycles in the lower portion of
the sequence and several conglomerate-BIF
cycles in the upper portion of the section. The
overall trend, however, is from peridotitic rocks
at the bottom to sedimentary rocks above. The
early geoclinal deposits in the area were metamorphosed to greenschist facies (Section 17.5)
and formed *greenstone belts*. These were
intruded by broad granite-gneiss batholiths
200–300 million years later (Fig. 18.1), most of
which originated from continued fractional
crystallization in the upper mantle, but some
resulted from ultrametamorphism of graywacke
to shale sediments.

Figure 18.1. Greenstone belts (dark) and granitic plutons (light) in the Pilbara block, northwestern Australia. The circular pluton at the top is about 50 km across. (Courtesy EROS Data Center, Sioux Falls, S.D.)

The continental nuclei where Archean (3.8–$2.7 \cdot 10^9$ y B.P.) and Proterozoic (2.7–0.6 $\cdot 10^9$ y B.P.) rocks outcrop are called *shields*. As in the Swaziland Sequence, granite-gneiss batholiths account for 80% of the shield area, and greenstone belts account for the remaining 20%. The major shields are

Canadian
Fennoscandian
Angaran (northeastern Siberia)
African
Brazilian
Western Australian
Eastern Antarctic

The shields are bordered by *stable platforms*, which are continental areas floored by shield extensions and covered mostly with a thin (< 2-km) layer of sediments (mainly clean sandstones, shales, and limestones). The sediments can thicken to 5–6 km in basins and thin out in domes. The shield + stable-platform areas of the world that have never subducted are called *cratons* (see Section 15.3).

18.3. MARINE TRANSGRESSIONS

The continental crust grades into the oceanic crust through a tapering zone whose upper surface is the continental slope. In the Pangean reconstruction, the continents fit against each other almost perfectly if the boundary between the continental and oceanic crust is taken to be halfway down the continental slope. This is then the geophysical boundary between continents and oceans. The geographic boundary (i.e., the shoreline) is 2 km higher up.

The shoreline is not notably stable. Sea level has oscillated through geological time by as much as 600 m. A rise in sea level is accompanied by an inland migration of the shoreline called an *ingression*. A drop in sea level will move the shoreline out to sea. This is called a *regression*. Marine ingressions and regressions are collectively called *transgressions*. If all the ice now present on Earth were to melt, the sea level would rise by 77 m, which would reduce to 54 m with complete isostatic adjustment. Table 18.3 lists the major transgressions that have occurred during geological time. The great marine ingressions of the Early Ordovician and Late Cretaceous covered 50% of the present land surface, while the great regression of the Late Oligocene added about 5% to the land surface.

Fig. 18.2 summarizes the changes in sea level since the beginning of the Cambrian. Superimposed upon a great cycle, with a wavelength of 400 million years (left), were numerous secondary cycles (right). The sea level change during ingressions is much slower (0.1 cm/1,000 y) than during regression (2 cm/1,000 y). This is elegantly explained by my brand-new *geosouffé theory*, which says that the Earth's internal heat causes a slow swelling of the mid-ocean-ridge system, producing ingression; swelling causes the sea-floor surface rocks

Table 18.3. Major marine transgressions

Geological period	Age (10^6 y)	Sea-level change (m)
Early Ordovician	505–480	+ 350
Early Jurassic	190	− 165
Late Cretaceous	90–70	+ 350
Late Oligocene	30	− 200
Middle Miocene	13	+ 135
Pleistocene ice ages	3–0	− 120

to crack, establishing hydrothermal circulation to a depth of 5 km or more; heat is rapidly lost, the mid-ocean ridges pancake (like your mother's soufflé), and a rapid regression ensues.

The more rapid marine regressions and ingressions associated with the waxing and waning of ice during periods of glaciation have opposite dynamics: the regressions are slow (80,000 years) and the ingressions are rapid (10,000 years). This is because it takes time to build up ice on land. In fact, heat exchange between the growing ice sheets and the ocean (which supplies the moisture) cools the surface water, which having become denser, sinks and is replaced by warmer water from below. The building of ice sheets therefore requires cooling of the entire seawater column. In contrast, deglaciation spreads a layer of lower-salinity seawater across the ocean, and the Sun warms it up, creating a stable stratification. Warming up of the ocean surface is rapid, and so are the melting of the ice and the concomitant marine ingression.

Ingressions and regressions related to sea-floor tectonism or to glaciation have been worldwide and synchronous. In addition, there have been regional or local ingressions and regressions caused by regional tectonism or local phenomena. An example of a regional regression is that caused by the postglacial uplift in eastern Canada and Scandinavia. When the ice started melting, about 16,000 y ago, the land began rebounding and rising, causing the sea to withdraw from coastal regions. The land has now risen 500 m, with an additional 250 m still to go.

An example of a local change is the Roman Temple of Jupiter Serapis at Pozzuoli, near Naples, Italy (Fig. 18.3). The temple was built

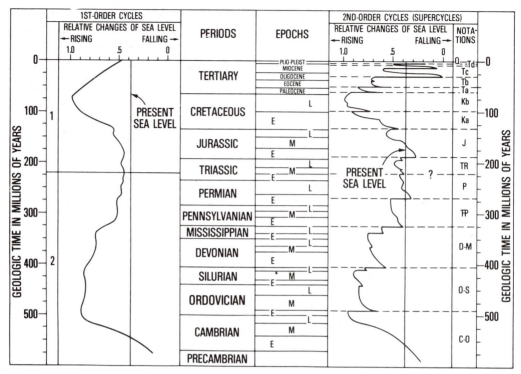

Figure 18.2. Phanerozoic transgressions: major cycle on the left, more detailed curve on the right. The maximum sea-level rise was +350 m (Early Ordovician, Late Cretaceous); the greatest drop was −250 m (Late Oligocene). The total change in sea level was 600 m, or about 15% (*E* = Early; *M* = Middle; *L* = Late). (Vail, Mitchum, and Thompson, 1977, p. 84, Fig. 1.)

close to the shore in the second century B.C.E. It then sank to about 4 m below sea level, as demonstrated by pholad borings. After that, it rose about 2 m, bringing the pholad borings to 2 m above sea level. (Pholads are a family— Pholadidae—of bivalve mollusks that bore holes in rocks, in which they live and from which they emerge for feeding.)

Marine ingression reduces the albedo of the Earth, because seawater is a better absorber of solar radiation than is dry land. As a result, the temperature gradient between low and high latitudes is reduced at times of ingression, with polar temperatures in the range of 20°C. Part of this reduction is attributable to increased CO_2 content in the atmosphere, from the increased volcanic activity along the mid-ocean-ridge system and from increased organic productivity. Marine ingression leads to widespread deposition of fossil-rich marine sediments on land. It is from the study of these sediments and fossils that geologists have been able to reconstruct the environmental evolution of our planet (see Part V).

Marine regressions interrupt marine sedimentation on land, and erosion may reduce the sedimentary deposits previously formed. As a result, the continuity of the marine record in any given region of a continent rarely exceeds a few tens of millions of years. A continuous record covering the entire Phanerozoic can be reconstructed only by correlating and patching together different sections. The succession of species in time is the cornerstone of stratigraphic correlation.

18.4. CORRELATION AND RELATIVE DATING

Different environments contain different types of organisms and produce different types of sediments. *Facies* is the sum total of the sedimentological and paleontological characteristics

Figure 18.3. The temple of Jupiter Serapis at Pozzuoli, Italy. Originally the temple was 40 m long and 34 m wide. Its roof was supported by 46 columns, each 12.5 m high and 1.5 m across. Three of these columns still stand. Pholad borings on the columns between 3.5 and 6 m above the base indicate that the temple sank rapidly to 6 m below high-tide level, and the floor was rapidly covered with a layer of mud 3.5 m thick, preventing pholad activity. The temple has since risen, and the mud has washed away, but the floor still remains under water. (Dana, 1896, p. 349, Fig. 321.)

Table 18.4. Taxonomic categories for *Homo sapiens*

Taxonomic Category	Appearance on Earth (10^6 y)
Superkingdom Eucaryota	1,500
Kingdom Animalia	1,000 (?)
Phylum Chordata	500
Class Mammalia	250
Order Primates	30
Family Hominidae	5
Genus *Homo*	2
Species *sapiens*	0.1

[Taxonomists classify living organisms in categories that range from superkingdom to species. Table 18.4 shows the principal taxonomic categories for *Homo sapiens*, together with their times of appearance. As may be seen, the higher the taxonomic category, the longer the life span of the category.]

Different environments are characterized by different sets of species, all of which evolve through time. Stratigraphy has three main purposes:

- To determine the age of a rock from its fossil content,
- To determine the environment in which that rock was formed, using both the physical and chemical characteristics of the rock and its fossil content,
- To correlate rock layers that may occur in different regions or even different continents.

The lifetimes of species of common marine invertebrates are on the order of 1 to 10 million years. It would seem that the error should be on the order of at least ± 1 million years. Species, however, follow each other through time, so that such criteria as the time of the first or last occurrence of a given species, or the concurrent presence of species that have little overlap in time, allow stratigraphers to increase the precision of age determination to perhaps $\pm 100,000$ years or even better in favorable cases (i.e., when the facies are similar and the fossils are abundant).

Marine deposits are easier to correlate than

of a layer or group of layers. Some environments (e.g., tropical coral reefs, tropical rain forests), have persisted on Earth since the Paleozoic. Other environments (e.g., glacial lakes) have reoccurred intermittently. Even persistent environments have evolved through time, however, because life has evolved and life profoundly interacts with the environment. The environment now represented by Florida Bay, for instance, was widespread in Paleozoic time in the North American midcontinent, as well as in Europe and elsewhere. There are corals now, and there were corals then, all of which belong to the same phylum (Cnidaria); the families and lower ranks, however, are different.

land deposits for two reasons: (1) The oceans mix rapidly (< 1,500 y), enabling even sessile marine animals, most of which have planktic larvae, to diffuse throughout the oceans (and be recorded in the sediments) on a time scale commensurate with the rate of oceanic mixing. (2) A wide variety of marine animals deposit shells of calcium carbonate that are preserved as fossils. By comparison, the most widespread group of land animals—the insects—have left a scanty fossil record, even though they have been around for 400 million years. Land deposits are more difficult to correlate in terms of their fossils, because land fossils are generally much more scarce, and land faunas exhibit much greater regionality than do marine faunas.

Magnetostratigraphy has recently become important in stratigraphic correlation and dating. Sediments exhibit magnetization because, when the particles settle, those that possess magnetic polarity orient themselves according to the ambient magnetic field. As a result, the sediment as a whole acquires a weak magnetization. Furthermore, the sediments contain magnetotactic bacteria—bacteria that contain open chains of microcrystals of magnetite, each about 0.1 μm across. A magnetotactic bacterium uses its magnetic chain (called *magnetosome*) to orient itself with respect to the vertical and remain close to the sediment surface, where organic matter is most abundant. Reworking of bottom sediments by benthic animals does not erase the magnetic signature contained in the sediment because, in resettling, the particles orient themselves again. The magnetization of a stratigraphic section is assessed in terms of both strength and direction. The section is sampled, and the orientation of the sample with respect to the present N–S direction is recorded. The magnetization of a rock sample includes the original

magnetization established when the sediment was deposited and, superimposed on it, an overprint of any change in the magnetic environment subsequent to the original deposition. Except in cases of a sediment subsequently cemented with iron oxides, the overprint usually is weaker than the original magnetization. The overprint is eliminated by heating the sample stepwise, until a stable magnetization, called *thermoremanent magnetization*, is achieved. The S–N direction of magnetization in the sample is determined by means of a *spinner magnetometer*: The sample is made to spin in three orthogonal directions in space within a sensing coil. The strength of the current induced in the sensing coil is a function of the orientation of the sample. The remanent S–N direction of the fossil magnetic field can thus be determined. The system includes a set of coils that cancel out the ambient magnetic field.

If a stratigraphic section is sampled at close stratigraphic intervals, the reversals revealed by the section can be compared with the standard polarity scale (Figs. 11.31 and 11.32). The section can thus be dated. Magnetostratigraphy has made it possible to correlate land deposits with marine deposits now outcropping on land and with deep-sea deposits still on the ocean floor.

The most accurate method of stratigraphic correlation, applicable mainly to deep-sea sediments, is by means of oxygen isotopes. The oxygen-isotopic composition of the calcite of planktic microfossils is related to both the oxygen-isotopic composition of the seawater and the temperature at which the calcite precipitated. If these parameters vary in a systematic way, as during the recent ice ages, variations in the oxygen-isotopic composition of the fossils can be used for correlation. Precision is within 3,000 y or so, limited by the bioturba-

Table 18.5. Chronological and Stratigraphic Subdivisions

Chronology	Example	Time range (10^6 y)	Stratigraphic Terms
era	Cenozoic	65–0	erathem
period	Neogene	23.3–1.64	system
epoch	Pliocene	5.2–1.64	series
age	Piacenzian	3.40–1.64	stage

A

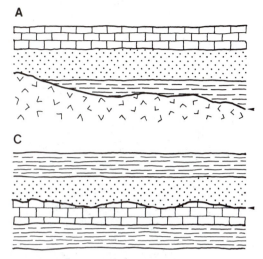

B

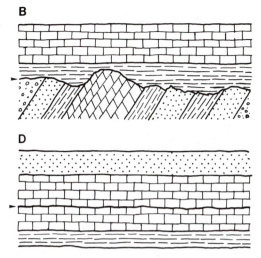

C

D

Figure 18.4. Unconformities: (A) nonconformity, (B) angular unconformity, (C) disconformity, (D) para- conformity (brickwork = limestone; dotted = sand- stone; dashed = shale; angles = granite).

tion of the sediment by benthic organisms (about 10 cm of sediment thickness). This method is particularly useful for the Late Cenozoic, when isotopic changes were both marked and periodic (see Chapter 24).

Geological time is divided into eras, periods, epochs, and ages. To each one of these subdivisions corresponds a stratigraphic term. Table 18.5 shows the subdivisions of geological time and their stratigraphic equivalents. The chronological units are subdivided into Early, Middle, and Late. The corresponding strati- graphic subdivisions are Lower, Middle, and Upper.

Portions of geological time have been divided into zones on the basis of particularly abundant and/or rapidly speciating taxa. The Cenozoic and much of the Mesozoic have been subdivided into zones based on coccolitho- phorids (planktic protophytes) and foramini- fera (planktic protozoans). These biostratigra- phic zones average in duration from $1 \cdot 10^6$ y (Neogene) to $3 \cdot 10^6$ y (Cretaceous). The Jurassic has been subdivided into 74 zones based on ammonites (an extinct order of cephalopods). Because the Jurassic lasted $62 \cdot 10^6$ y, the average duration of an ammonitic zone is 840,000 y. Pelagic organisms are particularly suitable for biostratigraphic zonation because of their wide dispersion and because their speciation and extinction generally were rapid and usually were synchronous worldwide.

18.5. UNCONFORMITIES

A break in the sedimentary record is called an *unconformity*. There are different types:

Nonconformity: an erosional surface separating igneous or metamorphic rocks below from sedimentary rocks above (Fig. 18.4A)
Angular unconformity: the bedding of sedimen- tary rocks below is at an angle to that of the sedimentary rocks above, indicating a period of diastrophism and erosion (Figs. 18.4B and 18.5)
Disconformity: the bedding of the sedimentary rocks below is parallel to that of the sedi- mentary rocks above, but a conspicuous ero- sional surface separates the two (Fig. 18.4C)
Paraconformity: the bedding of the sedimen- tary rocks below is parallel to that of the sedimentary rocks above, but the two are separated by a period of nondeposition (Fig. 18.4D)

Stratigraphic sections more nearly continuous than those on land occur on the deep-ocean floor. Sections of Globigerina-ooze sediment are rich in foraminiferal and coccolith fossils that reflect the environmental conditions at the sea surface where they once lived. Since the mid-1950s, isotope and micropaleontological studies of deep-sea sediment sections have clarified the history of the ice ages, which had confused geologists for more than 100 years.

Figure 18.5. Upper beds of the Old Red Sandstone formation (Permian) lying unconformably on vertical beds of Silurian shales and graywackes at Siccar Point, Cockburnspath, Berwickshire, on the North Sea coast of southeastern Scotland. It was here that the Scottish geologist James Hutton (1726–1797) realized that the crust of the Earth was not static. He concluded that all features that we see, including the loftiest mountain ranges, are the results of slow processes that are going on right under our eyes. He enunciated the principle of uniformitarianism, which says that the same processes active today were active in the past. He published his views in a famous book, *Theory of the Earth*, published in 1785. (Courtesy H. M. Geological Survey.)

Since 1967, studies of sediment sections retrieved from the deep-sea floor by the drilling vessels operated by the Deep-Sea Drilling Program, which was initiated in 1967, and its successor, the Ocean Drilling Program, have clarified the history of our planet during the past $180 \cdot 10^6$ y (Fig. 18.6). These studies have shown, for instance, that Australia separated from Antarctica $40 \cdot 10^6$ y ago, which permitted the development of the Antarctic Circumpolar Current and led to the formation of the Antarctic ice sheet. They have also shown that the Mediterranean dried up and was reflooded several times between $6 \cdot 10^6$ and $5 \cdot 10^6$ y ago and that North America and South America became joined by the Central American isthmus about $4 \cdot 10^6$ y ago.

Unfortunately, because of subduction, there is no portion of the sea floor older than about $180 \cdot 10^6$ y. For times earlier than that, stratigraphers still have to make use of the discontinuous records that marine ingressions left on land.

THINK

* In the Isua section of Greenland there are rocks that are $3.8 \cdot 10^9$ y old. If the present, total amount of ^{238}U on Earth is set arbitrarily as equal to 1, and considering that this amount can be viewed as growing back in time by $e^{\lambda t}$ (instead of decreasing with time by $e^{-\lambda t}$), what was the amount of ^{238}U when the Isua rocks were being formed?

* Marine ingressions and regressions are slow enough for isostatic equilibrium to be maintained. If so, a marine ingression will force the

Figure 18.6. (A) The drilling vessel *Submarex*, the first ship used for deep-sea drilling, was a converted World War II submarine chaser, 63 m long and 7 m wide and capable of doing 17 knots (17 nautical miles per hour). It was equipped with a skid-mounted Howard-Turner drilling rig and a 12.1-m mast. Drilling was done from a small rotary platform welded off the side of the ship. It was used by the author of this book to core sediments off the crest of a submarine ridge between Jamaica and Nicaragua. Deepest penetration achieved was 56 m below the sea floor. The ship was chartered from Global-Marine of Los Angeles, California (now in Houston, Texas). For details of this operation, see Cesare Emiliani, "A New Global Geology," in *The Oceanic Lithosphere*, ed. Cesare Emiliani (New York: John Wiley, 1981), pp. 1687–728. (B) The *Glomar-Challenger*, built and operated by Global-Marine especially for deep-sea drilling, initiated operations in the Gulf of Mexico in 1967 and was decommissioned in 1983. This ship, 122 m long and 18 m wide, had a drilling rig built amidship that operated through a well cut into the ship. During its 17 years of operations, it drilled the ocean floor at 624 sites worldwide and recovered 97,054 m of cores. Deepest penetration was 1,740 m below the sea floor at a site in eastern North Atlantic. The *Glomar-Challenger* was used in the famous expedition led by Arthur Maxwell in 1970 that proved the reality of sea-floor spreading (Section 12.1). (C) The *Joides-Resolution* is the new, larger, and better-equipped drilling ship that followed the *Glomar-Challenger*. This ship, 143 m long and 21 m wide, was commissioned in 1984 and is operated by SEDCO. By April 1992 it has recovered 69,693 m of cores. Deepest penetration achieved was 2,000.4 m, at a site in the eastern equatorial Pacific. (Courtesy Global Marine Drilling Co., Houston, Texas, and Ocean Drilling Program, College Station, Texas.)

land down and the sea floor up, whereas a marine regression will do the opposite. According to current estimates, the Ordovician and Late Cretaceous marine ingressions raised the sea level over the continents by 350 m (Table 18.3) with respect to the present sea level. Taking the density of the asthenosphere to be 3.3 g/cm³, calculate the actual sea-level rise during the Ordovician and Late Cretaceous ingressions, and, from the volume of the water on land, calculate how much the sea floor rose. Did the rise of the sea floor displace enough water to flood the continents by the observed amount? If not, what other processes might have entered the picture?

* Present at least two devastating arguments proving that my brand-new geosoufflé theory cannot possibly work.

V

Evolution of life and environment

P ART V discusses the evolution of life and environment, in six
chapters:

A historical approach is followed throughout, in the sense that organic
structures, systems, and taxa are introduced in the order in which they
appeared (or are believed to have appeared) in geological time. Thus,
the procaryotic cell is discussed Chapter 19, but the eucaryotic cell is
discussed in Chapter 20, after representative Early Cryptozoic stratigraphic sections have been introduced, so that the reader will have a
feeling for the environment where these cells evolved and lived. Similarly,
the taxa that appeared in the Late Proterozoic and the subsequent
Phanerozoic are introduced at the appropriate times. This approach
is used to emphasize the close relationship between organic evolution
and environmental evolution. It provides the reader with an orderly,
integrated picture of the true evolution of Planet Earth.

A key to Part V is the geological time scale (Table A8), which is
based on a variety of geochemical and geophysical methods (see Sections 3.4 and 11.8). We adopt, in this book, the time scale synthesized
by W. Brian Harland and coauthors and published in 1989 by Cambridge
University Press.

Chapter 19 begins with brief discussions of how life may have originated on Earth. The structures of the molecules key to living processes
are reviewed. Bacteria are treated in some detail, because those
organisms were the only forms of life on Earth for the first two-thirds
of its history.

Chapter 20 includes an analysis of some of the classic Cryptozoic
sections (Isua, North Pole, Swaziland, Gunflint), a discussion of the
eucaryotic cell and how it operates, a brief review of mitosis and meiosis,
the basic principles of genetics and evolution, a review of the taxa that
are believed to have appeared in Late Cryptozoic time, and a discussion
of the Cryptozoic–Phanerozoic transition.

Chapter 21 presents, in taxonomic order, the taxa that appeared
(or are believed to have appeared) in the earliest Phanerozoic.

Chapters 22–24 contain narratives of the Paleozoic, Mesozoic, and Cenozoic history of our planet, with emphasis on the major events that guided the organic and environmental evolutions up to and including the time when modern humans appeared on Earth and began interfering with the environment.

Chapter 24 ends with an exhortation to control the growth of the human population, to forgo interracial and intercultural strife, to foster religious tolerance, to take good care of our little planet, and to enjoy the rest of the Quaternary.

19 ORIGIN AND EARLY EVOLUTION OF LIFE

19.1. ORIGIN OF LIFE

People are awed by life. Some call it "unique". Many call it "a miracle". All are puzzled and bewildered by it. Indeed, there is excellent cause for bewilderment—life, as we know it today, is the end product of 15 billion years of cosmic and terrestrial evolution and therefore is utterly complex. We see a fantastic machine with billions of moving parts, a machine that has been fine-tuned by nature to the utmost degree—and we are bewildered. I hope, however, that this book will convince you that, far from being a miracle, life is a *necessary* and *inevitable* consequence of the way the world originated and evolved. Indeed, the origin and evolution of life on Earth were foregone conclusions right from the beginning, given the mass, the chemical composition, and the physical and astronomical parameters of our planet. Life must have originated and evolved also in many of the other Earthlike planets that almost certainly dot the universe. Life probably originated on Mars, too, but it did not go very far, because Mars lost most of its atmosphere to outer space. The Moon and all other planets in our solar system are too hot or too cold or chemically unsuitable for the development of life.

While life on Earth is not to be considered a miracle, what certainly would have to be considered a miracle, and one of the first magnitude to boot, would be just the opposite: the absence of life from a planet like the Earth, at a distance of about $150 \cdot 10^6$ km from a star like the Sun, $15 \cdot 10^9$ y after the Big Bang—*that* would be a miracle.

But what is life? There are countless definitions, of course. Here is one that I just cooked: "Life is the set of molecular systems each one of which is capable of reproducing itself from generation to generation." This definition applies to life in general. From it, we can readily derive the definition of a living organism: "A living organism is a molecular system capable of reproducing itself from generation to generation."

Life has had an important part in shaping the environmental evolution of Planet Earth. The Earth formed 4.6–$4.5 \cdot 10^9$ y B.P. Asteroidal bombardment continued between $4.5 \cdot 10^9$ and $3.8 \cdot 10^9$ y B.P. while the Earth differentiated into core and mantle and degassed, forming the early atmosphere. It was during that turbulent period that life on Earth originated. It probably originated not just once, but many times, and in many different places, only to be stressed or even wiped out by the next asteroidal impact. About $3.8 \cdot 10^9$ y ago the asteroidal bombardment petered out, and life took firm hold on the surface of the young Earth.

As discussed in Chapter 7, the dominant gases in the earliest atmosphere were (in order of abundance) CH_4 (methane), H_2O, N_2, NH_3 (ammonia), and H_2S. Life probably originated when methane was still the dominant gas in the atmosphere. That is suggested by the famous Urey-Miller experiment. Stanley Miller (b. 1930) was a graduate student at the University of Chicago in the early 1950s, working under the Nobel Prize chemist Harold Urey (1893–1981). Miller succeeded in synthesizing amino acids, the building blocks of proteins, under the primitive-Earth conditions suggested by the Russian scientist Alexander Oparin in 1926 and by Harod Urey in the early 1950s. Miller placed water in a flask, and above the water a gaseous mixture of molecular hydrogen (H_2, at a pressure of 10 cm of mercury), methane (CH_4, at a pressure of 20 cm), and ammonia (NH_3, at a pressure of 20 cm). He then passed an electric spark through the mixture, which also contained water vapor

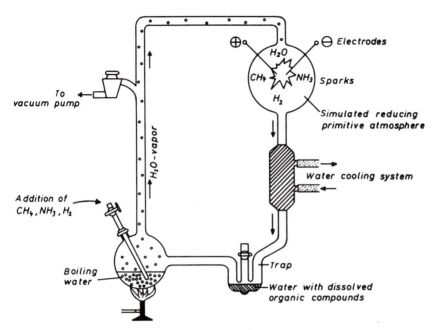

Figure 19.1. The Urey-Miller experiment. Water vapor and the added gases (CH_4, NH_3, and H_2) were cycled through an electric spark and cooled in a jacket. The condensates were collected in a trap, and the rest was recycled. Within a week at least four protein-forming amino acids had been manufactured and collected in the trap. (Schwemmler, 1984, p. 20, Fig. 1.)

from the liquid water below (Fig. 19.1). In one week the water had turned orange-red. Chemical analysis showed the presence, in abundance, of at least four amino acids. Subsequently, similar experiments succeeded in producing all the 20 protein-forming amino acids found in nature.

As soon as the early atmosphere of the Earth began forming, photolysis—the dissociation of molecules by light—began splitting H_2O into H_2 (which escaped) and O (which went to oxidize CH_4 to CO_2). In a short time, all CH_4 was transformed into CO_2. At the time, as mentioned in Section 7.4, the sky was red, and the Sun was bluish. The earliest photosynthesizing organisms had to use that kind of spectrum. The descendants of those earliest photosynthesizers may be the modern extreme-halophile archaebacteria, which use only chlorophyll a. Chlorophyll a uses red light (peak absorption at 680 nm) and blue light (peak absorption at 440 nm).

Earth, however, had liquid water. In the presence of water, CO_2 reacted with silicates to form carbonates. As CO_2 was removed from the atmosphere, the sky turned blue. Chloro-

phyll b was adopted, which uses blue light (peak absorption at 490 nm). The procaryotic Prochlorophyta and all higher photosynthesizers use mixtures of chlorophyll a (70%) and chlorophyll b (30%).

Stanley Miller showed that amino acids can be synthesized directly from the simplest molecules (CH_4. NH_3, H_2O, and H_2). But amino acids are only the building blocks for proteins. In order to have life, even the simplest form of life, a truly bewildering array of organic compounds must be formed, and conditions favorable to their interaction must develop. Therefore, before continuing our discourse on how life may have originated on Earth, we must learn the compositions and structures of at least some of the more important species of organic molecules that are involved in living processes.

THINK

* The mass of the Earth is 1.23 times that of Venus. The atmosphere of Venus, 96.4% of which is CO_2, has a pressure of 92 atm. Calculate the

total amount of CO_2 on Venus, the total amount expected in the atmosphere of the primitive Earth, the CO_2 pressure on Earth, and the total mass of carbonate that would form if all the CO_2 were precipitated as $CaCO_3$. Compare your result with the recent estimate of $276 \cdot 10^6 \, km^3$ of carbonate rock existing on Earth (assume an average density of $2.83 \, g/cm^3$ for the rock). If the two figures do not agree, how would you reconcile them?

19.2. THE MOLECULES OF LIFE

Life is based on the properties of carbon and nitrogen, each of which can assume eight oxidation states (from C^{4+} to C^{4-}; from N^{3+} to N^{5-}). The major component elements of living matter are, in order of abundance, O, C, H, N, Ca^+, P, K^+, S, Na^+, Cl^-, Mg^{2+}, and Fe. As an example, Table 19.1 shows the chemical composition of the human body (not including many other elements, mainly metals and metalloids, that are needed in trace amounts). The major molecular constituents of living systems are amino acids, proteins (including catalytic proteins called *enzymes*), nucleic acids, carbohydrates, and lipids. Pigments and a group of compounds involved in energy-transfer processes will be discussed in Section 19.3.

The *amino acids* are a family of relatively simple organic compounds characterized by a common, monovalent $NH_2-CH-COOH$ group, with side groups attached to the C of the CH segment. Amino acids range in mass from 75 u (glycine) to 240 u (tryptophan). Twenty amino acids partake in the formation of all animal and vegetal proteins (Fig. 19.2). Amino acids are readily formed abiotically by the action of UV radiation or electrical discharges through an atmosphere simulating that of the primitive Earth (Section 19.1). Some amino acids are synthesized more readily than others, however. Table 19.2 shows the yields of amino acids in an electrical-discharge experiment. It has been estimated that under primitive-Earth conditions the concentration of amino acids in the ocean may have reached 10^{-4} M (where M = molar concentration = 1 Av of solute per liter of solution).

Amino acids can link to each other by dehydration—one H^+ is removed from the

Table 19.1. Composition of the human body (cf. Table 2.1)

Element	Percentage	
	By Mass	By Number of Atoms
O	65	25.6
C	18	9.5
H	10	63.1
N	3	1.32
Ca	1.5	1.23
P	1.0	0.20
K	0.35	0.06
S	0.25	0.05
Na	0.15	0.04
Cl	0.15	0.03
Mg	0.05	0.01
Fe	0.004	0.00045

amino terminal, and an OH^- group from the carboxylic terminal. The two combine to form one molecule of water. In fact, dehydration is the main process by which polymerization generally occurs.

A chain consisting of 15 amino acids or fewer is called a *peptide chain*. Longer chains are called *polypeptide chains*. A protein is a chain of amino acids. The smallest protein consists of eight amino acids. Common protein chains range from 130 to 630 amino acids, corresponding to a molecular-mass range of 14,000–70,000 u.

Table 19.2. Yields of amino acids in an electrical-discharge experiment

Amino Acid	Concentration (10^{-6} Av/L)
alanine	790
glycine	440
aspartate	34
valine	20
leucine	11
glutamate	8
serine	5
isoleucine	5
proline	2
threonine	1

Source: Loomis, W. F. (1988), *Four Billion Years*. Sinauer Associates, Sunderland, Massachusetts, p. 8, Table 1.

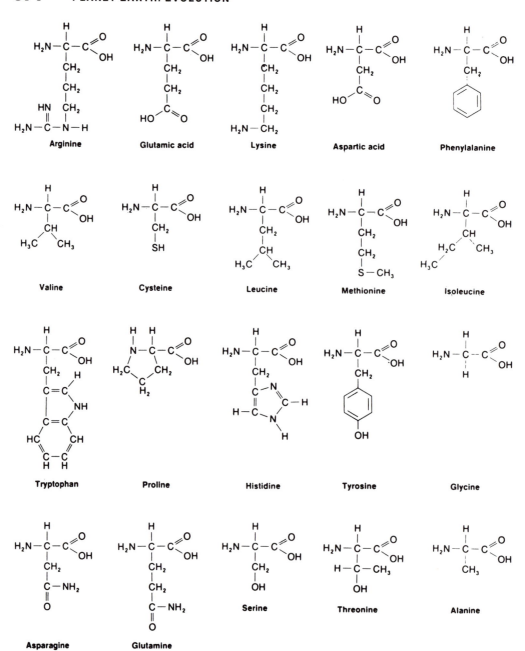

Figure 19.2. Amino acids. The 20 protein-forming amino acids have chemical structures that are particularly stable. All amino acids have the same NH_2–CH–COOH backbone. Proteins are formed when amino acids link to each other by the loss of one H from the NH_2 end of the backbone and one OH from the other end. The H and the OH combine to form one molecule of H_2O. Protein synthesis, therefore, requires dehydration. (Freifelder, 1987, p. 59, Fig. 3.1.)

A polypeptide chain commonly forms a spiral structure, called an α *helix*, that has positive helicity and contains 3.6 amino acids per turn. The spiral is stabilized by hydrogen bonds that link a peptide group to a peptide group two units behind and to another peptide group two units ahead. Another common protein structure is the β *structure*, in which two or more parallel or antiparallel chains are stabilized by hydrogen bonds between chains.

Proteins that have mainly α-helix or β-structure patterns are fibrous and form the structures of cells and tissues. Hair, for example, typically consists of fibrous protein chains staggered and held together by hydrogen bonds. In contrast, globular proteins, which include the vast group of catalytic proteins called *enzymes*, contain alternating α-helix and β-structure sections that are tightly folded. Globular proteins perform regulatory and enzymatic functions.

Enzymes are catalytic proteins that accelerate reaction rates by factors ranging from 10^4 to 10^{15} and make it possible for the chemical reactions on which life is based to occur at room temperature. Many enzymes are pure proteins. Other are proteins associated with a nonprotein component called a *coenzyme* or *prosthetic group*. At the core of the enzyme is an *active center* responsible for capturing the reactants and forcing them to react. Capture occurs because the electronic surface of the active center is complementary to that of the reactant (called *substrate*). This makes linkage possible—where one surface has a slight excess of negative charge, the opposite side has a slight deficiency, and vice versa.

In coenzymes, the core usually is a metal ion, but it is the structure of the peptide chain surrounding the metal ion that is important for recognizing and capturing a reactant. For example, UV radiation on water produces H_2O_2. The Fe^{3-} ion reduces H_2O_2 to H_2O with an efficiency of 10^{-5}. If the same Fe^{3-} ion is enclosed in a porphin ring to form a heme group (Fig. 19.3), the efficiency rises to 10^{-2}. If the heme group is enclosed in a polypeptide chain to form the enzyme catalase, the efficiency is raised to 10^5. The increase in efficiency is therefore by a factor of 10^{10}.

Enzymes are reaction-specific, which means that each reaction requires a specific enzyme.

The simplest bacterial cell may contain 2,000 different enzymes, while the human body may contain 100,000. The *cytochromes* are an important group of enzymes, each of which consists of a polypeptide chain (Fig. 19.4), with a heme group (Fig. 19.3) as coenzyme. Cytochromes are *chelates*, with an Fe at the center of the heme group.

[*Chelates* are compounds that can grab and hold for future release an atom or group of atoms; the word *chelate* derives from $\chi\eta\lambda\acute{\eta}$, which means *claw* or *pincher*.]

The Fe in the heme group can be oxidized to Fe^{3+} (by removal of one electron) or reduced to Fe^{2+} (by addition of one electron). Cytochromes can be arranged in order of decreasing electron-donor capacity. Electrons passing down the cytochrome chain release energy for chemical processes.

Artificial polymerization of amino acids to form chains 100 or so units long has been achieved in the laboratory under simulated primitive-Earth conditions. Polymerization of amino acids on the primitive Earth may have taken place on the beaches, where the tide would have caused alternating wetting and desiccation, and where UV radiation from the Sun would have provided the energy needed to break and form chemical bonds. It is believed that all sorts of proteins, including enzymatic proteins, formed and were carried to the ocean by the waning tide. The ocean thus became a sort of "primordial soup" in which the concentration of organic compounds may have reached 10^{-4} molar. Indeed, it seems that the ancient beaches, inlets, bays, and bayous of the world were the laboratories from which life emerged.

Today, the ordering of amino acids in protein chains in arranged by *nucleic acids*, the repositories of genetic information. The nucleic acids, RNA (ribonucleic acid) and DNA (deoxyribonucleic acid), are polymers consisting of *nucleotides*. A nucleotide consists of a pentose sugar with a PO_4 group attached to the 5'-C in the pentose ring (meaning the carbon in position 5, i.e., the carbon in the $-CH_2$ group attached to the fourth C atom starting from the 3 o'clock position in the pentose ring), and one of four *bases* (adenine, cytosine, guanine, and uracil in RNA; adenine, cytosine, guanine, and thymine in DNA)

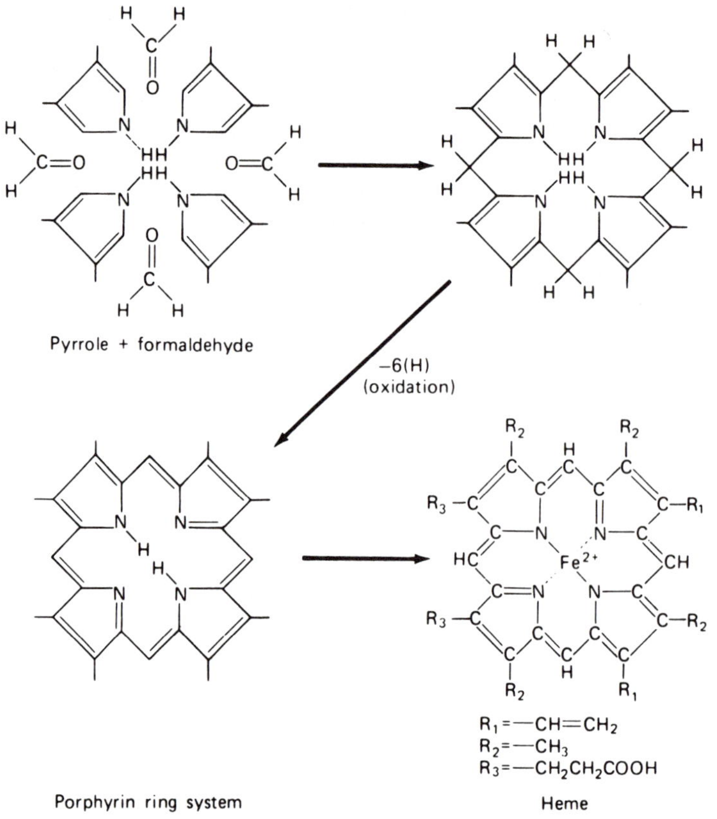

Pyrrole + formaldehyde

−6(H)
(oxidation)

Porphyrin ring system

$R_1 = -CH = CH_2$
$R_2 = -CH_3$
$R_3 = -CH_2CH_2COOH$

Heme

Figure 19.3. Chelates. Formation of a chelate (in this case, heme) from four pyrrole rings and four formal-dehyde molecules. (Emiliani, 1987, p. 39; adapted from Calvin, 1969, p. 147, Fig. 7.2.)

attached to the 1'-C (the carbon in position 1) (Fig. 19.5). The pentose sugar is *ribose* in RNA and *deoxyribose* in DNA. The difference is that the DNA pentose has an H attached to the carbon in position 2 instead of an OH as in RNA. The bases are ring structures that are either single (*pyrimidines*) or double (*purines*). The pyrimidine bases are *adenine* and *guanine*; the purine bases are *cytosine* and *uracil* in RNA, and *cytosine* and *thymine* in DNA. The bases range in molecular mass from 111 u to 151 u. A sugar-base unit is called a *nucleoside*; a sugar-phosphate-base segment is called a *nucleotide*. Nucleotides average 330 u in molecular mass.

The nucleotide components—sugars, bases, and phosphate—could have formed under primitive-Earth conditions. In fact:

1. Sugars are readily formed from the polymerization of formaldehyde (HCHO), a compound that in turn is readily formed in discharge experiments. Polymerization can be achieved by passing a formaldehyde solution over clays, which act as catalysts. Among the sugars that form are ribose and deoxyribose.

2. In a methane-rich atmosphere, hydrogen cyanide (HCN) would readily have formed and dissolved in the ocean, where it may have reached a molar concentration of 10^{-5}. The purine and pyrimidine hetero-cyclic ring structures (Fig. 19.6) that went to form the nucleic acid bases (adenine, guanine, cytosine, thymine, and uracil) may have formed from the polymerization of hydrogen cyanide (HCN). Cytosine spontaneously hydrolyzes to produce uracil.

3. Phosphorus for the phosphate groups

	1–8	9	10									20
Human	——	Gly–Asp–Val–Glu–Lys–Gly–Lys–Lys–Ile–Phe–Ile–Met–
Rhesus monkey	——	Gly–Asp–Val–Glu–Lys–Gly–Lys–Lys–Ile–Phe–Ile–Met–
Horse	——	Gly–Asp–Val–Glu–Lys–Gly–Lys–Lys–Ile–Phe–Val–Gln–

21   30   40
Lys–Cys–Ser–Gln–Cys–His–Thr–Val–Glu–Lys–Gly–Gly–Lys–His–Lys–Thr–Gly–Pro–Asn–Leu–
Lys–Cys–Ser–Gln–Cys–His–Thr–Val–Glu–Lys–Gly–Gly–Lys–His–Lys–Thr–Gly–Pro–Asn–Leu–
Lys–Cys–**Ala**–Gln–Cys–His–Thr–Val–Glu–Lys–Gly–Gly–Lys–His–Lys–Thr–Gly–Pro–Asn–Leu–

41   50   60
His–Gly–Leu–Phe–Gly–Arg–Lys–Thr–Gly–Gln–Ala–Pro–Gly–Tyr–Ser–Tyr–Thr–Ala–Ala–Asn–
His–Gly–Leu–Phe–Gly–Arg–Lys–Thr–Gly–Gln–Ala–Pro–Gly–Tyr–Ser–Tyr–Thr–Ala–Ala–Asn–
His–Gly–Leu–Phe–Gly–Arg–Lys–Thr–Gly–Gln–Ala–Pro–Gly–**Phe**–**Thr**–Tyr–Thr–**Asp**–Ala–Asn–

61   70   80
Lys–Asn–Lys–Gly–Ile–Ile–Trp–Gly–Glu–Asp–Thr–Leu–Met–Glu–Tyr–Leu–Glu–Asn–Pro–Lys–
Lys–Asn–Lys–Gly–Ile–Thr–Trp–Gly–Glu–Asp–Thr–Leu–Met–Glu–Tyr–Leu–Glu–Asn–Pro–Lys–
Lys–Asn–Lys–Gly–Ile–Thr–Trp–**Lys**–Glu–**Glu**–Thr–Leu–Met–Glu–Tyr–Leu–Glu–Asn→Pro–Lys–

81   90   100
Lys–Tyr–Ile–Pro–Gly–Thr–Lys–Met–Ile–Phe–Val–Gly–Ile–Lys–Lys–Lys–Glu–Glu–Arg–Ala–
Lys–Tyr–Ile–Pro–Gly–Thr–Lys–Met–Ile–Phe–Val–Gly–Ile–Lys–Lys–Lys–Glu–Glu–Arg–Ala–
Lys–Tyr–Ile–Pro–Gly–Thr–Lys–Met–Ile–Phe–**Ala**–Gly–Ile–Lys–Lys–Lys–**Thr**–Glu–Arg–**Glu**–

101   110   112
Asp–Leu–Ile–Ala–Tyr–Leu–Lys–Lys–Ala–Thr–Asn–Glu
Asp–Leu–Ile–Ala–Tyr–Leu–Lys–Lys–Ala–Thr–Asn–Glu
Asp–Leu–Ile–Ala–Tyr–Leu–Lys–Lys–Ala–Thr–Asn–Glu

Figure 19.4. Cytochrome c, an enzyme important in mitochondrial oxidation processes, consists of a polypeptide chain of 112 amino acids. The first 8 occur only in bacteria. In eucaryotes, the chain starts with the 9th amino acid. As shown in this illustration, humans differ from rhesus monkeys by only one amino acid in position 66 corresponding to a difference of 0.9%, and from horses by 11 amino acids, corresponding to a difference of 10.6%. They differ from chickens by 22 amino acids (21%), from rattlesnakes by 29 amino acids (32%), and from beer yeast by 62 amino acids (59%). (Ayala, 1982, p. 200, Fig. 7.9.)

occurs in nature in the mineral apatite [$Ca_5(PO_4)_3(F, Cl,OH)$], which is very stable. Under primitive-Earth conditions, however, reaction of apatite with oxalic acid ($HOOC–COOH$, probably derived from the hydrolysis of HCN) will form calcium oxalate (CaC_2O_4) and release the phosphate group. In the presence of urea ($H_2N–CO–NH_2$), a compound readily formed in discharge experiments, and at a temperature of a few tens of degrees Celsius, the attachment of the $–PO_3^{2-}$ group to the nucleo-side group proceeds readily. This process is called *phosphorylation*.

If the base is adenine, the nucleoside is called *adenosine*, and the nucleotide is called *adenosine monophosphate* (AMP). The addition of a second $–PO_3^{2-}$ group makes AMP into ADP (adenosine diphosphate), and the addition of a third $–PO_3^{2-}$ group makes ADP into ATP (adenosine triphosphate) (Fig. 19.7). Bond energy between the first and second phosphate groups is 0.33 eV, and that between

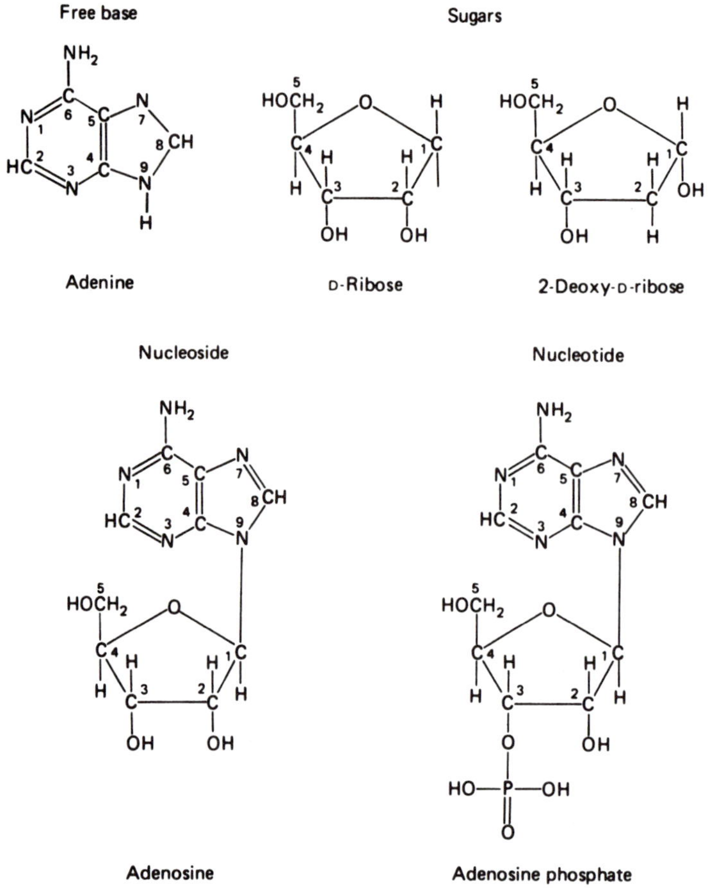

Free base

Sugars

Adenine

D-Ribose

2-Deoxy-D-ribose

Nucleoside

Nucleotide

Adenosine

Adenosine phosphate

Figure 19.5. An example (adenine) of base linkages to form a nucleotide. (Emiliani, 1987, p. 149.)

the second and third phosphate groups is 0.32 eV. These bonds are weak because the negative oxygen ion in each bound PO_3 group strongly repels its equivalent in the next group. Being weak, these bonds can be broken by simple hydrolysis. This is the reason that nature has chosen ATP as the molecule responsible for storing and releasing chemical energy.

Today, mononucleotides are polymerized to polynucleotide chains by enzymes, DNA polymerases, that are large enzymatic proteins. Polymerase precursors may have led to the formation of short DNA and RNA chains. The formation of primitive RNA is particularly critical, because RNA can be autocatalytic, which means that it can catalyze its own duplication without the help of enzymes.

Carbohydrates are molecules consisting of C, H, and O atoms. The fundamental carbohydrate is glucose ($C_6H_{12}O_6$), which is the most common organic substance on Earth. Glucose polymerizes to form cellulose (6,000 glucose units) and starch (60,000 glucose units).

Lipids are a family of substances that include carboxylic acids (also called *fatty acids*) and their complexes, glycolipids and phospholipids. Carboxylic acids consist of a –COOH head, with a long $-CH_2-CH_2-\cdots-CH_3$ tail attached (Fig. 19.8). Three carboxylic acids may combine, by dehydration, with a glycerol molecule to form a *triglyceride*. One of the carboxylic acid chains may be replaced by the group [$(CH_3)_3N^+-CH_2-CH_2-PO_4^-$], which is highly polar, forming a *phospholipid* (in this case *lecithin*, Fig. 19.9). Phospholipids thus

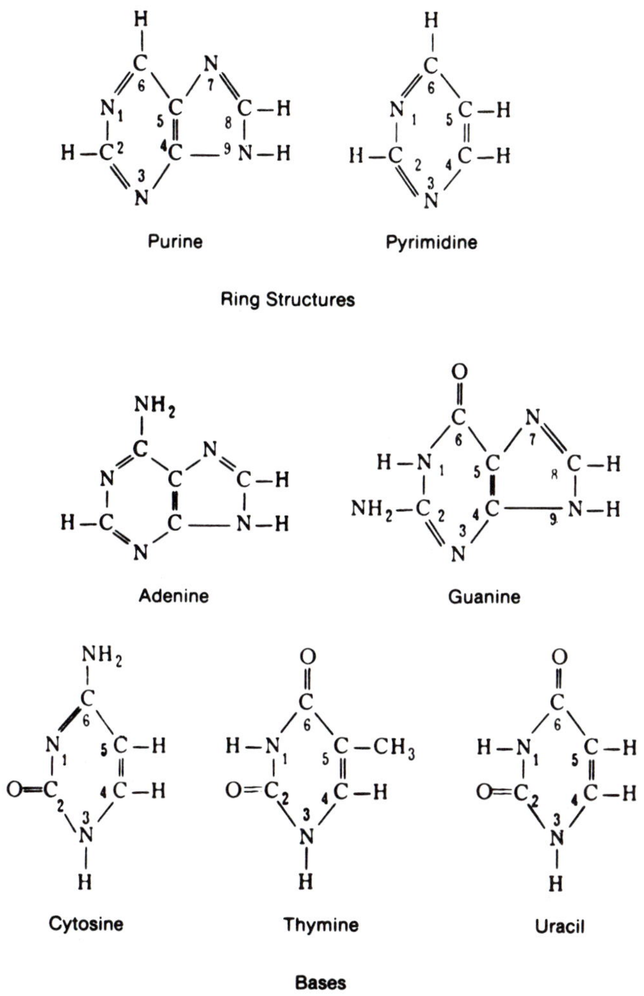

Purine　　　　　　　Pyrimidine

Ring Structures

Adenine　　　　　　　Guanine

Cytosine　　　　Thymine　　　　Uracil

Bases

Figure 19.6. Above: The ring structures of purine and pyrimidine. Below: The bases of RNA and DNA. (Emiliani, 1987, p. 22.)

have a polar head, which attracts water molecules and is thus *hydrophilic*, and a non-polar tail (the $-CH_3$ terminus) that does not bind to water and thus is *hydrophobic*. Phospholipids are important constituents of biological membranes.

Life needs energy for its processes. The Earth has two sources of energy available—external energy (i.e., the Sun) and internal energy. Solar light is by far the major source of energy for life today. Solar radiation quanta at wavelengths of about 450 and 650 nm, where chlorophylls absorb, have frequencies of $6.6 \cdot 10^{14}$ and $4.6 \cdot 10^{14}$ Hz, equivalent (by $E = h\nu$) to energies

of 2.7 and 1.9 eV. These energies are sufficient to move electrons to higher energy levels up to ionization in the complex molecular system where photosynthesis occurs. The energy released as the electrons return to the ground state is stored in chemical bonds that can be broken by hydrolysis, releasing energy for further processes.

The only internal source of energy utilized for living processes is the bacterial oxidation of compounds in vents along the mid-ocean-ridge system and in hot springs. Examples are the anaerobic oxidation of FeS to FeS_2, which releases 0.7 eV of energy, and the aerobic

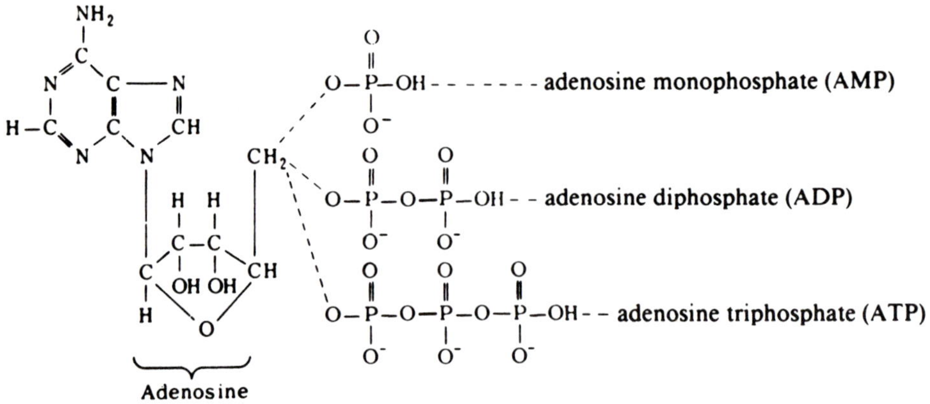

Figure 19.7. Adenosine phosphates. The bonds AMP-phosphate (forming ADP) and ADP-phosphate (forming ATP) are created using energy derived from photosynthesis or the oxidation of foodstuff. Their energy is about 0.3 eV. These bonds are easily breakable by hydrolysis, releasing 0.3 eV per bond. (King and Stansfield, 1985, p. 9.)

oxidation of H_2S to H_2O. which releases 2.05 eV of energy, and of S to H_2SO_4, which releases 7.2 eV of energy.

The primitive Earth was an immense laboratory rich in water, reduced gases, and a wealth of minerals with which liquid and gases were interacting and where energy for atomic rearrangements and the formation of complex

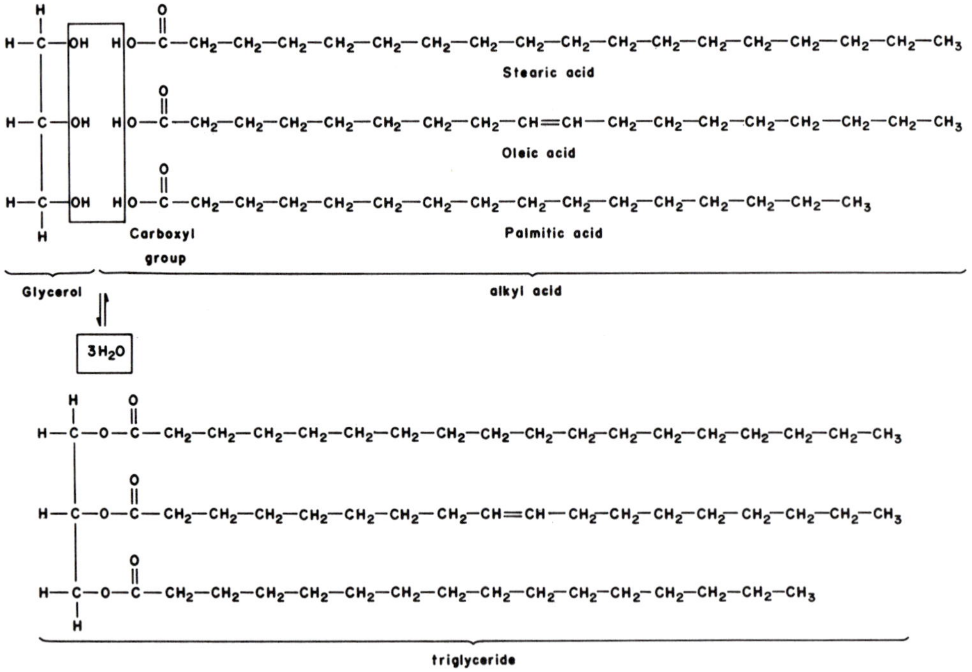

Figure 19.8. Three carboxylic acids combine with a glycerol to form a triglyceride by dehydration. (Curtis and Barnes, 1989, p. 68, Fig. 3.12.)

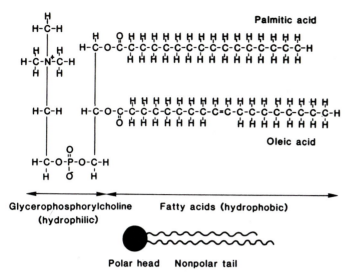

Palmitic acid

Oleic acid

Glycerophosphorylcholine   Fatty acids (hydrophobic)
(hydrophilic)

Polar head   Nonpolar tail

Figure 19.9. Lecithin, a typical phospholipid with a hydrophilic head and a hydrophobic tail. It is derived from the triglyceride by replacement of a carboxylic acid with the group $[(CH_3)_3N^+-CH_2-CH_2-PO_4-]$, which is highly polar. (Emiliani, 1988, p. 145, Fig. 9.10)

molecules—UV radiation from the Sun, lightning, and volcanic heat—was plentiful. Submarine vents were more numerous and widespread than today, possibly by a factor of 10 or more, contributing to the formation of organic compounds. Under those conditions it would have been virtually impossible that the ocean would not have become in fact a true primordial soup, with a molar concentration of organic compounds in steady-state equilibrium with the physicochemical environment amounting to perhaps 10^{-4}. In terms of salinity, that would have been about 3.5 mg of NaCl per liter of water (about the same as in fresh water). The primordial soup was therefore a very dilute solution.

In order for the organic compounds in the primordial soup to interact with each other, concentration was necessary. That could have occurred in a number of ways, including evaporation and the formation of brines in tidal pools. An efficient method of concentrating *and preserving* organics is foaming. Phospholipids in water form a film at the water surface, with the hydrophilic head in the water, and the hydrophobic tail in the air. Foaming rolls up these films into little spheres (called *micelles*, Fig. 19.10), with the hydrophilic heads

pointing out and the hydrophobic tails pointing in. Inside the micelles are other hydrophobic compounds that are captured by the hydrophobic tails during the process of foaming.

There was, of course, no dearth of foaming in the primitive ocean, both on wave crests in the open ocean and along the beaches of the world, and so organic compounds were concentrated in the open ocean and dumped on the beach. There, under the copious energy supplied by the Sun, anything that could, did happen.

The step from a micelle with a mixture of organic compounds inside to a functional living cell is an enormous leap, however. The biochemistry of even the simplest bacterium is so extraordinarily complex that it would seem impossible to believe that it came about by random steps. Here, however, we must remember three crucial facts:

1. The Earth's surface was an enormous laboratory, compared with the sizes of the molecular experiments in progress.

2. There was an enormous amount of time available—millions to hundreds of millions of years, possibly half a billion.

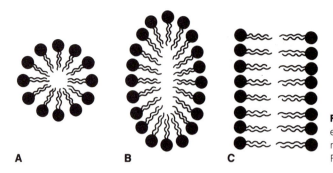

Figure 19.10. A spherical micelle (A), an ellipsoidal micelle (B), and a lipid bilayer membrane (C). (Emiliani 1988, p. 146, Fig. 9.11.)

3. Molecular experiments proceed with great rapidity—most enzymes can process thousands of molecules per second, with catalase holding the record of 100,000 molecules per second in the reduction of H_2O_2 to H_2O.

It is to be expected that anything that could, did in fact happen, and that life eventually emerged.

THINK

* If the Moon has been receding from the Earth at 4 cm/y for the past $250 \cdot 10^6$ y, and at an average of 2 cm/y for the preceding $4 \cdot 10^9$ y, how high were the lunar tides $4 \cdot 10^9$ y ago?

* The diffusion coefficients of chemicals in dilute solution range around $10^{-10} \, m^2 \, s^{-1}$. Assuming that two molecules beyond a distance of $(10^{-10} \, m)^{1/2}$ cannot interact within any one second, assuming that only one chemical reaction occurs each second within this distance, and further assuming that the surface of the ocean was the same as today and that the time available was 10^8 y, calculate how many reactions could take place on the surface of the primitive ocean.

* The structure of adenosine in Fig. 19.5 appears different from that in Fig. 19.7. Is it really different?

19.3. EARLY LIFE ON EARTH: THE BACTERIA

The Urey-Miller experiment indicated that life originated on Earth when the atmosphere still consisted of CH_4, H_2O, and NH_3. That atmosphere probably lasted only a short time, perhaps 10^8 y, because photolysis freed oxygen from H_2O that went to oxidize CH_4 to CO_2. It is likely that the earliest organisms had to endure harsh conditions, namely, high temperature ($90°C$), an atmosphere consisting predominantly of CH_4, H_2O, and NH_3, and an atmospheric pressure as high as 97 atm (Section 7.4). The earliest organisms may have been similar to some of the modern archaebacteria, which can endure such harsh conditions and which need only inorganic substances for their life processes.

Locally, life violates the second law of thermodynamics—it extracts energy from the environment and uses it to increase the local order. This happens because the system is open. The energy used by the earliest bacteria, the archaebacteria, was chemical energy. Archaebacteria differ from the other bacteria (called *eubacteria*) in a number of characteristics, the most significant of which is that their cell membrane does not consist of phospholipids but of glycerol, with attached hydrocarbons instead of carboxylic acids.

The most primitive archaebacteria use exclusively inorganic substances for their energy requirements and are found in fumaroles and submarine vents where hydrogen and sulfur gases provide the energy and where temperatures of 90–$100°C$ are available. Salty fumaroles emitting sulfurous gas off the island of Vulcano, Italy, and on the coast of the island of São Miguel, Azores, where the temperature reaches $110°C$, host thriving populations of, respectively, *Pyrodictium* and *Hyperthermus* (Fig. 19.11). These archaebacteria exhibit optimal growth at temperatures between $95°C$ and

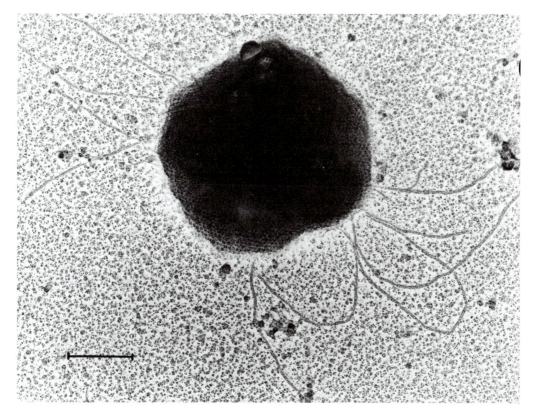

Figure 19.11. The archaebacterium *Hyperthermus butylicus*, found in a salty fumarole on the coast of the island of São Miguel, Azores (length of bar = 0.5 μm); where the salinity is about 17%, and temperatures range up to 112°C. *H. butylicus* grows best at temperatures between 95°C and 106°C. It derives its energy from the fermentation of peptides and by converting elemental sulfur and molecular hydrogen to H_2S. In the primitive Earth, the peptides may have formed abiotically in the fumarole itself. (Courtesy Wolfram Zillig, Max-Planck-Institut für Biochemie, Martinread, Germany.)

106°C and oxidize hydrogen with sulfur to form H_2S:

$$H_2 + S = H_2S - 0.3 \, eV$$

[the negative sign indicates that the energy (0.3 eV) has been lost by the system].

As soon as the atmosphere began to contain some CO_2, other primitive archaebacteria, such as the methanogens, may have evolved. These include *Methanothermus*, which grows in fumarole fields at temperatures around 90°C, and *Methanococcus* and *Methanopyrus*, which grow in deep-sea vents at temperatures between 90°C and 110°C. Methanogens reduce CO_2 to methane with hydrogen:

$$CO_2 + 4H_2 = CH_4 + 2H_2O - 1.43 \, eV$$

Methanogens are strictly anaerobic. The descendants of these primitive archaebacteria are found today in sewers, in bogs, and in the stomachs of ruminants, where they produce methane from the fermentation of organic compounds such as formic acid (HCOOH) and methyl alcohol (CH_3COOH).

[*Fermentation* is the breakdown of organic compounds—mainly sugars—into simpler compounds containing less energy.]

In the primitive Earth, archaebacteria probably were ubiquitous, because high temperatures (90–100°C) and pressures (97 atm) prevailed at the Earth's surface (Section 7.4). Nucleic acids, some amino acids, and many proteins are highly unstable at these high tem-

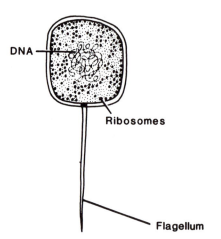

DNA —

Ribosomes

Flagellum

Figure 19.12. A modern procaryotic cell. There are no organelles, and the DNA forms a single loop centrally or subcentrally located. The flagellum, when present, is a simple whiplike structure.

peratures. The thermophilic archaebacteria, therefore, must be fast resynthesizers of these compounds to keep up with thermal decomposition.

Primitive archaebacteria like *Methanopyrus* are cells about 8–10 μm long and 0.5 μm across. The cells of archaebacteria are procaryotic; that is, the cell does not contain a nucleus (the term *procaryote* derives from the Greek πρό, meaning *before*, and κάρυον, meaning *nut* or *kernel*; a cell with a nucleus is called *eucaryote*). A procaryotic cell (Fig. 19.12) consists simply of a cell membrane, cytoplasm with some 10,000 ribosomes in it, and a loop of DNA. The ribosomes are exceedingly small bodies— 250 Å across—where protein synthesis occurs; they consist of two-thirds RNA (ribosomal RNA or rRNA) and one-third protein. At optimum growth temperature (98°C), the cells of *Methanopyrus* duplicate every 50 minutes. Cell duplication ceases at temperatures lower than 85°C or higher than 110°C.

DNA (Fig. 19.13) is always double-stranded, even in the most primitive archaebacteria, with the two strands wound around each other with positive helicity. In procaryotic cells, the DNA double helix forms a single closed loop, the *chromosome*. In addition to this loop, all procaryotes have *plasmids*, which are smaller loops dispersed in the protoplasm that contain 2 to 30 genes (versus the several thousand genes in

the main loop—see Section 20.4 for a discussion of genes). Bacterial cells may contain a few of the larger plasmids and 10–20 of the smaller ones.

DNA replicates by a complex process (Fig. 19.14) in which a section of the DNA double helix unwinds, starting from a given point, with the help of the enzyme *helicase*. A "bubble" with two forks is formed. The enzyme *DNA polymerase* adds nucleotides, beginning at each fork and proceeding in opposite directions, to form segments that are complementary to the original DNA chains. The segments, 1,000–2,000 bases long in procaryotic cells, are joined by the enzyme *DNA ligase*, until a chain is formed that is as long as the original DNA chain. The rate of fork movement is such that about 1,500 base pairs per second are assembled. Plasmid DNA replicates differently (see below).

The process of DNA replication is so fast that errors, on the order of 1 in 10^6, are made. For instance, an incorrect base (most frequently uracil instead of thymine) may be introduced; the C–N bond between base and sugar may break, resulting in the absence of a base; a base may be replaced by a spurious chemical; a break may be caused in a single strand by ionizing radiation; a double strand may break if two single-strand breaks happen to be close to each other; opposite bases may bond covalently because of radiation or antibiotics, with the result that the two strands of DNA become glued together, and replication is prevented.

Most of this damage, if not too intense, can be repaired. A base-pair mismatch arising during DNA replication, for instance, is detected by DNA polymerase, and the wrong base is replaced with the right one. If the damage persists, a mutation results. Mutation rates in bacteria range from 10^{-4} to 10^{-8}.

The *genome*, the genetic endowment of an individual or a species, consists of the total DNA content. The genome consists of 10^6 base pairs in the simplest bacteria and of $5.3 \cdot 10^9$ base pairs in *Homo sapiens*.

In some viruses, the genome consists of RNA, which is then double-stranded and forms a double helix like DNA. The double-RNA helix is held together by hydrogen bonds as in DNA (adenine to uracil, cytosine to guanine).

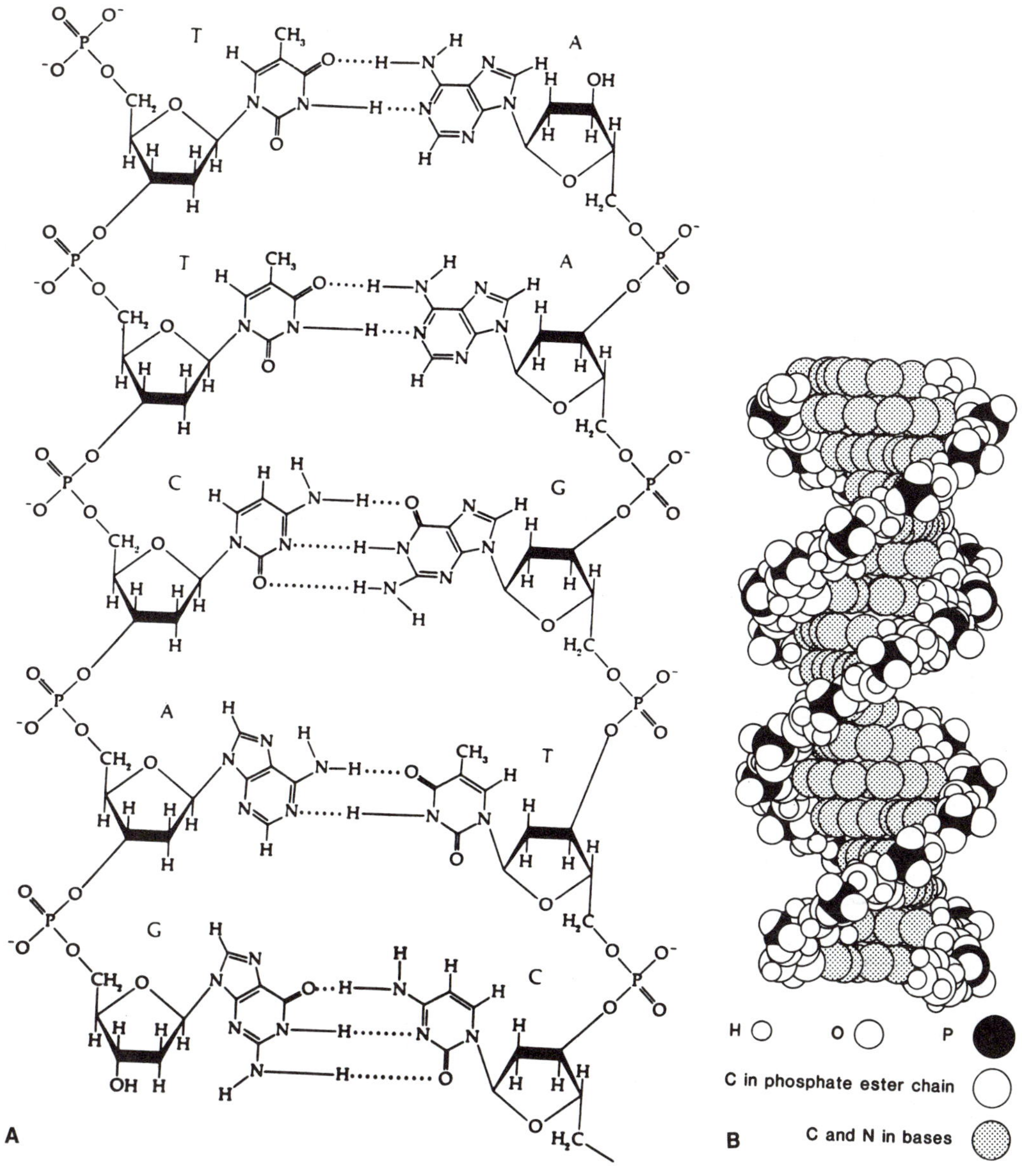

Figure 19.13. (A) A double strand of DNA unwound and stretched out. The dotted lines represent hydrogen bonds. Notice that the cross-chain links are between purine and pyrimidines (adenine to thymine, cytosine to guanine). (B) The double helix in its spiral structure. The pitch is $3.4 \cdot 10^{-10}$ m. (Emiliani, 1988, pp. 137–8, Figs. 9.3 and 9.4.)

Different species have DNA chains that differ in length and, more importantly, in the arrangement of the bases. The 1,500,000 nucleotides in the smallest DNA molecule can be arranged in $4^{1,500,000} = 10^{903,090}$ different ways (permutations of 4 elements 1,500,000 at a time). Therefore, the potential number of different species is essentially infinite.

[Permutations and combinations

$n!$, read n factorial, is equal to $1 \times 2 \times 3 \times \cdots \times (n-1) \times n$; thus, $3! = 1 \times 2 \times 3 = 6$.

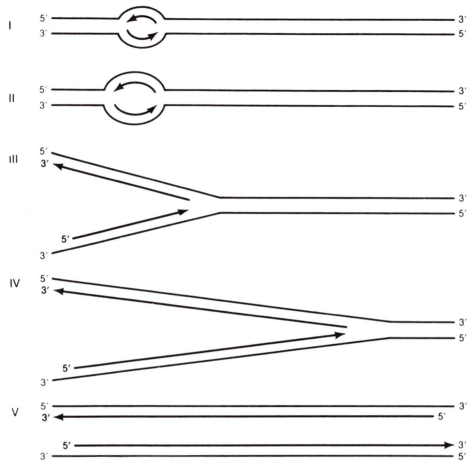

Figure 19.14. DNA replication in *Escherichia coli* phage T7. Replication is bidirectional, starting at a point about 17% from the left end. Because bidirectional replication proceeds at equal rates, replication to the left is finished first. Bidirectional replication is the major mode of chain growth among eucaryotes. The notations 3′ and 5′ refer to the two ends of a DNA strand. The 3′ end is the hydroxyl group attached to the carbon in position 3 around the sugar ring (carbon atoms are numbered 1 to 4 starting from 3 o'clock). The 5′ end refers to the phosphate group attached to the $=CH_2$ group, whose C is assigned position 5. This C is attached to the C in position 4 in the sugar ring. (Freifelder, 1987, p. 268, Fig. 9.40.)

A. *Permutations* consider the order of the elements:
  1. Permutation of n elements:
     a. without repetition:
        $$P_n = n!$$
     b. with repetition:
        $$\bar{P}_n = n^n$$
  2. Permutation of n elements taken k at a time:
     a. without repetition:
        $$P_{n,k} = P_n/P_{n-k} = n!/(n-k)!$$
     b. with repetition:
        $$\bar{P}_{n,k} = n^k$$
B. *Combinations* do not consider the order of elements:
  1. Combination of n elements:
     $$C_n = 1$$
  2. Combination of n elements taken k at a time $(k < n)$:
     $$C_{n,k} = P_{n,k}/P_k$$
     $$= [n!/(n-k)!]/k!$$
     $$= n!/k!(n-k)!]$$

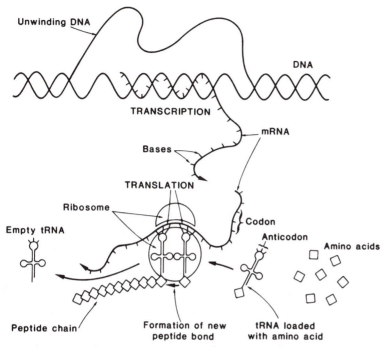

Figure 19.15. Synthesis of proteins. A section of DNA in the nucleus, 1,500 to 3,000 nucleotides long, unwinds and transcribes its sequence into a corresponding section of mRNA. The mRNA section leaves the nucleus, threads through a ribosome, and lines up a complementary triplet of tRNA. The tRNA molecule, consisting of 73 to 93 (average 76) nucleotides arranged in a cloverleaf pattern, carries attached to one end a specific amino acid. A second tRNA molecule carrying its own amino acid is brought into the ribosome next to the first tRNA molecule. The two amino acids form a bond (via a specific enzyme); the first tRNA molecule is released; and the ribosome moves along to pick up a third rRNA molecule with a third amino acid attached to it. Only two tRNA molecules are at the binding site inside the ribosome at any given time. The chain of amino acids (protein) reflects the sequence of mRNA base triplets, which in turn derives from the sequence of base triplets in the original DNA section. (Emiliani, 1988, p. 143, Fig. 9.8.)

The synthesis of proteins, which form much of the structures of archaebacteria and all higher organisms, is organized by DNA. DNA manufactures an RNA segment several hundred to a few thousand nucleotides long called *messenger RNA* (mRNA); mRNA threads through a ribosome (Fig. 19.15) where a segment of RNA folded in the shape of a cloverleaf (called *transfer RNA* or tRNA) transports a single amino acid. A triplet of bases at the top of the cloverleaf recognizes the complementary triplet in the mRNA and deposits its amino acid within the ribosome. There, the amino acid is linked to the preceding amino acid still in the ribosome and the process continues. The relationship between mRNA triplets of bases and specific amino acids is called the *genetic code* (Fig. 19.16). Viruses that use RNA as genetic material have their RNA protected by a sheath of specific proteins. The early association of nucleic acids with proteins may eventually have led to the development of the genetic code.

Bacteria come in three fundamental forms: spheroidal (*coccus*), rod-shaped (*bacillus*), and spiral (*spirillum*). Cocci and bacilli may form groupings or filaments that may be considered incipient colonies. Many bacteria are not motile, but many others are, by means of flagella that are embedded in the cell membrane. Modern bacteria may be aerobic, semi-anaerobic, or fully anaerobic (like the methanogens). The optimal temperatures for their growth and reproduction range from about

2nd →	U	C	A	G	3rd ↓
U	PHE PHE LEU LEU	SER SER SER SER	TYR TYR STOP STOP	CYS CYS STOP TRP	U C A G
C	LEU LEU LEU LEU	PRO PRO PRO PRO	HIS HIS GLN GLN	ARG ARG ARG ARG	U C A G
A	ILE ILE ILE MET	THR THR THR THR	ASN ASN LYS LYS	SER SER ARG ARG	U C A G
G	VAL VAL VAL VAL	ALA ALA ALA ALA	ASP ASP GLU GLU	GLY GLY GLY GLY	U C A G

Figure 19.16. The genetic code. The genetic code relates a triplet of mRNA bases to a specific amino acid. More than one triplet may code for the same amino acid. The triplet AUG (adenine-uracil-guanine) is used both to code methionine and to start an amino acid chain. Three triplets (UAA, UAG, and UGA) signal the end of the chain. (Emiliani, 1988, p. 144, Fig. 9.9.)

100°C (for *Methanococcus* and *Methanopyrus*) to less than 10°C. Thermophilic bacteria probably are the oldest.

19.4. PHOTOSYNTHESIS

Several groups of bacteria are capable of harnessing solar light to manufacture organic compounds—the complex process known as photosynthesis. A battery contains a chemical pump that separates charges, so that one terminal has more electrons and the other terminal has fewer. Electrical energy is released when electrons flow through a wire from the negative to the positive terminal. If the flow is through the filament of a light bulb, light is produced. Photosynthesis performs the same function as the battery—charge separation. Charge separation occurs when an atom is ionized, a process that requires energy.

Discharge experiments have produced pyrroles that combine to form a porphin ring (Fig. 19.3). The porphin ring readily binds metal ions. If the metal ion is Mg^{2+}, the ring forms the core of chlorophyll (which has a molecular mass about 900 u); if it is Fe^{2+}, the ring forms the core of hemoglobin (which has a molecular mass of 65,322 u). In chlorophyll, the porphin ring has a hydrophobic tail of

C and H groups ending with $-CH_3$. There are several types of chlorophyll: chlorophylls a, b, c, d, and e (Fig. 19.17), bacteriochlorophyll, and chlorobium chlorophyll, with slightly different structures. In addition, there are other pigments also capable of being excited by solar light, including phycoerythrin (in red algae), phycocyanin (in cyanobacteria), rhodopsin (in extreme-halophilic archaebacteria), and carotene (in bacteria and plants) (Table 19.3).

Procaryotes, except Prochlorophyta, use mainly chlorophyll a and, secondarily, phycobilins and carotenes. Prochlorophyta (single-celled or filamentous procaryotes, living as plankton in lakes or in the ocean or as symbionts on the surface of marine tunicates— see below) and all photosynthesizing eucaryotes use a mixture of 70% chlorophyll a and 30% chlorophyll b with some carotenoids (see absorption spectrum in Fig. 19.32).

The most common type of chlorophyll, chlorophyll a, consists of 72 H atoms, 55 C, 5 O, 4 N, and 1 Mg. The energies needed to ionize, from the ground state, H, C, O, and N range from 11.3 eV for C to 14.5 eV for N; that needed to ionize Mg is somewhat less, but still a high 7.6 eV. Photoionization would require highly energetic photons from the far-UV region of the spectrum. Chlorophylls can absorb only photons of longer wavelengths and therefore lower energies. However, an

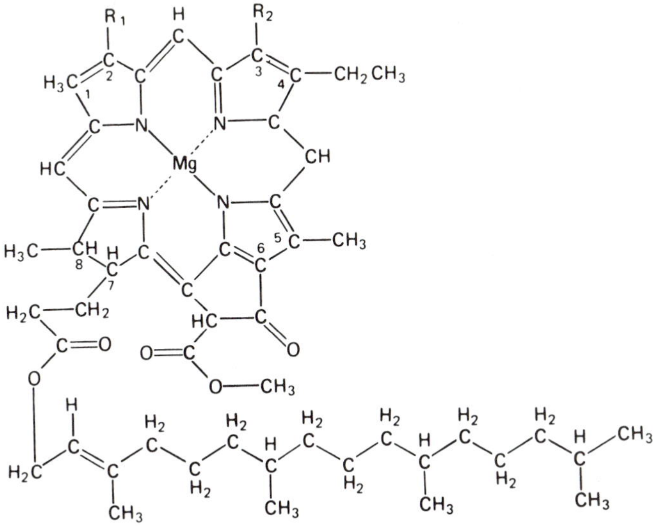

Figure 19.17. Chlorophylls. Chlorophylls are a group of comparatively large molecules (molecular mass 893.3 to 907.3 u), each with a hydrophilic head consisting of a magnesium atom at the center of a porphin ring, and a hydrophobic tail consisting of a carbon–hydrogen group that attaches itself to cell membranes. The most common are chlorophyll a ($R_1 =$ –CH=CH$_2$, $R_2 =$ –CH$_3$); chlorophyll b ($R_1 =$ –CH=CH$_2$, $R_2 =$ –CHO); chlorophyll c ($R_1 =$ –CH=CH$_2$, $R_2 =$ –CH$_3$, and the group –CH=CH–COO replacing the C in position 7); and chlorophyll d ($R_1 =$ –CHO, $R_2 =$ –CH$_3$). Green algae and all higher plants use a mixture of 70% chlorophyll a and 30% chlorophyll b; brown algae and diatoms (freshwater and marine protophytes) use chlorophyll c; red marine algae use chlorophyll d. Chlorophyll absorbs mainly red and blue light, but little green light, which is the reason that plants look green. (Emiliani, 1987, p. 40.)

Table 19.3. Pigments capable of being energized by solar light

Name	Absorption Band (nm)					Molecular Mass (u)
	UV (> 3.3 eV)	Blue (3.3–2.5 eV)	Green–Orange (2.5–2.0 eV)	Red (2.0–1.6 eV)	Infrared (< 1.6 eV)	
bacteriochlorophyll a	376		590		840, *870*	871.46
bacteriochlorophyll b		403	604	*760*		—
chlorobium chlorophyll a		446		730		—
chlorobium chlorophyll b		457		750		—
carotene	340	460				536.89
chlorophyll a		*436*		676		893.5
chlorophyll b		*480*		650		907.5
phycocyanin				620		273,000
phycoerythrin		498	*540–568*			290,000
rhodopsin		498				40,000

Italics identify strongest absorption peaks. Phycocyanin and phycoerythrin are called *phycobilins*.

atom or a molecule can give up an electron at an energy below that needed for ionization if the electron is already excited and if there is in the immediate vicinity another atom or molecule desirous of absorbing that electron. A *redox* potential forms between the two that facilitates the electron transfer. This facilitation decreases in inverse proportion to the sixth power of the distance, which means that donor and acceptor must be in the immediate vicinity of each other.

[*Redox* means *reduction–oxidation*, because one atom or molecule receives an electron (and is therefore reduced), while the other donates an electron (and is therefore oxidized). In chemistry, the redox potential (symbol Eh) is expressed relative to the hydrogen half-cell (symbol E_0) (i.e., to an electrolytic cell one electrode of which consists of the substance under study, and the other electrode consists of hydrogen at 1 atm bubbling over a Pt rod at 25°C). Eh is measured in volts or electron volts (the E in Eh stands for voltage, and the h stands for hydrogen half-cell). The element with the most negative Eh (strongest electron donor) is lithium, which is eager to donate its lone electron in the L shell. At the other extreme, the element with the highest positive Eh (strongest electron acceptor) is fluorine, which, with 7 electrons in its L shell, has a consuming desire to absorb one more electron. Elements and compounds can be arranged in an *electrochemical series* ranging from lithium (Eh $= -3.04$ V) through hydrogen (Eh $= 0$ V, by definition) to fluorine (Eh $= +2.87$ V). The electrochemical series deals with elements in solution, which makes Eh quite different from ionization potentials: An atom that, in vacuo, would require a certain amount of energy to get rid of a valence electron and become a postive ion will require less energy if it is in close proximity of an electron acceptor (or donor, if the atom wishes to become a negative ion). Table 19.4 lists some of the common elements in order of increasing Eh. For each element, the voltage shows how energetic the electron transfer is. In a Li/Pt cell, for instance, the reaction is $Li \rightarrow Li^+ + e^-$ on the Li side, and $H^+ + e^- \rightarrow H$ on the Pt side. The electrons donated by Li are transferred to the hydrogen side, with a voltage difference of 3.05 V (negative on the Li side). In a F/Pt cell, H donates electrons that are picked up by

Table 19.4.   **Electrochemical Series**

Element	E_0 (V)
Li	−3.04
Cs	−2.92
Rb	−2.92
K	−2.92
Ba	−2.90
Sr	−2.89
Ca	−2.87
Na	−2.71
Mg	−2.34
Be	−1.85
Al	−1.67
Mn	−1.18
Zn	−0.76
Cr	−0.74
Fe	−0.44
Cd	−0.40
Co	−0.28
Ni	−0.25
Sn	−0.14
Pb	−0.13
H	0.00
Cu	0.34
I	0.53
H_2O	0.69
Hg	0.80
Ag	0.80
Cl^-	1.36
Au	1.50
F^-	2.65

F, with a voltage difference of 2.65 V across the cell (negative on the H side). In a Zn/Cu cell (Fig. 19.18), Zn donates electrons, and Cu accepts them, with a voltage difference of 1.10 V (negative on the Zn side).

Notice that Eh complements pH: Eh measures the ability of the chemical environment to supply or remove electrons, and pH measures the ability of the environment to supply or remove protons (H^+).]

Photons captured by chlorophylls energize the transfer of an electron from a donor molecule to an acceptor molecule, a transaction that, if the two molecules are in very close proximity, requires 1.2 eV of energy or even less. Electron and donor are reunited not through a wire but through a series of molecules. The energy released is used to phosphorylate ADP to ATP and, in green bacteria and eucaryotes, to produce a reductant.

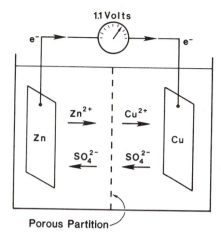

Porous Partition

Figure 19.18. The Daniell cell, a Zn/Cu electrolytic cell with a dilute solution of $ZnSO_4$ on the left side, and $CUSO_4$ on the right side. Cu is more electronegative than Zn (i.e., it has greater attraction for electrons than does Zn). Electrons are carried to the Zn electrode (the *anode*) by the SO_4^{2-} anion, and from there they pass through a conductor to the Cu electrode (the *cathode*). The Zn^{2+} and Cu^{2+} cations moving to the cathode pick up the electrons and complete the circuit. The porous partition prevents convective mixing of the two solutions, which might bring Cu^+ ions directly in contact with the Zn electrode—the Cu^{2-} would directly remove electrons from the zinc, thus reducing the flow of electrons through the external circuit.

There are two soluble electron carriers in nature, *nicotinamide adenine dinucleotide* (NAD), which becomes NADH when reduced, and *nicotinamide adenine dinucleotide phosphate* (NADP), which becomes NADPH when reduced (Fig. 19.19). Hydrolysis of ATP to ADP and oxidation of NADH to NAD and of NADPH to NADP produce the energy needed for the reduction of CO_2 and H_2O to carbohydrates.

The three major groups of photosynthesizing bacteria are the purple sulfur bacteria (*Chromatiaceae*), the purple nonsulfur bacteria (*Rhodospirillaceae*), and the green bacteria (*Chlorobiaceae*). In these bacteria, the pigment, usually bacteriochlorophyll a, with some carotene, is packed in minute spherules, about 40 nm across, called *chromatophores*. The chromatophores consist of 65% protein, 25% phospholipids, and 4.6% bacteriochlorophyll a or b, plus phycobilins. Each chromatophore contains approximately 40 *reaction centers*, each surrounded by 10 light-harvesting units, 25 ubiquinone molecules, and a few carotenoids.

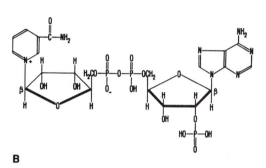

A

B

Figure 19.19. NAD (A) and NADP (B) are two coenzymes (the nonprotein portion of an enzyme) important for respiration and other processes. (Diem and Lentner, 1970, p. 344, Table 12.)

[*Quinone*, of which benzoquinone is a representative, consists of a benzene ring with two oxygens replacing two hydrogens; *ubiquinones* are a family of quinones in which the remaining hydrogens in the benzene ring are replaced by two $-COCH_3$ groups and by a $-CH_2-CH-C-CH_2-$ chain up to 10 units long, with $-CH_3$ side groups. The name *ubiquinone* derives from the fact that they are ubiquitous in biological systems.]

In *Rhodopseudomonas sphaeroides*, each reaction-center unit in the chromatophore contains 4 molecules of bacteriochlorophyll a, 2 phycobilin molecules, and 1 carotenoid, plus 2 ubiquinone molecules and one Fe^{2+} that function as electron receivers.

A light-harvesting unit consists about 50 bacteriochlorophyll molecules. Bacteriochlorophyll in the 10 light-harvesting units surrounding a reaction center capture photons and become excited (they do not become ionized because the energies of the photons in the main absorption band of bacteriochlorophyll—870 nm—is too low, only 1.42 eV). The excitation energy of the molecules is funneled to the

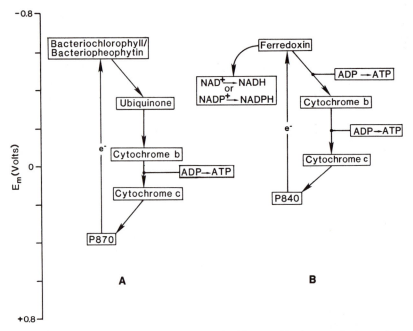

Figure 19.20. Photosynthesis in purple bacteria (A) and in green bacteria (B). In purple bacteria, capture of photons with $\lambda = 870$ nm energizes an electron in a molecule of bacteriochlorophyll in P_{870} with about 1.2 eV of energy. The electron returns to base through ubiquinone, cytochrome b, and cytochrome c, and in the process, one molecule of ADP is phosphorylated to ATP. In green bacteria the electron energy (about 0.8 eV) is used either to form ATP from ADP or to reduce NAD^+ to NADPH.

bacteriochlorophyll molecules in the reaction center by resonance (like energy transfer between adjacent tuning forks). Two of the four bacteriochlorophyll molecules in a reaction center form a dimer that absorbs intensely at 870 nm. The absorbing pigment is called P_{870}, where P stands for pigment. P_{870} is excited by the absorption of 1.42 eV of energy. This energy is used to boost the electron from $+0.4$ to -0.8 V, which is sufficient to transfer it to an adjacent acceptor. The donor (D) becomes a positive ion (D^+), and the acceptor (A) becomes a negative ion (A^-). Because A becomes reduced and D becomes oxidized, the D^+–A^- pair forms a *redox pair*. The potential difference between the two (redox potential or Eh), 1.2 eV (Fig. 19.20A), is taken relative to the D–A potential difference (if any) before the excitation event. In biochemistry, the redox potential usually is referred to a midpoint potential (symbol E_m), taken to be 0, referring to the potential difference of zero between two solutions that contain the same concentrations of positive and negative ions (or, respectively, oxidized and reduced forms). On the E_m scale,

the redox of D^+ is $+0.4$ V, and that of A^- is -0.8 V, for a total of 1.2 V.

In the purple bacteria, the donor is a molecule of bacteriochlorophyll, and the acceptor is another molecule of bacteriochlorophyll plus a molecule of bacteriophycobilin (Fig. 19.20A). The electron is passed from the acceptor to two ubiquinone molecules in succession, and from them to cytochromes b and c. Cytochrome c returns the errant electron to P^+, which is neutralized back to P. This process releases the original 1.2 eV of energy, some of which is used to phosphorylate ADP to ATP, and some is dissipated in increased molecular and atomic agitations (i.e., heat). This system is called *Photosystem I* (PSI).

In green bacteria (Fig. 19.20B), the electron source P absorbs at a wavelength that is a little shorter (840 nm). The donor is a molecule of bacteriochlorophyll, and the acceptor is Fe_2-S_2 ferredoxin. (Ferredoxins are protein chains with 25–40 amino acids and an Fe_2-S_2 group or an Fe_4-S_4 group bound to cysteine in the chain.) The ferredoxin molecule can do either of two things: It can pass the electron to an

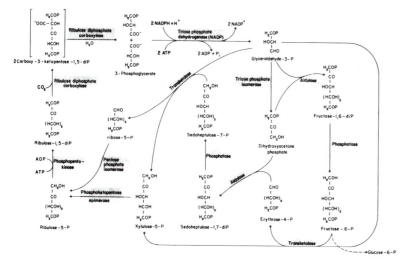

Figure 19.21. The Calvin cycle, reduced on purpose so as to be almost unreadable. This illustration is flashed here only to give a visual impression of how complicated this cycle is and how much work the Nobel prize winner, American biochemist Melvin Calvin, must have done to unravel it. (Moment and Habermann, 1973, p. 362. Fig. 15.26.)

NAD or NADP molecule, giving it a negative charge, which is then immediately neutralized by the capture of a H^+ ion from the environment to form the reduced NADH or NADPH; or the electron can be returned to P^+ via two cytochrome b molecules, phosphorylating ADP to ATP in the process. The system used by green bacteria is still called Photosystem I.

Photosynthesis in purple bacteria produces only phosphorylation, whereas in green bacteria it produces both phosphorylation and NADH or NADPH. Green bacteria, therefore, are more advanced than purple bacteria. This is also indicated by the fact that purple bacteria are strictly anaerobes, whereas green bacteria include a group, the *Chloroflexaceae* (gliding filamentous bacteria), that are *facultative anaerobes*; that is, they can take it or leave it (referring to oxygen, of course). Photosynthesis utilizes only 2% of the incident solar photons under full sunshine, but up to 20% in dimmer light.

Both ATP and NADPH are needed to synthesize carbohydrates from CO_2 and water. Purple bacteria produce NADPH from NADP using H_2S, or other compounds as reducing agents. Carbohydrate synthesis takes place in the *Calvin cycle* (Fig. 19.21), a cyclical set of reactions that do not need light, in which six CO_2 molecules and six water molecules combine to form glucose. Eight photons are needed to reduce one molecule of CO_2 and evolve one molecule of O_2.

Rhodopsin (Table 19.3) is a red pigment with peak absorption at 498 nm (blue-green) functioning as photoreceptor in extreme halophile archaebacteria and higher organisms (it is present in animal retinas). Rhodopsin consists of *retinal* (a chain of nine C atoms with –H and –CH$_3$ side groups), bound on one side to a hexagonal C ring (with –H's and –CH$_3$'s) and on the other to the protein *opsin* (molecular mass ~ 40,000 u) (Fig. 19.22).

Carotenes (Fig. 19.23) are yellow-to-red

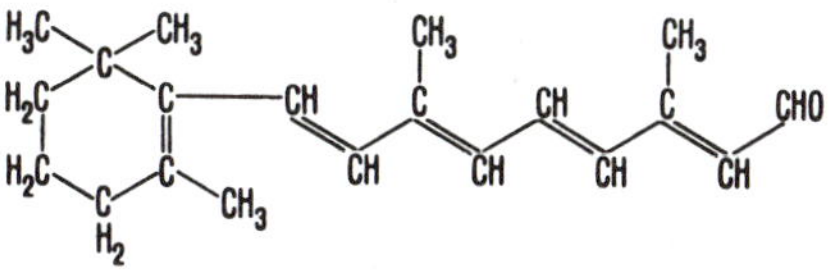

Figure 19.22. Retinal (aldehyde of vitamin A) consists of a chain of 9 carbon atoms attached on one side to a benzene ring and terminating on the other side with a –CHO group. Retinal loses an OH from the –CHO terminus, and the protein *opsin* (molecular mass ~ 40,000 u) loses an H from its NH$_2$– terminus. The two form a bond, creating rhodopsin, a pigment important in vision. The OH$^-$ and the H$^+$ join to form a molecule of water. The formation of rhodopsin requires, therefore, dehydration. (Diem and Lentner, 1970, p. 459.)

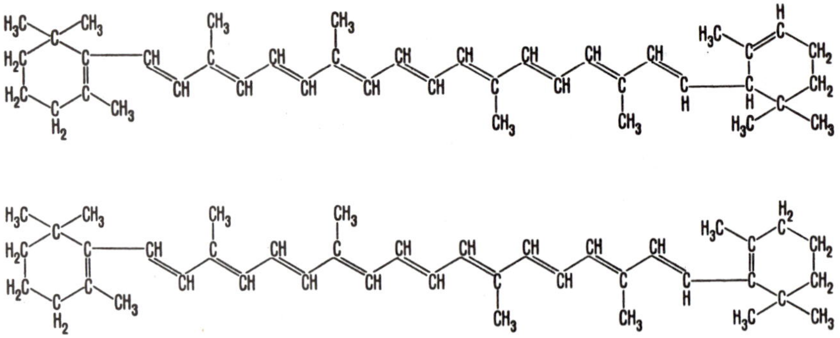

Figure 19.23. Carotenes are yellow-to-red pigments consisting of a chain of about 18 carbon atoms bound on either side by a benzene ring. (Diem and Lentner, 1970, p. 458.)

pigments consisting of chains of about 18 carbon atoms (α-carotene and β-carotene) with H and CH_3 side attachments, bound at either end by hexagonal C rings with $-H$'s and a $-CH_3$. Carotenes are accessory pigments in photosynthesis, with peak absorption in the blue and near-UV portions of the spectrum. In unicellular algae, carotenoproteins activate phototaxis (motion in response to light).

19.5. METABOLISM

Metabolism, the process of energy transferin living systems, consists of two phases: *anabolism*, which is the complexing of simpler molecules into more complex ones (thereby increasing the net bond energy), and *catabolism*, which is the reverse process. Photosynthesis is an example of anabolism, and *glycolysis* and *fermentation* are examples of catabolism. Metabolism may be summarized in a single redox equation.

$$H_2A + B + energy \rightleftarrows H_2B + A$$

where A is a hydrogen donor, and B is a hydrogen acceptor. The transfer of a hydrogen molecule means the transfer of two protons and two electrons. The reaction proceeds to the right in anabolism, and to the left in catabolism. Anabolic processes must supply all the components that form a bacterial cell. Table 19.5 shows the major components. Amino acids are synthesized in the bacterial cell by taking the carbon from glucose, the $-NH_2$ group from NH_4^+ in solution, and the sulfur from SO_4^{2-}, also in solution (both of mineral derivation). The energy for the synthesis is supplied by the

hydrolysis of ATP to ADP and by the reduction of NAD^+ or $NADP^+$ to NADH or NADPH. Needless to say, the reactions are mediated by specific enzymes.

The RNA and DNA nucleotides are synthesized by taking the carbon from glucose, the phosphate group from ATP, and the $-NH_2$ group from NH_4^+ in solution. Energy is supplied by ATP, and the reactions are mediated by enzymes.

Bacterial lipids include phospholipids and lipoproteins as part of the cell membrane (lipoproteins are *conjugated proteins*; a conjugate protein is a protein with attached a nonprotein component that in this case is a carboxylic acid). The lipids include triglycerides as energy-storage units. The glycerol head of a triglyceride is derived from glucose. The carboxylic acids are derived from acetyl co-

Table 19.5. Components of bacterial cells

Component	Number of Different Molecular Species	Percentage by Mass
water	1	70
proteins	3,000	15
DNA	1	2
RNA	3,000	6
carbohydrates	50	3
lipids	50	2
intermediates[a]	500	2
inorganic ions	15	1

[a] The intermediates are transient molecules along metabolic pathways.

enzyme A, a coenzyme consisting of adenosine, with attached two phosphate groups followed by a protein component. Acetyl coenzyme A is formed from pyruvic acid, with carboxylation and added CO_2.

In photosynthesizing bacteria, glucose is produced by energizing the Calvin cycle with solar energy. In chemosynthetic bacteria, the Calvin cycle still operates, but the energy is supplied by the oxidation of such elements or compounds as H_2, H_2S, Fe^{2+}, NH_4^+, and others. The energy needed for the polymerization of amino acids to form proteins, of nucleotides to form RNA and DNA, and of sugars to form polysaccharides, is provided by ATP.

The solar energy stored in chemical compounds is retrieved by breaking down the products that have been synthesized. In glycolysis, one molecule of glucose is degraded to two molecules of pyruvic acid (CH_3–CO–COOH). The energy of two ATP bonds is needed to initiate the process, degrading two ATP molecules to two ADP molecules. There are nine successive reactions, which are mediated by specific enzymes. The process produces four ATP molecules from four ADP molecules and two NADH molecules from two NAD^+ molecules. The net gain is two ATP molecules and two NADH molecules. The pyruvic acid, which consists of three groups, a methyl ($-CH_3$), a keto ($=CO$), and a carboxyl ($-COOH$), is highly reactive. Under anaerobic conditions, pyruvic acid is degraded by bacteria into acetaldehyde (CH_3CHO), which, with the addition of an H from NADH (which thus becomes NAD^+), is transformed into ethanol (CH_3CH_2OH). Bacteria can ferment alcohols, many acids (including amino acids), and purines, degrading them to simpler compounds that have less total bond energy.

19.6. REPRODUCTION AND FUNCTION

All procaryotes reproduce asexually by simple cell division, called *binary fission*, that is initiated by the duplication of DNA. Plasmid DNA duplicates at the same time as the DNA in the chromosome, with an error rate of 10^{-4}. Plasmids can exchange single-strand nucleotide segments with the chromosome, and vice versa. The remaining single strand and the

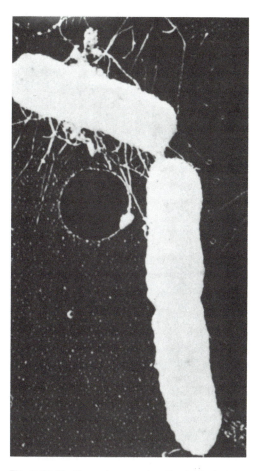

Figure 19.24. Young love: conjugation in *Escherichia coli* ($\times$ 20,000). Genetic material crosses from donor to acceptor through a *pilus* (pl. *pili*), a filamentous connection with a hollow core. (Anderson and Wollman, 1957, p. 450.)

transferred single strand are duplicated prior to insertion into the plasmid DNA and into the bacterial chromosome, respectively. Plasmids provide additional genes, which, among other things, may protect the bacteria against antibiotics. Although reproduction in bacteria is asexual, exchange of genetic material does occur. Exchange takes place through a protoplasmic bridge that forms between two bacterial cells in close proximity, a process called *conjugation* (Fig. 19.24).

The exchange of genetic material involves mainly plasmids. The DNA in the plasmid loop of the donor cell unwinds while duplicating, and a single strand passes through the bridge into the acceptor cell. There the strand rolls up while duplicating, forming a second plasmid. The result is two identical plasmids in the two cells. The acceptor may, in turn, become a donor, thus showing that it has become a male. Exchange of genetic material involving the chromosome also occurs. A section of DNA in the main loop of a donor (male) cell unwinds while duplicating, and a single-strand piece of DNA crosses the bridge and becomes incorporated into the chromosome of the acceptor (female) cell while duplicating itself. This process is called *recombination*, and the new DNA is called *recombinant DNA*.

In cell duplication, the chromosome attaches itself to the cell membrane and duplicates, and the two chromosomes are separated as the cell membrane elongates. The elongation of the membrane eventually causes the cell to fission in the middle, between the two separated chromosomes. Cell duplication takes minutes to hours.

Bacteria are ubiquitous in modern nature, ranging from the hot infernos of fumaroles and hot springs to endolithic habitats in Antarctic rocks. In the billions of years since they came into existence, bacteria have learned to metabolize an astonishing variety of substances. It is expected that even the toughest plastics will soon become prey to bacterial mutants. Nearly all bacteria have learned to produce spores within cells (called *endospores*) that have tough walls that can protect the genetic material under adverse conditions (e.g., a drought or temperature beyond tolerance limits).

Bacteria perform all kinds of functions within the biosphere. Although many bacteria cause diseases in both animals and plants, most are essential components of the biosphere. Among the most useful bacterial activities we may count nitrogen fixation, which means the reduction of nitrogen to ammonia (NH_3). Nitrogen is a fundamental component of amino acids, nucleotides, and many other organic substances. In the primitive earth, nitrogen was readily available as NH_3 both in the atmosphere and dissolved in water. Today, nitrogen forms 78% of the atmosphere (by mass), but the $N\equiv N$ bond is very strong (9.8 eV), and most organisms cannot crack it. Bacteria, however, soon learned how to do it, by using the enzyme *nitrogenase*.

Nitrogenase consists of two proteins, a larger one (molecular mass = 220,000 u) containing 1–2Mo, 18–36Fe, and 18–36S, and a smaller one (molecular mass = 60,000 u) containing a 4Fe-4S cluster. Both proteins are destroyed by contact with oxygen at room temperature. Hence, nitrogen fixation must take place under strict anaerobic conditions. Nitrogen fixation requires, in addition to nitrogenase, a source of energy (provided by ATP), a reductant (which usually is ferredoxin), and Mg^{2+} ions. One molecule of N_2 is reduced to two molecules of NH_3, with the degrading of 16 ATP to 16 ADP and the release of one molecule of H_2 (which then goes to reduce ferredoxin). Nitrogen fixation may be summarized as follows:

$$N_2 + 16ATP \rightarrow 2NH_3 + H_2 + 16ADP$$

The principal nitrogen-fixing bacteria are *Azobacter*, *Rhizobium*, and *Frankia*. *Azobacter* is a free-living bacterium that efficiently reduces N_2 to NH_3 under aerobic conditions by removing any oxygen that gets into the cell via oxidation of an iron–sulfur protein. *Rhizobium* lives symbiotically in the roots of leguminous plants, forming nodules within which nitrogen fixation occurs. *Frankia* lives similarly in the roots of *Alnus* (alder), *Casuarina* (Australian pine), and other nonleguminous plants.

We owe a great debt of gratitude to bacteria: They not only established the basic biochemistry by which life as we know it today can function but also continue to be helpful in a myriad of ways that range from providing us with nitrogen compounds all the way to decomposing garbage. Thanks, bacteria!

THINK

* Some bacteria use small crystals of magnetite packaged inside their cells as compasses. Why would a bacterium want to navigate? What happens during and after a magnetic reversal?

* Do you expect archaebacteria to exist on Venus? On Mars?

* Using $E = h\nu$, calculate the energy (in electron volts) of the frequencies absorbed by bacteriochlorophyll a (Table 19.3).

* A bacterial cell reproduces until the population has grown to the optimum size in equilibrium

with the environment. Why, then, continue repro-ducing, seeing that in an equilibrium situation half of the cells created in reproduction are wiped out by the environment? Why can a bacterial cell in an equilibrium population not live forever, just happily metabolizing?

• The earliest nitrogenase probably arose abiotic-ally in the "primordial soup" and was adopted by some bacteria for nitrogen fixation. What were the earliest bacteria using as a source of nitrogen to make amino acids before nitrogenase was adopted? What are archaebacteria in deep-sea vents using today as a source of nitrogen to make their amino acids?

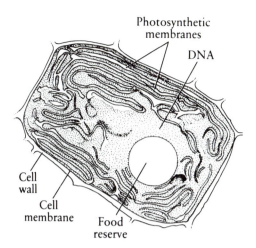

Figure 19.25. Cyanobacteria. *Anabaena cylindrica*, 5 × 3 μm. (Curtis and Barnes, 1989, p. 441, Fig. 21.18.)

19.7. CYANOBACTERIA

Cyanobacteria are photosynthesizing pro-caryotes that evolved $3.5 \cdot 10^9$ y ago and are still thriving today. The process of cyanobacterial photosynthesis, which, unlike bacterial photo-synthesis, produces free oxygen, may be sum-marized as follows:

$$6CO_2 + 6H_2O + energy$$
$$\rightleftarrows C_6H_{12}O_6 + 6O_2 \pm 7.1\,eV$$

The reaction goes to the right during photo-synthesis, and to the left during *respiration* (which here means *biooxidation of reduced organic substances*, not *breathing*). Energy is absorbed in forming glucose and is released when glucose is oxidized. Glucose ($C_6H_{12}O_6$) is the most abundant organic substance on Earth. The energy stored in glucose by photosynthesis is retrieved by the breakdown of glucose to simpler substances by the process of glycolysis.

Cyanobacteria are single celled or, more commonly, filamentous, with individual cell sizes in the filaments ranging from 1 to 60 μm (Fig. 19.25). Today, they are ubiquitous—they are found as plankton in the ocean and fresh-water lakes, on wood and vegetal litter, in soils, on rock and sediment surfaces, in microcracks just below rock surfaces (*endolithic environ-ment*), in layers within sediments, and even in hot springs at temperatures as high as 75°C.

Several genera of filamentous cyano-bacteria, most notably *Anabena* and *Nostoc*, can produce, along their filaments, special cells with thickened walls, called *heterocysts*, where

nitrogen fixation can take place under strictly anaerobic conditions even when the environ-ment contains free oxygen. Nitrogen-fixing cyanobacteria are symbionts with fungi to form lichens and with a number of algae and plants. The advantage to the host is clearly the cyanobacterial production of nitrogen compounds. Another type of cell developed by filamentous cyanobacteria is the *akinete*, a resting spore with thick walls that can survive even in the desert and is quickly brought back to life by a rainstorm. Cyanobacteria can move around by a gliding or even a snakelike motion. Coordination among the cells in the filaments to achieve these motions probably is by means of chemical signals from cell to cell. Transmission is very slow, but then so is the motion.

Cyanobacteria evolved in the Archean and were the dominant form of life during much of the Proterozoic. They perfected photo-synthesis by adding a second system, called *Photosystem II* (PSII), which splits the water molecule and releases oxygen to the atmosphere (Fig. 19.26). They also built layered structures, called *stromatolites* (see below), in which much of the world's iron resources are concentrated.

Cyanobacteria use chlorophyll a instead of bacteriochlorophyll. Their cells have a com-plex system of internal membranes, called *thylakoids*, to which the molecules of chloro-phyll a and other pigments are attached. In

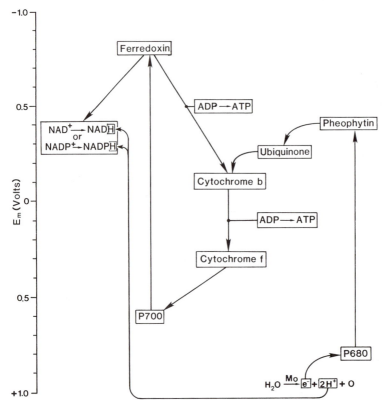

Figure 19.26. Photosynthesis in cyanobacteria. A photon is collected in PSII (P680) and an electron is energized by 0.8 eV. Half of the energy is used to phosphorylate ADP to ATP. The still-energized electron enters PSI (P700), where it is energized by an additional 0.9 eV. The energy is used to reduce NADP to NADPH (using ambient H$^+$) or, if there is already enough NADPH around, to phosphorylate a second ADP to ATP. The positively ionized chlorophyll molecule in P680 is neutralized by an electron from a water molecule, removed by a complex of four proteins that contain molybdenum.

PSII, photons are collected by an apparatus similar to that of PSI: The photon energy is funneled to a chlorophyll molecule at the reaction center called P$_{680}$ (because it absorbs mainly at $\lambda = 680$ nm). An electron is boosted from $+0.8$ to -0.4 V, acquiring 1.2 eV of energy. The electron is released to a pheophytin molecule (pheophytin has the same structure as chlorophyll, less the Mg atom at the center). The positively ionized chlorophyll molecule is neutralized by an electron from a water molecule, which, as a result, disintegrates. The breakup of the water molecule is caused by a complex of four proteins, with molecular masses ranging from 18,000 to 34,000 u, that contain molybdenum. The exact reaction remains unknown. Meanwhile, the pheophytin molecule passes the electron down through ubiquinone to cytochrome c and cytochrome f, a process that leads to the phosphorylation of one molecule of ADP to ATP and the expenditure of 0.3 eV. The electron is then passed along to a chlorophyll molecule in the reaction center P$_{700}$ of PSI. The chlorophyll molecule in P$_{700}$ boosts the electron from $+0.6$ to -0.8 V, with the help of photons harvested by its own light-harvesting system, and passes it, through two Fe-S centers, to a ferredoxin molecule. From there the electron either goes to reduce NAD$^+$ to NADH or NADP$^+$ to NADPH, or is cycled back to P$_{700}$ via cytochromes b and f, in which case another molecule of ADP is phosphorylated to ATP. The photosynthetic process in cyanobacteria is summarized in Fig. 19.26. As may be seen, cyanobacterial photosynthesis produces two

ATPs and one NADH or NADPH, instead of only ATP (or NHDH/NADPH) as in bacteria (Fig. 19.20). The total Eh is 2.6 eV, compared with 0.8–1 eV in the PSI of bacteria.

Cyanobacteria evolved when the atmosphere was still reducing, about $3.5 \cdot 10^9$ y ago. Because their photosynthesis liberates oxygen, cyanobacteria combine the superoxide ion O_2^- (a powerful oxidant formed in the cells) with hydrogen (from the breakdown of H_2O in PSII), using the enzyme *superoxide dismutase*:

$$2O_2^- + 2H^+ \xrightarrow{\text{superoxide dismutase}} H_2O_2 + O_2$$

The H_2O_2 (still a strong oxidant) is then reduced to water by means of the enzyme *catalase*:

$$2H_2O_2 \xrightarrow{\text{catalase}} 2H_2O + O_2$$

For some two billion years the oxygen was quickly removed from the water by the abundant iron that was in solution in the ocean (Fe^{2+} is soluble in water under anoxic conditions). It is to be expected that the concentration of iron was highest in coastal basins and lagoons, in proximity to iron sources (the rocks being weathered on land). There, the cyanobacteria flourished as long as iron quickly removed the oxygen (oxygen is bad news for cells, because it oxidizes organic matter). In nature it often happens that when two processes act in tandem, one overshoots the other, leading to a forced oscillation that continues through time. In this case, the two processes are oxygen production and oxygen removal. If production overshoots removal, too much oxygen accumulates in the water, which interferes with nitrogen fixation and leads to a decrease in the cyanobacterial population. Decreasing production results in increasing accumulation of iron, whereupon the water is again free of oxygen, and the cyanobacterial populations expand. This sinusoidal change in productivity is recorded in the layers of the stromatolites. Stromatolites are laminated sediment accumulations in which organic-rich layers alternate with organic-poor layers. The organic component consists of filamentous cyanobacteria. The inorganic component consists of $CaCO_3$ and/or SiO_2 particles. The laminae are less than 1 mm thick, each

consisting of a pair of lamellae, an organic-rich lamella formed during the day by growing cyanobacteria, and on organic-poor lamella formed during the night by sediment accumulation. Stromatolites occur as crusts or may exhibit a columnar or domal structure (Figs. 19.27 and 19.28).

The earliest cyanobacteria have been found in the Warrawoona Formation of northwestern Australia, dating from $3.5 \cdot 10^9$ y ago (Fig. 19.29). Stromatolites are present in Archean times (Warrawoona, Australia; Onverwacht, Swaziland; Insuzi, Natal) and more commonly in Proterozoic times (Gunflint, Ontario). Modern representatives occur in the Bahamas (Fig. 19.30) and in Shark Bay, Western Australia. Stromatolites include crusts that are centimeters to inches thick (Fig. 19.31), hemispheroids 10–20 cm thick, and columns up to 50 cm across and 1 m high.

Stromatolitic deposits form the major iron resources of the world. They are found in all continents where Cryptozoic formations occur. The more important ones are in Minnesota, Ontario, eastern Canada, central and southern

Figure 19.27. Stromatolite section from the Precambrian Denault Formation of Newfoundland. (Donaldson, 1963.)

Figure 19.28. Stromatolitic mound on Victoria Island, northern Canada. (Young, 1974, p. 32, Fig. 12.)

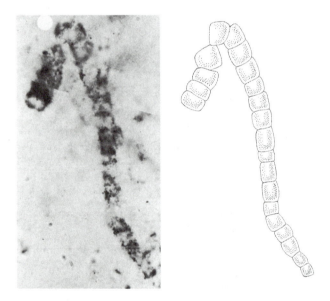

Figure 19.29. A 3.5-billion-year-old filamentous cyanobacterium from the Warrawoona Formation of northwestern Australia (left), and schematic representation of the same (right). The individual cells range in size from 4 to 7 μm. This cyanobacterium is similar to the modern *Nostoc*, one of the most common cyanobacteria. (Schopf, 1983, p. 288, Photo 9.8.)

Figure 19.30. Modern stromatolites in the Bahamas. (Courtesy E. Shinn, U.S. Geological Survey, St. Petersburg, Fla.)

Russia, India, Australia, South Africa, western Africa, Brazil, and Venezuela.

THINK

* Fig. 19.31 shows a definite stratification of different types of cyanobacteria. Some of these

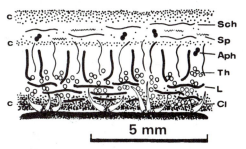

Figure 19.31. Cross section of an algal mat, 4.5 mm thick, from an intertidal pool in the upper Florida Keys. Oxygenation decreases from the surface down. Different species of cyanobacteria are adapted to different oxygen content: Sch, *Schizothrix*; Sp, *Spirulina*; Aph, *Aphanocapsa*; Th, thiobacteria (sulfur bacteria); L, *Lyngbya*; Cl, *Calothrix*. (Carr and Whitton, 1973, p. 458, Fig. 21.12.)

cyanobacteria are filamentous, and the filaments can move with coordinated motion. How do you expect the cells in the filament to communicate with each other so as to achieve coordinated motion? No nervous system, of course.

19.8. PROCHLOROPHYTA

The phylum Prochlorophyta includes the genera *Prochloron* (a single-celled symbiont on the body surfaces of colonial tunicates in the tropical Pacific), *Prochlotothrix* (single-celled, planktic, living in Dutch lakes), and *Prochlorococcus* (single-celled, planktic, marine). Prochlorophyta are procaryotes that, unlike all other photosynthesizing procaryotes, use for their photosynthesis a mixture of 70% chlorophyll a and 30% chlorophyll b plus some carotenoids. Chlorophyll b has the highest absorbance peak at 480 nm (Table 19.3), which is in the blue region of the spectrum (Fig. 19.32).

The use of chlorophyll b in addition to chlorophyll a indicates that prochlorophytes form a link between procaryotes and eucar-

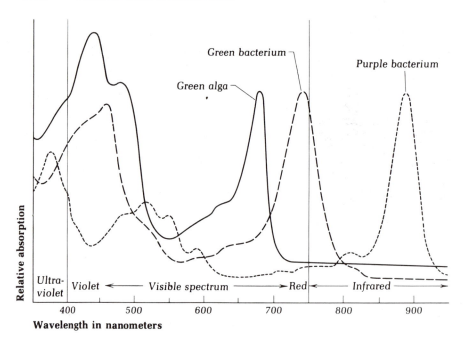

Ultra-violet | Violet ← Visible spectrum → Red ← Infrared →

Wavelength in nanometers

Figure 19.32. *Action spectrum* (rate of photosynthesis versus wavelength of photons harvested) for a purple bacterium, a green bacterium, and a green alga. The action spectrum is an index of the efficiency of the system to photosynthesize. (Stainer, Doudoroff, and Adelberg, 1986, p. 563, Fig. 17.3.)

yotes. Green algae (Section 20.6) and all plants use the mixture developed by prochlorophytes or their remote ancestors.

THINK

* If, in the beginning, when CO_2 was the dominant gas in the atmosphere, the sky was red and the sun was bluish, why was chlorophyll b not used, seeing that it absorbs in the blue region of the spectrum?

19.9. VIRUSES AND VIROIDS

Viruses are not living organisms or cells—they are "particles" about 0.1–$0.3\,\mu$m across that contain a single- or double-stranded DNA or RNA genome coiled with positive helicity. The genome is protected by a protein coating that may or may not be itself enclosed in a membrane (called *envelope*). A viral particle is called a *virion*. Viruses can reproduce only in-

side the living cells of bacteria, plants, or animals. Bacterial viruses are also called *bacteriophages*, or *phages* for short.

There are both tailed forms (Fig.19.33) and nontailed forms (Fig. 19.34). The nontailed forms usually are icosahedral particles 0.03–$0.1\,\mu$m across (the *icosahedron* is a polyhedron with 20 triangular faces). Most of the nontailed viruses have double-stranded DNA, but some have single-stranded DNA or RNA. The tailed forms (Fig. 19.33) have no envelope, a head $0.1\,\mu$m across (usually icosahedral), a tail $0.2\,\mu$m long, and double-stranded DNA.

Plant viruses have mostly single-stranded RNA contained in filaments $0.03 \times 1\,\mu$m. Some have single- or double-stranded DNA in icosahedral particles 0.03–$0.1\,\mu$m across. Animal viruses have double-stranded DNA, single- or double-stranded RNA, or, in some cases, single-stranded DNA.

Base pairs number from 5,000 to $2.5 \cdot 10^8$ in most families, corresponding to about 5 to 250,000 genes in most families.

Retroviruses are a family of spherical viral particles (diameter $0.1\,\mu$m) that contain a

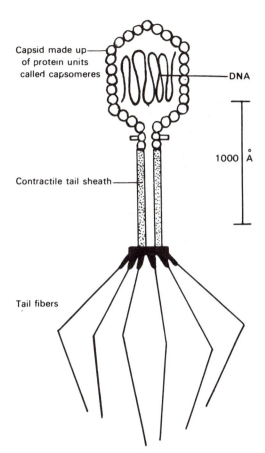

Capsid made up of protein units called capsomeres

DNA

1000 $\overset{\circ}{A}$

Contractile tail sheath

Tail fibers

Figure 19.33. A tailed virus form. The diameter of the capsid is about 1 μm. (Moment and Habermann, 1973, p. 96, Fig. 4.24.)

Poxviridae are a family of viruses in which the genome consists of $30 \cdot 10^9$ base pairs capable of coding a variety of enzymes. This allows the Poxviridae to reproduce inside the cytoplasm without interfering with the genome of the host cell.

Viral Reproduction

Viruses reproduce by two processes, the *lytic process* and the *transforming process*.

[*Lytic* derives from the Greek λύειν, meaning *to loosen* or *to dissolve*. It has the same root as Lysol, which dissolves dirt, and *analysis*, which means the separation of a complex concept into its constituent parts for easier understanding. Do not confuse *lytic* with *lithic*, which derives from λίθος (stone) and means *stony*. Notice that the Greek υ transliterates into the English *y*, and the Greek ι transliterates into the English *i*.]

In the lytic process, the viral particle lands on a receptor site on the host-cell membrane. The nontailed forms are absorbed into the cell by phagocytosis, and the tailed forms inject their genomes through their tails into the host cell. The viral genome reproduces either in the cytoplasm or in the nucleus of the host cell, depending on the type of virus. Viral mRNA is produced and threaded into the cell's ribosomes to form viral proteins. The daughter viral genome coats itself with the viral proteins, completing the formation of the new viral particle. The lytic process lasts 6–48 h, during which time 10^3–10^6 viral particles per cell are produced, and the host cell dies when it bursts or lyses.

In the transforming process the viral particle enters the host cell as in the lytic process. The viral genome is inserted into the genome of the host cell, or else it reproduces independently in the cytoplasm as a plasmid. The host cell is not killed, and it can reproduce, but it becomes different (e.g., cancer cells).

Viruses mutate at a rate 10^6 times greater than procaryotes or eucaryotes because they lack regulatory enzymes (like DNA polymerase) that can correct errors as they happen during the process of nucleic acid replication.

Viroids are small, uncoated, but functional, viral particles with RNA genomes (1.0–

genome consisting of two identical mRNA molecules. Upon entering the host cell, the virion makes a DNA copy of its mRNA, using an enzyme called *reverse transcriptase*. The DNA segment inserts itseft into the DNA of the host cell, which thus is made to produce viral mRNA. The viral mRNA is used by the host cell to make all viral-protein components (including reverse transcriptase). The components assemble themselves into a daughter virion, which leaves the host cell and invades another cell. This coding of DNA by mRNA is unique in the entire biosphere, where it's DNA that codes mRNA. Examples of retroviruses: oncoviruses, which produce tumors, and lentiviruses, which produce slowly ("lenti-") developing diseases, including human acquired immunodeficiency syndrome (AIDS).

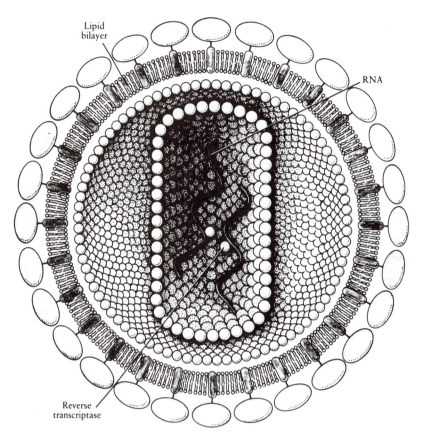

Lipid
bilayer

RNA

Reverse
transcriptase

Figure 19.34. Structure of the spherical HIV virus (human immunodeficiency virus) that causes AIDS in humans. (Curtis and Barnes, 1989, p. 814, Fig. 39.22.) The diameter of the HIV virion is about 0.1 μm. The HIV particle contains at the center two RNA molecules and two or more molecules of reverse transcriptase. These are surrounded by a protein envelope, a lipid bilayer, and an outer layer of glycoproteins whose structure is similar to that of the outer layer of T lymphocytes. The T lymphocyte (T stands for *thymus*, the organ where T lymphocytes originate) is one of the two types of cells that produce antibodies (the other is the B lymphocyte, where B stands for *bone*, because the B lymphocytes are made in bone marrow). The T lymphocyte does not recognize the viral particle as a foreign body and lets it in. Once in, the viral RNA emerges and, via reverse transcriptase, manufactures a piece of complementary DNA and inserts it into one of the chromosomes of a T lymphocyte. After a period, which may be years, replication of the viral particles begins and spreads like wildfire to other T lymphocytes. The immune system collapses, and the organism is left defenseless against viral and microbial infections.

$1.3 \cdot 10^4$ u = 15–20 genes) capable of reproducing within a host cell. They cause plant diseases.

Viruses and viroids probably have come and gone all along in the history of our planet. Viruses act as *antigens*; that is, they elicit the production of *antibodies* by the organism they invade. Specific sites on the bodies of pathogenic bacteria are also antigens, eliciting the production of specific antibodies. Both antigens and antibodies are protein complexes. Their molecular masses are on the order of 150,000 u to 900,000 u. The antibodies can neutralize the sites used by the viral particle for attachment to a cell, can agglomerate the viral particles in lumps that are then destroyed by phagocytes, or can altogether dissolve the viral (or bacterial) intruder.

The appearance of new viruses and viroids

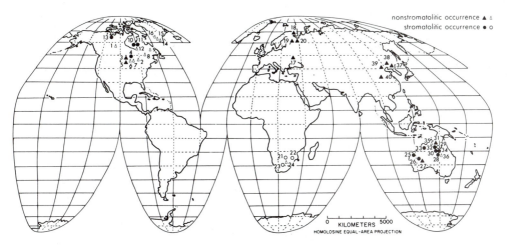

nonstromatolitic occurrence ▲ △
stromatolitic occurrence ● ○

0   KILOMETERS   5000
HOMOLOSINE EQUAL-AREA PROJECTION

Figure 19.35. This world map shows the distribution of 40 occurrences of microfossils reported from Early Proterozoic formations ($2.5–1.6 \cdot 19^9$ y B.P.). Remains of coccoid bacteria are often difficult or impossible to identify as microfossils as opposed to spheroidal inorganic structures of similar dimensions. As a result, only 24 of the 40 occurrences can be verified as containing true microfossils (black triangles and circles). In addition to these Proterozoic occurrences, earlier microfossils have been reported from the Archean upper Onverwacht ($3.5 \cdot 10^9$ y) and Fig Tree (ca. $3.2 \cdot 19^9$ y) groups of the Swaziland Sequence of SE Africa (close to no. 22 on the map; see also Fig. 20.4) and from the Warrawoona ($3.5 \cdot 10^9$ y) and Fortescue ($2.8 \cdot 10^9$ y) groups of NW Australia (close to no. 33 on the map; see also Fig. 19.29). (Hofmann and Schopf 1983, p. 322, Fig. 14.1.)

can be expected to be a common occurrence because of their high rates of mutation. In the beginning, the victims will be unable to produce the appropriate antibodies and may be defenseless against these new mutants. I have hypothesized that infestations by new viral mutants may have been responsible for many (perhaps most) of the extinctions of individual species that are documented in the geological record (see Chapter 20).

THINK

* Modern viruses, of which the virus that causes AIDS is an example, obviously are highly evolved. Viruses affecting bacteria (the phages) are much simpler. Would you expect viruses to have been common early on the primitive Earth? Would you expect interaction between viruses and primitive bacteria to have had an important function in the early evolution of life on Earth?

20
THE CRYPTOZOIC:
SECTIONS AND TAXA

20.1. OVERVIEW

The Cryptozoic covers the time from the formation of the solar system (end of the Gamowian, $4.7 \cdot 10^9$ y ago) to the beginning of the Cambrian ($570 \cdot 10^6$ y ago). It thus represents 87% of the life of our planet. The Cryptozoic is divided into three eons (the highest time rank, above era): Hadean ($4.7–3.8 \cdot 10^9$ y), Archean ($3.8–2.7 \cdot 10^9$ y), and Proterozoic ($2.7–0.57 \cdot 10^9$ y). The earliest rock materials yet discovered on Earth are grains of zircon (zirconium silicate, $ZrSiO_4$) from Mount Narrayer, northwestern Australia, dated at $4.2 \cdot 10^9$ y, and some rocks in the area of the Great Slave Lake in northwestern Canada, dated at $3.96 \cdot 10^9$ y. The Cryptozoic is also called the *Precambrian*, although, strictly speaking, the name *Precambrian* implies *all* the time before the Cambrian, including the Gamowian and the Planckian. Table 20.1 shows the major subdivision of the Cryptozoic.

The Earth's crust began forming in the early Cryptozoic. Magmatic differentiation and the *lithologic cycle* (erosion, sedimentation, subduction, magma formation, emplacement of new crustal rocks) produced nuclei of granodioritic rocks that grew in extension during the Cryptozoic to form the ancestral continents:

1. **North America:** present-day North America, less Florida, plus northern Mexico, northern Ireland, Scotland, western Norway, Spitzbergen, and eastern Siberia (Verkhoyansk Range and parts east)

2. **Europe:** Europe to the Urals, plus southern Newfoundland, New England, Nova Scotia, Turkey, Turkmenistan, and northern Iran, less southern Spain, Italy, and the Balkans

3. **Siberia:** Siberia between the Urals and the Verkhoyansk Range

4. **China:** China, plus Kazakhstan, Afghanistan, southeastern Asia, Indonesia, the Philippines, Japan, and Korea

5. **Gondwana:** Africa, plus Italy, southern Spain, Arabia, India, Australia, Antarctica, South America, Central America, and Florida

Two major types of terranes formed the continental crust during the Archean: greenstone belts, covering 30% of the land surface, and granite-gneiss batholiths, covering 70% of it (see Fig. 18.1). The greenstone belts include the oldest rock sections on Earth. They were formed at low temperatures (300–400°C) and at moderate pressure (1.5–15 kb = 5–50 km of depth). A typical greenstone belt consists, from bottom (a) to top (c), of the following sequence:

c. graded volcaniclastic sediments
(mainly graywackes)   10%
b. felsic volcanics (andesites
to rhyolites)   25%
a. ultramafic-to-mafic volcanics
(komatiites to basalts)   65%

[*Felsic* refers to rocks rich in feldspars, feldsparthoids, and silica; *mafic* refers to rocks with 60–80% Fe–Mg silicates; *ultramafic* refers to rocks with more than 80% Fe–Mg silicates. *Komatiite* is effusive peridotite. The percentage refer to the thickness within the sequence.]

Greenstone belts usually exhibit great thicknesses: 21 km in Swaziland, 18–30 km in India and northwestern Australia, 7–14 km in the Superior Province of Canada. The granite-gneiss batholiths were intruded into the greenstone belts $200–800 \cdot 10^6$ y after formation of the belts. The Cryptozoic batholiths are granodioritic, with more Na than K, while the younger once are granitic, with more K than Na.

Table 20.1.   Major subdivisions of geologic time

Age (10^9 y)	Eon	Event
4.7–3.8	Hadean	formation and differentiation of the Earth; evolution of bacteria
3.8–2.7	Archean	evolution of stromatolitic crusts and domal stromatolites
2.7–1.7	Proterozoic (Early)	cylindrical stromatolites
1.7–0.570	Proterozoic (Late)	branched stromatolites; evolution of eucaryota; appearance of the Ediacaran fauna
0.570–0	Phanerozoic	appearance of Archaeocyatha; radiation of metazoa and metaphyta.

During the Proterozoic, the continental crust expanded to cover almost 30% of the surface of the Earth. Little additional expansion, if any, has occurred during the Phanerozoic.

THINK

* Archean terranes are seen to consist of greenstone belts with pods of granite-gneiss. Why not granite-gneiss belts with pods of greenstones?

20.2. CRYPTOZOIC SECTIONS

The Banded Iron Formations

The Cryptozoic includes the Archean (3.8–2.7·10^9 y ago) and the Proterozoic (2.7–0.57·10^9 y ago). Cryptozoic sections reveal the early history of the continental crust and the early stages in the evolution of life on Earth. Many are characterized by the occurrence of banded ironstones, which are part of extensive *Banded Iron Formations*.

Banded Iron Formations (BIF's) formed commonly in Archean and Early Proterozoic time (3.8–1.7·10^9 y ago). These formations account for 15% of the total thickness of sediments deposited during that time. The standard sequence includes, from top to bottom:

Shale (highest)
Ironstones
Shale
Quartzite
Dolomite (lowest)

Volcanic deposits are also common in the BIF sections. The ironstones consist of laminae of mainly quartzite (formerly chert), alternating with laminae rich in Fe oxides [magnetite, Fe_3O_4 (27.6% oxygen); hematite, Fe_2O_3 (30.1% oxygen)] and, less commonly, iron carbonate [siderite, $FeCO_3$ (41.4% oxygen)].

The atmosphere was reducing when the BIFs were deposited. The concentration of oxygen in the atmosphere remained low for more than two billion years after the appearance of cyanobacteria, because the photosynthetic oxygen released was used up in the oxidation of surface rocks. Fe^{2+} (ferrous iron), which is soluble in water under anaerobic conditions, was carried by streams to the lagoons and ponds where cyanobacteria thrived and was quickly oxidized to magnetite and hematite by the photosynthetic oxygen produced by the cyanobacteria (Fig. 20.1). A sort of prey–predator relationship developed, with the cyanobacteria overshooting the oxygen scavenger (i.e., the iron) and poisoning themselves by introducing too much oxygen into the water. That, plus possible tidal and seasonal cycles, would explain the laminations. The Banded Iron Formations range from 100 m to 1,000 m in thickness. They are distributed in all ancestral continents and contain the world's major iron resources.

After the atmosphere became oxygen-rich, Fe^{2+} became transportable only by solutions rich in humic acids—the complex organic acids common in soils. Iron is normally precipitated in the B horizon of a soil as hematite or siderite ($FeCO_3$). Reducing conditions persist today in modern bogs, which are rich in decaying organic matter. There, oxygen released by the cyanobacteria oxidizes the iron

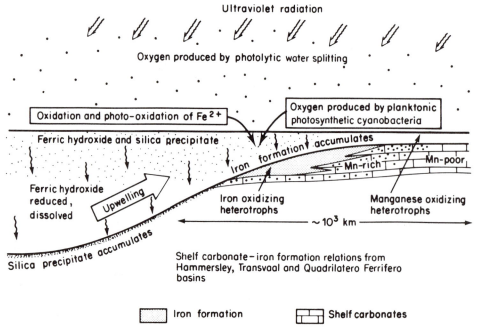

Figure 20.1. Depositional environment of a Banded Iron Formation. (Windley, 1984, p. 91, Fig. 6.12.)

as it did in the Cryptozoic. Characteristic BIF minerals and rocks are as follows:

Fe minerals

Hematite: Fe_2O_3

Magnetite: Fe_3O_4

Maghemite: Fe_2O_3 (structure of hematite, but with crystal imperfections)

Greenalite: a greenish, altered Fe silicate, $(Fe^{2+}, Fe^{3+})_{5-6}Si_4O_{10}(OH)_8$

Siderite: $FeCO_3$

Rocks

Chalcedony: cryptocrystalline quartz (crystal size $<5\,\mu m$)

Chert: microcrystalline quartz (crystal size $\sim 20\,\mu m$)

Jasper: chert with disseminted Fe oxides

Taconite: bedded ferruginous chert with at least 25% Fe, variously colored

Jaspilite: banded ferruginous chert with at least 25% Fe, reddish in color

Cryptozoic Sections

The following are some of the more important Cryptozoic sections.

ISUA, SOUTHWESTERN GREENLAND

The Isua section (Figure 20.2) consists of a greenstone belt dating from $3.86 \cdot 10^9$ y ago, intruded by granitic batholiths $3.6 \cdot 10^9$ and again $3.0 \cdot 10^9$ y ago. The greenstone belt consists of three successive formations (1 is the lowest unit; 3 is the highest):

3. *Banded ironstones*: quartzite with layers rich in Fe oxides (mainly magnetite); age, $3.76 \cdot 10^9$ y; percentage of extractable carbon, 0.5% (very low).

2. *"Schists"*: metamorphosed volcanic ash deposited below water, with a layer of metarhyolite broken into slabs.

1. Amphibolite: amphibole + plagioclase ($\frac{2}{3}$ hornblende + $\frac{1}{3}$ bytownite, which is $Ab_{30}An_{70}$ to $Ab_{10}An_{90}$), with layers of magnetite-bearing banded ironstones; it contains 10–15% Fe and up to 0.5% extractable (organic?) carbon; age; $3.8 \cdot 10^9$ y.

NORTH POLE, NORTHWESTERN AUSTRALIA

The North Pole area of northwestern Australia contains greenstone belts dating from $3.6 \cdot 10^9$ y ago that were intruded by granitic batholiths

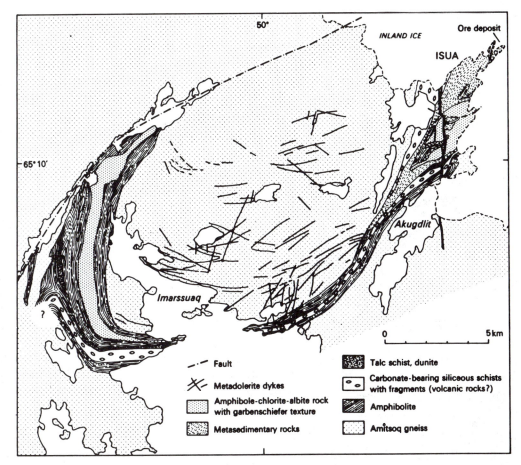

Figure 20.2. Isua, southwestern Greenland: geological map. The Amitsoq gneiss has been dated at $3.7 \cdot 10^9$ y of age. (Windley, 1984, p. 178, Fig. 1.)

$2.8 \cdot 10^9$ y ago. A typical greenstone-belt section includes the following units (1 is the lowest unit; 7 is the highest):

7. Pillow basalts

6. Interbedded sandstones (cross-bedded) and shales

5. Evaporites and possible stromatolites

4. Laminated mudstones and gyspum casts; intraformational breccias

3. Cross-bedded sandstone

2. Horizontally bedded sandstones

1. Pillow basalts

The pillow basalts are indicative of subaqueous extrusion, while the sedimentary formations are indicative of shallow-water deposition and even dryness.

SWAZILAND, SOUTHEASTERN AFRICA

The Swaziland Sequence (Fig. 20.3) is 22 km thick and consists of ultramafic-to-mafic rocks at the bottom, followed by calcalkaline volcanics, which are in turn followed by a sedimentary section at the top. The sequence may be summarized as follows (1 is the lowest unit; 3 is the highest):

3. Sedimentary group; repetition of:
   d. BIF
   c. Shales
   b. Graywackes
   a. Conglomerates

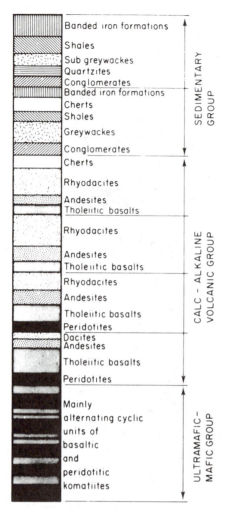

Figure 20.3. The Swaziland Sequence, southeastern Africa: generalized rock sequence. Total thickness = 22 km. (Windley, 1984, p. 30, Fig. 3.3.)

2. Calcalkaline group; repetition of:
   c. Dacites
   b. Andesites
   a. Tholeiitic basalts

   [*Tholeiite* is basalt enriched in Fe.]

1. Ultramafic–mafic group; repetition of:
   b. Basaltic komatiites
   a. Peridotitic komatiites

This section demonstrates increasingly acidic volcanicity, decreasing volcanic activity, increasing sediment formation and deposi-

tion, and the expansion of living organisms (Fig. 20.4).

THE CANADIAN SHIELD

The Canadian Shield, forming most of Canada east of the Rockies, is a classical area where Cryptozoic rocks are widely exposed and have been much studied by Canadian and American geologists. The generalized stratigraphic succession is summarized as follows (1 is the lowest unit; 4 is the highest):

4. Keeweenawan: lava flows with Cu, red siltstones (volcanic ash?)

3. Huronian
   Animikie:
   Rowe (argillite, > 2,000 ft)
   Gunflint (chert, 400 ft.) ($2.2 \cdot 10^9$ y)
   Kakabeka (quartzite above, conglomerate below) (lowest within Animikie)
   Cobalt: 500 ft of tillite (deposited at 60° latitude), followed by 12,000 ft of arkosic quartzite ($2.3 \cdot 10^9$ y)
   Bruce: basal conglomerate, followed by bedded quartzite, 5,000 ft total thickness

Major unconformity

2. Timiskaming ($2.7 \cdot 10^9$ y?): great thickness of conglomerates, with poorly sorted, thinly laminated graywakes (glacial-marine?)

1. Keewatin ($2.8 \cdot 10^9$ y): basaltic pillow lavas (subaqueous), and andesites (subaerial)

THE GUNFLINT IRON FORMATION

The Gunflint Iron Formation extends from Gunflint Lake (Minnesota) to the head of Thunder Bay (Ontario). It is about 5,000 ft thick and dates from about $1.9 \cdot 10^9$ y ago. The stratigraphy of the Gunflint Iron Formation is summarized as follows (1 is the lowest unit; 4 is the highest):

4. Upper limestone: 5–20 ft

3. Upper Gunflint
   taconite: 150–180 ft
   tuffaceous shale: 5–16 ft
   stromatolitic chert: 48–86 ft
   taconite: 150–210 ft

2. Lower Gunflint
   tuffaceous shale: 4–22 ft
   stromatolitic chert: 2–15 ft

1. Basal conglomerate: 1–5 ft

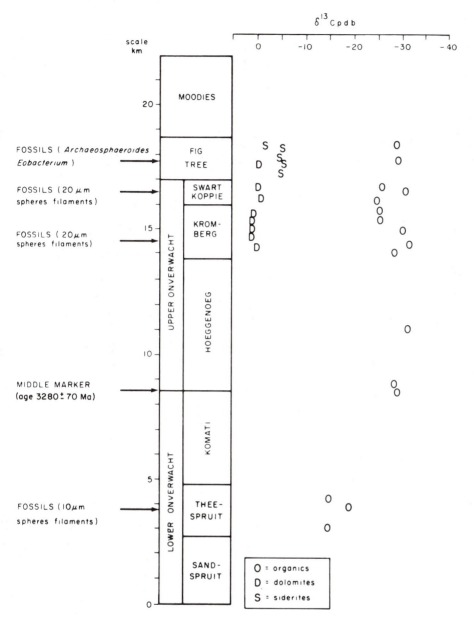

Figure 20.4. The Swaziland Sequence: formations, microfossils (?), and carbon-isotopic compositions of organic compounds, dolomites, and siderites. There are two stable isotopes of carbon, ^{12}C (98.9%) and ^{13}C (1.1%). Organic matter is enriched in ^{12}C and depleted of ^{13}C. The difference is about 2%. A shift of about 2% toward the lighter carbon is seen to have occurred between the Lower and Upper Onverwacht. It is believed that the organic matter in the Lower Onverwacht was produced abiotically, and that of the Upper Onverwacht was produced by organisms (bacteria). (Windley, 1984, p. 46, Fig. 32.12.)

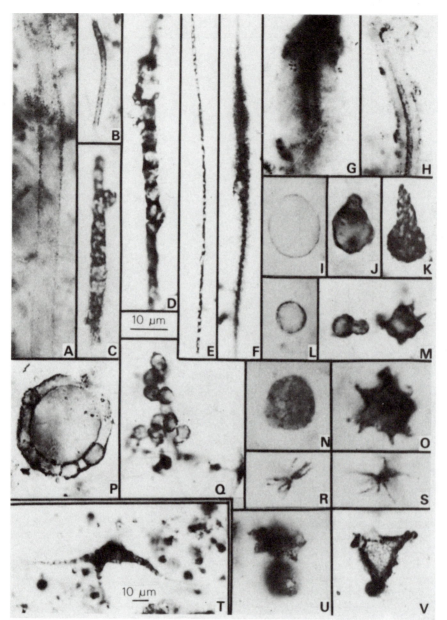

Figure 20.5. Microorganisms from the Gunflint Formation: (A) *Animikiea septata*; (B) *Gunflintia minuta*; (C,D) *Gunflintia grandis*; (E,F) degraded filament of *G. minuta*; (G,H) filaments surrounded by sheaths; (I) *Huroniospora*; (J,K) budding *Huroniospora*; (L) *Huroniospora microreticulata*; (M) budding *H. microreti-* *culata*; (N) *Huroniospora macroreticulata*; (O) *Eomichrystridium (?) barghoorni*; (P) *Eosphaera tyleri*; (Q) Huroniosporas in a cluster; (R,S) *Eoastrion simplex*; (T) *Veryhachium (?)*; (U) *Kakabekia umbellata*; (V) *Exochobrachium triangulum*. (Hofmann and Schopf, 1983, Photo 14.1.)

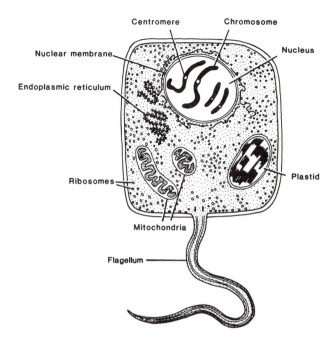

Centromere   Chromosome

Nuclear membrane

Nucleus

Endoplasmic reticulum

Ribosomes

Plastid

Mitochondria

Flagellum

Figure 20.6. A modern eucaryotic cell with its various organelles. DNA is segregated inside a nucleus bound by a nuclear membrane. DNA is partitioned into chromosomes that number from two to several hundred [fruit fly, 8; rhesus monkey, 42; human, 46; Atlantic salmon, 60; radiolarian (marine protozoan) 800] and range in length from 0.5 μm to 50 μm. The flagellum, when present, consists of a ring of nine pairs of microtubules plus two microtubules along the axis. Microtubules consist of fibrous proteins. (Emiliani, 1988, p. 133, Fig. 9.1.)

[Taconite has laminae enriched in iron oxides and, more rarely, iron carbonates.]

A rich assemblage of procaryotes (both bacteria and cyanobacteria) has been found in the cherts of the Gunflint Iron Formation (Fig. 20.5).

Procaryotes were the only form of life from the beginning of the Archean to mid-Proterozoic time, about $1.7 \cdot 10^9$ y ago, when the eucaryotes evolved. The cyanobacteria, which reigned supreme during that long period of time, left an abundant record of life and transformed the hydroatmosphere from a reducing one to an oxidizing one, leading to the evolution of the eucaryotes.

THINK

* Why is there a time lag in prey–predator situations? Why is equilibrium not maintained at all times?

* Suppose you land on Mars and find ancient stromatolites and even evidence of strong tides for a period. What is your conclusion?

20.3. THE EUCARYOTIC CELL

Eucaryotic cells evolved about $1.7 \cdot 10^9$ y ago. In contrast to procaryotic cells, they have a number of internal structures called *organelles* (Fig. 20.6). These include the nucleus, the endoplasmic reticulum, ribosomes, the Golgi apparatus, mitochondria, chloroplasts, and flagella (optional).

The *nucleus* is enclosed within a nuclear membrane and contains either *chromatin*, which is the ensemble of nucleic acids and nucleoproteins (proteins associated with nucleic acids) during the cell's resting stage, or *chromosomes*, if DNA has replicated and cell division is about to occur. DNA replication involves the formation of numerous bubbles (5,000 in *Drosophila*, separated by 30,000 base pairs), instead of just one as in procaryotes. As in procaryotes, DNA polymerase adds nucleotides, beginning at each fork and proceeding in opposite directions, to form segments that are complementary to the original DNA chain. The segments, 100–200 bases long in eucaryotic cells, are joined by DNA ligase to form a chain as long as the original chain.

[The DNA double helix (see Fig. 19.13) inside

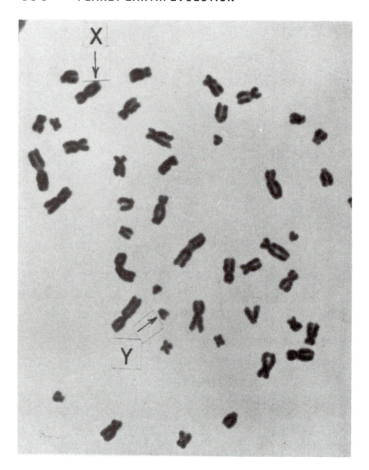

Figure 20.7. The 46 chromosomes of humans. (A) Male chromosomes, characterized by the XY pair. (Tijo and Puck, 1958, p. 1229.)

the nucleus is chopped into segments that are folded into chromosomes at the time of cell duplication. The human genome consists of $5.3 \cdot 10^9$ base pairs partitioned into 46 chromosomes (Fig. 20.7). Each chromosome averages, therefore, $1.1 \cdot 10^8$ base pairs. This long string is folded around successive groups of basic proteins called *histones* (8 histones form a core, with 140 base pairs around it and link to the next core; this forms a unit called the *nucleosome*, which contains 150–240 base pairs, Fig. 20.8). The string of nucleosomes is then folded again and again. In the double helix, the distance along the axis between one base pair and the next is $3.4 \cdot 10^{-10}$ m (see Fig. 19.13B). The total length of the human genome is therefore $5.3 \cdot 10^9 \times 3.4 \cdot 10^{-10} =$

1.8 m. Humans have 46 chromosomes, and so the average length of a chromosome should be $1.8/46 = 3.9$ cm. The histone folding reduces this length to $3.9/140 = 280$ μm, and further folding reduces it to the observed 5–10 μm. The fruit fly has 8 chromosomes, and Radiolaria have more than 800.]

The *nucleolus* is a cluster of chromatin, often from different chromosomes, that appears during duplication; it manufactures rRNA (ribosomal RNA) from chromosomal DNA.

The *ribosomes*, the sites of protein synthesis, are exceedingly small bodies ($2.5 \cdot 10^{-8}$ m across, equal to 250 atomic diameters) consisting of rRNA and protein (manufactured in situ by rRNA). There are 10^6–10^7 ribosomes

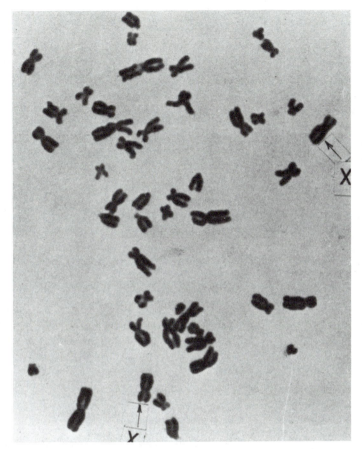

Figure 20.7 (*cont.*). (B) Female chromosomes, characterized by the XX pair. (Tijo and Puck, 1958, p. 1229.)

per cell. They are the sites of protein synthesis via mRNA and tRNA.

The *endoplasmic reticulum* is a highly folded extension of the nuclear membrane into the cytoplasm to which ribosomes are attached. It is the site of lipid and membrane manufacture, in addition to the proteins manufactured by the ribosomes.

The *Golgi apparatus* is a stack of vesicles whose function is to transfer proteins from the endoplasmic reticulum to the cell membrane to make cell walls and for export. The Golgi

Figure 20.8. The folding of the double helix around histone cores and DNA linkage from histone core to histone core.

apparatus, named after the Italian histologist Camillo Golgi (1843–1926), who discovered it in 1898, is found in nearly all eucaryotic cells (the mammalian red blood cells are an exception).

Mitochondria, 2–3 μm across, are the sites where food is oxidized. The mitochondria have their own DNA and may have derived from symbiotic procaryotes.

Chloroplasts, about 10 μm across, occur only in photosynthesizing protists and plants, not in bacteria or animals. They contain chlorophyll in flattened, folded membranes stacked in 20–50 layers, called *thylacoids*. Thylacoids form *grana*, several of which form a chloroplast. These may be one to several thousand chloroplasts per cell. Chloroplasts have their own DNA and, like mitochondria, may have derived from symbiotic procaryotes.

Flagella are used for locomotion. Each consists of a ring of nine pairs of microtubules surrounding a pair of microtubules. Microtubules consist of fibrous proteins. Flagella arise from a *basal body*, which has its own DNA. They too may have derived from symbiotic procaryotes.

In eucaryotes, the genome is too long to be coiled into a single molecule, and therefore it is broken up into chromosomes. In sexually reproducing organisms, there are two homologous chromosomes, one from one parent and one from the other. The two genes for the same character in homologous chromosomes are called *alleles*. The process of cell reproduction is far more complicated in eucaryotes than in procaryotes. There are two different processes: one, called *mitosis*, for body cells; the other, called *meiosis*, for sex cells.

In the simple cell division (binary fission) of procaryotes, the chromosome attaches itself to the cell membrane and duplicates, and the two chromosomes are sequestered on the two sides of the cell as the cell elongates and divides (Section 19.6). Among the dinoflagellates (Section 20.6), which are eucaryotic, the chromosomes are permanently attached to the nuclear membrane. During mitosis, the chromosomes duplicate and are partitioned in the two halves of the nucleus as the nuclear membrane elongates and the cell fissions. This kind of duplication is transitional between that of procaryotes and that of more advanced eucaryotes (green algae and up). Fully developed mitosis and meiosis are summarized as follows:

Mitosis (Fig. 20.9)

1. Chromosomes uncoil.

2. The DNA in each chromosome uncoils, replicates, and forms a new, double chromosome consisting of two identical parts (*chromatids*) attached to each other by a specialized chromatin granule called a *centromere*.

3. At the same time, the *centrioles* duplicate, and each pair of daughter centrioles moves toward the opposite poles of the cell (the centrioles are short—0.2 μm—cylindrical, extranuclear organelles consisting of a bundle of nine triplets of microtubules around a central cavity; they form during cell replication).

4. The nuclear membrane dissolves, and the chromatid pairs, joined at the centromere, move to the equatorial plane of the cell.

5. A set of microtubules forms a *spindle* radiating from the centrioles (the cells of higher plants have no centrioles, but the spindle forms anyway).

6. Some of the microtubules bind to the centromeres.

7. The opposite chromatids are pulled to the opposite poles of the cell.

8. A constriction appears at the equator.

9. The cell divides.

The result of mitosis is the production of two identical daughter cells. The process takes 1 to 20 hours.

Meiosis (Fig. 20.10)

1. Same as in mitosis.

2. Same as in mitosis.

3. Chromatid pairs from homologous chromosomes pair to form *tetrads*.

4. DNA sections are exchanged between homologous chromatids in a process called *crossing over* or *meiotic recombination*; sections of specific length are exchanged,

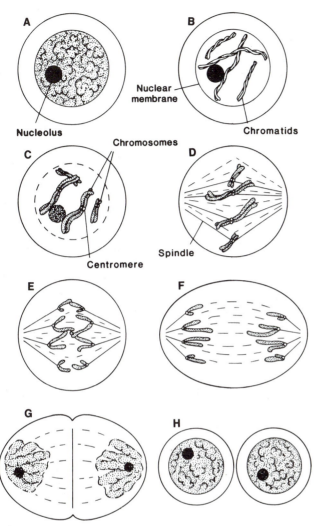

A

B

Nuclear
membrane

Nucleolus

Chromatids

Chromosomes

C

D

Centromere

Spindle

E

F

G

H

Figure 20.9. The process of mitosis (see text).
(Emiliani, 1988, pp. 152–3, Fig. 9.17.)

leaving the gene sequence between breaks intact (this explains *gene linkage*); the hybrid chromatids are called *recombinants*.

5. The nuclear membrane dissolves, and the tetrads move to the equatorial plane of the cell.

6. The spindle forms (as in step 5 of mitosis).

7. Some of the spindle microtubules bind to the centromeres.

8. The opposite tetrads are pulled toward the poles of the cell.

9. A constriction appears along the equator.

10. The cell divides.

11. The chromosomes in each cell again align themselves along the equator of the cell.

12. A new spindle forms.

13. Some of the spindle fibers bind to the centromeres.

14. The centromeres divide.

15. The opposite chromatids are pulled toward the poles of the cell.

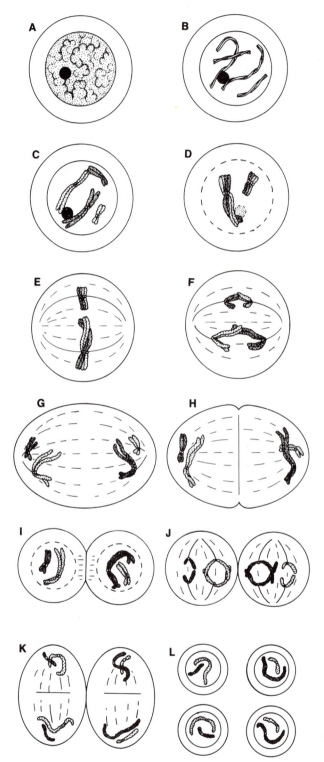

Figure 20.10. The process of meiosis (see text). (Emiliani, 1988, pp. 154–5, Fig. 9.18.)

Figure 20.11. Early eucaryotes from the Late Proterozolc of the Grand Canyon. (Courtesy Trevor Ford, University of Leicester, England.)

16. A constriction appears along the equator.

17. The cell divides.

In meiosis, the two genomes (one from father and one from mother) have been mixed and split into four by two successive cell divisions. Four *haploid* gametes are produced (i.e., gametes with a single set of alleles). Sexual reproduction mates two haploid gametes (one from father and one from mother) to form a *diploid* individual (i.e., an individual with a double set of alleles).

Protophyte and protozoan cells are mostly 10 μm to 2 mm in size, while metazoan cells are mostly 0.1–10 μm. The early eucaryotes from the Late Proterozoic deposits of the Grand Canyon shown in Fig. 20.11 were once globose, single-called planktic algae 1.5–2 mm across. A very attractive theory advanced by Lynn Margulis (b. 1938) says that the eucaryotic cell evolved by symbiotic assimilation of procaryotes that went to form chloroplasts, mitochondria, and flagella, all of which have their own DNA.

THINK

* There is little doubt that the evolutionary step that brought about the eucaryotic cell from pro-caryotic cells was the biggest step in the long history of life on earth. Obviously, it was not a single step, but a long series of steps with many flase starts. Assuming that the symbiotic theory of Lynn Margulis is correct, can you arrange all organelles in the order in which they may have appeared?

20.4. GENETICS IN A CAPSULE

All the information needed to produce an organism is stored in its DNA.

The *gene* is the unit of genetic information, responsible for a given physical or chemical characteristic of the phenotype. It is a specific section of a DNA molecule, called the *cistron*, which corresponds to a specific protein. Common proteins consist of 130–630 amino acids. The cistrons for these proteins consist, therefore, of 390–1,890 nucleotides (3 per amino acid). mRNA may consist of a single cistron (*monocistronic mRNA*), to encode a single protein, or several cistrons (*polycistronic mRNA*), to encode several proteins, which often belong to a specific metabolic pathway. The genes responsible for the synthesis of proteins are called *structural genes*. There are also *regulatory genes*, which are responsible for regulating the synthesis of proteins.

Procaryotes reproduce asexually, but occasionally some genetic material is exchanged in the process of conjugation (Section 19.6). Eucaryotes reproduce mainly sexually, but many reproduce asexually or alternate between sexual and asexual reproduction. Sex was invented by the green algae some $1.7 \cdot 10^9$ y ago (thank you, algae!). Sexual reproduction is governed by Mendel's laws.

Gregor Mendel (1822–1884) was an Austrian botanist and a monk in the Abbey of Saint Thomas in Brünn (now Brno) in Moravia. He conducted breeding experiments on pea plants between 1856 and 1864 and discovered that physical characteristics caused by single genes were transmitted from generation to generation *without blending*. Specifically, Mendel discovered that for some characteristics (e.g., the height of pea plants), one character (tallness) is dominant over the other (shortness), so that an organism carrying both will exhibit only the dominant one. For other characteristics (e.g., the color of the flower of the snapdragon plant) be found that one character (red color) was not dominant over the other (white color), thus producing an organism with an intermediate characteristic (pink flowers). If two organisms with intermediate characters were interbred, however, the original characters would reappear intact in their offspring. Those discoveries led Mendel to formulate his two laws:

Law of Segregation: "There are two alleles for each characteristic, each segregating in a different gamete."

Law of Independent Assortment: "Alleles segregate in different gametes independently of each other" (except for gene linkage).

Mendel's first paper, *Versuche über Pflanzen Hybriden*, was published in 1865 in volume 4, pp. 3–47, of the *Verhandlungen* of the *Naturforschenden Verein* of Brünn. A second paper appeared in the same journal in 1869. In spite of the fact that the journal was distributed in Europe and in the United States, Mendel's work went unnoticed until 1900, when it was simultaneously and independently discovered by three botanists, the Dutch Hugo de Vries (1848–1935), the German Karl Correns (1864–1933), and the Austrian Erich Tschermak (1871–1962). It was then realized that Mendel's

discovery solved one of the two major problems besetting Darwin's theory of evolution—how the environment could select the best "variety" if the "varieties" kept interbreeding and producing intermediate types that would produce even more intermediate types, and so on. (The other problem—where are the missing links?—is still unsolved; see Section 20.5.)

Sex

Sex is the exchange of DNA material between two cells and the incorporation of the exchanged material into the cell's genome. It is a mechanism for increasing genetic variability among the offspring produced by one generation and is in turn responsible for producing the next generation. Consider one individual having two identical alleles *AA* for a given characteristic and another individual having the two alleles *aa* for the same characteristic. Asexual reproduction would continue producing *AA* and *aa* individuals. Sexual reproduction, on the other hand, will create a third type, namely *Aa*. The human genome consists $5.3 \cdot 10^9$ nucleotide pairs. A gene averages 1,200 nucleotide pairs (for an average protein consisting of 400 amino acids), which could be taken to indicate that the human genome should contain about $4.4 \cdot 10^6$ genes. That is not so, however, because long sequences in the DNA chain within a gene, called *introns*, are not translated into mRNA and therefore do not function as carriers of genetic information. It is estimated that humans have about 10^5 genes (pairs of alleles). Even so, it is obvious that sex can produce an enormous number of genetic types, and in fact, except for monozygotic twins, there are no two individuals who are genetically identical. (Monozygotic twins are twins arising from a single *zygote*, which is the name for the fertilized egg.)

Sexual reproduction is never perfect. Errors of varying magnitudes are made, creating mutations. The smaller ones, called *point mutations*, often are reversible. An example is base-pair mismatch that can be self-corrected by means of the enzyme DNA polymerase (see Section 19.3).

Chromosomal mutations are changes in chromosomal structure or in chromosomal

Table 20.2. Mutation rates

Organism	Factor	Mutation rate/gene
Bacteria	sensitivity to antibiotics	10^{-6} to 10^{-8}
Chlamydomonas	sensitivity to streptomycin	10^{-6}
Corn	point mutations	$5 \cdot 10^{-4}$ or less
Drosophila	eye color	$4 \cdot 10^{-5}$
Humans	various genetic defects (dwarfism, hemophilia, etc.)	10^{-4} to 10^{-5}

number. Changes in chromosomal structure involve duplications, deletions, inversions, or translocations of DNA segments. Changes in chromosomal number usually involve loss of an entire chromosome, giving rise to severe, usually lethal, mutations.

Haploid organisms have one set *n* of chromosomes. Diploid organisms, deriving from two parents by sexual reproduction, have a double set (*2n*) of chromosomes. Sometimes the chromosomes in the spindle phase of either mitosis or meiosis do not separate properly. Organisms with *3n*, *4n*, *5n*, or even *6n* sets of chromosomes can form, a state called *polyploidy*. The cells and the bodies of polyploid organisms usually are larger than those of the normal, *2n* individuals. Polyploidy is induced artificially in plants (corn, etc.) in order to profit from the larger size.

Mutations that affect the phenotype are called *somatic*. *Germinal mutations* are mutations that affect the sex cells and therefore all subsequent generations. Mutation rates vary by a factor of 10,000, depending upon the type of organism and the particular gene or set of genes involved (Table 20.2). Because the human genome consists of $\sim 10^5$ genes, each human carries 1–10 mutations within himself or herself. Most of these are hardly noticeable point mutations.

Genetic Variability

Mutation and recombination result in the production and maintenance of a standing crop of variant genomes within a given species. Highly variable species are called *polytypic*. Variability ensures the survival of a species should the physical, chemical, or trophic environment change (within reasonable limits).

Monotypic species are species with similar genomes and therefore uniform phenotypes (e.g., blind fish in caves). Monotypic species are susceptible to extinction should the environment change.

THINK

* Listen to what Adolf Hitler (1889–1945), who incorporated his version of genetics into government policy, had to say:

> Jegliche Rassenkreuzung führt zwangsläufig früher oder später zum Untergang des Mischproduktes, solange der höherstehende Teil dieser Kreuzung selbst noch in einer reinen irgendwie rassenmässigen Einheit vorhanden ist. Die Gefahr für das Mischprodukt ist erst beseitigt im Augenblick der Bastardierung des letzten höherstehenden Rassenreinen. Darin liegt ein, wenn auch langsamer natürlicher Regenerationprozess begründet, der rassische Vergiftungen allmählich wieder ausscheidet, solange noch ein Grundstock rassisch reiner Elemente vorhanden ist und eine weitere Bastardierung nich mehr stattfindet. (*Mein Kampf*, Part II, Chapter 2)

Anything wrong with this piece? (If you do not know any German, grab a German–English dictionary and translate the piece—it will be a good exercise for you.)

20.5. EVOLUTION AND EXTINCTION

The fundamental unit in taxonomy is the species. There are many definitions of *species*, all of them necessarily loose. Carl von Linné (1707–1778), who placed taxonomy on a solid scientific foundation, defined species (in 1738) as follows: "*Species tot sunt quot diversas formas ab initio produxit infinitum Ens,"* meaning "the species are all those diverse forms that the Infinite Being produced initially."

Ernst Mayr (b. 1904), professor at Harvard, defined species (in 1942) as follows: *"Species are groups of actually or potentially interbreeding natural populations, which are reproductively separated from other such groups."* This definition does not apply to asexually reproducing organisms.

I now propose the following definition: *"A species is the set of organisms whose genomes are sufficiently similar that exchange of genes or sets of genes could take place without changing the aspect of the species."* This definition applies to both asexually and sexually reproducing organisms.

The definition of species is necessarily vague, though. In fact, in asexually reproducing organisms, the identification of a species depends on the stability of its genome through time. As point mutations accumulate, the genome drifts. At what point is the drift sufficient to produce a new genome? In sexually reproducing organisms, the identification of a species depends upon the chemical compatibility of two different genomes. Compatibility is in fact the foundation of the modern definitions of species.

Ernst Mayr, in his famous book *Systematics and the Origin of Species*, published in 1942 by the Columbia University Press, illustrates the inevitable vagueness of the concept of species using the warbler species *Phylloscopus trochiloides*, which occupies a ring of habitats all around Tibet. Beginning on the west and proceeding counterclockwise around Tibet, one encounters the following interbreeding subspecies: *viridianus → ludlowi → trochiloides → obscuratus → plumbeitarsus* and back to *viridianus*. However, there is no interbreeding between *viridianus* and *plumbeitarsus*. In other words, starting with *viridianus*, *Phylloscopus trochiloides* is one species going counterclockwise, but two species going clockwise.

Speciation is the process by which new species come into existence. According to Darwin, species come into existence because (1) natural populations are variable; (2) the environment continuously and slowly changes; (3) the changing environment selects the most suitable variety; and (4) continuing change

eventually produces a variety sufficiently different from the original to be called a new species.

The first point is true. However, in Darwin's time it was generally believed that sexual reproduction would produce intermediates between the parents. A Scottish engineer attempted to demonstrate to Darwin that variable, randomly mating natural populations would reduce to uniformity within a finite number of generations. Darwin had no answer to that argument. Mendel's work, showing that genetic traits are transmitted undiluted from generation to generation, would have invalidated the argument of the Scottish engineer. Unfortunately, Darwin, like everybody else, failed to take notice.

The second and third points are also true, but because environmental change is generally slow, it would take a long time for a series of intermediate "varieties" to be produced, ending finally with a new species. Those intermediate types, the famous *missing links*, were not to be seen in the fossil record. Where were they? Darwin's answer was "keep digging": If we dig up enough fossils, we shall find the missing links. Now, 130 years after Darwin, the missing links are still missing. Yet, because the environment is continuously changing and because its change is slow, when a species changes from A to B there should be a lot of intermediate types; in fact, there should be a lot more intermediate types than there are boundary types A and B. Indeed, this is clearly seen across space: There are comparatively few pure white Scandinavians and comparatively few pure black Nilotics, but there are lots of intermediate types, stretching from Scandinavia to Sudan through Germany, Switzerland and Austria, Italy, Sicily, and Egypt.

Neo-Darwinism is the "modern" theory of evolution that incorporates modern genetics and a few accruements. It runs as follows:

1. Natural populations are variable (because mutations keep them so).

2. Environmental changes occur on time scales of a century or less (secondary climate cycles), millennia to tens of millennia (ice ages), or tens of millions to hundreds of millions of years (tectonism, plate motions).

3. The environment selects the fittest mutant.

4. A new species may appear in a very short time ($< 10,000$ y) if a small population becomes isolated from its stock; the new species may then return to the territory occupied by the stock and may (a) wipe out the stock; (b) be wiped out by the stock; or (c) coexist with the stock if not competing with it.

Neo-Darwinism solves the missing-link problem by claiming that speciation occurs in small populations in limited, isolated areas, greatly reducing the chances of leaving a detectable or discoverable fossil record.

Recently, two new theories of evolution have been proposed: *punctuated equilibrium* and *extinctive evolution*. The punctuated-equilibrium model maintains that a species keeps its identity during a long period of stasis, and then its genome suddenly scrambles into a new type (thereby forming a new species).

The extinctive-evolution model maintains that established populations exhibit considerable inertia to change and change only if environmental threshold conditions are violated. Peripheral to the established populations are marginal populations living under conditions of continued environmental stress. These populations are the storehouse of genetic variability. Occasionally the evolution of a virus that fits the genome of an established species will lead to extinction of that species and its replacement by a suitable, marginal species if one is available. The new species will not necessarily be better adapted than the extinct species. If no suitable species is available from the marginal environment, the niche vacated by the extinct species remains vacant, and diversity is reduced. Extinction is abrupt ($< 1,500$ y for marine plankton). Speciation is not abrupt, but the appearance of a new species in the stratigraphic record is abrupt. According to this model, the "missing links" have never existed in the environments occupied by established populations.

The extinctive-evolution model requires the chance evolution of a virus fitting a specific genome. That should be easy, for viruses mutate a million times faster than cellular organisms—procaryotes or eucaryotes—because viruses lack mending enzymes to correct errors made during nucleic acid duplication. In addition, the virus must be able to spread rapidly throughout the range of the genome under attack.

Localized exterminations, (i.e., the disappearance of a species from a local habitat) are common events, but the species survives elsewhere and eventually returns. The broader the extermination, the rarer the event is. Worldwide extermination (true extinction) is the rarest. The scale of extermination thus follows a Poisson distribution.

Extinctive evolution predicts that the greater the number of individuals in a species, the greater the probability that a suitable virus will make contact and invade the species' genome. Indeed, the record of marine plankton demonstrates that a species disappears abruptly, worldwide, just when the species is at the peak of its success and without even a trace of discomfort to the dozens of similar species living in the same environment. The grim reaper is clearly species-specific, which viruses are. The most abundant species living today in the world ocean is the coccolithophorid *Emiliania huxleyi* (Fig. 23.10), with a standing crop of about 1 avogadro of cells. The total, worldwide number of living human cells is similar, also about 1 avogadro. Extinctive evolution predicts that both *Emiliania huxleyi* and *Homo sapiens* are candidates for extinction. Perhaps the current AIDS crisis is an example of the process of extinctive evolution at work (see Section 24.3).

The four models discussed above refer to the standard mode of evolution. Whatever model one adopts, the process of evolution is extremely effective. In four billion years it has made possible the writing of this book.

Occasionally, strong environmental upsets occur that scramble a major portion of the biosphere (or all of it). These events are the famous *mass extinctions*, characterized by simultaneous extinctions of a number of ecologically and trophically unrelated taxa. Mass extinctions have occurred from time to time in the past. In fact, the boundaries between successive geological periods are based on the disappearance of older species and the appearance of new ones. The largest mass extinctions occurred at or near the end of the Cambrian ($515 \cdot 10^6$ y ago), at the end of the Ordovician ($439 \cdot 10^6$ y), in the late Devonian

$(367 \cdot 10^6 \text{ y})$, at the end of the Permian $(245 \cdot 10^6 \text{ y})$, at the end of the Cretaceous $(65 \cdot 10^6 \text{ y})$, and during the late Quaternary (the past 50,000 y). In each case, except the Permian and Quaternary, mass extinction occurred when marine ingression was at a peak. This means that the midocean-ridge systems were very active and were spewing gases, which commonly include CO_2. A vast outpouring of basalt in northeastern Siberia at the end of the Permian (see Table 10.10) also produced a lot of CO_2.

All mass extinctions, except the Permian and Quaternary, affected mainly tropical marine faunas and floras of shallow habitat. It is possible that a runaway greenhouse effect overcame the ability of the biosphere to absorb excess CO_2, and temperature rose a few degrees. That would have been enough to cause mass extinctions in the tropical biosphere, because most of the tropical biosphere is already living close to the lethal threshold. The American alligator, for instance, has a body temperature of $36°C$. A temperature rise of only $2°C$ is sufficient to kill it. Our own body temperature is $37°C$, and we would die if our body temperature were raised by $2°C$ for any length of time (for example, by being accidentally locked inside a sauna). The enzymatic processes on which life rests work best just below the temperature at which the system will collapse.

The mass extinction at the end of the Cretaceous was exacerbated by an asteroidal impact (Section 23.6). The Earth was enveloped in steam, dust, and smoke. Water that splashed into the troposphere rained out the dust. The water that splashed into the stratosphere, together with the CO_2 produced by the burning of forests and soils and carbonate bedrock, added to the ongoing greenhouse effect. That may have been the coup de grace for those brave taxa that were attempting to survive the greenhouse effect.

The most recent mass extinction started 50,000 years ago in the Old World and spread to the Americas 10,000 years ago. More than 200 genera of large animals were exterminated. The cause? Humans. Today, human activities are still leading to the extinction of increasing numbers of taxa. It is likely that the ongoing mass extinction, marking the end of the Cenozoic, will be the largest of all.

THINK

* *Homo sapiens* evidently is a single species, because it can interbreed across the entire racial spectrum. Yet it obviously consists of a rich ensemble of subspecies, each one identifiable by specific physical characteristics. The type subspecies from which all others apparently originated is the African subspecies, which one could call *Homo sapiens africanus*. Using Latin names, can you name the other major subspecies, the Caucasian, the Oriental, the Australian, and the American Indian?

* *Subspecies* and *race* refer to the same concept — genomic identity of a population within a species. Yet the name *race* has a bad connotation, while *subspecies* doesn't. Why?

* *Psychospecies* and *psychosubspecies* (two names that I just invented) need to be defined. See if you can do it, and provide examples.

20.6. DINOFLAGELLATA, CHLOROPHYTA, AND EUGLENOPHYTA

The dinoflagellates, known since the Paleozoic, are single-cell algae $5\,\mu m$ to $2{,}000\,\mu m$ in size, with the body usually covered with a stiff cellulose wall. There are two flagella wrapped around an equatorial furrow that divides the body into two parts. The dinoflagellates have a vegetative stage and a resting stage in which the cell is enclosed in a cyst with a resistant, fossilizable wall. More than 2,000 modern species have been described, most of them marine. Dinoflagellates and the taxa discussed in this section and subsequent sections (20.6 through 20.13) are known, or are likely to have originated in the Cryptozoic.

In dinoflagellates, the chromosomes are permanently attached to the internal surface of the nuclear membrane. In cell division, the chromosomes duplicate and are pulled apart and separated when the nuclear membrane elongates and fissions. This type of cell division may be considered intermediate between the simple cell division of procaryotes and mitosis (Section 20.3). It is possible that dinoflagellates were the earliest eucaryotes, although there is no Proterozoic fossil record.

The chlorophytes (green algae) were also early eucaryotes. They evolved about $1.5 \cdot 10^9$ y ago and have remained abundant through geological time. They introduced a number of major innovations, including "vision", mitosis, meiosis (i.e., sex), and respiration.

Eucaryotes have cells with organelles, including a nucleas with the DNA partitioned into chromosomes. Reproduction is by mitosis or meiosis, and respiration is aerobic in most taxa. Eucaryotes use enzymes (catalases, peroxidases, and superoxide dismutases) that reduce dangerous oxygen compounds (e.g., H_2O_2) to H_2O.

A typical modern green alga is *Chlamydomonas*, with a broadly ellipsoidal body and two flagella (Fig. 20.13). *Chlamydomonas* has an eyespot, a red pigmented spot that is sensitive to light. It has a chloroplast that occupies 50% of its internal space; the chloroplast has embedded in it a spherical body, the *pyrenoid*, that contains enzymes that catalyze the conversion of sugar to starch, which can be stored. Other organelles are the nucleus (with DNA packaged into chromosomes), the nucleolus, mitochondria, the endoplasmic reticulum with its ribosomes, several Golgi bodies, and a contracting vacuole at the base of each flagellum. Reproduction is by alternation of generations—asexual reproduction by mitosis, and sexual reproduction by meiosis. In sexual reproduction, two cells make contact via flagella, which carry sex-distinguishing factors and then fuse into a single, larger cell with four flagella, two sets of chromosomes, and two sets of organelles. Immediately after fusion, the zygote absorbs the flagella, turns black on the outside, and enters a resting stage (zygospore) that lasts four days during which meiosis takes place. After that the zygospore opens, and four haploid baby cells emerge.

Green algae are aerobic. Aerobic respiration occurs in three stages. First there is glycolysis, a process by which glucose ($C_6H_{12}O_6$) is broken down to two molecules of pyruvic acid ($CH_3-CH-COOH$), with the net production of 2 ATP from 2 ADP, and of 2 NADH from 2 NAD (Section 19.4). Glycolysis is followed by the *Krebs cycle* (also called *citric acid cycle*). A molecule of pyruvic acid is oxidized to the acetyl radical CH_3-CO-, which is then attached to coenzyme A (CoA) to form acetyl coenzyme A (acetyl CoA). CoA lets the acetyl group into the Krebs cycle, in the course of which the two carbons in the pyruvic acid molecule are oxidized to CO_2, and the energy is used to form 1 ATP, 3 NADH, and 1 FADH (FADH is reduced FAD, *flavin adenine dinucleotide*, a compound with a function similar to those of NAD and NADP; see Section 19.4). Because two molecules of pyruvic acid are produced by the glycolysis of one molecule of glucose, one molecule of glucose ends up producing, via the Krebs cycle, 2 ATP, 6 NADH, and 2 FADH.

The Krebs cycle is followed by an electron-transfer chain in which the 3 NADH and the 1 FADH release $2e^-$ each down a chain of cytochromes, forming up to 36 ATP from 36 ADP and terminating by reducing O to O^{2-}, which captures $2H^{2+}$ to form H_2O. Adding the 2 ATP formed by the glycolysis preceding the Krebs cycle brings the total of ATP formed to 38. The 38 ATP produced from the 38 ADP by the respiration of 1 glucose molecule correspond to a total of 12 eV of stored energy (0.3 eV per ATP bond).

Green algae have been recorded from several Cryptozoic formations (Table 20.3). Green algae include both unicellular and multicellular species, as well as transitional species. Today, green algae number more than 9,000 species, most of them in marine or freshwater hatitats. Other habitats include soil, damp wood, deserts (as cysts deposited by wind and waiting for the rain), and the surfaces of glaciers and snowfields. Most species of green algae are microscopic, consisting of single cells or thin filaments. Some are quite large. *Valonia*, a tropical species, consists of an ovoidal, multi-nucleated cell a few centimeters across. *Ulva*, another tropical species, has a broad, leafy thallus, with folia a few centimeters across. *Codium magnum* (Gulf of Mexico) may reach a width of 20 cm and a length of several meters. *Acetabularia* is a stalked, unicellular alga several centimeters long anchored to the substrate by a rootlike portion of the stalk that contains the nucleus (Fig. 20.12).

The Volvocales are an order of single-cell or colonial mononucleated green algae that are normally flagellated and motile. The order includes the single-cell *Chlamydomonas* (Fig. 20.13) and the colonial *Gonium, Pandorina*,

Table 20.3.   The Proterozoic record of Chlorophyta

Morphology	Location	Age (y)
single cells with remnants of organelles	Bungle-Bungle dolomite, northern Australia	$1.5 \cdot 10^9$
tetrahedral groups of cells (mitotic division?)	Amelia dolomite, northern Australia	$1.5 \cdot 10^9$
multicellular filaments with large cells	Beck Spring dolomite, eastern California	$1.3 \cdot 10^9$
multicellular algae with stems and branches	Belt Supergroup, Montana	$1.3 \cdot 10^9$
very large (2 mm across) spheroidal unicells	Utah shales	$950 \cdot 10^6$
unicells with remnants of organelles	Bitter Springs formation, central Australia	$850 \cdot 10^6$
flask-shaped microfossils	Kwagunt formation, eastern Grand Canyon	$800 \cdot 10^6$
large (100 μm) ellipsoidal unicells	Grand Canyon shales	$800 \cdot 10^6$
multicellular filaments	Olkhin formation, Siberia	$725 \cdot 10^6$

and *Volvox*. These colonial types form a transition between unicellular and multicellular organization.

Gonium consists of only 4 to 16 biflagellated cells (depending upon the species), arranged in a plane. Flagellar motion is coordinated. Each cell divides and produces a new colony.

Pandorina (Fig. 20.14A) consists of 16 or 32 biflagellated cells in ellipsoidal shape. Flagellar motion is coordinated, making the colony roll through the water. The colony is polar, with larger eyespots at one end of the colony. Upon aging, the colony sinks to the bottom. Each cell divides and forms a new colony inside the mother colony, which then breaks apart, releasing the new colonies. *Eudorina* (Fig. 20.14B) differs from *Pandorina* in having the cells separated from each other (but still connected by protoplasmic filaments) inside a common gelatinous cytoplasmic envelope called *coenosarc* (from the Greek κοινός, meaning *common* and σάρξ, meaning *flesh, tissue*).

Volvox (Fig. 20.14C) is a hollow sphere 0.5–1.5 mm across, consisting of 500–50,000 *Chlamydomonas*-like biflagellate cells 20 μm across embedded in a coenosarc. The flagella beat in unison so as to spin the sphere around. Each cell has a photoreceptor-eyespot combination. Eyespot size decreases from the an-

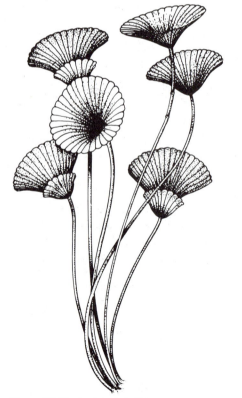

Figure 20.12. *Acetabularia.* (Moment and Habermann, 1973, p. 255, Fig. 10.7.)

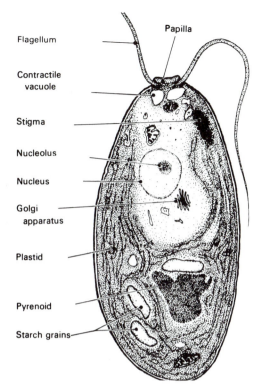

Flagellum

Papilla

Contractile
vacuole

Stigma

Nucleolus

Nucleus

Golgi
apparatus

Plastid

Pyrenoid

Starch grains

Figure 20.13. *Chlamydomonas.* (Moment and Habermann, 1973, p. 251, Fig. 10.2.)

and function and thus are intermediate between protophytes and metaphytes.

The evolution of protophyta to protozoa is exemplified by the modern protist *Euglena* (Fig. 20.15), which is a protophyte with protozoan characteristics. *Euglena* has a fusiform body 10 μm long, with a single anterior flagellum 5 μm long (plus a rudimentary one that does not emerge from the cell). The flagellum has a contracting vacuole at its base. The cell

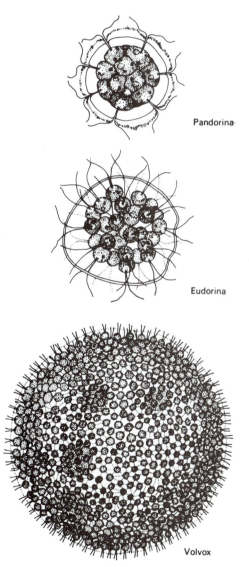

Pandorina

Eudorina

Volvox

Figure 20.14. The Volvocales: *Pandorina* (45 μm across), and *Volvox* (0.3 mm across). (Moment and Habermann, 1973, p. 252, Fig. 10.3.)

terior to the posterior side of the colony, and the colony moves as whole toward light (or away from it if the light is too intense). Most cells specialize in nutrition. As the colony matures, 2–50 posterior cells enlarge, lose their eyespots and flagella, and multiply, forming an invagination toward the interior, with daughter-cell flagella pointing toward the inside of the invagination; the invagination is everted (flagella pointing toward the outside of the invagination) and then turned loose to form a daughter colony inside the mother sphere. Other cells release male or female gametes that mate inside the mother sphere and form a zygote that develops a thick, spinose wall. After a few days, during which meiosis takes place, a single zoospore is released. The zoospore multiplies and forms a new colony inside the mother sphere. Eventually the mother sphere breaks apart, and the daughter spheres are released. The Volvocales show incipient cell differentiation in both form

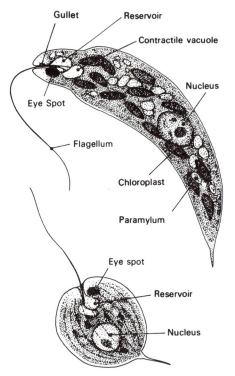

Figure 20.15. Euglenoids: *Euglena* (above, length 70 μm) and *Phacus* (below, diameter 25 μm). (Moment and Habermann, 1973, p. 257, Fig. 10.11.)

has an outer membrane, under which is a *pellicle* composed of proteinaceous stripes that allow peristaltic motion of the whole cell (*peristaltic* means *wavelike*). The cell wall contains no cellulose. Inside the cell there is one nucleus, with nucleolus and numerous chloroplasts with chlorophyll a and b, carotene, and xanthophylls (yellow carotenoids).

Euglena has a photoreceptor with an eyespot above—the eyespot acts as a shield that casts a shadow on the photoreceptor unless *Euglena* is facing toward light. The products of photosynthesis are stored as *paramylum* (a sugar, $C_5H_{10}O_5$) instead of starch. Reproduction is asexual, by mitosis. Under unfavorable conditions, *Euglena* forms thick-walled resting cysts.

The absense of cellulose, the presence of paramylum, the (practically) single flagellum, and the peristaltic motion are protozoan characteristics. In addition, *Euglena* loses its chloroplasts if placed in darkness and does not regain

them if returned to light. Some species can live without chloroplast by absorbing nutrients from the environment, like protozoans. These species become protozoans simply by experiencing a period of darkness. Apparently, protozoans can readily evolve from protophytes.

THINK

* Imagine that you are a little *Chlamydomonas*-like green alga happily swimming around and eying your Proterozoic world. What would you see?

* Which political system do the colonial Volvocales best represent?

20.7. PORIFERA

The Porifera (sponges) are intermediate between protozoa and metazoa. They are sessile, benthic filter-feeders. Sponges can be sessile or encrusting. The body structure of a free sponge is essentially a sac with an opening on top (*osculum*) leading into a hollow cavity (*paragaster*) (Fig. 20.16). The outer wall has pores (*ostia*) either leading directly into a choanocyte-lined paragaster (*ascon* type) or leading into simple (*sycon* type) or complex (*leucon* type) choanocyte-lined chambers within the wall that empty into the paragaster. A choanocyte is a cell with a collar from which a flagellum protrudes. Most sponge species, except those belonging to the Stromatoporoidea, reinforce their tissues with spicules, which can be calcareous or siliceous.

Encrusting sponges include the extinct Stromatoporoidea and the modern Sclerospongiae. The Stromatoporoidea ranged from the middle Ordovician to the late Devonian. They reappeared in the Jurassic and became extinct at the end of the Cretaceous. The Stromatoporoidea formed crusts of calcium carbonate consisting of layers of minute cells, about 0.3 mm across (Fig. 20.17). In Devonian time, stromatoporoid reefs up to 20 m thick and 200 m long were built in high-energy coastal environments. The stromatoporoid surface is perforated by tiny inhalant pores (*ostia*, 0.4 mm across) and by larger, exhalant

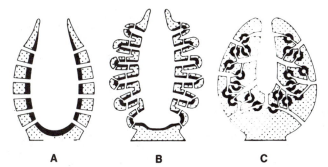

Figure 20.16. Structure of solitary sponges: (A) ascon, (B) sycon, (C) leucon. (Lehmann and Hillmer, 1983, p. 6, Fig. 27.)

pores (*oscula*, 0.6 mm across) at the centers of star-shaped systems (*astrorhizae*) of surface grooves that extend 5 mm across.

The Sclerospongiae form mound-shaped aragonitic masses 1–50 cm across in cavities within reefs. Their surface is similar to that of the Stromatoporoidea, including astrorhizae. Water is pulled in by choanocytes lining the ostia and is expelled through the oscula (filter-feeding).

The living tissue of encrusting sponges occupies the outer few millimeters of the skeletal structure. Sponges exhibit significant cell differentiation. The major cell types are as follows:

Pinacocytes: epithelial cells lining the outer body surface in simple sponges (ascon type), or both the outer surface and the paragaster in larger, more complex sponges
Archaeocytes: amoebocytes and other free cells between the inner and outer wall surfaces that carry food from choanocytes to other cells or secrete sponge spicules of $CaCO_3$ or $SiO_2 \cdot nH_2O$
Choanocytes: flagellated cells lining the inner

body surface (ascon type), simple chambers (sycon type), or complex chambers (leucon type); choanocytes not only move water from the outside in but also absorb the food particles which are then passed around to the archaeocytes and the epithelial cells
Porocytes: cylindrical cells forming pores
Sclerocytes: cells that form spicules

The skeleton of solitary sponges consists of *spongin* (a fibrous protein) with or without spicules; that of encrusting sponges consists of calcite or aragonite.

Some sponges have the ability to rebuild the sponge structure from small pieces of sponge tissue or even from disaggregated cells. The disaggregated cells can live independently of each other. If dispersed in a fluid, they will reaggregate and reform the sponge.

Sponges can reproduce either asexually or sexually. Asexual reproduction takes place by budding or by forming clumps of archaeocytes with a hardened covering that can survive desiccation (freshwater sponges). In sexual reproduction, archaeocytes produce sperms and eggs that form free larvae with choanocytes

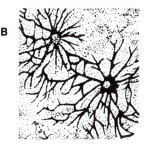

Figure 20.17. *Stromatopora*. (A) encrusting mass (5 cm across). (B) Astrorhizae (0.5 mm across). (C) Internal structure, showing laminae and pillars (× 15). (Lehmann and Hillmer, 1983, p. 60, Figs. 41 and 42.)

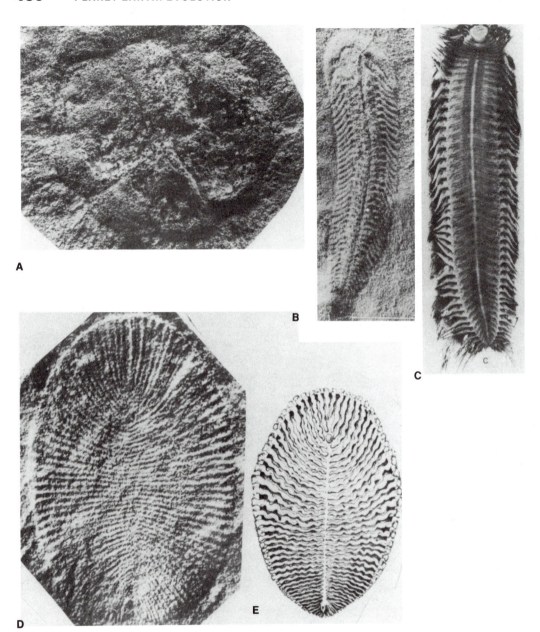

Figure 20.18 Ediacaran fauna (with modern comparisons): (A) *Conomedusites lobatus* (medusoid, Ediacara, ×0.75). Polychaete worms: (B) *Spriggina floundersi* (Ediacara, ×1.75) and (C) *Laetmonice producta* (modern, ×1). (D) *Dickinsonia costata* (Ediacara, ×0.8). (E) *Spinther alaskensis* (modern, ×0.8). (F) and (G) *Parvancorina* (Ediacara, ×1) and (G) *Tribrachidium* (Ediacara, ×1.5) (both of unknown affinities). (Glaessner, 1984, p. 63, Fig. 2.5, p. 59, Fig. 2.4, p. 58, Fig. 2.4.)

F   G

Figure 20.18 *(cont.)* (F) and (G)

that enable them to swim. The habitat of sponges is mainly marine, but also fresh water. There are five classes:

1. Calcarea: with calcareous spicules (Cambrian–Holocene)

2. Hexactinellida: with six-ray siliceous spicules (Cambrian–Holocene)

3. Demospongiae: skeleton consisting of spongin with or without siliceous spicules (Cambrian–Holocene); Demospongiae are classified into four orders:
   Monaxonida: single siliceous needles (e.g., *Cliona*)
   Tetractinellida: siliceous tetraxon spicules
   Keratosa: spongin skeleton (e.g., *Euspongia*, the bath sponge)
   Myxospongiae: no skeleton

4. Stromatoporoidea: encrusting, without spicules (Ordovician–Devonian; Jurassic–Cretaceous)

5. Sclerospongiae: encrusting (Holocene)

The sponges exhibit much greater cell differentiation than do the colonial Volvocales. Nevertheless, the ability of individual sponge cells to live independently suggests that sponges are highly differentiated colonial protozoa. They are assigned to the subkingdom Parazoa, intermediate between protozoa and metazoa.

THINK

* Modern coral reefs provide ample shelter for fish, which is the reason that reef fishes are brightly colored and highly visible—when a big, bad predator shows up, the reef fishes dash into nooks and crannies in the reef and easily escape capture. By comparison, open-water fishes cut a low profile in terms of body coloring—they try to look as inconspicuous as possible, which is an excellent idea, because there is no way to hide in the open ocean. Do you think that stromatoporoid reefs in Devonian time provided as good a shelter for the fish as modern reefs do?

20.8. THE EDIACARAN FAUNA

The Ediacaran fauna was discovered in 1947 by Reginald Sprigg in the Ediacara Hills, 300 miles north of Adelaide, South Australia. It has subsequently been found in 20 locations around the world (central Australia, China, Siberia, Russia, Norway, England, Newfoundland, North Carolina, northwestern Canada, Mato Grosso, Namibia). It consisted exclusively of soft-bodied metazoa (no mineralized tissues) living in shallow-water coastal environments. The age of the Ediacaran fauna is latest Proterozoic (590–570·10^6 B.P.).

About 1,500 specimens have been collected

Table 20.4.   Ediacaran fauna: composition

Type	Percentage
Scyphozoa (medusoids—class of phylum Cnidaria)	63
Polychaeta (class of phylum Annelida)	22
Pennatulacea (order of subclass Alcyonaria, class Anthozoa, phylum Cnidaria)	5.5
Parvancorina (dorsal shield of arthropod carapace?)	5
Tribrachidium (threefold symmetry—unknown affinity)	4
Praecambridium (a chitinous arthropod carapace?)	0.5

in the Ediacara Hills along a 3–6 m-thick sand-stone layer 45 m below the earliest Cambrian. Examples of Ediacaran organisms are shown in Fig. 20.18. The composition of the Ediacaran fauna is shown in Table 20.4. Additional types (rare) are Hydrozoa and Conulata, which are two classes of the phylum Cnidaria (see Section 20.9).

The Ediacaran fossils actually are only imprints of the soft bodies flattened between sandstone or shale beds. The beds indicate oxidizing conditions, although the oxygen content of the atmosphere probably was only 2% (equal to 10% of the present amount). The burial and preservation of the soft bodies indicate that there were neither predators that would feed on them nor bacteria yet adapted to those oxidizing conditions.

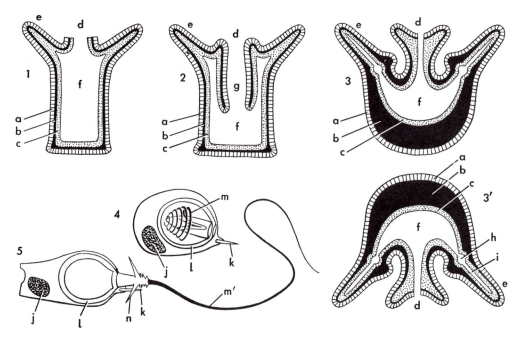

Figure 20.19. Cnidaria: (1) simple polyp (Hydrozoa); (2) more advanced polyp (Anthozoa) with gullet (g); (3) medusa (inverted above, to show similarity to polyp); (4) nematocyte with nematocyst withdrawn and (5) extended; (a) ectoderm; (b) mesoglea; (c) endoderm; (d) mouth/anus combination; (e) tentacles; (f) gastric cavity; (g) gullet; (h) peripheral circulatory canal in medusa that extends into tentacles (i); (j) nucleus of nematocyte cell; (k) sensor hair, which, if touched, causes contraction of the vacuole (l) and discharge of the nematocyst (m); (n) barb at base of filament. (Simpson, Pettigrew, and Tiffany, 1957, p. 526, Fig. 22.3.)

THINK

* If an Ediacaran fauna existed today, in which sedimentary environment could it be preserved? Where can this sedimentary environment be found?

20.9. CNIDARIA (COELENTERATES)

The phylum Cnidaria (or coelenterates) (Fig. 20.19) includes four classes:

1. Hydrozoa (Ediacaran–Recent) (e.g., *Hydra* which has the ability to regenerate its full body even if cut in half)

2. Scyphozoa (Ediacaran–Recent): jellyfish (limited regenerative ability)

3. Conulata (Ediacaran–Triassic): inverted pyramidal, chitinous shells attached at the apex, with a marked quadrilateral symmetry (Fig. 20.20)

4. Anthozoa (Ediacaran–Recent): stony corals, sea anemones, sea fans

In Cnidaria, the mouth opens into a blind body cavity (there is no anus); the mouth functions both as mouth and as anus (rather disgusting, I should think). Cnidaria have *nematocysts*, which are independent organelles inside stinging cells called *nematocytes*. A nematocyst consists of a capsule, one end of which is invaginated, forming a long, coiled tube. The tube shoots out when triggered by touching a sensor on the surface of the cell. There are four types of tubes:

1. Sticky tubes: to hold the coelenterate when moving on a substrate

2. Tubes with barbs: same function as the sticky tubes

3. Grasping tubes: they wrap themselves around a prey

4. Stinging tubes: with spines and toxins, to capture a prey.

Nematocytes occur on the ectoderm of Hydrozoa and on both the ectoderm and endoderm of Scyphozoa and Anthozoa.

Cnidaria have a gelatinous tissue, with no cells between ectoderm and endoderm. This tissue, called *mesoglea*, consists of *collagen*, a

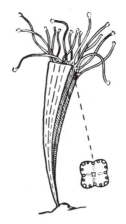

Figure 20.20. Cnidaria (Conulata). Reconstruction of a living adult conularid (× 0.7). (Shrock and Twenhofel, 1953, p. 119, Fig. 4.10.)

protein rich in glycin that is the major constituent of connective tissues and bones (see Section 22.3). Cnidaria have a strong muscular system and a well-developed nerve system. Many cnidarians have pigmented, light-sensitive cells located in structures called *ocelli* (plural of *ocellus*, which means *little eye* in Latin) at the bases of the tentacles.

Hydrozoa and Scyphozoa reproduce by alternation of generations: Polyps produce other polyps or medusae by budding; medusae (which can be male or female) form gametes, which mate and form a fertilized egg (zygote). The zygote undergoes successive cleavages, producing 2, 4, 8, 16, ... cells of smaller and smaller dimensions until a solid sphere of cells is formed. The cells move away from the center of the sphere, forming a hollow sphere of cells (called *blastula*, about 100 μm across), with larger cells on one side (vegetal pole) and smaller cells on the other (animal pole). These cells surround the hollow cavity (*blastocoel*). The blastula fills with cells from the surface layer (ectoderm) and becomes a ciliated *planula*, an ellipsoidal body 200 μm in length that swims around for a while. The planula becomes attached by its front end and develops mouth and tentacles at its rear end, becoming a polyp. The polyp, in turn, produces medusae by budding. In most Hydrozoa the polyp is the dominant form, but in Scyphozoa the medusa is the dominant form.

The life cycle of *Obelia*, a modern hydro-

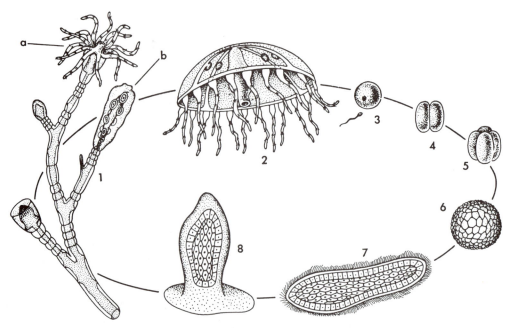

Figure 20.21. Cnidaria (Hydrozoa). Life cycle of *Obelia* (the polyp is the principal stage). A polyp (1) forms a colony by budding. The colony has feeding polyps (1a) and reproductive polyps (1b) that shed medusae (2). Medusae produce gametes (3), which fuse and grow (4–6) into a planula (7). The planula becomes attached and forms a new polyp (1). (Simpson, Pittendrigh, and Tiffany, 1957, p. 527, Fig. 22.4.)

zoan, is shown in Fig. 20.21. Hydrozoa include *Millepora* (Fig. 20.22), a common modern reef-builder. The life cycle of *Aurelia*, a modern scyphozoan, is shown in Fig. 20.23.

In Anthozoa there is no medusa stage. The polyps reproduce asexually by budding or sexually by means of gametes produced in gametocytes embedded in the endoderm. The zygote develops into a planula that becomes attached and develops into a new polyp.

The Anthozoa can be solitary or colonial. Colonial Anthozoa are reef-builders. The polyps deposit an aragonitic cup (the *corallite*) ranging in size from a few millimeters to more than 1 cm. The cup has septa extending from the wall toward the center, numbering multi-

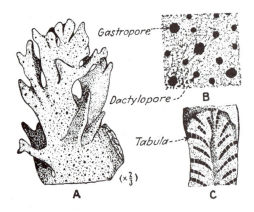

Figure 20.22. Cnidaria (Hydrozoa). *Millepora*, a modern reef-builder. (A) colony; (B) detail of surface of colony, showing gastropores, housing the larger feeding polyps (gastrozooids), and dactylopores, housing the protective polyps (dactylozoids); (C) sagittal section showing tabulae (living floor of polyp). (Shrock and Twenhofel, 1953, p. 108, Fig. 4.4.)

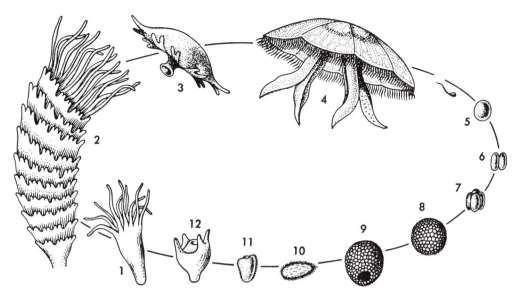

Figure 20.23. Cnidaria (Scyphozoa). Life cycle of *Aurelia* (the medusa is the principal stage). The polyp is small (1); it continuously sheds medusae (2–4), which produce gametes (5), which fuse and grow (6–9) into a planula (10); the planula becomes fixed (11) and grows into a new polyp (12–1). (Simpson, Pittendrigh, and Tiffany, 1957, p. 527, Fig. 22.4.)

ples of 4 in the Paleozoic or multiples of 6 in the Mesozoic and Cenozoic. The septa cause the body of the polyp to form invaginations, thus increasing the surface of the food-absorbing gastric cavity. A colony may reach a height of 2 m and may contain hundreds of thousands or even millions of corallites. *Corallum* is the name applied to the skeletal framework of the colony.

The Tabulata were an abundant subclass of anthozoan corals that had a rather open structure and ranged from the Ordovician through the Permian (Fig. 20.24). More massive were the Rugosa (Fig. 20.25). The subclass Zoantharia (Fig. 20.26) includes all modern stony corals, the builders of modern coral reefs. The subclass Alcyonaria includes the horny corals (Gorgonacea) and the Pennatulacea (Fig. 20.27). In the Pennatulacea the colonies consist of an axial polyp, with secondary polyps branching from it. The skeleton is horny, and the colony is fixed to the substrate, reaching a height commonly around 15–20 cm, but occasionally up to 2–3 m. The axial stem has four longitudinal canals around a horny axis. The cavity in secondary polyps is connected to an axial canal. Pennatulacea have strong muscular and nerve systems and are well represented in the Ediacaran fauna. The occurrence of pennatulaceans in the Ediacaran fauna indicates that the Alcyonaria, and therefore the Cnidaria, were among the first metazoan taxa to evolve.

THINK

* What is the advantage of the polyp–medusa cycle?
* Which famous scientists would best qualify to be called nematocysts?

20.10. PLATYHELMINTHES

The coelenterates have only two dermal layers (ectoderm and endoderm—the mesoglea is not dermal) and are termed, therefore, *diblastic*. The Platyhelminthes (flatworms) are the lowest animals with three dermal layers: ectoderm, mesoderm, and endoderm (Fig. 20.28). They are termed *triploblastic*.

Flatworms have no *pseudocoelom* nor *coelom*, a body cavity developed in animals

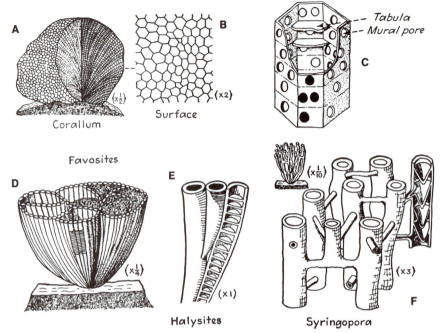

Figure 20.24. Cnidaria (Anthozoa), subclass Tabulata (Ordovician–Permian), characterized by nonseptate corallite floored with tabula. (A–C) *Favosites*: colony (A), detail of surface showing corallites (B), detail of corallite (C). (D–E) *Halysites*: colony (D) and detail (E). (F) *Syringopora*: colony and detail. (Shrock and Twenhofel, 1953, p. 162, Fig. 4.41.)

higher than flatworms either between mesoderm and endoderm (*pseudocoelom*, in nematodes, etc.) or within the mesoderm (*coelom*, in annelids, mollusks, arthropods, chordates, etc.). The digestive tract is a separate, central cavity bound by the endoderm. Flatworms are the most primitive organisms with organs formed by two or more types of tissues.

Flatworms have bilateral symmetry and range in size from centimeters to inches. They move by coordinated motion of protozoan-like cilia on the ventral side. The muscles are longitudinal and transversal. The nervous system consists of two rows of nerve cells along either side of the body, joining into a ganglion in the head, which may be connected to light-sensitive organs (ocelli). The digestive system includes a mouth that protrudes from a cavity at midbody on the ventral side, followed by an elongated blind gut with side branches. The excretory system consists of tubules running along either side of the body, with lateral extensions ending in *flame cells*, which are cells with cilia that push the water and waste products out through an excretory canal. The reproductive system is hermaphroditic, with an erectile penis and a copulatory sac (vagina) in each individual, leading to double intercourse as the standard mode of fertilization. There is no circulatory system.

The phylum Platyhelminthes consists of three classes:

1. Turbellaria: marine and freshwater forms (Fig. 20.29)

2. Trematoda: parasitic (flukes)

3. Cestoda: parasitic (tapeworms, of which *Taenia* is an example)

Whereas Trematoda and Cestoda obviously are late adaptations, the Turbellaria probably evolved in Late Proterozoic time. They have a larval stage consisting of a pear-shaped, free-swimming larva called *trochophore*, a type

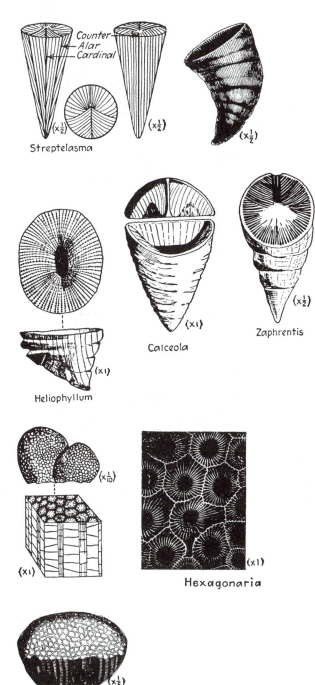

Streptelasma

Heliophyllum

Calceola

Zaphrentis

Hexagonaria

Favistella

Figure 20.25. Cnidaria (Anthozoa), subclass Rugosa (Tetracorallia) (Ordovician–Permian). Corallite septation follows a fourfold symmetry. Notice *Calceola* with operculum. (Shrock and Twenhofel, 1953, p. 160, Fig. 4.39.)

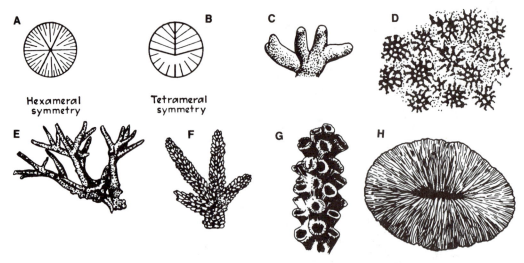

Figure 20.26. Cnidaria (Anthozoa), subclass Zoantharia (Hexacorallia) (Triassic–Holocene). The corallite septation follows a sixfold symmetry. (A) Hexameral addition of septa (longest ones first) compared to fourfold addition in Rugosa (B); (C–D) *Porites* (C, × 0.2; D, × 6);

(E–G) *Acropora cervicornis* (E, × 0.125; F, × 0.5; G, × 3); (H) *Fungia*, a solitary coral of the tropical Indo-Pacific (× 0.33). (A and B: Shrock and Twenhofel, 1953, p. 150, Figs. 4.34N and 4.34Q; C–H: Lehmann and Hillmer, 1983, p. 76, Fig. 54A.)

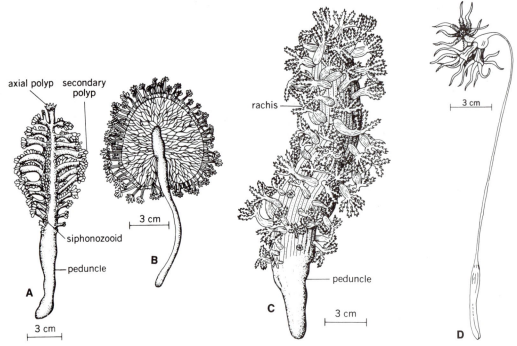

Figure 20.27. Cnidaria (Anthozoa), subclass Alcyonaria, order Pennatulacea (modern): (a) *Pennatula*; (b) *Renilla*; (c) *Veretillum*; (d) *Umbellula*. (*McGraw-*

Hill Encyclopedia of Science and Technology, 6th ed., 1987, vol. 13, p. 186, Figs. 1 and 4.)

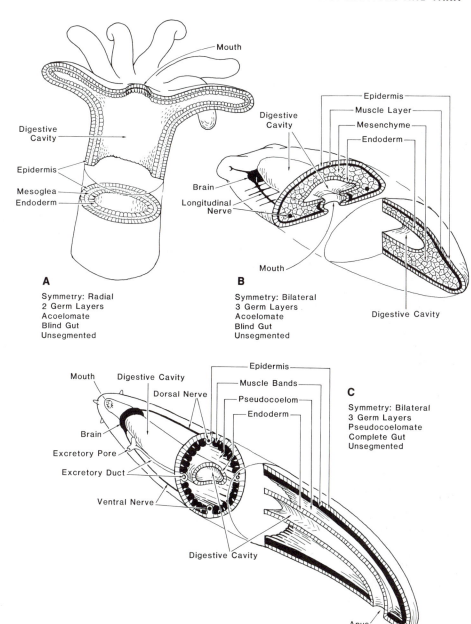

A

Symmetry: Radial
2 Germ Layers
Acoelomate
Blind Gut
Unsegmented

B

Symmetry: Bilateral
3 Germ Layers
Acoelomate
Blind Gut
Unsegmented

C

Symmetry: Bilateral
3 Germ Layers
Pseudocoelomate
Complete Gut
Unsegmented

Figure 20.28. Body organization of (A) Cnidaria (coelenterates), (B) Platyhelminthes (flatworms), and (C) Nematoda. (*Continued*)

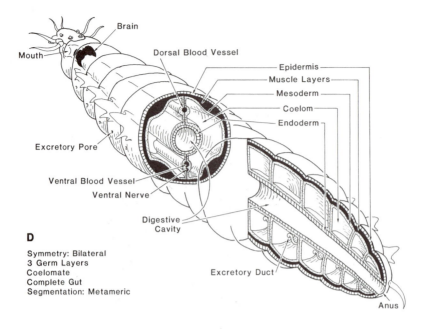

Brain
Mouth
Dorsal Blood Vessel
Epidermis
Muscle Layers
Mesoderm
Coelom
Endoderm
Excretory Pore
Ventral Blood Vessel
Ventral Nerve
Digestive Cavity
Excretory Duct
Anus

D

Symmetry: Bilateral
3 Germ Layers
Coelomate
Complete Gut
Segmentation: Metameric

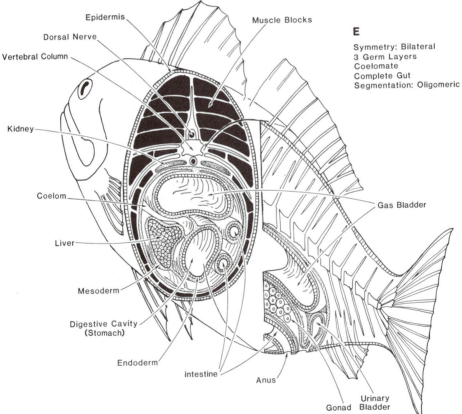

Epidermis
Muscle Blocks
Dorsal Nerve
Vertebral Column
Kidney
Coelom
Liver
Mesoderm
Digestive Cavity
(Stomach)
Endoderm
Intestine
Anus
Gonad
Urinary Bladder
Gas Bladder

E

Symmetry: Bilateral
3 Germ Layers
Coelomate
Complete Gut
Segmentation: Oligomeric

Figure 20.28 (*cont.*). Body organization of (D) Annelida and (E) fish.

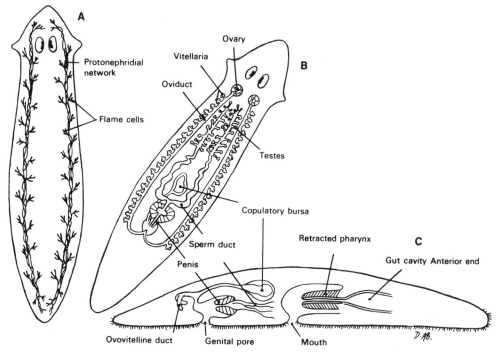

Figure 20.29. Platyhelminthes. *Planaria* (class Turbellaria). Internal structure: (A) excretory system; (B) reproductive system; (C) sagittal section. (Moment and Habermann, 1973, p. 127, Fig. 6.14.)

of larva that is common to other taxa (see Fig. 21.19). They have the ability to regenerate a complete body even if cut in half.

THINK

* Investigate the biology of *Taenia* and see if it could be used as an effective weight controller for humans.

20.11. NEMATODA

Nematodes (Fig. 20.30) are mainly microscopic organisms living in soils in great numbers (on the order of a million per kilogram of soil). Some are parasitic, including *Ascaris*, which lives in human and animal intestines and can grow to 30 cm in length.

Nematodes have an unsegmented, triploblastic body with a pseudocoelom (between mesoderm and endoderm). The body is covered by a tough cuticle made of scleroproteins.

The cuticle cannot grow, and as a result, growth is by molting. There are four or more longitudinal bands of muscles, but no cross-muscles. The nerve system consists of a peripharingeal nerve ring connected to a dorsal cord and a more important ventral nerve cord. More than 50% of the body cells of nematodes are nerve cells. The digestive system is graced with an anus. Sexes are separate. The fertilized eggs produce baby nematodes that reach adulthood through four molts. There is no circulatory system.

20.12. ANNELIDA

Annelida (Fig. 20.31) are triploblastic coelomates with segmented bodies. The phylum includes three classes:

1. Polychaeta: mainly marine
   a. free-living: swimming or crawling on mud on sea floor or coral reefs
   b. sedentary: in tubes with protruding gills or tentacles (e.g., *Sabella*, *Sabellaria*)

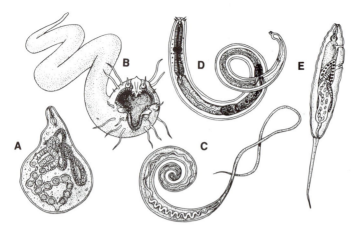

Figure 20.30. Nematoda (mostly millimeter size or less): (A) female of the rootknot nematode; (B) a carnivorous nematode; (C) intestinal whipworm; (D) soil nematode; (E) tailed nematode. (Moment and Habermann, 1973, p. 437, Fig. 19.1.)

2. Oligochaeta: terrestrial (*Lumbricus*); some freshwater or marine forms

3. Clitellata (Hirudinea): leeches

Polychaetes are by far the most important class of Annelida from the paleontological point of view. (*Polychaeta* derives from the Greek πολύς, which means *many*, and χαίτη, which means *hair*, referring to the many hairs on the surfaces of the worms in the class; in contrast, the Oligochaeta have few hairs—from the Greek ὀλίγος, for *few*.) The exterior of the polychaete body is covered with a collagen cuticle (see Section 22.3 for collagen). The muscles include a layer of circular muscles just below the epithelial cell layer

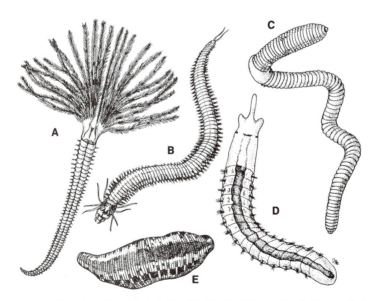

Figure 20.31. Annelida: (A) *Sabella*; (B) *Nereis*; (C) *Lumbricus* (the common earthworm); (D) *Stylaria*; (E) *Hirudo* (the medical leech). (Moment and Habermann, 1973, p. 444, Fig. 20.1.)

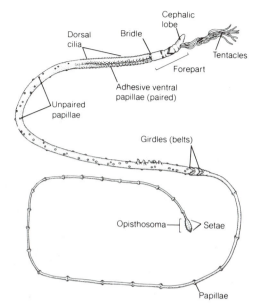

Figure 20.32. Brachiata, class Pogonophora: diagrammatic and shortened view of a pogonophoran removed from its tube. (Margulis and Schwartz, p. 239, Fig. C.)

(which secretes the cuticle), with, below it, a set of longitudinal muscles. The nervous system includes a dorsal brain, a circumpharingeal collar, and a ventral nerve cord. The excretory system includes two nephridia per segment connected to the outside via excretory pores.

Respiration is by diffusion of oxygen through the body surface. The circulatory system consists of a dorsal blood vessel with several hearts near the head and a return ventral vessel.

Some swimming polychaetes have well-developed eyes, with lens and retina. Most have various appendages for foraging or sensory functions. Many polychaetes have two thin appendages in each segment that are highly vascularized and are used for both gas exchange and swimming.

Sexes are separate, and fertilization is external. The zygote may develop into a trochophore larva, which allows dispersion. This is especially important for the many species of polychaetes that live in holes in the sediment on the ocean floor or in tubes encrusting hard rocks or reefs in shallow water.

THINK

* Why do polychaetes have more hair than oligocaetes?

* Sabellariid worms (i.e., worms belonging to the genus *Sabellaria*) are tropical, encrusting worms that form tubes by cementing together sand grains and other sedimentary particles. The tubes are cemented to each other to form strong crusts or even reefs capable of resisting strong breakers. Could sabellariid worms be used to stem beach erosion?

* Considering the pertinent internal anatomies, as well as the groove patterns of different types of tires, write a learned essay demonstrating that a flatworm could evolve (by punctuated equilibrium) from a fishing worm (Fig. 20.31C) that was run over by a car.

20.13. BRACHIATA

The phylum Brachiata includes two classes, the Pogonophora (Yudomian–Holocene) and the Vestimentifera (Holocene). They have been described only since the 1950s.

The Pogonophora (Fig. 20.32) have long, slender bodies (mostly 1 mm across and 10 cm long). At the upper end of a tube there are a few to 200 tentacles. There is a nervous system, consisting of a brain with a ventral cord; there is a circulatory system consisting of a heart with a ventral vessel that continues in a dorsal vessel; and there are muscles (not yet well studied).

The Pogonophora are pseudocoelomate: They have their body cavities lined with muscle tissues, not with dermal layers. There is no mouth, no digestive tract, no anus. Nutrition is by absorption of food particles through the tentacles. Digestion is by means of symbiotic bacteria. The sexes are separate. Fertilization occurs inside the female tube, and ciliated larvae emerge. The Pogonophora live mainly in burrows in deep (300–4,000 m), cold marine sediments. *Paleolina* occurs in the upper Yudomian (latest Cryptozoic) of Siberia.

The Vestimentifera also have slender bodies, but reaching a length of 1 m. Vestimentifera live in hot vents along the East Pacific Rise. Inside the body there is a vascularized sac, which occupies most of the trunk and is filled

Table 20.5.   The Infracambrian glaciations

State	Age (10^6 y B.P.)	Areas affected
Infracambrian 3	650–600	Europe, North America, Australia
Infracambrian 2	800–720	North America, Europe, Siberia, China, Africa, Australia
Infracambrian 1	950	Africa, Siberia, China

with sulfur bacteria (Section 19.3). These bacteria are the primary energy producers for the Vestimentifera. *Riftia* is an example.

THINK

* What happens to *Riftia* and its sulfur bacteria when the vent goes dead?

20.14. THE CRYPTOZOIC– PHANEROZOIC TRANSITION

The boundary between Cryptozoic and Phanerozoic is set at the first appearance of Archaeocyatha (Section 21.2). This boundary is often called the Precambrian–Cambrian boundary, but, as noted earlier, the name *Precambrian* implies all the time before the Cambrian, all the way back to the Big Bang. The name *Cryptozoic* is therefore preferable.

The Cambrian is characterized by the appearance of the following groups of animals:

Archaeocyatha
Hyolitha
Mollusca
Brachiopoda
Conodonta
Arthropoda
Echinodermata
Foraminifera
Radiolaria
Hemichordata

These will be discussed in Chapter 21.

An excellent section showing the transition between Cryptozoic and Phanerozoic exists along the Lena River valley in northeastern Siberia. The section begins with the tillites of the Infracambrian glaciations. The Infracam-

brian glaciations ranged from 900 to 650 million years ago and affected the entire globe. The evidence indicates that even continents at low latitudes were covered with ice at one time or another. The glaciation occurred in three stages (Table 20.5). Glaciations occurred also during the Phanerozoic. In fact, we are right now in a glaciation that started several million years ago. As recently as 20,000 y ago, all lands north of a front running from Seattle to New York and from London to Moscow and beyond were covered with enormous ice sheets. Judging from the pattern of glaciation during the past few million years, ice does not simply form, stay some time, and then disappear: It comes and goes with periodicities of 100,000 and 400,000 y (Chapter 22). This means that, periodically, the environment changes radically. The response of the biosphere is fragmentation and dispersal of populations, genetic isolation, and rapid speciation. In other words, a period of glaciation is a period of rapid evolution.

Before the Infracambrian glaciations started, there were only protophytes (and perhaps protozoans) on Earth. At the end of that glaciation, complex metazoa (the Ediacaran fauna) were already in existence.

The cause of the Infracambrian glaciations may have been a reduction in volcanic activity (which would have reduced the CO_2 input into the atmosphere), coupled with an increase in carbonate deposition. That would have resulted in a decrease in the CO_2 concentration in the atmosphere, which may have been sufficient to generate a worldwide glaciation because the Sun was then radiating less energy than today.

[After the T Tauri stage (Section 6.3), solar radiation stabilized, and the Sun began converting hydrogen into helium in its core in an orderly way. During the following $4.5 \cdot 10^9$ y,

Table 20.6. **The Siberian section**

3. Adtabanian: limestones and dolomites (highest)
   With Olenellida (trilobites)
   Corresponds to the Holmia zone (basal Cambrian)
   of Europe
2. Tommotian: reddish argillaceous limestones
   With archaeocyathids, sponges, hyolithids, gastropods, brachiopods (no trilobites)

Cryptozoic–Phanerozoic boundary
1. Yudomian:
   Upper: dolomites with
   Anabarites (worm)
   Paleolina (Pogonophora, single class of phylum
   Brachiata)
   Girvanella (filamentous cyanobacterium depositing $CaCO_3$; Cryptozoic–Jurassic)
   Lower: dolomites with
   c. Ediacaran fauna
   b. stromatolites
   a. tillites

the concentration of helium in the Sun's core has increased, from an initial 26% to the present 62% (see Table 7.2). As the hydrogen concentration in the core decreases, the core contracts, its temperature rises, and the Sun's energy output increases. The increase has amounted to 25% during the past $4.5 \cdot 10^9$ y, which means that during the Infracambrian glaciation the solar output was about 4% less than it is today.]

An Earth largely covered with ice would have a very dry atmosphere, which would contribute to maintaining low temperatures. In fact,

if both CO_2 and H_2O were absent from the modern atmosphere, the Earth today would have a temperature of about $-15°C$.

The Siberian section is summarized in Table 20.6. The section begins with the tillites of the Infracambrian glaciations and terminates with Cambrian limestones and dolomites. The Archaeocyatha are the first widespread phylum to deposit calcium carbonate structures. Together with the Archaeocyatha, other groups appeared that also were capable of mineralizing their tissues and forming protective hard parts (Table 20.6). Apparently, the increase in oxygen, which fostered greater alertness and faster movements, also led to the development of predation.

THINK

• The temperature of a blackbody radiator is calculated using the Wien displacement law, which says that T (K) $= 2,897.8/\lambda$, where λ (in μm) is the wavelength at which the blackbody radiates the greatest amount of power. Fig. 13.2 shows that the peak power of the Earth's backradiation into space is at $10\,\mu$m. Calculate the average temperature of the Earth using the Wien displacement law. Next deduce the average temperature of the Earth in the Late Proterozoic, when solar radiation was 4% lower than today. Was that temperature low enough to cause widespread glaciation? If it was, why did we not have much larger and perhaps continuous glaciations in the earlier Proterozoic and the Archean, when solar radiation was even lower?

21

EARLY PHANEROZOIC PHYLA AND RELATED TAXA

There was an enormous radiation in the biosphere during the Late Proterozoic, perhaps related to the glaciations that took place during that time. A number of the phyla that had evolved during the Late Proterozoic developed the ability to deposit or aggregate hard parts to form protective integuments or supporting structures. In most cases the hard parts consisted of calcium carbonate. In other cases, calcium phosphate, chitin (a polysaccharide), cemented clastic particles, opaline silica, or fibrous proteins were used. Widespread tissue mineralization is taken to mark the beginning of the Phanerozoic and thus of the Cambrian.

This chapter briefly describes the major phyla that began leaving a fossil record in the Cambrian (and related taxa). Important taxa left an abundant record dating from as early as the first two stages of the Cambrian, the Tommotian ($570-560 \cdot 10^6$ y ago) and the Adtabanian ($560-554 \cdot 10^6$ y ago). This chapter is necessarily largely descriptive and therefore rather boring. Nevertheless, it is essential, for a proper understanding of the evolution of life and environment, that the structures of the various taxa be clearly understood. It is particularly important that the reader develop a feeling for the enormous diversity of these early taxa, which indicates a long pre-Phanerozoic history and which shows that the course of life was set early in the Phanerozoic. The evolving taxa contributed importantly to the evolution of the environment—the subject matter of the remaining chapters of this book—and form the chronological frame of reference for the Phanerozoic history of our planet.

21.1. RHIZOPODA

Most Protozoan groups are naked, but some are shelled. The phylum Rhizopoda in-

Figure 21.1. *Amoeba proteus*, a common naked protozoan (× 100). (Borradaile and Potts, 1961, p. 76, Fig. 42.)

cludes three classes, Lobosia, Actinopoda, and Foraminifera, two of which (Actinopoda and Foraminifera) include most of the shelled marine protozoans.

Lobosia (Eocene–Holocene)

The Lobosia include the orders Amoebida (naked) and Arcellida (shelled). Representative of the Amoebida is *Amoeba proteus* (Fig. 21.1), a protozoan that moves around by means of cytoplasm flowing into lobate extensions (*pseudopodia*) and pulling the cell along. *Amoeba proteus* is 0.5 mm across, but other species range from 0.01 mm to 3 mm. The Amoebida reproduce asexually, by mitosis. The Arcellida construct shells consisting of secreted silica elements or agglutinated sand grains. Representative genera are *Arcella* and *Difflugia*, the earliest specimens of which have been found in the Green River Formation of Eocene age (see Section 16.1).

Actinopoda (Cambrian–Holocene)

The Actinopoda are not only exclusively marine but also exclusively planktic. The shells are small (0.1–1 mm) and lacy. The cytoplasm

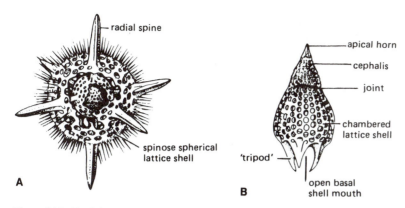

Figure 21.2. Radiolaria: (A) a spumellarian (radial symmetry); (B) a nassellarian (axial symmetry). (Lehmann and Hillmer, 1983, p. 33, Fig. 23.)

is partitioned by a perforated membrane into an inner part (which contains the cell nucleus) and an outer part. Reproduction probably is by alternation of generations, with asexual reproduction much more common than sexual reproduction. The life span of an individual generally is several weeks. Actinopoda have anastomosing pseudopodia, a characteristic they share with the Foraminifera. Living "species" number over 1,000, and thousands of "species" have been described.

[The word *species* is here in quotation marks because the concept of species becomes nebulous when dealing with protozoa and protophyta that frequently multiply asexually. It becomes even more nebulous when one considers these "species" through time. What micropaleontologists do is to assign a shell (or other remains) to a given species if it looks sufficiently similar to the shell (or other remains) for which that species was named. This is, of course, entirely subjective, but there is not much else one can do when dealing with fossils, because generally it is not possible to examine fossil genetic material.]

Actinopoda usually are solitary, but some species are colonial in the sense that they form agglomerations of hundreds of individuals held together by a common gelatinous mass. They live from the polar seas to the equator, different assemblages being characteristic of different water masses.

The class is divided into three subclasses, Polycystinea, Acantharia, and Phaeodaria. The Polycystinea deposit shells of opaline silica ($SiO_2 \cdot nH_2O$) and include two orders, the Spumellaria and the Nassellaria. The Spumellaria (Fig. 21.2A) have radial symmetry, and the Nassellaria have axial symmetry (Fig. 21.2B). The Polycystinea are distinguished by having an extraordinary number of chromosomes, up to 1,500 per nucleus (as compared with 8 in the fruit fly and 46 in humans). The Acantharia deposit shells with radial symmetry, consisting of strontium sulfate ($SrSO_4$); they are rarely preserved as fossils. The Phaeodaria (Fig. 21.3) deposit shells with radial symmetry, consisting of silica and organic matter. Polycystinea and Phaeodaria are collectively called Radiolaria. Radiolaria live in the open tropical ocean, from the surface to depths of a few hundred meters. If the sea floor is sufficiently deep (> 4,500 m) for calcium carbonate to dissolve, the radiolarian

Figure 21.3. Phaeodaria: a modern phaeodarian. (Tasch, 1973, p. 70, Fig. 3.13A.)

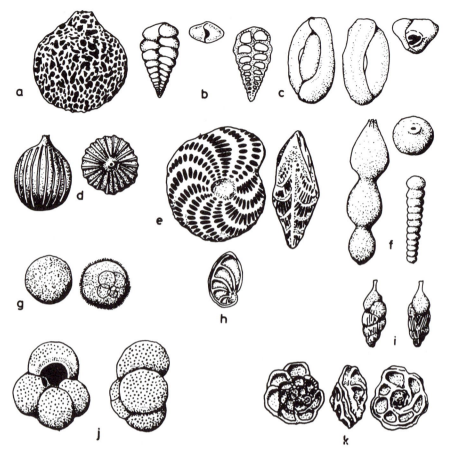

Figure 21.4. Modern foraminifera: (a) *Saccammina* (× 30); (b) *Textularia* (× 13); (c) *Quinqueloculina* (× 25); (d) *Lagena* (× 50); (e) *Elphidium* (× 40); (f) *Nodosaria* (× 40); (g) *Orbulina* (× 10); (h) *Lenticulina* (× 20); (i) *Uvigerina* (× 50); (j) *Globigerina*; (k) *Globotruncana*, a Cretaceous genus; g, j, and k are planktic; all others are benthic. (Lehmann and Hillmer, 1983, p. 28, Fig. 16A.)

shells are concentrated, forming a deposit known as *radiolarian ooze*. A great belt of radiolarian ooze stretches from east to west along the equatorial Pacific. Radiolarite, found scattered in the geological record, is a deep-sea radiolarian ooze that has recrystallized to compact chert.

Foraminifera (Cambrian–Holocene)

Foraminifera are a large class of exclusively marine protozoans that includes several thousand living benthic species (Fig. 21.4) and about 40 living planktic species (Fig. 21.5). Fossil species number tens of thousands.

[*Benthic* derives from the Greek βένθος mean-ing *the deep*; *planktic* derives from πλαγτός, meaning *floating around*; *nektic* derives from νήκτος, meaning *swimming*.]

The shell morphology may vary from a single vaselike structure to a highly complex structure with hundreds of chambers. The material may be mineral grains glued together, magnesian calcite (with up to 15% Mg substituting for Ca), virtually pure calcite, or aragonite (in one group). The walls of carbonate shells may be perforated or imperforated. Generally, the shells of benthic species are thicker and heavier than those of planktic species. Most foraminiferal shells are 0.2–2 mm across, but some range up to a few centimeters.

Foraminiferal cells often are multinucle-

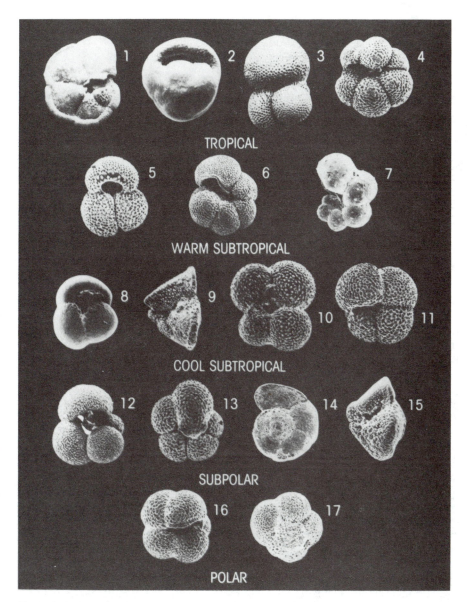

Figure 21.5. Some of the more abundant species of modern planktic foraminifera, typical of different latitudes: (1) *Globorotalia menardii*; (2) *Pulleniatina obliquiloculata*; (3) *Globigerinoides sacculifera*; (4) and (6) *Neogloboquadrina dutertrei*; (5) *Globigerinoides rubra*; (7) *Hastigerina pelagica*; (8) *Globorotalia inflata*; (9) *Globorotalia truncatulinoides*; (10) *Globigerina falconensis*; (11) *Neogloboquadrina pachyderma* (positive helicity); (12) *Globigerina bulloides* (13) *Globigerina quinqueloba*; (14) and (15) *Globorotalia truncatulinoides*; (16) and (17) *Neogloboquadrina pachyderma* (negative helicity) (all × 16). (Kennett, 1982, p. 540, Fig. 16.1.)

ated. The cytoplasm extends outside the shell through one or more openings and/or through pores in the shell wall. Outside the shell, the cytoplasm forms pseudopodia that are anastomosed (different from *Amoeba*, but similar to the Actinopoda). The pseudopodia are used to capture food particles and (in benthic Foraminifera) also for locomotion (similar to *Amoeba*).

Frequently, the protozoan reproduces

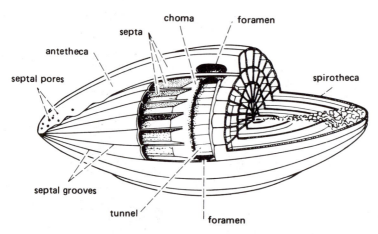

Figure 21.6. *Fusulina* (× 10), internal structure. The chambers, separated by longitudinal septa, run from pole to pole; the septa, which do not reach the floor, were straight in the earliest species, but become progressively more sinuous with time. (Lehmann and Hillmer, 1983, p. 24, Fig. 16A.)

asexually—each nucleus grabs a piece of protoplasm and scrambles out of the shell (as fast as possible in planktic Foraminifera, because the shell sinks fast once the protoplasm is gone). Occasionally, especially if environmental conditions have taken a turn for the worse, meiosis occurs, gametes are produced, and sexual reproduction ensues. In some species, three individuals partake in a kind of sexual orgy. The life span of an individual is on the order of weeks.

A major foraminiferal group, the fusulinids (Fig. 21.6), evolved in Mississippian time and became widespread during the Pennsylvanian and Permian periods. The fusulinid shell is planispiral, elongated along the axis of coiling, and multichambered. Fusulinids were abundant from the Upper Mississippian to Permian time in shallow, carbonate environments. They were distributed across the tropical belt, which, in the Late Paleozoic, stretched from the northwestern coast of South America across the North American platform and eastward through Europe to southeastern Asia.

The fusulinids are called "larger foraminifera" to distinguish them from most other foraminiferal taxa that have smaller and simpler shells. Larger foraminiferal taxa (Orbitoidiadae, Discocyclinidae, Camerinidae, Alveolinellidae) evolved again in Upper Cretaceous times, were particularly abundant during the Eocene and Oligocene ($56.5-23.3 \cdot 10^6$ y ago), and again

were (and some of them still are) distributed in carbonate environments around the tropics.

Different species of benthic foraminifera are adapted to different environments, from rocky shores to mangrove swamps, sandy beaches, open shelves, continental and island slopes, the deep-sea floor, and even deep-sea trenches. Foraminifera usually are abundant in marine sediments. They provide important information about the environment in which the sediment was deposited. The larger foraminifera are so complex that their genetic makeup does not last long through time (the more complex the genome is, the easier it is for significant mutations to occur); they speciate relatively rapidly (a species may last less than 10^6 y) and therefore are good guide fossils. The smaller benthic foraminifera, which account for most of the species, do not make good time markers because their species last a long time (usually more than 10^7 y).

During the Jurassic, a group of foraminifera adapted to the planktic life style. Their shells became thinner and lighter, and in most species the shell wall became abundantly perforated. Today there are about 40 species of planktic foraminifera—very few compared with the number of benthic species. Planktic foraminifera show latitudinal zonation, with the number of species decreasing from the tropics to the polar seas. The species of planktic foraminifera last for shorter times than do the

Figure 21.7. Archaeocyatha. A typical archaeocyathid, 10 cm tall, 2 cm across, attached to the bottom. The body presumably occupied the space between the inner and outer cones, drawing water through the perforations in the outer cone, removing nutrients, and expelling processed water and waste products through the central cavity. Archaeocyathids constructed reefs up to 10 m thick along shallow Cambrian shelves. (Emiliani, 1988, p. 198, Fig. 11.1.)

benthic species—mostly $1-8 \cdot 10^6$ y. They are, therefore, important time markers from the Cretaceous (when they became abundant) to the present.

21.2. ARCHAEOCYATHA (Tommotian to early Middle Cambrian)

The Archaeocyatha were exclusively marine organisms that constructed a calcareous skeleton consisting of an inverted double cone, with the inner cone connected to the outer cone by vertical partitions (Fig. 21.7). The outer cone was attached to the bottom by an apical holdfast. These structures averaged 10–15 cm in length and 1–3 cm across at the top. Some were bowl-shaped, and some reached a length of 30 cm. Some archaeocyathids were solitary, and others were colonial. The colonies, which grew by branching or by budding, formed conspicuous reefs on the stable platforms of the Early Cambrian. The extinction of the Archaeocyatha coincided with the flourishing of the sponges, which may have found the archaeocyathid reefs an excellent substrate on which to settle.

21.3. HYOLITHA (Tommotian to end of Permian)

The Hyolitha had a conical, probably aragonitic, shell, with a trianguloid cross section (the widest side was the ventral side), commonly 1–3 cm long, but some up to 20 cm long. The shell had an operculum (covering lid) that could be opened and closed. They were marine mud-feeders. The Hyolitha had a ventral mouth near the shell opening. The digestive system consisted of a highly folded gut continuing into a straight, dorsal segment at the aboral end, and terminating with an anus above the head. The respiratory system consisted of paired gills above the head and below the anus. The class contains two orders:

1. Hyolithida (Fig. 21.8). The shells are gently concave in lateral profile. The lower wall of the shell extends forward, forming a small shelf called *ligula*. Two curved, carbonate rods, called *helens*, extend laterally from the junction of the ligula to the shell.

2. Orthothecida. The Orthothecida have

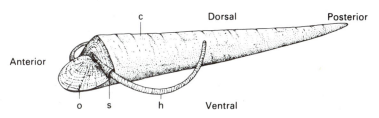

Figure 21.8. Hyolithida. *Hyolithes*: c, conch; h, helen; o, operculum; s, slot between operculum and conch where the helens emerge. (Boardman, Cheetham, and Rowell, 1987, p. 437, Fig. 1A.)

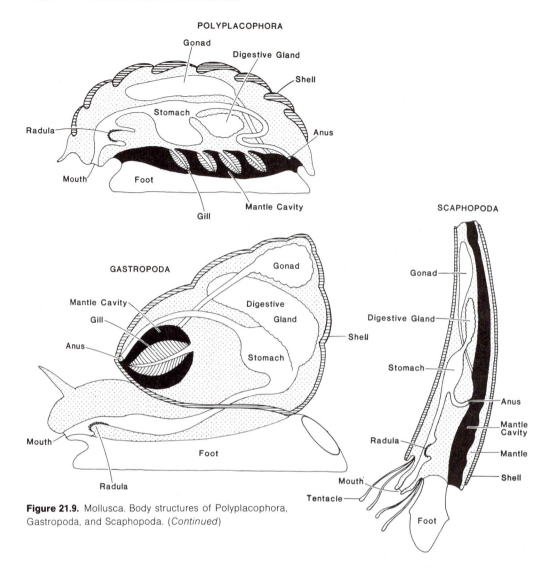

Figure 21.9. Mollusca. Body structures of Polyplacophora, Gastropoda, and Scaphopoda. (*Continued*)

straight shells, with no ligula and no helens. They range from the Tommotian to the Middle Devonian.

The Hyolitha were *not* ancestral to gastropods or other mollusks.

THINK

* Doesn't the location of the gills, just below the anus, look mighty peculiar? Why would one place the intake (gills) in such close proximity of the discharge?

21.4. MOLLUSCA

The ancestral mollusk had a mouth followed by a gut, ending in a posterior anus; there were also a ventral foot and a dorsal shell. There are eight classes of mollusks. The body structures of the dominant classes are shown in Fig. 21.9.

Aplacophora (Holocene)

The Aplacophora have a wormlike body 0.5–4 mm long, without a shell. The body is covered with a cuticle with embedded aragonitic

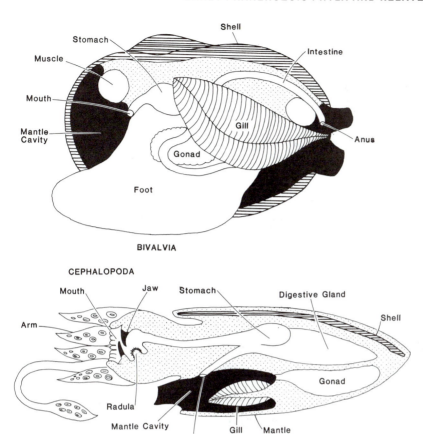

Figure 21.9 (*cont.*). Mollusca. Body structures of Bivalvia and Cephalopoda.

spicules. They are benthic marine animals living from the shore to 9,000 m of depth. Although known only from the present time, they probably originated in the Late Proterozoic.

Monoplacophora (Tommotian to Devonian, Holocene)

The Monoplacophora have a univalve shell, not torted (twisted), without operculum (Fig. 21.10). They are benthic algal grazers. They were thought to be extinct until a living representative, *Neopilina*, was dredged from the bottom of the Mid-America Trench off Costa Rica in 1952. *Neopilina* has a dorsal shell, a head with paired sensory organs (tentacles) above and a mouth below with a strong *radula*. The radula, common in mollusks, is a tongue with teeth; it is used to scrape up microalgae and also to per-

forate shells and get to the meat inside. The respiratory and circulatory system includes five pairs of gills along either side of the body, with vessels leading to five paired hearts. There are five paired nephridia (kidneys) along each side of the body. The digestive system includes a mouth extending into a straight gut that terminates in a dorsal anus at the other end. There are five pairs of retractor muscles, plus three pairs of muscles for the head and one circular muscle to control the single large foot. Sexes are separate. There are no copulatory organs—the gonads empty in seawater.

Polyplacophora (Late Cambrian to Holocene)

Polyplacophora have a dorsal shell composed of eight segments and a mantle girdle allowing

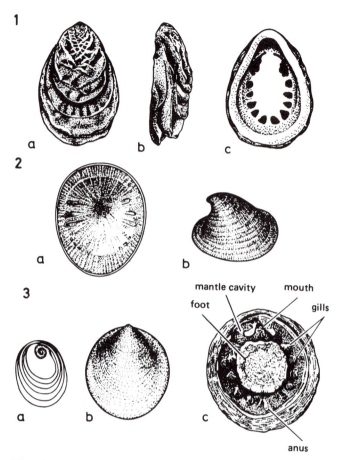

Figure 21.10. Monoplacophora: (1) *Tryblidium* (× 1); (2) *Scenella* (2a × 2.5; 2b × 2); (3) *Neopilina* (3a, larval shell, × 2.5; 3b, dorsal view, × 0.8; 3c, ventral view, × 1). (Lehmann and Hillmar, 1983, p. 85, Fig. 60.)

them to cling to rocky substrates along high-energy coastlines. They are benthic algal grazers. An example is *Chiton* (Fig. 21.11).

Gastropoda (Tommotian–Holocene)

Gastropoda have a univalve shell that usually is torted and has an operculum (Fig. 21.12). Torsion is caused by the left side of the visceral mass growing faster than the right side, so that the intestinal tract and other organs are rotated 180° counterclockwise as seen from above. The anus comes to occupy a position above the head, with the gills above it.

The shells of most gastropod genera and species have positive helicity (Fig. 21.12). One family (the Triphoridae), one genus (*Contra-*

conus), and some species of the genus *Busycon* (e.g., *Busycon perversum*) have negative helicity. The earliest gastropod (*Aldanella* of Tommotian time) already shows fully developed positive helicity. The class Gastropoda includes three subclasses:

1. Prosobranchia (Tommotian–Holocene). The Prosobranchia, the dominant gastropod group, are mainly marine, with fully torted shells. As a result of torsion, most Prosobranchia have one anterior gill, the right gill having been lost.

2. Opistobranchia (Mississippian–Holocene). The Opistobranchia are strongly detorted, with a posterior gill. The shell is commonly concealed in the mantle or is absent.

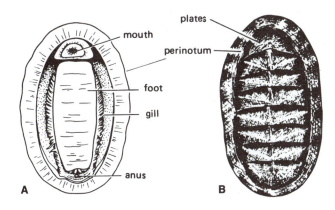

Figure 21.11. Polyplacophora. *Chiton*: (A) ventral view; (B) dorsal view (× 1). (Lehmann and Hillmer, 1983, p. 83, Fig. 59.)

Included in the group are the planktic Pteropoda (e.g., *Cavolinia*).

3. Pulmonata (Pennsylvanian–Holocene). The Pulmonata are mainly terrestrial. The gill has been lost, and the mantle cavity has become vascularized and altered into a lung. This group includes the terrestrial snails and slugs.

Rostroconchia (Tommotian–Permian)

In the Rostroconchia, the shell consisted of two mirror sides joined by continuous shell material across the dorsum. There was no hinge—the whole structure was rigid. There was an anterior gap between the two sides for protrusion of the foot. The body structure probably was similar to that of the Bivalvia, to which Rostroconchia probably were ancestral.

Bivalvia (Early Cambrian–Holocene)

The Bivalvia (Fig. 21.13), previously called Pelecypoda or Lamellibranchia, are mainly *infaunal*; that is, they live in burrows that they quickly dig into the sediment using the foot. They have no head. The two valves (right and left) are held together by a horny ligament that tends to keep them open. The valves are brought together by the action of two adductor muscles, a smaller one anterior and a larger one posterior, or by a single posterior muscle. The inner surfaces of valves are lined by the mantle. The body is centrally located, with the gills in two mantle cavities on either side of the body, which extends into a foot. Water is drawn into the mantle cavities, and expelled from there, usually by means of a posterior

siphon (a double tube with the incurrent canal below and the excurrent canal above).

A bivalve shell consists of three layers, the *periostracum*, the *prismatic layer*, and the *ostracum*. The periostracum, also present in Gastropoda, is the outer layer, which consists of conchiolin (a fibrous protein). The prismatic layer is the main layer, consisting of calcite prisms perpendicular to the shell surface. The nacreous layer is the inner layer. It consists of aragonitic laminae with intervening layers of conchiolin. The aragonitic laminae and the conchiolin layers do not have perfectly smooth and parallel surfaces. As a result, light is diffracted and iridescence (color separation) is produced.

The bivalve shells have a hinge structure consisting of alternating teeth and sockets. The beak on the dorsal margin of the shell, called *umbo*, usually is slanted toward the anterior. The muscle attachments leave internal imprints on the nacreous layer. In many genera the imprints are of about the same size. In many others a smaller, anterior imprint marks the attachment of the anterior muscle, and a larger, posterior imprint marks the attachment of the posterior muscle. The siphon also leaves an imprint, called *pallial sinus*. Most bivalves are marine, but some live in fresh water.

Cephalopoda (Late Cambrian–Holocene)

The Cephalopoda are exclusively marine and nektic (free-swimming). Swimming is done by jet propulsion: Water enters from a frontal opening, bathes four gills, and enters a cavity called the *hyponome*. The hyponome is lined with powerful muscles that squirt the water

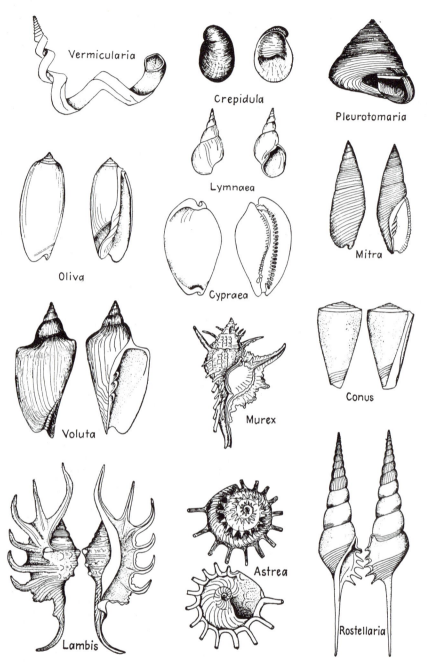

Figure 21.12. Gastropoda: different types of shells.
(Shrock and Twenhofel, p. 412, Fig. 10–40.)

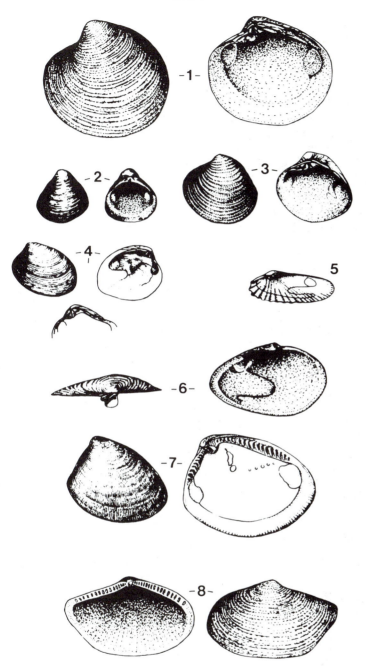

Figure 21.13. Bivalvia: (1) *Arctica* (× 0.7); (2) *Corbicula* (× 0.7); (3) *Venus* (× 0.4); (4) *Tapes* (× 0.4); (5) *Petricola* (× 0.6); (6) *Mya* (× 0.4); (7) *Nucula* (× 3); (8) *Yoldia* (× 1). (Lehmann and Hillmer, 1983, p. 128, Fig. 87, p. 112, Fig. 77.)

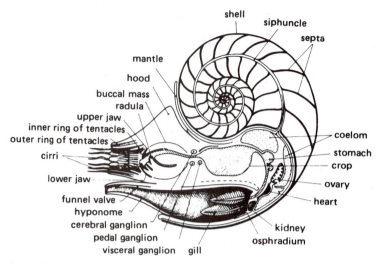

Figure 21.14. *Nautilus*, cross section. (Lehmann and Hillmer, 1983, p. 134, Fig. 91.)

out. In squids, the hyponome contains an ink sac.

The shells can be straight (*orthoconic*), gently curved (*cyrtoconic*), or planispiral. They are partitioned into chambers by septa that are concave toward the aperture and are connected by a tube (*siphuncle*) that is centrally or dorsally located. The sutures marking the connections of the partitions to the outer wall of the shell exhibit a highly characteristic morphology. They can be straight (*orthoceratitic*) or with undulation. The portions of the undulation that are concave toward the aperture are called *lobes*; those that are convex are called *saddles*. The undulations can be smooth and gentle (*agoniatitic* type) or smooth and more pronounced (*goniatitic*). The lobes can be smooth, with the saddles having secondary undulations (*ceratitic*); or both lobes and saddles may have secondary undulations (*ammonitic*).

The head of a cephalopod has a well-developed brain (highest IQ among all mollusks), with eyes and other sensory organs. The mouth is surrounded by tentacles. The anal opening is between the gills, at the end of the mantle cavity below the head (not a bright arrangement, it seems to me). The class Cephalopoda is divided into three subclasses:

1. Nautiloidea (Late Cambrian–Holocene). The Nautiloidea have four gills, straight

to planispiral shells, orthoceratitic sutures, and a central siphuncle. The modern *Nautilus* (Fig. 21.14) has a planispiral shell and lives in the western Indo-Pacific at 200–700 m of depth; it comes inshore to deposit eggs on seagrass clumps, which are then squirted over by the male. *Ellesmoceras* (Late Cambrian–Ordovician) is the earliest cephalopod, with a small, orthoconic shell with closely spaced septa. The Endoceratida (Ordovician–Silurian) include the longest orthoceratid shells on record (up to 4.5 m).

2. Ammonoidea (Early Devonian–Late Cretaceous). The Ammonoidea had four gills and planispiral shells with agoniatitic, goniatitic, ceratitic, or ammonitic sutures (Fig. 21.15). More than 2,000 genera have been described. The highly complex ammonitic sutures (extreme in Fig. 23.11) required a complex genome, and as a result, a species did not last a long time. The average species duration among Ammonoidea was about 840,000 y, making the Ammonoidea useful for high-resolution biostratigraphy.

3. Coleoidea (Early Devonian–Holocene). The Coleoidea have two gills and, in most species, an internal orthoconic or cyrtoconic shell. Some species have no shell. The Belemnoidea (Fig. 21.16) have a small,

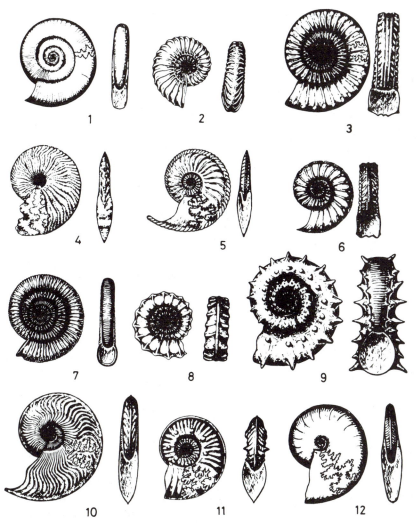

Figure 21.15. Ammonoidea: (1) *Psiloceras* (× 0.3); (2) *Schlotheimia* (× 0.3); (3) *Arietites* (× 0.5); (4) *Oxynoticeras* (× 0.35); (5) *Amaltheus* (× 0.2); (6) *Pleuroceras* (× 0.3); (7) *Dactylioceras* (× 0.35); (8) *Echioceras* (× 0.5); (9) *Eoderoceras* (× 0.4); (10) *Harpoceras* (× 0.5); (11) *Sonninia* (× 0.5) (12) *Oppelia* (× 0.5) (1–10, Lower Jurassic; 11, Middle Jurassic; 12, Upper Jurassic–Lower Cretaceous). (Lehmann and Hillmer, 1983, p. 162, Fig. 11.7.)

internal orthoconic shell (called *phragmocone*) surrounded by a thick calcitic deposit (called *rostrum*) and a forward extension called *proostracum*. The Sepioidea, of which *Sepia* (Fig. 21.17) is a representative, have an internal shell with a minute rostrum and a large phragmocone (the common cuttlebone). In the Teuthoidea the rostrum and the phragmocone are absent, and the proostracum is shaped like a lancet (e.g., *Loligo*). The Coleoidea include the Octo-poda (octopuses). Octopuses have no hard parts, but the female of *Argonauta*, which is many times larger than the male, secretes a thin, ribbed, nonchambered, planispiral shell that she uses as an egg sac.

Scaphopoda (Middle Ordovician–Holocene)

The scaphopoda (Fig. 21.18) are marine and benthic. They have a gently curved, conical shell

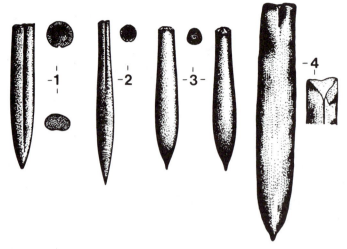

Figure 21.16. Belemnoidea: (1) *Belemnopsis* (Middle Jurassic, × 0.5); (2) *Hibolites* (Middle–Upper Jurassic, × 0.3); (3) *Actinocamax* (Upper Cretaceous, × 0.5); (4) *Gonioteuthis* (Upper Cretaceous, × 0.75). (Lehmann and Hillmer, 1983, p. 178, Fig. 132.)

open at both ends, wider at the anterior end and narrower at the posterior end. Scaphopods spend their time with two-thirds of the body embedded in bottom sediment, head down, feeding on benthic foraminifera. The rear end remains above the sediment surface to allow water to escape (during digging) or to return (during withdrawal).

The Scaphopoda have an anterior mouth at the end of an extensible proboscis surrounded by numerous (up to 150) prehensile feeding tentacles called *captacula*. There is a simplified circulatory system, with no heart. There are no gills—the oxygen intake is through the mantle. The nervous system consists of a cerebral ganglion, a pedal ganglion, and a visceral ganglion. The foot is shaped like a wedge to dig into the sediment (the name Scaphopoda derives from the Greek σκάφος, which means *digging*, and ποδός, the genitive of πούς, which means *foot*). A representative genus is *Dentalium*.

Fertilization in mollusks may be internal or external. Internal fertilization produces no larval stages. In external fertilization the zygote may develop into a trochophore larva (Fig. 21.19), which grows into the *veliger* larva with an enlarged girdle of ciliated cells called

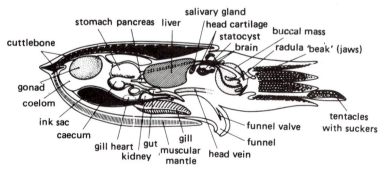

Figure 21.17. *Sepia*, cross section. (Lehmann and Hillmer, 1983, p. 171, Fig. 124.)

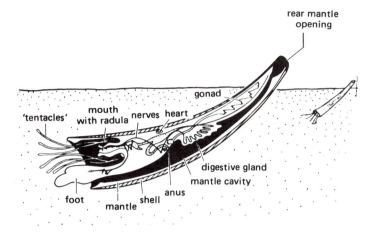

Figure 21.18. Scaphopoda, body structure. (Lehmann and Hillmer, 1983, p. 105, Fig. 71.)

velum. In addition to mollusks, the trochophore larval stage is found among Annelida, Brachiopoda, and Bryozoa.

21.5. BRACHIOPODA
(Tommotian–Holocene)

The Brachiopoda have two valves like the Bivalvia, but that is where the similarity ends. The brachiopods are totally different animals. Not only are the shells dorsal (or *brachial*) and ventral (or *pedical*), instead of left and right as in Bivalvia, but the body structure has nothing in common with that of the Bivalvia (Fig. 21.20). The valves are opened and closed by, respectively, adductor and diductor muscles.

Brachiopods are characterized by a *lophophore* (from the Greek λόφος, for *crest*), an extensible organ consisting of two *brachia* bearing thin, ciliated tentacles (*cirri*) on either side of a *brachial groove* that leads to the mouth. The lophophore, when retracted, is coiled around a supporting carbonate ribbon (*brachidium*) that hangs from the dorsal valve. Brachiopods have paired nephridia to remove nitrogenous waste. Excretion is through the mouth. The circulatory system consists of a longitudinal, dorsal tube with a contractile

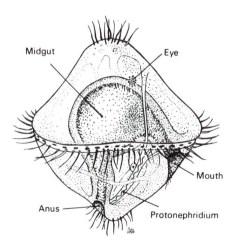

Figure 21.19. The trochophore larva. (Moment and Habermann, 1973, p. 450, Fig. 21.2.)

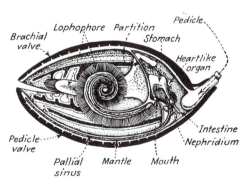

Figure 21.20. Brachiopod, cross section. (Shrock and Twenhofel, 1953, p. 265, Fig. 9.4.)

vesicle, which functions as a heart, and side branches. The nervous system includes two frontal ganglia (dorsal and ventral) with nerve fibers running to each of the cirri.

In some genera the organism lies directly on the substrate. In most species, however, the body is anchored to the substrate through a fleshy stalk called *pedicle*. The pedicle exits through the *delthyrium*, an apical opening between the two valves (earliest forms) or in the ventral valve (later forms); the delthyrium is partly closed, limiting the opening, by *deltidial plates* of shell material. In free or cemented forms, which have no pedicle, the deltidial plates completely close the delthyrium.

Sexes are separate. The zygote develops into a free-swimming, ciliated trochophore larva. The phylum is divided into two classes, Inarticulata and Articulata.

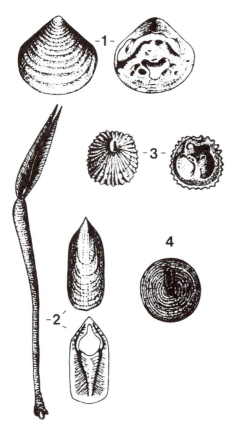

Figure 21.21. Brachiopoda (Inarticulata): (1) *Obolus* (Cambrian–Ordovician, × 3); (2) *Lingula* (Ordovician–Holocene, × 1); (3) *Crania* (Ordovician–Holocene, × 1); (4) *Orbiculoida* (Ordovician–Cretaceous, × 1.2). (Lehmann and Hillmer, 1983, p. 244, Fig. 198.)

Inarticulata (Tommotian–Holocene)

In the Inarticulata (Fig. 21.21), the two valves are not hinged by teeth and sockets. Most genera have chitinophosphatic shells, but some have calcareous shells; most have a fleshy pedicle, but some genera have lost the pedicle and rest directly on the sea floor or cement themselves to it. A representative genus is *Lingula*, one of the longest-lived genera among animals or plants (Ordovician–Holocene). Its longevity is in part due to the fact that the shell has a very simple morphology, implying a rather simple (and therefore durable) genome. In contrast to most brachiopods, *Lingula* is infaunal, living in vertical burrows into which it retracts by action of the pedicle anchored to the bottom of the burrow.

Articulata (Early Cambrian–Holocene)

In the Articulata (Fig. 21.22), the valves are calcareous and hinged, with a teeth-and-socket arrangement. There is a horny pedicle, which is atrophied in some genera. The Articulata are more advanced and much more diversified than the Inarticulata. However, the digestive tract ends in a blind intestine (excretion is through the mouth).

The habitat of the brachiopods is benthic and sessile (via pedicle) or free, with the ventral shell resting on the bottom, often supported by spines. Half of the species live between 100 and 400 m of depth; the other half live deeper. One species has been dredged from 6,179 m of depth. The distribution is worldwide.

The diversity of brachiopods increased from one genus in the Tommotian to 900 genera in the Devonian, and then decreased to about 100 genera today.

21.6. ARTHROPODA (Cambrian–Holocene)

Arthropoda are the largest phylum of living eucaryotes in terms of both the number of species (10^6–10^7?) and the number of individuals. They are found in practically all environments on Earth. Characteristics common to all arthropods are bilateral symmetry, a segmented body, jointed legs, a tendency of the

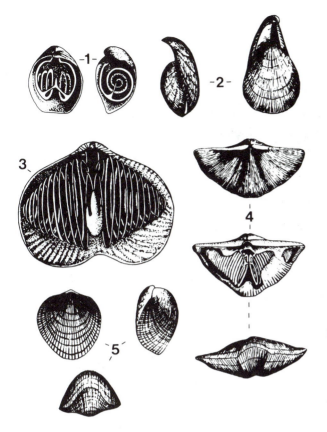

Figure 21.22. Brachiopoda (Articulata): (1) *Dayia* (Upper Silurian, × 1); (2) *Uncites* (Middle Devonian, × 0.5); (3) *Athyris* (Devonian–Permian, × 2); (4) *Spirifer* (Devonian–Carboniferous, × 3); (5) *Atrypa* Silurian–Devonian, × 2). (Lehmann and Hillmer, 1983, p. 251, Fig. 204.)

anterior segments to fuse and form a head, lateral jaws (left–right rather than up–down as in humans), and elaborate sensory organs (antennae, eyes). Arthropoda are classified into four subphyla:

1. Trilobita (Cambrian–Permian)

2. Chelicerata
   Class Merostomata
   Order Aglaspida (Cambrian–Ordovician)
   Order Eurypterida (Ordovician–Permian) (mainly brackish to freshwater environments)
   Order Xiphosura (Ordovician–Holocene) (*Limulus*, the horseshoe crab)
   Class Arachnida (Silurian–Holocene) (spiders, scorpions)
   Class Pycnogonida (Devonian–Holocene) (sea spiders)

3. Crustacea (Cambrian–Holocene)
   Class Branchiopoda (*Daphnia*)
   Class Ostracoda (Cambrian–Holocene)
   Class Cirripedia (barnacles)

Class Malacostraca (crabs, crayfishes, lobsters, shrimps)
Class Copepoda

4. Uniramia
   Class Myriapoda (Devonian–Holocene) (centipedes, millipedes: 30–300 legs, actually)
   Class Insecta (Middle Devonian–Holocene)

Four of these classes, Arachnida, Pycnogonida, Myriapoda, and Insecta, have left very scanty and discontinuous paleontological records. The taxa that have left important records are discussed here briefly.

Trilobita (Cambrian–Permian)

Trilobites (Figs. 21.23 and 21.24) had a segmented body, usually 1–5 cm long, but some up to 40–60 cm long. The Devonian *Tetraspis* had a record length of 67.5 cm. The trilobite body consisted of three parts: the head (cephalon), the thorax, and the *pygidium*. The cephalon consisted of a few fused segments. It

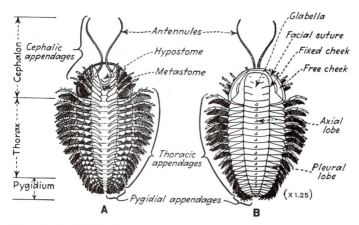

Figure 21.23. Trilobita: Ventral (A) and dorsal (B) views of the Ordovician trilobite *Triarthrus becki*. (Shrock and Twenhofel, 1953, p. 1953, p. 580, Fig. 13.26.)

included a braincase (*glabella*), paired composite eyes with 15 to 15,000 elements (but some species that were mud-dwelling were blind), and paired antennae. The thorax consisted of 2 to 44 segments. Each segment consisted of an axial portion, on either side of which was a *pleuron* that extended laterally into an *endopodite* (for swimming or crawling) and an *exopodite* with a gill attached to it. The

posterior part of the body, called *pygidium*, consists of 2 to 29 fused segments.

The dorsum of a trilobite was protected by an *integument* consisting of chitin mineralized with apatite [calcium phosphate, $Ca_5(PO_4)_3$ (F, Cl, OH)] and calcite $(CaCO_3)$. (*Integument*, which means *external cover*, derives from the Latin *integere*, which means *to cover up*. *Tegere*, to cover, is the root for deck, thatch,

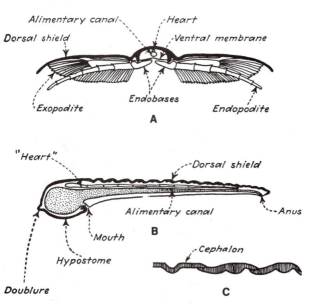

Figure 21.24. Trilobita: (A) Cross section and (B) logitudinal section of the Ordovician trilobite *Ceraurus* *pleurexanthemus*. (Shrock and Twenhofel, 1953, p. 581, Fig. 13.27.)

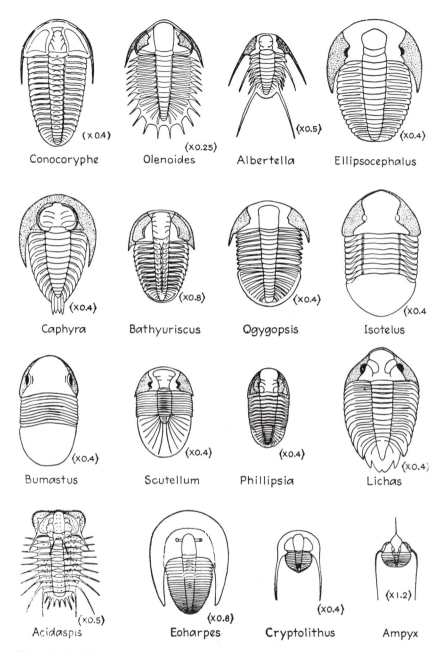

Figure 21.25. Trilobita, various genera. (Shrock and Twenhofel, 1953, p. 595, Fig. 13.36.)

and tectonics.) The ventral side of the trilobite body was unprotected: The trilobites rolled up to protect their undersides.

Trilobites had a complex morphology (Fig. 21.25) and therefore a complex genome. The life span of most species was rather short, ranging from less than one million years to a few million years. Trilobites, therefore, can be used as a basis for a rather detailed biostratigraphy.

$(×1\frac{1}{2})$

Figure 21.26. Aglaspida: dorsal view of *Aglaspella eatoni*. (Shrock and Twenhofel, 1953, p. 567, Fig. 13.17.)

Trilobites were exclusively marine, living on shallow bottoms. They fed on algae, protozoans, and mollusks. When fishes evolved and spread (Silurian–Devonian), trilobites underwent a marked decline, until they became extinct at the end of the Permian.

The first appearance of trilobites on Earth marked the beginning of the Adtabanian, the second age of the Cambrian period.

Aglaspida (Cambrian–Ordovician)

Aglaspida (Fig. 21.26) had a small body, 2–6 cm long, consisting of a semicircular cephalon with compound eyes and paired

appendages; a thorax with 12 segments; and a *telson* (a caudal segment). The carapace was chitinophosphatic.

Eurypterida (Ordovician–Permian)

Eurypterids (Fig. 21.27) were the largest known arthropods, attaining a length of 3 m (*Pterigotus*). The head had paired eyes on the dorsal side, and strong jaws and teeth on the ventral side. The thorax consisted of 13 segments, ending with paired legs (for crawling) and a pair of paddles (for swimming). The thorax ended in a telson. The Eurypterids were at first marine, but later they invaded brackish and freshwater environments. They undoubtedly were fierce predators.

Xiphosura (Ordovician–Holocene)

Xiphosura (Fig. 21.28) evolved in the Ordovician, but they have survived to this day, apparently with little change. A modern representative is *Limulus*, the horseshoe crab common from Maine to Yucatán and from Japan to Indonesia.

Limulus has a large, chitinous carapace with a strong telson. It has two medial simple eyes and two lateral compound eyes. There are six paired legs on the cephalothorax and six paired abdominal appendages, on the last five of

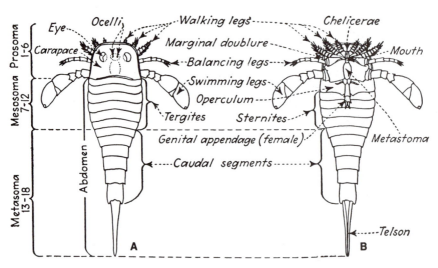

Figure 21.27. Eurypterida. *Eurypterus*: (A) dorsal view; (B) ventral view. (Shrock and Twenhofel, 1953, p. 565, Fig. 13.15.)

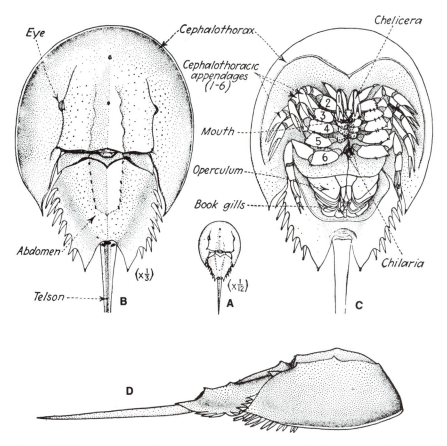

Figure 21.28. Xiphosura. *Limulus polyphemus*, the horseshoe crab: (A and B) dorsal views; (C) ventral view; (D) lateral view. (Shrock and Twenhofel, 1953, p. 562, Fig. 13.13.)

which are 150–200 leafy gills. *Limulus* lives in shallow coastal waters. It buries itself in the sediment at night, but during the day it crawls around and forages. If it is overturned by a wave or something, it can right itself using the telson as a lever. It swims upside down, propelling itself with its appendages.

Crustacea

Of the five classes of Crustacea listed earlier, only the Ostracoda have left an abundant paleontological record (Late Cambrian–Holocene).

Ostracoda (Fig. 21.29) have a poorly segmented body and a bivalve shell of calcium carbonate. The shell was up to 3 cm in size in Paleozoic ostracods, but is smaller (1–5 mm) in Mesozoic and Cenozoic ostracods. Ostra-

cods live in aquatic environments, both marine and fresh water. They swim or crawl on the bottom. More than 2,000 species are living today. Their abundance and the common preservation of the shell make ostracods very useful tools for biostratigraphy. The other orders of Crustacea have left very scanty paleontological records or none at all.

21.7. ECHINODERMATA (Cambrian–Holocene)

The echinoderms (Fig. 21.30) are fully marine, benthic animals (but there are a few swimming holothurians and crinoids). The body has an apparent pentaradial symmetry in many, but bilateral symmetry in others because of the

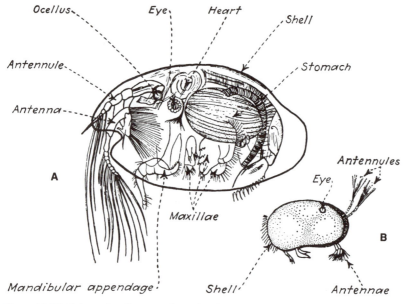

Figure 21.29. Ostracoda. Cross section showing a modern ostracod: (A) animal; (B) shell. The shell is closed by a pair of adductor muscles. (Shrock and Twenhofel, 1953, p. 549, Fig. 13.5.)

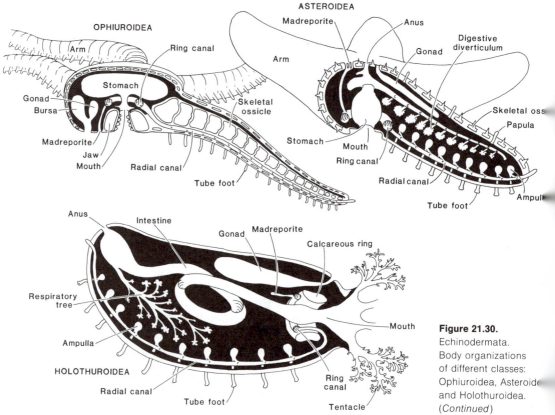

Figure 21.30. Echinodermata. Body organizations of different classes: Ophiuroidea, Asteroidea and Holothuroidea. (*Continued*)

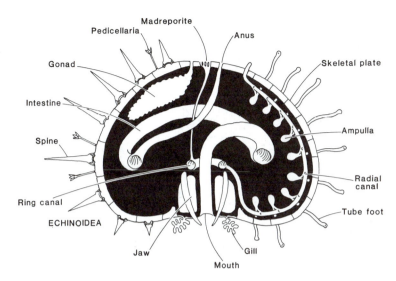

Figure 21.30. (*Cont.*). Echinoidea and Crinoidea.

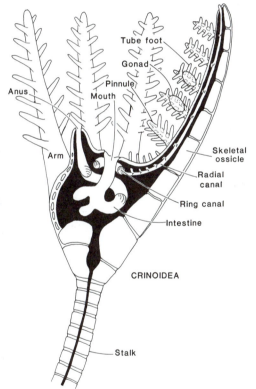

eccentric position of the anus. Body length ranges from a few millimeters to 20 m (giant stalked crinoids). The skeleton consists of calcitic plates with 3–15% Mg (the percentage of Mg increases with temperature).

Echinoderms have a unique water-vascular system with 5 rays radiating from a circular inner canal. Water intake is through a *sieve plate* called *madreporite*; the water is pumped through radial canals in and out of *tube feet*, hollow finger-like extensions that become alternately turgid and limp, thereby allowing the animal to move water and particles toward the mouth and also to move around. The tube feet are implanted on five rows of plates (*ambulacra*) extending radially from the mouth, either on the free arms or attached to the body. In many taxa, a groove (*ambulacral groove* or *food groove*) extends along the middle of each ambulacrum. Sexes are separate, with the zygote developing into a free-swimming, ciliated larva called *dipleurula* that has bilateral symmetry. The dipleurula attaches itself by the oral end and develops into an adult. The Echinodermata are divided into five subphyla and a total of 11 classes:

Subphylum Homalozoa

CLASS CARPOIDEA (CAMBRIAN–DEVONIAN)

In the Carpoidea (Fig. 21.31) the body consisted of a stem, a theca a few centimeters across walled by plates irregularly distributed (no symmetry), and one or two arms, possibly with food grooves. The mouth was at the top of the chamber, with the anus at the bottom.

Subphylum Echinozoa

CLASS EDRIOASTEROIDEA (EARLY CAMBRIAN–PENNSYLVANIAN)

The Edrioasteroidea (Fig. 21.32) had a discoidal body consisting of a variable number of irregular plates, with attached to it 5 ambulacra radiating from the centrally located mouth. The anus was to one side between arms 1 and 5.

CLASS HOLOTHUROIDEA (ORDOVICIAN–HOLOCENE)

The Holothuroidea (sea cucumbers) (Fig. 21.33) have an orally–aborally elongated, muscular body, with the mouth surrounded by tentacles. The body wall contains free calcareous sclerites, 10–100 μm long. Holothuroidea are mud-feeders, living from the low-tide zone to the bottom of the Marianas Trench (−10,915 m).

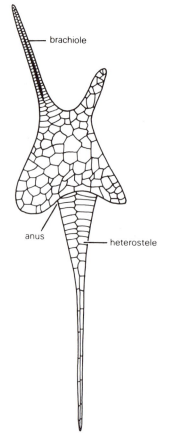

Figure 21.31. Homalozoa (Carpoidea). *Dendrocystites* (Ordovician, × 1) (Clarkson, p. 267, Fig. 9.48a.)

CLASS ECHINOIDEA (MIDDLE ORDOVICIAN–HOLOCENE)

Echinoidea (sea urchins) (Fig. 21.34) have a subspherical-to-discoidal theca consisting of firmly linked calcitic plates. The mouth is ventral, with a strong masticatory organ called

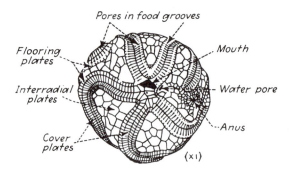

Figure 21.32. Edrioasteroidea. *Edrioaster bigsbyi* (Middle Ordovician, × 0.8), oral view. (Shrock and Twenhofel, 1953, p. 691, Fig. 14.27.)

Figure 21.33. Holothuroidea. *Holothuria.* (Brusca and Brusca, 1990, p. 823, Fig. 12H.)

Aristotle's lantern. The anus is aboral, lateral-posterior, or on the oral side. Spines and pedicellariae are mounted on the aboral surface. *Pedicellariae* are tiny jaws, mounted on a tiny stem, that snap open and shut when disturbed to keep the dorsal surface free of intruders, such as settling larvae (cf. the avicularia of Bryozoa, Section 22.2). Five ambulacra radiate from the mouth, with rows of tube-feet. There is an ocular plate at the end of each ambulacrum. The habitat is subtidal to hadal.

Subphylum Blastozoa

CLASS EOCRINOIDEA
(EARLY CAMBRIAN–SILURIAN)
Eocrinoidea were primitive echinoderms with pores along the sutures between plates.

CLASS CYSTOIDEA (ORDOVICIAN–DEVONIAN)
The Cystoidea (Fig. 21.35) had a globular theca with or without stem, consisting of irregularly distributed plates ranging in number from 13 to more than 200. The mouth was at the top of the chamber, followed (displaced to one side) by the *genital pore* (for the discharge of gametes), the hydropore (for water intake for the water-vascular system), and the anus.

CLASS BLASTOIDEA (SILURIAN–PERMIAN)
The Blastoidea (Fig. 21.36) were stalked forms, with pentameral symmetry. There were five food grooves along the dorsal side of the theca, with a row of *brachioles* (free filaments to move water) on either side. Blastoidea had a complex system of folds, called *hydrospires*, hanging from the thecal roof into the thecal cavity and connected to the outside through a system of five holes (*spiracles*) surrounding the mouth.

Subphylum Crinozoa

CLASS CRINOIDEA (ORDOVICIAN–HOLOCENE)
The Crinoidea (Fig. 21.37) are mostly stalked and sessile benthic forms. A few are free-swimming. The body consists of a cup and a stalk. The cup, called *calyx*, consists of rings of calcareous plates arranged with pentameral symmetry. Five arms (*brachia*) radiate from the

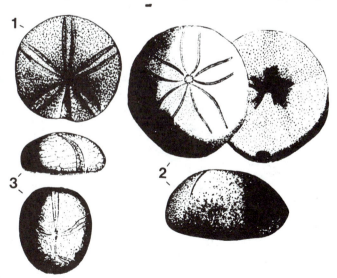

Figure 21.34. Echinoidea: (1) *Clypeus* (Middle–Upper Jurassic, × 0.25); (2) *Echinolampas* (Eocene–Holocene, × 0.7); (3) *Collyrites* (Middle–Upper Jurassic, × 0.5). (Lehmann and Hillmer, 1983, p. 304, Fig. 250.)

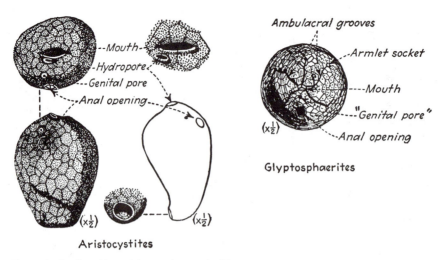

Figure 21.35. Cystoidea. *Aristocystites* and *Glypto-sphaerites* (Ordovician). (Shrock and Twenhofel, 1953, p. 657, Fig. 14.6.)

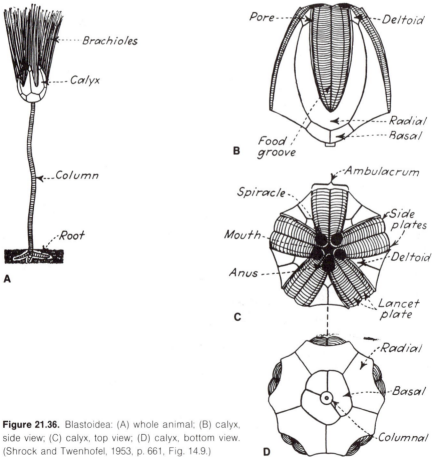

Figure 21.36. Blastoidea: (A) whole animal; (B) calyx, side view; (C) calyx, top view; (D) calyx, bottom view. (Shrock and Twenhofel, 1953, p. 661, Fig. 14.9.)

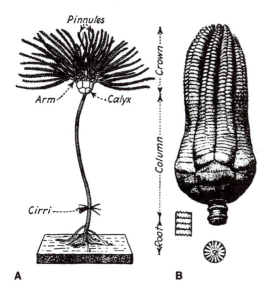

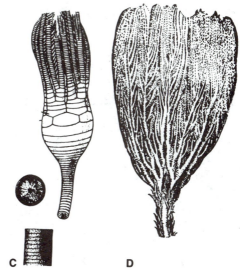

Figure 21.37. Crinoidea: (A) entire animal; (B) *Encrinus* (Triassic, × 0.5); (C) *Apiocrinus* (Jurassic, × 0.5, × 1); (D) *Seirocrinus* (Jurassic, × 0.6). Some species of *Seirocrinus* have crowns (calyx + arms) with up to 1,400 arms, and stems up to 18 m long. (Shrock and Twenhofel, 1953, p. 672, Fig. 14.17A; and Lehmann and Hillmer, 1983, p. 280, Fig. 229.)

rim of the calyx. The brachia consist of dorsal calcitic plates (up to hundreds) lined ventrally with a food grove, on either side of which are cilia that guide the water with food particles (mainly dinoflagellates and tintinnids) toward the mouth. The food groove continues across the soft roof of the cup to the centrally located mouth. The five arms may branch and sub-branch, with the total number of arms remaining a multiple of 5. The stalk consists of hollow cylindrical, elliptical, quadrangular, or polygonal calcitic segments called *entrochites*. It may have short side branches called *cirri*. The stalk is short in all living species, but it was longer in fossil species. In many species the stalk is missing, and the cup rests directly on the substrate. The eggs are produced by oocytes in the arms and are fertilized while attached to the arms. The zygote develops into a dipleurula that swims around and settles by the anterior (oral) end and develops into an adult.

Crinoids are abundant in the modern ocean. Stemless forms live at depths less than 100 m; stemmed forms live at depths greater than 100 m down to the bottom of deep-sea trenches.

Subphylum Asterozoa

CLASS SOMASTEROIDEA (EARLY ORDOVICIAN)

Somasteroidea include the earliest starfish, *Villebrunaster* (Fig. 21.38), with no ambulacral grooves. They probably were ancestral to both Asteroidea and Ophiuroidea.

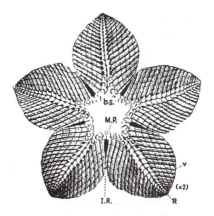

Figure 21.38. Somasteroidea. *Villebrunaster thorali* (Early Ordovician, France, × 2). (Shrock and Twenhofel, 1953, p. 700, Fig. 14.35A.)

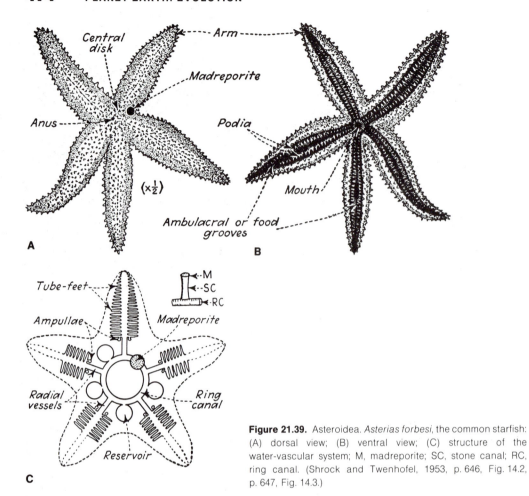

Figure 21.39. Asteroidea. *Asterias forbesi*, the common starfish: (A) dorsal view; (B) ventral view; (C) structure of the water-vascular system; M, madreporite; SC, stone canal; RC, ring canal. (Shrock and Twenhofel, 1953, p. 646, Fig. 14.2, p. 647, Fig. 14.3.)

CLASS ASTEROIDEA (ORDOVICIAN–HOLOCENE)

The Asteroidea (Fig. 21.39) have a star-shaped or pentagonal, flattened body, with a central, ventral mouth and 5 to 40 ambulacra radiating from it. The anus is lateral between rays 1 and 5 (giving true bilateral symmetry). The water-vascular system radiates from the mouth to the tip of each ambulacrum and branches into the ventral tube-feet. Water intake is through an aboral madreporite. The dorsal side often has calcareous spines, and in many species it has also pedicellariae. At the end of each arm there is a light-sensitive plate called *ocular plate*.

CLASS OPHIUROIDEA (MISSISSIPPIAN–RECENT)

Ophiuroidea (brittle stars) (Fig. 21.40) have a body consisting of a central, flattened disc from which radiate five or more thin, long (up to

60 cm) arms covered by a column of vertebrae, each consisting of four pairs of *ossicles*. The vertebrae are articulated so that the arms can move rapidly in snakelike fashion.

Echinoderms are exclusively marine, living from near shore to the bottom of deep-sea trenches. Most feed on microscopic organisms, including trochophore larvae. The holothurians are mud-feeders—they eat the mud on the sea floor and extract from it bacteria, foraminifera, and particulate organic matter. Their activity reworks the bottom sediment to an average depth of several centimeters.

21.8. CHAETOGNATHA [Cambrian(?)–Holocene]

The Chaetognatha (arrow worms) (Fig. 21.41) form a phylum consisting of about 40 modern

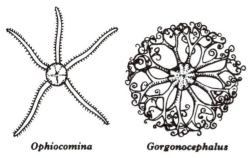

Ophiocomina   **Gorgonocephalus**

Figure 21.40. Ophiuroidea. *Ophiocomina* and *Gorgonocephalus* (modern, × 0.5). (Borradaile and Potts, 1961, p. 687, Fig. 479.)

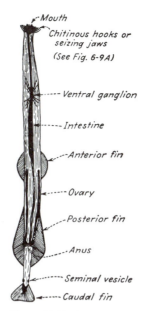

Figure 21.41. Chaetognatha. *Sagitta*, a modern chaetognathan about 5 cm long. (Shrock and Twenhofel, 1953, p. 193, Fig. 6.8A.)

species of marine, mostly planktic worms (a few benthic), with a slender body, lateral fins, a caudal fin, and sets of grasping teeth around the mouth. *Sagitta*, abundant in the modern seas, is 2–7 cm long. Chaetognatha can swim by moving the caudal fin in short spurts, with long periods of rest between spurts. During these resting periods, the Chaetognatha drift passively with the water. For this reason, they are generally considered planktic rather than nektic. Chaetognatha are hermaphroditic, with internal fertilization.

A single fossil specimen from the Middle Cambrian Burgess shale of British Columbia has been assigned to the Chaetognatha.

21.9. CONODONTA
(Early Cambrian–Triassic)

The Conodonta form an extinct phylum of slender marine animals with grasping teeth around the mouth that are similar to chaetognathan teeth. The teeth were supported on a common bar, blade, or platform (Fig. 21.42). Single, conical teeth also occurred. These structures were of millimeter size and consisted of microcrystalline apatite. They are found in a variety of Paleozoic sediments, ranging from conglomerates to black shales, which indicates a nektic–planktic life style similar to that of the modern Chaetognatha.

Although conodont teeth are common in many Paleozoic settings, only one specimen of the entire animal has been discovered (in 1983, in a Lower Carboniferous shale in Scotland). That animal was about 3 cm long and 2 mm

wide; it had a head with conodont teeth, a slender body, lateral fins, and a caudal fin. In contrast to the Chaetognatha, the body of the conodont animal appears to have been finely segmented longitudinally. In that, it resembles a hemichordate.

21.10. HEMICHORDATA
(Middle Cambrian–Holocene)

The Hemichordata are intermediate between the invertebrates and the chordates. There are three classes:

Enteropneusta

The Enteropneusta have no mineralized parts and therefore have left no fossils. A modern representative is *Balanoglossus gigas* (Fig. 21.43), with a body up to 2 cm wide and 2.5 m long. *Balanoglossus* is marine benthic, living in burrows cemented with slime. It has a dorsal nerve chord, underneath which is the *notochord*, a tough, flexible rod consisting of a fibrous sheath filled with gelatinous cells. The function of the notochord is to prevent the telescoping of the body.

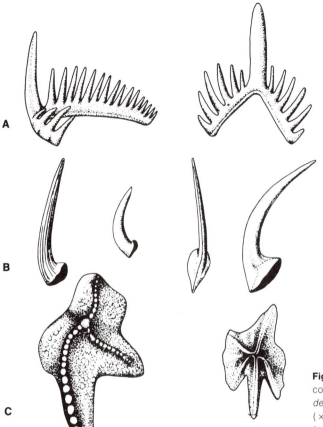

Figure 21.42. Conodonta: (A) compound conodonts: *Lingonodina* (left) and *Hibbardella* (right) (× 75); (B) simple conodonts (× 37); (C) platform conodont; *Palmatolepis* (× 20). (Lehmann and Hillmer, 1983, p. 258, Fig. 209.)

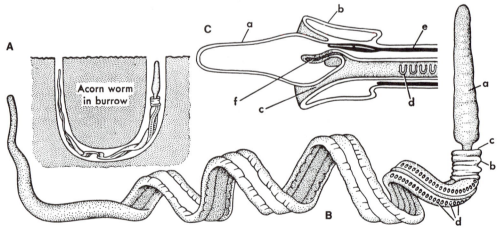

Figure 21.43. Hemichordata (Enteropneusta). *Balanoglossus* (acorn worm): (A) in burrow (× 0.1); (B) a, burrowing proboscis; b, collar; c, mouth; d, gill slits (× 1.5); (C) detail of head region: a, burrowing proboscis; b, collar; c, mouth; d, gill slits; e, nerve chord; f, notochord (× 1.5). (Simpson, Pittendrigh, and Tiffany, 1957, p. 543, Fig. 22.14.)

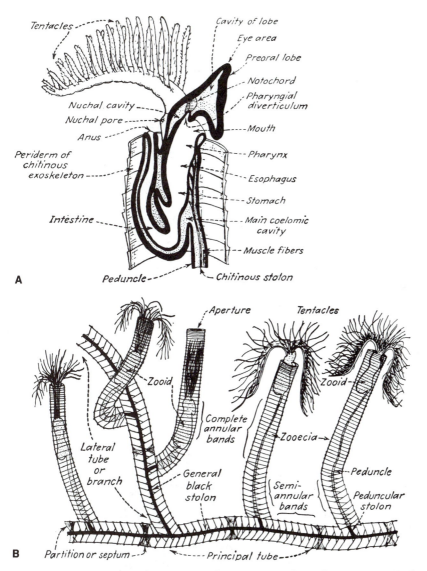

Figure 21.44. Hemichordata (Pterobranchia). *Rhabdopleura*: (A) sagittal (along length of body) section (× 22.5); (B) colony (× 4.5). (Shrock and Twenhofel, 1953, p. 739, Fig. 15.4.)

Pterobranchia

The Pterobranchia also have no mineralized parts and therefore have left no fossils. A modern representative, *Rhabdopleura* (Fig. 21.44), is a marine epibenthic animal 2–3 mm to 1.5 cm high living in solid tubes consisting of scleroprotein secreted by glands in the animal's head. *Rhabdopleura* can retract inside the tube by means of a peduncle anchored to the bottom of the tube. The tubes are connected by stolons and form branches by budding.

Graptolithina (Middle Cambrian–Lower Mississippian)

Graptolithina also had no hard parts, but they have left a good fossil record in the mud flooring anaerobic basins. The Graptolithina were

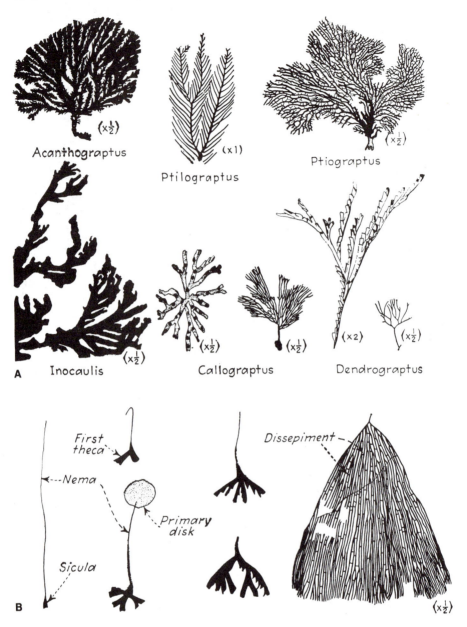

Figure 21.45. Graptolithina (Dendroidea): (A) benthic colonies; (B) *Dictyonema* (hemiplanktic): development of colony. (Shrock and Twenhofel, 1953, p. 751, Fig. 15.14, p. 749, Fig. 15.12F.)

colonial animals living in rows of conical chambers (*thecae*) attached to a rod (*stipe*). The colonies consisted of one, two, or more stipes branching from a common filament (*nema*) attached to the bottom or connected to a float on the sea surface. The colonies originated by budding from a sexually pro-

duced initial conical chamber (*sicula*). Adjacent nemata could be connected through *dissepiments*. All structures were made of scleroproteins (fibrous proteins). The size of an individual theca was about 1 mm. There are six orders, of which two, the Dendroidea and the Graptoloidea, are by far the most important.

DENDROIDEA (MIDDLE CAMBRIAN–EARLY MISSISSIPPIAN

Dendroidea (Fig. 21.45) were the more primitive of the Graptolithina. They were sessile, rising from the sea floor like trees with many branches. There was one exception. *Dictyonema*, which was hemiplanktic (attached to floating seaweeds). The height of the colonies was 2–10 cm.

GRAPTOLOIDEA (EARLY ORDOVICIAN–LATE SILURIAN)

The colonies (Fig. 21.46) were exclusively planktic (hanging from a float via a nema) or hemiplanktic. The thecae opened downward (Fig. 21.47). The colonies were uniserial (*Monograptus*), biserial (*Glossograptus*) or (rarely) quadriserial (*Phyllograptus*) (Fig. 21.46). The lengths of the colonies ranged around 1–2 cm.

The four minor orders of Graptolithina (the Camaroidea, Crustoidea, and Stolonoidea of the Ordovician, and the Tuboidea of the Ordovician–Silurian) consisted of forms adapted to unusual environments (unusual for the graptolites, that is). The Tuboidea were similar to the Dendroidea, but branching was irregular (an indication of life in some kind of marginal environment); the other three orders were encrusting forms.

The scleroproteins of Graptolithina were subject to oxidation and destruction in aerobic environments. Preservation was largely limited to anaerobic environments (black shales).

21.11. CHORDATA (Cambrian–Holocene)

The phylum Chordata includes the subphyla Urochordata (tunicates), Cephalochordata (amphioxus), and Craniata (vertebrates, from agnathous fishes to humans) (see Chapter 22). The phylum includes about 40,000 living species (17,000 fishes, 2,200 amphibians, 6,000 reptiles, 9,000 birds, and 4,400 mammals).

Chordates are characterized by a dorsal nerve cord, the notochord, and paired gills or lungs. The notochord is a flexible cord consisting of a fibrous sheath filled with gelatinous cells. In the Craniata, the notochord is replaced by a segmented vertebral column, and the brain is protected by a skull.

The Urochordata (tunicates) have left no paleontological record. Today, there are about 1,300 species distributed throughout the oceans. The tunicate larva (Fig. 21.48) bears a close resemblance to a fish larva, exhibiting a dorsal nerve cord, a notochord, and gill slits.

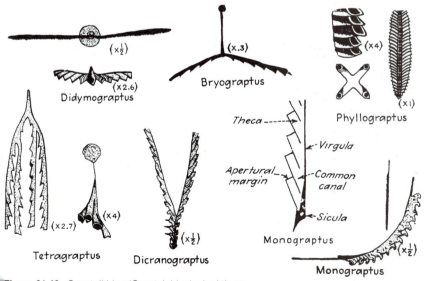

Figure 21.46. Graptolithina (Graptoloidea): planktic or hemiplanktic colonies. (Shrock and Twenhofel, 1953, p. 760, Fig. 15.25.)

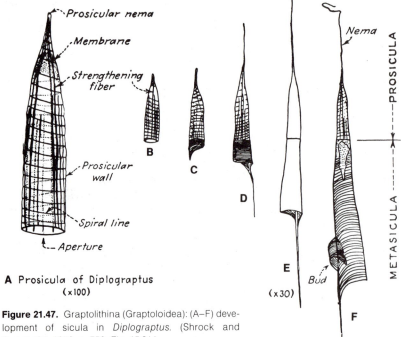

A Prosicula of Diplograptus
(×100)

(×30)

Figure 21.47. Graptolithina (Graptoloidea): (A–F) development of sicula in *Diplograptus*. (Shrock and Twenhofel, 1953, p. 756, Fig. 15.21.)

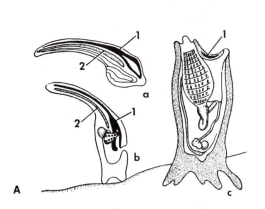

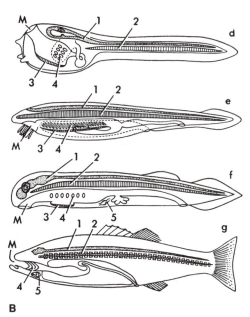

Figure 21.48. Urochordata (tunicates): (A) tunicate larva (a) settling (b) and growing into adult (c) about 2 cm tall; (B) larval tunicate (d) compared to Branchiostoma (e), an agnathous fish (f), and an advanced fish (g); 1, nerve cord; 2, notochord or vertebral column; 3, endostyle (a groove in the pharynx that evolved into the thyroid gland); 4, gill slits; 5, heart. (Simpson, Pittendrigh, and Tiffany, 1957, p. 559, Fig. 22.23.)

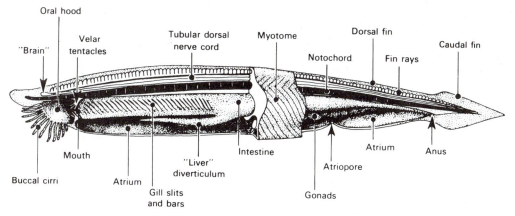

Figure 21.49. Cephalochordata. *Branchiostoma* (*Amphioxus*): internal structure (× 3). (Moment and Habermann, 1973, p. 478, Fig. 23.6.)

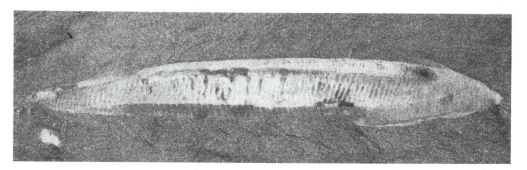

Figure 21.50. Chordata: the earliest chordate, a *Branchiostoma*-like organism from the Middle Cambrian Burgess Shale. The head is to the left. (Courtesy S. Conway Morris, Cambridge, England.)

These features do not survive the larval stage, however. When the larva settles and attaches itself by the oral end, the body undergoes a profound modification, in the course of which the chordate characteristics are lost.

The Cephalochordata also have left no paleontological record. They are represented today by *Branchiostoma*, which used to be called *Amphioxus* and is informally called *lancelet*. Lancelets are fishlike animals of small size (a few centimeters in length). They have a well-developed dorsal nerve cord and an equally well-developed notochord (Fig. 21.49). The mouth is surrounded by a crown of cirri that draw water into the mouth. The water is filtered through the gills, and particulate matter and microorganisms are caught in a mucous membrane and from there directed by ciliated cells to the gut.

The earliest chordate was found in the Middle Cambrian Burgess Shale (Fig. 21.50). It bears a remarkable resemblance to the modern *Branchiostoma*. Aside from this fossil, there is little evidence of chordate life in the Cambrian. The next chordate fossil is a fish, *Arandaspis*, represented by bone fragments found in the Middle Ordovician of Australia. Thus the third subphylum of the Chordata, the Craniata, may have made its appearance after the end of the Cambrian.

22

THE PALEOZOIC

The name *Paleozoic* derives from παλαιός, which means *ancient*, and ζῷον, which means *animal*. It is the first era of the Phanerozoic, ranging from 570 to 245 million years ago. Its beginning is set at the time when fossil remains became abundant. By the end of the Paleozoic, most of the major phyla, except mammals and birds, appear to have been in place. One major phylum, the Archaeocyatha, whose appearance in the fossil record marks the very beginning of the Paleozoic, became extinct shortly thereafter, in Middle Cambrian time (thus lasting "only" some 50 million years).

The Paleozoic era comprises six periods: Cambrian, Ordovician, Silurian, Devonian, Carboniferous (split into Mississippian and Pennsylvanian in the United States), and Permian. The type sections for the first four periods are in Wales and in southwestern England. (A *type section* is the sequence of strata originally described as representing a given stratigraphic unit.)

The Paleozoic saw the rapid expansion of life and also two great mass extinctions, toward the end of the Devonian ($367 \cdot 10^6$ y ago) and at the end of the Permian. In both cases, life recuperated rapidly, with new taxa replacing the old ones that had not survived.

22.1. CAMBRIAN (570–510·10^6 y B.P.)

The Cambrian was first described from a section just east of Harlech Castle in northern Wales and was named after Cambria, the Roman name for Wales (Fig. 22.1).

The map of the world in Cambrian time looked very different from that of today. Africa lay across the South Pole and its north coast barely reached 50°S. So Africa was cold. In contrast, North America lay right across the equator, with what is now its Arctic coastline facing east, and what is now Mexico facing west. About 5,000 km away to the east lay Siberia, and to the southeast was Europe. Beyond Siberia and Europe lay a great landmass that included China, India, Australia, Antarctica and South America, all bound to Africa.

The planetary wind system has remained essentially the same through geological time. The two high-pressure belts at 30°N and 30°S, produced by the convergence of the Hadley and Ferrel cells, were in place also in Cambrian time. However, North America and Siberia lay between the two belts, and Europe lay between 30°S and 60°S. These lands had abundant rain. The climate was tropical-equatorial in North America and Siberia, and temperate in Europe. Only the northern portion of the great landmass farther east, which included northern South America and Australia, lay directly under the convergence and had little or no rain.

At the beginning of the Cambrian a new system of mid-ocean ridges came into existence. The sea floor swelled along the ridges, and seawater spilled over the continents. A major marine ingression took place during the Cambrian. At the end of the Cambrian and at the beginning of the Ordovician, between 510 and 490 million years ago, sea level stood very high, about 350 m higher than today. About 50% of the land was covered with shallow seas in which marine animals and plants thrived and speciated.

In the shallow seas of North America, Siberia, and Europe, widespread but thin (~ 300 m) deposits of sandstone, shale, and limestone formed. Some of the sandstones are cross-bedded (which indicates tidal effects). At the peak of the marine ingression, in Late Cambrian and Early Ordovician time, dry

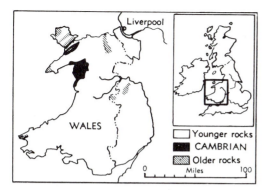

Figure 22.1. The Cambrian outcrops in Wales. (Moore, 1958, p. 108, Fig. 5.1.)

land in North America was restricted to the Canadian shield to the east and, farther west, to the Black Hills of North Dakota, the Ozarks in Missouri, the Ouachita, Arbuckle, and Wichita areas of Oklahoma, and the Llano of Texas. These areas appeared as islands on the vast, shallow sea.

Some 5,000 km to the east, across an ocean as wide as the Atlantic, lay Siberia. As sea level rose during the Cambrian, western Siberia was flooded. Only its eastern portion, the Angara shield, remained above water.

Europe lay southeast of North America and southwest of Siberia, in the temperate belt between 30°S and 60°S. Southern Europe, at 60°S and surrounded by the frigid southern ocean, had a rather miserable climate. Northern Europe, located at 30°S, was faring a lot better. The Cambrian ingression covered most of Europe, except the Fennoscandian shield. Shallow-water deposits formed all around the shield. In the Oslo region, above the crystalline basement rocks of the shield are thin sandstones that are overlain by shales and glauconitic sandstones with *Olenellus* (one of the earliest trilobite genera), for a total thickness of 30–50 m. Above these sediments there are 75 m of shales with *Paradoxides* (a trilobite characteristic of the Middle Cambrian), followed above by *Olenus* (a trilobite characteristic of the Late Cambrian). The shales containing *Olenus* are overlain with more shales, which, however, contain the graptolite *Dictyonema flabelliforme* instead of *Olenus*. The first appearance of this graptolite has been chosen by geologists to mark the

beginning of the next geological period, the Ordovician.

Along the borders of the continental masses, thick sedimentary piles of geoclinal facies were deposited. Typical are the sediments deposited along the northern border of North America. The coastline was running west to east and was facing north. It lay in the region of the trade winds that, as today, blew from the northeast. There a thick pile of sandstones accumulated first, as the sea advanced over the land. With the shoreline moving farther inland, the seawater became progressively clearer, and thick accumulations of carbonate rocks formed. The advance of the sea onshore was not smoothly continuous; it underwent numerous stops and even reversals. These are reflected in characteristic limestone–shale cycles, repeated several times. The thick pile of Cambrian sediments deposited along the northern coastline of North America now forms the backbone of the Canadian Rockies (Fig. 22.2). The majestic peaks of Mounts Eisenhower, Assiniboine, Temple, Robson, and so forth, consist mainly of flat-lying carbonate–shale sequences of Cambrian sediments.

A famous deposit in the Canadian Rockies is the Burgess Shale of Middle Cambrian age. This shale, 2 m thick, accumulated in a small, anaerobic basin on the ancient continental slope, 200 m below the rim of the continental shelf. It was the product of repeated turbidity currents. The animals swept by the turbidites and buried on the basin's floor were preserved because of the anoxic conditions. Geologists have recovered 44 genera of arthropods, 19 of worms, 18 of sponges, 8 of brachiopods, 7 of priapulids, 6 of annelids, and 5 or fewer of hemichordates and chordates, echinoderms, coelenterates, and mollusks. Many have not been found anywhere else. The sad conclusion is that standard fossil deposits represent only a small part of the total living fauna.

Fig. 22.3 shows the succession of trilobite species that lived in the shallow Cambrian sea of North America. The letters are the initials of successive faunal zones characterized by specific trilobite assemblages. This illustration is representative of many similar illustrations showing the succession of species of plants or animals through geological time. The illustration is highly intriguing on several counts:

Figure 22.2. Cambrian. Nearly all rocks in this illustration are Middle–Upper Cambrian carbonates over 3 km thick. Mount Robson (3,954 m high, the highest peak in the Canadian Rockies) consists entirely of Cambrian strata. (Photograph by K. T. Hyde, Calgary; from Holland, 1971.)

1. As can be seen by comparing the ranges of the species with the time scale, some species existed for only 100,000 y or less, and others lived as long as $3 \cdot 10^6$ y; we do not know how long an individual trilobite lived, but assuming that it was one year, a trilobite species would have lasted 100,000 to $3 \cdot 10^6$ generations.

2. At times there were only two or three species in the ecosystem, and at other times the number of species present was a dozen; in other words, the *diversity* of the trilobite fauna ranged from 2 to about 12.

3. Sometimes the disappearance of a species occurred at about the same time that another species appeared; at other times a species disappeared without replacement; altogether, it seems that most species disappeared for no apparent reason.

4. Three times, at the boundaries between C and A, between E and T, and between S and the Ordovician, eight to nine species disapppeared at the same time, indicating the temporary establishment of conditions adverse to the whole group.

5. Two to three species managed to survive across the boundaries, but in $2 \cdot 10^6$ y or so a robust diversity was reestablished.

6. The time intervals between the onsets of adverse conditions averaged $5 \cdot 10^6$ y.

The illustration shows the actual record. As recently as the middle of the nineteenth century this record would have been explained as the result of successive creations (148, in this case). Today scientists believe in evolution. But which model? Neo-Darwinism, the "punctuated equilibria" of Stephen Jay Gould and Niles Eldredge, or my own "extinctive evolution"? You decide.

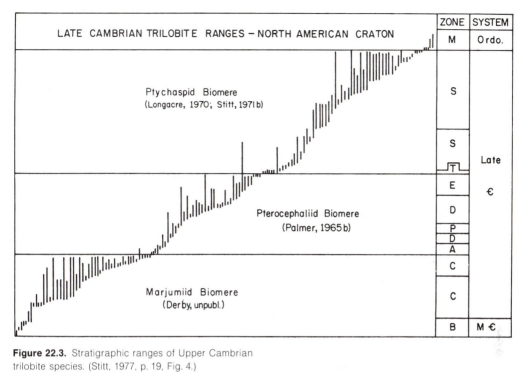

Figure 22.3. Stratigraphic ranges of Upper Cambrian trilobite species. (Stitt, 1977, p. 19, Fig. 4.)

Example of a Type Section: the Cambrian of Wales

The beginning of the Cambrian and the boundaries between all successive geological periods and their subdivisions (epochs, ages, etc.) are based, by international agreement, on specific changes in the marine invertebrate fauna. Thus, the beginning of the Cambrian is defined as the time when the first Archaeocyatha appeared on Earth, and the beginning of the next period, the Ordovician, is defined as the time when the graptolith *Dictyonema flabelliforme* appeared. In addition, for each geological period or subdivision, a type section is chosen that is supposed to be representative of that time interval. However, the sedimentological environment depends on a number of factors that change from place to place (climate, geomorphology, type of sediments being supplied, etc.). As a result, no single type section can be truly representative.

The type section for the Cambrian is around Harlech Castle in Wales (Fig. 22.4). There, in a basin not more than 20 km across, clastic sediments reaching a maximum thick-

ness of 3,630 m in the center were deposited. Tectonism has transformed this basin into a dome that has been eroded, exposing the oldest sediments in the center. The Cambrian sediments at Harlech are clastics that were deposited by turbidity currents at middle-southern latitudes. During the same time, the great carbonate deposits that now form the backbone of the Canadian Rockies were deposited 8,000 km away, at low northern latitudes. The two facies are completely different, and the fossils they contain are also different. In fact, the trilobite faunas of the Canadian Rockies have hardly any species in common with those in Wales.

Because the Harlech section is the type section for the Cambrian, it is presented here in some detail, as an example. As is customary in stratigraphy, the section is described from the bottom up. No. 1 represents the earliest Cambrian, outcropping at the center of the dome. No. 12 represents the transition to the Tremadocian, the first epoch of the next geological period, the Ordovician. The Tremadocian outcrops along the margin of the

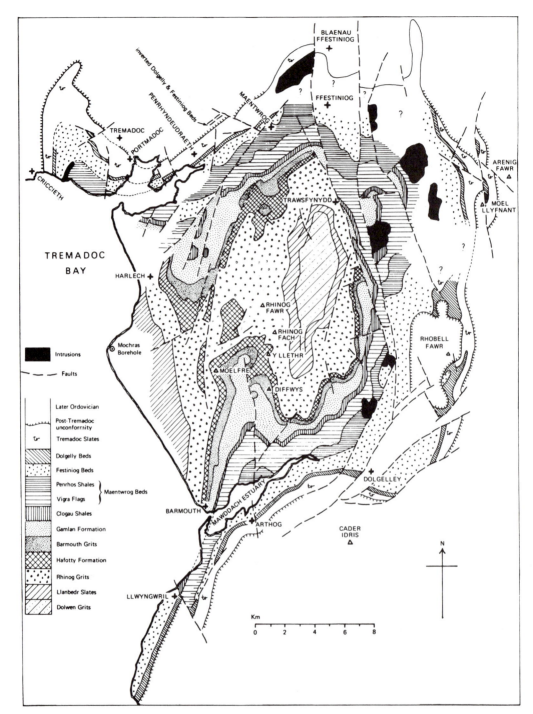

Figure 22.4. Cambrian. The Harlech dome (northern Wales). (Holland, 1974.)

basin. To properly follow the succession of facies changes and of events, one should read this section from the bottom up, that is, starting with No. 1 and ending with No. 12.

Stratigraphic Section Across the Harlech Dome, Wales

12. Transition to light-gray slates (Tremadocian).

11. Dolgelly beds (60–180 m thick): dark-blue laminated mudstones and slates with pyrite, indicating continued anaerobic conditions.

10. Festiniog beds (520–800 m thick): thinly laminated bluish-gray to yellowish-gray slates; anaerobic conditions.

9. Penrhos shales (310 m thick): dark-blue shales with pyrite (12.48%) and organic matter (3.9%), indicating anaerobic conditions.

8. Vigra flags (340 m thick): fine-grained, light-gray, ungraded graywacke in 10–20-cm-thick beds called "ringers" because of their sound when struck with a hammer. The beds exhibit ripple marks and convolute bedding, and they alternate abruptly with shales. The Vigra flags accumulated under aerobic conditions in very shallow water. The section has more ringers below, more slate above. The upper, slaty section accumulated in deeper water. It contains trilobites, which preferred deeper water.

7. Clogau shales (75–100 m thick): uniform, dark-blue, laminated mudstones with pyrite (21.9%) and organic matter (5.8%). Following the rapid accumulation of the Gamlan turbidites, the Clogau shales indicate quiet deposition of muds on an anaerobic bottom.

6. Gamlan formation (230–300 m thick): The Gamlan formation is all turbidites, with graded-bedded graywacke and shale sequences characterized by flow marks, load casts, and convolute bedding.

5. Barmouth grits (100–200 m thick): The Barmouth formation starts with shallow-water pebbly layers and cross-bedded clastics and continues with deeper-water turbidites:

> graded-bedded graywackes (highest)
> coarse, cross-bedded shales (shallow water)
> pebbly layers (lowest)

4. Hagfotty formation (up to 250 m thick): The Hagfotty formation includes the famous manganese ore bed, together with blue and green mudstones and blue grits (the green color is due to reduced iron, and the blue color is due to finely disseminated manganese). The formation includes the following:

grits and shales (highest)	10–15 m
blue grit	1 m
laminated blue mudstone	2 m
Mn ore bed	0.5 m
green mudstone	5 cm
blue-gray mudstone (lowest)	10 m

The Mn ore bed contains rhodochrosite ($MnCO_3$) and spessartine (Mn garnet, $Mn_2Al_2Si_3O_{12}$).

3. Rhinog grits (400–800 m): graded graywackes, medium-to-coarse-grained, cross-bedded, with interbedded shales; pebbly layers are common.

2. Llanbedr slates (90–200 m): bluish slates with few sandy beds.

1. Dolwen grits (150+ m): cross-bedded, gray-green, fine-to-medium-grained graywackes with pebbly layers (grit is a fine-grained sandstone in which the sediment particles are angular). Crossbedding shows that at the beginning the Harlech basin was very shallow. As subsidence began, more sediments flowed in. The Rhinog grits are largely turbidite deposits. Cross bedding and pebbly layers indicate temporary infilling of the basin and shallow-water deposition.

The section terminates with a transition to light-gray slates that contain *Dictyonema flabelliforme* and are therefore of earliest Ordovician age.

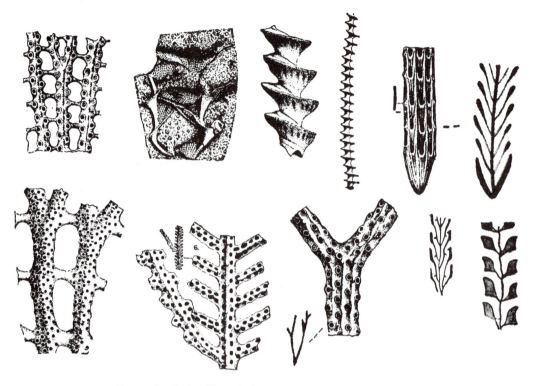

Figure 22.5. Bryozoa. Forms of colonies. The spiral form is *Archimedes* (Mississippian–Permian). (Shrock and Twenhofel, 1953, p. 246, Fig. 7.44.)

22.2. BRYOZOA (Ordovician–Holocene)

Bryozoa are minute, exclusively colonial, mainly marine animals depositing calcium carbonate *zooecia* (zooecia are the enclosures in which the animal lives). A colony is started by a sexually produced trochophore larva settling and developing into an *ancestrula*, which forms a *stolon* consisting of modified zooids. The colony grows by budding from the stolon. The body is triploblastic coelomate.

The Bryozoa have a circular (Stenolaemata, Gymnolaemata) or horseshoe-shaped (Phylactolaemata) lophophore for sweeping food particles, small crustaceans, and so forth, into the mouth. The mouth is followed by a U-shaped digestive tract that ends with an anus outside the lophophore. The muscles include longitudinal and transversal muscles, retractor muscles from the base to the tentacles to retract the tentacles, and a circular muscle (sphincter) just below a collar surrounding the base of the tentacles to close the mouth. The retracted tentacles are invaginated, with the cilia facing inside; they are extruded by pressure when the transversal muscles contract. There is no respiratory, vascular, or excretory system.

In the process of excretion, the entire body disintegrates and forms a brown, compact mass. The individual reforms, and the brown mass comes to rest in the stomach. It is then extruded through the anus.

Bryozoa are hermaphrodites: Each individual contains ovaries and testes. The zygote is kept inside tentacle sheaths until it grows into the trochophore larva. Bryozoa are subdivided into three classes:

The Stenolaemata (Middle Ordovician–Holocene) are entirely marine bryozoans depositing calcitic zooecia. Stenolaemata were the dominant marine Bryozoa from the Ordovician to the Cretaceous, and they

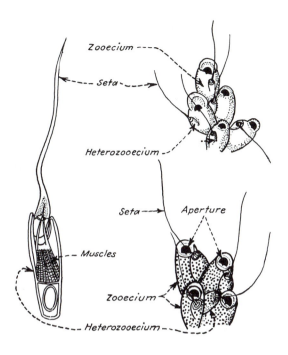

Figure 22.6. Bryozoa. Zooecia (cavities occupied by the bryozoans) and heterozooecia with vibracula (× 11). (Shrock and Twenhofel, 1953, p. 239, Fig. 7.38.)

are still common today. An example is *Archimedes*, a colony that grows spirally (Fig. 22.5).

The Gymnolaemata (Late Ordovician–Holocene) are mainly marine, but can be found also in brackish and fresh water. The Gymnolaemata became the dominant Bryozoa during the Cenozoic. The basal and side walls of the zooecia usually are calcified, but the frontal wall often is not completely calcified. The Cheilostomata (an order of Gymnolaemata) have *vibracula*, which are modified cells with a strong muscle and a whip (Fig. 22.6); they also have *avicularia*, snapping, bird-beak-like structures mounted on the sides of the colony to wave away or crush intruders (especially larvae of encrusting organisms). An example is *Bugula* (Fig. 22.7).

The Phylactolaemata (Miocene–Holocene) are exclusively freshwater Bryozoa with uncalcified zooecia.

Bryozoa are a well-diversified group of mostly marine animals numbering 4,000 modern species. Bryozoa live mainly in shallow (0–100 m), tropical, marine waters, but have been found even in deep-sea trenches. The shapes of the bryozoan colonies depend in part on the dynamics of the waters in which they grow. This may create difficulties for the taxonomist, but on the other hand it provides a way to assess the environmental conditions.

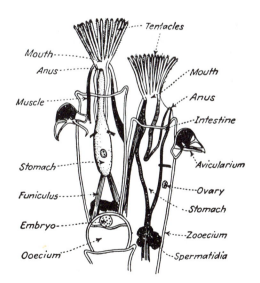

Figure 22.7. Bryozoa. *Bugula avicularia*, a modern bryozoan. (Shrock and Twenhofel, 1953, p. 198, Fig. 7.2.)

More than 16,000 fossil species have been described.

22.3. ORDOVICIAN (510–439·10⁶ y B.P.)

The Ordovician began with a whisper. Hardly anybody noticed. The only difference was some *Dictyonema* floating hither and thither. The type section for the Ordovician is again in Wales (Fig. 22.8). The name Ordovician derives from the Ordovices, the Roman name for a Celtic tribe living in northern Wales. The first epoch of the Ordovician, the Tremadocian, is named after the town of Tremadoc, about 10 km north of Harlech Castle. It was during the Tremadocian that the sea reached its greatest extent inland everywhere in the world. On shallow, well-aerated, shelf areas carbonate deposits formed that were rich in corals, bryozoa, brachiopods, and trilobites. In basins and troughs, turbidites acccumulated. In the deeper bottoms offshore, radiolarian muds and shales rich in graptolites were deposited. The radiolarian muds diagenized into chert. The source of the clastic sediments in Wales and in the Lake District to the north was a mountain range to the northeast.

North America still lay across the equator. In fact, the equator ran from San Francisco to Labrador. The vast interior of the continent was covered with a shallow sea. An elevated ridge, the Transcontinental Arch, not all of it reaching above sea level, ran from Los Angeles to Minnesota and beyond. It ran parallel to the equator, but about 800 km to the south. In the shallow sea north of the Transcontinental Arch, shallow-water limestones rich in brachiopods, crinoids, cephalopods, trilobites, and bryozoa accumulated. To the south of the arch, salinity was very high, and dolomites formed. On the borders of the continent, in what is now the northwest and where the Gulf states now are, cherts and black shales rich in graptolites accumulated. A vast sheet of extremely well rounded, well-sorted sand, the St. Peter Sandstone, derived from the Canadian shield to the east and moved by the trade winds, covered the upper central states, from Missouri to Indiana and southern Wisconsin.

Although some fragments of fish bones have been reported from Cambrian deposits in Wyoming, undisputable fish remains are

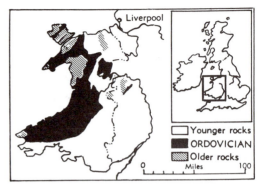

Figure 22.8. The Ordovician outcrops in Wales. (Moore, 1958, p. 136, Fig. 6.1.)

found in Ordovician deposits. What makes a fish are a couple of distinctive characteristics— a vertebral column and paired gills. Fish evolved from hemichordates. The vertebral column evolved from the notochord. The purpose of the gills is to extract oxygen dissolved in water, stick it to the iron ion at the center of the hemoglobin molecule in the fish's blood, and have the blood carry it to metabolic sites (the mitochondria), where it is used to oxidize food and make ATP bonds. The trick is to expose blood to water without hemorrhaging. That is what the gill does.

The earliest fishes, belonging to a group called Agnatha (Fig. 22.9), had a tough life. For a start, they did not know how to swim. Second, they had no jaws, their teeth being implanted directly in the flesh around the mouth as in the modern lampreys. Third, they were considered a delicacy by a host of other animals that had been around for a while and were therefore quite streetwise. So the poor early fishes, perhaps only a few centimeters long, found it necessary to cover their bodies with bony plates. (Translation: Those mutations that had bony plates covering the body had a better chance of survival.) That, of course, ruled out swimming entirely. The early fishes were condemned to squiggle through the mud and eat whatever meager leftovers they found there. Eventually they developed jaws from an anterior pair of gill bars, reduced their armor to the front half of the body, and learned some awkward swimming. That took 100 million years. Just when things were looking up, the sharks evolved. The sharks were

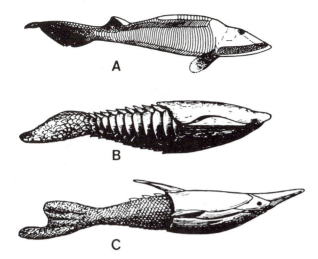

Figure 22.9. Chordata (Agnatha): (A) *Hemicyclaspis* (Upper Silurian–Lower Devonian, × 0.3); (B) *Anglaspis* (Lower Devonian × 0.5); (C) *Pteraspis* (Lower Devonian, × 1). (Romer, 1966, p. 17, Fig. 17, p. 20, Fig. 22.)

excellent swimmers (they still are); they could catch anything they wanted (they still can); and they had formidable teeth. They couldn't care less about bony protections. And so the armored fishes were wiped out. That's the way the ball bounces. We owe the early fishes an eternal debt of gratitude, however, because they invented *bone*.

Bone is an extraordinarily useful substance, without which you and I would look like disgusting mounds of jelly. Bone is mineralized cartilage. Cartilage consists of collagen fibers strung together. Collagen is a protein molecule containing about 1,000 amino acids, the most common of which (about 30% of the total) is glycine. A collagen molecule is coiled with negative helicity and a pitch of 3 amino acids (~ 0.8 nm). It is then twisted in the opposite direction (positive helicity), with a pitch of 11 amino acids (2.9 nm). Three chains are entwined with positive helicity (the *triple helix*) and bound to each other by hydrogen bonds, forming a *tropocollagen* strand about 280 nm long. Tropocollagen strands, staggered by 7 nm and bound to each other by hydrogen bonds, form a *fibril* about 120 nm thick. Fibrils are bundled into *fibers* about 0.7 μm across and 10 μm long to form a network. In bone, the network is mineralized with minute crystals of hydroxylapatite [$Ca_5(PO_4)_3OH$] and calcite ($CaCO_3$), giving bone its strength. Compact bone is almost as strong as cast iron, but is more flexible. The hydroxylapatite makes up 60–80% of the dry mass of the bone. In the human body, the bones contain 99% of the body's calcium and 88% of the phosphorus.

Collagen exists in many other tissues, such as epithelia, ligaments, and cartilage. It is, in fact, the main constituent of connective tissues and one of the most common proteins in animal bodies.

During the last few million years of the Ordovician, a major glaciation developed in the Southern Hemisphere. Thick ice sheets covered North Africa and adjacent regions of South America. Sea level dropped, and the vast inland seas that had covered much of the continents withdrew. Dry land replaced shallow seas. The rich marine faunas that populated the vast, shallow Ordovician seas were devastated. Some 50% of the genera disappeared.

An island arc off the southern coast of North America (which is now the eastern coast) dumped sediments in a trough along the coast. Subduction from the south pushed that island arc against the continent and folded the sediments into a mountain range (the Taconic range, the roots of which are visible in the Taconic Mountains of eastern New York state). Sediments eroded from that range formed a huge delta, the Queenston delta, that extended from New England to Alabama. That delta formed while glaciation was in progress in North Africa. A glaciation consists of a sequence of glacial and interglacial ages, during which ice advances and retreats, and as a consequence, sea level goes up and down. The surface of the Queenston delta was exposed during each ice age, and its surface desiccated. The climate was hot and dry, and so the sediments were well oxidized and red in color. (The red muds that created havoc in the Juniata and Susquehanna valleys when hurricane Agnes passed over that area in 1972 were part of the ancient Queenston delta.)

But the glaciations ended, and at the beginning of the Silurian the sea advanced inland once more, and the old environments were re-established. Rapid speciation repopulated the world, so that in perhaps 10^7 y the diversity of life on Earth had returned to the level it had had during the Late Cambrian and the Tremadocian. Indeed, life has a remarkable ability to bounce right back after a major disaster.

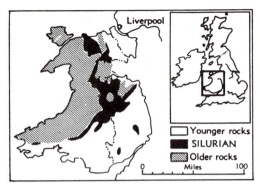

Figure 22.10. The Silurian outcrops in Wales. (Moore, 1958, p. 164, Fig. 7.1.)

22.4. SILURIAN (439–408.5·10^6 y B.P.)

The Silurian was named after the Silures, a fierce Celtic tribe living in southern Wales that resisted the Romans for 30 years (48–78 C.E.). Its type section is again in Wales (Fig. 22.10).

The series of marine regressions at the close of the Ordovician, caused by glaciation in North Africa, exposed much of the land. A significant amount of erosion took place while tectonic motions tilted the Ordovician sediments in Wales. The transgressive Silurian sea deposited sediments with an angular unconformity on the eroded and tilted surface of the underlying Ordovician. Above the basal conglomerates and sandstones, on the east there are shallow-water carbonates and shales with rich faunas of trilobites, brachiopods, and corals, and on the west and northwest (Lake District) there are deep-water deposits of shales with graptolites reaching a thickness of 4,000 m.

In North America, Silurian sediments grade from sandstones (in a belt running from New York State to Alabama) to shales along a parallel belt farther inland, and then to limestones that floored a vast area from Michigan and Wisconsin to Texas. Michigan, exclusive of the upper peninsula, formed a basin in which sediments accumulated for 300 million years (Cambrian to Pennsylvanian), reaching a thickness of 5.5 km at the center. Coral reefs, from less than 1 km to up to 10 km across and 3 to 300 m high, developed and

flourished in the shallow sea around the rim of the basin and along a belt stretching to Illinois and Wisconsin on one side and to Labrador on the other. The reef-builders were stromatoporoids, corals, calcareous algae, and sponges; the reef-dwellers were brachiopods, gastropods, pelecypods, and crinoids. Their skeletal fragments, tossed around by waves and storms, form aprons of loose debris around the ancient reefs. The reefs stuck out only a few meters above the surrounding bottom.

Corals and calcareous algae tended to occupy the windward side of a reef, as they do today, while to the lee communities of trilobites, brachiopods, crinoids, gastropods, and bivalves flourished. Heinz Lowenstam (b. 1916) studied the distribution of these organisms and found that the prevailing ancient winds were from the southwest. Between the Late Cambrian and Early Silurian, North America had rotated counter-clockwise, so that the Arctic coast was facing north when the reefs developed (Middle Silurian). The broad belt of reefs was closely following the equator. The prevailing winds, therefore, should have been from the east. Europe and Siberia, however, had been approaching North America from the east. By middle Silurian time those two landmasses were only a few hundred kilometers offshore, Europe to the east and Siberia to the northeast. So a monsoon circulation is likely to have developed, with strong winds crossing the equator from southwest to northeast during the summer, and a reverse, weaker circulation during the

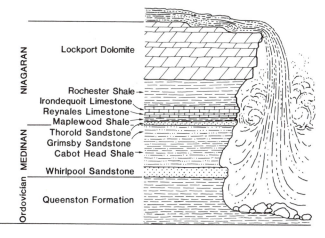

NIAGARAN

Lockport Dolomite

Rochester Shale →
Irondequoit Limestone
Reynales Limestone
Maplewood Shale

MEDINAN

Thorold Sandstone
Grimsby Sandstone
Cabot Head Shale →

Whirlpool Sandstone

Ordovician

Queenston Formation

Figure 22.11. Niagara Falls (section): dashes, shale; dots, sandstone; straight brickwork, limestone; slanted brickwork, dolomite (these graphic symbols have been officially adopted for geological sections). Medinan = Early Silurian; Niagaran = Middle Silurian. (Adapted from Moore, 1958, p. 174, Fig. 7.12.)

winter. The strong southwesterly summer winds had an important part in shaping the Silurian reefs in terms of structure, morphology, and the distribution of reefbuilding and reef-dwelling organisms.

At the end of the Silurian came a big crunch: Europe hit North America from the east. The thick Cambrian, Ordovician, and Silurian sediments that had been deposited for 200 million years off the east coast of North America and the west coast of Europe were folded into a great mountain range, the Caledonian mountain range, that stretched from New York northeastward to Norway. Its roots remain exposed in Scotland. Somewhat later, the crunch folded sediments deposited in a trough across Europe and formed a mountain range that ran from Cornwall to Bohemia. The sea level dropped sharply at the end of the Silurian, and the mountain ranges that had formed were rapidly eroded. (At a rate of 1 mm/year, a 3,000-m-high mountain will be flattened out in $3 \cdot 10^6$ y, neglecting isostatic adjustment, or in 10^7 y with complete isostatic adjustment.)

A major marine regression accompanied

mountain building at the end of the Silurian. The sea level dropped 200 m, and again vast areas of the continents became dry land. In that process of drying, salt deposits of gypsum, anhydrite, and halite were formed in Michigan (reaching a thickness of 550 m at the center of the basin), as well as in Ohio, West Virginia, Pennsylvania, and New York. At the same time, the limestone of the reefs became dolomite. The Niagara Falls exist because a hard layer of dolomite, the Lockport Dolomite, 25 m thick, overlies weaker shale layers, the lowest of which floors the gorge and belongs to the rim of the Queenston delta (Fig. 22.11). The shales are eroded by the turbulent water at the foot of the falls, and the Lockport Dolomite is undermined.

The first Osteichthyes (the bony fishes, characterized by a bony skeleton) appeared in the Silurian. They developed the jaw, apparently from two or three frontal gill arches (Fig. 22.12). The earliest group of bony fishes, the Acanthodii, were small (~ 10 cm long) and had their bodies covered with bony scales (Fig. 22.13). The scales consisted of an inner layer of lamellar bone grading into a layer of spongy

Figure 22.12. Osteichthyes. Development of the jaw from the supports for the two to three frontal gill arches. (Romer, 1966, p. 7, Fig. 6.)

Figure 22.13. Osteichthyes (Acanthodii). *Climatius,* a Lower Devonian acanthodian about 8 cm long. Among the ventral, paired fins, the front pair is the pectoral, the rear pair is the pelvic, and the ones in between are accessory. (Romer, 1966, p. 34, Fig. 46.)

bone that was covered with enamel. Bony fishes have reduced the thickness and calcification of their scales across geological time. That reduction coincided with increasing dexterity in swimming. The Acanthodii were the first heterocercal fishes (caudal fin with the upper lobe larger, into which the spinal cord extends); they had one or two dorsal fins and several pairs of ventral fins. All fins, except the caudal fin, were reinforced anteriorly by spines.

The Silurian saw the evolution of the first land plants. In order to become adapted to life on dry land, the plants had to develop stronger stems and roots, tough cell walls to conserve water, and a vascular system capable of moving water and dissolved minerals upward and organic substances downward. In modern trees, the upward flow is through an inner layer (*xylem*) of elongated, spindly cells called *tracheids* communicating end-to-end through perforations in their walls. The core of the stem consists of pith (a storage tissue) in many plants. The xylem forms a cylindrical layer around the pith, or, if there is no pith, it forms the core of the stem. The downward flow is through a cylindrical layer (*phloem*) surrounding the xylem and consisting of the cylindrical walls of cells joined end-to-end through perforated walls (*sieve elements*).

The earliest plants had no true roots, trunks, or leaves. They were flexible structures with spore-bearing organs. An example is *Psylophyton*, which was 0.5–1 m high and had a stem 1–2 cm across (Fig 22.20). It had no leaves, no stomata (it breathed through pores distributed along the stem and branches), and no differentiated root system, but it was vascular. It had both xylem and phloem, but no pith (xylem formed the core of the stem). *Psylophyton* had branches with curved tips that bore sporangia. It reproduced by means

of spores (as ferns do) that were contained in the sporangia. Spores are haploid and grow into a haploid *gametophyte*. The gametophyte produces gametes (sperms and eggs, from different organs). Fertilization produces a diploid plant (*sporophyte*) with diploid sporangia that, by meiosis, produce haploid spores. The cycle then repeats itself. *Psylophyton* has a distant descendant, *Psilotum nudum*, known in Florida as whiskfern.

The first air-breathing animals were scorpions. Their gills adapted to absorb oxygen from the air, rather than from water, and became lungs. There are 5.2 ml of oxygen gas dissolved in 1 liter of seawater at 20°C, but there are 209 mL dissolved in 1 liter of air. Adaptation to air breathing means reducing the intake of oxygen. That is what the early scorpions did. Some fish also developed lungs, a little later.

22.5. DEVONIAN (408.5–362.5·10^6 y B.P.)

The Devonian was named after Devon, a region in southwest England where the type section is found (Fig. 22.14). The great Caledonian mountain range that had been pushed up at the close of the Silurian was rapidly eroded during the Devonian, and a thick (up to 4 km) deltaic deposit, the Old Red Sandstone, formed

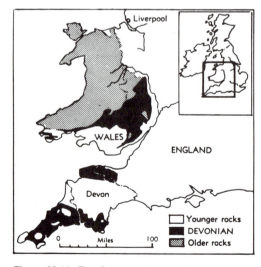

Figure 22.14. The Devonian outcrops in Wales and southwestern England. (Moore, 1958, p. 188, Fig. 8.1.)

Figure 22.15. Chondrichthyes (Cladoselacii). *Chlado-selache*, a Devonian shark (0.5–1.2 m long). (Romer, p. 38, Fig. 50.)

over northeastern North America, Ireland, England (except Cornwall), Scotland, and Scandinavia on both sides of the Caledonian mountain range. The stretch along which the Old Red Sandstone was deposited was across 30°N, where the convergence of the Hadley and Ferrel cells lay. As a result, the climate was dry, and the sediments were red in color because of oxidized iron particles (hematite, Fe_2O_3). Such sediments are called *red beds*.

Away from the Old Red Sandstone areas, toward the North American interior on one side and Cornwall, France, and central Europe on the other, shallow-water limestones were deposited, with rich assemblages of brachiopods, bryozoa, and crinoids together with many reefs made by stromatoporoids and tabulate corals. Trilobites were much less abundant than they had been in earlier Paleozoic times. In the deeper waters around the continental margins, sandy and muddy deposits formed, with many gastropods, bivalves, and cephalopods.

The Devonian is called the "age of fishes" because of the rapid radiation of the Chondrichthyes (Fig. 22.15) and the Osteichthyes (Fig. 22.16). The class Chondrichthyes includes the subclass Elasmobranchii (Silurian–Holocene), which includes the sharks (Selachii) and the rays and skates, and the subclass Holocephali (Jurassic–Holocene), which includes the chimaeras. Their skeleta consist of cartilage (with some mineralization in the

chimaeras). Both subclasses enjoy internal fertilization: The male has a pair of entering organs (*mixopterygia*) that are appendages of the pelvic fins and have a canal through their length originating from a sperm-producing gland at their base.

The class Osteichthyes includes the subclass Actinoptergii (ray-finned fishes, the largest group of modern fishes) and the freshwater subclass Choanichthyes (Fig. 22.17), which are also called Sarcopterygii because of their muscular fins. In the Choanichthyes the nostrils connect with the throat, but in all other bony fishes the nostrils are blind sacs lined with olfactory epithelium. The subclass Choanichthyes includes two orders, the Crossopterygii (Devonian–Holocene), which includes the coelacanth, and the Dipnoi (also Devonian–Holocene), which includes the lungfishes of Australia (Fig. 22.18). These two orders were much closer to each other in the Devonian than they are today, indicating an origin from a common stock. During the Devonian, the Crossopterygii gave rise to the amphibians (Fig. 22.19).

Amphibia start their lives essentially as fishes, with gills absorbing oxygen dissolved in

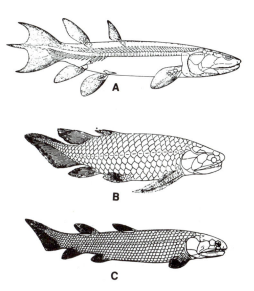

Figure 22.17. Choanichthyes (Crossopterygii), from the Devonian: (A) *Eusthenopteron*, 30–50 cm long; (B) *Holoptycus*, 50–70 cm long; (C) *Osteolepis*, about 20 cm long. (Romer, 1966, p. 68, Fig. 99, p. 71, Fig. 100.)

Figure 22.16. Osteichthyes (Chondrostei). *Cheirolepis* (× 0.33). (Romer, 1966, p. 55, Fig. 74.)

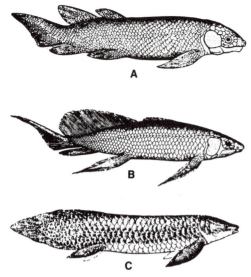

Figure 22.18. Osteichthyes (Dipnoi): (A) *Dipterus* (Devonian); (B) *Scaumenacia* (Upper Devonian); (C) *Neoceratodus* (Holocene) with all median fins fused into a homocercal fin. (Romer, 1966, p. 75, Fig. 105.)

water. As they grow, they develop lungs and become air-breathing animals. Amphibian eggs are unprotected (no shell), and so they must be laid in water (in some modern frog species the eggs are retained in pouches on the back of the female until hatched). A larva emerges from the egg, closely resembling a fish (e.g., tadpoles). The larva undergoes a metamorphosis: The gills disappear, lungs appear, and the embryonic paired fins develop into limbs. In frogs, oxygen absorption is not only through lungs but also through the skin of the body.

The welding of Europe to North America at the end of the Silurian eliminated many shallow-water environments. Furthermore, those two landmasses, welded into one, moved under the Hadley-Ferrel cell convergence, so that desertic conditions were widespread. There was much desiccation. Many freshwater fishes had to get out of shrinking bodies of water and quickly find other ones, preferably larger and not too far away. Early mutants that could breath air and could use their fins to propel themselves on dry land had a better chance of survival. Thus the fins became limbs, and better lungs were developed. Many varied

types of amphibia dominated the land from the Middle Devonian to the end of the Triassic ($380–208 \cdot 10^6$ y ago). The group suffered a severe decline during the Jurassic and Cretaceous ($208–65 \cdot 10^6$ y ago), but picked up again during the Cenozoic (the past 65 million years) and is now represented by the Anura (frogs), Urodela (salamanders and newts), and Apoda (a small group of wormlike, burrowing amphibians).

The Devonian also saw the diversification of land plants (Fig. 22.20). The Devonian forests included Lycopodiophyta and Sphenophyta (horsetails), in addition to Psilophyta. A typical representative of the Lycopodiophyta was *Protolepidodendron*, about 2 cm across and some 50 cm tall, with the stem covered with tiny leaves. An example of the Sphenophyta was *Archaeocalamites*. It was a small plant, with a stem about 1 cm across and nodes about 3 cm apart, from which a whorl of branches or leaves originated.

22.6. MISSISSIPPIAN ($362.5–322.8 \cdot 10^6$ y B.P.)

The beginning of the Mississippian was marked by a broad ingression of the ocean, with shallow seas expanding across both North America and Europe. Only the Canadian and Fennoscandian shields remained above sea level. Sedimentary regimes were established similar to those of the preceding Devonian— deltaic and lagoonal facies around the shield, shallow carbonates farther offshore, across the North American and European platforms, and deeper facies, with cephalopods, around the margins of the continent. Blastoids and crinoids reached their apex in the Mississippian seas. The Mississippian was a nice, quiet period, and so there is not much to say about it. Everybody seems to have had a good time.

22.7. PENNSYLVANIAN ($322.8–290 \cdot 10^6$ y B.P.)

Like all good things, the Mississippian did not last long: a mere 40 million years. The end came when the seawater withdrew from the continental platforms, creating general

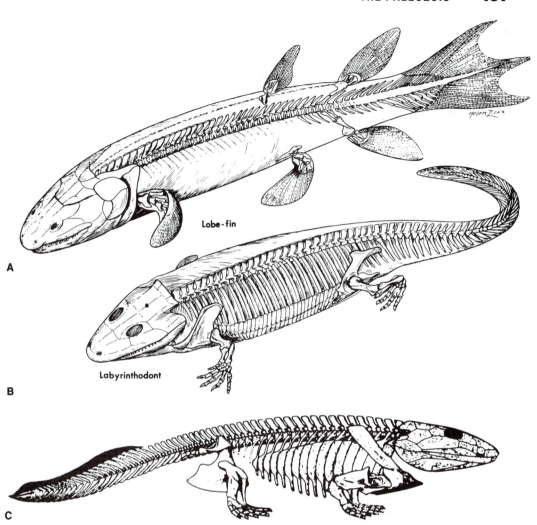

A

Lobe-fin

B

Labyrinthodont

C

Figure 22.19. Fish to amphibian: (A) crossopterigian fish. (B) labyrinthodont amphibian. (McAlester 1968, p. 79, Fig. 4.9.) (C) *Ichthyostega*, the oldest amphibian (about 1 m long). (Romer, 1966, p. 87, Fig. 118.)

consternation: Environments were destroyed, biota were exterminated. What happened? What happened was the first pulse of glaciation in the Southern Hemisphere.

If the Earth did not have an atmosphere, or if it had an atmosphere without water vapor and CO_2 in it, the average surface temperature would be the same as that of the Moon and everything would be frozen. Fortunately, water vapor and CO_2 molecules in the atmosphere keep the Earth warm.

The amount of CO_2 in the atmosphere depends on the balance between input (from volcanoes) and output (absorption by organic matter and by precipitates, followed by storage as coal, fossil fuels, or limestone). If this balance is shifted toward a higher concentration of CO_2 (e.g., too much volcanic activity), the temperature rises, the ocean warms up, and more H_2O is added to the atmosphere, which makes the atmosphere even warmer (positive feedback, producing a *greenhouse effect*).

If ice forms on the land, the Earth's albedo increases, the temperature drops, the oceans cool, and more CO_2 and H_2O return to the oceans, which makes the temperature drop

Figure 22.20. A Middle Devonian forest: (A) *Archaeo-calamites*; (B) *Protolepidodendron*; (C) *Psilophyton*; (D) *Eospermatopteris*, an early tree fern that grew to a height of 12 m. (Courtesy Chicago Museum of Natural History.)

even more (positive feedback, this time producing a glaciation). A glaciation, once started, grows all by itself.

But what starts a glaciation? The first requirement is that a significant continental mass occupy a polar or near-polar position; the second is that volcanic activity be low; the third is that carbonate precipitation and the storage of organic matter in sediments be able to reduce the concentration of CO_2 in the atmosphere. These conditions apparently were met in Ordovician time, when ice developed in North Africa, and again at the beginning of the Pennsylvanian, when ice developed in South Africa and adjacent lands.

During the Silurian and Devonian the landmass that occupied the Southern Hemisphere, consisting of Africa, South America, Antarctica, India, and Australia, all welded to each other, had been moving northward. During the Late Pennsylvanian this landmass hit

North America and Europe. The collision created mountain ranges in southern and eastern North America (the Ouachitas and the Alleghenies), in northwest Africa (the Mauritanids), and across central Europe (the Hercynids). Meanwhile, Siberia moved westward and collided with Europe from the east, forming the Urals. By the end of the Pennsylvanian, all landmasses, except China, formed a single supercontinent that stretched from pole to pole. In 1912 this supercontinent was named *Pangea* by the German meteorologist Alfred Wegener (1880–1930). Pangea was surrounded by a single ocean, called *Panthalassa*. On the eastern side of Pangea, Panthalassa extended westward into a broad gulf, called *Tethys*, bounded on the north by China and on the south by Australia.

Glaciation started at the beginning of the Pennsylvanian around the South Pole and extended northward to cover eastern South

America, South Africa, Madagascar, India, and Western Australia. A much smaller ice sheet probably developed over northeastern Siberia, which was located close to the North Pole. Meanwhile, North America and Europe enjoyed a tropical climate.

Ice advanced and retreated many times during the Pennsylvanian glaciation. Each ice advance caused a drop in sea level (because the water to make the ice comes from the ocean), and each time the ice melted, the sea level rose and advanced inland. When sea level was high, vast areas of North America, Europe, and Asia were covered by shallow seas, and marine shales and limestones were deposited. When sea level was low, continental sandstones and shales were deposited on top of the marine sediments. Soils developed and forests grew, only to be destroyed and buried under marine shales and limestones when the sea returned.

A regime of rhythmic sedimentation thus became established during the Pennsylvanian. These rhythmic deposits, called cyclothems (see Fig. 17.2), consist of continental sandstones and shales below, a coal layer in the middle (formed from the remains of forests), and marine shales and limestones above. More than 100 cyclothems have been identified, indicating that sea level (and therefore the ice in the Southern Hemisphere) advanced and retreated more than 100 times. Many of the coal reserves of the world are contained in Pennsylvanian cyclothems.

What caused ice to expand and retreat rhythmically? Ice is very sensitive to insolation, the amount of solar radiation it receives. If insolation is too high, the ice melts; if it is too low, the ice expands. The sun is a very stable nuclear reactor, and all indications are that the solar output does not change significantly over a period of millions or even tens of millions of years. The partition of solar radiation between the Northern Hemisphere and Southern Hemisphere, however, changes periodically.

We all know that when it is summer in the Northern Hemisphere it is winter in the Southern Hemisphere, and vice versa. This is due to the inclination of the Earth's axis from the normal to the orbital plane, which now stands at $23°28'$. This inclination, however, does not remain constant, but oscillates between $21°39'$ and $24°36'$, with a periodicity of $41,000$ y

(Chapter 8). When the inclination is larger, the summers are hotter and the winters are colder; when the inclination is smaller, the summers are cooler and the winters are warmer.

The eccentricity of the Earth's orbit is not constant either. It oscillates between 0.004–0.020 and 0.040–0.050, with a periodicity of $100,000$ y. The difference averages about 0.03. Because of the inverse square law, this translates into a power oscillation in seasonality amounting to about 6%.

When eccentricity and inclination are high, and when summer occurs at perihelion, the summers will be particularly hot (and winters particularly cold). The opposite (cool summers and warm winters) occurs when eccentricity is low, inclination is low, and the summer occurs at aphelion. In the 1920s the German climatologist Wladimir Köppen (who, incidentally, was the father-in-law of Alfred Wegener) figured out that glaciation could expand only if there were a stretch of cool summers (even if accompanied by warm winters), because the important thing for ice to expand is for at least some ice to survive the summer. A cold winter may produce more snow, but that will not matter if it all melts during a hot summer.

Because of the combined effects of precession of the Earth's axis and changes in orbital eccentricity, periods of cool summers occur every 100,000 years or so. Superimposed on the 100,000-year cycle, eccentricity exhibits a "supercycle" of 400,000 years. This means that every 400,000 years, the eccentricity maxima and minima are particularly pronounced.

The changes in insolation, opposite in the Northern Hemisphere and Southern Hemisphere, had no effect whatsoever on climate when the continents were distributed around the low latitudes, as in Cambrian time. But when there are significant continental masses at or near the poles, when the volcanoes are quiet, and when the sea has largely withdrawn from the continents, that is, when a glaciation is about to start anyway, a period of cool summers will precipitate it. So the astronomical motions of the Earth set the pace of the ice ages.

The Pennsylvanian glaciation was largely limited to the Southern Hemisphere. Its periodicity apparently was controlled by the major ellipticity cycle, the 400,000-year cycle. The 100 + cyclothems that have been recorded

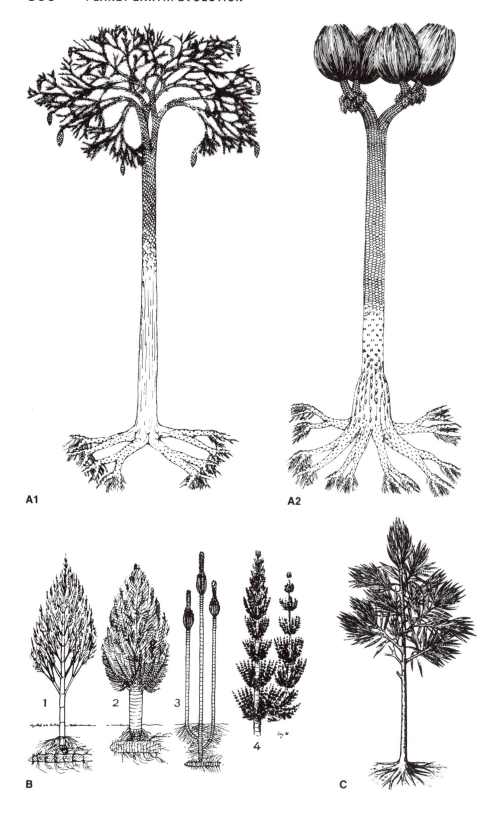

A1

A2

B

C

represent $100 \times 400,000 = > 40 \cdot 10^6$ y. Glaciation continued during the next geological period, the Permian.

The plants that made up the Pennsylvanian forests included species that reached heights of 40 m. The more common were the Lycopodiophyta (e.g., *Lepidodendron* and *Sigillaria*), the Sphenophyta (*Calamites*), the Polypodiophyta (ferns), the Cycadophyta (Lyginopteridales, seed ferns), and the Coniferophyta (Cordaitales) (Fig. 22.21). The Lycopodiophyta, Sphenophyta, and Polypodiophyta reproduced by spores that were formed by the conjugation of gametes. Plant spores often are covered with a hardy substance and can remain viable for a period of time even under dry conditions. The gametes, on the other hand, are delicate. As a result, plants that reproduce by spores are limited to swamps and humid soils, where the gametes can survive for the brief time it takes them to meet and mate. Swamps and humid soils were common during the Pennsylvanian interglacials.

There were, however, also continental areas (e.g., Canadian and Fennoscandian shields). In addition, the continental surface greatly expanded during glacial intervals. Seed plants made their appearance. In seed plants, the gametophytes (*pollen grain*, male; *ovule*, female) are small structures developed on the sporophyte. The ovule ends up by producing a single egg. When a pollen grain comes into contact with the ovule, it grows one or more gametes and a *pollen tube*, through which the gametes reach the egg in the ovule. After fertilization, the zygote grows into an *embryo*. The ovule produces a tough outer covering that encloses not only the embryo but also the female gametophyte. The seed is a capsule containing a germinated zygote (*embryo*). The embryo is surrounded by protective layers, and a seed can survive many years of adverse conditions (the record is held by some seeds of *Lupinus arcticus* from the Arctic permafrost that were

found to be able to germinate after 10,000 y in deep freeze). When conditions are right, the embryo resumes its growth and germinates, using the female gametophyte in the seed as food.

On dry land, seed ferns (Lyginopteridales) and early seed-bearing conifers (Cordaitales) were abundant. *Glossopteris*, a seed fern, was widespread in the cool areas of the Southern Hemisphere and of Siberia.

Insects, which first appeared in the Devonian, expanded greatly in the Pennsylvanian, when land forests became widespread. Insects developed a system of internal canals (*tracheae*) for the distribution of oxygen to all parts of the body. Many insects can close their tracheal openings to the outside and therefore can survive for a brief period of time under water. That ability probably was developed in order to survive sudden outpourings or even flooding. Birds and mammals inherited their lungs from lungfishes via amphibia and reptiles. Many birds and mammals can stay under water for a while because they have learned to hold their breath.

22.8. PERMIAN (290–245·10⁶ B.P.)

The Permian was named for exposures of marine strata in the region around the city of Perm in Russia. During the Early Permian, glaciers continued expanding and contracting in the Southern Hemisphere as they had during the Pennsylvanian. But then, about $270 \cdot 10^6$ y ago, glaciation expanded more than before, and a marked drop in sea level occurred. Continental redbeds were deposited in northern Europe (the New Red Sandstone of England and Scotland, and the Rothliegende of Germany) and in eastern North America.

The vast, shallow sea that was extending across the North American midcontinent broke into several broad basins that where connected with the open ocean only intermittently. When the connection was reduced or cut, evaporation would cause precipitation of salts on the bottom and formation of evaporites. Salt layers, each up to 8–10 m thick, formed across vast areas of the midcontinent and the southwest. A similar situation existed in Europe. The sea advanced from the north and formed the Zechstein sea, an inland

Figure 22.21. (Facing page) Pennsylvanian forest trees. (A) Lycopodiophyta: 1, *Lepidodendron* (Taylor 1981, p. 132, Fig. 8.7); 2, *Sigillaria* (Taylor, 1981, p. 155, Fig. 8.28). (B) Sphenophyta: 1, *Calamites carinatus*; 2, *Calamites multiramis*; 3, *Calamites schulzi*; 4, *Calamites sachsei* (Moret, 1943, p. 85, Fig. 37). (C) *Cordaites* (Cordaitales) (Stewart, 1983, p. 328, Fig. 25.4).

sea that covered much of Germany. Another inland sea covered the Russian platform. These, too, broke up into separate basins, on the bottoms of which evaporites formed. The aggregate thickness of the salt layers ranged from 100 to 1,000 m. During the Middle and Late Permian, evaporites alternated with redbeds in both North America and Europe. Altogether, the Middle and Late Permian were times of increasing marine regression, decreasing sea level, and widespread aridity.

The stratigraphy of Gondwana, the southern half of Pangea, was makedly different from that of the northern half. From the beginning of the Carboniferous to the end of the Triassic, thick continental deposits accumulated in South America, South Africa, Madagascar, India, Australia, and Antarctica. These sediments are largely fluvial, lacustrine, and deltaic, formed by the rivers that emerged from the huge ice cap that covered the south polar regions. These rivers had a long way to go before finally reaching the ocean. To the south there was the ice cap, and the continental margins to the southwest and the southeast were active margin and therefore were raised. The rivers could flow only toward the northeast, away from the active margins, filling intervening basins and emptying into the Tethyan Gulf.

The Gondwana deposits are characterized nearly everywhere by tillites at the bottom (up to 1 km thick in Brazil). These are followed above by sandstones and shales several kilometers thick, with coal beds. They are capped by basaltic flows of Jurassic age up to 1 km thick, the result of the breakup of Pangea (Fig. 23.13). The Gondwana deposits attain an aggregate thickness of up to 10 km. They form the Gondwana system of India and the Karroo sytem of Africa. In Madagascar and Australia there are marine intercalations within the thick continental deposits, and there is no basalt at the top.

The sandstones and shales contain a very characteristic flora, the *Glossopteris* flora (Fig. 22.22), which consisted of seed ferns (*Glossopteris, Gangamopteris, Sphenopteris, Sphenophyllum*). Elements of that flora are also found in Angara (northeastern Siberia), which was (and still is) at high northern latitudes.

The continental deposits of the Southern Hemisphere contain remains of amphibians and reptiles, but not in abundance. A characteristic reptile is *Mesosaurus*, which lived in South Africa and Brazil in the Lower Permian and was adapted to swimming in lakes.

In North America and Europe the fragmentation of shallow epicontinental seas into separate basins led to the rapid diversification of amphibians (Fig. 22.23) and reptiles (Fig. 22.24). The reptiles differ from the amphibians in many characteristics, the most important of which is the amniotic egg. The amniotic egg, of which the chicken egg is a perfectly good example, is a kind of "space capsule" in which the embryo can develop to a stage sufficiently advanced for the newborn to more or less fend for itself.

From inside out, the amniotic egg (Fig. 22.25) consists of the embryo, wrapped in a protective membrane called *amnion*. Connected with the embryo, but outside the amnion, are the *yolk sac*, a sac full of yolk that provides the food for the developing embryo, and the *allantois*, a sac in which the waste products of the developing embryo are stored. Amnion, yolk sac, and allantois are immersed in *albumen*, which consists largely of *albumin*, a nutritious glycoprotein. All this is enclosed in the *chorion*, a membrane that allows oxygen in and CO_2 out, which in turn is enclosed in a shell. The shell consists mainly of porous $CaCO_3$.

The advantage of the amniotic egg over the bare eggs of amphibia and fishes is obvious: Reptiles are freed from the aquatic (or at least humid) environment and can roam the land as far as they please.

The earliest reptiles appeared in the Early Pennsylvanian. From the skeletal point of view, they were very similar to the amphibians of that time. A subclass of reptiles, the Synapsida, which also evolved in the Pennsylvanian, developed mammalian characteristics in the Permian. Most synapsids were carnivorous. To improve mastication, tooth differentiation was introduced. Tooth differentiation is a major mammalian characteristic. In addition, the synapsids improved quadrupedal motion. Mammals evolved from synapsids during the Triassic.

A significant skeletal difference was the development of the lacrimal duct, which allowed the early reptiles to keep their eyes moist even in the desert. In the Permian, the

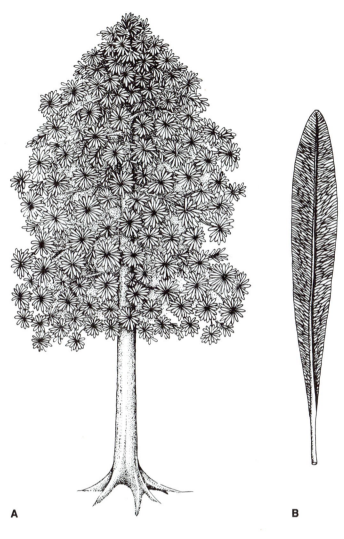

A

B

C

D

Figure 22.22. Gondwana flora. (A) *Glossopteris* tree (× 0.05). (Stewart, 1983, p. 306, Fig. 23.5.) (B) Leaves of *Glossopteris decipiens* (× 0.4). (C) *Gangamopteris cyclopteroides* (× 0.4). (D) *Schizoneura gondwanensis.* (Kummel 1970, p. 396, p. 396, Fig. 11.11.)

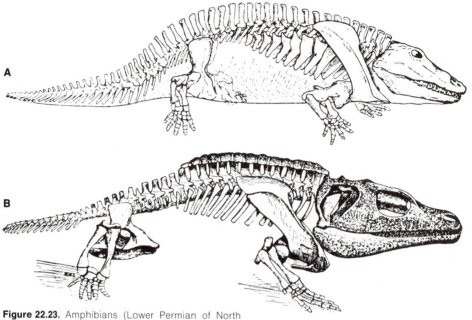

Figure 22.23. Amphibians (Lower Permian of North America): (A) *Eryops* (× 0.1); (B) *Cacops* (× 0.3). (Romer, 1966, p. 91, Fig. 122.)

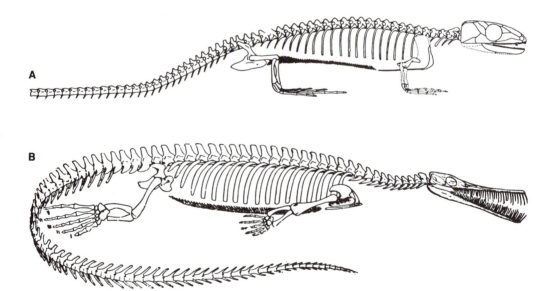

Figure 22.24. (A) *Hylonomus*, the oldest known reptile (Lower Pennsylvanian of North America) (× 0.2). (B) *Mesosaurus*, an aquatic reptile from the Lower Permian of South Africa and South America (× 0.5). Although reptiles had evolved to occupy the land, some had readapted to aquatic life as early as the Early Permian. (Romer, 1966, p. 103, Fig. 147, p. 117, Fig. 170.)

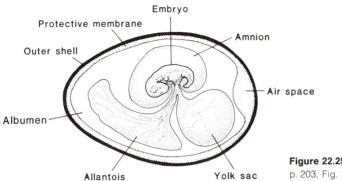

Figure 22.25. The amniotic egg. (Emiliani, 1988, p. 203, Fig. 11.7.)

reptiles had a lot of crying to do, because things were rapidly deteriorating, with all that drying up and whatnot. But the worst was yet to come. At the end of the Permian, the sea dropped even more, and practically everybody was dying of thirst. As though that was not enough, a great outpourng of basalt surfaced in Siberia, throwing a lot of water vapor and CO_2 into the atmosphere and creating a large greenhouse effect. That was the last straw: About 95% of the existing species became extinct, high-latitude as well as low-latitude ones. It was a great disaster, marking the end of the Paleozoic era. Here is a list of the major taxa that disappeared for good:

Heteractinid sponges
Fusulinid foraminifera
Tabulate corals
Rugose corals
Hyolithids
Trilobites
Eurypterids
Cystoids
Blastoids
Conodonts
Acanthodians

May they rest in peace.

THINK

* Could cyclothems form during the current sequence of glacial–interglacial ages? If so, where? If not, why not?

* Calculate the depth of the water in a marine basin that becomes isolated from the ocean and, following desiccation, leaves a salt deposit 10 m thick. Assume the basin to have vertical sides and flat bottom.

* The weight of the fertilized human egg is practically zero. The average weight of a baby at birth is 3.4 kg. Although growth of the fetus is modestly accelerated, rather than linear with time, assume nevertheless that it is linear with time and that the average weight of a fetus is thus 1.7 kg. Nutritional requirements for a fetus are about 45 cal/g of weight (*cal* here and in the following refers to a food calorie: 1 food calorie equals 1 kcal). Human milk rates at 0.7 cal/g. Calculate the mass of food (assuming that it is equivalent to milk) that a fetus absorbs from the mother in its lifetime in the mother's uterus. Calculate the mass of waste products produced in the same time (= total intake minus mass of baby at birth). Assuming that the density of organic matter is equal to 1, calculate the volume of an egg that the mother would have to lay if humans laid eggs like reptiles and birds.

The Mesozoic was the "Middle age" of the biosphere. It was the age not only of the reptiles but also of the cephalopods and of the calcareous marine plankton—Coccolithophoridae and planktic foraminifera. The Mesozoic started with the breakup of Pangea and ended with a bang—a giant outpouring of basalt and an asteroidal impact that caused a runaway greenhouse effect and devastated the biosphere.

23.1. THE BREAKUP OF PANGEA

The basaltic flows that cap the Gondwana formations in the Southern Hemisphere resulted from rifting associated with the breakup of Pangea. The lithosphere of Pangea was a hot lid (because it was rich in U, Th, and ^{40}K), trapping mantle heat. When too much heat accumulates, rocks melt. Molten rocks are 10% less dense than solid rocks. A mass of molten rock at depth will produce a welt on the surface. But because the lithosphere is rigid, a welt means a break. If the welt is circular, radial breaks will form, increasing in width away from the center. These radial breaks are called *aulacogens* (from the Greek αὐλαξ, for *furrow*). If the welt is elongated, a rift will form along its crest, and radial aulacogens will form around the ends. Aulacogens will also form at bends in the welt. Continental sediments accumulate in these depressions, followed by marine sediments. The process, from incipient welting to the establishment of marine conditions, may take from 10 to 40 million years. An example of an aulacogen being formed today is the Danakil depression in Africa, where the extension of the Carlsberg Ridge that penetrates the Gulf of Aden makes a narrow turn.

The breakup of Pangea (Fig. 23.1) started when an elongated welt, running northeast–southwest, began forming in Late Permian time at the junction between Connecticut–New Jersey on one side and Morocco on the other. Aulacogens were formed radiating from its northeastern end (Morocco) and southwestern end (Connecticut–New Jersey). The base sediments of these aulacogens are about $225 \cdot 10^6$ y old (Late Triassic). Aside from the evaporites in the Moroccan aulacogens, truly marine sediments in the area date from much later, about $180 \cdot 10^6$ y ago.

The aulacogen cutting across Connecticut and extending southwestward to New Jersey accumulated the thick (6,000 m) Newark series, which consists of conglomerates at the bottom, followed by sandstones and shales above, with additional conglomerate layers. Occasionally, basalt would rise through dikes and form horizontal sills. One of these sills, cut by the Hudson River, forms the Palisades along the river opposite New York City (Fig. 10.5).

The Moroccan aulacogens on the eastern side of the rift were periodically flooded by the western terminal of Tethys. There, thick continental deposits alternate with evaporitic layers.

The rift that formed between northeastern North America and northwestern Africa soon began extending southward, forming aulacogens indenting the northern coast of South America. It then turned east, forming the large Benue aulacogen of Cameroon, and then turned once more south, forming aulacogens in Patagonia. The South American and African aulacogens were formed in the Early Cretaceous and date from 145 to 125 million years ago. Other rifts formed in other portions of Pangea, eventually breaking it into the continents that we see today. Rifting ages in the various areas are shown in Table 23.1.

The new oceans created by rifting were

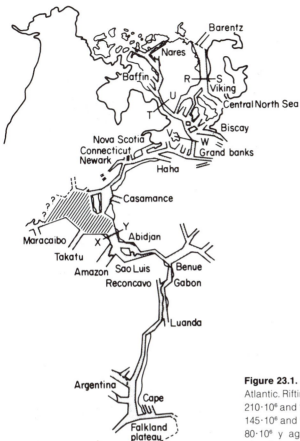

Figure 23.1. The breakup of Pangea and the opening of the Atlantic. Rifting between lines V–W and X–Y formed between $210 \cdot 10^6$ and $170 \cdot 10^6$ y ago; rifts south of X–Y formed between $145 \cdot 10^6$ and $125 \cdot 10^6$ y ago; those north of T–U formed about $80 \cdot 10^6$ y ago; and those north of R–S formed between $80 \cdot 10^6$ y and $60 \cdot 10^6$ ago. (Burke, 1976.)

shallower than old Panthalassa because they were floored by younger, hotter, and therefore more elevated rocks. As the fragmentation proceeded and new ridges were formed between the separating landmasses, the average depth of the ocean as a whole decreased, and the ocean began spilling over the land. A great marine ingression started at the beginning of the Jurassic ($208 \cdot 10^6$ y ago). By the end of the Cretaceous ($65 \cdot 10^6$ y ago), the Americas, Eurasia, Africa, India, and Australia–Antarctica were independent landmasses separated by new mid-ocean ridges. New oceans had been created at the expense of Panthalassa. The marine ingression reached its peak at the end of the Cretaceous, when Greenland broke away from Norway. The Late Cretaceous ingression was as large as the great ingression that had taken place in Late Cambrian–Early Ordovician time.

Plates separated by a rift are tilted away from it and slide off on the asthenosphere. Plates do not move away from a spreading axis because they are "pulled" by a subducting limb or "pushed" by the dikes surfacing along the spreading axis, because rocks have no resistance to either pull or push (they break or crumble); gravity is what makes the plates move the way they do.

We are now on the way to reconstructing a new Pangea. The Americas are moving northwestward, and Africa and Australia are moving straight north. In less than 50 million years, Africa will be stuck against Europe, and Australia against southeastern Asia. North America will hit eastern Asia in 200 million years, and south America will hit Australia 200 million years later. The new Pangea will be a northern supercontinent, with a big ice sheet in the north, huge deserts in its middle belt,

Figure 23.2. The Gruppo Sella in the Dolomites, northeastern Italy. (Courtesy Joeroen A. M. Kenter.)

and lowlands toward the equator, where the sea will advance and retreat as the ice in the north waxes and wanes. Antarctica will remain where it is because it is encircled by a continuous spreading axis. It may develop high-latitude floras and faunas similar to those in the north. It will function just like Angara in Carboniferous and Permian times. In fact, the whole setup will be similar to the old Pangea, only upside down—most of the land will be around the North Pole.

THINK

* Consider a lithospheric plate 100 km thick and 5,000 km across. Assume that it is perfectly bonded by covalent and ionic bonds. You want to break it into two pieces by pulling on the opposite sides. Estimate the bond strength and calculate the energy needed.

23.2. THE TRIASSIC (245–208 · 10⁶y B.P.)

The mass extinction that marked the end of the Permian left an impoverished biosphere. But the biosphere is resilient. For instance, only two genera of ammonoids survived the end of the Permian, but four million years later, at the end of the Early Triassic ($241 \cdot 10^6$ y ago), there were 100 new genera. When opportunity arises, the biosphere goes into a frenzy of evolutionary activity.

The name *Triassic* derives from the three parts into which the Triassic is divided on the European platform:

3. Keuper (highest) $235–208 \cdot 10^6$ y B.P.
2. Muschelkalk $241–235 \cdot 10^6$ y B.P.
1. Bunter (lowest) $245–241 \cdot 10^6$ y B.P.

Bunter deposits indicate arid conditions similar to those of the Permian. The sediments consist of variegated red-to-purple sandstones and shales with conglomerate beds, with a total thickness of 300–600 m. These sediments, derived from the Hercynian mountains of the Massif Central to the west and the Bohemian massif to the east, filled the Zechstein basin, where evaporites had been deposited during the Permian.

In the Middle Triassic (Muschelkalk), the Tethyan Gulf began advancing westward, and the sea spread over the European platform. The waters of the Muschelkalk sea had salinities higher than normal. The marine species that could tolerate those conditions multiplied greatly and formed sediments 200–400 m thick that were rich in molluscan and brachiopod shells and in entrochites (the segments of crinoidal stems).

The lower part of the Muschelkalk sediments (Wellenkalk) consists of shallow-water limestones, 70–80 m thick, with ripple marks. This is followed above by 30–100 m of dolomites with gypsum and anhydrite (Anhydritgruppe), indicative of hypersaline conditions. Above that is the main fossiliferous lime-

Table 23.1. Ages of Pangea's rifts

Rifting Between Landmasses (X/Y)	Approximate Age of Earliest Marine Ingression (10^6 y)
North America/Africa	180
South America–Africa/ Madagascar– Seychelles–India–Australia– Antarctica	160
Madagascar–Seychelles–India/ Australia–Antarctica	140
Africa/South America	125
Seychelles/India	100
Northeastern North America/ Europe	80
New Zealand–Campbell Plateau/Australia	80
Greenland/Norway	65
Australia/Antarctica	55
South America/Antarctica	30
Africa/Arabia	10
Baja California/Mexico	5
Palestine/Jordan	<0

stone (Hauptmuschelkalk), with abundant fossils. The top layer of the Muschelkalk, 10–50 m thick, consists of alternating marine and lagoonal deposits, some with carbonized remains of plants (Lettenkohle).

The third, uppermost subdivision of the Germanic Trias, the Keuper, consists of varicolored marls and porous dolomites, 450 m thick, indicative of lagoonal conditions. The Muschelkalk sea did not cover the Russian platform, but it extended northwestward to the North Sea area.

The southern border of the European platform fronted the deep waters of the Tethyan Gulf. Shallow-water limestones, including reefs, were deposited on the shelf facing the open ocean. The reefs, containing a new type of coral, the Hexacorals, were separated by basins in which marls with cephalopod shells were deposited. The limestones have all been transformed into dolomite and form the mountain range in northeastern Italy called Dolomiti. The main dolomite (Hauptdolomit), up to 1,000 m thick, was formed in the Late Triassic.

[The name *dolomite*, which is applied to both the mineral $CaMg(CO_3)_2$ and the rock formed largely or totally by this mineral, derives from the name of the French geologist Déodat Guy de Dolomieu (1750–1801), who was the first to describe the mineral and the rock.]

Dolomitic platforms extend to central and southern Italy, Sicily, and Yugoslavia. The Bahamas provide a modern example of these platforms with their intervening basins.

Further west in Europe, toward the head of the Tethyan Gulf, the salinity was too high for corals to form reefs. There, the Triassic is represented by shales, with gypsum below, followed by limestones that were largely formed by the encrusting green alga *Diplopora* (Permian–Triassic, Fig. 23.3) and other Dasycladaceae (a family of Chlorophyta).

Further offshore along the southern margin of Europe, fine sandstones and shales accumulated that contain abundant cephalopod shells. This sequence—a band of neritic sediments bordering the continent to the south, paralleled by a band of bathyal sediments farther offshore—is seen also in the Balkans, the Caucasus, and the Himalayas.

In North America, the Triassic is represented by the thick Newark series along the east coast, by widespread red-bed and fluvial sediments in the Great Plains, and by marine sediments in scattered basins in the western states.

The dominant Triassic land plants were the Cycadales, a class of Cycadophyta that had evolved from the Lycopodiophyta in Late Permian time. The Cycadales, in contrast to the Lycopodiophyta, have specialized pollen and seed cones (*strobili*) that are borne on different plants (they are, therefore, *dioecious*). The ovules are borne on the scales of female cones (as in Coniferophyta). In gross appearance the cycads resemble palm trees, with a bulbous or columnar trunk and a crown of fronds on top. Heights range from 1 to 18 m.

Gaining importance, but not dominance, were the Ginkgophyta, dioecious plants of which *Ginkgo biloba* is the only extant genus and species. The pollen is borne on catkinlike structures [a catkin is a type of inflorescence (cluster of flowers on a common stalk) in which the flowers are attached directly to the common stalk without an individual stem]. The ovules, in pairs, are borne on stalks and form seeds upon pollination and fertilization. These reproductive organs are intermediate between

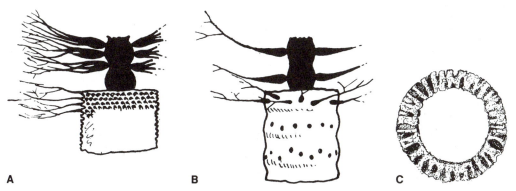

Figure 23.3. Diploporidae (Dasycladaceae). The Diploporidae (Triassic–Oligocene) are an extinct family of Dasycladaceae (phylum Chlorophyta) characterized by a stem from which whorls of branches radiated. The Diploporidae deposited around their stems cylinders of calcium carbonate 1–2 mm across, with perforations for the branches to get through: (A) *Diplopora phanerospora* (Triassic); (B) *Oligoporella pilosa* (Middle Triassic); (C) cross section of a diploporid segment from the Middle Triassic of northern Italy (× 25). (Moret, 1943, p. 40, Figs. 15.7 and 15.9, p. 45, Fig. 17.2.)

those of the Cycadophyta and those of the Cycadeoidales. Pollination is by wind.

The Cycadeoidales (Bennettitales) are an extinct (Triassic–Cretaceous) class of plants that were common in the Mesozoic forests. They closely resembled the Cycadales in the shapes of their leaves and trunks. They had, however, stalked fruiting structures containing both pollen and seeds and resembling primitive flowers. The Angiospermophyta (flowering plants) may have evolved from this group (Cretaceous).

In Early Triassic time a remarkable group of reptiles, the thecodonts, evolved, of which *Euparkeria* is a typical example. *Euparkeria* (Fig. 23.4) evidently was built for running.

Although *Euparkeria* could walk on all fours, its hindlimbs were twice as large as its forelimbs and were built for speed. *Euparkeria* needed a lot of energy for all that running. A palate evolved that separated the mouth from the nostrils, so that food could be chewed without interfering with respiration. *Euparkeria* could walk and chew gum at the same time.

During the Triassic, the thecodonts evolved into the saurischian (Fig. 23.5) and ornithischian (Fig. 23.6) dinosaurs. Both groups appeared in Late Triassic time. The two differ in the structure of the pelvis (Fig. 23.7). The pelvis consists of three bones (*ilium*, *ischium*, and *pubis*), with an opening at their triple junction (the *acetabulum*) in which the head of the femur

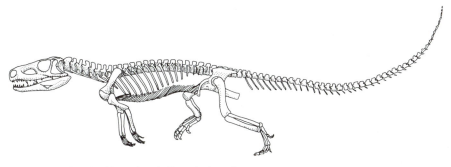

Figure 23.4. Reptiles (thecodonts). *Euparkeria*, a thecodont reptilian from the Lower Triassic, about 1 m long. (Adapted from Paul, 1987.)

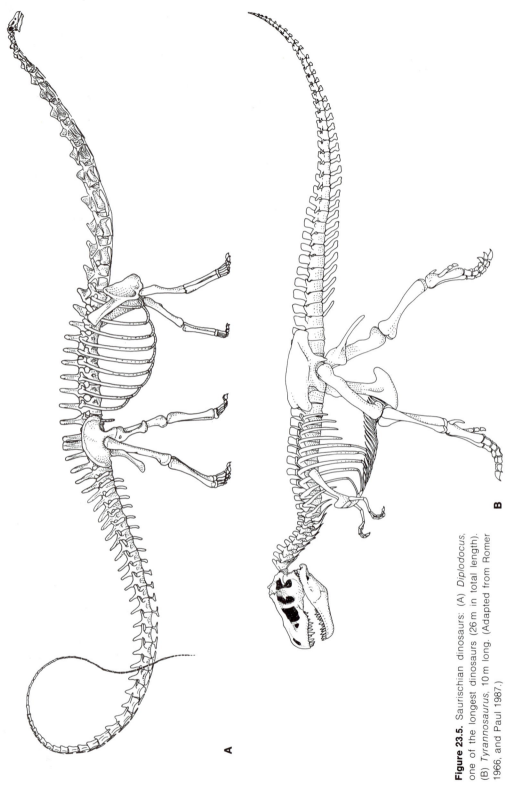

Figure 23.5. Saurischian dinosaurs: (A) *Diplodocus*, one of the longest dinosaurs (26 m in total length). (B) *Tyrannosaurus*, 10 m long. (Adapted from Romer 1966, and Paul 1987.)

A

B

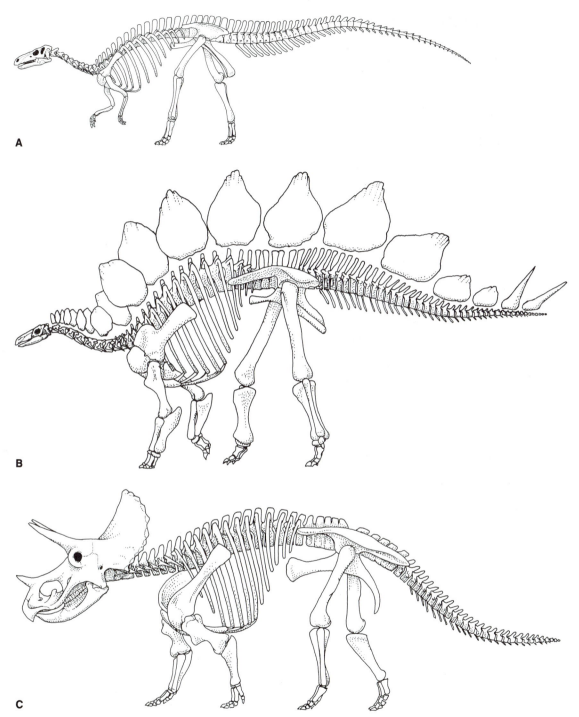

Figure 23.6. Ornithischian dinosaurs: (A) *Campto-saurus*, 5 m long. (B) *Stegosaurus*, about 6 m long. (Adapted from Czerkas 1987 and Paul 1987.) (C) Triceratops, up to 6 m long. (Adapted from Bakker 1987.)

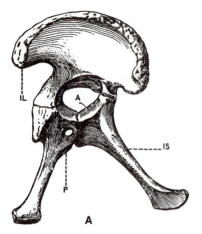

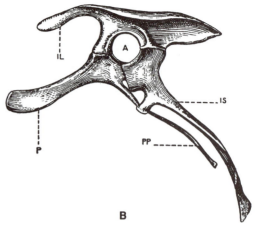

Figure 23.7. Reptiles (dinosaurs). Pelvic girdles showing ilium (*IL*), ischium (*IS*), pubis (*P*) with postpubis extension (*PP*), and acetabulum (*A*). (A) Pelvis of the saurischian *Diplodocus*; (B) pelvis of the ornithischian dinosaur *Iguanodon*. Front is to the left. (Colbert, 1968, p. 100, Fig. 15.)

fits. In saurischian dinosaurs, the three bones radiate from the acetabular area. In the ornithischian dinosaurs, the pubic bone is elongated in opposite directions from the acetabular area, forming a quadriradiate pelvis. The dinosaur classification and the principal characteristics of each major group are shown in Table 23.2.

Many dinosaurs had large bodies. Among

Table 23.2. Classification and characteristics of dinosaurs

Order Saurischia
    Suborder Sauropodomorpha: partly bipedal in Late Triassic, evolving into large, quadrupedal, herbivorous types in the Jurassic and Cretaceous [e.g., *Aptosaurus* and *Diplodocus* (Fig. 23.5A), Late Jurassic]
    Suborder Therapoda: fully bipedal, carnivorous [e.g.,*Tyrannosaurus* (Fig. 23.5B), Late Cretaceous]

Order Ornithischia
    Suborder Ornithopoda: bipedal [e.g., *Camptosaurus* (Fig. 23.6A), Late Jurassic to Early Cretaceous]
    Suborder Stegosauria: bipedal/quadrupedal, with a double row of plates along the back (vascularized heat-exchange devices?) and horns on the tail [e.g., *Stegosaurus* (Fig. 23.6B), Late Jurassic]
    Suborder Ankylosauria: quadrupedal, with bony plates on back (e.g., *Ankylosaurus*, Late Cretaceous)
    Suborder Ceratopsia: horned dinosaurs [e.g., *Triceratops* (Fig. 23.6C), Late Cretaceous]

the largest are *Diplodocus*, 27 m long (Fig. 23.5A), and the record-holder, *Seismosaurus*, 43 m long, reaching 18 m above ground at the shoulder and weighing 90 tons. They were very active and generated a lot of body heat that had to be dissipated. It is likely that some of the more bizarre structures exhibited by the dinosaurs, such as dorsal plates and the canalization of the skull bones, had the purpose of dissipating body heat. The dinosaurs reached their peak diversity in the Late Cretaceous, but became extinct at the end of the Cretaceous.

The discovery, for some dinosaur species, of egg nests with an average of 12 eggs carefully half-buried in sand and carefully laid on a spiral arrangement, with the narrow end of the egg pointing toward the center, suggests careful manipulation by the mother (or perchance by the father?). Furthermore, the nests are laid out in close proximity of each other, suggesting that the dinosaurs carefully tended their hatcheries. This implies a rather advanced social organization, with town meetings and an appropriate set of pecking orders, like the penguins of Antarctica.

Other important groups of reptiles that evolved in the Triassic are the lizards (Lacertilia, Early Triassic–Holocene), the snakes (Serpentes, early Triassic–Holocene), the crocodiles (Crocodylia, Late Triassic–Holocene), the turtles (Chelonia, Late Triassic–Holocene) (Fig. 23.8), and the marine reptiles (Ichthyosauria and

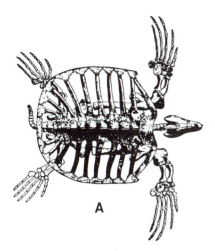

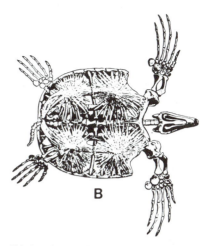

A

B

Figure 23.8. Turtles. *Archelon*, a marine turtle from the Upper Cretaceous of North America (2.5 m long): (A) dorsal view: (B) ventral view. (Romer, 1966, p. 115, Fig. 168.)

Plesiosauria, Early Triassic–Late Cretaceous). The turtles are a "primitive" group of reptiles (i.e., they have not evolved much since they first appeared). Most species were and are freshwater dwellers (rivers, lakes, marshes), but some have become adapted to land (the tortoises, Eocene–Holocene), and others became fully marine (Chelonidae, Dermochelyidae; Late Cretaceous–Holocene).

A typical marine reptile was *Ichthyosaurus* (Early Jurassic–Early Cretaceous) (Fig. 23.9), which had a fishlike body with paired ventral fins, a dorsal fin, and a homocercal caudal fin. The paired fins were derived from the limbs, often with increased number of phalanges. *Ichthyosaurus* reproduced by internal fertilization and was viviparous: Fossils have been found with the baby inside the mother or in the process of being born (in which case the mother apparently died during childbirth).

The mammals also evolved in Triassic time. The difference between dropping an egg and giving birth to a baby may seem quite large at first, but intermediate steps are easily seen, even today. The monotremes of Australia are mammals (they have teats), but produce eggs like reptiles. *Ornithorhynchus* (the duck-billed platypus) is a semiaquatic animal. The female drops two eggs in a nest made of damp vegetation and returns ten days later, when the eggs are hatched, to feed the babies. In the more advanced echidnas (spiny anteaters), which are terrestrial, one egg is laid in the marsupium of the mother. An immature baby hatches in

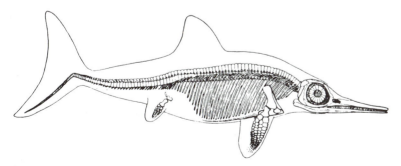

Figure 23.9. *Ichthyosaurus*, a marine reptile (0.80 m long). (Romer, 1966, p. 118, Fig. 172.)

10 days and stays there for another 2 weeks, feeding on the milk from teats that are in the pouch. In the still more advanced marsupials, the baby is born alive, but immature; it crawls into the pouch, attaches itself to one of the teats (which range in number from 2 to 27), and remains there feeding for 10 to 100 days, depending on the species. In montremes there is no connection between the mother's circulatory and excretory system and that of the infant. In marsupials and all higher mammals, the egg is not supplied with sufficient yolk for adequate nourishment of the embryo. Also inadequate are the provisions for storage and elimination of waste products. This situation is remedied by the mother's circulatory and excretory systems taking over through the placenta, a sophisticated, highly vascularized exchange system connecting the egg's chorion with the epithelium of the uterus. In marsupials, placentation is primitive. In the more advanced mammals (Eutheria), placentation is fully developed.

A major advantage that the mammals have over the reptiles is in the growth of the long bones. In mammals, the epiphyses are ossified early in life and coated with cartilage for smooth articulation. Bone growth occurs between epiphyses and shaft, where there is a second layer of cartilage that is continuously replaced by bone. At the end of growth, this second layer disappears, being completely replaced by bone. In reptiles, the cartilage occurs only at the two ends of the long bones, and bone growth occurs by continued replacement of that cartilage. As a result, articulation remains rather sloppy, as anybody who has seen a crocodile or an alligator running around can readily testify. The only gracefully moving reptiles are the snakes, mainly because they have no legs.

Another major advantage of the mammalian life-style is the nurturing of the babies, which means education or learning the tricks needed to survive and, possibly, to prosper. In primitive mammals, education lasts a couple of weeks. In humans it may take more than 25 years (until the kid gets a Ph.D.).

Throughout the Triassic and the remainder of the Mesozoic, the mammals played a secondary role with respect to the dinosaurs. It was only after the end of the Cretaceous, when the dinosaurs had disappeared, that the mammals finally got their chance.

THINK

* Some of the largest dinosaurs were swamp-dwellers, feeding on plants. How could they grow so large? Why did the meat eaters also grow very large?

23.3. HAPTOPHYTA (Earliest Jurassic–Holocene)

The Haptophyta are a phylum of single-cell, biflagellated, mostly planktic marine algae up to a few tens of micrometers across in size. In addition to the two flagella, they have a haptonema, a specialized flagellumlike structure used by the cell to collect food particles. Haptophyta undergo a coccolithophorid stage (which gave them the name Coccolithophoridae), during which the cell deposits calcitic platelets called *coccoliths* (Fig. 23.10). Once formed, the coccoliths are transported to the surface of the cell and eventually shed. They reach the ocean floor probably as part of fecal pellets. They are a major component of Globigerina ooze.

Coccoliths are manufactured in a great variety of forms, which makes it possible to distinguish a large number of "species"—living species number about 150, and more than 2,000 fossil species have been described. Some species range from the Arctic to the Antarctic across the equator. Other species prefer middle or low latitudes. Haptophyta reproduce 1 to 5 times a day; they speciate rather rapidly (a species generally lasts a few hundred thousand to a few million years) and have wide distribution. They are, therefore, excellent guide fossils. They range in time from the earliest Jurassic to the Holocene.

The coccoliths usually are less than $10 \mu m$ across, and yet they are exquisitely sculpted. Visible light has a wavelength of $0.5 \mu m$, which is greater than the details of the sculptings. As a result, these details must be studied with an electron microscope, which has a resolution several hundred times higher than that of optical microscopes. Electron microscopes use electrons accelerated through a voltage drop

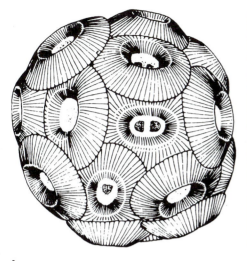

A

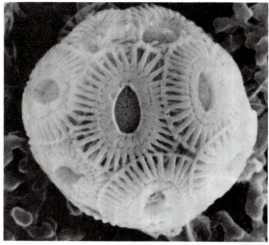

B

Figure 23.10. Coccolithophoridae: (A) *Coccolithus pelagicus* (× 625) (Lehmann and Hillmer 1983, p. 17, Fig. 11); (B) *Emiliania hyxleyi* (× 6,000) (This genus was named after me by the authors W. W. Hay and H. P. Mohler because I translated into Latin their description of the genus—Coccolithophoridae are considered "plants," and botanical rules require that new taxa be described in Latin. That's class!) (Courtesy Pat Blackwelder, University of Miami.)

of 50,000–100,000 V. The electrons therefore acquire an energy of 50,000–100,000 eV. The frequency of electrons with 100,000 eV of energy is obtained from $E = h\nu$ and the wavelength from $\lambda = c/\nu$.

Naked protoplasm has a density similar to that of seawater. So a nonflagellated, naked cell would float at the surface of the ocean. This is not a good place to be, because the sun is too strong and the rain dilutes the surface water (which plays havoc with the cell's osmotic balance). To escape the surface layer, nonflagellated planktic organisms need to be weighted down. The shell does the weighing.

In Haptophyta, the cell is happily swimming around, but an insidious force of nature is growing inside the cell into a tremendous urge to reproduce. So the cell starts manufacturing coccoliths and coating itself with them—no more swimming. The coccoliths are of course heavy, because they consist of calcite, which has a density of $2.7\,\text{g/cm}^3$. So the cell starts sinking. The surface layer of the ocean is highly turbulent, however, and water motion may force the cell up many times against its wishes. The cell responds by making more coccoliths. Eventually, sinking wins. But, of course, sinking is the knell of death for a planktic, photosynthesizing cell. So, at the last moment, just when things are getting desperate, the cell divides into two, and the two daughter cells escape the sinking casket. There is no sex, of course, because haptophytes are innocent little things. Altogether, it would seem that coccolith production is the way by which the Haptophyta convince themselves to reproduce. What remains to be seen is why they sculpt their coccoliths so exquisitely. Could it be that they just want to look pretty?

THINK

* What is the wavelength of an electron accelerated through a voltage difference of 100,000 V? How much shorter is it than the wavelength of visible light?

23.4. THE JURASSIC
(208–145.6·10^6 y B.P.)

The Jurassic is divided into three epochs: the Lias (Early Jurassic, 208–178·10^6 y ago), the Dogger (Middle Jurassic, $178-157.1 \cdot 10^6$ y ago), and the Malm (Late Jurassic, $157.1-145.6 \cdot 10^6$ y ago). The Middle and Late Jurassic saw the widening of the North Atlantic rift and the creation and widening of a rift between western Gondwana (South America–Africa) and eastern Gondwana (Madagascar–Seychelles–India–Australia–Antarctica). As a result, a significant marine ingression took place.

In Europe, the sea advanced to cover a broad area extending from England through central Europe to Russia. The Fennoscandian shield to the north and the Hercynian massifs of France and central Europe remained, of course, above sea level. During the Lias, clays and marls were the dominant deposits. The Dogger sediments were predominantly fossiliferous limestones, with abundant molluscan and brachiopod shells and layers of calcareous or ferruginous oolites. In fact, the ferruginous oolites are so abundant in the Dogger of Lorraine, Luxembourg, and adjacent areas of Belgium and Germany that they form the major iron resource of western Europe.

Ferruginous oolites consisting of tiny (0.5 mm across) ooids or spheroids of hematite (Fe_2O_3) or goethite [$FeO(OH)$] are the dominant type of Phanerozoic iron ore. One of the largest deposits is the Clinton formation (Silurian) in the eastern United States, which outcrops discontinuously from New York State to Alabama and reaches a thickness of 8 m in places. Whereas in the Clinton the iron oxides form coatings on quartz and other mineral nuclei, in western Europe the iron oxides coat predominantly calcareous nuclei. The composition of the sediment where the ferruginous oolites occur indicates a fully marine, well-aerated environment. Because iron cannot be in solution under these conditions, the formation of the iron oxide coatings must have occurred after deposition, under reducing conditions. One should notice that reducing conditions may be found beneath only a few centimeters of sediment in bays and bayous, and sometimes even along open shorelines.

The Malm sediments consist again predominantly of clays, but with intercalated layers of limestones and calcareous oolites. Jurassic sediments on the European platform are only a few hundred meters thick. South of the European platform, in the deeper water of Tethys, siliceous clays were deposited that have undergone low-grade metamorphism. They now contain chlorite and muscovite, which give them a shine and have caused geologists to call them *schistes lustrés*. The schistes lustrés contain intercalations of radiolarites and numerous intrusions of gabbro, diabase, and peridotite. This assemblage is typical of the eugeoclinal facies. During the Jurassic, the push of Africa from the south was beginning to fold the sediments of the Alpine trough. As a result, a series of ridges separated by deep basins came into existence. Toward the end of the Jurassic, the tops of the ridges were already above sea level.

The ammonites (see Fig. 21.15) diversified enormously during the Jurassic. Because the ammonites have septal structures that can be very complicated (Fig. 23.11), requiring a lot of DNA information, specific patterns did not last long. Furthermore, ammonites, being nektic, had wide distribution in the Jurassic seas. They are, therefore, ideal guide fossils. Stratigraphers have been able to subdivide the Jurassic into a large number of stages using ammonites.

The North Atlantic rift began widening $180 \cdot 10^6$ y ago (Middle Jurassic) at a rate of 20 km per million years (2 cm/y). At the end of the Jurassic the young Atlantic was about 700 km wide and 2,000–3,000 m deep. The circulation was relatively poor in that narrow, elongated, sinuous basin, and black shales were deposited during the Middle Jurassic and again during the Middle Cretaceous. The organic matter in the black shales includes the remains of phytoplankton and zooplankton, as well as vegetal matter from the bordering continents transported to the sea floor by turbidity currents.

West of the Mississippi, widespread red beds, similar to those of the Triassic, were deposited during the Early Jurassic. The cross-bedded Navajo sandstone (see Fig. 16.2) was deposited at that time. In the Middle Jurassic, a broad, shallow sea advancing from the Arctic covered the western interior from Wyoming to New Mexico. Sandstones and shales were deposited in the north, and limestones and evaporites in the south. That sea, called the Sundance Sea, was replaced, in the Late Jurassic, by the fluvial sands and clays of the Morrison formation. The Morrison formation is only about 100 m thick, but it is famous because of the number of large dinosaur skeletons that have been recovered from it.

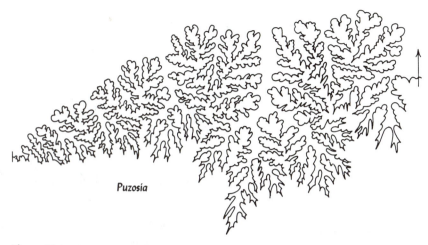

Puzosia

Figure 23.11. Ammonitic septum. The illustration depicts half a septum, from the periphery (arrow) to the left side of a *Puzosia* shell of Cretaceous age. A mirror-image septum extends from the arrow to the right side. (Weller, 1969, p. 229, Fig. 229.)

In Jurassic time the Pacific Ocean was much wider than today, because the Atlantic Ocean and Indian Ocean were practically nonexistent. The Pacific was floored by four plates, the Pacific plate to the southwest, the Kula plate to the northwest, the Farallon plate to the northeast, and the Phoenix plate to the southeast. Fig. 23.12 shows these plates at a somewhat later time. The Pacific, Kula, and Farallon plates met at a triple junction, and so did the Pacific, Farallon, and Phoenix plates. There was a major spreading line between the Pacific plate and the other three plates. The subduction of the Kula, Farallon, and Nazca plates continued during the Cretaceous and the Cenozoic, producing intense volcanism and mineralizations all around the Pacific margin. It also produced, by magmatic differentiation, extensive batholiths, especially in western North America (see Fig. 10.6). At the same time, the Pacific plate expanded, so that it now occupies most of the Pacific.

The separation between west and east Gondwana, which started at the beginning of the Malm, about 160 million years ago, produced a vast outpouring of flood basalts up to 1.5 km thick that covered $3 \cdot 10^6$ km^2 of land in South America and South Africa (Table 10.10). These basalts form the top of the Karroo system of South Africa (Fig. 23.13).

The Jurassic–Early Cretaceous ocean saw the flourishing of important groups of marine plankton (Table 23.3). Dinoflagellates (Fig. 23.14A) are a phylum of unicellular, biflagellated algae (see Section 20.6). They have been recorded from sediments as early as Late Silurian, but they have been abundant since the Jurassic. The Calpionellids (Fig. 23.14B) are planktic protozoa belonging to the phylum Ciliophora, with shells shaped like little vases 0.05–0.2 mm in size. These shells resemble those of modern tintinnids, except that the latter are composed of organic matter. The silicoflagellates (Fig. 23.14C) are a class of small (20–50 μm), unicellular, uniflagellated algae belonging to the phylum Chrysophyta. They form lacy shells of opaline silica.

Vast areas of the ocean surface that previously had been essentially void of life became quickly populated. Primary production in the surface waters by the newly evolved phytoplankton made it possible for the planktic foraminifera to live in the open ocean. In turn, bathyal and abyssal fauna found a food supply in the organic debris raining down from the open oceanic surface. So the bathyal and abyssal fauna thrived too.

The Pterosauria (flying reptiles) appeared in the Early Jurassic, and the birds in the Late Jurassic. It takes a lot of energy to fly, so a number of adaptations were necessary to make flight possible. Basically, one needs wings, a tail, very strong pectoral muscles, a body cover to conserve body heat, lightweight bones, and good eyes to spot food from high up. *Rhampho-*

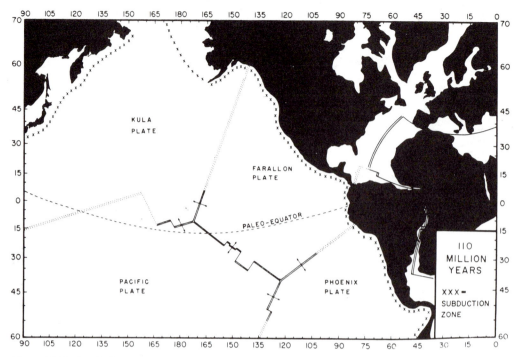

Figure 23.12. Pacific Ocean. Lithospheric plates in the Pacific Ocean $110 \cdot 10^6$ y ago. (Kennett, 1982, p. 184, Fig. 6.2.)

Figure 23.13. The 1.5-km-thick basaltic lavas topping the Karroo system in the Natal National Park, South Africa. (Kummel 1970, p. 391, Fig. 11.8.)

Table 23.3. Groups of marine plankton that flourished during the Jurassic and Early Cretaceous

Group	Type of Cell Covering	Stratigraphic Range
Haptophyta (coccolithophorid)	calcitic platelets	Earliest Jurassic–Holocene
Dinoflagellata	organic cell wall	Paleozoic (scarce); Early Jurassic–Holocene
Foraminifera (planktic)	calcitic shell	Late Jurassic–Holocene
Calpionellida	calcitic shell	Late Jurassic–Cretaceous
Silicoflagellata	opaline silica shell	Early Cretaceous–Holocene

rhynchus (Fig. 23.15A), an early flying reptile, had regular bones and no body cover, but had all the rest. Its wings were made out of a membrane stretched between the fourth finger and an extraordinarily elongated fifth finger. That portion of its brain dealing with vision was much enlarged.

Most remains of Jurassic flying reptiles have been found in marine sediments, suggesting that they were diving for fish. *Rhamphorhyncus* was no exception. It probably lived on cliffs facing the ocean. Judging from its skeletal structure, *Rhamphorhyncus* must have found it impossible to take off from the ground. It probably let itself fall off a cliff.

The earliest bird on record is *Archaeopteryx*, from the Upper Jurassic of Germany. *Archaeopteryx* (Fig. 23.15B) had feathers like a bird

and teeth like a reptile. People give more importance to feathers than to teeth, and so *Archaeopteryx* is considered a bird. In terms of skeletal structure, anatomy, and mode of reproduction, reptiles and birds are very similar.

There have been no convincing theories on the origin of flight. My theory is that those reptiles that liked to sun themselves on cliffs facing the ocean, like German tourists on the French Riviera, must have noticed sooner or later that in the sea below there were plenty of fish. So why not dive and try to catch some? Good idea, said everybody. In the beginning, of course, there was nothing but belly flops, to the great delight of the sharks waiting below (a powerful selective pressure). But after a while some reptiles may have become quite good at

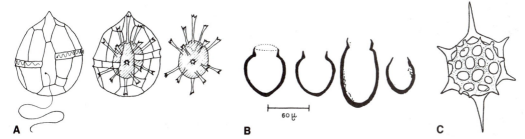

Figure 23.14. Planktic microorganisms. (A) *Gonyaulax*, a dinoflagellate. The theca of the vegetative stage is shown on the left. Two flagella emerge from two pores in the theca; one is wrapped around the equatorial depression (called *cingulum*), and the other, extending along a longitudinal depression (called *sulcus*), propels the organism forward. The resting stage is formed inside the vegetative stage (center), and it emerges following the destruction of the theca (right). The resting stage bears no morphological resemblance to the vegetative stage. In the fossil the two stages cannot be recognized as part of the same organism. Fossil resting stages, called *histrichospherids*, are treated, therefore, as an independent taxon and are named accordingly. The resting stage on the right, for instance, is called *Oligosphaeridium*. (Williams, 1978, p. 294, Fig. 1.) (B) Cross sections of Calpionellids from the Upper Jurassic of the Alps (× 300). (Moret, 1940, p. 78, Fig. 26.) (C) Shell of *Cannopilus hemisphaericus*, a silicoflagellate (× 800). (Haq, 1978, p. 270, Fig. 7.)

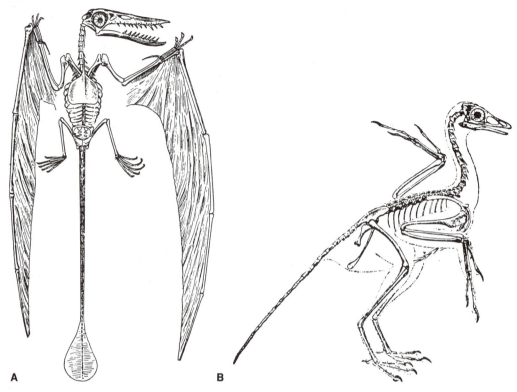

Figure 23.15. Flying reptiles and birds. (A) *Rham-phorhynchus*, a flying reptile from the Upper Jurassic of Europe and eastern Africa (50 cm long). (Romer, 1966, p. 114, Fig. 222.) (B) *Archaeopteryx*, the earliest bird, from the Upper Jurassic of Europe, about 30 cm long. (Romer 1966, p. 166, Fig. 235.)

it (survival of the fittest). Even the best divers had a problem, though: By the time they hit the water, more often than not the fish was gone. So wings and a rudder were developed for directional flight. With wings and rudder, a flying reptile could home in on a fish like a fish-seeking missile. Feathers, and the ability to close down skin capillaries to preserve body heat, quickly followed when the flying reptiles realized that they had to spend a lot of time sunning themselves between flights in order to warm up. Well, such is my theory, which is undoubtedly absolutely 100% correct because I just invented it while I was typing this page. You are the first to know.

THINK

* Why were ferruginous oolites not formed in the Archean, when so much iron was in solution and oxygen was being produced by the cyano-bacteria?

23.5. BACILLARIOPHYTA

Bacillariophyta is the phylum of the diatoms (Fig. 23.16), single-cell, nonflagellated algae that, like Radiolaria, form shells (*frustules*) of opal-ine silica ($SiO_2 \cdot nH_2O$). The shell consists of two valves, one larger and one smaller. The lip of the larger valve fits around the smaller one. Reproduction is by simple cell division. The two valves separate (hence the name *diatom*), and each constructs a new valve that fits *inside* the other. As a result, size decreases with each generation. At a certain point, a larger cell (*auxospore*) forms that returns the size to the original.

Bacillariophyta are divided into two classes: Centrales and Pennales. In the Centrales, the

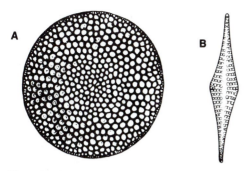

Figure 23.16. Diatoms: (A) *Coscinodiscus* (Centrales) (× 250); (B) *Nitzchia* (Pennales) (× 250). (Lehmann and Hillmer, 1983, p. 16, Figs. 9 and 10.)

frustule, which usually is circular, triequilateral, or quadriequilateral, has radial symmetry. In the Pennales, the frustule, which usually is ellipsoidal, isoscelian, or fusiform, has bilateral symmetry. Shell size ranges from 0.01 to 1 mm.

Diatoms are abundant in both seawater and fresh water. They live on the bottom free or attached by one valve. They may be attached to floating organisms (from *Sargassum* to the skin of whales). Or they may be freely floating in the water. Marine diatoms appeared during the Middle Cretaceous ($\sim 10^8$ y ago), and freshwater diatoms appeared during the Miocene ($\sim 20 \cdot 10^6$ y ago). More than 12,000 living "species" have been described.

Diatomite, a soft, highly porous sediment consisting of diatom frustules, is widely used in industry to make filters and as an additive to paints, rubber, and cement.

23.6. THE CRETACEOUS (145.6–65·10⁶ y B.P.)

At the beginning of the Cretaceous, the Madagascar-Seychelles-India landmass separated from the Australia-Antarctica landmass. Later in the Early Cretaceous, $125 \cdot 10^6$ y ago, the South Atlantic began opening. The North Atlantic kept widening throughout the Cretaceous, and at the end of that period, $65 \cdot 10^6$ y ago, it reached a width of 4,000 km. In contrast, the nose of Brazil was still only 700 km from Liberia, although the South Atlantic was considerably broader.

The creation of additional mid-ocean ridges,

in the South Atlantic and in the Indian Ocean, decreased the mean depth of the ocean, which forced the ocean onto the land. The sea advanced on all continents. Maximum ingression occurred in Late Cretaceous time. The sea covered most of Europe (except Scandinavia), most of North Africa, most of North America (except the Canadian shield), and a 500-km-wide strip of land in Australia running from the Gulf of Carpentaria to the Great Australian Bight. Emergent remained South America (except for the western geocline), Asia (except for the Himalayan geocline), and Africa south of the equator.

In the Anglo-Parisian basin, the type area for the European Cretaceous, the Lower Cretaceous consists of continental and lagoonal sands and clays. The Middle Cretaceous (Albian-Aptian stages) saw a marine ingression that deposited neritic limestones with oysters, followed by glauconitic sands 10 m thick and a 100-m-thick layer of bluish clay called the Gault Clay.

[*Glauconite* is a mineral, $(K, Na)(Al, Fe, Mg)_2$-$(Al, Si)_4O_{10}(OH)_2$, that forms on the sea floor either by alteration of fecal pellets or of detrital illite or biotite, or by direct precipitation from seawater in the presence of organic matter. It is one of the few authigenic minerals formed in seawater (*authigenic* means *generated in place*, from the Greek αυθι, which means *there, on the spot*). Because it contains potassium (which includes ^{40}K), it can be used to determine the age of the deposit in which it occurs. Detrital minerals cannot be used for dating. Although biotite and most clays also contain potassium, the age they reflect is the age in which the biotite or the clay minerals were formed, usually tens or hundreds of millions of years earlier than the time when the minerals are deposited on the sea floor to form a sediment.]

In the Late Cretaceous, the Anglo-Parisian basin had less terrigenous sedimentation. Above a second layer of glauconitic sand, chalk 500 m thick was deposited. This chalk consists of coccoliths, planktic and benthic foraminifera, and fragments of brachiopod and molluscan shells and of bryozoa, corals, and polychaete worm tubes. In addition to these carbonate elements, the chalk contains an appreciable fraction of clay, 1% or less of detrital quartz, and scattered layers of chert. The water depth under which the chalk was deposited is estimated to have been about 300 m, which, considering that the sea level was higher, would place the chalk deposition at the outer edge of the continental shelf. This chalk forms the White Cliffs of Dover and gives England its Latin name Albion.

In the region of the Alps and the Apennines, the Late Cretaceous is characterized by shallow-water limestones adjacent to the European platform, grading into deep-water clays, which, farther offshore, grade into pelagic limestones. In the Tertiary, these deep-water clays were intruded by gabbros, diabases, and serpentinites and were squeezed out of their trough to form a chaotic mass, the *argille scagliose* of the northern Apennines. The Cretaceous–Tertiary pelagic limestones, containing a rich assemblage of coccoliths and planktic foraminifera, were also uplifted and can now be seen in central and southern Italy.

In North America, the Cretaceous sea covered the east coast, the Gulf states, and a broad section from Texas to the Arctic. The east-coast deposits consist of sandstones, shales, and marls that are in part continental and, generally, do not exceed 300 m in thickness. The border between the Appalachians to the west and the Cretaceous deposits to the east forms the fall line, a line along which waterfalls or rapids in the rivers mark the margin between hard pre-Cretaceous river bedrock and the softer Cretaceous rocks. Many cities are located along the fall line, including Trenton, Philadelphia, Wilmington, Baltimore, Washington, Richmond, Columbia, and Augusta. In earlier times, they used the waterfalls to run mills.

Limestones, marls, and chalk were deposited in the Gulf states, especially in the Late Cretaceous. In the western seaway, sedimentation was mainly shales. The emplacement of batholiths farther west (Fig. 10.6) was accompanied by explosive volcanism, which produced ash layers. These ash layers, up to 8 m thick, have devitrified and converted to montmorillonite and colloidal silica, forming a deposit called *bentonite*. Montmorillonite and colloidal silica can absorb water. The maximum three-dimensional volume increase of bentonite is 11-fold. Bentonite is the main ingredient of muds for oil-well drilling.

Orogeny on the west, associated with the subduction of the Farallon plate, created a mountain range similar in size to the modern Andes. This mountain range shed huge volumes of clastics that accumulated in a trough running from British Columbia to northern Mexico. The sediments attain a thickness of more than 5 km in places.

The Middle Cretaceous was characterized by widespread anoxic conditions in the world ocean. It appears that many times between $125 \cdot 10^6$ and $85 \cdot 10^6$ y ago the ocean had no dissolved oxygen below a depth of 300 m or so. During anoxic episodes, the ocean simply was not overturning. What happened can be traced once more to activity along mid-ocean-ridge crests. The huge marine ingression then in progress indicates that the ridges were very active. If so, CO_2 probably was pouring into the atmosphere in unprecedented amounts from the volcanic vents. That created a large greenhouse effect. Surface temperatures at the equator probably rose only a couple of degrees, but the temperature in polar areas (and therefore also on the ocean floor) rose from an estimated 0°C in Early Triassic time to an estimated 20°C in Middle Cretaceous time. There is now 58 times more CO_2 in the ocean than in the atmosphere, but the modern ocean is very cold at depth. The Middle Cretaceous ocean was warm, and so a lot of the CO_2 it contained was released to the atmosphere, exacerbating the greenhouse effect. The biosphere responded with a vigorous expansion in productivity, which resulted in a much greater flux of organic matter to the ocean floor and the establishment of anoxic conditions.

The storage of CO_2 as organic matter and carbonate oozes in the deep sea, and in shallow seas as organic carbonates in reefs, shell beds, and so forth, was the response of the biosphere to the excess of CO_2 in the atmosphere. The anoxic sediments of the eastern Central Atlantic basin are unusual in that they contain abundant plant remains. The major plant groups of the Early Cretaceous were the Cycadophyta and the Coniferophyta. The Early Cretaceous saw the appearance of the Angiospermophyta, which diversified rapidly during the Middle and Late Cretaceous. Unlike the Cycadophyta and the Coniferophyta, which have no herbs, the angiosperms have numerous taxa of annual herbaceous plants that are the basis of modern agriculture.

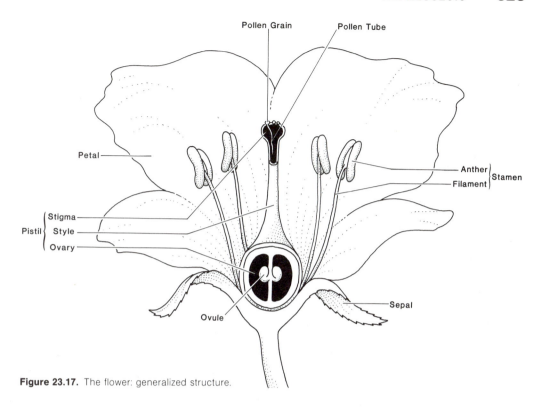

Figure 23.17. The flower: generalized structure.

The single most important feature characterizing the angiosperms is the reproductive system.

[Angiosperms have true flowers (Fig. 23.17), which consist, from outside in, of a crown of green sepals, a crown of petals, a crown of stamina (each of which consists of a filament carrying a pollen-producing *anther*), and a central *pistil*, a vaselike structure with an ovary below and a neck (the *style*) above that terminates in a lip (the *stigma*). The ovary carries ovules attached to its wall (the *placental wall*). Within each ovule, a megaspore divides meiotically to produce four haploid megaspores, three of which disintegrate; the fourth grows mitotically into a haploid female gametophyte called the *embryo sac*, consisting of 8 nuclei partitioned among 7 cells, one of which contains 2 nuclei, and another one is the egg cell. Pollen is formed in the anther. The anther (diploid) contains four microsporangia that contain a cluster of *microspore mother cells* (also diploid). Each microspore mother cell divides meiotically to produce four haploid microspores. Each microspore divides mitotically and grows into a *pollen grain* (the male gametophyte). The pollen grain contains two male gametes.]

When the pollen grain lands on the stigma, it grows a haploid *pollen tube* (still male gametophyte) that penetrates through the neck of the pistil (style) and reaches the embryo sac. Of the two male gametes, one fertilizes the egg, and the other joins the cell in the embryo sac that has two nuclei and forms a 3n (triploid) cell. This cell grows into a 3n endosperm that fills the embryo sac and is used as food by the embryo. In some plants (e.g., peas) the endosperm is used up by the embryo while the seed is being formed; in other plants (wheat, corn) the endosperm is kept in the seed to be used by the embryo when the seed germinates. The endosperms of nongerminated seeds of rice, wheat, and corn feed most of the people on Earth.

The angiosperms diversified and spread

Table 23.4. Terminal Cretaceous extinctions

Who got wiped out
Ammonites
Belemnites
Dinosaurs
Ichthyosaurs
Pterosaurs

Who almost got wiped out
Coccolithophorid haptophytes
Planktic foraminifera
Reef corals
Neritic gastropods, bivalves, crinoids, echinoids
Bony fishes

Who came through in reasonably good shape
Amphibians
Mammals
Sharks
Benthic foraminifera
Calcareous algae
Diatoms
Radiolaria
Plants

---

vigorously during the Late Cretaceous. They contributed greatly to fixing carbon and therefore to reducing the CO_2 excess in the atmosphere.

Just when the mid-ocean ridges were beginning to relax and the biosphere seemed to have managed to control the CO_2 problem, India split from the Seychelles block, and a great outpouring of basalt emerged from the rift. That outpouring, dated at $66 \cdot 10^6$ B.P., forms the *Deccan traps*, which cover 500,000 km^2 of territory in western India. The thickness ranges up to 2 km along the west coast, but tapers off to zero at the eastern margin. Individual basaltic flows range in thickness from 1 to 40 m. That vast outpouring of basalt lasted more than one million years and was accompanied by a release of CO_2 that raised even higher the already dangerously high surface temperature of the Earth.

It has been known for a long time that a great mass extinction occurred at the end of the Cretaceous. Perhaps 90% of the Cretaceous species were wiped out. Extinction was not indiscriminate, however, as Table 23.4 shows. The organisms that became extinct or almost so were particularly susceptible to a temper-

ature increase because (1) they were already living in high-temperature environments or (2) their eggs were laid in such environments or (3) they had large bodies that could not efficiently dissipate body heat.

Ammonites and belemnites were cephalopods. They lived in the open, tropical ocean at various depths, but probably, like the modern *Nautilus*, came inshore to deposit their eggs in shallow water. Haptophyta and planktic foraminifera live mainly in the open, tropical ocean at shallow depths. Reef corals and neritic mollusks and echinoderms live, again, mainly in shallow, tropical waters. All these organisms were highly susceptible to a temperature rise. Also susceptible were the bony fishes, which produce eggs that float to the surface.

Most dinosaurs had large bodies. Their big problem was the dissipation of body heat, because production of body heat is proportional to body mass, but dissipation is proportional to body surface. Pterosaurs were pretty big too. The wingspan of *Rhamphorhynchus* was more than 1 m, and that of *Pteranodon* was fully 8 m. Failure to dissipate body heat quickly leads to heat stroke, and that's fatal.

The organisms that suffered the least—the amphibians, mammals, birds, and calcareous algae—either were tolerant of high temperatures or did not live in high-temperature environments. Mammals and birds, for instance, can tolerate temperatures up to 45–50°C for periods of time because they keep cool by evaporating water (through the skin and/or the buccal mucosa). In addition, the mammals and birds of the Late Cretaceous were small in size and did not need to dissipate much body heat. The amphibia can tolerate temperatures up to 40°C. By comparison, the American alligator, *Alligator mississippiensis*, drops dead when, if forced to stay in the Sun, its body temperature rises above 38°C (from a normal 36°C). As for the sharks, they reproduce by internal fertilization—no eggs floating to the surface only to be cooked.

Most benthic foraminifera are *eurybathyal*; that is, they have a considerable depth range and could survive simply by moving to deeper water. Radiolaria are planktic, but they too can live at depth. Diatoms are abundant at high latitudes, where an increase in temperature would not be expected to create a hardship. As for the plants, they produce seeds precisely for the purpose of weathering periods of adverse conditions.

Judging from who died and who survived, it would appear that the mass extinction that terminated the Cretaceous was due to a temperature rise. In fact, recent paleobotanical studies indicate that the surface temperature in the North American midcontinent rose by 5–10°C at the close of the Cretaceous.

But there is more: If the temperature rise were due to a gradually increasing greenhouse effect produced by volcanism alone during the Late Cretaceous, one would expect that low-latitude species would have had enough sense to move to cooler, higher latitudes (like Miamians who go to North Carolina for the summer) or would have become adapted to the changing environment, following the classical tenets of Darwinian evolution. They certainly had the time to do it, because a big basaltic outflow like the one that created the Deccan traps would not be expected to be an overnight affair. Indeed, recent radiometric measurements indi-

cate that the basaltic flow in India lasted more than one million years. Yet, whenever a continuous Cretaceous–Tertiary stratigraphic record is available, we do not see a pattern of slow adaptation, but a pattern of practically instantaneous extinctions and replacements.

At Gubbio in central Italy (Fig. 23.18), for instance, there is a monotonous series of limestone layers with a rich tropical assemblage of Cretaceous coccoliths and planktic foraminifera. This series is abruptly interrupted by a clay layer about 2–3 cm thick, above which the limestone series continues. But there is a big difference. The limestone layers immediately above the clay contain an assemblage totally different from that below. There is only one foraminiferal species, and one coccolithophorid species, both quite scarce and very small in size. These two species resemble high-latitude species.

In 1978, Louis and Walter Alvarez analyzed the concentration of iridium in the clay layer at Gubbio and found that it was 30 times greater than in the layers above and below

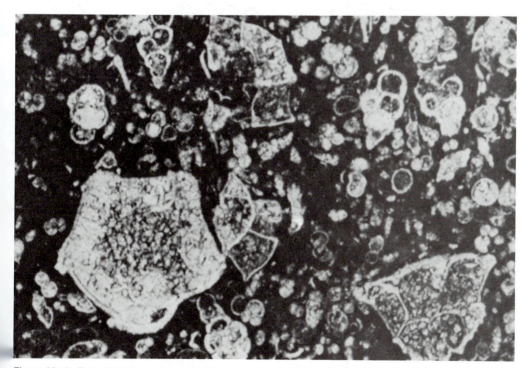

Figure 23.18. The Gubbio section. (A) The rich, tropical fauna of planktic foraminifera in the limestone layer immediately below the clay layer.

(Luterbacher and Premoli–Silva 1964, Pl. 4, Fig. 2.)
(*Continued*)

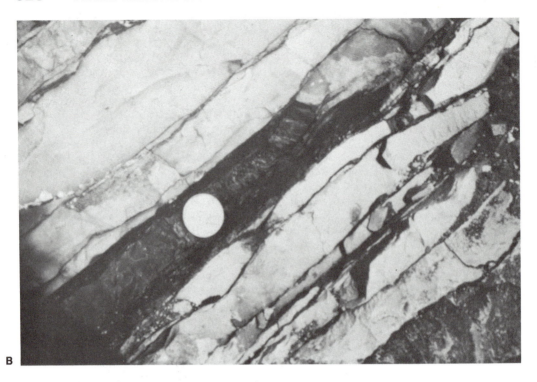

B

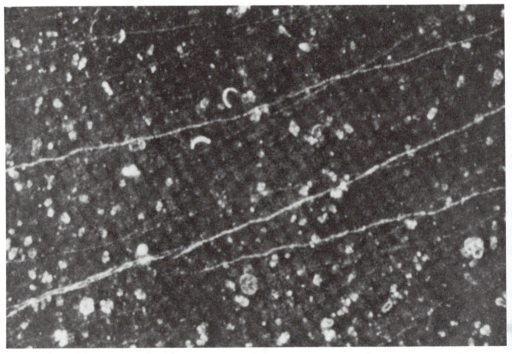

C

Figure 23.18 (*cont.*). (B) The clay layer where iridium is enriched 30 times; the coin across the clay layer is 25 mm across. (Courtesy Walter Alvarez, University of California, Berkeley, Calif.) (C) The impoverished planktic foraminiferal fauna in the limestone layer overlying the clay layer. (Luterbacher and Premoli–Silva 1964, Pl. 3, Fig. 1.)

Table 23.5. Platinum metals concentrations

Element	Abundance (ppb)		
	Earth's Crust	Chondrites	Siderites
Ru	1	1,000	7,300
Rh	1	200	2,300
Pd	10	1,200	3,700
Os	1	700	4,300
Ir	1	500	2,700
Pt	5	1,500	10,100

(after the carbonate fraction had been removed by acidification).

Iridium (Ir) is one of the platinum metals, the others being ruthenium (Ru), rhodium (Rh), palladium (Pd), osmium (Os), and platinum (Pt). These elements are rare in the Earth's crust because they were scavenged and carried to the core at the time when the Earth's core formed. They were more abundant in the original material, which is presumed to have been similar in composition to the chondritic meteorites. The concentrations of the platinum metals in the Earth's crust, in chondrites (stony meteorites), and in siderites (iron meteorites) are shown in Table 23.5 in parts per billion (ppb) by mass. Louis and Walter Alvarez chose iridium as representative of the platinum metals because it is easier to analyze. They explained the enrichment in iridium they had discovered at Gubbio (which they and others subsequently discovered at many other locations around the world) as being due to an asteroid 10 km in diameter hitting the Earth. The size of the asteroid was estimated from the concentration of iridium in the Gubbio clay and other similar clays.

The asteroid seems to have broken up into two unequal pieces before hitting ground. The smaller piece fell in Iowa and excavated a crater 32 km across (the Manson crater). The larger piece fell in northern Yucatan and the adjacent Gulf of Mexico and excavated a crater 180 km across (the Chicxulub crater). The asteroidal fragments vaporized instantly, and giant clouds of debris, water vapor from the Gulf of Mexico, and CO_2 from the vaporization of limestone rocks were thrown into the atmosphere and the stratosphere. There was total darkness for weeks, but soon the water vapor cleaned the atmosphere by condensing around the dust particles and removing them in a sort of worldwide muddy rainstorm. (See also Section 7.5.)

Let's see if we can reconstruct what actually happened (Fig. 23.19). Let us assume that you and I are two dinosaurs, lazily lounging around on a riverbank in Montana (where in fact the last dinosaur bones were found—presumably yours and mine):

YOU: Nice day, eh?

I: Nice day indeed.

YOU: Perhaps a bit too hot. But then again, it's summer.

I (*yawning*): Relax. The worst is over. The mid-ocean ridges are cooling, and the sea has begun withdrawing. The next few million years will be like the good old times of the Early Cretaceous. No more stinky, anoxic sediments on the ocean floor.

YOU (*pointing to the sky*): Look! A ball of fire! It's big! It's coming this way!

I (*paying sudden attention*): !!!!! It's a !!!!!!! asteroid! [Expletives expurgated by the publisher—available on request.]

YOU: Good God! What will happen?

I (*hurriedly*): The most probable angle of entry is 30°, and the most probable velocity is 25 km/s. The asteroid will cross the atmosphere in less than 10 seconds and will dig a crater 100 km deep and 200 km wide.

YOU: And then what?

I: There will be a great big dust cloud engulfing the whole Earth. If the asteroid falls into the ocean, there will also be lots of steam and of CO_2 (from the burning of carbonates on the sea floor), which will create a large greenhouse affect and raise the temperature by 5 or 10 degrees. There will be darkness and hot rains for months.

YOU: But that's awful.... Look out! There are stones falling all over the place!

I: (*pointing south*): There comes the cloud!

YOU: Let's duck!

I: Let's duck where? There are no holes big enough for us!

[*The cloud rolls in*]

YOU (*coughing*): I can't see a thing! Jesus, it's hot! It rains hot water!

Figure 23.19. The demise of the dinosaurs. The illustration describes what really happened. Forefront: Little mammals seeking refuge in a hole. Left: *Triceratops.* Right: *Tyrannosaurus.* In the air, an unidentified flying reptile in the process of losing its teeth and therefore becoming a bird. (Emiliani, 1980, p. 505, Fig. 1.)

I (*panting*): Aw, shut up!

[*Pause, while we cough and pant*]

YOU: Hey! Get your paws off me!

I: Oh pardon me! I couldn't see you. I was trying to catch some of these darned little mammals that are scurrying between my feet, probably trying to find a hole in the ground to hide in. Devious little bastards! I bet they're having fun now that we can't see a thing! But wait until it clears up—we'll teach them a lesson!

[*Next morning. A little mammal named Ralph to his wife, Edith:*]

RALPH: You can come out now, Edith.
EDITH: But Ralph, it's too dark!
RALPH: I want to show you something.
EDITH (*emerging from the hole*): Where are you?
RALPH: Up here!
EDITH: Eeeek! What are you doing on top of a dinosaur? Aren't you afraid?
RALPH: Naw! He's dead. They are all dead—look around and see for yourself.
EDITH (*looking around*): What happened?

RALPH: An asteroid hit. Lots of dust and steam, with a trace of iridium. They all died of a heat stroke, those big fat slobs. Asteroid in, dinosaurs out. Edith, my gal, we are in the *Cenozoic*, the age of mammals!
EDITH: Oh, Ralph!
RALPH: Look at all this space—it's all ours!
EDITH: Ooh, Ralph!
RALPH: We will have a lot of speciating to do.
EDITH: Oooh, Ralph! Let's!

[*And they kissed passionately and quickly disappeared into the hole.*]

THINK

* Suppose that an asteroid comparable to the one that hit the Earth at the end of the Cretaceous were to hit the Earth now. Carefully choose the impact site so as to produce *minimum* damage, and then write a gripping script for a film to be entitled *Asteroid!* Sell the script for an astronomical sum, and retire to a villa overlooking the impact site.

24

THE CENOZOIC

We divide the Cenozoic era into three periods: the Paleogene (which includes Paleocene, Eocene, and Oligocene), the Neogene (which includes Miocene and Pliocene), and the Quaternary (which includes Pleistocene and Holocene). Informally, Paleogene and Neogene often are grouped into a single chronological unit called the Tertiary (a name introduced by Arduino—see Chapter 18).

The name *Cenozoic* derives from καινός, meaning *new*, and ζῷον, meaning *animal*. The names *Paleogene* and *Neogene* derive from παλαιός (old) and νέος (new), and from γένος (race). The names *Eocene*, *Miocene*, *Pliocene*, and *Pleistocene*, introduced by Charles Lyell, derive from the Greek καινός, for *recent*, ἠώς, for *dawn*, μεῖον, for *lesser*, πλεῖος, for *full*, and πλεῖστος, for *fullest*, referring to the percentage of extant molluscan species in Cenozoic fossil faunas (πλεῖος and πλεῖστος are the comparative and superlative of πολύς, which means *many*). *Oligocene*, added in 1854 by the German Heinrich von Beyrich (1815–1896), derives from ὀλίγος, which means *few*.

24.1. THE PALEOGENE
(65–23.3·10^6 y B.P.)

The terminal Cretaceous extinctions left a vast number of ecological niches empty. At the beginning of the Paleogene, перестройка was the word of the day: The biosphere had to be reconstructed. The empty niches had to be filled. Life met that challenge head-on.

The Paleogene is divided into three epochs: the Paleocene (65–56.5·10^6 y B.P.), the Eocene (56.5–35.4·10^6 y B.P.), and the Oligocene (35.4–23.3·10^6 y B.P.). The biosphere diversified vigorously during the Paleogene. Let us consider the planktic foraminifera, for instance,

for which we have a continuous record thanks to the Deep-Sea Drilling Project (Fig 24.1). During the final epoch of the Cretaceous, the Maestrichtian, about 70 million years ago, there were 34 species of planktic foraminifera in the world ocean. Only a few species survived the holocaust. From that stock, planktic foraminifera diversified into 14 species in 5 million years, increasing to 18 species in another 5 million years, that is, by the end of the Paleocene (56.5·10^6 y ago).

The number of species rose to a Cenozoic peak of 66 in the middle Eocene (about 45 million years ago), corresponding to a speciation rate of 3.3 species per million years during the first 20 million years of Cenozoic time. Following that peak, diversity dropped to a minimum of 15 species during the late Oligocene (about 25 million years ago), corresponding to an extinction rate of 2.5 species per million years. Diversity then rose again to the modern number of 43 species, corresponding to a speciation rate of 1.5 species per million years. The number of planktic foraminiferal species has averaged 35. The excess number of species in the Eocene and the deficiency in the Oligocene were due to the paleogeography of the Earth during those periods.

On the European platform, Paleocene, Eocene, and Oligocene were ingressive cycles separated by minor regressions. The sea covered the Anglo-Parisian basin and a stretch of northern Europe from eastern England to the Baltic states. Scandinavia, as usual, was not covered.

The earliest ingression (Paleocene or Danian) covered only portions of Denmark and deposited a bryozoan limestone and glauconitic sands above it. The Eocene began with a marine ingression that covered the Anglo-Parisian basin, Denmark, and northern Germany. In the center of the Anglo-Parisian

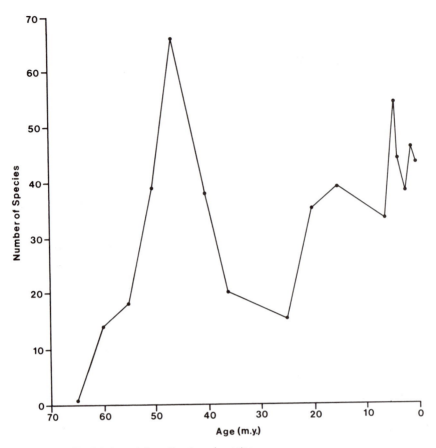

Figure 24.1. Planktic foraminifera. Number of species versus time.

basin, mainly marine sands, limestones, and marls were deposited, while on the edges the sedimentation consisted mainly of continental freshwater limestones and clays. These opposite conditions, with intermediate lagoonal facies, alternated with each other during much of the Eocene over large areas of the basin. There were at least seven such oscillations during the Eocene. The marine sediments are sands and marls, with shallow-water limestones in the Middle Eocene (Lutetian). The continental sediments are freshwater limestones and clays.

During the Middle Oligocene (Stampian), the marine Fontainebleau sands were deposited. These are very pure quartz sands, up to 70 m thick, which, because of their purity, are used to make common glass as well as quartz glass.

[Quartz glass is pure, fused silica. It is difficult to work because of its high melting point (1,610°C). Common glass is quartz fused with sodium carbonate (Na_2CO_3) to lower the quartz's melting point and viscosity, and CaO (which forms calcium silicate) to stabilize the glass and give it strength. Crown glass used for windows consists of 72% SiO_2, 15% Na_2O, and 13% CaO. Lead improves the optical properties of glass, and boron reduces its thermal expansivity (e.g., Pyrex). The green color of glass bottles is due to iron impurities. There are more than 50,000 formulas for different glasses.]

Following the deposition of the marine Fontainebleau sands, the sea left the Paris Basin, and only lagoonal deposits, including freshwater limestones, were formed during the rest of the Oligocene. The total thickness of the Paleogene sediments around Paris is about

170 m, whereas in southeastern England it reaches almost 300 m, including 100 m of the London Clay deposited in early Eocene time. Southeastern England was not covered by the Oligocene sea.

The Oligocene ingression spread from the North Sea eastward, reaching across northern Germany to the Baltic republics and depositing mainly sandstones and clays.

The Alpine trough system was filled up by the end of the Eocene with thick layers of sands and clays from the emerging Alpine ridges. A narrow marine passage persisted along the northern border of the Alpine arc. That channel connected the western Mediterranean with the eastern Mediterranean via the Austrian, Pannonic, and Dacian basins of eastern Europe. Farther south, the main Alpine troughs filled with *flysch* during the Eocene. Flysch is a sediment consisting of deep-water clays, with thin layers of sandstone and limestone.

During the Oligocene, the continued northward push of Africa folded the Alpine troughs into a large mountain range, the Alps. Thick banks of sandstones (*molasse*) from the young mountain range filled the perialpine trough and interrupted the passage between western and eastern Mediterranean north of the Alps.

In North America, Paleocene, Eocene, and Oligocene deposits occur along the east coast and in the Gulf states. In Texas and Louisiana, the classic formations Midway (mainly shales), Wilcox (sandstones and shales), Claiborne (mainly calcarenites), and Jackson (marls) succeed each other through the Eocene. They are followed in the Oligocene by the Vicksburg Limestone.

In the western interior, continental sediments accumulated, including the frequently cross-bedded Fort Union sandstone (Paleocene) and the lacustrine Green River Formation (Eocene) (see Fig. 16.5). The Green River Formation (see Sections 16.1 and 17.4) was deposited in a lake in southwest Wyoming and northwest Colorado. It is 600 m thick, it covers an area of 125,000 km^2, and it consists of finely laminated oil shale. The minerals forming the inorganic part of the Green River shales are quartz, dolomite, feldspars, illite, and pyrite. The presence of pyrite clearly indicates that the bottom was anaerobic, and the presence of dolomite and trona ($Na_2CO_3 \cdot NaHCO_3 \cdot 2H_2O$) indicates evaporitic conditions. The fossil plants found in the Green River Formation indicate a temperature range of 15°C to 20°C and a precipitation range of 60–75 cm/year.

The laminations, with alternating layers of organic-rich and organic-poor sediments, are related to seasonal changes in organic productivity: During the rainy season, the lake was flooded and algae bloomed; during the dry season, the lake shrank, the algae died, and a thin layer of precipitates formed on the bottom. Varve counts indicate that the Green River Formation took $4 \cdot 10^6$ y to accumulate (Section 16.1).

The shallow-marine carbonates of the Paleogene are characterized by several species of larger benthic foraminifera with planispirally coiled shells that have numerous chambers per coil. Among the more common genera are *Camerina* (Eocene–Oligocene) (formerly called *Nummulites*, because it resembles a coin, which is *nummus* in Latin), *Discocyclina* (Paleocene–Eocene), and *Lepidocyclina* (Middle Eocene–Early Miocene). Limestones consisting largely of the shells of these foraminifera are common in the Paleogene of Europe and North Africa.

The Paleogene saw the diversification of the placental mammals. Some of the major groups that differentiated in the Paleogene became extinct later (the Condylartha, ancestral ungulates; the Notoungulata of South America; and the Creodonta, which were early carnivores), but no new ones have evolved since the Eocene. Table 24.1 lists the living mammalian orders, with the number of species in each.

A classical example of evolution is that of the horses, which progressed from being four-toed in the Paleocene to being one-toed today (Fig. 24.2). Some people will tell you that that evolutionary trend resulted from the need to increase speed. I say, fat chance! Try cutting off all your toes, except the middle one, and see if you can run faster. I have a much better theory. Have you noticed that lions in zoos and in TV nature shows seem to be fast asleep most of the time? The reason is that lions roam at night and sleep during the day (just like your cat). Horses, on the other hand, roam during the day and sleep at night. Now, what happens if a roaming herd of horses stumbles on a pride of sleeping lions? Naturally, they *tiptoe* away! I can assure you that nothing is more conducive to preserving the species than tiptoeing

Table 24.1. **Living mammalian groups and their numbers of species**

Subclass Monotremata (ornithorhynchus, echidna)	6
Subclass Marsupialia (kangaroos, etc.)	248
Subclass Eutheria	
Order Edentata (anteaters, armadillos)	32
Order Pholidota (pangolins)	7
Order Tubulidentata (aardvark)	1
Order Pinnipedia (seals)	32
Order Sirenia (dugongs, manatees)	4
Order Cetacea (porpoises, whales)	92
Order Insectivora (hedgehogs, moles, shrews)	374
Order Chiroptera (bats)	981
Order Rodentia (squirrels, beavers, rats)	1,729
Order Hyracoidea (hyraxes)	6
Order Lagomorpha (rabbits, hares)	66
Order Artiodactyla (even-toed ungulates)	194
Order Perissodactyla (odd-toed ungulates)	16
Order Proboscidea (elephants)	2
Order Carnivora (wolves, bears, lions)	252
Order Primates (monkeys, apes, humans)	195
Total	4,237

rupted flow. That current effectively cut off Antarctica from the warmer oceans to the north and led to the glaciation of Antarctica. The first ice formed around the South Pole and expanded toward the ocean, which apparently was reached about 15 million years ago. Long before that, strong catabic winds streaming out from the growing Antarctic ice sheet had begun cooling the surrounding ocean, which led to accelerated formation of cold bottom water. The entire mass of the ocean cooled off, and the bottom-water temperature slowly dropped to its present value around 2°C. Obviously the bottom and midwater faunas were affected, but the change was slow enough to allow adaptation. Increasing glaciation in Antarctica during the Cenozoic led to the recent ice ages.

THINK

* Considering that the quartz in most marine sands is mixed with varying amounts of other minerals, shell fragments, and so forth, how could such pure quartz sands as the Fontainebleau sands have formed?

* Arrange the taxa shown in Table 24.1 in order of decreasing (or increasing) numbers of species, and explain why the order is the way it is.

away from a pride of sleeping lions. The better tiptoers had a better chance to get away, and so the habit of tiptoeing spread, and the horses became single-toed. Another brilliant theory.

Modern horses, belonging to the genus *Equus*, evolved from more primitive horses about 3.5 million years ago in North America, and from there they spread rapidly to all continents (except Australia and Antarctica). They became extinct in the Americas about 10,000 years ago, apparently victims of overkill by the rapidly expanding human population.

While the mammals were diversifying and evolving, the Atlantic kept widening: Australia separated from Antarctica 55 million years ago, at the boundary between Paleocene and Eocene, and South America separated from Antarctica 30 million years ago, in mid-Oligocene time. The creation of the Drake Passage made it possible for the largest ocean current in the world, the Antarctic Circumpolar Current, to come into existence as an uninter-

24.2. THE NEOGENE (23.3–1.6·10⁶ y ago)

The Neogene is divided into two epochs: the Miocene (23.3–5.2·10⁶ y ago) and the Pliocene (5.2–1.6·10⁶ y ago).

The Miocene is characterized by a transgressive cycle in northwestern Europe. The North Sea extended over Belgium, Holland, and northwestern Germany, and deposited sandstone and clays. Most of Italy remained below sea level during the Miocene, accumulating mainly sands and clays. The type section for the Miocene is in fact in northwest Italy, represented by bluish marls rich in foraminifera and mollusks.

The Mediterranean extended northward up the Rhone Valley and into Switzerland, reestablishing the perialpine passage with the eastern Vienna, Pannonic, Dacian, Euxinic, and Caspian basins, where brackish (Sarmatian)

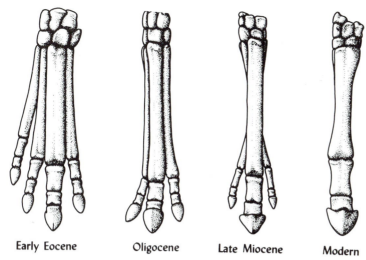

Early Eocene   Oligocene   Late Miocene   Modern

Figure 24.2. Horses. Evolution of the lower part of the limb during the Cenozoic. In Eocene times, the forelimbs had four toes, and the hindlimbs had three. In Oligocene and Miocene time, both forelimbs and hindlimbs had three toes. These were further reduced to one during the Pliocene. There was a concurrent fivefold increase in size (not shown in the illustration) and a 30-fold increase in brain capacity. (Weller, 1969, p. 552, Fig. 552.)

and freshwater (Pontic) sediments accumulated. In the perialpine depression, thick, poorly sorted sandstone beds (molasse) accumulated. In Middle Miocene time, about 15 million years ago, the Antarctic ice sheet reached the shoreline of Antarctica, and ice shelves began extending into the sea. Ice buildup continued during the rest of the Miocene, causing the sea level to drop. The end of the Miocene was characterized by a worldwide regression. The perialpine passage was rapidly filled with thick beds of molasse. The Vienna, Pannonic, and Dacian basins also filled up.

Until toward the end of the Miocene, the Mediterranean was connected to the Atlantic via a passage through southern Spain. That passage became interrupted $6.2 \cdot 10^6$ y ago. The interruption was not complete, however, and the passage was reestablished from time to time. The overall reduction of water flow from the Atlantic, combined with the excess evaporation characteristic of the Mediterranean, was sufficient to reduce the amount of water in the Mediterranean tenfold. Residual basins filled with hypersaline brines occupied the deeper areas of the Mediterranean (western Mediterranean, Tyrrhenian Sea, Ionian Sea, and eastern Mediterranean). Those basins were isolated from each other and may have become totally desiccated in spite of runoff from the Black Sea, the Nile, and the Italian, French, and Spanish rivers.

If the depth of the Mediterranean were uniform, total desiccation would produce a salt layer 26 m thick. The combined surface of the residual basins was much smaller than the surface of the Mediterranean, however, so that the thickness of the salt would be greater, perhaps as much as 100–200 m. Operations in the Mediterranean by the Deep-Sea Drilling Project have revealed that the thickness of the salt beds is much greater, up to 2–3 km in places. It is clear that the Mediterranean was periodically refilled, which is best explained by periodic oscillations in Antarctic ice. If each refill produced 100–200 m of salt, some 10 or 20 refills would be needed to account for the thickness of the salt. Assuming a periodicity of 100,000 y, similar to that of the more recent ice ages, the deposition of the salt would have taken on the order of 10^6 y. That was in fact the case. Salt deposition began $6.2 \cdot 10^6$ y ago and terminated $5.2 \cdot 10^6$ y ago.

Salt deposition terminated because a new passage opened, the Strait of Gibraltar, through which enough Atlantic water enters to replenish the excess Mediterranean water that evaporates each year.

The establishment of a passage between Atlantic and Mediterranean at Gibraltar brought back marine conditions in the Mediterranean and marked the beginning of the Pliocene ($5.2–1.64 \cdot 10^6$ y B.P.). Marine sediments, mainly sands and clays, were deposited on the evaporites, not only in the deep Mediterranean floor but also in basins bordering the Mediterranean. Type sections for the Pliocene are in Italy and Sicily. They are above sea level and available for inspection because Italy has been (and is being) rapidly uplifted by the African push. Marine Pliocene deposits are found at altitudes of 500 m in Tuscany and more than 1,000 m in Calabria.

In America, marine Neogene formations extend along the Atlantic coast and across Florida. They consist of marls and shales for the Miocene, and mainly sandstones for the Pliocene. The total thickness is on the order of 150 m. Along the Gulf Coast, the Miocene and Pliocene sediments, mostly sands and clays, attain great thicknesses (3–4 km). In the Great Plains, which extend from the foot of the Rockies to the Mississippi, a broad apron of fluviatile sediments, averaging 50 m in thickness, was formed during the Neogene.

West of the Rockies are the Colorado Plateau and the Basin and Range Province. The Colorado Plateau, now averaging about 1,700 m in altitude, was a lowland until about 10 million years ago (Middle Miocene). Uplift took place from 10 to 5 million years ago. In those 5 million years, the Colorado River, which before was meandering through the lowlands, incised its own bed and formed the Grand Canyon. The average rate of uplift (and incision) was 0.3 mm/y.

The Basin and Range Province is bordered by the Colorado Plateau to the east and by the Sierra Nevada to the west. This province was a back-arc basin overlying the subducting Farallon plate. But the last of the Farallon plate disappeared 40 million years ago. Subduction ceased, and transform motion (right-lateral) between the North American and Pacific plates took over. No more subduction meant no more downpulling of the back-arc basin. The back-arc basin bounced back and updomed, with block faulting as a result. The thickness of the continental crust is 35–40 km under the Colorado Plateau and 45–50 km under the Sierra Nevada, but only 25–30 km

under the Basin and Range Province. Block faulting has created a series of grabens (basins) and horsts (ranges). The basins have been rapidly filled with sediments (up to 3 km thick) during the Pliocene and the Quaternary, forming the very flat playas.

Extensive volcanism accompanied the uplift of the Colorado Plateau, the updoming of the Basin and Range Province, and the renewed uplift of the Sierra Nevada. In addition, flood basalt poured out from fissures in the continental crust $16 \cdot 10^6$ y ago and in three million years covered 500,000 km² of land in Washington, Oregon, Idaho, and northern California, and formed the Columbia Plateau. Individual basaltic flows are 30–150 m thick and may be 150 km long, indicating that the basalt was very hot and very fluid. The total thickness reaches 3.5 km. It is clear that the underlying mantle is very hot. In fact, the entire region is believed to be underlain by a huge hot spot, the Yellowstone hot spot.

Farther south, in southern California, the continantal borderland was block-faulted in Miocene time, and several basins were formed that accumulated great thicknesses of sediments rich in organic matter. The Los Angeles basin has 8 km of Miocene sediments, 2.7 km of Pliocene sediments, and 500 m of Pleistocene sediments. The sediments are sandstones and shales, with some conglomerate beds. They were rich in organic matter when formed because the basin was located in an area of upwelling and therefore of high productivity. Since then, the organic matter has become petroleum, the sediments have been gently folded, and the petroleum has accumulated in the anticlines. At Rancho La Brea in Los Angeles, the petroleum has seeped to the surface, the lighter hydrocarbons have evaporated, and a pool of heavy oil and tar has remained.

The expansion of ice in Antarctica continued to lower global temperatures during the Pliocene. In middle Pliocene time, about $4 \cdot 10^6$ y ago, volcanic activity closed the seaway between North America and South America that had existed throughout the Mesozoic and Cenozoic. That event had a profound effect on oceanic circulation. The Antilles Current, which until then had continued westward into the Pacific, was deflected into the Gulf of Mexico and became part of the Gulf Stream. The Equatorial Counter Current in the Pacific,

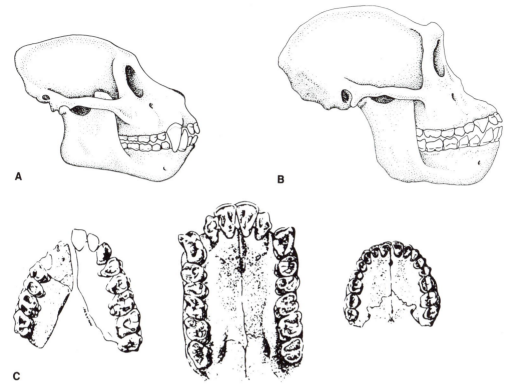

A   B

C

Figure 24.3. (A) *Aegyptopithecus*. (Emiliani, 1988, p. 217, Fig. 11.22.) (B) *Ramapithecus*. (Emiliani, 1988, p. 218, Fig. 11.23.) (C) Upper jaw dentition of *Ramapithecus* (left); a modern chimpanzee (center); and *Homo sapiens* (right). (Buettner-Janusch, 1966, p. 126, Fig. 8.1.)

which until then had crossed into the Atlantic, had to turn back after bouncing against the west coast of Central America.

There are indications that when the seaway between North and South America was open, a counterclockwise current circled North America and kept the Arctic warm. When the passage was closed, that current ceased, the Arctic Ocean cooled, and ice began forming. Arctic deep-sea sediment cores indicate that the Arctic pack ice formed $3 \cdot 10^6$ y ago and that Greenland became glaciated at about the same time. Shortly after, in the late Pliocene, ice formed over Canada and over Scandinavia. The great ice ages that mark the Quaternary were under way.

The Neogene saw the major evolutionary steps that led to the appearance of *Homo sapiens* on Earth. The lineage that produced modern humans can be traced to African primates. The earliest was *Aegyptopithecus* (Fig. 24.3a), a small tree-dwelling primate that lived in Egypt 28 million years ago (Late Oligocene). Next came *Dryopithecus*, which lived in Africa from 20 to 10 million years ago (Early to Middle Miocene). *Dryopithecus* expanded to Eurasia about 14 million years ago when the African-Arabian landmass collided with Asia. The foot of *Dryopithecus* indicates that *Dryopithecus* did not hang from trees, but walked on land and probably was already bipedal. *Kenyapithecus* is another genus of advanced primates that lived between 15 and 10 million years ago. It may have been the first systematic tool user, because remains of *Kenyapithecus* dating from 12 million years ago have been found together with piles of bones that had been crushed with stones that had been brought in from somewhere else. A branch of *Kenyapithecus* that migrated to Asia, known as *Ramapithecus* (Fig. 24.3B), had its teeth arranged in an arc, as in humans (Fig.

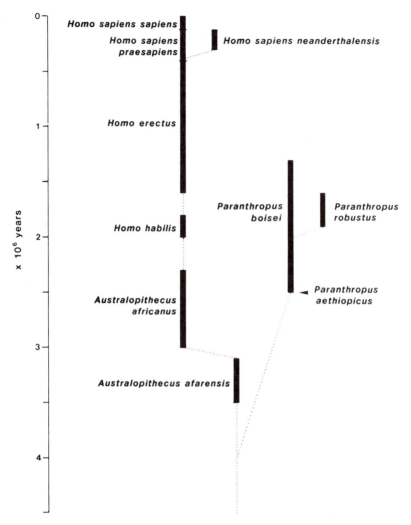

Figure 24.4. The hominid lineage. (Emiliani, 1988, p. 222, Fig. 11.26.)

24.3C), indicating that *Kenyapithecus* did too. In contrast, apes have their teeth arranged in a very different, rectangular pattern.

The Hominoidea is the superfamily that includes the families Pongidae (apes) and Hominidae. The Pongidae include the living genera *Hylobates* (gibbon), *Pongo* (orangutan), *Gorilla*, and *Pan* (chimpanzee), plus several fossil genera. The family Hominidae includes the living genus *Homo* and two fossil genera, *Australopithecus* and *Paranthropus*.

There is a gap of 6 million years in the fossil record between *Kenyapithecus-Ramapithecus* and the earliest Hominidae. Fig. 24.4 shows

the genealogical tree of the Hominidae. The earliest hominid genus was *Australopithecus*, whose fossil record ranges from 3.7 to 2.3 million years ago. The earliest species, *Australopithecus afarensis* (Fig. 24.4A), lived in eastern Africa between 4 and 3 million years ago. It was 1 m high, it weighed about 30 kg, it walked upright, and its ratio of brain size (in cubic centimeters) to body weight (in kilograms) was 13 (compared with 3.5 for a modern gorilla, 8.9 for a chimpanzee, and 20.8 for a modern human). *Australopithecus africanus* (Fig. 24.5B) was a more advanced species that lived between 3 and 2 million years ago in eastern and southern

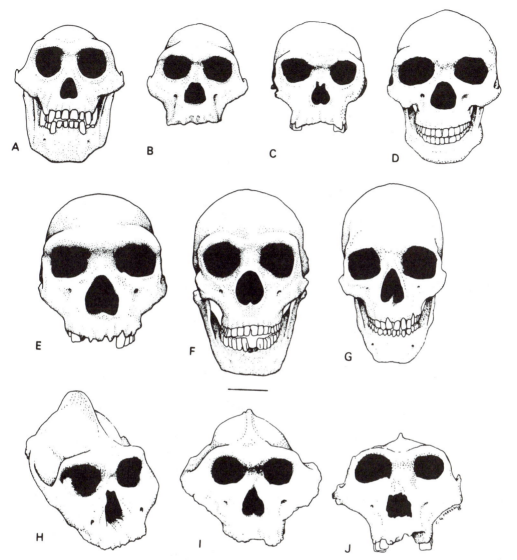

Figure 24.5. Human ancestors: (A) *Australopithecus afarensis*; (B) *Australopithecus africanus;* (C) *Homo habilis*; (D) *Homo erectus*; (E) *Homo sapiens* "*praesapiens*"; (F) *Homo sapiens neanderthalensis*; (G) *Homo sapiens sapiens*; (H) *Paranthropus aethiopicus*; (I) *Paranthropus boisei*; (J) *Paranthropus. rubustus*. (Emiliani, 1988, p. 221, Fig. 11.25.)

Africa. It had a brain-size/body-weight ratio of 14. Next came the first *Homo, Homo habilis* (Fig. 24.5C), which lived in Africa around 2 million years ago, at the close of the Pliocene. The beginning of the Quaternary, $1.64 \cdot 10^6$ y ago, saw the appearance of *Homo erectus* (Fig. 24.5D). Remains of *Homo erectus* have been found throughout the Old World at sites ranging in age from $1.6 \cdot 10^6$ to 400,000 y ago. *Homo erectus* had a brain-size/body-weight ratio of 17, was a capable hunter, and knew the use of fire. *Homo erectus* evolved into *Homo sapiens* "*praesapiens*" (Fig. 24.5E) about 250,000 years ago. About 125,000 y ago *Homo sapiens* "*praesapiens*" branched into *Homo sapiens neanderthalensis* (the Neanderthals) (Fig. 24.5F) and *Homo sapiens sapiens* (modern humans) (Fig. 24.5G).

Parallel to the evolution of the *Australopithecus–Homo* lineage, another genus appeared: *Paranthropus* (Fig. 24.5H–J). *Paranthropus* was much heavier than *Australopithecus* (50 kg vs. 30 kg), but had a comparatively small brain (brain-size/body-weight ratio = 9–10) and was a vegetarian. It lived in Africa, together with *Australopithecus* and *Homo*, between 2.5 and 1.5 million years ago. It left no descendants. Too bad, because, compared with *Australopithecus*, who, judging from its descendants, must have had a nasty disposition, *Paranthropus* probably was a jolly good fellow and nice to play with (though a bit weak on intellectual conversation).

THINK

* The surface of the Mediterranean is $2.5 \cdot 10^6$ km². Assuming that the deep Mediterranean basins where salt accumulated in great thickness had a combined surface equal to 10% of that of the Mediterranean, assuming that the average thickness of salt is 2 km, knowing that the density of halite is 2.16 g/cm³, knowing that the average salinity of the ocean is 34.72‰ (meaning 34.72 g of salts in 1 kg of seawater), knowing that the average temperature of the ocean water is 3.5°C, and knowing that the average density of seawater with 34.72‰ salinity and 3.5°C temperature is 1.027 g/cm³, calculate the average salinity of the ocean water before the Mediterranean desiccation events.

24.3. THE QUATERNARY ($1.64 \cdot 10^6$ y ago to the present)

The Quaternary, the last period of the Cenozoic, is defined as the time since the appearance of the benthic foraminiferal species *Hyalinea baltica* in the Mediterranean. This event occurred $1.64 \cdot 10^6$ y ago and is recorded in marine silts and clays in Calabria, southern Italy.

In addition to *Hyalinea baltica*, the molluscan species *Arctica islandica* and a dozen other North Atlantic species are found in Calabria. These species are living not only in the North Atlantic but also in the Baltic Sea.

Species living in the Baltic must be capable of tolerating the low salinities prevailing there (< 16‰). It is possible that the species that entered the Mediterranean did so through a temporary passage running from Bordeaux to Narbonne and acting as a salinity filter. The appearance of the northern species in the Mediterranean was not an event of great significance per se. It marks, nevertheless, a specific point in time that, by international agreement, defines the beginning of the Quaternary.

The Quarternary is divided into two epochs: the Pleistocene ($1.64 \cdot 10^6$ y ago to 10,000 y ago) and the Holocene (the past 10,000 y). Paleomagnetic research has shown that during the Pliocene and Quaternary the magnetic field of the earth reversed itself several times (see Fig. 11.32). Four polarity epochs are distinguished: the Gilbert epoch, predominantly reversed ($5.4–3.4 \cdot 10^6$ y); the Gauss epoch, predominantly normal ($3.40–2.48 \cdot 10^6$ y); the Matuyama epoch, predominantly reversed ($2.48–0.73 \cdot 10^6$ y); and the Brunhes epoch, predominantly normal ($0.73 \cdot 10^6$ y to the present). Within each epoch, except the Brunhes, there are reversed events (see Fig. 11.32). Within the reversed Matuyama epoch, for instance, there are the normal events called Réunion ($2.1–1.9 \cdot 10^6$ y), Olduvai ($1.83–1.62 \cdot 10^6$ y), and Jaramillo ($0.97–0.91 \cdot 10^6$ y).

Hyalinea baltica appeared near the top of the Olduvai-event section in Calabria and therefore dates from $1.64 \cdot 10^6$ y ago. *Hyalinea baltica* apparently speciated earlier, and but it rapidly spread at that time not only to the Mediterranean but also to the Atlantic and the Caribbean.

The Quaternary is characterized by extensive glaciations in the Northern Hemisphere (Fig. 24.6). Every 100,000 years or so, ice expanded southward from the Canadian Arctic and from Scandinavia to a front running from Seattle to New York and from London to Berlin, Moscow, and eastward into Siberia. The North American ice sheet was larger than the present Antarctic sheet, and the Scandinavian–Siberian ice sheet was not much smaller. The water that went to build up the ice sheets came from the ocean, of course. At the peak of each ice age, the sea level was about 120 m lower than it is today.

The closing of the seaway between North America and South America about 3 million

Figure 24.6. The ice ages. Extent of ice in the Northern Hemisphere during the last ice age, about 20,000 y ago. In addition to the major ice sheets, mountain glaciers covered all high mountain ranges of the world. (Flint, 1952, p. 375.)

years ago apparently created conditions suitable for the development of major glaciations in the Northern Hemisphere. The first major glaciation occurred about $2.36 \cdot 10^6$ y ago. From then on, the astronomical motions of the Earth (see Section 22.7) have been timing the succession of ice ages.

Let us follow the development of a typical ice age: Given the present world physiography, and a land surface about 200 m higher than now in both the Canadian Arctic and Fennoscandia, the first combination of low eccentricity and low obliquity causes the ice to expand southward in both the Canadian Arctic and Scandinavia. Expanding ice increases the global albedo; temperature drops, and more ice forms. More ice means greater albedo, lower temperature, and even more ice. This positive-feedback loop brings the ice front all the way to the mid-latitudes. As ice accumulates, isostatic adjustment takes place. For an ice thickness of 2 km, similar to that of the ice in Greenland today, the ground sinks about 700 m, because the density of ice is one-third that of mantle rocks.

But as the northern ice sheets expand, the ocean surface cools, less moisture evaporates, and less precipitation falls on the ice sheets. This leads to thermal equilibrium between ice and ocean. Equilibrium is soon broken, however, when the northern areas of the North Pacific and North Atlantic become covered with pack ice. At present, pack ice covers the Arctic to an average thickness of 3 m. Evaporation from an ice surface is negligible compared with that from an open water surface. When pack ice forms over the North Pacific and North Atlantic, evaporation from the ocean, which fed the ice sheets, stops, and the ice sheets are starved.

A glacier is like a stream: It exists only so long as it flows (i.e., as long as accumulation matches ablation). If ablation overtakes accumulation, an ice sheet does not shrink, but keeps flowing out and thinning. As long as ice flow continues, the surface of the ice does not shrink, albedo remains high, and temperature remains low (in part also because of heat absorption by ice melting). More than 50% of the ice is returned to the sea as ice water before the ice starts shrinking. Eventually it does, but by that time the surface of the land is so depressed that the ice-sheet surface is below the snowline. When pack ice disappears, renewed evaporation produces rains that literally wash away the residual ice. It takes an estimated 15,000 y for the land surface to bounce back to the original altitude and reestablish conditions suitable for another ice age. After that, a new ice age will start when the astronomical conditions are again favorable.

The detailed history of the ice ages has been reconstructed by analyzing deep-sea sediments. Columns of deep-sea sediments up to 20 m length can be retrieved by piston coring. Much longer sediment columns can now be retrieved using drilling vessels. In Globigerina-ooze sediment, the fossil faunas of coccoliths and planktic foraminiferal shells reflect the ecological conditions at the surface at the time when they formed. A tropical assemblage is replaced by a temperate assemblage if the surface water

cools, and vice versa if the surface water warms up. So, simple micropaleontological analysis reveals the succession of the ice ages.

A more precise picture can be obtained by determining the oxygen-isotopic composition of the foraminiferal shells or coccoliths. There are three stable isotopes of oxygen: ^{16}O (99.7%), ^{17}O (0.04%), and ^{18}O (0.2%). The ratio of ^{18}O to ^{16}O is 1:499 in nature. This ratio is not constant, however. There is about 7‰ more ^{18}O in water vapor than in the water from which the vapor came. This is so because the $H_2^{16}O$ molecule is lighter than the $H_2^{18}O$ molecule (18 u vs. 20 u). In the liquid phase, the lighter molecule ($H_2^{16}O$) travels faster (Section 2.2) and therefore will hit the underside of the water–air interface more often than the heavier molecule. As a result, the lighter molecule has a better chance to escape into the air. When atmospheric water vapor precipitates as rain or snow, the heavier molecule, $H_2^{18}O$, tends to condense in preference to the lighter molecule. An air mass moving from the Gulf of Mexico to Canada and dropping rain all along the way will become progressively isotopically lighter. In fact, Greenland ice, which is the last fraction to precipitate, is about 30‰ isotopically lighter than seawater.

When ice builds up during an ice age, the preferential removal of lighter water leaves behind an ocean that is enriched (by 1‰) in the heavier isotope. If nothing else happened, the oxygen in the $CaCO_3$ of foraminiferal shells growing in the open ocean during an ice age would be enriched in ^{18}O by 1‰. But there is also a temperature effect. The lower the temperature at which the shell forms, the greater the concentration of ^{18}O in the shell with respect to that in the water. So there are two effects adding to each other: During an ice age, the seawater is 1‰ heavier; it is also colder, which causes the foraminiferal shells to become even heavier (by another 1‰). Altogether, foraminiferal shells contain 2‰ more ^{18}O during an ice age than during an interglacial age.

This difference can easily be measured by means of a mass spectrometer, and so the succession of the ice ages can be accurately determined by measuring the oxygen-isotopic composition of foraminiferal shells downcore. At the normal rate of sedimentation of 2.5 cm per 1,000 y, a piston core 20 m long represents

800,000 years of time. The earlier history of the ice ages can be studied by analyzing drill cores.

We now have a continuous isotopic record for the past 60 million years (Fig. 24.7A). This record shows that seawater became progressively heavier, and temperature progressively dropped, as ice built up in Antarctica. Strong drops occurred 55, 38, 12, and 5 million years ago (Fig. 24.7A). A more detailed deep-sea core record ranging from 4 to 1.5 million years ago (Fig. 24.7B) shows a definite change about 3.2 million years ago, perhaps signifying the glaciation of Greenland.

The more recent ice ages are shown in great detail by the isotopic record of the past 750,000 years (Fig. 24.7C). The record shows that it takes some 50,000 to 100,000 years to make an ice age, but only 10,000 years to destroy it. The reason is that as ice grows, the sea surface cools. But cooler water sinks and is replaced by warmer water from below. The building of an ice sheet, therefore, cools off not only surface water but also deeper water. In contrast, when ice melts, a surface layer of low-salinity water forms over the ocean and remains at the surface because of its lower density. As the sun warms it up, it becomes even less dense. Melting of the ice does not require a warming of the entire ocean mass, only of the surface layer. As a result, deglaciation is much faster than glaciation.

During the last ice age, ice reached its maximum extension 20,000 years ago. Deglaciation started almost immediately and progressed rapidly (Fig. 24.8). Sometimes ice meltwater would pile up behind an ice dam, and when the dam collapsed a huge flow would follow. One such great flood occurred in the American northwest 13,500 years ago when an ice dam holding back about 2,000 km^3 of ice meltwater (Lake Missoula) collapsed. A huge mass of muddy water and debris rushed across the area into the Columbia River, cutting broad channels called *coulees* and forming the so-called Channeled Scablands. A similar terrane exists on Mars, especially in the equatorial region between 20°W and 40°W. Evidently there were floods on Mars, too, when there was still water there, more than 3 billion years ago. As a result of the flood that formed the Scablands, the sea level rose very rapidly, from −100 to −80 m

(Fig. 24.9). By 12,000 years ago, more than 50% of the ice had returned to the ocean, and the sea level had risen to −60 m. At that point, another major meltdown occurred, and a great flood took place down the Mississippi Valley. Pebbles, which now are confined to the upper reaches of the river, were carried all the way down the lower Mississippi Valley. The sea level rose very rapidly from −60 to −40 m (Fig. 24.9). The minor peak between 12 and 11 cm in Fig. 24.8 is the isotopic signature of this event.

The oxygen-isotopic record of deep-sea cores shows that both minima and maxima had very short durations, an indication of climatic instability. Interglacial conditions similar to those prevailing today lasted only a short time during earlier interglacials, perhaps 5,000 years or so. This causes considerable concern, because the present interglacial has already lasted more than 5,000 years. We should soon be heading for another glacial age (barring human effects on climate). In fact, one could argue that the next glacial age was already under way in the sixteenth through eighteenth centuries, a period when temperatures were so low in Europe that the period came to be known as the Little Ice Age. One could further argue that progress toward the next ice age was stopped by the industrial revolution, which began adding CO_2 to the atmosphere. Today, the main concern is that the CO_2 we are adding to the atmosphere may create a runaway greenhouse effect. Also of concern is the addition of reduced carbon compounds that will decrease the amount of ozone in the stratosphere and destroy the UV protective shield that we now enjoy.

Not all ice ages were of equal severity, and not all interglacial ages were equally warm. The isotopic oscillations have been numbered with odd, numbers for the interglacials and even numbers for the glacials, increasing with age. No. 1 is the present interglacial, the Holocene, and no. 2 is the last ice age, which peaked 20,000 years ago. That ice age was one of the most severe. Studies of plant remains (especially pollen) in lake deposits indicate that a cold, dry steppe extended from the ice margin almost to the tropics. Florida and the Bahamas were dry, windswept plateaus rising about 120 m above the glacial sea. A frigid gap

δ¹⁸O, ‰ w.r.t. P.D.B.

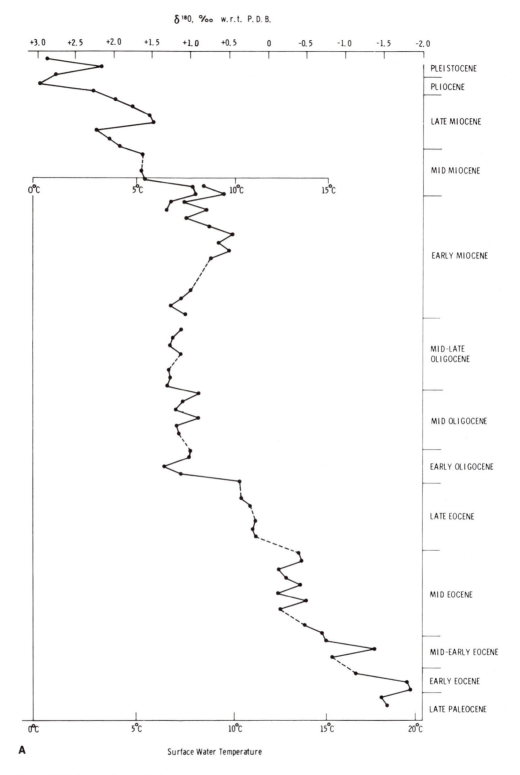

A

Surface Water Temperature

Figure 24.7. See caption on facing page.

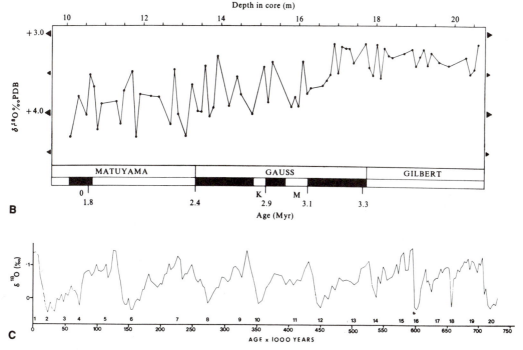

Figure 24.7. The oxygen-isotopic record of planktic foraminifera. All values refer to the Chicago standard PDB-1. (A) Isotopic record, 60·10[6] y long, showing the increasing amount of ice on Earth and the accompanying temperature decrease. Approximate temperatures are shown only for Cenozoic time before the Middle Miocene, because after that time the Antarctic ice significantly expanded, and northern glaciations began. The isotopic composition of seawater was drastically altered, making temperature estimates uncertain. (Shackleton and Kennett, 1975.) (B) Isotopic record from 4·10[6] to 1.6·10[6] y B.P. from the equatorial Atlantic. The isotopic minimum at 2.36· 10[6] y represents the first major northern glaciation. (Shackleton and Opdyke, 1977.) (C) Isotopic record of the past 750,000 y from the tropical Atlantic and the Caribbean, showing the alternation of glacial and interglacial ages. (Emiliani, 1978, p. 353, Fig. 2.)

300 km wide existed across southern Germany between the Scandinavian ice sheet to the north and the Alpine ice cap to the south. A dry climate extended across the Mediterranean into central Asia.

When sea level was 120 m below its present level, the continental shelves of the world were exposed. Rivers incised the shelves and created drainage system that are now flooded by the oceans. Among the major areas that were exposed were the Bahama Banks, the Florida, Yucatán, and Nicaragua shelves, the Persian Gulf, the southern portion of the South China Sea, the Java Sea, and the Arafura Sea. The broad Arctic shelves were largely covered with ice, and so was the North Sea and the shallow banks off New England, Nova Scotia, and Newfoundland. In addition, innumerable straits were dry land and were used as passages by the land fauna (including humans). The most famous is the Bering Strait, through which humans reached North America, apparently more than 30,000 years ago.

The environmental conditions during the interglacial high-temperature intervals (*hypsithermals*) strongly contrast with those during the glacial temperature minima (*bathythermals*). During the last hypsithermal, 125,000 years ago, global temperatures were significantly higher than today, especially in middle and high latitudes. In fact, red soils developed in the North American midcontinent as well as in southern Europe, whereas only brown soils have developed during the Holocene. The

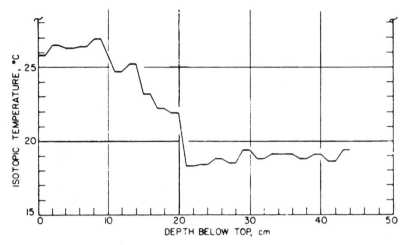

Figure 24.8. Detailed oxygen-isotopic record of the past 30,000 y. This record, from planktic foraminifera from an equatorial Atlantic core, shows that we are past the last hypsithermal. The secondary isotopic peak at 11–12 cm below the top of the core reflects the meltdown of the ice sheets at the end of the last ice age, about 11,000 y ago. (Emiliani, 1955, p. 554, Fig. 8.)

oxygen-isotopic record from an ice core drilled through the ice sheet of eastern Antarctica by Soviet scientists (Fig. 24.10) shows a higher peak for the last interglacial (G) than for the present interglacial (A), indicating that about $2.2 \cdot 10^6 \, \text{km}^3$ of ice melted. That apparently was the cause for the sea level rising 6 m above its present level (see Fig. 16.10). The amount of ice that melted was equivalent to the ice mass now stored in Greenland or in West Antarctica.

Oxygen-isotopic analysis of ice cores sampling the entire thickness of the Greenland and Antarctic ice sheets has shown that ice formed during the last interglacial age still exists at the bottoms of the ice sheets of eastern Antarctica and of Greenland. Evidently these ice sheets did not melt during the last interglacial. It appears, therefore, that the higher sea level of 125,000 y ago was due to melting of the West Antarctic ice. That could happen again. Warmer temperatures and higher sea

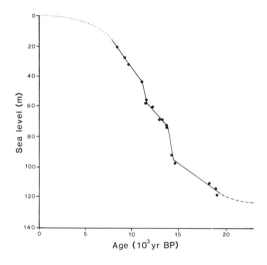

Figure 24.9. Sea-level rise from 20,000 y ago to the present. At 20,000 y ago the sea level stood about 120 m below the modern level. The sea-level rise was determined by dating specimens of *Acropora palmata* from fossil reefs now below sea level off the west coast of Barbados. The coral species grows within 2 m of sea level and therefore is a very good indicator of sea level. Dating was done by the $^{230}\text{Th}/^{234}\text{U}$ method (see Section 3.4). (Data from Bard et al., 1990, Fig. 1.)

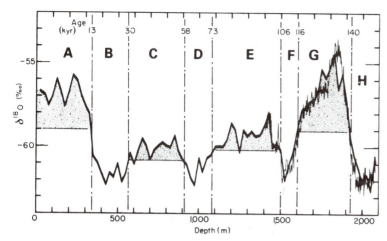

Figure 24.10. Oxygen-isotopic record from the Vostok ice core in Antarctica (78°28'S, 106°48'E, 2,083 m long), subdivided into warmer (A, C, E, G) and colder (B, D, F, H) stages. The record is continuous from the penultimate glaciation (H) to the present. The bottom scale gives the depth in meters below the core top. The top scale gives the age in thousands of years. Because the ice below compresses as more ice accumulates on top, stage G has a thickness similar to that of stage A, but it represents almost twice the time. The isotope peak of the last interglacial age (G) is higher than that of the present interglacial (A), reflecting the additional amount of ice that melted at that time and the concomitant high sea level of +6 m (see Fig. 16.10). The isotopic trend of the present interglacial (A) indicates that we have passed the peak. (Lorius et al., 1985, p. 592, Fig. 1.)

levels could be just ahead, before the next glaciation begins to develop in earnest.

Significant volcanic activity took place during the Quaternary along all subduction margins. In addition, volcanic activity was intense along an arc extending from the French Riviera through Tuscany and Latium into Naples, Calabria, and Sicily. Volcanoes still active in this region are Vesuvius, Etna, Stromboli, and Vulcano.

The Holocene is the current geological epoch. Its beginning is set arbitrarily at 10,000 years ago (8000 B.C.E.). At that time, residual ice still covered areas of Canada and Scandinavia, and the sea level stood at about −30 m.

The Holocene is divided into five ages, based on European pollen sequences. During the first age, the Preboreal (10,000–9,500 y B.P.), birch and pine forests replaced tundra in western Europe. Spruce forests and, farther south, deciduous forests covered the North American midcontinent. During the Boreal (9,500–8,200 y B.P.), ice retreated to the Scandinavian mountains and northern Canada, forests expanded, and the sea level rose to −20 m. During the Atlantic (8,200–5,900 y B.P.), the ice disappeared almost completely from the northern continents (except for Greenland), the sea level reached its present height, and deciduous forests dominated the northern continents. The temperature at the peak of the Atlantic was perhaps 1°C higher than today (Fig. 24.8). Rainfall over the Sahara and eastern Africa was much higher than today, and African lake levels, which had dropped several hundred meters during the last ice age, were higher than today. The Subboreal (5,900–2,250 y B.P.) was characterized by a cooler and more humid climate. Lastly, the Subatlantic (2,250–0 y B.P.) represents the establishment of modern climate. Within the Subatlantic, however, significant climatic oscillationss have occurred. There was a warm interval from 950 to 1250 C.E., when Greenland was green and Scandinavian settlers arrived. That warm period was followed by the Little Ice Age, which peaked in the seventeenth century and waned during the eighteenth century. The nineteenth and twentieth centuries saw the establishment of today's climate. Harsh winters in the early and late parts of the

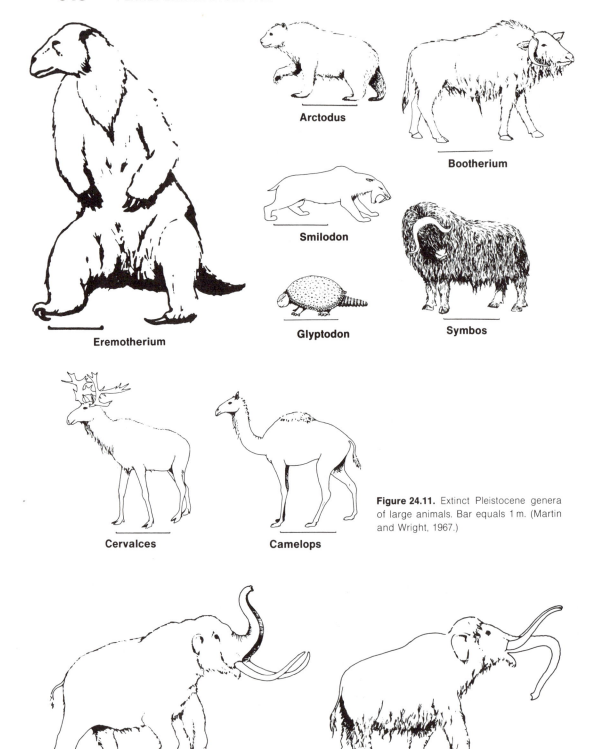

Eremotherium

Arctodus

Bootherium

Smilodon

Glyptodon

Symbos

Cervalces

Camelops

Figure 24.11. Extinct Pleistocene genera of large animals. Bar equals 1 m. (Martin and Wright, 1967.)

Mammut

Mammathus

nineteenth century were replaced by a period of rising temperature between 1900 and 1940. Temperatures remained stable between 1940 and 1980, but a dramatic increase has occurred during the 1980s.

Asteroidal and cometary impacts have occurred with considerable frequency during the Quaternary. One formed Meteor Crater in Arizona, 1,200 m across and 180 m deep, 50,000 years ago (Section 7.6). A major impact occurred 700,000 years ago and produced a vast number of tektites that spread from Australia through Indonesia to southeastern Asia (see Section 7.3); no crater has been found, however. Other major craters that produced tektite fields are the Zhamanshin crater (600 km east of the Aral Sea), which is 10–15 km across and dates from $1.1 \cdot 10^6$ y ago, and the Bosumtwi crater in Ghana, which is 10.5 km across and dates from $1.3 \cdot 10^6$ y ago. The Bosumtwi impact also produced a field of microtektites that covers a portion of the eastern equatorial Atlantic. A recent impact was the Tunguska impact in eastern Siberia on June 30, 1908, when a small comet exploded 8.5 km above ground (see Section 7.3). The trees in the forest below were flattened to a distance of 15 km, but no crater was excavated (see Fig. 7.3).

Asteroidal and cometary impacts pose an increasing threat as the human population expands across the globe. Today, an impact almost anywhere on Earth can be guaranteed to have serious repercussions, to say the least. An "asteroidal watch" is in effect. There are even preliminary plans to intercept an Earth-bound asteroid and redirect it away from the Earth.

The enormous environmental changes that took place every 100,000 years during the Quaternary were forceful enough to fragment, disperse, and isolate all kinds of populations. That is a prerequisite for speciation and evolution—genetic isolation of small groups. In fact, the Quaternary has been a time of rapid organic evolution, most notably evidenced by the genus *Homo*, which produced three species (*habilis, erectus,* and *sapiens*) and three subspecies (*sapiens "praesapiens", sapiens sapiens,* and *sapiens neanderthalensis*) in two million years.

The mammalian fauna was abundant and varied at the beginning of the Quaternary, but it was devastated by *Homo sapiens*, especially during the last ice age and after. More than 200 genera of large animals were driven to extinction, including mastodons, mammoths, ground sloths, saber-tooth cats, and many species of llamas, peccaries, camels, oxen, horses, bears, antelopes, caribou, and armadilloes (Fig. 24.11). Ice-age civilization reached its peak during the last deglaciation, when people had enough leisure time to paint caves. The overkill led to famine, however, which was not alleviated until agriculture was invented about 10,000 years ago.

The ice ages of the Quaternary both caused and made possible the fragmentation and migration of *Homo* from a parental stock in Africa. Some of those migratory groups remained by and large genetically isolated for thousands or tens of thousands of years, leading to subspeciation and the formation of races. Following the end of the last ice age and the advent of modern civilization, genetic recombination is proceeding rapidly throughout the world, enriching the human genome and possibly leading to superior physical types capable of, for instance, better resisting diseases.

24.4. WHAT NEXT?

If humans had not evolved and developed technologies that can deeply affect climate and the environment, the Quaternary, with its procession of glacial and interglacial ages, might have continued for millions or tens of millions of years, coming to an end only in the distant future when a new system of convection cells in the outer core and the mantle might move the northern continents away from the North Pole.

Instead of a new ice age, we may be facing rising temperatures. Humans are now adding about $5 \cdot 10^9$ tons of CO_2 and $500 \cdot 10^6$ tons of CH_4 to the atmosphere each year. The volume concentration of CO_2 in the atmosphere has risen from 280 ppm to 345 ppm in 100 years. During the past 10 years it has risen by an average of 1.3 ppm per year. Both CO_2 and CH_4 absorb infrared radiation from the Earth's surface, which means that much of it is reradiated downward, increasing the tem-

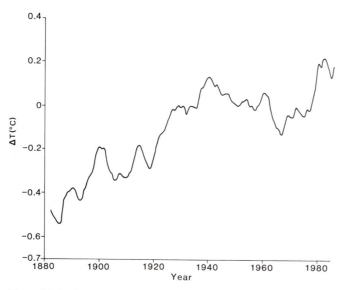

Figure 24.12. Global temperature rise, 1880–1990 (5-y averages). (Hansen and Lebedeff, 1988.)

perature of the troposphere and of the Earth's surface. This is the greenhouse effect about which everybody talks nowadays. Fig. 24.12 shows the trend of global temperature during the past 110 years. An overall temperature increase amounting to about 0.8°C can be seen. As a result, glaciers have been retreating, and the sea level has risen by about 10 cm since 1880 (Fig. 24.13). If the current rate of growth in the consumption of fossil fuels is projected into the future, a doubling of the 1880 concentration of CO_2 in the atmosphere may be predicted for the year 2100.

A rising temperature has, of course, a number of concomitant effects. Warmer seawater means an expanded ocean and an even higher sea level. It also means a more humid atmosphere. Because the H_2O molecule is also an efficient absorber (and reemitter) of infrared radiation, a further temperature rise may be expected. But more water vapor in the atmosphere also means more clouds, which reflect away solar radiation and tend to lower surface temperatures. And then there is the Sun, whose emission may not be absolutely constant, even on a time scale of decades to centuries, as is generally assumed. Fig. 24.12 shows that between 1940 and 1965 the global temperature did not increase, but decreased appreciably. There is no clear explanation for this event. In

fact, the climate system, involving the ocean, the atmosphere, the surface of the land, and volcanic activity, is so complex that no clear-cut climate prediction over a time interval longer than a few days can yet be made, in spite of the huge computers that are now being used. It is possible that human activities during the past century have balanced out—some causing a temperature increase, and others (such as deforestation) causing a decrease. Not knowing exactly how the system works, we should try to control *all* the activities that we expect to have an impact on climate.

At the root of our climatic and environmental problems, including atmospheric and oceanic pollution, is of course, the population explosion. In the 1930s the United States had 120 million people, and the per capita consumption of energy was 5,000 W. Now the population has doubled, and the per capita consumption of energy has also doubled, to 10,000 W. In the industrialized world, doubling the population obviously means a fourfold increase in waste products. Nearly all the problems that confront us at the present time would disappear if the world population were to be stabilized to a level that would place no stress on the resources of our planet. Unfortunately, the political and religious leaders of the world provide no leadership whatsoever

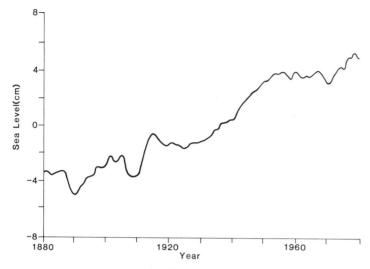

Figure 24.13. Sea-level rise, 1880–1980 (5-y averages).
(Gornitz, Lebedeff, and Hansen, 1982.)

on this number-one problem. They should be ashamed of themselves.

What may possibly save the world from ruin is education. Unfortunately, it takes one generation to educate a cadre of knowledgeable and expert teachers, and another generation for the teachers to do their work. So where do we go from here? We go to infinity, obviously. Fig. 24.14 shows the *doomsday curve*, originally (1960) calculated by three engineers from the University of Illinois, and more recently (1988) adjusted by myself to the more recent demographic data. This curve, based on the demographic data from 1650 on, predicts that the human population will reach 8 billion people in the year 2000, 14 billion people in 2010, 60 billion in the year 2020, and infinity in the year 2023. That, of course, will not happen, because environmental devastation, famine, disease, and strife will grow at even higher exponential rates. There is little doubt, however, that the next few decades will be the most critical in the long history of our planet.

If you have read this book in its entirety, you will realize that you have traveled a very long distance, on a journey that started with the moment of creation and saw the evolution of stars, planets, the Earth, and the life on it. The journey has not been smooth, but the result—humans—is indeed remarkable. Even more

remarkable is the spectacle of modern humans, wrapped in different cultures and speaking different languages, who, instead of congratulating each other for the wonderful variety of human life and experience, spend their time fighting each other in the name of their pet social and religious systems. Perhaps humans ought to recognize that they are *primarily and foremost animals*, and as such they belong to one and the same species. The fact that one such animal speaks Chinese and considers himself or herself a Buddhist, whereas another such animal speaks Hebrew and considers himself or herself a Jew, is really quite irrelevant, seeing that a child can be raised to become with equal aplomb either, or, for that matter, anything else. Those of us who feel a need to believe in one god or another should adopt the relaxed attitude of religious tolerance so admirably expressed in a beautiful hymn in the Koran (Surah 109):

Say: "O you unbelievers,
I do not worship what you worship.
Nor do you worship
whom I worship.
Nor will I worship
what you worship.
Nor will you worship
whom I worship.
To you your way,
to me my way."

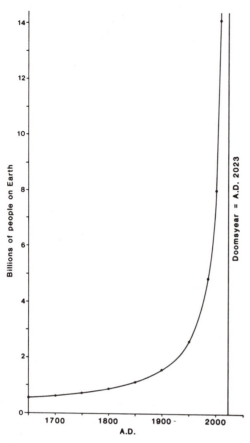

Figure 24.14. The *doomsday curve*: The human population growth since 1650, with projection into the future, indicating that the human population should go to infinity in the year 2023. That will not happen, of course, because of skyrocketing pollution, disease, and starvation. (Emiliani, 1988, p. 224, Fig. 11.28.)

Wasting time fighting each other on religious or any other ground is not a very good idea, especially now when we need to do some serious thinking about our future. Not only is our planet at risk, but something else, quite ominous, may be looming ahead. *Homo sapiens* appeared on earth at the same time as the coccolithophorid *Emiliania huxleyi*. Both species have become very abundant. Today, there are $5 \cdot 10^9$ people, each one with $6 \cdot 10^{13}$ cells in his or her body, giving a total of $3 \cdot 10^{23}$ living human cells. At sea, it is estimated that there are 10^{24} living cells of *Emiliania huxleyi*. These two numbers are ominously close. The deep-sea sediment record shows that ordinarily a coccolithophorid species becomes abruptly extinct when it becomes that abundant. I have suggested before that the normal mode of extinction is by a lethal virus that has accidentally evolved to fit the genome of a given species and is offered the opportunity to (1) come in contact with a cell of that species and (2) spread throughout the population. So contact and mixing are essential factors for extinction. If a virus evolved in the Caspian Sea that would be suitable for the kiwi bird, forget it. The virus would look around for a while and, finding nothing attractive, would shrug its shoulders and either mutate or drop dead. But both *Homo sapiens* and *Emiliania huxleyi* are dense, worldwide, well-mixed populations. Each human or *huxleyi* cell makes an attractive target. The chances of lethal viral infection would seem excellent for both. For humans, HIV (Fig. 19.34) may be just such a virus.

So, instead of quarreling, let us take care of each other and of our planet, and let us enjoy the rest of the Quaternary.

THINK

* Anaxagoras (ca. 500–428 B.C.E) said that human intelligence resulted from the power of manipulation when the forelimbs were freed from the task of locomotion, that is, when vertical posture was achieved. Today, most scientists hold this view, supported by the observation that the early Australopithecines had a modern pelvis (making vertical posture possible) but a primitive brain. What was the selective advantage that made vertical posture possible, seeing that it comes with enormous disadvantage of fully exposing to attack and injury the softest and most delicate parts of the human anatomy? I suggested in the magazine *Discovery* (March 1991) that it was the desired ability to stomp on grapes and make wine, leading the better stompers to win over the most desirable mates. Considering that vertical posture is indeed enormously disadvantageous, can you think of any better theory? No? I thought you couldn't.

VI The historical perspective

T HE cultured person not only has some idea about how the present world came to be and how it works (the subject matter of this book) but also knows how this information was accumulated and who the principal contributors were. The historical perspective is an essential part of our cultural heritage. In this Part VI, I list (by order of birth) the people most responsible for producing the view of the world we now hold, together with their achievements. An alphabetical index is given at the end of this Part VI.

Before the Greeks arrived from the north, there was the Minoan civilization in Crete (2500–1645 B.C.E., Bronze Age). It was the first advanced (i.e., mercantile) civilization in Europe. Commerce, of course, involves coinage, writing, arithmetic, and so forth, and so the Minoan civilization was quite advanced for its time. The capital was Knossos on the north coast. The Minoans manufactured pottery and objects in copper and bronze.

[Bronze is an alloy of 70–90% copper (melting point 1,063°C) and 10–20% tin (melting point 231°C). In contrast, iron has a much higher melting point (1,535°C), requiring a *forge* to be worked. In a forge, air is forced through the burning coal by means of a bellows or a compressor.]

Minoan settlements on the mainland were destroyed by Greek invaders from the north. Then came the giant explosion of Thera (1645 B.C.E.), which covered Crete and the entire eastern Mediterranean with ashes. A tsunami destroyed Minoan harbors on the north coast. And that was the end of the Minoans.

Next came the Mycenaean civilization (1400-1250 B.C.E., Iron Age), which extended from Ithaca to Rhodes and included the Peloponnesus, Attica, Miletus, and Crete. The capital was Mycenae in northern Pelopon-

nesus. They were Greek, writing in a syllabic script called *Linear B*.

The war with Troy was fought for 10 years around 1250 B.C.E. It marked the end of the Mycenaean civilization, because it drained the resources of Greece and left the door open for more Greek invaders from the north.

The Greeks

The Greeks came in three nations:

The Ionians, who settled in Attica, Euboea, the Cyclades, Thrace, and Ionia (the central portion of the western coast of Asia Minor, i.e., modern Turkey),

The Aeolians, who settled in northwestern Greece, northwestern Peloponnesus, Troas (district around Troy), Lesbos, and Aeolis (the district between Troas and Ionia, opposite Lesbos),

The Dorians, who settled in southeastern Peloponnesus, Crete, the southern Cyclades, southern Asia Minor, and Rhodes.

The war with Troy was the subject of Homer's *Iliad*. Homer (fl. 900 B.C.E.) also wrote the *Odyssey*, which records the travels and travails of Odysseus (Ulysses) on his return from Troy to Ithaca. It makes a wonderful read, because it shows the Mediterranean through the eyes of the early Greeks. People didn't travel much in those days, and so stories were floating around by word of mouth, properly embellished and magnified at each relay. The *Odyssey* reports the first observations in marine geology, made by Odysseus himself in the Strait of Messina (*Odyssey* XII, 242–243).

The key to Greek civilization was the πόλις (city), often with a citadel (ἀκρόπολις), a separate and independent political unit. The different city-states were often at war with each other, like the Italian states during the Renaissance. A kind of national unity, however, was fostered by the common language, by the oracle of Delphi, dedicated to Apollo, and by the Olympic Games, which were celebrated at Olympia in northwestern Peloponnesus.

What really started Western civilization on its track was the Ionian school of philosophy in Miletus, a coastal town 75 km south of modern Smyrna, Turkey. The school was founded by Thales.

[Philosophy means *love of knowledge*, an attempt to explain nature in rational rather than theistic terms.]

Beginning with Thales of Miletus, and ending with scientists still living today, we shall now review the major contributions of these people to our understanding of the natural world.

Thales of Miletus (ca. 640–548 B.C.E.), the "father of philosophy," learned practical geometry in Egypt and transformed it into theoretical geometry. The Egyptians used geometry to build pyramids and plot land. Thales concluded that a point has no dimensions, that a line has length but no width, and that a surface has length and width but no thickness. He also discovered that the opposite angles formed by two intersecting lines are equal (Fig. VI.1) and that the sum of the internal angles of a triangle is equal to 2 right angles (Fig. VI.2). Thales was very interested in astronomy. He stated that the Sun and the stars were not gods, but balls of fire, and determined that the diameter of the solar disc was 1/720 of the zodiac (or $360°/720 = 0.5°$).

Thales also divided the year into 365 days and the month into 30 days and predicted the solar eclipse of May 28, 585 B.C.E. He believed that all substances originated from water and

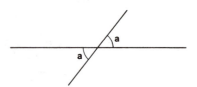

Figure VI.1. The opposite angles formed by two intersecting lines are equal.

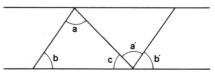

Figure VI.2. Two sets of parallel lines intersecting each other demonstrate that the sum of the internal angles of a triangle is equal to 180°. In fact, $a' = a, b' = b$, and $a' + b' + c = 180°$.

that the Earth was a disc resting on water (Fig. VI.3).

Thales' most important contribution to the evolution of Western thought was his use of theoretical thinking. It is obvious that his definitions of points, lines, and surfaces are theoretical, but more important, his conclusion that all substances derived from water indicates that he was on the same track as modern physicists, who are still trying to find the "primordial" particle out of which all matter—

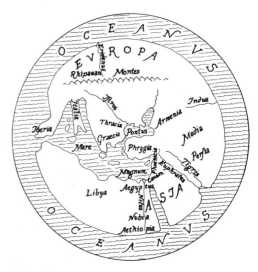

Figure VI.3. The world according to Hecateos of Miletus (ca. 550–476 B.C.E.). This was the world of Thales.

energy is made. Perhaps Thales chose water because, in his common experience, water existed in all three phases: solid (as ice or snow), liquid (as water), and gaseous (as water vapor). In contrast, stone is only solid, wine is only liquid, and air is only gas. The mere fact that Thales asked himself about a primordial substance was of tremendous significance. It was that kind of theoretical thinking that set Western civilization on its course.

Anaximander of Miletus (ca. 610–540 B.C.E.) was a pupil of Thales. He rejected water as the primordial substance, probably because he thought water was only another substance among a great many. He chose instead the ἄπειρον, which means *indefinite*. Anaximander believed that the Earth was at rest in space, kept there by a balance of "internal forces," and that all planets were born fluid, but had been dried up by the Sun (indeed, during the formation of the solar system the Sun blew the gases away from the inner planets and in fact "dried them up"). Anaximander also believed that life originated in the sea and that land animals derived from marine animals that had been left stranded on the beach by retreating seas and had learned to breathe air—these animals then generated all later land life. Finally, Anaximander believed that humans did not appear as such, because their helpless infancy would have made their survival impossible. It would seem that the concept of evolution was in fact invented by Anaximander.

Anaximenes of Miletus (fl. 525 B.C.E.) was a pupil of Anaximander. According to him, air, wind, clouds, water, soil, and stone are progressive condensations (by decreasing temperature) of a primordial gas that must be, therefore, air. Anaximenes also believed that earthquakes were due to the solidification of an originally fluid Earth.

[We must pause here a moment and emphasize that all we know about these early Greek philosophers derives from quotations by later authors—none of the original manuscripts have survived.]

Pythagoras (ca. 582–507 B.C.E.) was born in Samos, went to Miletus, spent 25 years in Egypt. When Cambyses (king of Persia from 529 to 522 B.C.E.) conquered Egypt (525 B.C.E.), he followed Cambyses back to Babylon, where

he spent 10 years, and finally returned to Samos at age 56. Soon, however, he was kicked out of Samos because of his unorthodox views. He went to Croton, in southern Italy, where he founded the Pythagorean School. That school was in fact a sect that preached a life-style based on notions that Pythagoras had picked up in the East. These notions included a belief in the transmigration of souls, the practice of asceticism, avoidance of food or clothing derived from animals, wearing sandals or going bare-footed and so forth. Women were accepted.

But Pythagoras was not just another mystic—he was also a mathematician and a scientist. He discovered the *triangular numbers* (numbers generated by a triangular array of dots, Fig. VI.4A), the *square numbers* (numbers generated by a square array of dots, Fig. VI.4B), and the fact that the *alternate* and the *corresponding* angles formed by a line crossing two parallel lines are equivalent (Fig. VI.5). This means that the sum of the internal angles of a triangle is equal to 180°, a fact that had already been discovered by Thales (Fig. VI.2). Pythagoras also introduced the Pythagorean table (Fig. VI.6) and discovered the Pythagorean theorem (see Fig. 5.1)

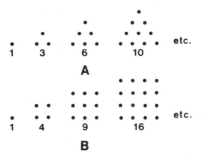

Figure VI.4. (A) Triangular numbers. (B) Square numbers.

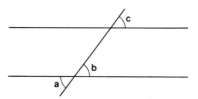

Figure VI.5. The alternate angles (*a* and *c*) and the corresponding angles (*b* and *c*) formed by a line crossing two parallel lines are equivalent.

1	2	3	4	5	6	7	8	9	10	11	12
2	4	6	8	10	12	14	16	18	20	22	24
3	6	9	12	15	18	21	24	27	30	33	36
4	8	12	16	20	24	28	32	36	40	44	48
5	10	15	20	25	30	35	40	45	50	55	60
6	12	18	24	30	36	42	48	54	60	66	72
7	14	21	28	35	42	49	56	63	70	77	84
8	16	24	32	40	48	56	64	72	80	88	96
9	18	27	36	45	54	63	72	81	90	99	108
10	20	30	40	50	60	70	80	90	100	110	120
11	22	33	44	55	66	77	88	99	110	121	132
12	24	36	48	60	72	84	96	108	120	132	144

Figure VI.6. The Pythagorean table.

In the physical sciences, Pythagoras discovered that the musical intervals depend upon the arithmetic ratio of the lengths of different strings under the same tension. For instance:

2:1 = octave
3:2 = fifth
4:3 = fourth
etc.

That strengthened his opinion that numbers rule the world. In particular, the transparent "celestial spheres," to which he believed the celestial bodies to be attached, by turning as they did at different rates with specific ratios, would give off harmonious sounds because they turned at different rates and the rates had specific ratios (unfortunately, the harmonious sounds were never heard by anybody).

The Pythagorean system of astronomy not only placed the Earth in space but also had it orbiting. At the center, according to Pythagoras, there was the *central fire*, around which were orbiting, in succession, the Antiearth, the Earth, the Moon, the Sun, Mercury, Venus, Mars, Jupiter, Saturn, and the fixed stars. Pythagoras introduced the central fire and the Antiearth to explain the difference in frequency between lunar and solar eclipses. He explained why we never see either of them by claiming that those two bodies were on the other side of the Earth, facing the Southern Hemisphere, and that the Antiearth always accompanied the Earth. He also explained the retrograde motion of the outer planets by means of *epicycles* (see Fig. 5.18).

Pythagoras' system may seem rather strange, but in fact it was a colossal advance because it was the first time that the Earth was considered to move in space.

Xenophanes (ca. 570–480 B.C.E.) was born in Colophon, a town 45 km south of modern Smyrna. He moved to Elea, a coastal town 105 km southeast of Naples, Italy, and founded the Eleatic School. There he preached monotheism: "There is one God, neither in shape nor in thought like mortals. ... He is all sight, all mind, all ear. ... He resides motionless in one place and does not move. ... Yet mortals imagine gods to be born and to have clothes and body and voice like themselves. ... The gods of the Ethiopians are swarthy and flat-nosed and the gods of the Thracians are blond and blue-eyed. ... Homer and Hesiod attributed to the gods the faults of humans—theft, adultery, deceit, and other lawless acts.... Oxen, lions, and horses would fashion gods after their own shapes." Xenophanes believed that all things come from the Earth and return to Earth. Having noticed that shells are sometimes found on mountains, he concluded that such mountains must once have been below sea level.

Heraclitus (ca. 540–475 B.C.E.) was born and lived in Ephesus, a coastal city in Asia Minor 60 km south of modern Smyrna. He claimed that the primary substance was fire, that war was the master of everything, that peace led to stagnation (like still water rotting in a pond), and that everything changes (πάντα ῥεῖ, meaning *everything flows*).

Anaxagoras (ca. 500–428 B.C.E.) was the last of the Ionian school. He was born in Clazomene in Asia Minor (100 km north of Miletus), moved to Athens, and taught Pericles, the Athenian general and statesman (ca. 495–429 B.C.E). Anaxagoras explained solar and lunar eclipses, discovered respiration in plants and fishes, and believed that life rained down on Earth with condensing water. He explained human intelligence by the powers of manipulation when the forelimbs were freed from the task of locomotion (this is entirely modern!).

Empedocles of Agrigentum (ca. 490–430 B.C.E.) believed that there are four ultimate, unchangeable elements: fire, air, water, and earth, which are joined by love into a single sphere and separated by strife into different proportions, forming the various substances. Aside from the notion of love and strife, the idea of four elements instead of the ἄπειρον of Anaximander seems a big step backward. Empedocles believed that fossils were remains of *extinct* species and that in the early stages of a "strife" era, organs (heads, limbs, etc.) were made and combined haphazardly: Only those combinations adapted to the environment survive and thrive; the rest disappear. What we have here is a pale notion of mutation and a pretty good idea of survival of the fittest.

Herodotus (ca. 484–425 B.C.E.) was Greece's first true historian. He was born in Halicarnassus (40 km south of Miletus), moved to Samos, then to Athens, and then to Thurii, an Athenian colony in Calabria, 100 km northwest of Croton. Herodotus traveled extensively in Egypt, Persia, Armenia, Sicily, and southern Italy. While in Egypt he saw fossil echinoderms in the desert and concluded that they were the remains of ancient marine organisms that had lived there when Egypt was below the sea. Like Anaximander before him, he believed that the sea could move across the land.

Hippocrates (ca. 460–370 B.C.E.) was born in Kos, an island 100 km northwest of Rhodes, where he founded a school of medicine. Hippocrates believed that disease was due to natural causes. Treatment was limited to rest, diet, and exercise. Hippocrates, in a way, invented preventive medicine by prescribing cleanliness and dietary moderation. The Hippocratic books, written by Hippo-crates and his disciples across more than 100 years of time, contain carefully recorded case histories and the results of treatments.

Democritus (ca. 460–360 B.C.E.) lived in Abdera (Thrace) and was a disciple of Leucippus (fl. 445 B.C.E.), also of Abdera. Nothing is known about Leucippus beyond his name. We know more about Democritus, who probably represented Leucippus' ideas. Democritus believed that the universe consisted of *atoms* and *void*. The atoms were the smallest possible parcels of matter, entirely indivisible (ἄτομος means *uncuttable*) (see Section 1.1 for a hypothesis on how Democritus may have reached that conclusion). Democritus believed that the atoms had different shapes: The atoms of air and water were very smooth and slippery; those of iron were hard and jugged; and so forth. Democritus did not believe in gods, but he believed that the human soul existed and consisted of "special atoms."

Aristotle (384–322 B.C.E.) was born in Stagira, Macedonia. His father was court physician. King Philip of Macedonia appointed him tutor to his son Alexander (356–323 B.C.E.), then 13 years old (343). Alexander became king in 336 at the age of 20, and at the time of his death in 323 B.C.E. at the age of 31, he had conquered a vast empire stretching from Greece and Egypt to India.

Aristotle studied for 20 years (367–347 B.C.E.) under Plato in Plato's Academy, but left when he couldn't succeed Plato. He founded his own school in Athens (the Lyceum), which included a library and a museum. When Alexander became emperor, Aristotle remained in Athens until Alexander's death (323). Feeling that the Macedonian party in Athens was becoming unpopular, he withdrew to his mother's house in Chalcis, on the island of Euboea, and died there the following year.

Aristotle was independently wealthy. In addition, Alexander gave him a grant of 800 talents of silver to pursue his researches (the talent was a unit of weight equal to 26.04 kg = 57.5 lb = 920 oz; at the current price of $4.21/oz (Feb. 7, 1992), 800 talents of silver are equal to $3,098,560, a nice bundle).

Whereas Plato had strong inclinations toward mathematics (but hated the natural sciences), Aristotle preferred the natural sciences, especially biology. He wrote 400 books.

Aristotle invented *logic*, which is based on the *syllogism*. An example: *Man is rational* ("major premise"), *Socrates is a man* ("minor premise"), and therefore *Socrates is rational* ("conclusion"). Aristotle believed that god must exist, a πρῶτον κινέων, a "prime mover" that started the world, although in Aristotle's opinion the world operates on its own by its own internal energy. [The prime mover was later amplified by Thomas Aquinas (ca. 1225–1274) into a *primum movens immobile*, a prime mover that does not itself move—see *Summa Theologica*, 1st Part, Question 3, Art. 1c.]

According to Aristotle, the soul is the sum of the powers of any organism. In plants, the soul is nutritive and reproductive; in animals it is nutritive, reproductive, locomotive, and sensitive; in humans it is the same as in animals, plus reason and thought. A part of the soul (the "personality") dies with the body, and another part ("pure thought") survives.

In ethics, Aristotle expounded the "golden mean" as the best, condemning both extremes as vices: "between cowardice and rashness is courage". In government he supported aristocracy as the best form, provided it was not hereditary. The wealthy middle class (i.e., the "golden mean") should rule by constitutional means. Public office should be open to anybody who has "prepared himself".

In astronomy, Aristotle accepted the geocentric system. He described accurately the process of evaporation of water from the ocean, precipitation, erosion, runoff, and sedimentation, stating that "Egypt is the work of the Nile". He believed that new continents arose and old continents disappeared—the face of the Earth was continuously changing. He proposed the pyramid of nature, ranging from the lowest organisms at the base through minute gradations to humans at the top. Aristotle discovered that birds and reptiles are close in structure, that monkeys are intermediate between quadrupeds and humans, and that characters defining a "genus" appear in the development of an individual before the characters typical of the "species". This is akin to the nineteenth-century dictum that "ontogeny recapitulates phylogeny". Aristotle believed that prominent variations (e.g., genius in humans) are diluted in mating and lost in successive generations. He described a woman

in Elis (northwest Peloponnesus) married to a black man and reported that the children were white, but some of the grandchildren were black. He asked himself, "Where was blackness hidden?" (Mendel probably asked himself the same question about pea-plant characteristics when he began his work.) Aristotle also studied embryology, by cracking chicken eggs at different stages of development.

Aristotle was mainly an observer and classifier. He did not develop any theory (e.g., evolution), and some of his thinking and many of his observations were faulty. For instance, he thought that the male element in reproduction was only a "stimulant" (but, if so, why do many children resemble their father?). He also thought that the brain was an organ to cool blood, that the male skull had more sutures than the female skull, that males had eight ribs on each side, and that females had fewer teeth than males. He believed that sea birds were born from mussels on pilings by the seashore, that fossils were born in mud by a *vis formativa* ("a forming force"), and that comets were atmospheric phenomena because they were too irregular to be "celestial".

Aristotle was adopted as the official philosopher of the Christian church. His authority in the Middle Ages was undisputed: Saying *ipse dixit* ("he himself said so") would silence any argument.

Euclid of Alexandria (ca. 325–270 B.C.E.) authored a 13-volume textbook on mathematics (Τὰ Στοιχεῖα, meaning *The Elements*) that is still in use (στοῖχος means *row* or *line*).

Aristarchus of Samos (ca. 310–230 B.C.E.) was the first to propose the heliocentric system, a system in which the Earth and all other planets revolve around the Sun. Unfortunately nobody believed him.

Archimedes (ca. 287–212 B.C.E.) was a mathematician and an engineer living in Syracuse, Sicily. He is famous for having discovered the principle of buoyancy that says that "a body floating or totally immersed in a fluid experiences an upward force equal to the weight of the fluid displaced".

Eratosthenes (276–194 B.C.E.) was a mathematician and a geographer (Fig. VI.7). He was born in Cyrene (Libya), was educated in Athens, and became head librarian at the great

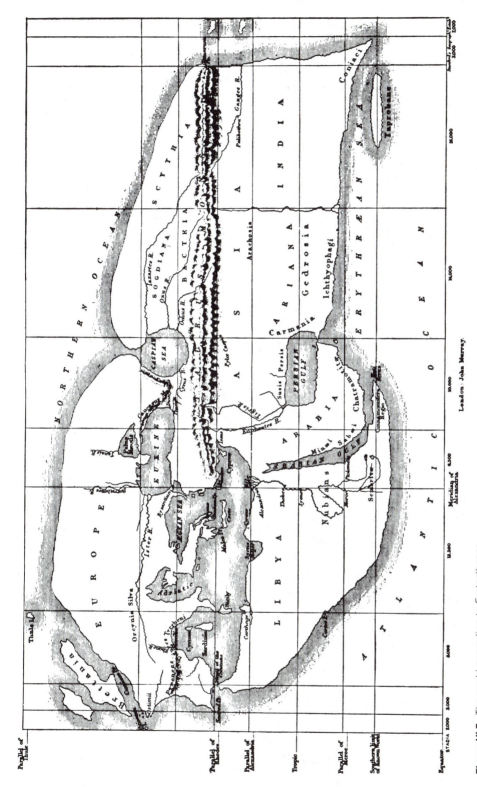

Figure VI.7. The world according to Eratosthenes.

559

library in Alexandria. He is famous for having determined the size of the Earth (see Section 5.2).

Apollonius of Alexandria (ca. 260–190 B.C.E.) wrote a treatise (which survives) on conic sections.

Hipparchus (ca. 190–120 B.C.E.), the great Greek astronomer, was born in Nicea (the modern Iznik, 90 km southeast of Istanbul) but set up his observatory in Rhodes. He determined the distance and size of the Moon (see Section 5.2). He discovered that the celestial north pole was not fixed but moved around a circle counterclockwise. The motion was not less than 1° per century, requiring 36,000 years to complete a circle (we now know that this motion is due to the precession of the Earth's axis and takes 25,800 years). Hipparchus made a catalogue of more than 1,000 stars, giving their precise positions in terms of celestial longitude and latitude. He also ranked the stars in terms of their visual magnitude, assigning magnitude 1 to the brightest stars and magnitude 6 to the weakest stars still visible with the naked eye. His system of celestial coordinates and stellar magnitudes is still used today.

Hero (fl. 60 C.E.), an engineer from Alexandria, invented the steam engine, a hollow sphere with two nozzles facing in opposite directions. When water was boiled inside the sphere, the steam escaping through the nozzles would make the sphere turn. Hero constructed many mechanical devices and wrote two books: *Pneumatics* and *Mechanics*.

Ptolemy (Claudius Ptolemaeus) (ca. 100–170) was probably an Egyptian, living in Alexandria. He was an excellent mathematician. He "proved" the geocentric system using mathematical arguments. In his system, the Earth was at the center of the universe. Around it orbited, at increasing distances, the Moon, Mercury, Venus, the Sun, Mars, Jupiter, and Saturn. The outer planets (i.e., the planets beyond the Sun) performed *epicycles* along their *deferents*. The deferent was the planetary orbit; the epicycle was a secondary cycle performed by the planet as it progressed along its orbit (see Fig. 5.18). Because his system placed the Earth at the center of the universe, it was adopted by the Christian church. Anybody

who wanted to place the Sun at the center would be in serious trouble. It was only after Galileo's discovery that Venus had phases like those of the Moon that the Ptolemaic system was reluctantly dropped.

Galen (ca. A.D. 130–200) was born in Pergamum (the modern Bergama, in western Turkey), traveled extensively, and settled in Rome. He dissected a great many animals and applied his anatomical discoveries to humans (dissection of human cadavers was not in favor at the time). Galen wrote about 500 books, including philosophical commentaries. Of the medical works that are Galen's or are attributed to him, 162 survive. They remained the standard medical texts in Arab and European medical schools until Vesalius (1514–1564).

The Romans

Rome was founded by three tribes, the Aemilii, the Cornelii, and the Manlii, about 750 B.C.E. They annoyed the Etrurians to the north enough for the Etrurians to subjugate them and place an Etrurian king in charge. The Romans revolted in 509 B.C.E., and decided that there should be no more kings and in fact no more single rulers. They set up a legislative Senate and an executive branch consisting of two consuls, who were changed every year. The consuls were replaced by a single emperor only after Julius Caesar, in 44 B.C.E.

The Romans were great engineers and excellent administrators. They built their empire on two foundations: the Roman law ("everybody is equal under the law") and racial equality. They did not think themselves superior, for the simple reason that everywhere they went (Etruria, southern Italy, Greece, Egypt) they found more advanced civilizations than theirs. Their bag was law and order—they had little sense of humor and little imagination, and they contributed little to scientific or philosophical advancement.

Titus Livius (59 B.C.E. to 19 C.E.), a Roman historian, believed that fossils were the remains of marine organisms living where the ocean once was. He therefore agreed with Herodotus. Seneca (55 B.C.E. to 39 C.E.), a philosopher, and Pliny the Elder (23–79 C.E.), the closest thing the Romans had to a scientist, also believed that fossils were the remains of marine organisms. Pliny the Elder died while watching the eruption of Mount Vesuvius that buried Pompeii in 79 C.E. His nephew, Pliny the Younger, recorded the event.

The Arabs

The Arabs performed a great service for science and learning during the darkest times for Europe—from 600 to 1200 C.E. They saved a lot of ancient Greek texts and developed a great civilization that spread westward to Spain and eastward to Iran, India, and central Asia.

Alhazen (ca. 965–1039), Arab physicist, was born in Basra, Iraq, and died in Cairo, Egypt. He studied the principles of light reflection and refraction; he made lenses and explained their properties and magnifying power. We wonder how he could have failed to invent the telescope.

Avicenna (979–1037), born in Bukhara (Uzbekistan), was the greatest Arab philosopher in the East. He was also a famous physician: His *Canon of Medicine* was translated into Latin and remained the standard medical textbook in European universities until 1650. He believed, however, that fossils were formed by a *vis plastica*, a plastic force that could act only imperfectly. According to him, fossils were abortions—organisms that failed to become alive.

Medieval Science

Leonardo Fibonacci (1170–1240), of Pisa, learned math from Arab tutors in North Africa and popularized the Arabic system of numeration. He also invented the Fibonacci series (0, 1, 1, 2, 3, 5, 8, 13, 21, ...) in which each number is the sum of the two preceding numbers.

Roger Bacon (1220–1292), English scholar, was interested in many aspects of science and was a strong proponent of experimentation. He, like Alhazen, made lenses, but he too failed to invent the telescope.

Brunetto Latini (ca. 1210–1294) was born in Florence, took refuge in France from 1261 to 1268 for political reasons, and then returned to Florence, where he held various public offices until his death. While in France, he wrote, in French, *Li Livers dou Tresor*, an encyclopedic summary of the knowledge of the time. In addition to history, biography, philosophy, and even etiquette, the book covers a variety of subjects in the natural sciences, including astronomy, geology, and zoology. That book was one of the foundations of Dante's scientific knowledge.

Dante Alighieri (1265–1321), Florentine politician and poet, wrote the *Divine Comedy*, a poem in which he describes his travel through hell, purgatory, and paradise. Dante understood gravity—he recounts how, having reached the bottom of hell, at the center of the Earth, he had to turn himself upside down to climb up on the other side. Dante disseminated through his poem his view of nature, which was largely gleaned from Latini's *Tresor*. Because the *Divine Comedy* was an instant success, interest in the natural sciences was kindled.

Peregrinus (Peter de Maricourt) (1240–?) was a French engineer who became interested in the properties of the *lodestone*, an iron oxide mineral or rock naturally magnetized in the magnetic field of the Earth (*lodestone* derives from *lode*, an English word meaning *mineral deposit*). Peregrinus showed that a magnetic needle cut out of lodestone has a north pole and a south pole, that similar poles repel each other (and opposite poles attract), and that one could not separate an individual pole by cutting a magnet. He believed that the north pole of a magnetized needle was attracted by the north pole of the celestial sphere. Peregrinus perfected the compass (which had been invented by the Chinese and brought to Europe by Arab seafarers) by placing a magnetized needle on a pivot with a graduated circle around it.

Giovanni Boccaccio (1313–1375) was born in Paris, out of wedlock, to an Italian father and a French mother, but was raised from infancy in Certaldo, near Florence. He wrote the *Decameron*, a collection of 100 short stories set around the great plague of 1348. He was a great admirer of Dante Alighieri. He wrote a *Vita di Dante* and a commentary to the *Divine Comedy*. In addition to that, Boccaccio wrote a treatise on natural science entitled *De*

Montibus, Silvis, Fontibus, etc. In it he concluded that fossils were the remains of marine organisms, just like Livy and Herodotus. But those views on fossils were forgotten, and from 1300 to 1500 there were three incredible schools of thought regarding fossils: One school maintained that they were the results of the *vis plastica* advocated by Aristotle; a second school said that they were *lusi naturae* (which means *jokes of nature*); the third school claimed that they were leftovers of the Biblical flood.

Nicholas of Cusa (1401–1464), a German scholar, believed that the Earth rotated around its axis, was not at the center of the universe, and moved in space. He also believed that the stars were other suns placed at various distances in infinite space and had planets like the Earth. Good boy.

Modern Science

Leonardo da Vinci (1452–1519), painter, scientist, engineer, and many other things, is regarded by many as one of the few "superbrains" that the human race has produced. He understood inertia, that falling bodies accelerate, and that perpetual motion is impossible. He believed that the Earth is *not* at the center of the universe and that it spins on its own axis. He read Boccaccio and drew conclusions about long-term geological change. Like Boccaccio and Herodotus, he regarded fossils as organisms left behind by a retreating sea. Leonardo wrote his observations with his left hand from right to left, in notebooks that nobody could read. Shortly before his death he ordered a large number of lenses and mirrors; it is suspected that he was on the way to discovering the telescope. He never published anything.

Nicolaus Copernicus (1473–1543) was born in Thorn (Prussian Poland) of a Cracovian father. He went to the University of Bologna in 1496 to study canon law (i.e., church law). While there, he attended, in the year 1500, some lectures by Domenico Maria Novara (1454–1504), the resident astronomer, who believed in the heliocentric system. Copernicus found that one could predict more precisely the motions of the planets if one assumed that the planetary orbits were circular and that the Sun was at the center. His system revived that proposed by Aristarchus 1700 years before: The Sun was at the center, followed, at increasing distances, by Mercury, Venus, the Earth (with the Moon revolving around it), Mars, Jupiter, Saturn, and the fixed stars. Copernicus published his work, *De Revolutionibus Orbium Coelestium*, in the year he died (1543), because he was afraid that the pope would lean on him.

Georgius Agricola (which means *George the farmer* in Latin, his real name being Georg Bauer, which means *George the farmer* in German) was born in 1494 and died in 1555. He studied philology at the universities of Leipzig, Bologna, and Padua, earned a degree in medicine at the University of Ferrara, and became city physician at Joachimstahl (now Jachimov) in the Erzgebirge, Bohemia (1527–1533). Joachimstahl has rich deposits of uraninite (UO_2) and pitchblende (fine-grained colloforn uraninite), from which Marie Curie isolated radium in 1898. There are also silver deposits, from which silver for the Austrian Empire was mined (*thaler*, the name of the silver coin of the Austrian Empire, derives from *Joachimsthal*, and the name *dollar* derives from *thaler*).

Agricola left Joachimsthal in 1533 and became city physician at Chemnitz in the Erzgebirge. He remained there from 1534 to his death in 1555 and wrote *De Re Metallica*, a great treatise on mining, minerals, and the extraction of metals from metal ores. He also wrote the book *De Natura Fossilium*, which deals mainly with minerals, but in which he claims that fossils are mineral concretions formed by "special processes". Agricola is considered the "father of mineralogy".

Andreas Vesalius (1514–1564), Flemish anatomist, migrated to Italy in order to carry out dissections on human bodies. In 1543, when he was 29 year old, he published his great work on human anatomy, *De Corporis Humani Fabrica*. The book is illustrated with magnificent and extremely accurate drawings of human anatomy.

Joseph Justus Scaliger (1540–1609) was perhaps the foremost scholar of the Renaissance. He was born in France of an Italian father. He converted to Protestantism and moved first to Geneva and then to Holland. He was fluent in Latin, Greek, Hebrew, and Arabic. He established a single historical chronology not only for Roman and Greek history but also for Oriental histories. In order to fix the dates of lunar and solar eclipses, of planetary motions, and of any other celestial phenomenon that might be observed, Scaliger devised the Julian Date (JD), a continuous chronology beginning on January 1, 4713 B.C.E. and simply reckoned in consecutive days (see Section 8.2). This system is in use to this day.

William Gilbert (1544–1603) was an English physician and physicist. In the year 1600 he published the book *De Magnete*, in which he discussed the magnetic properties of matter. He discovered that a piece of iron became magnetized when rubbed with a lodestone and that a spherical piece of lodestone also exhibited north and south magnetic poles. He then suggested that the Earth itself was a giant magnet and that the end of a magnetized needle pointing north was attracted by the Earth's north pole, not by the celestial north pole, as Peregrinus had suggested.

Tycho Brahe (1546–1601) was a Danish astronomer (see Chapter 5) who latinized his Danish name Tyge into Tycho. He was the protégé of King Frederick II, who built him an observatory, called Uraniborg, on the island of Hveen (now Ven) off Copenhagen. For seventeen years (1580–1597) Tycho recorded the exact positions of planets with respect to the backdrop of the fixed stars. He observed a supernova in 1572—the first time a change was observed in the realm of the fixed stars (which were supposed to be perfect and immutable). Tycho also observed a comet (1577) that crossed the orbits of the planets. That demonstrated that there were no such things as the "celestial spheres". These observations made him quite famous. Tycho quarreled with the new king, Christian IV, and moved to Prague, a guest of Emperor Rudolph II of Hapsburg. There he met Kepler, who became his assistant. Tycho's observations supported Copernicus, but Tycho was a devout Christian and steadfastly maintained that the Earth was at the center of the universe. To accommodate his observations, he claimed that whereas all other planets revolved around

the Sun, the Sun itself revolved around the Earth! That arrangement, called the *Tychonian system*, convinced no one.

John Napier (1550–1617) was a Scottish nobleman who invented logarithms. He used e (= 2.71828…) as a base and spent twenty years constructing his logarithmic tables. (*Logarithm* derives from λόγος, meaning *word, ratio*, or *reason*, and ἀριθμός meaning *number*.) See *Briggs*.

Henry Briggs (1561–1630), English mathematician, developed the common logarithms, using 10 as a base instead of e. In 1624 he published his tables of logarithms, which he had calculated to 14 decimal places.

Galileo Galilei (1564–1642), born in Pisa, was a mathematician and a physicist (see Section 5.4). He discovered the laws of dynamics (pendulum, free fall, projectiles) and rediscovered the telescope (1609). Using his telescope, Galileo discovered mountains on the Moon, the four major satellites of Jupiter, and observed that the Milky Way is an agglomeration of stars. He also discovered that Venus had phases like the Moon and that crescent Venus was 7 times larger than full Venus. This observation demonstrated that Venus revolves around the Sun and vindicated both Aristarchus and Copernicus. According to Galileo, the road to truth rests on experiment plus mathematical analysis, because the senses can fool you (how true! Who could believe that the Earth is not flat, that heavy bodies do not fall faster than lighter bodies, and that time is relative???).

In 1632 Galileo published his *Dialogo Sopra i Due Massimi Sistemi del Mondo*, in which the Ptolemaic and Copernican systems are debated (the defender of the Ptolemaic system was called Simplicio and was made to look like a fool). The church had condemned Copernicus in 1616, and Galileo was forced to recant in 1633. Eventually those condemnations would haunt the church, with the result that the church learned the lesson of not interfering with science.

Hans Lippershey (1570–1619) was a German-Dutch optician credited with having invented the telescope (see Section 5.4). It has variously been said that the basic discovery—that two lenses placed in front of each other create a magnified image—was made by one of his assistants, by two children playing with lenses, or by Lippershey himself. Also, it is not clear whether both lenses were positive (giving an inverted image) or the ocular was negative and the objective positive (giving a right-side-up image).

Johannes Kepler (1571–1630) was a young German mathematician who went to Prague in 1597 to study under Tycho Brahe and inherited Tycho's notebooks when Tycho died (1601). He discovered the three laws of planetary motions (see Sections 5.4 and 7.4). Kepler was very much interested in astrology. His motivation for discovering how planets really move was largely to improve the precision of horoscopes.

William Harvey (1578–1657) was the founder of modern physiology. He made numerous experiments on the circulation of blood and published his conclusions in 1628. It was the first advance in this area since Galen.

Willebrord Snel (or **Snell**) (1580–1626), Dutch physicist, formulated Snell's law of refraction.

René Descartes (1596–1650). Descartes invented the Cartesian coordinates, a way of locating a point on a plane with respect to two mutually orthogonal axes intersecting at an origin 0, or a point in space with respect to three mutually orthogonal axes, also intersecting at an origin 0. *Cartesian* derives from *Cartesius*, the latinized name of Descartes.

Otto von Guericke (1602–1686), German physicist, invented a water pump that could remove nearly all the air from a container, thus demonstrating the existence of vacuum (denied by Aristotle). He showed that, in vacuo, a bell was muffled and a flame extinguished. Most dramatically, he demonstrated that two large hemispheres fitting together could not be separated by two teams of eight horses if the air inside had been pumped out.

Evangelista Torricelli (1608–1647), Italian physicist, was secretary and companion to Galileo during the last three months of Galileo's life. At the suggestion of Galileo, Torricelli measured the height of a mercury column in a vertical tube filled with mercury, closed at the top end, and with the other end dipping into a basin full of mercury. He thus measured atmospheric pressure. The fact that air had a definite weight meant that the atmosphere had a finite height and did not extend throughout the universe (as was then generally believed).

William Gascoigne (1612–1644), English astronomer, added cross hairs on the focal plane of his telescope, which enabled him to center a star precisely (see Section 5.5) a micrometer allowed him to make accurate measurements of the angular distances between stars within the field of view of the telescope and thus to locate stars very accurately.

Blaise Pascal (1623–1662), French mathematician, studied the mechanics of fluids. He sent his brother-in-law up the Puy-de-Dôme, a mountain in central France, with two barometers to see if pressure diminished with height, as Torricelli maintained. It did.

Giovanni Domenico Cassini (1625–1721) was an Italian astronomer who made important observations on the planets (rotational periods of Mars and Jupiter, the precise motions of Jupiter's moons, etc.) and taught at the University of Bologna. In 1669 he moved to Paris, and there, in 1672, he was able to determine the parallax of Mars and thus to establish the size of the solar system. His value for the astronomical unit (average radius of the Earth's orbit) was $138 \cdot 10^6$ km, which is close to the modern value of 149,597,871 km.

Francesco Redi (1626–1697), Italian physician and poet, demonstrated that maggots do not develop on rotting meat if the meat is behind a screen to keep the flies out. That was the first demonstration refuting the spontaneous generation of life.

Robert Boyle (1627–1691), Irish physicist and chemist, discovered the inverse relationship between pressure and volume in a gas, which means that gases consist of atoms separated by vacuum. He made the distinction between elements and compounds, as well as that between acids and bases. He is considered the father of modern chemistry.

Marcello Malpighi (1628–1694), Italian biologist and microscopist, studied the structure of human and animal tissues. He discovered capillary blood vessels and showed that they link the arterial system to the venous system.

Christian Huygens (1629–1695), Dutch physicist and astronomer, invented the pendulum clock, the first precise timekeeper. He maintained that light is a wave, not a bunch of corpuscles as Newton believed.

Antony van Leeuwenhoek (1632–1723), Dutch microscopist, built microscopes consisting of a single tiny, highly biconvex lens. Because the lens was ground to perfection, Leeuwenhoek could see with a magnification of up to 200. He reported his numerous observations (on bacteria, protozoa, red blood cells, rotifers, *Hydra, Volvox,* and canine spermatozoa) in 375 illustrated letters to the Royal Society of London. He discovered bacteria in 1683.

Robert Hooke (1635–1703), English physicist, discovered Hooke's law, which says that the strain on a solid is proportional to the applied stress.

Niels Stensen (Nicolaus Steno) of Copenhagen (1638–1686) was a physician, anatomist, and geologist. He converted to Catholicism in 1667, befriended the Medici family in Florence, spent time at their court, and was buried in San Lorenzo, Florence. Steno wrote important works on human anatomy. He also wrote *De solido* (1669), a basic work on geology and paleontology based on his observations in Tuscany. He showed that the angle between corresponding faces of crystals of a given substance remains the same regardless of the size of the crystal. Steno established the *principle of superposition*, which says that younger rock layers lie above older layers.

Isaac Newton (1642–1727), Lucasian Professor of Mathematics at Cambridge University, invented calculus (simultaneously with Gottfried Wilhelm Leibnitz, 1646–1716).

[Calculus is the branch of mathematics that deals with the relationship between or among continuously changing quantities when the relationships are themselves continuously changing. For instance, the relationship between the radius r and the surface $S = 4\pi r^2$ of a sphere is variable ($dS/dr = 2\pi r$), depending upon the value of r. In contrast, the relationship between the radius r and the circumference of $C = 2\pi r$ of a circle is constant ($dC/dr = 2\pi$).]

Newton studied the refraction of light through prisms, invented the reflecting telescope (see Fig. 5.31), and discovered the three laws of motion and the law of universal gravitation, which he published in his *Philosophiae Naturalis Principia Mathematica* (1687).

Olaus Roemer (1644–1710) was a Danish astronomer who moved to Paris in 1672. There, he noticed that the disappearance of Jupiter's satellites behind Jupiter, as seen from the Earth, occurred progressively earlier when the Earth, in its journey around the Sun, was approaching Jupiter, and occurred progressively later when the Earth was moving away from Jupiter. Roemer concluded that those changes were due to the fact that the speed of light was not infinite, as believed by most at the time, but finite. The maximum change was 16.5 minutes, which Roemer correctly thought was the time needed for light to cross the Earth's orbit. From the radius of Earth's orbit, as determined by Cassini in 1672, Roemer calculated (in 1676) that the speed of light was 277,000 km/s. If Cassini's value had not been 7% too small, Roemer would have obtained a value 7% greater for the speed of light, or 296,390 km/s. This is very close to the modern value of 299,792.458 km/s.

Edmund Halley (1656–1742), English astronomer, established an observatory on the island of St. Helena and produced the first catalog of stars in the Southern Hemisphere. He recognized that a comet that had appeared in 1682, which he had observed, had the same path as comets that had appeared previously, in 1456, 1531, and 1607. He concluded that it was the same comet coming back, and he predicted a return in the year 1758. This comet, now known as Halley's comet, did indeed return in 1758 and, since then, in 1835, 1910, and 1986. In 1686, Halley explained the origin of the trade winds as resulting from the rising of hot air at the equator and its sinking at high latitudes (see *Hadley*). In 1718, Halley pointed out that several prominent stars had changed their positions since Greek times, and even since the time of Tycho, indicating that the "fixed" stars are not fixed, but have their own peculiar motions.

Georg Ernst Stahl (1660–1724), German chemist, invented the theory of "phlogiston", according to which a substance that burned or became oxidized lost phlogiston. The resulting "calx" (= oxide) or "ashes" contained no phlogiston (see *Lavoisier*).

Johann Bartholomaeus Adam Beringer (1670–1740), professor of natural history at the University of Würzburg, believed that fossils were "capricious fabrications of God". In 1728 he published *Lithographiae Wirceburgensis*, with 21 plates illustrating clay tablets with Hebrew, Babylonian, Syriac, and Arabic inscriptions, as well as stars, moons, suns, comets, and so forth, also in clay, which he thought were genuine fossils. Actually they had been manufactured by his students and "salted" in a hill where Beringer led field trips. When the students, feeling that they had gone too far, confessed, he refused to believe them and went ahead with his publication. Finally the students planted a tablet that said "VIVAT BERINGERIUS—GOTT". Beringer finally realized that he had been taken. He ruined himself trying to buy back all the copies of his book and died bankrupt and heartbroken. Hobard, a bookseller in Hamburg, bought all available copies from Beringer's heirs, resold them, and even printed a second edition.

George Hadley (1685–1768), English meteorologist, explained the direction of the trade winds as a result of the rotation of the Earth.

Daniel Bernoulli (1700–1782), Swiss mathematician, pioneered the study of hydrodynamics.

Carl von Linné (Carolus Linnaeus) (1707–1778), Swedish botanist and professor at the University of Uppsala, published in 1735 his *Systema Naturae*, in which he introduced the modern system of classification and nomenclature for all organisms. He defined species and grouped species into genera, genera into classes, and classes into orders. He fit humans into his classification scheme and named humans *Homo sapiens.*

George-Louis Leclerc, compte de Buffon (1707–1778), French naturalist and founder of the Musée d'Histoire Naturelle, wrote *Histoire de la Terre* (1749), *Histoire des Quadrupeds* (1750–1755, 10 vols.), and *Histoire des Oiseaux* (1770–1783, 9 vols.), among other works. He believed in the separate creation of all species, but also that the environment could modify or even kill species. He suggested that the Earth

could be 75,000 y old and that life might have originated as early as 30,000 y ago. Those were colossal underestimates, of course, but it was the first attempt to break through the Biblical 6,000 y.

Giovanni Arduino (1714–1795), Italian naturalist, taught at the University of Padua. He founded the science of stratigraphy (see Section 18.1). Arduino's stratigraphic system is as follows:

Old formations
1. Primary: mountains formed by crystalline rocks

2. Secondary: mountain formed of sandstone, limestone, and shale, with fossils

3. Tertiary: gravels, sands, clays, marls

Modern formations
1. sediments deposited by floods

2. volcanoes and lavas (the Colli Euganei, near Padua, are Quaternary volcanoes piercing through Tertiary formations)

Immanuel Kant (1724–1804), German philosopher, thought that the solar system had condensed from a rotating, flattened nebula; that the Milky Way was but one of many "island universes" (i.e., galaxies); and that tidal friction must slow down the Earth.

James Hutton (1726–1797), Scottish geologist, the "father of geology", recognized the difference between igneous and sedimentary rocks and believed that the Earth was continuously changing by slow processes currently in progress. He believed that the same processes had operated in the past and formulated the principle of *uniformitarianism*. Hutton thought that the internal heat of the Earth was the primary source of energy for the geological processes (this theory was called *vulcanism*, cf. *Werner*). In 1785 he published *Theory of the Earth with Proofs and Illustrations*, the first comprehensive treatise on how the Earth works. There he stated that "there is no vestige of a beginning, no prospect of an end" (which placed him in conflict with the Bible).

Lazzaro Spallanzani (1729–1799), Italian biologist, proved that broth that had been boiled and then sealed never spoiled (i.e., microorganisms would not grow in it). That was a further proof against spontaneous generation (see *Redi* and *Pasteur*).

Henry Cavendish (1731–1810), English physicist and chemist, discovered hydrogen and argon and showed that water is a combination of hydrogen and oxygen. He was able to determine the gravitational force of attraction between two lead balls in close proximity (see Fig. 11.1). That enabled him to determine the universal gravitational constant G.

Joseph Priestley (1733–1804), English chemist, isolated many gases, including nitrogen, ammonia, nitrogen oxides, carbon monoxide, hydrogen chloride, and sulfur dioxide.

Charles Augustin Coulomb (1736–1806), French physicist, discovered the inverse square law of electrical and magnetic attraction (or repulsion).

Joseph Louis Lagrange (1736–1813), French mathematician, worked on celestial mechanics, especially the motions of the Sun-Earth-Moon system and the system of Jupiter and its four moons.

James Watt (1736–1819), Scottish engineer, invented the modern steam engine (see *Hero*).

William Herschel (1738–1822), German-English astronomer, discovered Uranus with a telescope that he and his sister Caroline had made. He also determined the proper motions of a number of stars and concluded that the Sun itself was moving toward a point in the constellation Hercules near the star Vega, (the *solar apex*).

Carl Wilhelm Scheele (1742–1786), Swedish chemist, discovered oxygen and chlorine.

Antoine Laurent Lavoisier (1743–1794), French chemist, killed the idea of phlogiston (see *Stahl*) by showing that burning is the acquisition of oxygen, not the loss of phlogiston. He also established the principles of chemical nomenclature.

René Juste Haüy (1743–1822), French mineralogist, founded the science of crystallography by showing that cleaving a crystal produces fragments that replicate the form of the original crystal.

Jean Baptiste Pierre Lamarck (1744–1829), French naturalist, proposed evolution for the first time since the ancient Greeks. He claimed, erroneously, that acquired physical traits can be inherited.

Alessandro Volta (1745–1827), Italian physicist, invented the electric battery, which tremendously enhanced the study of electricity.

Giuseppe Piazzi (1746–1826), Italian astronomer, established an astronomical observatory in Palermo, Sicily. There, on January 1, 1801, he discovered the first asteroid and named it Ceres after the Roman goddess of agriculture who was especially worshipped in Sicily. He also discovered that 61 Cygni had the fastest proper motion.

Pierre Simon, marquis de Laplace (1749–1827), French astronomer and mathematician, worked on celestial mechanics, planetary perturbations, and so forth. He wrote the treatise *Mécanique Céleste* (1799–1825, 5 vols.), in which he started the dastardly practice of presenting an infernal equation followed by the phrase "from which immediately follows", in turn followed by an even more infernal equation, when in effect it takes days to work out the derivation.

Abraham Werner (1750–1817), German geologist who never traveled out of Saxony, believed that all rocks were sedimentary and had been laid down under a primordial worldwide ocean. This theory came to be called *neptunism*. For a period, geologists were divided between the vulcanists who believed in Hutton (see *Hutton*) and the neptunists who believed in Werner. Werner believed that volcanoes were recent phenomena produced by the underground burning of coal seams. In stratigraphy he had a scheme with wonderful German names:

1. *Urgebirge*: original mountains, crystalline (*Ur* means *original, primordial*; *Gebirge* means *mountain range*)

2. *Übergangsgebirge* (*Gang* means *pace, march*; *Übergang* means *passage, transition*; and so *Übergangsgebirge* means *transitional mountains*, those made of quartzite or limestone)

3. *Flötzgebirge* (*Flötz* means *horizontal seam*); these are the sedimentary mountains

4. *Aufgeschwemmtgebirge* (*schwemmen* means *to float*); these are the alluvial mountains, consisting of loose rocks

Heinrich Wilhelm Matthäus Olbers (1758–1840), German physician and amateur astronomer, discovered the asteroids Pallas and Vesta and five comets. He is famous as the author of the Olbers paradox (see Section 5.5).

John Dalton (1766–1844), English chemist, maintained that all matter consists of atoms (thus resuscitating Democritus) and developed a quantitative atomic theory.

William Nicol (1768–1851), British geologist, invented the Nicol prism for the study of minerals and rocks in polarized light.

Georges Léopold Chrétien Frédéric Dagobert Cuvier (1769–1832), French comparative anatomist, believed in the rigid concept of species. Noting that there were no intermediate types between species, he explained the succession of fossil faunas by means of successive creations, the last creation being that of humans.

William ("Strata") Smith (1769–1839), English canal engineer, recognized that sedimentary layers are characterized by their fossil content, and developed the concept of stratigraphic correlation. He formulated the *principle of faunal successions*, which says that different faunas succeed each other in time, and the *principle of superposition*, which says that younger strata overlie older strata. This principle had been previously formulated by Steno.

Friedrich Wilhelm von Humboldt (1769–1859), German naturalist, was independently wealthy and a whirlwind of activity. He traveled extensively in Europe, South America, Russia, and Siberia. He discovered the Humboldt current and the fertilizing properties of guano, and he climbed Mount Chimborazo in Ecuador (6,267 m), whose top is the place on Earth most distant from the center of the Earth (2,151 m farther away than the top of Mount Everest). He wrote *Kosmos*, a comprehensive treatise on natural history.

Amadeo Avogadro, count of Quaregna (1776–1856), Italian physicist, reasoned that equal volumes of gases at the same pressure and temperature must contain the same number of atoms or molecules (but he did not know how many). This is Avogadro's law.

Hans Christian Oersted (1777–1851), Danish physicist, demonstrated the connection between electricity and magnetism by passing a current through a wire aligned with a compass needle and showing that the needle swings by 90° (1819).

Karl Friedrich Gauss (1777–1855), German mathematician, is considered one of the greatest mathematicians of all time, equal to Archimedes and Newton. Gauss contributed to many field of mathematics. He was also interested in the Earth, particularly its shape, and in terrestrial magnetism. He invented the *heliotrope*, an instrument that reflected solar light over great distances and was particularly useful in surveying. He accurately determined the positions of the magnetic poles and created an observatory specifically dedicated to geomagnetic studies.

Joseph Louis Gay-Lussac (1778–1850), French chemist, discovered that gases in specific proportions combine without residue. That most fundamental observation set chemistry on a firm, quantitative base.

Friedrich Wilhelm Bessel (1784–1846), German astronomer, determined that the distance to 61 Cygni, a star with the largest proper motion then known (see *Piazzi*), was 6 light-years. That was the first determination of the distance to a fixed star. Notice that Kepler thought that stars were 0.1 l.y. away, and Newton thought they were 2 l.y. away.

William Prout (1785–1850), English chemist and physiologist, discovered HCl in the human stomach and hypothesized that all elements consist of combinations of hydrogen atoms.

Adam Sedgwick (1785–1873), British geologist, established the Cambrian period.

Joseph von Fraunhofer (1787–1826), German optician, discovered (1814) the absorption lines in the solar spectrum.

Georg Simon Ohm (1787–1854), German physicist, formulated *Ohm's law*, the relationship $V = IR$ among voltage (V), current (I), and resistance (R).

Michael Faraday (1791–1867), English chemist and physicist, established the nomenclature and laws of electrolysis, discovered benzene, and invented the transformer and the generator. One Avogadro number of electrons is called a faraday, and the SI unit of capacitance is called a farad.

Gustave Gaspard de Coriolis (1792–1843), French physicist, discovered the Coriolis acceleration and the Coriolis effect.

Roderick Impey Murchison (1792–1871), British geologist, established the Silurian, Devonian, and Permian periods.

Sadi Carnot (1796–1832), French physicist, studied heat engines and created the science of thermodynamics. He showed that the maximum amount of work that can be obtained from a perfect (frictionless) heat engine with a heat source (e.g., steam) at absolute temperature T_1 and a heat sink (e.g., cooling water) at absolute temperature T_2 is equal to $(T_1 - T_2)/T_2$.

Charles Lyell (1797–1875), Scottish geologist, popularized the principle of uniformitarianism, formulated by Hutton. He wrote a comprehensive and influential treatise entitled *The Principles of Geology* (3 vols., 1830–1833). He added the Eocene, Miocene, Pliocene, and Pleistocene to the stratigraphic column.

Friedrich Wöhler (1800–1882), German chemist, synthesized urea in 1828, the first organic substance to be synthesized (until then, people had thought that organic substances could be made only by living processes).

Heinrich Georg Bronn (1800–1852) of Heidelberg demonstrated that many species pass from one geological period to the next.

George Biddell Airy (1801–1892), English geophysicist and astronomer, developed the principle of *isostasy*, according to which mountains are blocks of different thicknesses, reaching to different heights while floating isostatically on a fluid mantle.

Alcide Dessalines d'Orbigny (1802–1857), French specialist on foraminifera, recognized 27 successive faunas and proposed 27 successive creations.

Christian Johann Doppler (1803–1853), Austrian physicist, discovered the Doppler effect and suggested its application to light (Doppler shift).

Justus von Liebig (1803–1873), German organic chemist, discovered *isomerism*, the property of many organic compounds to have the same chemical formula but different structures and properties.

Matthew Fontaine Maury (1806–1873), American oceanographer, director of the U.S. Naval Observatory and Hydrographic Office, organized the first systematic survey of the ocean from information supplied by merchant ships.

Louis Agassiz (1807–1873), Swiss naturalist, worked with Cuvier in Paris and then became professor of natural history at Neuchâtel in Switzerland. He convinced himself that the Alpine glaciers had once been much more extended, and he proposed the Great Ice Age. In traveling through northern Europe and England, he discovered more evidence of former glaciation. In 1847 he became professor at Harvard and found even more evidence of glaciation in North America. His Great Ice Age caught the popular imagination, and he became very famous. He stated, in opposition to Bronn, that no species passes from one period to the next, and he steadfastly refused to believe in evolution.

John Henry Pratt (1809–1871), British geophysicist, proposed in 1854 his principle of *isostasy*, according to which mountains rise high because they consist of less dense rocks. Pratt visualized the mountains as standing on a flat lithospheric surface (cf. *Airy*).

Charles Darwin (1809–1882), British naturalist, was independently wealthy (his maternal grandfather was Josiah Wedgwood, maker of the Wedgwood porcelains). He circumnavigated the globe on the H.M.S. *Beagle* (December 27, 1831, to October 2, 1836). The small but significant differences between finches occupying different islands in the Galapagos convinced him that the environment can modify a species (he recognized 14 species). Darwin was elected a member of the Geological Society upon his return. On September 28, 1838, he read Malthus' work *An Essay on the Principle of Population*. That led him to formulate his theory of evolution. In 1839 he published *A Naturalist's Voyage on the Beagle*, a report that made him famous. During the next 20 years (1838–1858) Darwin gathered evidence in favor of evolution. In 1858, just when he was getting ready to publish his theory of evolution, he received for review a paper by Alfred Russel Wallace (1823–1913) in which the principles of evolution were laid out. At Lyell's insistence, Darwin and Wallace published a joint paper in the journal of the Linnean Society (presented July 1, 1858). A year later, on November 24, 1859, Darwin's eagerly awaited book on evolution appeared: *On the Origin of Species by Means of Natural Selection.* The first edition of 1,250 copies was exhausted on the first day of publication. The book produced a storm of controversy. Thomas Huxley (1825–1895), Darwin's close friend, took it upon himself to defend and propagandize evolution. In spite of grave problems (the mixing of genetic characteristics that should have wiped out variability, and the fact that the missing links apparently were missing), Darwin's work was so meticulous and so well documented that his theory of evolution won over the majority of natural scientists in a relatively short time.

Urbain Jean Joseph Leverrier (1811–1877), French astronomer and mathematician, calculated the position of an outer planet that had been suspected from the anomalies of Uranus' orbit (see *Adams*). He told the German astronomer Galle to look at a certain position in the sky, and Galle discovered Neptune (September 23, 1846).

Robert Wilhelm Bunsen (1811–1899), German chemist, invented the Bunsen burner and pioneered the use of spectroscopy for chemical analysis.

Johann Gottfried Galle (1812–1910), German astronomer, discovered Neptune (September 23, 1846) at a location in the sky predicted by Leverrier.

Julius Robert Mayer (1814–1878), German physicist, enunciated the principle of conservation of energy.

Daniel Kirkwood (1814–1895), American astronomer, discovered the Kirkwood gaps in the asteroidal belt and showed that an asteroid within a gap would have an orbital period that would be a simple fraction of Jupiter's orbital period. Perturbation by Jupiter would accelerate or decelerate the asteroid, forcing it into a more distant or a closer orbit, thus maintaining the gap.

James Joule (1818–1889), English physicist, studied in great detail the equivalence of mechanical energy and heat. In 1846 he discovered the phenomenon of magnetostriction—the deformation of ferromagnetic materials subjected to a magnetic field.

Heinrich Ernst von Beyrich (1815–1896), German paleontologist, wrote a treatise on the Tertiary mollusks of northern Germany and added the Oligocene to the stratigraphic column.

Jean Foucault (1819–1868), French physicist, demonstrated the rotation of the Earth in 1851 by suspending a 28-kg iron ball from 60-m wire hanging from the cupola of the Pantheon in Paris and made to swing along the meridian. A Foucault pendulum swings on a plane that is fixed relative to the stars. A Foucault pendulum suspended at the pole would make one apparent turn in one sidereal day (clockwise in the Northern Hemisphere, counterclockwise in the Southern Hemisphere). A Foucault pendulum suspended at the equator along the equatorial plane would not rotate at all. At intermediate latitudes, the pendulum turns $15°/$sidereal hour $\times \sin \varphi$ ($\varphi =$ latitude). Foucault invented the gyroscope and a method to silver glass surfaces that revolutionized the manufacture of reflecting telescopes.

John Couch Adams (1819–1892), English astronomer, calculated the position of an outer planet that had been suspected from the anomalies of Uranus' orbit (see *Leverrier*), but his calculations, which were similar to those of Leverrier, could not be used because of the poor celestial maps available in England at the time.

George Gabriel Stokes (1819–1903), British physicist, developed *Stokes' law*, describing the motion of a sphere through a viscous fluid.

John Tyndall (1820–1893), British physicist, studied the scattering of light by particles.

Hermann Helmholtz (1821–1894), German physicist and physiologist, studied the physics and physiology of vision and hearing.

Gregor Mendel (1822–1884), Austrian monk and geneticist, worked at the Saint Thomas monastery in Brünn (now Brno), Moravia. He published the results of his experiments in two papers (1866 and 1869). These papers were ignored by the scientific community until they were discovered in the year 1900. They contain Mendel's laws, which wipe out one of the fundamental objections to Darwin's theory of evolution, namely, the reduction of variability from generation to generation. In fact, Mendel showed that genes are transmitted intact from generation to generation without mixing.

Rudolf Clausius (1822–1888), German physicist, promulgated the second law of thermodynamics, which says that, in a closed system, any process must increase the entropy of the system.

Louis Pasteur (1822–1895), French chemist, proved that broth boiled in a flask with a long, twisted neck would not spoil even if the neck were open to air. This finally proved that there was no spontaneous generation of life (cf. *Redi* and *Spallanzani*).

Alfred Russel Wallace (1823–1913), English naturalist, worked in the Indonesian archipelago and discovered the *Wallace line*, a line between Borneo and Bali on one side, and Celebes and Lombok on the other, that separates the Asiatic fauna from the Australian fauna. While sick with malaria in Borneo, he wrote, in two days, a paper describing evolution as a result of natural selection. He sent his paper to Darwin for review (see *Darwin*). He also wrote a book, entitled *Island Life*, in which he discussed the ice ages of Mars. Later in life he became a socialist, a feminist, and a mystic.

Gustav Robert Kirchhoff (1824–1887), German physicist, formulated (in 1845) the Kirchhoff laws of electrical circuitry and pioneered spectroscopy. He developed the concept of blackbody emission, which became fundamental in the formulation of quantum physics.

William Thomson, Lord Kelvin (1824–1907), Scottish physicist, proposed $-273°C$ as the absolute zero, where molecular motions cease. He also proposed that a new temperature scale be established with that point as its zero (this is now the Kelvin temperature scale). Lord Kelvin attempted to estimate the age of the Earth by assuming that it came from the Sun, that at the beginning it had the same temperature as the Sun, and that it cooled gradually. The result was an age between $20 \cdot 10^6$ and $40 \cdot 10^6$ y. Lord Kelvin refused to accept the concept of radioactivity, and he died leaving no children.

Johann Jakob Balmer (1825–1898), Swiss mathematician, discovered the numerical relationship, which bears his name, between the spectral lines of the hydrogen series pertaining to visible light.

Henry Clifton Soddy (1826–1908), British geologist, developed a technique for making thin sections of rocks and minerals and discovered the crystalline structure of steel.

George Johnstone Stoney (1826–1911), Irish physicist, proposed the name *electron* for the smallest electrical charge (1891).

Friedrich August Kekulé (1829–1896), German organic chemist, elucidated the structure of the benzene ring.

James Clerk Maxwell (1831–1879), Scottish mathematician and physicist, worked out the distribution of molecular speeds (1860) (see *Boltzmann*). His major work (1864–1873) was the proof that an oscillating electric charge produces an electromagnetic field that radiates outward at a constant speed. This speed, which Maxwell could calculate from the ratio of electromagnetic to electrostatic units, turned out to be the speed of light. Thirty years later, Einstein may have used this "coincidence" to conclude that the speed of light (in vacuo) is constant, which led directly to relativity (see *Einstein*).

Orthniel Charles Marsh (1831–1899), American vertebrate paleontologist, led four major expeditions to the American West and brought back a wealth of fossil reptiles and mammals. The fossils showed the evolution of horses toward monodactility (see Fig. 24.2) and the enlargement of the vertebrate brain through time.

Eduard Suess (1831–1914), Austrian geologist, suggested, on the basis of geological and paleontological similarities, that the southern continents were once grouped into a single supercontinent. He named that supercontinent *Gondwanaland*, which actually means *land of the land of the Gonds* (the name is now simplified to *Gondwana*, which means *land of the Gonds*).

Alfred Nobel (1833–1896), Swedish chemist, invented dynamite (a mixture of nitroglycerin and diatomite) and established the Nobel Prizes in five fields (chemistry, physics, physiology/medicine, literature, and peace; a sixth field, economics, was added in 1869).

Dmitri Ivanovich Mendeléev (1834–1907), Russian chemist, developed the periodic table of the elements (1869).

Josef Stefan (1835–1893), Austrian physicist, showed that the amount of radiation from a hot body is proportional to the fourth power of the temperature (Stefan-Boltzmann law—Boltzmann was Stefan's student assistant).

Johannes van der Waals (1837–1923), Dutch physicist, devised the equation of state for real gases (1873).

Ernst Mach (1838–1916), Austrian physicist, worked on airflow and noticed the sharp change that occurred when the relative speed of air and an object immersed in it reached the speed of sound. The *Mach number* reflects the speed of an object in a medium relative to the speed of sound in that medium (Mach 2 means twice the speed of sound, etc.). Mach also suggested that the properties of space depend on the mass distribution within it. This is called *Mach's principle.*

Josiah Willard Gibbs (1839–1903), American physical chemist, discovered the *phase rule*, which says that in a heterogeneous system at equilibrium, the *degrees of freedom* (= number of independent variables) are equal to the number of components plus 2 minus the number of phases. In the ice/water/water-vapor system, for instance, there is one component (H_2O) in three phases. The degrees of freedom are, therefore, $1 + 2 - 3 = 0$.

Ludwig Edward Boltzmann (1844–1906), Austrian physicist, related entropy to disorder. He demonstrated that $E = \frac{3}{2}kT$ (where $E =$ energy, $k =$ Boltzmann constant, $T =$ temperature) and worked out, independently of Maxwell, the distribution of molecular speeds as a function of temperature.

Wilhelm Konrad Roentgen (1845–1923), German physicist, discovered x-rays (1895).

Hugo de Vries (1848–1935) was a Dutch botanist. Having discovered (1886) in a field of primroses some that looked different and bred true, he proposed the theory of mutation (1900). (Actually, the sudden appearance of

"freaks" had been noticed since antiquity by herdsmen; a short-legged sheep that appeared in 1791 had been bred because it could not jump fences.) De Vries worked out the laws of inheritance, after which he searched the literature to make sure that nobody had done it before. It was in doing so that he discovered Mendel's publications. In that same year (1900), the German botanist Karl Correns (1864–1933) and the Austrian botanist Erich Tschermak (1871–1962) had independently worked out the laws of inheritance and, upon searching the literature, had also discovered Mendel's work.

George Francis FitzGerald (1851–1901), Irish physicist. In 1895, FitzGerald explained the failure of the Michelson-Morley experiment by stating that the length of a moving body contracts in the direction of motion. He developed a simple formula that was elaborated by Lorentz.

Edward Walter Maunder (1851–1928), British astronomer, discovered that between 1645 and 1715 there had been little solar sunspot activity. That period, called the *Maunder minimum,* coincided with a period of particularly cool summers and cold winters called the Little Ice Age (see Section 24.3).

Antoine Henri Becquerel (1852–1908), French physicist, discovered radioactivity (1896). He demonstrated that electrons are emitted by uranium (1900) (they are actually emitted by ^{234}Th and ^{231}Pa when ^{238}U decays; and by ^{231}Th when ^{235}U decays).

Jacobus Henrikus van't Hoff (1852–1911), Dutch physical chemist, studied the rotation of polarized light by chemical compounds and founded the science of stereochemistry.

Albert Abraham Michelson (1852–1931), German-American physicist, ran the famous Michelson-Morley experiment of 1887 at the Case School of Applied Science (now Case Western Reserve University) in Cleveland, Ohio. That experiment, done in cooperation with Edward Morley (1838–1923), showed that there was no "ether", the intangible "fluid" through which light waves were supposed to propagate.

Hendrik Antoon Lorentz (1853–1928), Dutch physicist, developed the *Lorentz transformation*, showing that bodies in motion slightly contract in the direction of motion: $L_v = L_0(1 - v^2/c^2)^{1/2}$ (where L_v = length of body in motion, L_0 = length of body at rest, v = velocity of body, c = speed of light). This is called the *Lorentz-FitzGerald contraction*.

Friedrich Wilhelm Ostwald (1853–1932), German physical chemist, established the *Ostwald law*, the law relating the degree of ionization to concentration in a dilute electrolytic solution.

Johannes Rydberg (1854–1919), Swedish spectroscopist, generalized the Balmer formula into a formula that describes the wavelengths of lines in all spectral series of the hydrogen atom.

Nikola Tesla (1856–1943), Croatian-American electrical engineer, developed the alternating-current induction motor and worked on transmission networks. He also developed the *Tesla coil*, a transformer with a high-frequency, high-voltage output.

Joseph John Thomson (1856–1940), English physicist, discovered the electron (1897), which had already been named by Stoney.

Heinrich Rudolph Hertz (1857–1894), German physicist, discovered radio waves.

Andrija Mohorovičić (1857–1936), Serbian geologist, studied a 1909 earthquake and showed that the speed of the seismic waves increased abruptly from crust to mantle. That discontinuity is called the Mohorovičić discontinuity.

Eugène Dubois (1858–1940), Dutch naturalist, discovered *Pithecanthropus erectus* (now called *Homo erectus*) in Java in 1891.

Max Karl Ernst Ludwig Planck (1858–1947), German physicist, assumed that energy is not infinitely divisible but comes in *quanta*; he further assumed that energy is proportional to frequency, the proportionality constant (h) being the energy of 1 Hz. He then showed that those two assumptions were correct by developing a formula that fits the radiative spectrum of a blackbody at any temperature.

Svante Arrhenius (1859–1927), Swedish physical chemist, established that the conductivity of an electrolytic solution is due to ionic dissociation.

William Henry Bragg (1862–1942), British physicist, developed x-ray crystallography, together with his son William Lawrence, and determined interatomic distances in crystals. The two shared a Nobel Prize in 1915.

Walther Nernst (1864–1941), German physical chemist, established the third law of thermodynamics, which says that a perfect crystal has zero entropy at absolute zero.

Thomas Hunt Morgan (1866–1945), American geneticist, developed the chromosome theory of heredity and showed that genes are strung in linear fashion along the chromosomes.

Marie Sklodowska Curie (1867–1934), Polish-French chemist, separated radium from uranium ore (with husband Pierre, 1898).

Henrietta Swan Leavitt (1868–1921), American astronomer, discovered the relationship between period and brightness (power received) in Cepheids in the Small Magellanic Cloud (1912).

Fritz Haber (1868–1934), German chemist, devised a method, in 1913, for making ammonia from atmospheric nitrogen by combining nitrogen with hydrogen at 400°C, under pressure, with an iron catalyst ($N_2 + 3H_2 \rightarrow 2NH_3$). Until that time, the only sources of ammonia were natural nitrate deposits, especially those of the Atacama desert.

Robert Andrew Millikan (1868–1953), American physicist, measured the charge of a single electron by means of his famous oil-droplet experiments (1913–1917). He determined how the velocity of single-, double-, or multiple-charged oil droplets falling under the influence of gravity was countereffected by a charged plate above. The velocities were shown to reduce to whole numbers, the smallest being related to the elementary electrical charge—the charge of a single electron or proton. Millikan succeeded in calculating this charge in CGS_{esu} units and was awarded the Nobel Prize in physics in 1923.

Jean Brunhes (1869–1930), French geographer, discovered reversed magnetization in a lava flow in the Massif Central in France (1906).

Charles Thompson Rees Wilson (1869–1959), Scottish physicist, devised the Wilson cloud chamber for observing atomic particles in flight. In 1917, Wilson managed to carry out the first artificial radioactive reaction by bombarding nitrogen atoms with alpha particles and stripping protons from their nuclei.

Ernest Rutherford (1871–1937), British physicist, demonstrated that alpha particles are helium atoms without the electrons (1909) and concluded that the fundamental positive charge is hydrogen stripped of its electron (he called that *proton*). He showed that whereas most alpha particles passed through gold foil, some were sharply deflected. That demonstrated that atoms are mainly empty space, with all the positive charge concentrated in a minute space in the center, which Wilson called *nucleus* (1911). Rutherford developed a model of the atom as a microscopic "solar system" consisting of a nucleus and orbiting electrons (1911).

Karl Schwarzschild (1873–1916), German astronomer, introduced the concept of the black hole and worked out its radius in terms of its mass.

Ejnar Hertzsprung (1873–1967), Danish astronomer, discovered in 1905 the relationship between spectral type and luminosity. It was independently rediscovered in 1913 by the American astronomer Henry Norris Russell (1877–1957). The relationship is expressed in the Hertzsprung-Russell diagram.

Guglielmo Marconi (1874–1937), Italian engineer, sent radio waves (which were known to be like light, and therefore to travel in a straight line) to a distance of 100 m in 1895, to a distance of 15 km in 1896, from shore to a warship 20 km away in 1897, to a distance of 29 km in 1898, and from England to Newfoundland in 1901. Arthur Edwin Kennelly, British-American electrical engineer (1861–1939), and Oliver Heaviside, English physicist (1850–1925), suggested (1902) that the radio waves were reflected by a layer of ionized molecules in the upper atmosphere. That layer, called *ionosphere*, was discovered by Appleton in 1924.

Francis William Aston (1877–1945), English physicist, invented the mass spectrometer and discovered the nuclear binding energy.

Frederick Soddy (1877–1956), English radiochemist, formulated the laws of radioactive decay in 1900, together with Rutherford. He coined the name *isotope*.

Henry Norris Russell (1877–1957) developed

the relationship between stellar luminosity and spectral type (see *Hertzsprung*).

Lise Meitner (1878–1968), Austrian physicist, worked with Otto Hahn on neutron bombardment of uranium and, together with her nephew Otto Frisch, developed the concept of nuclear fission. She had to overcome many prejudices in her long life, being Austrian, a woman, Jewish, and Protestant. She declined to work on the atom bomb project.

Albert Einstein (1879–1955), German physicist, was a high-school dropout (Munich). Nevertheless, he went to college, in Switzerland, became a Swiss citizen, and landed a job at the patent office in Berne (1901). In 1905, the year he received his Ph.D., he published three fundamental papers: (1) application of Planck's work to the photoelectric effect (for which he received the Nobel Prize in 1921); (2) mathematical analysis of Brownian motion; and (3) relativity of time, based on the assumption that the speed of light is constant regardless of the relative speed of systems moving with uniform motion relative to each other. (That assumption probably was based on the observation that electrostatic and electromagnetic systems of measurement are related to each other via a constant that is equal to the speed of light — see *Maxwell.*) In 1916 Einstein published his general theory of relativity for accelerated systems. Some consequences are the precession of perihelion in planetary orbits, the gravitational redshift, and the deflection of light by strong gravitational fields (which was observed during the solar eclipse of March 29, 1919).

Milutin Milankovitch (1879–1958), Serbian mathematician, calculated the changes in insolation at different latitudes for the past 10^6 y (1920–1938), and proposed that these changes were the cause of the ice ages (this was confirmed by isotope analysis of deep-sea cores — see Section 24.3).

Max von Laue (1879–1960), German physicist, passed a beam of x-rays through a crystal in 1912 and showed that x-rays diffracted. That opened the way to study the structure of minerals by x-ray analysis (see *Bragg*).

Otto Hahn (1879–1968), German physical chemist, fissioned uranium by bombarding it with neutrons. That work was done in association with the Austrian physicist Lise Meitner.

Alfred Lothar Wegener (1880–1930), German meteorologist, proposed continental drift in 1912 and reconstructed Pangea. He died on his fourth expedition to Greenland.

Arthur Eddington (1882–1944), English astrophysicist, discovered the mass–luminosity relationship in stars. He also led the expedition that proved, during the solar eclipse of 1919, that the light from stars is curved while passing by the Sun. That was the first experimental verification of relativity.

Hans Wilhelm Geiger (1882–1945), German physicist, invented the Geiger counter and applied it to the study of radioactivity.

Max Born (1882–1970), German physicist, was the founder of quantum mechanics.

Motonori Matuyama (1884–1956), Japanese geologist, discovered that the Earth's magnetic field was reversed in the Early Pleistocene (1929).

Niels Henrik David Bohr (1885–1962), Danish physicist, in 1913 improved the atomic model of Rutherford by introducing the concept that the electrons are in stationary "orbits" and that energy is emitted or absorbed in "quantum jumps" when electrons move between orbits. "Quantum reality" requires that electrons be indeterminate physical quantities that can interact as masses or waves (1927).

Harlow Shapley (1885–1972), American astronomer, demonstrated that the Sun is not at the center of the Galaxy, but is 30,000 l.y. away. He showed that globular clusters are distributed in a spherical halo 50,000 l.y. in radius, centered on the center of the Galaxy. He determined the distances to some Galactic Cepheids, which made it possible to replace the period/brightness graph of Leavitt with a period/luminosity graph (luminosity is the power emitted), thus leading to the determination of the distance to a Cepheid by determining its period and by measuring the power received.

Erwin Schrödinger (1887–1961), Austrian physicist, showed that electrons bound in atoms can be interpreted as standing waves and that the electronic orbits are the loci where a whole number of waves can fit.

Felix Andries Vening Meinesz (1887–1966), Dutch geophysicist, used submarines for gravity measurements at sea and discovered the belt of negative gravity anomalies accompanying the deep-sea trenches.

Hans Pattersson (1888–1966), Swedish oceanographer, organized and led the Swedish Deep-Sea Expedition of 1947–1948, which opened the postwar era of oceanographic exploration.

Chandrasekhara Raman (1888–1970), Indian physicist, discovered the *Raman effect*, the scattering and frequency changes of light passing through a transparent medium due to interaction of the photons with the vibrational and rotational motions of the molecules of the medium.

Inge Lehmann (b. 1888), Danish seismologist, discovered the solid inner core of the Earth (1936).

Beno Gutenberg (1889–1960), German-American geophysicist, explained the seismic "shadow zone" as due to a liquid core (1913) and discovered the asthenosphere (1948).

Edwin Powell Hubble (1889–1953), American astronomer, used the 100-in. Mount Wilson telescope and demonstrated that most "nebulae" are extragalactic (1924). He also demonstrated that the light received from the galaxies is redshifted in proportion to their distances. The proportionality constant is the Hubble constant (~ 12–18 km/s per 10^6 l.y.) (1929).

William Lawrence Bragg (1890–1971). See *William Henry Bragg* (1862–1942).

Arthur Holmes (1890–1965), English geologist, showed that convection must occur in the Earth's mantle even though the mantle is solid. He also developed methods for radioactive dating of rocks and minerals.

James Chadwick (1891–1974), English physicist, discovered the neutron (1932).

Edward Victor Appleton (1892–1965), English physicist, discovered the ionosphere (1924).

George Paget Thomson (1892–1975), English physicist, son of J. J. Thomson, demonstrated that electrons can behave as waves (1927). Like his father, he received a Nobel Prize.

Louis Victor Pierre Raymond de Broglie (1893–1987), French physicist, demonstrated (by combining $e = mc^2$ with $E = h\nu$) that each particle has an associated "matter-wave" with wavelength inversely proportional to momentum.

Arthur Holly Compton (1892–1962), American physicist, discovered that the frequency of x-rays decreases when they are scattered by matter. He concluded that light particles (which he named *photons*) lose momentum in collision with electrons. That proved that light can act as a particle.

Harold Clayton Urey (1893–1981), American chemist, discovered deuterium. He also discovered a method for measuring paleotemperatures by oxygen-isotopic analysis of marine carbonates (see Section 24.3).

Raymond Dart (1893–1988), Australian anthropologist, discovered *Australopithecus africanus* in 1924.

Satyendranath Bose (1894–1974), Indian physicist, laid the foundation for the Bose-Einstein statistics, which describe particles with spin equal to an integral multiple of $h/2\pi$ ($h =$ Plank's constant).

Alexander Ivanovich Oparin (1894–1980), Russian biochemist, published a book (*Origin of Life on Earth*, 1936) in which he suggested that life originated when the Earth's atmosphere consisted of a mixture of methane and ammonia.

Louis Leakey (1895–1972), British-Kenyan paleoanthropologist and Quaternary geologist, worked in East Africa with his wife Mary Leakey (b. 1913) and son Richard. The Leakeys discovered *Paranthropus boisei* (1959) and *Homo habilis* (1960).

Patrick Maynard Blackett (1897–1974), English physicist, bombarded lead with gamma rays and discovered *pair production*, the production of matter from energy by the reaction $\gamma \rightarrow e^+ + e^-$. That was the first demonstration that $e = mc^2$.

Carl-Gustaf Rossby (1898–1957), Swedish-American meteorologist, discovered the jet stream and the Rossby waves (see Section 13.4).

Leo Szilard (1898–1964), Hungarian-American physicist, recognized the possibility of nuclear fission in 1934 and contributed to the Manhattan Project during World War II.

Wolfgang Pauli (1900–1958), Austrian physicist, formulated the *exclusion principle*, which says that only two electrons can occupy the same energy level, and only if they have opposite spins. That made it possible to arrange the electrons in shells (characterized by the principal quantum number n), subshells (characterized by the orbital quantum number l), and orbitals (characterized by the orbital magnetic quantum number m_l). The same orbital can be occupied by a maximum of two electrons provided they have opposite magnetic spin numbers m_s (which can have the value of $+\frac{1}{2}h/2\pi$ or $-\frac{1}{2}h/2\pi$).

Hans Adolf Krebs (1900–1981), German-British biochemist, discovered the Krebs cycle, the cycle in mitochondria by which glucose is broken down to CO_2 and H_2O, with the release of bond energy.

Jan Hendrick Oort (b. 1900), Dutch astronomer, described the rotation of the Galaxy and proposed the *Oort cloud*, a spherical cloud of comets surrounding the solar system to a distance of about 1 l.y.

Charles Richter (b. 1900), American seismologist, creator of the Richter scale of earthquake intensities.

Ernest Lawrence (1901–1958), American physicist, invented the cyclotron.

Enrico Fermi (1901–1954), Italian-American physicist, bombarded uranium with neutrons and discovered the importance of moderators to slow down neutrons to thermal velocities (1934). During those experiments, he fissioned ^{235}U, but (fortunately) thought that he had made element 93 instead (if fission had been discovered in 1934, Hitler might have had nuclear weapons at the start of World War II) (see *Hahn*). Fermi moved to the United States in 1938 and built the first nuclear reactor, which went critical at 3:45 P.M. on December 2, 1942, at the University of Chicago.

Robert Van de Graaff (1901–1967), American physicist, invented the Van de Graaff generator, which can routinely produce potentials as high as 10^7 V.

Werner Karl Heisenberg (1901–1976), German physicist, enunciated the *uncertainty principle* (1927), which says that $\Delta p \Delta r \geqslant h/2\pi$ or $\Delta E \Delta t \geqslant h/2\pi$ (p = momentum; r = position; E = energy; t = time). The two expressions are equivalent (see Section 2.5).

Linus Pauling (b. 1901), American chemist, did estensive work on the chemical bond and elucidated the structures of proteins.

Paul Adrien Maurice Dirac (1902–1984), English physicist, proposed the existence of antiparticles and antimatter (1930).

Robert Oppenheimer (1904–1967), American physicist, investigated cosmic-ray showers and became a leader of the Manhattan Project.

George Gamow (1904–1968), Russian-American physicist, proposed the Big Bang. In his model, all elements were formed within half an hour. This model was modified by Fowler.

Otto Frisch (1904–1979), Austrian physicist, developed the idea of nuclear fission, together with his aunt, Lise Meitner, and worked on uranium fission.

Karl Guthe Jansky (1905–1950), American radio engineer, discovered radio-wave emissions from the center of the Galaxy (1932).

Emilio Segrè (1905–1989), Italian-American physicist, bombarded molybdenum (element 42) with deuterons, which made element 43. He named it technetium (1937). That was the first element made artificially.

Carl David Anderson (1905–1989), American physicist, discovered the first antiparticle, the positive electron, which he named *positron* (1932) (see *Dirac*).

Harry H. Hess (1906–1969), American geologist, discovered guyots (1946) and proposed the theory of sea-floor spreading (1962).

Maria Goeppert-Mayer (1906–1972), Austrian-American physicist, developed a model explaining the pattern of nuclear magic numbers—the particular stability of nuclei with 2, 8, 20, 28, 50, 82, or 126 protons or neutrons.

William Maurice Ewing (1906–1974), American geophysicist, discovered that the crust under the Atlantic Ocean is fully oceanic, not sialic (1950).

Kurt Gödel (1906–1978), Austrian-American mathematician, proved that any system of axioms will contain within itself statements that cannot be proved or disproved on the basis of those axioms. It is impossible, therefore, to construct a self-consistent mathematical system.

Sin-Itiro Tomonaga (1906–1979), Japanese physicist, developed the theory of quantum electrodynamics independently of Feynman and Schwinger.

Hans Albrecht Bethe (b. 1906), German-American physicist, discovered the carbon cycle (1938) (see Section 6.2).

Clyde William Tombaugh (b. 1906), American astronomer, discovered Pluto on February 18, 1930. The discovery was announced on March 13, 1930.

Hideki Yukawa (1907–1981), Japanese physicist, predicted the existence of mesons (1935).

Willard Frank Libby (1908–1980), American chemist, developed the radiocarbon dating method at the University of Chicago (1948–1950).

Hannes Alfvén (b. 1908), Swedish physicist, studied the dynamics of plasmas in magnetic fields, and explained the momentum transfer from Sun to planets during the formation of the solar system (see Section 7.1).

John Tuzo Wilson (b. 1908), Canadian geophysicist, developed the concept of transform faults and hot spots.

Roger Revelle (b. 1909), American oceanographer, discovered, together with Maxwell, that the oceanic heat flow is similar to the continental heat flow (1952).

William Bradford Shockley (1910–1989), English-American physicist, invented the transistor (1948).

Subrahmanian Chandrasekhar (b. 1910), Indian-American astronomer, discovered the Chandrasekhar limit (see Section 6.3).

Carl Keenan Seyfert (1911–1960), American astronomer, discovered the Seyfert galaxies.

Luis Walter Alvarez (1911–1988), American physicist, discovered, together with his son Walter, the iridium excess at the Cretaceous–Tertiary boundary at Gubbio (Italy) and developed a theory of asteroidal impact (1978) to explain the observed extinctions at the end of the Cretaceous (see Section 23.6).

William Fowler (b. 1911), American physicist. Together with Geoffrey and Margaret Burbidge and Fred Hoyle, Fowler showed how elements are made inside stars. He received a Nobel Prize for his work.

Melvin Calvin (b. 1911), American biochemist, discovered the cycle (which bears his name) by which CO_2 molecules are incorporated into organic compounds during the process of photosynthesis.

Alfred O. Nier (b. 1911), American chemist, improved the mass spectrometer, measured the isotopic compositions of many elements, and developed several dating methods.

Grote Reber (b. 1911), American radio astronomer, discovered the first radio sources in the Galaxy (1938–1942).

Glenn Theodore Seaborg (b. 1912), American physicist, discovered elements 95–102.

Karl Friedrich von Weizsächer (b. 1912), German astronomer, discovered the carbon cycle (1938) independently of Bethe. He also developed the details of the nebular hypothesis for the origin of the solar system, taking into account turbulence, conservation of angular momentum, and so forth.

James Alfred Van Allen (b. 1914), American physicist, discovered the Van Allen belts (1958).

Max Perutz (b. 1914), Austrian-British molecular biologist, elucidated the structure of hemoglobin.

Fred Hoyle (b. 1915), British physicist, proposed that elements are formed inside stars (see *Fowler*).

Francis Crick (b. 1916), English biochemist, elucidated (1953) the structure of DNA (the "double helix"), together with Watson.

R. G. Mason (b. 1916), British geophysicist, discovered magnetic stripes in the oceanic crust (1958).

Walter Heinrich Munk (b. 1917), American geophysicist, studied the effects of atmospheric and oceanic motions on the rotation of the Earth.

Richard Feynman (1918–1988), American physicist, developed the theory of quantum electrodynamics, independently of Schwinger and Tomonaga.

Julian Seymour Schwinger (b. 1918), American physicist, developed the theory of quantum electrodynamics, independently of Feynman and Tomonaga.

Margaret Burbidge (b. 1919), British-American astrophysicist, worked on nucleosynthesis, on the chemical evolution of the universe, and, together with Fowler and Hoyle and her husband Geoffrey, on the formation of the elements inside stars.

Sam Epstein (b. 1919), Canadian-American geochemist, developed the isotopic method of paleotemperature analysis and pioneered the study of stable isotopes in nature.

Cesare Emiliani (b. 1922), Italian-American marine geologist, applied oxygen-isotopic analysis to deep-sea sediments and demonstrated the cyclic nature of the ice ages (1955) (see the Note at the end of this list).

Claire Patterson (b. 1922), American geochemist, dated meteorites at $4.5 \cdot 10^9$ y.

S. Keith Runcorn (b. 1922), British geophysicist, demonstrated continental drift (1962).

Bruce Heezen (1924–1977), American marine geologist, discovered magnetic anomalies at sea (1953). He also discovered the rift along the crest of the Mid-Atlantic Ridge (1956).

Frank Press (b. 1924), American geophysicist, demonstrated that the asthenosphere is continuous beneath continents and oceans (1959).

Geoffrey Burbidge (b. 1925), British astrophysicist, worked on element formation inside stars, together with Fowler, Hoyle, and Margaret Burbidge.

Arthur E. Maxwell (b. 1925), American oceanographer, discovered, together with Roger Revelle, that the oceanic heat flow is similar to the continental heat flow (1952).

Harmon Craig (b. 1926), American geochemist, pioneered the application of stable-isotope analysis to the hydrosphere, to the biosphere, and to statuary.

W. Jason Morgan (b. 1926), American geologist, one of the founders of the theory of plate tectonics (1968).

Abdus Salam (b. 1926), Pakistani physicist, developed a unified theory of the electromagnetic force and the weak force, independently of Glashow and Weinberg (the three shared a Nobel Prize in 1979).

Gerald J. Wasserburg (b. 1927), American geochemist, pioneered the application of potassium-argon dating to meteorites and terrestrial rocks.

James Dewey Watson (b. 1928), American biochemist, worked out the DNA double-helix structure with the bases *inside* doing the connecting. He shared the Nobel Prize with Crick (see *Crick*).

Murray Gell-Mann (b. 1929), American physicist, developed the quark model (1961).

Rudolph Ludwig Mössbauer (b. 1929), German physicist, discovered the *Mössbauer effect*, the recoilless emission of a gamma-ray photon by a nucleus strongly bound within a crystalline nucleus. This effect can produce a gamma-ray beam of sharply narrow energy that can then be used to study a wide range of phenomena.

Robert F. Schmalz (b. (1929), American geologist, proposed the theory of sea-floor spreading, independently of Hess (1961).

Maarten Schmidt (b. 1929), Dutch-American astronomer, discovered quasars (1963).

Stanley Lloyd Miller (b. 1930), American chemist, designed and carried out the Urey-Miller experiment at the University of Chicago (1953).

Sheldon Lee Glashow (b. 1932), American physicist, produced a theory unifying the electromagnetic and weak interaction (see *Salam*).

Neil Opdyke (b. 1933), American geophysicist, established a firm chronology for the Quaternary, in collaboration with Shackleton (1973).

Arno Allan Penzias (b. 1933), German-American physicist, discovered the microwave background radiation, together with Wilson (1964).

Steve Weinberg (b. 1933), American physicist, produced a theory combining the electromagnetic and weak interactions (see also *Glashow* and *Salam*).

Christopher G. A. Harrison (b. 1936), British-American geophysicist, discovered magnetic reversals in deep-sea sediments (1964).

Robert Woodrow Wilson (b. 1936), American physicist, codiscoverer of the microwave background radiation (see *Penzias*).

Nicholas John Shackleton (b. 1937), English physicist, established a firm chronology for the Quaternary, in collaboration with Opdyke (1973).

David Baltimore (b. 1938), American molecular biologist, discovered the enzyme *reverse transcriptase*, which transcribes RNA into DNA.

Stephen Jay Gould (b. 1941), American paleontologist and essayist, proposed the theory of punctuated equilibrium, jointly with Niles Eldredge.

Stephen William Hawking (b. 1942), English physicist, has theorized that "mini black holes" could have formed at the time of the Big Bang and that (because of the uncertainty principle) black holes may evaporate at a rate inversely proportional to their masses.

Dan P. McKenzie (b. 1942), British geophysicist, proposed the theory of plate tectonics (1967).

Jocelyn Bell (b. 1943), English astronomer, discovered pulsars (1967).

Niles Eldredge (b. 1943), American paleontologist, proposed the theory of punctuated equilibrium, jointly with Stephen Jay Gould.

Note: After a brief battle between modesty and thoroughness, I decided to include my own name in this list, taking the description of accomplishments from Millar D., Millar I., Millar J., and Millar M. 1989. *Concise Dictionary of Scientists*. Chambers/Cambridge University Press, Cambridge, England, 461 p. This work profiles what those authors deem to be the 1,000 most prominent scientists of all time. Buy a copy of the book and see if your name is there. If it is not, you may complain to Millar, Millar, Millar, and/or Millar.

INDEX OF NAMES

PHYSICAL AND CHEMICAL DATA

A1. CONSTANTS

Table A1 shows the latest values of the fundamental constants of nature, presented in alphabetical order. Anyone claiming to possess a modicum of scientific literacy should memorize the most important constants to one or two decimal places. Going down the list, I recommend the following 10 constants:

Avogadro number ($= g/u =$ number of atomic mass units in 1 g) $= 6.02 \cdot 10^{23}$
Bohr radius ($=$ radius of the H atom in the ground state) $= 0.5 \cdot 10^{-10}$ m
  ($\approx$ average atomic radius if multiplied by 2 or 3).
Boltzmann constant $= 1.38 \cdot 10^{-23}$ J/K
electron rest mass $= 1/1,836$ of the proton rest mass
Faraday constant $= 96,485$ coulombs (C) per avogadro (Av) of electrons
gravitational constant* $= 6.67 \cdot 10^{-11}$ N m^2 kg^{-2}
light speed in vacuo $= 299,792,458$ m/s (exactly)
ideal-gas volume $= 22.4$ liters/Av
Planck's constant $= 6.62 \cdot 10^{-34}$ J/Hz
Proton or neutron rest mass (in grams) $\approx 1/$Avogadro number

Every night, before going to bed, you should recite these 10 constants. They represent the wisdom of the world.

* Do not try to memorize the units of G. Remember, instead, Newton's law of gravitations $[F = G(m_1 m_2/r^2)]$, solve for G, and figure out the units.

Table A1. Constants

Quantity	Symbol	Value	Error	SI	cgs	Others
atomic mass unit	u	1.6605402	10	10^{-27} kg	10^{-24} g	—
		931.49432	28	—	—	MeV
Avogadro number	Av	6.0221367	36	10^{23} mol^{-1}	10^{23} mol^{-1}	—
Bohr magneton	μ_B	9.2740154	31	10^{-24} J T^{-1}	10^{-21} erg G^{-1}	—
Bohr radius	a_0	0.529177249	24	10^{-10} m	10^{-8} cm	—
Boltzmann constant, R/Av	k	1.3806513	4	10^{-23} J K^{-1}	10^{-16} erg K^{-1}	—
		8.6173442	25	—	—	10^{-5} eV K^{-1}
electron charge	e	1.60217733	15	10^{-19} C	10^{-20} aC	—
		4.80320680	15	—	10^{-10} statC	—

(cont.)

Table A1. Constants (continued)

Quantity	Symbol	Value	Error	SI	cgs	Others
electron charge/ mass ratio	e/m_e	1.75881962	53	$10^{11}\,C\,kg^{-1}$	$10^7\,aC\,g^{-1}$	—
		5.27280856	53	—	$10^{17}\,statC\,g^{-1}$	—
electron Compton wavelength	λ_C	2.42631058	22	$10^{-12}\,m$	$10^{-10}\,cm$	—
electron magnetic moment	μ_e	9.2847701	31	$10^{-24}\,J\,T^{-1}$	$10^{-21}\,erg\,G^{-1}$	—
electron magnetic moment in Bohr magnetons	μ_e/μ_B	1.001159652193	10	—	—	—
electron magnetic moment/proton magnetic moment	μ_e/μ_p	658.21068801	66	—	—	—
electron rest mass	m_e	0.91093897	54	$10^{-30}\,kg$	$10^{-27}\,g$	—
		5.48579903	13	—	—	$10^{-4}\,u$
		0.51099906	15	—	—	MeV
electron rest mass/ proton rest mass	m_e/m_p	5.44617013	11	10^{-4}	10^{-4}	10^{-4}
		1/1836.152702	11	—	—	—
Faraday constant	eAv	9.6485309	29	$10^4\,C\,Av^{-1}$	$10^3\,aC\,Av^{-1}$	—
fine-structure constant $(e^2/\hbar c)$	α	0.00729735308	49	—	—	—
	α^{-1}	137.0359895	61	—	—	—
gas constant	R	8.3144710	14	$J\,mol^{-1}\,K^{-1}$	$10^7\,erg\,mol^{-1}\,K^{-1}$	
		1.9858773	4	$cal\,mol^{-1}\,K^{-1}$	$cal\,mol^{-1}\,K^{-1}$	—
		82.057449	14	—	$atm\,cm^3\,Av^{-1}\,K^{-1}$	
gravitational constant	G	6.67206	8	$10^{-11}\,N\,m^2\,kg^{-2}$	$10^{-8}\,dyn\,cm^2\,g^{-2}$	
impedance of vacuum $(\mu_0/\varepsilon_0)^{1/2}$	Z_0	3.767303134	—	$10^2\,\Omega$	$10^2\,\Omega$	—
light speed in vacuo	c	299792458	0	$m\,s^{-1}$	$10^2\,cm\,s^{-1}$	—
light speed in vacuo squared	c^2	89875517873681764	0	$m^2\,s^{-2}$	$10^4\,cm^2\,s^{-2}$	—
magnetic-flux quantum	$h/2e$	2.06783461	61	$10^{-15}\,Wb$	$10^{-7}\,Mx$	
molar ideal-gas volume (STP)	V_m	22.41410	19	$10^{-3}\,m^3\,Av^{-1}$	$10^3\,cm^3\,Av^{-1}$	—
muon magnetic moment	μ_μ	4.4904514	15	$10^{-26}\,J\,T^{-1}$	$10^{-23}\,erg\,G^{-1}$	—
muon rest mass	m_μ	1.8835327	11	$10^{-28}\,kg$	$10^{-25}\,g$	—
		0.113428913	17	—	—	u
neutron Compton wavelength $(h/m_n c)$	$\lambda_{C,n}$	1.31959110	12	$10^{-15}\,m$	$10^{-13}\,cm$	—
neutron rest mass	m_n	1.6749286	10	$10^{-27}\,kg$	$10^{-24}\,g$	—
		1.008664904	14	—	—	u
		939.56563	28	—	—	MeV
nuclear magneton	μ_M	5.0507866	17	$10^{-27}\,J\,T^{-1}$	$10^{-24}\,erg\,G^{-1}$	
permeability constant $(4\pi \cdot 10^{-7})$	μ_0	12.5663706144...	0	$10^{-7}\,H\,m^{-1}$	—	—
		1	0	—	(Dimensionless)	—

Quantity	Symbol	Value	Error	SI	cgs	Others
permittivity constant $(1/\mu_0 c^2)$	ε_0	8.854187817	0	$10^{-12}\,\mathrm{F\,m^{-1}}$	$10^{-10}\,\mathrm{F\,cm^{-1}}$	
Plank's constant	h	6.6260755	40	$10^{-34}\,\mathrm{J\,Hz^{-1}}$	$10^{-27}\,\mathrm{erg\,Hz^{-1}}$	—
		4.1356692	12	—	—	$10^{-15}\,\mathrm{eV\,Hz^{-1}}$
$(h/2\pi)$	$\hbar$	1.0545726	6	$10^{-34}\,\mathrm{J\,Hz^{-}}$	$10^{-27}\,\mathrm{erg\,Hz^{-1}}$	—
proton Compton wavelength	$\lambda_{\mathrm{C,p}}$	1.32141002	12	$10^{-15}\,\mathrm{m}$	$10^{-13}\,\mathrm{cm}$	—
proton magnetic moment	μ_p	1.41060761	47	$10^{-26}\,\mathrm{J\,T^{-1}}$	$10^{-23}\,\mathrm{erg\,G^{-1}}$	—
proton magnetic moment in Bohr magnetons	$\mu_\mathrm{p}/\mu_\mathrm{B}$	1.521032202	15	10^{-3}	10^{-3}	—
proton mass/electron mass	$m_\mathrm{p}/m_\mathrm{e}$	1836.152701	37	—	—	—
proton rest mass	m_p	1.6726231	10	$10^{-27}\,\mathrm{kg}$	$10^{-24}\,\mathrm{g}$	—
		1.007276470	12	—	—	u
		938.27231	28	—	—	MeV
Rydberg constant	R_∞	1.0973731534	13	$10^{7}\,\mathrm{m^{-1}}$	$10^{5}\,\mathrm{cm^{-1}}$	—
Rydberg energy	hcR_∞	13.6056981	40	—	—	eV
Stefan-Boltzmann constant	σ	5.67051	19	$10^{-8}\,\mathrm{W\,m^{-2}\,K^{-4}}$	$10^{-5}\,\mathrm{erg\,cm^{-2}\cdot s^{-1}\,K^{-4}}$	

Note: The uncertainty in the figures (Error) is the standard deviation in the last digits. The atomic mass unit, for instance, should be read as being equal to $1.6605402 \pm 0.0000010\,\mathrm{u}$.

Source: Emiliani (1987, pp. 241–2), updated 1992.

A2. CONVERSION FACTORS

Neglecting units of measure that are no longer in use [the barleycorn ($\frac{1}{3}$ of one inch), the rod ($16\frac{1}{2}$ feet), the chain (66 feet), the league (3 miles), the parasang (6,000 guz), the momme ($= 10$ fun), the fusz ($= 12$ zolls), the pharoagh ($= 10$ parmaks), the batman, candy, gin, guz, immi, nin, ock, pik, pick, pot, pu, tan, to, tu, zolotnik, etc.—such wonderful names!], we still have a bewildering number of SI units and others in common use. This conversion table is one of the most extensive that you will find anywhere. The conversion factors are given to the last known decimal place. The old cgs units are included. Factors relating electromagnetic SI units to the corresponding cgs units are also given in terms of the speed of light, so that the reader may see how they are derived. For instance: coulomb $= c/10$ statcoulomb, farad $= 10^{-9}c^2$ statfarad, weber $= 10^8/c$ statweber. Similarly, factors involving π are given in terms of π. For instance: ampere-turn $= 4\pi/10$ gilbert, stilb $= \pi$ lambert, lumen $= 1/4\pi$ candela, oersted $= 10^3/4\pi$ ampere/meter.

To convert a unit on the left side to the corresponding unit on the right side, simply multiply by the conversion factor; to convert a unit on the right side to the corresponding one on the left side, simply divide by the conversion factor. For instance: meter $= 100$ centimeters; centimeter $= 1/100$ meter. Simple enough, right?

Table A2. Conversion Factors ($A = XB$; $B = A/X$; $c =$ *speed of light in vacuo in centimeters per second; an asterisk identifies an exact value*)

A	X	B
abampere	10*	ampere
	$2.997\,924\,58 \times 10^{10}$ ($= c$)	statampere
abampere-turn	10*	ampere-turn
	$12.566\,371$ ($= 4\pi$)	gilbert
abcoulomb	$0.002\,777\,8$	ampere-hour
	10*	coulomb
	$6.241\,506 \times 10^{19}$	electron charge
	$2.997\,924\,58 \times 10^{10}$ ($= c$)	statcoulomb
abfarad	10^{9*}	farad
	10^{15*}	microfarad
	$8.987\,552 \times 10^{20}$ ($= c^2$)	statfarad
abhenry	10^{-9*}	henry
	$1.112\,650 \times 10^{-21}$ ($= 1/c^2$)	stathenry
abmho	10^{9*}	mho
	$8.987\,552 \times 10^{20}$ ($= c^2$)	statmho
abohm	10^{-9*}	ohm
	$1.112\,650 \times 10^{-21}$ ($= 1/c^2$)	statohm
abtesla	10^{4*}	tesla
abvolt	$3.335\,641 \times 10^{-11}$ ($= 1/c$)	statvolt
	10^{-8*}	volt
abweber	1*	line
	1*	maxwell
	10^{-8*}	weber
acre	$4\,046.856\,423$	m^2
	$0.404\,685\,642\,3$	hectare
aeon	10^{9*}	year
albert	1*	einstein/second
ampere	0.1*	abampere
	1*	coulomb/second
	$2.997\,924\,58 \times 10^9$ ($= c/10$)	statampere
ampere-hour	360*	abcoulomb
	3600*	coulomb
	$0.037\,311\,7$	faraday
ampere-turn	0.1	abampere-turn
	$3.767\,303\,1 \times 10^{10}$	statgilbert
	$1.256\,637\,1$	gilbert
angstrom	10^{-8*}	cm
	10^{-10*}	m
	10^{-4*}	μm
are	$1\,076.391\,042$	ft^2
	100*	m^2
astronomical unit	$149\,597\,870.7$	km
	$8.316\,746$	light-minute
	$499.004\,784$	light-second
	$1.581\,284 \times 10^{-5}$	light-year
atmosphere	$1.013\,250*$	bar
	76*	cmHg (0°C)
	$1.013\,250 \times 10^6$	dyne/cm^2
	$1\,033.227$	g/cm^2
	$1.033\,227$	kg/cm^2
	$14.695\,95$	lb/in^2
	$10.332\,27$	mH_2O (3.98°C)
	760*	mmHg (0°C)

A	X	B
	$1.013\,250 \times 10^5$	N/m^2
	$1.013\,250 \times 10^5$	Pa
	760*	torr
atomic mass unit (u)	$931.494\,32 \times 10^6$	eV
	$1.660\,540\,2 \times 10^{-24}$	g
	$1.660\,540\,2 \times 10^{-27}$	kg
	931.494 32	MeV
avogram ($= 1/Av$)	$1.660\,540\,2 \times 10^{-24}$	g
bar	0.986 923 3	atmosphere
	10^{6*}	barye
	750.006 201 114	cmHg
	10^{6*}	$dyne/cm^2$
	1 019.716	g/cm^2
	10 197.16	kg/m^2
	14.503 77	lb/in^2
	10^{3*}	millibar
	750.062 011 14	mmHg
	1*	Nm^{-2}
	10^{5*}	Pa
barleycorn	$\frac{1*}{3}$	inch
	0.847	cm
barn	10^{-24*}	cm^2
	10^{-28*}	m^2
barrel	42*	gallon (U.S.)
	35*	gallon (British)
	158.987 295	liter
barye	$9.869\,233 \times 10^{-7}$	atmosphere
	10^{-6*}	bar
	1*	$dyne/cm^2$
	0.001 019 716	g/cm^2
	0.001*	millibar
	0.1*	Pa
bel	10*	decibel
becquerel	1*	dps (disintegrations per second)
	2.7×10^{-11}	curie
BeV	1*	GeV
Btu_{IT}	251.995 75	cal_{IT}
	1 055.0555 852 62*	J
bubnoff	1*	$m/10^6\,y$
	1*	$mm/1,000\,y$
bushel	8*	gallon (U.S., dry)
	35.239 072	liter
calorie (gram)	4.185 5	J
$calorie_{IT}$	$4.186\,8 \times 10^{7*}$	erg
	4.186 8*	J
candela	12.566 370 61 ($= 4\pi$)	lumen (at $\lambda = 555\,nm$)
	0.001 464 129 ($= 1/683$)	watt/steradian
	0.018 398 787	watt
candela cm^{-2} (stilb)	$3.141\,592\,653\ldots\,(= \pi)$	lambert
candela m^{-2} (nit)	1*	nit
carat	3.086 472 2	grain
	0.2*	g
Celsius	1*	centigrade
	1.8*	Fahrenheit
centigrade	1*	Celsius
	1.8*	Fahrenheit

(cont.)

Table A2. Conversion Factors ($A = XB$; $B = A/X$; $c =$ *speed of light in vacuo in centimeters per second; an asterisk identifies an exact value*)(*continued*)

A	X	B
centimeter	10^{8*}	angstrom
	0.1^*	decimeter
	0.393 700 79	inch
	0.01^*	m
	10^{4*}	μm
	10^*	mm
	10^{7*}	nm
cm s^{-1}	0.036^*	km/h
	36^*	m/h
circumference	$6.283\,185\,306\ldots(=2\pi)$	radian
cord	128^*	ft^3
	3.624 556	m^3
coulomb	0.1^*	abcoulomb
	$6.241\,506\,4 \times 10^{18}$	electron charge
	$1.036\,427 \times 10^{-5}$	faraday
	$2.997\,924\,58 \times 10^9\ (=c/10)$	statcoulomb
cubic centimeter	10^{-15*}	km^3
	0.001^*	liter
	10^{-6*}	m^3
cubic decimeter	1^*	liter
	0.001^*	m^3
cubic foot	$28\,316.846\,592^*$	cm^3
	$28.316\,846\,592^*$	liter
	0.028 316 846	m^3
cubic inch	$16.387\,064^*$	cm^3
	$0.016\,387\,064^*$	liter
	$1.638\,706\,4^*$	m^3
cubic kilometer	10^{15*}	cm^3
	10^{9*}	m^3
cubic meter	10^{6*}	cm^3
	35.314 666 72	ft^3
	10^{-9*}	km^3
	$1,000^*$	liter
	10^{9*}	mm^3
cubic millimeter	10^{-3*}	cm^3
	10^{-9*}	m^3
cubit (Egyptian)	52.4	cm
cubit (Greek)	52.67	cm
curie	$3.7 \times 10^{10*}$	dps (disintegrations per second)
dalton	1^*	atomic mass unit (u)
day (mean solar)	1.002 737 91	day (sidereal)
day (sidereal)	0.997 269 57	day (mean solar)
debye	10^{-18*}	statcoulomb-centimeter
decibel	0.115 129 255	neper
decimeter	10^*	cm
	0.1^*	m
degree	0.017 453 293	radian
dekameter	10^*	m
digit (Greek)	1.86	cm
	$1/40^*$	$2\,(\text{cubit}^2)^{1/2}$
digit (Roman)	1.84	cm
dram	1.771 845	gram
	$0.062\,5^*\ (=1/16)$	ounce (avdp.)

A	X	B
dyne	$10^{-5}*$	newton
dyne·cm	$1*$	erg
	$10^{-7}*$	N·m
dyne/cm^2	$9.869\,233 \times 10^{-7}$	atmosphere
	$10^{-6}*$	bar
	$1*$	barye
	$7.500\,620\,114 \times 10^{-4}$	mmHg
	$0.1*$	N/m^2
	$0.1*$	Pa
einstein	$1*$	Av hν
electron charge	$1.602\,177\,3 \times 10^{-20}$	abcoulomb
	$1.602\,177\,3 \times 10^{-19}$	coulomb
	$4.803\,206\,7 \times 10^{-10}$	statcoulomb
electron volt (eV)	$1.602\,177\,3 \times 10^{-12}$	erg
	$2.417\,988\,2 \times 10^{14}$	hertz
	$1.602\,177\,3 \times 10^{-19}$	J
	$10^{-6}*$	MeV
	$123.984\,25$	nm (λ)
	$1.073\,535\,4 \times 10^{-9}$	u
eon (see aeon)		
erg	$2.388\,459 \times 10^{-8}$	cal$_{IT}$
	$6.241\,506\,5 \times 10^{11}$	eV
	$2.389\,201 \times 10^{-8}$	g-cal
	$10^{-7}*$	J
	$2.389\,201 \times 10^{-11}$	kcal (g-cal)
	$2.388\,459 \times 10^{-11}$	kcal$_{IT}$
	$6.241\,506\,5 \times 10^{5}$	MeV
	$6.700\,531 \times 10^{2}$	u
erg/s	$10^{-7}*$	watt
erg·s or erg/Hz	$1.509\,189 \times 10^{26}$	Planck's constant
Fahrenheit	$0.555\,555\,555$	Celsius
	$0.555\,555\,555$	centigrade
farad	$10^{-9}*$	abfarad
	$1.000\,495$	farad (Int.)
	$10^{6}*$	microfarad
	$8.987\,552 \times 10^{11}\,(=10^{-9}c^2)$	statfarad
farad (Int.)	$0.999\,505$	farad
faraday	$96\,485.309$	coulomb
fathom	$6*$	foot
	$1.828\,8*$	m
femtometer	$10^{-13}*$	cm
	$1*$	fermi
	$10^{-15}*$	m
femtosecond	$10^{-15}*$	second
fermi	$10^{-13}*$	cm
	$10^{-15}*$	m
foot	$30.48*$	cm
	$0.304\,8*$	m
foot (Greek)	31.6	cm
	$3/5*$	cubit
foot (Roman)	29.49	cm
	$4*$	palm
foot-pound	$1.355\,818$	J
franklin	$3.335\,640\,952\ (=10/c)$	coulomb
	$1*$	statcoulomb

(cont.)

Table A2. Conversion Factors ($A = XB$; $B = A/X$; $c =$ *speed of light in vacuo in centimeters per second; an asterisk identifies an exact value*)(continued)

A	X	B
furlong	660*	ft
	201.168*	m
	0.125*	mile (statute)
	220*	yard
gal	1*	cm/s^2
	0.01*	m/s^2
	1 000*	mgal
gallon (British)	4.546 090	liter
	4*	quart (British)
	4.201	quart (U.S.)
gallon (U.S.)	3.785 411 784	liter
	8*	pint
	4*	quart
gallon (U.S., dry)	4.404 884	liter
gamma	10^{-5}*	gauss
	10^{-5}*	oersted
	10^{-9}*	tesla
gauss	1*	abtesla
	0.999 670	gauss (Int.)
	1*	line/cm^2
	1*	maxwell/cm^2
	1*	oersted
	$3.335 640 952 \times 10^{-11}$ ($= 1/c$)	stattesla
	10^{-4}*	tesla
	10^{-8}*	weber/cm^2
	10^{-4}*	weber/m^2
g·cal (see gram-calorie)		
GeV (see giga electron volts)		
giga electron volt (GeV)	10^9*	electron volt
gilbert	0.079 577 472 ($= 1/4\pi$)	abampere-turn
	0.795 774 72 ($= 10/4\pi$)	ampere-turn
	$2.997 924 58 \times 10^{10}$ ($= c$)	statgilbert
gilbert/cm	1*	oersted
grain	0.064 798 91	g
	64.798 91	mg
gram	$6.022 137 \times 10^{23}$	avogram
	5*	carat
	0.564 383	dram
	10^{-3}*	kg
	10^3*	mg
	10^{-6}*	ton (metric)
gram-calorie	4.185 5	J
gram/cm^2	0.000 967 841	atmosphere
	0.000 980 665	bar
	10*	kg/m^2
	0.014 223 343	lb/inch^2
	0.735 559 14	mmHg
hectare	2.471 053 815	acre
	100*	are
	10^4*	m^2
hectogram	100*	g
	0.1*	kg
hectoliter	100*	liter

A	X	B
hectometer	100*	m
henry	10^{9*}	abhenry
	$1.112\,650 \times 10^{-12}\ (=10^9/c^2)$	stathenry
horsepower (mechanical)	550.0*	foot-pound per second
	0.745\,700	kilowatt
	745.700	watt
hour (mean solar)	1/24*	day (mean solar)
	60*	minute
	3\,600*	second
	1.002\,737\,90	hour (sidereal)
hour (sidereal)	0.997\,269\,58	hour (mean solar)
	59.836\,175	minute (mean solar)
	3\,590.170\,5	second (mean solar)
inch	2.54*	cm
	0.083\,333\ldots	foot
	0.0254*	m
	1\,000*	mil
inch of Hg	25.4*	mmHg
jansky	10^{-26*}	$\text{W m}^{-2}\,\text{Hz}^{-1}$
joule	$9.478\,171\,209 \times 10^{-4}$	BTU_{IT}
	0.238\,845\,9	cal_{IT}
	10^{7*}	erg
	$6.241\,506\,48 \times 10^{18}$	eV
	0.238\,920\,1	g-cal
	$2.777\,78 \times 10^{-7}$	kilowatt-hour
	$6.241\,506\,48 \times 10^{12}$	MeV
karat (1/24 gold)	41.667	mg/g
kayser	$123.976\,6 \times 10^{-6}$	eV
	1*	wave/cm
kelvin	1*	Celsius
	1*	centigrade
kilocalorie/mole	$4.339\,313 \times 10^{-2}$	eV/particle
kilogram	1\,000*	g
	15\,432.361	grain
	35.273\,962	ounce (avdp.)
	2.204\,622\,6	pound (avdp.)
	10^{-3*}	ton (metric)
kilogram/cm^2	0.967\,841\,1	atmosphere
	0.980\,665	bar
	735.559\,14	mmHg (standard)
	28.959\,021	inch Hg (standard)
	14.223\,343	lb/inch^2
	98\,066.5	Pa
kilogram/m^2	$9.678\,41 \times 10^{-5}$	atmosphere
	$9.806\,65 \times 10^{-5}$	bar
	0.1*	g/cm^2
	0.001\,422\,334\,3	lb/inch^2
	0.073\,555\,914	mmHg (standard)
	9.806\,65	Pa
kilojoule/mole	$1.036\,427 \times 10^{-2}$	eV/particle
kilometer	10^{5*}	cm
	1\,000*	m
	0.539\,956\,80	mile (nautical)
	0.621\,371\,19	mile (statute)
kilometer/hour	27.777\,778	cm/s

(cont.)

Table A2. Conversion Factors ($A = XB$; $B = A/X$; c = speed of light in vacuo in centimeters per second; an asterisk identifies an exact value)(continued)

A	X	B
	0.277 777 78	m/s
	0.539 956 80	mile (nautical)/hour
	0.621 371 19	mile (statute)/hour
kiloton	$4.1868 \times 10^{12*}$	J
kilowatt	3 412.141 8	Btu_{IT}/hour
	$8.598 452 \times 10^5$	cal_{IT}/hour
	$8.601 123 \times 10^5$	g-cal/hour
	10^{10*}	erg/s
	1.341 02	horsepower (mechanical)
	$3.6 \times 10^{6*}$	J/h
	1 000*	watt
kilowatt-hour	3 412.141 8	Btu_{IT}
	$8.598 452 \times 10^5$	cal_{IT}
	$8.601 123 \times 10^5$	g-cal
	$3.6 \times 10^{6*}$	J
knot	51.444 444	cm/s
	1.687 809 9	ft/s
	1.852*	km/h
	0.514 444 444	m/s
	1*	mile (nautical)/hour
lambert (luminance)	0.318 309 886 ($= 1/\pi$)	candela/cm^2
	$0.318 309 886 \times 10^4$	nit
	1*	lumen/cm^2
langley	1*	g-cal/cm^2
league (statute)	4.828 032*	km
	3*	mile (statute)
light-year	63 239.727	astronomical unit
	$9.460 528 4 \times 10^{12}$	km
	0.306 594 89	parsec
line	1*	maxwell
	10^{-8*}	weber
line/cm^2	1*	gauss
	10^{-4*}	tesla
liter	1 000*	cm^3
	0.001*	m^3
	33.814 022	ounces (U.S., fluid)
	0.879 877 0	quart (British)
	1.056 688 2	quart (U.S., liquid)
liter/second	2.118 883 5	ft^3/min
	15.850 324	gallon (U.S., liquid)/min
lumen ($\lambda = 555$ nm)	0.079 577 472 ($= 1/4\pi$)	candela
	0.001 470 588 2	watt
lumen/cm^2	1*	lambert (luminance)
	1*	phot (illumination)
lux (illumination)	1*	lumen/m^2
	10^{-4*}	phot
maxwell	1*	abweber
	1*	gauss-cm^2
	1*	line
	$3.335 641 \times 10^{-11}$ ($= 1/c$)	statweber
	10^{-8*}	volt-second
	10^{-8*}	weber
maxwell/cm^2	1*	gauss

A	X	B
	$3.335\,641 \times 10^{-11}$ ($= 1/c$)	stattesla
megaton	$4.186\,8 \times 10^{15*}$	J
megamho/cm	0.001^*	abmho/cm
megohm	10^{6*}	ohm
	$1.112\,650 \times 10^{-6}$ ($= 10^{15}/c^2$)	statohm
meter	10^{10*}	angstrom
	100^*	cm
	$0.546\,806\,65$	fathom
	$3.280\,839\,9$	ft
	0.001^*	km
	$1\,000^*$	mm
meter/second	3.6^*	km/h
	$1.943\,844\,5$	knot
	$1.943\,844\,5$	mile (geographic)/hour
	$2.236\,936\,3$	mile (statute)/hour
metric ton (see ton)		
MeV (see million electron volts)		
mho	10^{-9*}	abmho
	1^*	ohm^{-1}
	$8.987\,552 \times 10^{11}$ ($= 10^{-9}c^2$)	statmho
microfarad	10^{-15*}	abfarad
	10^{-6*}	farad
	$8.987\,552 \times 10^5$ ($= 10^{-15}c^2$)	statfarad
microgram	10^{-6*}	g
	0.001^*	mg
micrometer	10^{4*}	angstrom
	10^{-4*}	cm
	10^{-6*}	m
	10^{-3*}	mm
micromicrofarad	10^{-12*}	farad
mil	10^{-3*}	inch
	$2.54 \times 10^{-3*}$	cm
mile (geographic)	1.000	min of latitude at 45°latitude
	$6\,076.115\,5$	ft
	1.852^*	km
	$1\,852^*$	m
	$1.150\,779$	mile (statute)
	1^*	nautical (Int.)
mile (Roman)	$1\,000^*$	step (Roman)
	$1\,475.0$	m
mile (U.S., statute)	$5\,280^*$	ft
	8^*	furlong
	$1.609\,344^*$	km
	$1\,609.344^*$	m
	$1\,760^*$	yard
mile (U.S., statute)/hour	44.704^*	cm/s
	88^*	ft/min
	$1.466\,666\,7$	ft/s
	$1.609\,344^*$	km/h
	$0.868\,976\,24$	knot
	$26.822\,4^*$	m/min
	$0.447\,04^*$	m/s
millibar	10^{-2*}	bar
	$1\,000^*$	barye
	10^{2*}	Pa
milligal	10^{-3*}	gal

(cont.)

Table A2. Conversion Factors ($A = XB$; $B = A/X$; $c =$ speed of light in vacuo in centimeters per second; an asterisk identifies an exact value)(continued)

A	X	B
milligram	0.005*	carat
	0.015 432 358	grain
	10^{-3}*	g
milligram/liter	1*	ppm (parts per million)
milliliter	1*	cm^3
	10^{-3}*	liter
millimeter	0.1*	cm
	0.001*	m
million electron volts (MeV)	1.602 177 3	J
mmHg	0.001 315 789 5	atm
	0.001 333 223 1	bar
	1 333.223 1	dyn/cm^2
	1.359 51	g/cm^2
	13.595 1	kg/m^2
	$7.500 617 \times 10^{-3}$	Pa
	1*	torr
month (lunar, synodic)	29.530 604 2	day (mean solar)
myriagram	10^4*	g
	10*	kg
nanometer	10*	angstrom
	10^{-7}*	cm
	10^{-9}*	m
	10^{-3}*	μm
	10^{-6}*	mm
neper	8.685 890	decibel
newton	10^5*	dyne
	7.233 013 871	poundal
	0.224 808 943	pound force
newton-meter	10^7*	dyn-cm
	1*	joule
	0.737 562 15	lb-ft
nit (luminance)	1*	candela/m^2
	$\pi \times 10^{-4}$*	lambert
oersted	79.577 472 ($= 10^3/4\pi$)	ampere/meter
	1*	emu
	$2.997 924 58 \times 10^{10}$ ($= c$)	esu
	1*	gauss
	1*	gilbert/cm
ohm	10^9*	abohm
	$1.112 650 \times 10^{-12}$ ($= 10^9/c^2$)	statohm
ounce (apoth., troy)	480*	grain
	31.103 476 8	gram
	1.097 142 9	ounce (avdp.)
	20*	pennyweight
	24*	scruple
ounce (avdp.)	16*	dram (avdp.)
	28.349 523	g
	0.911 458 3	ounce (apoth., troy)
ounce (U.S., fluid)	29.573 530	cm^3
	8*	dram (U.S., fluid)
palm (Roman)	4*	digit (Roman)
	7.37	cm
parsec	206 264.806	astronomical unit

A	X	B
	$30.856\,772 \times 10^{12}$	km
	$3.261\,633$	light-year
pascal	$9.869\,233 \times 10^{-6}$	atmosphere
	$10^{-5}*$	bar
	$10*$	barye
	$10*$	dyn/cm^2
	$1.019\,716 \times 10^{-5}$	kg/cm^2
	$1.450\,377 \times 10^{-4}$	$lb/in.^2$
	$7.500\,617 \times 10^{-3}$	mmHg
	$1*$	N/m^2
	$7.500\,617 \times 10^{-3}$	torr
peck	$0.25*$	bushel
	$8*$	quart (U.S., dry)
phot (illumination)	$1*$	$lumen/cm^2$
	10^4*	lux
pint (U.S., dry)	$0.5*$	quart (U.S., dry)
	$0.550\,610$	liter
pint (U.S., liquid)	$473.176\,48$	cm^3
	$8*$	gallon (U.S.)
	$0.473\,176\,48$	liter
	$16*$	ounce (U.S., fluid)
	$0.5*$	quart (U.S., fluid)
poise	$1*$	$g\,cm^{-1}s^{-1}$
	$0.1*$	pascal-second
pound (avdp.)	$256*$	dram (avdp.)
	$453.592\,37*$	g
	$0.453\,592\,37*$	kg
	$16*$	ounce (avdp.)
	$0.031\,081$	slug
pound (force)	$4.448\,221\,615\,260\,5$	newton
pound/inch^2	$0.068\,045\,96$	atm
	$70.306\,958$	g/cm^2
	$703.069\,58$	kg/m^2
	$6\,894.757$	Pa
poundal	$13\,825.495\,437\,6*$	dyn
	$1*$	$lb\text{-}ft/s^2$
	$0.138\,254\,954\,376*$	newton
quart (U.S., dry)	$1.101\,221$	liter
	$2*$	pint (dry)
quart (U.S., liquid)	$946.352\,946$	cm^3
	$256*$	dram (U.S., liquid)
	$0.25*$	gallon (U.S., liquid)
	$0.946\,352\,946$	liter
	$32*$	ounce (U.S., liquid)
	$2*$	pint (U.S., liquid)
quintal	10^5*	g
	$100*$	kg
radian	$57.295\,78$	degree
radiation dose	$10^{-2}*$	J/kg of tissue
roentgen	$2.58 \times 10^{-4}*$	coulomb/kg of air
second	$1*$	atomic second
	$1*$	ephemeris second
	$2.777\,777\,777\ldots \times 10^{-4}$	hour
slug	$14.593\,903$	kg
	$32.174\,0$	pound (advp.)

(*cont.*)

Table A2. Conversion Factors ($A = XB$; $B = A/X$; $c =$ *speed of light in vacuo in centimeters per second; an asterisk identifies an exact value*)(continued)

A	X	B
stadion (Greek)	600*	foot (Greek)
	189.60	m
stadium (Roman)	184.31	m
	125*	step
statampere	$3.335\,641 \times 10^{-11}$ $(=1/c)$	abampere
	$3.335\,641 \times 10^{-10}$ $(=10/c)$	ampere
statcoulomb	$3.335\,641 \times 10^{-11}$ $(=1/c)$	abcoulomb
	$3.335\,641 \times 10^{-10}$ $(=10/c)$	coulomb
	$2.081\,942 \times 10^{9}$	electron charge
statfarad	$1.112\,650 \times 10^{-21}$ $(=1/c^2)$	abfarad
	$1.112\,650 \times 10^{-12}$ $(=10^9/c^2)$	farad
stathenry	$8.987\,552 \times 10^{20}$ $(=c^2)$	abhenry
	$8.987\,552 \times 10^{11}$ $(=10^{-9}c^2)$	henry
statohm	$8.987\,552 \times 10^{20}$ $(=c^2)$	abohm
	$8.987\,552 \times 10^{11}$ $(=10^{-9}c^2)$	ohm
stattesla	$2.997\,924\,58 \times 10^{10}$ $(=c)$	gauss
	$2.997\,924\,58 \times 10^{6}$ $(=10^{-4}c)$	tesla
statvolt	$2.997\,924\,58 \times 10^{10}$ $(=c)$	abvolt
	$2.997\,924\,58 \times 10^{2}$ $(=10^{-8}c)$	volt
statweber	$2.997\,924\,58 \times 10^{10}$ $(=c)$	maxwell
	$2.997\,924\,58 \times 10^{2}$ $(=10^{-8}c)$	weber
step (Roman)	5*	foot (Roman)
	1.475	m
stere	1*	m^3
stilb	1*	candela/cm^2
stoke	1*	cm^2/s
sverdrup	10^{6*}	m^3/s
tesla	10^{4*}	abtesla
	10^{4*}	gauss
	10^{4*}	line/cm^2
	10^{4*}	maxwell/cm^2
	10^{4*}	oersted
	10^{-4}	weber/cm^2
	1*	weber/m^2
ton (long)	1 016.047	kg
	2 240*	pound (avdp.)
	1.016 047	ton (metric)
ton (metric)	10^{6*}	g
	1 000*	kg
ton (of refrig., U.S.)	72 574.8	kcal/day
ton (short)	907.185	kg
	2 000*	pound (avdp.)
	0.907 185	ton (metric)
torr	1*	mmHg
	133.322 4	Pa
volt	10^{8*}	abvolt
	0.003 335 641 $(=10^8/c)$	statvolt
watt	3.412 142	Btu$_{IT}$/hour
	860.112 29	g-cal/hour
	10^{7*}	erg/s
	1*	J/s
	0.001*	kilowatt
watt-hour	3 600*	joule

A	X	B
	0.860 112 29	kcal (g-cal)
	0.859 845 23	kcal$_{IT}$
weber	10^{8*}	abweber
	10^{8*}	line
	10^{8*}	maxwell
	0.003 335 641 ($= 10^8/c$)	statweber
	1*	volt-second
weber/cm^2	10^{8*}	gauss
	10^{8*}	oersted
	10^{4*}	tesla
yard (U.S.)	91.44*	cm
	0.5*	fathom
	3*	ft
	0.914 4*	m
year (sidereal) (1992)	365.256 365 66	day (mean solar)
	366.256 78	day (sidereal)
	31.558 149 993 × 10^6	second (ephemeris)
	1.000 037 721	year (tropical)
year (tropical) (1992)	365.242 193 42	day (mean solar)
	366.242 58	day (sidereal)
	31.556 926 464 × 10^6	second (ephemeris)
	0.999 961 229	year (sidereal)

Source: Emiliani (1988, pp. 236–47), updated 1992.

A3. EARTH: ASTRONOMICAL AND GEOPHYSICAL DATA

Mother Earth deserves special treatment, namely, an extensive table of astronomical and geophysical data. Here it is, a table that summarizes the data that are scattered through the text, and more. The values given are the most recent and are given to the last decimal place available.

Table A3. Earth: Astronomical and Geophysical Data

semimajor axis (= mean Sun–Earth distance)	= 1 astronomical unit (AU)
	= 149,597,870.7 km
	= 499.005 light-seconds
semiminor axis	= 149,576,881.1 km
	= 0.9998597 AU
eccentricity ($(a^2 - b^2)^{1/2}/a$ (= distance of either focus from center/ semimajor axis = distance of either focus from center in AU)	= 0.01675104
perihelion distance	= 0.98324896 AU
aphelion distance difference [(aphelion − perihelion)/mean]	= 1.01675104 AU
	= 3.35%
orbital velocity (mean)	= 29.784 km/s^{-1}
inclination of axis from normal to plane of orbit	= 23°26′28″
	= latitude of tropics
	= colatitude of polar circles

(cont.)

Table A3. Earth: Astronomical and Geophysical Data (*continued*)

inclination angle secular change	$= \sim 21°39'$ to $\sim 24°36'$
precessional angle	$= 2 \times$ inclination of axis from normal to orbital plane
	$= 46°52'28''$
precessional period	
general	$= 25{,}800\,\text{y}$
climatic (summer at perihelion to summer at perihelion)	$= 21{,}000\,\text{y}$
time	
sidereal year	$= 31{,}558{,}150\,\text{s}$
	$= 365.256\,366\,\text{d}_E$
tropical year	$= 31{,}556{,}926\,\text{s}$
	$= 365.242\,199\,\text{d}_E$
ephemeris day (d_E)	$= 86{,}400\,\text{s}$
ephemeris second (s_E)	$= 1/31{,}556{,}925.9747$ of tropical year 1900
atomic second (s)	$= 9{,}192{,}631{,}770$ periods of radiation emitted or absorbed by transition between the two hyperfine levels of the ground state of ^{133}Cs
	$=$ ephemeris second
radius	
polar (c)	$= 6{,}356.779\,\text{km}$
equatorial (mean) (a)	$= 6{,}378.139\,\text{km}$
mean $(a^2 c)^{1/3}$	$= 6{,}371.011\,\text{km}$
flattening	$= 21.360\,\text{km}$
	$= 0.00033$
	$= 0.33\%$
dimensions	
equator	$= 40{,}075.24\,\text{km}$
meridional quadrant	$= 10{,}002.02\,\text{km}$
polar circumference	$= 40{,}008.08\,\text{km}$
length of 1° of latitude at 45° N	$= 111.132\,\text{km}$
	$= 60.006$ nautical miles
length of 1′ of latitude at 45°	$= 1.00017$ nautical miles
	$= 1{,}852.31\,\text{m}$
[nautical mile now defined as equal to 1,852 m exactly]	
area	
land	$= 148.017 \cdot 10^6\,\text{km}^2$
ocean	$= 362.033 \cdot 10^6\,\text{km}^2$
total	$= 510.050 \cdot 10^6\,\text{km}^2$
mean elevation	
land	$= +840\,\text{m}$
ocean	$= -3{,}729\,\text{m}$
ocean basins	$= -4{,}500\,\text{m}$
mass	
solid Earth	$= 5.9737 \cdot 10^{24}\,\text{kg}$
ocean	$= 1.4 \cdot 10^{21}\,\text{kg}$
atmosphere	$= 5.1 \cdot 10^{18}\,\text{kg}$
total	$= 5.976 \cdot 10^{24}\,\text{kg}$
volume	
ocean	$= 1{,}349.929 \cdot 10^6\,\text{km}^3$
total	$= 1.0831 \cdot 10^{12}\,\text{km}^3$
density (mean)	$= 5.518\,\text{g}\,\text{cm}^{-3}$
	$= 5.518\,\text{kg}\,\text{dm}^{-3}$

internal structure	
continental-crust thickness	
mean	$= 35\,km$
range	$= 20–80\,km$
oceanic-crust thickness	$= 7\,km$
base of crust (Mohorovičić discontinuity)	
oceanic (mean)	$= -12\,km$
continental (mean)	$= -35\,km$
base of upper mantle	$= -670\,km$
mantle–core boundary	$= -2,885\,km$
outer-core/inner-core boundary	$= -5,170\,km$
gravitational acceleration (mean, at sea level) (g)	$= 9.81260\,m/s^2$
gravitational acceleration (standard, g_o)	$= 9.80665\,m/s^2$
magnetic field (mean)	$= 0.5$ gauss
	$= 0.5 \cdot 10^{-4}$ tesla
geothermal flux	$= 6.142 \cdot 10^{-2}\,W/m^2$ (cf. with $\sim 1\,kW/m^2$ at sea level with vertical Sun)
age	$= 4.6 \cdot 10^9\,y$
artificial satellites	
velocity to attain circular orbit (minimum for orbiting)	$= 7.91\,km/s$
velocities to attain elliptical orbits	$> 7.91\,km/s,\ < 11.19\,km/s$
velocity to attain parabolic orbit (minimum for escape)	$= 11.19\,km/s$
velocity for hyperbolic orbits	$> 11.19\,km/s$

Source: Emiliani (1987, pp. 260–3).

A4. ELEMENTARY PARTICLES

There is no life without elementary particles. Everybody should know something about elementary particles. There are now more than 100 "elementary particles," that is, particles that have their own peculiar personalities—just like humans. Table A4 presents the more important ones and their characteristics. When you meet your friends, you recognize them as individuals and treat them as such. You know, of course, that they are made out of cells, but you do not address yourself to those cells. Similarly, most elementary particles are individuals with specific characteristics, but they consist of subparticles that are even more elementary—the famous quarks, for instance. Other particles, such as the field particles and the leptons, are thought to be truly elementary. The fact is that the ultimate particle, the fundamental building block out of which all other particles can be made, has not yet been found, and in fact it may not even exist.

Table A4. Elementary Particles

Particle Name	Symbol	Mass			Mean Life (seconds)	Decay	
		MeV	u	Electron = 1		Principal Mode(s)	Percentage
Classons							
graviton	—	0	0	0	infinite	—	—
photon	γ	0	0	0	infinite	—	—
Gauge bosons							
—	$W^{\pm}$	80,800	86.7	158,120	—	$e^+ e^-$	—
—	Z^0	92,900	99.7	181,800	—	$\mu^+ \mu^-$	—
Leptons							
electron	e^-	0.51099906	0.005485799	1	infinite	—	—
positron	e^+	0.51099906	0.005485799	1	infinite	—	—
e neutrino	ν_e	0	0	0	infinite	—	—
e antineutrino	$\bar{\nu}_e$	0	0	0	infinite	—	—
muon	$\mu^{\pm}$	105.65932	0.11342892	206.76833	$2.19709 \cdot 10^{-6}$	$e^{\pm} \nu \bar{\nu}$	—
μ neutrino	ν_{μ}	0	0	0	infinite	—	—
μ antineutrino	$\bar{\nu}_{\mu}$	0	0	0	infinite	—	—
tauon	$\tau^{\pm}$	1784.2	1.91540	3491.5619	$3.4 \cdot 10^{-13}$	$\mu^{\pm} \nu \bar{\nu}$	18.5
						$e^{\pm} \nu \bar{\nu}$	16.2
						etc.	
τ neutrino		0	0	0	infinite	—	—
τ antineutrino		0	0	0	infinite	—	—
Nonstrange mesons							
pion	$\pi^{\pm}$	139.5673	0.1498304	273.12401	$2.6030 \cdot 10^{-8}$	$\mu^{\pm} \nu$	100
	π^0	134.9630	0.1448876	264.1137	$0.83 \cdot 10^{-16}$	$\gamma \gamma$	98.802
						$e^+ e^- \gamma$	1.198
eta	η	548.8	0.589156	1074	$0.75 \cdot 10^{-18}$	$\gamma \gamma$	39.0
						$\pi^0 \pi^0 \pi^0$	31.8
						$\pi^+ \pi^- \pi^0$	23.7
						$\pi^+ \pi^- \gamma$	4.9

Category / Name	Symbol	Mass (MeV)	Mass (amu)	Mass (m_e)	Mean life (s)	Decay modes	%
Strange mesons							
kaon	$K^\pm$	493.667	0.529969	966.0738	$1.2371\cdot10^{-8}$	$\mu^\pm\nu$	63.51
						$\pi^\pm\pi^0$	21.17
						etc.	
	$K^0, \bar{K}^0$	497.67	0.534266	973.9074		consists of 50% K^0_S + 50% K^0_L	
	K^0_S	497.67	0.534266	973.9074	$0.8923\cdot10^{-10}$	$\pi^+\pi^-$	68.61
						$\pi^0\pi^0$	31.39
						etc.	
	K^0_L	497.67	0.534266	973.9074	$5.183\cdot10^{-8}$	$e^\pm\pi^\mp\nu$	38.7
						$\pi^\pm\mu^\mp\nu$	27.1
						$\pi^0\pi^0\pi^0$	21.5
						$\pi^+\pi^-\pi^0$	12.4
						etc.	
Charmed nonstrange mesons							
—	$D^\pm$	1869.4	2.006867	3658.29	$0.92\cdot10^{-12}$	$K^0\bar{K}^0\ldots$	48
						$e^\pm\ldots$	19
						$K^-\ldots$	16
						$K^+\ldots$	6
						etc.	
—	$D^0\bar{D}^0$	1864.7	2.001821	3649.09	$4.4\cdot10^{-13}$	$K^-\ldots$	44
						$K^0\bar{K}^0\ldots$	33
						$K^+\ldots$	8
						etc.	
Charmed strange mesons							
—	$F^\pm$	1971	2.11594	3857	$1.9\cdot10^{-13}$	$\eta\pi^\pm$	?
						etc.	
Bottom mesons							
—	$B^\pm$	5270.8	5.6584	10314.6	$1.4\cdot10^{-12}$	$D^0\pi^\pm$	4.2
						etc.	
—	B^0	5274.2	5.6620	10321.3	$1.4\cdot10^{-12}$	$D^0\ldots$	80
						etc.	
Nonstrange baryons							
proton	$p^\pm$	938.2796	1.00727647	1836.1515	infinite(?)	—	—
neutron	n^0	939.5731	1.00866490	1838.6827	914 ± 6	$pe^-\bar{\nu}$	100
Strangeness-1 baryons							
lambda	Λ	1115.60	1.197636	2183.1557	$2.632\cdot10^{-10}$	$p\pi^-$	64.2
						$n\pi^0$	35.8

(cont.)

Table A4. Elementary Particles (continued)

Particle Name	Symbol	Mass			Mean Life (seconds)	Decay	
		MeV	u	Electron = 1		Principal Mode(s)	Percentage
Strangeness-1 baryons (cont.)							
sigma	Σ^+	1189.36	1.276820	2327.4992	$0.800 \cdot 10^{-10}$	$p\pi^0$	51.64
						$n\pi^+$	48.36
	Σ^0	1192.46	1.280148	2333.5657	$5.8 \cdot 10^{-20}$	$\Lambda\gamma$	100
	Σ^-	1197.34	1.285387	2343.1155	$1.482 \cdot 10^{-10}$	$n\pi^-$	100
Strangeness-2 baryons							
Xi	Ξ^0	1314.9	1.411592	2573.1727	$2.90 \cdot 10^{-10}$	$\Lambda\pi^0$	100
	Ξ^-	1321.32	1.418484	2585.7362	$1.641 \cdot 10^{-10}$	$\Lambda\pi^-$	100
Strangeness-3 baryons							
omega	Ω^-	1672.45	1.795434	3272.8745	$0.819 \cdot 10^{-10}$	ΛK^-	68.6
						$\Xi^0\pi^-$	23.4
						$\Xi^-\pi^0$	8.0
						etc.	
Nonstrange charmed baryons							
—	Λ_c^+	2282.0	2.4498	4465.7	$2.3 \cdot 10^{-13}$	$e^+ \ldots$	4.5
						$pK^-\pi^+$	2.2
						etc.	

Source: Emiliani (1987, pp. 266–7), updated 1992.

A5. ELEMENTS

The ancient Greek physicists, Thales, Anaximander, Anaximenes, Heraclitus, and Xenophanes, claimed that there was only one element. They argued only about which one. For Thales it was water, for Anaximander it was the ἄπειρον, for Anaximenes it was air, for Heraclitus it was fire, and for Xenophanes it was earth. Enter Empedocles. Forgetting the ἄπειρον, Empedocles opted for all of the above and concluded that were four elements: air, water, fire, and earth. These ancient scientists had one thing in common—they were all wrong. Strangely enough, nine elements were known in antiquity, but not recognized as such. They were carbon, copper, gold, iron, lead, mercury, and silver. A tenth element, platinum, was known to pre-Columbian South American Indians. The identification of elements started in earnest in the seventeenth century. Three elements, antimony, arsenic, and phosphorus, were discovered in that century. Sixteen elements were discovered in the eighteenth century, 51 in the nineteenth century, and 12 in the twentieth century, making the total of 92. Two of these elements, promethium and technetium, have no stable isotopes and do not occur on Earth—they were made by artificial bombardment in 1937 and 1941, respectively. In addition, 16 transuranic elements have been manufactured in accelerators, beginning with neptunium and plutonium in 1940 and continuing to this day. The heaviest element made thus far is element number 109. Elements 104 to 109 have not yet been officially named.

Table A5 lists the elements in alphabetical order, together with the element number (= number of protons in the nucleus), its symbol (in parentheses), and the origin of the name.

Table A5. **Elements**

No.	Element and Symbol	Origin of Name
89	actinium (Ac)	From ἀκτίς, Greek for *ray.*
13	aluminum (Al)	From *alumen.* Latin for *aluminum, sodium,* or *potassium sulfate* [(Na or K)Al(SO$_4$)$_2$·12H$_2$O].
95	americium (Am)	From *America.*
51	antimony (Sb)	From the late Latin *antimonium.* The symbol Sb derives from the earlier Latin name *stibium,* for *antimony.*
18	argon (Ar)	ἀργόν, Greek for *inert.*
33	arsenic (As)	From ἀρσενικός, Greek for *orpiment* (arsenic sulfide, As$_2$S$_3$).
85	astatine (At)	From ἄστατος, Greek for unstable.
56	barium (Ba)	From βαρύς, Greek for *heavy.*
97	berkelium (Bk)	Named after Berkeley, California.
4	beryllium (Be)	From the Greek βήρυλλος, for *beryl,* a beryllium aluminosilicate mineral [Be$_3$Al$_2$(Si$_6$O$_{18}$)]
83	bismuth (Bi)	From the Latin *bisemutum,* derived from the German *Wismut* (bismuth).
5	boron (B)	From the Latin *borax,* derived from the Persian *burah* [Na$_2$B$_4$O$_5$(OH)$_4$·8H$_2$O].
35	bromine (Br)	From βρῶμος, Greek for *stench.*
48	cadmium (Cd)	From the Latin *cadmia* (zinc ore).
20	calcium (Ca)	From *calx,* Latin for *lime* (CaO).
98	californium (Cf)	Named after California.
6	carbon (C)	From *carbo,* Latin for *charcoal.*
58	cerium (Ce)	From Ceres, Latin goddess of agriculture.
55	cesium (Cs)	From the Latin *caesium,* for *bluish gray,* the color of cesium's spectral lines.

(cont.)

Table A5. **Elements** (*continued*)

No.	Element and Symbol	Origin of Name
17	chlorine (Cl)	From χλωρός, Greek for *greenish yellow*.
24	chromium (Cr)	From χρῶμα, Greek for *color*.
27	cobalt (Co)	From the German name *kobalt*, for the metal.
29	copper (Cu)	From the Greek κύπρος, for Cyprus, the island where copper was mined.
96	curium (Cm)	Named after Marie Curie (1867–1934) and her husband Pierre Curie (1858–1906), French chemists.
66	dysprosium (Dy)	From the Greek δυσπρόσιτος, for *unapproachable*.
99	einsteinium (Es)	Named after Albert Einstein, German-American physicist (1879–1955).
68	erbium (Er)	From Ytterby (which means Outer Village), a Swedish village on a small island east of Stockholm where erbium-containing minerals were found.
63	europium (Eu)	From *Europa*.
100	fermium (Fm)	Named after Enrico Fermi, Italian-American physicist (1900–1954).
9	fluorine (F)	From the Latin *fluere*, for *to flow*, because fluorine minerals are used as fluxes in metallurgy.
87	francium (Fr)	From *France*.
64	gadolinium (Gd)	Named after Johan Gadolin, Finnish chemist (1760–1852).
31	gallium (Ga)	From *gallia*, Latin for *France*.
32	germanium (Ge)	From *Germania*, Latin for *Germany*.
79	gold (Au)	Old English. The symbol Au derives from the Latin *aurum*, for *gold*.
72	hafnium (Hf)	From *Hafnia*, the Latin name for Copenhagen.
2	helium (He)	From the Greek Ἥλιος, for *Sun*, as helium was discovered on the Sun before being detected on Earth.
67	holmium (Ho)	From *Holmia*, the Latin name for Stockholm.
1	hydrogen (H)	From the Greek ὕδωρ, for *water*, and γεννάειν, for *to generate*; hydrogen thus means *producer of water*, referring to its forming water with oxygen.
49	indium (In)	From the Latin *indigum*, for *indigo*, referring to the color of the spectral lines.
53	iodine (I)	From the Greek ἰοειδής, for *violet*, referring to the color of the substance.
77	iridium (Ir)	From the Latin *iris*, for *rainbow*, referring to the many colors that iridium gives in solution.
26	iron (Fe)	Old English. The symbol Fe derives from the Latin *ferrum*, for iron.
36	krypton (Kr)	From κρυπτόν, Greek for *hidden*.
57	lanthanum (La)	From λανθάνειν, Greek for *to elude*, referring to the difficulty of separating out the element.
103	lawrencium (Lr)	Named after Ernest Lawrence, American physicist, (1901–1958).
82	lead (Pb)	Old English. The symbol Pb derives from the Latin *plumbum* for *lead*.
3	lithium (Li)	From the Greek λίθος, for *stone*, referring to the mineral petalite from which it was isolated in 1817.
71	lutetium (Lu)	From *Lutetia*, the Latin name for Paris.
12	magnesium (Mg)	From Μαγνησία, ancient Greek city in western Turkey.
25	manganese (Mn)	From *manganese*, the Italian name for the element.
101	mendelevium (Md)	Named after Dmitri Mendeleév, Russian chemist (1834–1907).
80	mercury (Hg)	From the Latin god, Mercurius (the Greek Hermes) famous for his speed and slipperiness. The symbol Hg derives from the Latin *hydrargyrum*, for *liquid silver*.
42	molybdenum (Mo)	From the Greek μόλυβδος, for lead, referring to the high density (10.2 g/cm^3 at 20°C) of molybdenum.
60	neodymium (Nd)	From the Greek νέος, for *new*, and δίδυμος, for *twin*. Originally, neodymium and praseodymium were believed to be a single element called didymium (symbol: Di).

No.	Element and Symbol	Origin of Name
10	neon (Ne)	From $\nu\acute{\varepsilon}o\nu$, Greek for *new*.
93	neptunium (Np)	So named because it follows uranium, as the planet Neptune follows the planet Uranus.
28	nickel (Ni)	From the German *Kupfernickel*, for *Nickolaus's copper*, referring to niccolite (NiAs), which appeared to contain copper but did not.
41	niobium (Nb)	Named after Niobe, the daughter of Tantalus, who turned to stone at the loss of her children (see tantalum).
7	nitrogen (N)	From the Greek $\nu\acute{\iota}\tau\rho o\nu$, for *soda* (carbonate of soda, Na_2CO_3), and $\gamma\varepsilon\nu\nu\acute{\alpha}\varepsilon\iota\nu$, for *to generate*.
102	nobelium (Nb)	Named after Alfred Nobel (1833–1896), Swedish chemist and inventor, who deeded his fortune to finance the Nobel Prizes.
76	osmium (Os)	From the Greek $\partial\sigma\mu\acute{\eta}$ for *odor*, referring to the odor of osmium tetroxide (OsO_4).
8	oxygen (O)	From the Greek $\partial\xi\acute{\upsilon}\varsigma$, for *sharp* (meaning acid), and $\gamma\varepsilon\nu\nu\acute{\alpha}\varepsilon\iota\nu$, *to generate*, because oxygen combines with hydrogen and nonmetals to produce acids [for example, carbonic acid (H_2CO_3), nitric acid (HNO_3), silicic acid (H_4SiO_4), phosphoric acid (H_3PO_4), sulfuric acid (H_2SO_4)].
46	palladium (Pd)	From the asteroid Pallas, discovered in 1801, shortly before the discovery of palladium (1803).
15	phosphorus (P)	From the Greek $\varphi\tilde{\omega}\varsigma$, for *light*, and $\varphi o\rho\varepsilon\acute{\upsilon}\varsigma$, for *bearer*.
78	platinum (Pt)	From the Spanish *platina*, meaning *small silver* (*plata* is Spanish for *silver*).
94	plutonium (Pu)	So named because it follows neptunium, as the planet Pluto follows the planet Neptune.
84	polonium (Po)	Named after *Polonia*, Latin for *Poland*.
19	potassium (K)	From the English potash or pot ash, an ash residue derived by evaporating the liquid obtained by leaching wood ashes in iron pots. The symbol K derives from the Latin *kalium*, for *potassium* (from the Arabic *qily*, for *saltwort*, a plant from whose ashes potassium can be leached out).
59	praseodymium (Pr)	From the Greek $\pi\rho\acute{\alpha}\sigma\iota o\varsigma$, for *leek green*, referring to the color of the salts, and $\delta\acute{\iota}\delta\upsilon\mu o\varsigma$, for *twin*. Originally, neodymium and praseodymium were believed to be a single element called didymium (symbol: Di).
61	promethium (Pm)	Named after Prometheus, who stole fire from Olympus and gave it to humankind.
91	protactinium (Pa)	So named because it decays into actinium by α decay.
88	radium (Ra)	From *radius*, Latin for *ray*.
86	radon (Rn)	From the root *rad-* of radium plus the ending *-on* characteristic of inert gases.
75	rhenium (Re)	From *Rhenus*, Latin for the Rhine River. Rhenium was discovered by the German chemist Walter Noddack (1893–1960) and associates in 1925.
45	rhodium (Rh)	From the Greek $\acute{\rho}\acute{o}\delta o\nu$, for *rose*, after the color of the compound.
37	rubidium (Rb)	From the Latin *rubidus*, for *red*, referring to the color of the spectral lines.
44	ruthenium (Ru)	From the Latin *Ruthenia*, for *Russia*, because the element was discovered in minerals from the Urals.
62	samarium (Sm)	Named after Vasilii Erafovich Samarski-Bykhovets (1803–1870), Russian mining engineer.
21	scandium (Sc)	From *Scandia*, Latin for *Scandinavia*.
34	selenium (Se)	From the Greek $\sigma\varepsilon\lambda\acute{\eta}\nu\eta$, for *moon*, so named to contrast with the related element tellurium, named after *tellus*, Latin for *earth*.
14	silicon (Si)	From *silex*, Latin for *flint*.
47	silver (Ag)	Old English. The symbol Ag derives from *argentum*, Latin for *silver*.

(cont.)

Table A5. Elements (*continued*)

No.	Element and Symbol	Origin of Name
11	sodium (Na)	From *soda*, Latin for *sodium carbonate* (Na_2CO_3). The symbol Na derives from the Arabic *natrun*, for *soda*.
38	strontium (Sr)	Named after Strontian, a mining village in Scotland where the element was discovered.
16	sulfur (S)	From the Latin *sulphur*.
73	tantalum (Ta)	From the mythical king Tantalus, son of Zeus and father of Niobe, who was condemned to stand in water up to his chin without being able to drink it (the water would recede if he tried), and with fruits dangling above his head, which also would recede if he tried to reach them.
43	technetium (Tc)	From τεχνητός Greek, for *artificial*.
52	tellurium (Te)	From *Tellus*, Latin for *earth*.
65	terbium (Tb)	From Ytterby (see erbium) where terbium-containing minerals were found.
81	thallium (Tl)	From the Latin *thallus*, referring to the green color of the spectral lines of the element.
90	thorium (Th)	Named after Thor, the god of thunder in Norse mythology.
69	thulium (Tm)	Named after Thule, the ancient name of Scandinavia.
50	tin (Sn)	Old English. The chemical symbol Sn derives from the Latin *stannum*, for *tin*.
22	titanium (Ti)	From the Greek Τιτάν, any of the children of Uranus (Heaven) and Gaia (Earth) in Greek mythology.
74	tungsten (W)	From the Swedish *tung*, for *heavy*, and *stem*, for *stone*. The symbol W derives from the German name *wolfram*, for *tungsten*.
92	uranium (U)	Named after the Greek god Uranus, father of Cronos and the other Titans and grandfather of Zeus.
23	vanadium (V)	Named after Vanadis, Norse goddess of love and beauty.
54	xenon (Xe)	From the Greek ξένον, for *stranger*.
70	ytterbium (Yb)	From Ytterby (see erbium) where ytterbium-containing minerals were found.
39	yttrium (Y)	From Ytterby (see erbium) where yttrium-containing minerals were found.
30	zinc (Zn)	From the German *Zink*, for *zinc*.
40	zirconium (Zr)	From the Persian *zargun*, for *gold-colored*.

Source: Emiliani (1988, pp. 248–53).

A6. ELEMENTS: ELECTRON STRUCTURE

Table A6 lists the elements in order of atomic number and shows the electron structure of each element when in the ground state. The first two lines at the top identify the shell, by letter (first line) and by the principal quantum number n (second line). The third line gives the numbers of l quantum states. The fourth line gives the numbers of m_1 quantum states (orbitals), and the fifth line gives the corresponding chemical notation. Note that the lanthanides, a string of elements between lanthanum (no. 57) and lutetium (no. 71), have similar numbers of electrons in the O and P shells, while still filling the N shell. Also note that the actinides (from thorium, no. 90, to lawrencium, no. 103) have similar numbers of electrons in the P and Q shells, while still filling the O shell. Because these two groups of elements have the same numbers of electrons in their outer shells, the elements in either group have very similar chemical properties. In the ground state, an orbital is left unfilled when its energy is higher than the energy of a higher orbital. If you are looking for an element in this table, first find its atomic number in Table A5.

Table A6. Elements: Electron Structure

Shell		K	L		M			N				O				P				Q			
n quantum no.		1	2		3			4				5				6				7			
Subshell (l quantum no.)		0	0	1	0	1	2	0	1	2	3	0	1	2	3	0	1	2	3	0	1	2	3
Orbital (m_l quantum no.)		1	1	3	1	3	5	1	3	5	7	1	3	5	7	1	3	5	7	1	3	5	7
Chemical notation		s	s	p	s	p	d	s	p	d	f	s	p	d	f	s	p	d	f	s	p	d	f
Atomic No.	Element Symbol																						
1	H	1																					
2	He	2																					
3	Li	2	1																				
4	Be	2	2																				
5	B	2	2	1																			
6	C	2	2	2																			
7	N	2	2	3																			
8	O	2	2	4																			
9	F	2	2	5																			
10	Ne	2	2	6																			
11	Na	2	2	6	1																		
12	Mg	2	2	6	2																		
13	Al	2	2	6	2	1																	
14	Si	2	2	6	2	2																	
15	P	2	2	6	2	3																	
16	S	2	2	6	2	4																	
17	Cl	2	2	6	2	5																	
18	Ar	2	2	6	2	6																	
19	K	2	2	6	2	6		1															
20	Ca	2	2	6	2	6		2															
21	Sc	2	2	6	2	6	1	2															
22	Ti	2	2	6	2	6	2	2															
23	V	2	2	6	2	6	3	2															
24	Cr	2	2	6	2	6	5	1															
25	Mn	2	2	6	2	6	5	2															
26	Fe	2	2	6	2	6	6	2															
27	Co	2	2	6	2	6	7	2															
28	Ni	2	2	6	2	6	8	2															
29	Cu	2	2	6	2	6	10	1															
30	Zn	2	2	6	2	6	10	2															
31	Ga	2	2	6	2	6	10	2	1														
32	Ge	2	2	6	2	6	10	2	2														
33	As	2	2	6	2	6	10	2	3														
34	Se	2	2	6	2	6	10	2	4														
35	Br	2	2	6	2	6	10	2	5														
36	Kr	2	2	6	2	6	10	2	6														
37	Rb	2	2	6	2	6	10	2	6			1											
38	Sr	2	2	6	2	6	10	2	6			2											
39	Y	2	2	6	2	6	10	2	6	1		2											
40	Zr	2	2	6	2	6	10	2	6	2		2											
41	Nb	2	2	6	2	6	10	2	6	4		1											
42	Mo	2	2	6	2	6	10	2	6	5		1											

(cont.)

Table A6. Elements: Electron Structure (*continued*)

Shell	K	L		M			N				O				P				Q			
n quantum no.	1	2		3			4				5				6				7			
Subshell (*l* quantum no.)	0	0	1	0	1	2	0	1	2	3	0	1	2	3	0	1	2	3	0	1	2	3
Orbital (*m$_l$* quantum no.)	1	1	3	1	3	5	1	3	5	7	1	3	5	7	1	3	5	7	1	3	5	7
Chemical notation	s	s	p	s	p	d	s	p	d	f	s	p	d	f	s	p	d	f	s	p	d	f
Atomic No. / Element Symbol																						
43 Tc	2	2	6	2	6	10	2	6	6		1											
44 Ru	2	2	6	2	6	10	2	6	7		1											
45 Rh	2	2	6	2	6	10	2	6	8		1											
46 Pd	2	2	6	2	6	10	2	6	10													
47 Ag	2	2	6	2	6	10	2	6	10		1											
48 Cd	2	2	6	2	6	10	2	6	10		2											
49 In	2	2	6	2	6	10	2	6	10		2	1										
50 Sn	2	2	6	2	6	10	2	6	10		2	2										
51 Sb	2	2	6	2	6	10	2	6	10		2	3										
52 Te	2	2	6	2	6	10	2	6	10		2	4										
53 I	2	2	6	2	6	10	2	6	10		2	5										
54 Xe	2	2	6	2	6	10	2	6	10		2	6										
55 Cs	2	2	6	2	6	10	2	6	10		2	6			1							
56 Ba	2	2	6	2	6	10	2	6	10		2	6			2							
57 La	2	2	6	2	6	10	2	6	10		2	6	1		2							
58 Ce	2	2	6	2	6	10	2	6	10	2	2	6			2							
59 Pr	2	2	6	2	6	10	2	6	10	3	2	6			2							
60 Nd	2	2	6	2	6	10	2	6	10	4	2	6			2							
61 Pm	2	2	6	2	6	10	2	6	10	5	2	6			2							
62 Sm	2	2	6	2	6	10	2	6	10	6	2	6			2							
63 Eu	2	2	6	2	6	10	2	6	10	7	2	6			2							
64 Gd	2	2	6	2	6	10	2	6	10	7	2	6	1		2							
65 Tb	2	2	6	2	6	10	2	6	10	9	2	6			2							
66 Dy	2	2	6	2	6	10	2	6	10	10	2	6			2							
67 Ho	2	2	6	2	6	10	2	6	10	11	2	6			2							
68 Er	2	2	6	2	6	10	2	6	10	12	2	6			2							
69 Tm	2	2	6	2	6	10	2	6	10	13	2	6			2							
70 Yb	2	2	6	2	6	10	2	6	10	14	2	6			2							
71 Lu	2	2	6	2	6	10	2	6	10	14	2	6	1		2							
72 Hf	2	2	6	2	6	10	2	6	10	14	2	6	2		2							
73 Ta	2	2	6	2	6	10	2	6	10	14	2	6	3		2							
74 W	2	2	6	2	6	10	2	6	10	14	2	6	4		2							
75 Re	2	2	6	2	6	10	2	6	10	14	2	6	5		2							
76 Os	2	2	6	2	6	10	2	6	10	14	2	6	6		2							
77 Ir	2	2	6	2	6	10	2	6	10	14	2	6	7		2							
78 Pt	2	2	6	2	6	10	2	6	10	14	2	6	9		1							
79 Au	2	2	6	2	6	10	2	6	10	14	2	6	10		1							
80 Hg	2	2	6	2	6	10	2	6	10	14	2	6	10		2							
81 Tl	2	2	6	2	6	10	2	6	10	14	2	6	10		2	1						
82 Pb	2	2	6	2	6	10	2	6	10	14	2	6	10		2	2						
83 Bi	2	2	6	2	6	10	2	6	10	14	2	6	10		2	3						
84 Po	2	2	6	2	6	10	2	6	10	14	2	6	10		2	4						
85 Al	2	2	6	2	6	10	2	6	10	14	2	6	10		2	5						
86 Rn	2	2	6	2	6	10	2	6	10	14	2	6	10		2	6						
87 Fr	2	2	6	2	6	10	2	6	10	14	2	6	10		2	6			1			

Shell	K	L		M			N				O				P				Q			
n quantum no.	1	2		3			4				5				6				7			
Subshell (*l* quantum no.)	0	0	1	0	1	2	0	1	2	3	0	1	2	3	0	1	2	3	0	1	2	3
Orbital (m_l quantum no.)	1	1	3	1	3	5	1	3	5	7	1	3	5	7	1	3	5	7	1	3	5	7
Chemical notation	s	s	p	s	p	d	s	p	d	f	s	p	d	f	s	p	d	f	s	p	d	f

Atomic No.	Symbol	K s	L s	L p	M s	M p	M d	N s	N p	N d	N f	O s	O p	O d	O f	P s	P p	P d	P f	Q s	Q p	Q d	Q f
88	Ra	2	2	6	2	6	10	2	6	10	14	2	6	10		2	6			2			
89	Ac	2	2	6	2	6	10	2	6	10	14	2	6	10		2	6	1		2			
90	Th	2	2	6	2	6	10	2	6	10	14	2	6	10		2	6	2		2			
91	Pa	2	2	6	2	6	10	2	6	10	14	2	6	10	2	2	6	1		2			
92	U	2	2	6	2	6	10	2	6	10	14	2	6	10	3	2	6	1		2			
93	Np	2	2	6	2	6	10	2	6	10	14	2	6	10	4	2	6	1		2			
94	Pu	2	2	6	2	6	10	2	6	10	14	2	6	10	6	2	6			2			
95	Am	2	2	6	2	6	10	2	6	10	14	2	6	10	7	2	6			2			
96	Cm	2	2	6	2	6	10	2	6	10	14	2	6	10	7	2	6	1		2			
97	Bk	2	2	6	2	6	10	2	6	10	14	2	6	10	9	2	6			2			
98	Cf	2	2	6	2	6	10	2	6	10	14	2	6	10	10	2	6			2			
99	Es	2	2	6	2	6	10	2	6	10	14	2	6	10	11	2	6			2			
100	Fm	2	2	6	2	6	10	2	6	10	14	2	6	10	12	2	6			2			
101	Md	2	2	6	2	6	10	2	6	10	14	2	6	10	13	2	6			2			
102	No	2	2	6	2	6	10	2	6	10	14	2	6	10	14	2	6			2			
103	Lr	2	2	6	2	6	10	2	6	10	14	2	6	10	14	2	6	1		2			
104	—	2	2	6	2	6	10	2	6	10	14	2	6	10	14	2	6	2		2			

Source: Emiliani (1987, pp. 273–5).

A7. GEMS

Many people have killed for gems, and others have spent long stretches in prison. The human fascination with gems dates from prehistory. Table A7 lists the gems in common use, together with chemical composition, mineral properties, color, and coloring agent. This table will be useful to you if you plan to get engaged or get married. It will clarify the confusion that exists, for instance, between emerald and oriental emerald and between topaz and oriental topaz (not to mention Brazilian emerald, which is something else again).

People relate certain gems to the months in which they were born. These are the generally accepted birthstones:

Month	Gem	Month	Gem
January	Garnet	July	Ruby
February	Amethyst	August	Peridot
March	Aquamarine	September	Sapphire
April	Diamond	October	Tourmaline
May	Emerald	November	Topaz
June	Alexandrite	December	Turquoise

Gems, of course, also have magic properties. If you plan to drink and drive, be sure to wear an amethyst. The name *amethyst*, in fact, derives from the Greek ἀμέθυος, which means *not drunk*. Tell that to the officer while he drags you to jail.

Table A7. Gems

Name	Composition	Crystal System[a]	G[b]	H[c]	n[d]	color	Coloring Agent
Amber	resin	A	2–2.5	1.05	1.54	yellow–brown	organics
Beryl	$Be_3Al_2(Si_6O_{18})$	H	2.64–2.8	$7\frac{1}{2}$–8	1.57–1.61		Fe^{2+}, Fe^{3+}
aquamarine	—	—	—	—	—	pale blue	Fe^{3+}
emerald	—	—	—	—	—	green	Cr^{3+}
heliodore	—	—	—	—	—	yellow	Fe^{3+}
morganite	—	—	—	—	—	pink	Mn^{2+}
Chrysoberyl	$BeAl_2O_4$	O	3.65–3.8	$8\frac{1}{2}$	1.75		
alexandrite	—	—	—	—	—	green–red	Cr^{3+}
cat's eye	—	—	—	—	—	opalescent	—
Corundum	Al_2O_3	T	3.99	9	1.77		—
oriental amethyst	—	—	—	—	—	purple	Fe^{3+}
oriental emerald	—	—	—	—	—	green	Fe^{2+}, Ti^{4+}
oriental topaz	—	—	—	—	—	yellow	Fe^{2+}, Fe^{3+}
ruby	—	—	—	—	—	red	Cr^{3+}
sapphire	—	—	—	—	—	blue	Fe^{2+}, Ti^{4+}
Diamond	C	C	3.51	10	2.42	white	—
	—	—	—	—	—	blue	—
	—	—	—	—	—	yellow	—
Garnets							
almandine	$Fe_3Al_2Si_3O_{12}$	C	4.32	7	1.83	red	Fe^{2+}
andradite	$Ca_3Fe_2Si_3O_{12}$	C	3.86	7	1.89	reddish	Fe^{3+}
grossularite	$Ca_3Al_2Si_3O_{12}$	C	3.59	$6\frac{1}{2}$	1.73	white–brown	V^{3+}
pyrope	$Mg_3Al_2Si_3O_{12}$	C	3.58	7	1.71	red	—
spessartine	$Mn_2Al_2Si_3O_{12}$	C	4.19	7	1.80	dark red	Mn^{2+}
uvarovite	$Ca_3Cr_2Si_3O_{12}$	C	3.90	$7\frac{1}{2}$	1.87	green	—
Jade							
jadeite	$NaAl(SiO_3)_2$	M	3.3	$6\frac{1}{2}$–7	—	green	—
nephrite							
actinolite	$Ca_2(Mg,Fe)_5Si_8O_{22}(OH)_2$	M	3.1–3.3	5–6	1.65	green	—
tremolite	$Ca_2Mg_5Si_8O_{22}(OH)_2$	—	3.0–3.2	5–6	1.61	white–gray	—
Lazurite (lapis lazuli)	$(Na,Ca)_8(AlSiO_4)_6\cdot$ $(SO_4, S, Cl)_2$	C	2.40–2.45	5–$5\frac{1}{2}$	1.5	blue	S_3

			Density	Hardness	R.I.	Color	Coloring agent
Malachite	$Cu_2CO_3(OH)_2$	M	3.90–4.03	$3\frac{1}{2}$–4	1.88	green	—
Mother of pearl	$CaCO_3$ (aragonite)	O	2.95	$3\frac{1}{2}$–4	1.68	iridescent	—
Olivine	$(Mg, Fe)_2SiO_4$	O	3.27–4.37	$6\frac{1}{2}$–7	1.69		
chrysolite	—	—	—	—	—	yellow–green	Fe^{2+}
peridot						green	—
Opal	$SiO_2 \cdot nH_2O$	A	2.0–2.25	5–6	1.44	iridescent	organics
Pearl	$CaCO_3$ (aragonite)	O	2.95	$3\frac{1}{2}$–4	1.68	iridescent	—
Quartz	SiO_2	T,H	2.65	7	1.54		
eucrystalline							
amethyst	—	—	—	—	—	purple	Fe^{3+}
citrine	—	—	—	—	—	yellow	Fe^{3+}
rock crystal	—	—	—	—	—	colorless	—
rose quartz	—	—	—	—	—	pink	—
smoky	—	—	—	—	—	brown	Al^{3+}
microcrystalline							
agate	—	—	—	—	—	variegated	—
chalcedony	—	—	—	—	—	translucent	—
cornelian	—	—	—	—	—	red	Fe oxides
flint	—	—	—	—	—	gray	—
jasper	—	—	—	—	—	red–green	Fe oxides
onyx	—	—	—	—	—	banded	—
Spinel	$MgAl_2O_4$	C	3.5–4.1	8	1.72	blue	Co^{2+}
	—	—	—	—	—	green	—
	—	—	—	—	—	red	—
	—	—	—	—	—	yellow	Co^{2+}
Topaz	$Al_2SiO_4(OH, F)_2$	O	3.4–3.6	8	1.61–1.63	blue	—
	—	—	—	—	—	colorless	—
	—	—	—	—	—	green	—
	—	—	—	—	—	red	—
	—	—	—	—	—	yellow	—
Tourmaline	$(Na, Ca)(Li, Mg, Al)$ $\cdot (Al, Fe, Mn)_6(BO_3)_3$ $\cdot Si_6O_{18}(OH)_4$	T	3.0–3.25	7–$7\frac{1}{2}$	1.64–1.68		
achroite	—	—	—	—	—	colorless	—
Brazilian							
emerald	—	—	—	—	—	green	—
indicolite	—	—	—	—	—	blue	—

(cont.)

Table A7. Gems (*continued*)

Name	Composition	Crystal System^a	G^b	H^c	n^d	color	Coloring Agent
Tourmaline (*cont.*)							
peridot of Ceylon	—	—	—	—	—	yellow	—
rubellite	—	—	—	—	—	red	Mn^{3+}
Turquoise	$CuAl_6(PO_4)_4(OH)_8 \cdot 5H_2O$	Tc	2.6–2.8	6	1.62	azure	Cu^{2+}
Zircon	$ZrSiO_4$	Tt	4.68	$7\frac{1}{2}$	1.92–1.96	colorless	—
	—	—	—	—	—	blue	—

^{a} Crystal systems; C, cubic; H, hexagonal; M, monoclinic; O, orthorhombic; T, trigonal; Tc, triclinic; Tt, tetragonal; A, amorphous.

^{b} Density relative to that of water at 3.98°C.

^{c} Hardness (Mohs scale).

^{d} Refractive index.

A8. GEOLOGICAL TIME SCALE

Treat Table A8 with awe and respect. It has taken a lot of effort by many people from many disciplines to produce it. Until well into the nineteenth century most scientists took for granted the biblical story of creation and believed that the Earth was 6,000 years old. As the century progressed, so did the study of geology. Darwin, in his famous book *The Origin of Species*, published in 1859, estimated the rate of denudation of the Weald anticline in southeastern England and concluded that more than 300 million years had elapsed since the Cretaceous.

A much different figure was obtained from the total thickness of rapidly accumulating sediments (sandstones and graywackes) deposited since the Cambrian, which was estimated to be about 150,000 m. The rate of silting in harbors close to large rivers, such as at Ravenna in northern Italy, suggested an average rate of sedimentation of 1 m per 1,000 y. The time elapsed since the beginning of the Cambrian would then be $150 \cdot 10^6$ y if sedimentation had been continuous.

An estimate for the age of the ocean (and therefore of the Earth itself) was based on the salinity of the ocean. Sodium and chlorine elutriated from continental rocks by rain and percolating waters, and brought to the ocean by rivers and groundwater, accumulate in the ocean because they are not removed by biological processes, as calcium is, for instance. The sodium content of sea water is $10.8 \, \text{g/L} = 10.8 \, \text{kg/m}^3$. The volume of the ocean is $1.356 \cdot 10^9 \, \text{km}^3 = 1.356 \cdot 10^{18} \, \text{m}^3$, which means that the ocean contains $1.5 \cdot 10^{19} \, \text{kg}$ of sodium. The flux of fresh water to the ocean is $3 \cdot 10^{16} \, \text{kg/y}$ (see Fig. 13.22). River water contains 6.4 mg of sodium per liter (world average). The flux of sodium is therefore $3 \cdot 10^{16} \times 6.3 \times 10^{-6} = 1.9 \cdot 10^{11} \, \text{kg/y}$, and the age of the ocean is $1.5 \cdot 10^{19}/1.9 \cdot 10^{11} = 0.8 \cdot 10^8$ y. A similar argument for chlorine (total in ocean $= 2.63 \cdot 10^{19} \, \text{kg}$; world average concentration in river water $= 7.8 \, \text{mg/L}$; supply to ocean $= 2.34 \cdot 10^{11} \, \text{kg/y}$) yields an age of $1.1 \cdot 10^8$ y for the ocean.

In the second half of the nineteenth century, Lord Kelvin applied himself to the problem of the age of the Earth by assuming that the Earth had originated as a molten sphere. He calculated how long it would take for the heat flux from the interior to decrease to the present value as the crust cooled off and solidified from the outside in. He concluded, in 1899, that the age of the Earth was 20 to 40 million years. That age happened to be close to the estimate of $20 \cdot 10^6$ y for the age of the Sun, based on the notion, then prevalent, that the observed solar energy output derived from gravitational shrinking. The glaring discrepancy between Lord Kelvin's estimate and the geological estimates remained unexplained.

The discovery of radioactivity finally produced methods by which one could actually *measure* the ages of geological formations and deduce the age of the Earth. The estimates grew longer and longer: from about $2 \cdot 10^9$ y before World War I to $3.5 \cdot 10^9$ y in the 1930s. Eventually, a firm age, $4.6 \cdot 10^9$ y, was established by Clair Patterson at the University of Chicago in 1956. It appears that all previous estimates were far off, and most of them on the low side. The reason that estimates based on the salinity of the ocean were far too low is that there are thick salt deposits buried in the continents and, as we know today, also under the Mediterranean. In addition, again as we know today, seawater subducts in subduction zones and is recycled. Large quantities of sodium and chlorine are trapped at any given time in the mantle. A steady-state equilibrium between subduction along subduction zones and resurfacing in volcanic vents is likely to have been established long before the Cambrian, maintaining the salinity of the ocean more or less constant from then on.

Clair Patterson's figure of $4.6 \cdot 10^9$ y for the age of the Earth has not changed. What has changed is the estimate for the age of the universe itself. The Big Bang theory was proposed by George Gamow in 1948, but it was not confirmed until 1964, with the discovery of the microwave background radiation by Arno Penzias and Robert Wilson. It was only then that scientists came to believe that the universe indeed had had a beginning. Because of the uncertainties surrounding the value of the Hubble constant, the age of the universe can be estimated only within broad limits—between 10 and 20 billion years. Assuming a probable value for the Hubble constant of 18 km/s per 10^6 l.y., we get an age of $16.5 \cdot 10^9$ y. The decimal figure is illusory, however, because the error is plus or minus a few billion years. The time scale referring to the origin and evolution of the Earth is much more firm—the ages listed in Table A8 carry an error of only a few percent or even less.

Cosmologic time t ranging from the beginning ($t = 0$) to $t = 5.390 \cdot 10^{-44}$ seconds is the Planckian era. It is the time interval defined by the Planck length, which is equal to $(Gh/2\pi c^3)^{1/2}$ divided by the speed of light. It is therefore equal to $(Gh/2\pi c^5)^{1/2}$. The Planckian is the time interval during which space, time, and energy came into existence. It is the moment of creation, about which we know nothing. In contrast, the Gamowian is the longest era, ranging from the end of the Planckian to the formation of the solar system. During the Gamowian, matter formed out of the primordial radiation as it cooled and condensed; stars were formed and were grouped into galaxies. More than 100 generations of massive stars came and went, continuously enriching interstellar matter with heavy elements.

Table A8.   Geological Time Scale

Era	Period	Epoch	Age (y B.P.)	Major Events
			0	
			50	Beginning of the Atomic Age (December 2, 1942)
		Holocene	3,000	Beginning of the Iron Age
			10,000	
			11,600	Intense deglaciation; giant floods down the Mississippi Valley
	Quaternary		20,000	Maximum of the last ice age
			125,000	Temperature maximum of the last interglacial: appearance of *Homo sapiens sapiens* and *Homo sapiens neanderthalensis*
		Pleistocene	250,000	Disappearance of *Homo erectus*; appearance of *Homo sapiens "praesapiens"*
			$1.5 \cdot 10^6$	Disappearance of *Paranthropus*; appearance of *Homo erectus*
			$1.64 \cdot 10^6$	Appearance of *Hyalinea baltica*
			$3.0 \cdot 10^6$	Appearance of *Australopithecus africanus*
			$3.2 \cdot 10^6$	Closing of the Central American isthmus; beginning of extensive northern glaciation
		Pliocene	$3.6 \cdot 10^6$	Appearance of *Australopithecus afarensis*
Cenzoic	**Neogene**		$5.2 \cdot 10^6$	Opening of Gibraltar passage
			$6.2 \cdot 10^6$	Isolation of the Mediterranean; salt deposits on its floor
		Miocene	$14 \cdot 10^6$	Arrival of Antarctic sheet at ocean; *Ramapithecus*
			$23.3 \cdot 10^6$	Alpine orogenesis apex
		Oligocene	$30 \cdot 10^6$	*Aegyptopithecus*
			$35.4 \cdot 10^6$	Sudden expansion of the Antarctic ice sheet; separation of Australia from Antarctica; apearance of artiodactyls, perissodactyls, and apes
	Paleogene	Eocene		
			$56.5 \cdot 10^6$	
		Paleocene		Appearance of globorotalids; radiation of placental mammals; first primates; flowering plants spread
			$65.0 \cdot 10^6$	
				Giant asteroidal impact; extinction of cycadeoidales, globotruncanids, ammonoids, belemnoids, ichthyosaurs, plesiosaurs, dinosaurs
	Cretaceous			First angiosperm and marsupials; opening of the South Atlantic; the White Cliffs of Dover
Mesozoic			$145.6 \cdot 10^6$	
	Jurassic			Opening of the North Atlantic; first coccoliths and planktic Foraminifera; first birds
			$208.0 \cdot 10^6$	
	Triassic			Palisades sill, New York; Carrara marble; appearance of dinosaurs, lizards, turtles; first mammals
			$245.0 \cdot 10^6$	

(*cont.*)

Table A8. Geological Time Scale (*continued*)

Era	Period	Epoch	Age (y B.P.)	Major Events
	Permian			Appalachian-Alleghanian/Hercynian-Variscan orogenesis; New Red Sandstone in Europe; Glaciation in the Southern Hemisphere; extinction of tetracorals, cystoids, placoderms
			$290.0 \cdot 10^6$	
	Carboniferous			Widespread formation of coal; cyclothems; first reptiles, winged insects
			$362.5 \cdot 10^6$	
Paleozoic	Devonian			Old Red Sandstone; Queenston-Juniata red beds; first sharks; first amphibia
			$408.5 \cdot 10^6$	Taconic–Caledonian orogeny
	Silurian			Lockport dolostone; first bony fishes; first trees
			$439 \cdot 10^6$	
	Ordovician			First corals; first vertebrates (jawless fishes)
			$510.0 \cdot 10^6$	
	Cambrian			Appearance of trilobites, brachiopods, echinoderms, and shelled mollusks
			$570 \cdot 10^6$	Appearance of Archaeocyatha
	Ediacaran			Ediacaran fauna
			$590 \cdot 10^6$	Appearance of metazoa
Proterozoic			$1.7 \cdot 10^9$	Increasing O_2 in atmosphere; appearance of eucaryota
			$2.7 \cdot 10^9$	Oldest stromatolites
Archean			$3.5 \cdot 10^9$	Earliest bacteria (heterotrophs)
			$4.0 \cdot 10^9$	Oldest terrestrial rocks
Hadean			$4.6 \cdot 10^9$	Age of the meteorites and oldest lunar rocks
			$4.7 \cdot 10^9$	Formation of the solar system
			$4.7 \cdot 10^9$	Nucleogenesis in the region of the solar system
				Formation and evolution of stars, quasars, and galaxies; general increase in the relative abundances of the heavier elements (continuing to the present)
			$16.5 \cdot 10^9$	$3 = 800,000$ y: electrons captured by nuclei; formation of H and He atoms and of H_2 molecules; the universe becomes transparent
Gamowian				3.8 m: stabilization of ^2H, ^3He, and ^4He nuclei; relative abundances: 74% H, 26% He (by mass) or 92% H, 8% He (by number of atoms)
				10 s: stabilization of electrons
				10^{-5}–10^{-3} s: stabilization of protons and neutrons
				10^{-8}–10^{-5} s: quark–antiquark annihilation
				10^{-10}–10^{-8} s: stabilization of quarks
				10^{-10} s: electromagnetic and weak forces separate
				10^{-33}–10^{-32} s: inflation
				10^{-35} s: strong force and electroweak force separate
				10^{-43} s: gravity separates
			$16.5 \cdot 10^9$	
Planckian				0–$5.390 \cdot 10^{-44}$ s: appearance of space, time, energy, and superforce
			$16.5 \cdot 10^9$	

A9. ISOTOPE CHART

Table A9 plots all naturally occurring isotopes in terms of the number of their protons (ordinates) and neutrons (abscissa). All figures are the latest (1992) and are given to the last decimal place. Also included are isotopes that are important in the process of formation of the elements (e.g., ^{5}He) and in supernova explosions (e.g., the transuranic elements); short-lived isotopes bracketed by, or in the neighborhood of, stable isotopes; and short-lived isotopes that occur in nature because they either are formed by the decay of long-lived isotopes (e.g., ^{234}U, ^{231}Pa, or ^{228}Ra) or are continuously formed by galactic protons bombarding upper atmosphere gases (e.g., ^{3}H and ^{14}C).

The following data, from top to bottom in each square, are given for each isotope:

- Element symbol and isotopic number (also called *mass number*, the total number of nucleons—protons and neutrons—in the nucleus)
- Mass of the neutral atom (nucleus plus electrons) in atomic mass units (u)
- Relative natural abundance (percent) in common terrestrial matter
- Decay mode, if any, with its relative frequency if more than one mode is present [α = alpha decay, with the daughter isotope two squares down and two to the left (loss of 2 protons and 2 neutrons); β^+ decay or k-capture, with the daughter isotope one square down and one to the right (transformation of a proton into a neutron); β^- decay, with the daughter isotope one square up and one to the left (transformation of a neutron into a proton); ε = K-capture (capture by the nucleus of an electron from the K shell, transforming a proton into a neutron); (n, p) = neutron in, proton out; (p, n) = proton in, neutron out; IT = isomeric transition (change in the internal energy of a nucleus); SF = spontaneous fission].
- Half life (y = years; d = days; h = hours; m = minutes; s = seconds; ms = milliseconds; μs = microseconds).

The illustration on page 620 summarizes the different modes of decay. Notice that the (n, p) reaction has the same effect of β^+ or K-capture, and the (p, n) reaction has the same effect as β^- decay.

Notice that the mass of an isotope expressed in atomic mass units is very close to the number of nucleons in the nucleus (mass number). The mass of oxygen-16 (^{16}O), for instance, is 15.9949146 u, that of iron-56 (^{56}Fe) is 55.934939 u, and that of uranium-238 (^{238}U) is 238.050785 u. This is because the mass of a nucleon inside the nucleus is almost exactly 1 u. Free protons and free neutrons have greater masses, 1.00727647 u for the proton and 1.00866490 u for the neutron. Protons and neutrons lose about 0.008 u of mass when entering to form a nucleus. The mass is transformed into energy that appears as gamma radiation emitted by the nucleus. The amount of energy lost by the nucleus is exactly the same amount you would have to put in to free again the protons and neutrons.

Notice also that for elements 1 to 7 (hydrogen to nitrogen), the actual mass is slightly larger than the number of nucleons (mass number); for elements 8 to 83 and 84 (oxygen to bismuth and polonium), it is slightly smaller, reaching a minimum at tin-118 (^{118}Sn); and for elements above bismuth, it is again slightly larger.

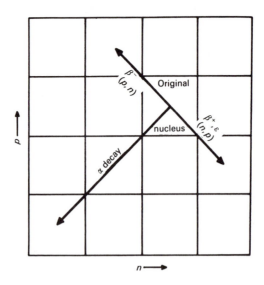

The exact masses of the proton, neutron, and electron are given below. Also given is the energy equivalent (in million electron volts, MeV) of 1 atomic mass unit.

mass of proton: 1.00727647 u
mass of neutron: 1.00866490 u
mass of electron: 0.00054858 u
1 u = 931.49432 MeV

If one calculates the mass that the nucleus of any isotope should have, by multiplying the mass of the protons by the number of protons in that nucleus and adding the mass of the neutron times the number of neutrons, the mass obtained is larger than the actual mass. The actual mass of the nucleus of any given isotope can be calculated from the table by subtracting the mass contributed by the electrons. The nucleus of helium-4, for example, consists of two protons and two neutrons, for a total mass of 4.03188274 u. The mass of the neutral atom of helium-4 listed in the table is 4.0026032 u. Subtracting the mass of the two electrons (0.0010972 u) yields a mass of 4.0015060 u for the nucleus, which is significantly (0.75%) smaller than the mass previously calculated.

What has happened is that in assembling the helium atom from its constituents, there has been a mass loss of 0.0303767 u, equivalent to 28.295723 MeV of energy. Mass has been transformed into energy according to Einstein's equation $e = mc^2$, and the energy has been radiated away by the nucleus in the form of gamma radiation. The sun produces its energy by this very process, the synthesis of helium from hydrogen, as explained in Section 6.2.

The energy lost by the nucleus when it formed from its individual constituents (in steps, of course) is what keeps the nucleus together. You would have to inject into it the same amount of energy in order to pry loose the protons and neutrons. The largest mass loss per nucleon is shown by the iron isotopes, indicating that the iron isotopes are particularly tightly bound. This is the reason why iron is an abundant element.

Some isotopes (e.g., chlorine-36) can decay in two different ways, resulting in two different products. One isotope, actinium-226, decays in three different ways. The different decays have different probabilities, the percentages of which are given in parentheses. Especially in the region of the heavier elements, a decay product itself may be radioactive and undergo further decay. You can follow the successive decays starting from uranium-238, uranium-235, or thorium-232. After a number of intermediate steps, all radioactive, you will end up with the stable isotopes lead-206, lead-207, and lead-208 (in that order).

Notice that the lighter elements have similar numbers of protons and neutrons. As one progresses through the table, the number of neutrons increases faster than the number of protons. Uranium-238, toward the end of the table, has 92 protons and 146 neutrons. An excess of neutron in the heavier nuclei is needed for the strong force to overcome the electrostatic repulsion among protons and keep the nuclei together. Beyond ^{238}U, however, no truly long-lived isotopes exist, because neutrons on opposite sides of large nuclei are too far apart to be linked by the strong force, while protons in the same situation still repel each other. As the excess of neutrons increases, so does the number of isotopes. The lighter elements tend to have only a few stable isotopes, but the heavier elements may have more. Tin for instance, has 10 stable isotopes and xenon has 9.

Elements with odd numbers of protons tend to have many fewer isotopes than those with even numbers of protons. Twenty-one elements have only one stable or long-lived radioactve isotope. Among these are beryllium, fluorine, sodium, aluminum, phosphorus, cobalt, arsenic, iodine, cesium, gold, bismuth, and thorium. Two elements of moderate mass, technetium (no. 43) and promethium (no. 61), and seven heavy elements, between bismuth and uranium, have no stable or long-lived radioactive isotopes at all. Technetium and promethium do not occur on Earth, but have been made artificially and have been observed in remnants of supernova explosions. The seven unstable elements between bismuth and uranium occur on Earth because they are continuously produced by the decay of the long-lived isotopes of uranium and thorium.

Table A9.   Isotope Chart

| | | | | | | | Ne20 19.992436 90.51 — \| | Ne21 20.993843 0.27 — \| | Ne22 21.991383 9.22 — \| |

Individual isotope box contents (arranged as in the chart):

Ne22 21.991383 9.22 — |

Ne21 20.993843 0.27 — |

| Ne20 19.992436 90.51 — \| | F19 18.998403 100 — \| | O18 17.999160 0.200 — \| |

| O17 16.999131 0.038 — \| |

| O16 15.9949146 99.762 — \| | N15 15.000109 0.37 — \| | C14 14.0032419 β^- 5715 y |

| O15 15.003065 β^+ 122.2 s | N14 14.0030740 99.63 — \| | C13 13.0033548 1.10 — \| |

| N13 13.005739 β^+ 9.965 m | C12 12.0000000 98.90 — \| | B11 11.009305 80.1 — \| | Be10 10.013534 β^- $1.51 \cdot 10^6$ y |

| B10 10.012937 19.9 — \| | Be9 9.012182 100.0 — \| | Li8 8.022485 β^- 0.838 s |

| Be8 8.005305 2α $\sim 1 \cdot 10^{-16}$ s | Li7 7.016003 92.5 — \| | He6 6.018886 β^- 0.807 s |

| Be7 7.016928 β^+ 53.29 d | Li6 6.015121 7.5 — \| | He5 5.01222 n,α $2 \cdot 10^{-21}$ s |

| Li5 5.01254 p,α $\sim 10^{-21}$ s | He4 4.0026032 99.999862 — \| | H3 3.0160493 β^- 12.33 y |

| He3 3.0160293 0.000138 — \| | H2 2.0141077 0.015 — \| |

| H1 1.00782503 99.985 — \| |

| n 1.00866490 |

Chart of the nuclides (Ne through Ca).

Nuclide	Mass	Abundance / —	Decay	Half-life
Ne20	19.992436	90.51	—	—
Ne21	20.993843	0.27	—	—
Ne22	21.991383	9.22	—	—
Na22	21.994434	—	β^+	2.602 y
Na23	22.989768	100.0	—	—
Na24	23.990961	—	β^-	15.02 h
Mg24	23.985042	78.99	—	—
Mg25	24.985837	10.00	—	—
Mg26	25.982593	11.01	—	—
Mg27	26.984341	—	β^-	9.462 m
Mg28	27.983877	—	β^-	20.90 h
A126	25.986892	—	β^+	750,000 y
A127	26.981539	100.0	—	—
A128	27.981910	—	β^-	2.241 m
Si28	27.976927	92.23	—	—
Si29	28.976494	4.67	—	—
Si30	29.973770	3.10	—	—
Si31	30.975362	—	β^-	2.62 h
Si32	31.974148	—	β^-	172 y
P31	30.973762	100	—	—
P32	31.973907	—	β^-	14.26 d
P33	32.971725	—	β^-	25.3 d
S32	31.972071	95.02	—	—
S33	32.971456	0.75	—	—
S34	33.9678667	4.21	—	—
S35	34.969032	—	β^-	87.2 d
S36	35.9670806	0.02	—	—
S37	36.9711255	—	β^-	5.05 m
S38	37.971162	—	β^-	2.84 h
Cl34	33.9737629	—	β^+	32.23 m
Cl35	34.9608527	75.77	—	—
Cl36	35.9683069	—	β^- (98.10) ε (1.90)	301,000 y
Cl37	36.9659026	24.23	—	—
Cl38	37.9680105	—	β^-	37.2 m
Cl39	38.968005	—	β^-	55.6 m
Ar36	35.9675455	0.337	—	—
Ar37	36.966776	—	ε	35.04 d
Ar38	37.962732	0.063	—	—
Ar39	38.964314	—	β^-	269 y
Ar40	39.962384	99.600	—	—
K39	38.963707	93.2581	—	—
K40	39.963999	0.0117	β^- (89.30) ε (10.70)	$1.277 \cdot 10^9$ y
K41	40.961825	6.7302	—	—
K42	41.962402	—	β^-	12.36 h
Ca40	39.962591	96.941	—	—
Ca41	40.962278	—	ε	103,000 y
Ca42	41.958618	0.647	—	—
Ca43	42.958766	0.135	—	—
Ca44	43.955481	2.086	—	—
Ca45	44.956185	—	β^-	162.7 d

(cont.)

Table A9. Isotope Chart (*continued*)

Zn
Zn63	Zn64	Zn65	Zn66	Zn67	Zn68	Zn69
62.933214	63.929145	64.929243	65.926035	66.927129	67.924846	68.926552
—	48.6	ε	27.9	4.1	18.8	IT (99.97)
β^+	—	—	—	—	—	β^- (0.03)
38.50 m	—	243.9 d	—	—	—	13.76 h

Cu
Cu63	Cu64	Cu65
62.929599	63.929766	64.927793
69.17	β^+ (62.90)	30.83
—	β^- (37.10)	—
—	12.701 h	—

Ni
Ni58	Ni59	Ni60	Ni61	Ni62	Ni63	Ni64
57.935346	58.934349	59.930788	60.931058	61.928346	62.929670	63.927968
68.077	ε	26.223	1.140	3.634	β^-	0.928
—	75,000 y	—	—	—	100 y	—

Co
Co59	Co60
58.933198	59.933820
100	β^-
—	5.271 y

Fe
Fe54	Fe55	Fe56	Fe57	Fe58	Fe59	Fe60
53.939613	54.938296	55.934939	56.935396	57.93277	58.934877	59.934078
5.8	ε	91.72	2.2	0.28	β^-	β^-
—	2.73 y	—	—	—	44.49 d	$1.49 \cdot 10^6$

Mn
Mn53	Mn54	Mn55
52.941291	53.940361	54.938047
ε	ε	100
$3.7 \cdot 10^6$ y	312.1 d	—

Cr
Cr50	Cr51	Cr52	Cr53	Cr54
49.946046	50.944768	51.940510	52.940651	53.938882
4.345	ε	83.79	9.50	2.36
—	27.70 d	—	—	—

V
V50	V51
49.947161	50.943962
0.250	99.750
ε (> 70)	—
β^- (< 30)	—
$\sim 1.5 \cdot 10^{17}$ y	—

Ti
Ti46	Ti47	Ti48	Ti49	Ti50
45.952629	46.951764	47.947947	48.947871	49.944792
8.0	7.3	73.8	5.5	5.4
—	—	—	—	—

Sc
Sc45	Sc46	Sc47
44.955910	45.955170	46.952409
100	β^-	β^-
—	83.83 d	3.349 d

Ca
Ca44	Ca45	Ca46	Ca47	Ca48
43.955481	44.956185	45.953689	46.954543	47.952533
2.086	β^-	0.004	β^-	0.187
—	—	—	—	—

Nuclide	Atomic mass	Abundance / decay	Decay mode	Half-life
Zr93	92.906474	—	β−	1.5·10^6 y
Zr92	91.905039	17.15	—	—
Zr91	90.905644	11.22	—	—
Zr90	89.904703	51.45	—	—
Y90	89.907152	—	β−	64.1 h
Y89	88.905849	100	—	—
Sr90	89.907738	—	β−	29.1 y
Sr89	88.907450	—	β−	50.52 d
Sr88	87.905619	82.58	—	—
Sr87	86.908884	7.00	—	—
Sr86	85.909267	9.86	—	—
Sr85	84.912937	—	ε	64.84 d
Sr84	83.913430	0.56	—	—
Rb87	86.909187	27.83	β−	48.8 10^9 y
Rb86	85.911172	—	β−	18.66 d
Rb85	84.911794	72.17	—	—
Kr86	85.910616	17.3	—	—
Kr85	84.912531	—	β−	10.72 y
Kr84	83.911507	57.0	—	—
Kr83	82.914135	11.5	—	—
Kr82	81.913482	11.6	—	—
Kr81	80.916590	—	ε	210,000 y
Kr80	79.916380	2.25	—	—
Kr79	78.920084	—	ε	35.0 h
Kr78	77.92040	0.35	—	—
Br81	80.916289	49.31	—	—
Br80	79.918528	—	β− (91.7), ε (8.3)	17.68 m
Br79	78.918336	50.69	—	—
Se82	81.916698	8.74	β−	1.4 10^20 y
Se81	80.917990	—	β−	57.25 m
Se80	79.916520	49.61	—	—
Se79	78.918498	—	β−	≤65.000 y
Se78	77.917308	23.77	—	—
Se77	76.919912	7.63	—	—
Se76	75.919212	9.36	—	—
Se75	74.922521	—	ε	119.77 d
Se74	73.922475	0.89	—	—
As75	74.921594	100	—	—
Ge76	75.921402	7.44	—	—
Ge75	74.922858	—	β−	82.8 m
Ge74	73.921177	35.94	—	—
Ge73	72.923463	7.72	—	—
Ge72	71.922079	27.66	—	—
Ge71	70.924954	—	ε	11.2 d
Ge70	69.924250	21.24	—	—
Ga71	70.924701	39.9	—	—
Ga70	69.926028	—	β− (99.8), ε (0.2)	21.15 m
Ga69	68.925580	60.1	—	—
Zn70	69.925325	0.6	—	—
Zn69	68.926652	—	β− (99.97), IT (0.03)	13.76 h
Zn68	67.924846	18.8	—	—

Nuclide	Atomic mass	Abundance (%)	Decay	Half-life
Cd106	105.90646	1.25	—	—
Cd107	106.90661	—	ε	6.52 h
Cd108	107.90418	0.89	—	—
Cd109	108.904953	—	ε	462.0 d
Cd110	109.903005	12.49	—	—
Cd111	110.904182	12.80	—	—
Cd112	111.902757	24.13	—	—
Cd113	112.904400	12.22	β^-	$9.3 \cdot 10^{15}$ y
Ag107	106.90509	51.839	—	—
Ag108	107.905952	— (97.15)	β^- (2.85) ε	2.39 m
Ag109	108.904756	48.161	—	—
Pd102	101.905634	1.02	—	—
Pd103	102.906114	—	ε	16.99 d
Pd104	103.904029	11.14	—	—
Pd105	104.905079	22.33	—	—
Pd106	105.903478	27.33	—	—
Pd107	106.905127	—	β^-	$6.5 \cdot 10^6$ y
Pd108	107.903895	26.46	—	—
Pd109	108.905954	—	β^-	13.7 h
Pd110	109.905167	11.72	—	—
Rh103	102.905500	100	—	—
Ru96	95.907599	5.54	—	—
Ru97	96.907556	—	ε	2.89 d
Ru98	97.90529	1.86	—	—
Ru99	98.905939	12.7	—	—
Ru100	99.904219	12.6	—	—
Ru101	100.905582	17.0	—	—
Ru102	101.904348	31.6	—	—
Ru103	102.906323	—	β^-	39.26 d
Ru104	103.905424	18.7	—	—
Tc97	96.906364	—	ε	$2.6 \cdot 10^6$ y
Tc98	97.907215	—	β^-	$4.2 \cdot 10^6$ y
Tc99	98.906254	—	β^-	213.000 y
Mo92	91.906807	14.84	—	—
Mo93	92.906813	—	ε	3500 y
Mo94	93.905085	9.25	—	—
Mo95	94.905841	15.92	—	—
Mo96	95.904678	16.68	—	—
Mo97	96.906020	9.55	—	—
Mo98	97.905407	24.13	—	—
Mo99	98.907711	—	β^-	65.942 h
Mo100	99.90748	9.63	—	—
Nb93	92.906377	100	—	—
Nb94	93.907281	—	β^-	24.000 y
Nb95	94.906835	—	β^-	34.97 d
Zr90	89.904703	51.45	—	—
Zr91	90.905644	11.22	—	—
Zr92	91.905039	17.15	—	—
Zr93	92.906474	—	β^-	$1.5 \cdot 10^6$ y
Zr94	93.906315	17.38	—	—
Zr95	94.908042	—	β^-	64.02 d
Zr96	95.908275	2.80	—	—

Nuclide	Atomic mass	Abundance/%	Decay mode	Half-life
Te120	119.904048	0.099	—	—
Te121	120.904947	—	ε	16.78 d
Te122	121.903054	2.60	—	—
Te123	122.904271	0.91	ε	1.3·10^{13} y
Te124	123.902818	4.79	—	—
Te125	124.904433	7.12	—	—
Te126	125.903314	18.93	—	—
Te127	126.905227	—	β$^-$	9.35 h
Sb121	120.903821	57.3	—	—
Sb122	121.905179	—	β$^-$ (97.62), ε (2.38)	2.72 d
Sb123	122.904216	42.7	—	—
Sn112	111.904827	0.97	—	—
Sn113	112.905176	—	ε	115.1 d
Sn114	113.902784	0.65	—	—
Sn115	114.903348	0.36	—	—
Sn116	115.901747	14.53	—	—
Sn117	116.902956	7.68	—	—
Sn118	117.901605	24.22	—	—
Sn119	118.903311	8.58	—	—
Sn120	119.902199	32.59	—	—
Sn121	120.904239	—	β$^-$	27.06 h
Sn122	121.903440	4.63	—	—
Sn123	122.905722	—	β$^-$	129.2 d
Sn124	123.905274	5.79	—	—
In113	112.904061	4.3	—	—
In114	113.904916	—	β$^-$ (99.5), ε (0.5)	71.9 s
In115	114.903880	95.7	β$^-$	441·10^{12} y
Cd108	107.90418	0.89	—	—
Cd109	108.904953	—	ε	462.0 d
Cd110	109.903005	12.49	—	—
Cd111	110.904182	12.80	—	—
Cd112	111.902757	24.13	—	—
Cd113	112.904400	12.22	β$^-$	9.3·10^{15} y
Cd114	113.903357	28.73	—	—
Cd115	114.905430	—	β$^-$	44.6 d
Cd116	115.904755	7.49	—	—

(cont.)

627

Table A9. Isotope Chart (continued)

Isotope	Mass	Abundance / %	Decay	Half-life
Xe124	123.905894	0.10	—	—
Xe125	124.906397	—	ε	17.1 h
Xe126	125.904281	0.09	—	—
Xe127	126.905182	—	ε	36.4 d
Xe128	127.903531	1.91	—	—
Xe129	128.904780	26.4	—	—
Xe130	129.903509	4.1	—	—
Xe131	130.905072	21.2	—	—
Xe132	131.904144	26.9	—	—
Xe133	132.905888	—	β^-	5.243 d
Xe134	133.905395	10.4	—	—
Xe135	134.907130	—	β^-	9.09 h
Xe136	135.907214	8.9	—	—

Isotope	Mass	Abundance / %	Decay	Half-life
Cs131	130.905444	—	ε	9.69 d
Cs132	131.906431	—	ε (98), β^- (2)	6.47 d
Cs133	132.905429	100	—	—
Cs134	133.906696	—	β^-	2.062 y
Cs135	134.905885	—	β^-	$3 \cdot 10^6$ y

Isotope	Mass	Abundance / %	Decay	Half-life
Ba130	129.906281	0.106	—	—
Ba131	130.906902	—	ε	11.8 d
Ba132	131.905043	0.101	—	—
Ba133	132.905988	—	ε	10.74 y
Ba134	133.904485	2.417	—	—
Ba135	134.905665	6.592	—	—
Ba136	135.904553	7.854	—	—
Ba137	136.905812	11.23	—	—
Ba138	137.905233	71.70	—	—

Isotope	Mass	Abundance / %	Decay	Half-life
I125	124.904620	—	ε	59.4 d
I126	125.905624	—	ε (56.3), β^- (43.7)	13.0 d
I127	126.904473	100	—	—
I128	127.905810	—	β^- (93.1), ε (6.9)	25.0 m
I129	128.904986	—	β^-	$15.7 \cdot 10^6$ y
I130	129.906713	—	β^-	12.36 h

Isotope	Mass	Abundance / %	Decay	Half-life
Te122	122.903054	2.60	—	—
Te123	122.904271	0.91	ε	$1.3 \cdot 10^{13}$ y
Te124	123.902818	4.79	—	—
Te125	124.904433	7.12	—	—
Te126	125.903314	18.93	—	—
Te127	126.905227	—	β^-	9.35 h
Te128	127.904463	31.69	—	$> 8 \cdot 10^{24}$ y
Te129	128.906594	—	β^-	69.6 m
Te130	129.906229	33.80	—	$2.5 \cdot 10^{21}$ y

↑

Nuclide	Mass	Abundance/%	Decay	Half-life
Gd152	151.919786	0.20	α	1.1·10¹⁴ y
Gd153	152.921745	—	ε	241.6 d
Gd154	153.920861	2.18	—	—
Gd155	154.922618	14.80	—	—
Gd156	155.922118	20.47	—	—
Gd157	156.923956	15.65	—	—
Eu151	150.919847	47.8	—	—
Eu152	151.921742	—	ε (72.08) β⁻ (27.92)	13.483 y
Eu153	152.921225	52.2	—	—
Sm144	143.911998	3.1	—	—
Sm145	144.913409	—	ε	340 d
Sm146	145.913053	—	α	103·10⁶ y
Sm147	146.914894	15.0	α	106·10⁹ y
Sm148	147.914819	11.3	α	7·10¹⁵ y
Sm149	148.917180	13.8	α	10¹⁶ y
Sm150	149.917273	7.4	—	—
Sm151	150.919929	—	β⁻	90 y
Sm152	151.919728	26.7	—	—
Sm153	152.922094	—	β⁻	46.30 h
Sm154	153.922205	22.7	—	—
Pm143	142.910930	—	ε	265 d
Pm144	143.912588	—	ε⁻	363 d
Pm145	144.912743	—	ε	17.7 y
Pm146	145.914708	—	ε (66.1) β⁻	—
Pm147	146.915135	—	β⁻	2.6234 y
Pm148	147.917473	—	β⁻	41.3 d
Pm149	148.918332	—	β⁻	53.1 h
Nd142	141.907719	27.13	—	—
Nd143	142.909810	12.18	—	—
Nd144	143.910083	23.80	α	2.1·10¹⁵ y
Nd145	144.912570	8.30	—	>6·10¹⁶ y
Nd146	145.913113	17.19	—	—
Nd147	146.916097	—	β⁻	10.98 d
Nd148	147.916889	5.76	—	—
Nd149	148.920145	—	β⁻	103.5 m
Nd150	149.920887	5.64	—	—
Pr141	140.907647	100	—	—
Ce136	135.907140	0.19	—	—
Ce137	136.907780	—	IT (99.22) ε (0.78)	34.4 h
Ce138	137.905985	0.25	—	—
Ce139	138.906631	—	ε	137.7 d
Ce140	139.905433	88.43	—	—
Ce141	140.908271	—	β⁻	32.501 d
Ce142	141.909241	11.13	α	>5·10¹⁶ y
La137	136.906460	—	ε	60.000 y
La138	137.907105	0.0902	ε (66.7) β⁻ (33.3)	106·10⁹ y
La139	138.906347	99.9098	—	—
Ba134	133.904485	2.417	—	—
Ba135	134.905665	6.592	—	—
Ba136	135.904553	7.854	—	—
Ba137	136.905812	11.23	—	—
Ba138	137.905233	71.70	—	—

(cont.)

Table A9. Isotope Chart (*continued*)

Nuclide	Mass	Abundance / —	Decay	Half-life
Yb168	167.933894	0.13	—	—
Yb169	168.935186	—	ε	32.03 d
Yb170	169.934759	3.05	—	—
Yb171	170.936323	14.3	—	—
Yb172	171.936378	21.9	—	—
Yb173	172.938208	16.12	—	—
Yb174	173.938859	31.8	—	—
Yb175	174.941273	—	β^-	4.19 d
Tm169	168.934212	100	—	—
Er162	161.928775	0.14	—	—
Er163	162.930030	—	ε	75.0 m
Er164	163.929198	1.61	—	—
Er165	164.930723	—	ε	10.36 h
Er166	165.930290	33.6	—	—
Er167	166.932046	22.95	—	—
Er168	167.932368	26.8	—	—
Er169	168.934588	—	β^-	9.40 d
Er170	169.935461	14.9	—	—
Ho163	162.928731	—	ε	4570 y
Ho164	163.930285	—	ε (58) β^- (42)	29 m
Ho165	164.930319	100	—	—
Dy156	155.925277	0.06	—	—
Dy157	156.925460	—	ε	8.1 h
Dy158	157.924403	0.10	—	—
Dy159	158.925735	—	ε	144.4 d
Dy160	159.925193	2.34	—	—
Dy161	160.926930	18.9	—	—
Dy162	161.926795	25.5	—	—
Dy163	162.928728	24.9	—	—
Dy164	163.929171	28.2	—	—
Tb157	156.924023	—	ε	110 y
Tb158	157.925411	—	ε (82) β^- (18)	180 y
Tb159	158.925342	100	—	—
Gd154	153.920861	2.18	—	—
Gd155	154.922618	14.80	—	—
Gd156	155.922118	20.47	—	—
Gd157	156.923956	15.65	—	—
Gd158	157.924099	24.84	—	—
Gd159	158.926384	—	β^-	18.56 h
Gd160	159.927049	21.86	—	—

Nuclide	Mass	Abundance / comment	Decay	Half-life
Pt190	189.959917	0.01	α	6·10^11 y
Pt191	190.961665	—	ε	2.96 d
Pt192	191.961019	0.79	—	—
Pt193	192.962977	—	ε	60 y
Pt194	193.962655	32.9	—	—
Pt195	194.964766	33.8	—	—
Ir191	190.960584	37.3	—	—
Ir192	191.962580	—	β⁻ (95.4), ε (4.6)	73.831 d
Ir193	192.962917	62.7	—	—
Os184	183.952488	0.02	—	> 1·10^17 y
Os185	184.954041	—	ε	93.6 d
Os186	185.953830	1.58	α	2.0·10^15 y
Os187	186.955741	1.6	—	—
Os188	187.955834	13.3	—	—
Os189	188.958137	16.1	—	—
Os190	189.958436	26.4	—	—
Os191	190.960920	—	β⁻	15.4 d
Os192	191.961467	41.0	—	—
Re185	184.952951	37.40	—	—
Re186	185.954984	—	β⁻ (93.5), ε (6.5)	3.78 d
Re187	186.955744	62.60	β⁻	42·10^9 y
W180	179.946701	0.12	—	> 1.1·10^15 y
W181	180.948192	—	ε	121.0 d
W182	181.948202	26.3	—	—
W183	182.950220	14.3	—	—
W184	183.950928	30.7	—	—
W185	184.953416	—	β⁻	74.8 d
W186	185.954357	28.6	—	—
Ta180	179.947462	0.012	ε	> 1.2·10^15 y
Ta181	180.947992	99.988	—	—
Hf174	173.940044	0.162	α	2.0·10^15 y
Hf175	174.941507	—	ε	70 d
Hf176	175.941406	5.206	—	—
Hf177	176.943217	18.606	—	—
Hf178	177.943696	27.297	—	—
Hf179	178.945812	13.629	—	—
Hf180	179.946546	35.100	—	—
Lu175	174.940770	97.41	—	—
Lu176	175.942679	2.59	β⁻	38·10^9 y
Yb172	171.936378	21.9	—	—
Yb173	172.938208	16.12	—	—
Yb174	173.938859	31.8	—	—
Yb175	174.941273	—	β⁻	4.19 d
Yb176	175.942564	12.7	—	—

(cont.)

Table A9. Isotope Chart (continued)

Each box lists: isotope, atomic mass, abundance (%) or decay mode, and half-life.

Bi (Z = 83)

Isotope	Mass	Abundance / Decay	Half-life
Bi206	205.978478	ε	6.243 d
Bi207	206.978446	ε	32.2 y
Bi208	207.979717	ε	368.000 y
Bi209	208.980374	100	—
Bi210	209.984095	β⁻	5.013 d
Bi211	210.987255	α (99.72) β⁻ (0.28)	2.14 m
Bi212	211.991255	β⁻ (64.08) α (35.94)	60.55 m
Bi213	212.994351	β⁻ (97.84) α (2.16)	45.6 m
Bi214	213.998691	β⁻ (99.93) α (0.02)	19.9 m

Pb (Z = 82)

Isotope	Mass	Abundance / Decay	Half-life
Pb203	202.973365	ε	2.1615 d
Pb204	203.973020	1.4 / ε	$\geq 1.4 \cdot 10^{17}$ y
Pb205	204.974458	ε	$15.2 \cdot 10^{6}$ y
Pb206	205.974440	24.1	—
Pb207	206.975872	22.1	—
Pb208	207.976627	52.4	—
Pb209	208.981065	β⁻	3.25 h
Pb210	209.984163	β⁻	22.6 y
Pb211	210.988735	β⁻	36.1 m
Pb212	211.991871	β⁻	10.64 h
Pb213	212.996510	β⁻	10.2 m

Tl (Z = 81)

Isotope	Mass	Abundance / Decay	Half-life
Tl203	202.972320	29.524	—
Tl204	203.973839	β⁻ (97.45) ε (2.55)	3.78 y
Tl205	204.974401	70.476	—
Tl206	205.976084	β⁻	4.20 m
Tl207	206.977404	β⁻	4.77 m
Tl208	207.981998	β⁻	3.053 m
Tl209	208.985444	β⁻	2.2 m
Tl210	209.990057	β⁻	1.30 m

Hg (Z = 80)

Isotope	Mass	Abundance / Decay	Half-life
Hg196	195.965807	0.15	—
Hg197	196.967187	ε	64.13 h
Hg198	197.966743	9.97	—
Hg199	198.968254	16.87	—
Hg200	199.968300	23.10	—
Hg201	200.970277	13.18	—
Hg202	201.970617	29.86	—
Hg203	202.972848	β⁻	48.61 d
Hg204	203.973467	6.87	—
Hg205	204.976047	β⁻	5.2 m
Hg206	206.977489	β⁻	8.15 m

Au (Z = 79)

Isotope	Mass	Abundance / Decay	Half-life
Au197	196.966543	100	—
Au198	197.968217	β⁻	2.694 d

Pt (Z = 78)

Isotope	Mass	Abundance / Decay	Half-life
Pt194	193.962655	32.9	—
Pt195	194.964766	33.8	—
Pt196	195.964926	25.3	—
Pt197	196.967315	β⁻	18.3 h
Pt198	197.967869	7.2	—

Chart of the nuclides (continued).

Nuclide	Mass	Decay mode(s)	Half-life
Rn222	222.017571	α; —	3.8235 d
Rn221	221.015470	β⁻ (78); α (22)	25 m
Rn220	220.011368	α; —	55.6 s
Rn219	219.009479	α; —	3.96 s
Rn218	218.005580	α; —	35 ms
At219	219.0113	α (97); β⁻ (3)	0.9 m
At218	218.008684	α (99.98); β⁻ (0.02)	1.6 s
At217	217.004694	α; —	32.3 ms
At216	216.002390	α; —	0.30 ms
At215	214.998638	α; —	0.10 ms
Po218	218.008966	α (99.98); β⁻ (0.02)	3.04 m
Po217	217.006260	α; —	< 10 s
Po216	216.001889	α; —	0.145 s
Po215	214.999419	α; —	1.780 ms
Po214	213.995176	α; —	163.7 μs
Po213	212.992833	α; —	4.2 μs
Po212	211.988842	α; —	45 s
Po211	210.986627	α; —	0.516 s
Po210	209.982848	α; —	138.376 d
Po209	208.982404	α (99.74); ε (0.26)	102 y
Po208	207.981222	α; —	2.898 y
Bi215	215.001930	β⁻; —	7.7 m
Bi214	213.998691	β⁻ (99.98); α (0.02)	19.9 m
Bi213	212.994359	β⁻ (97.84); α (2.16)	45.6 m
Bi212	211.991255	β⁻ (64.06); α (35.94)	60.55 m
Bi211	210.987255	α (99.72); β⁻ (0.28)	2.14 m
Bi210	209.984095	β⁻; —	5.013 d
Bi209	208.980374	—; —	100
Bi208	207.979717	ε; —	368.000 y
Bi207	206.978471	ε; —	32.2 y
Bi206	205.978478	ε; —	6.243 d
Pb214	213.999798	β⁻; —	26.8 m
Pb213	212.996510	β⁻; —	10.2 m
Pb212	211.991871	β⁻; —	10.64 h
Pb211	210.988735	β⁻; —	36.1 m
Pb210	209.984163	β⁻; —	22.6 y
Pb209	208.981065	β⁻; —	3.25 h
Pb208	207.976627	—; —	52.4
Pb207	206.975872	—; —	22.1
Pb206	205.974440	—; —	24.1
Pb205	204.974458	ε; —	15.2·10⁶ y

(cont.)

Table A9. Isotope Chart (continued)

Nuclide	Atomic mass	Decay / abundance	Half-life
Np237	237.048168	α	$2.14 \cdot 10^6$ y
Np238	238.050941	β⁻	2.117 d
Np239	239.052932	β⁻	2.355 d
Np240	240.056169	β⁻	61.9 m
U236	236.045563	α	$23.42 \cdot 10^6$ y
U237	237.048725	β⁻	6.75 d
U238	238.050785	99.2745; α	$4.468 \cdot 10^9$ y
U239	239.054290	β⁻	23.5 m
U232	232.037130	α	68.9 y
U233	233.039628	α	159,200 y
U234	234.040947	0.0055; α	245,000 y
U235	235.043924	0.7200; α	$704 \cdot 10^6$ y
Pa231	231.035880	α	32.760 y
Pa232	232.038565	β⁻	1.31 d
Pa233	233.040242	β⁻	27.0 d
Pa234	234.043300	β⁻	6.69 h
Th230	230.033128	α	75,380 y
Th231	231.036298	β⁻	25.52 h
Th232	232.038054	100; α	$14.05 \cdot 10^9$ y
Th233	233.041577	β⁻	22.3 m
Th234	234.043593	β⁻	24.10 d
Ac228	228.031015	β⁻	6.15 h
Ra228	228.031064	β⁻	5.76 y
Th226	226.024888	α	30.6 m
Th227	227.027703	α	18.718 d
Th228	228.028715	α	1.913 y
Th229	229.031755	α	7340 y
Ac225	225.023205	α	10.0 d
Ac226	226.026084	β⁻ (82.79), ε (17.20), α (0.01)	29.4 h
Ac227	227.027750	β⁻ (98.62), α (1.38)	21.773 y
Ra223	223.018501	α	11.434 d
Ra224	224.020186	α	3.66 d
Ra225	225.023604	β⁻	14.9 d
Ra226	226.025403	α	1599 y
Ra227	227.029171	β⁻	42.2 m
Fr221	221.014230	α	4.9 m
Fr222	222.017585	β⁻	14.4 m
Fr223	223.019733	β⁻ (99.99), α (0.01)	21.8 m
Ra222	222.015353	α	38 s
Rn219	219.009479	α	3.96 s
Rn220	220.011368	α	55.6 s
Rn221	221.015470	β⁻ (78), α (22)	25 m
Rn222	222.017571	α	3.8235 d
Rn218	218.005580	α	35 ms
At217	217.004694	α	32.3 ms
At218	218.00868	α (99.98), β⁻ (0.02)	2 s
At219	219.0113	α (97), β⁻ (3)	0.9 m

Md258
258.098594
—
α
—
55 d

Fm257
257.095100
—
α (99.79)
SF (0.21)
100.5 d

Fm256
256.091769
—
SF (91.9)
α (8.1)
2.63 h

Es255
255.090267
—
β⁻
α (8)
40 d

Fm255
255.089941
—
α
—
20.07 h

Es254
254.088024
—
—
β⁻
275.7 d

Fm254
254.086850
—
α (99.94)
SF (0.06)
3.240 h

Es253
253.084817
—
α
—
20.47 d

Fm253
253.085174
—
ε (88)
α (12)
3.0 d

Cf254
254.087320
—
SF (99.69)
α (0.31)
60.5 d

Cf253
253.085128
—
β⁻ (99.69)
α (0.31)
17.81 d

Cm250
250.078316
—
SF (~80)
α (~11)
β⁻ (~9)
9700 y

Cf252
252.081621
—
α (96.91)
SF (3.09)
2.645 y

Cm249
249.075949
—
β⁻
—
64.15 m

Bk250
250.078316
—
β⁻
—
3.217 h

Cf251
251.079581
—
α
—
888 y

Cm248
248.072344
—
α (91.74)
SD (8.26)
340,000 y

Bk249
249.074978
—
β⁻
—
320 d

Cf250
250.076401
—
α (99.92)
SF (0.08)
13.08 y

Cm247
247.070347
—
α
—
15.6·10⁶ y

Bk248
248.073116
—
α (>70)
β⁻ (<30)
>9 y

Pu246
246.070200
—
β⁻
—
10.84 d

Cf249
249.074844
—
α
—
351 y

Cm246
246.067219
—
α (99.97)
SF (0.03)
4730 y

Bk247
247.070300
—
α
—
1380 y

Am246
246.069770
—
β⁻
—
39 m

Pu245
245.067821
—
β⁻
—
10.5 h

Cf248
248.072182
—
α
—
333.5 d

Cm245
245.065484
—
α
—
8500 y

Bk246
246.068825
—
ε
—
1.80 d

Am245
245.066443
—
β⁻
—
2.05 h

Pu244
244.064200
—
α (99.88)
SF (0.12)
80.8·10⁶ y

Cf247
247.070993
—
ε (99.97)
α (0.03)
3.11 h

Cm244
244.062747
—
α
—
18.10 y

Bk245
245.066355
—
ε (99.88)
α (0.12)
4.94 d

Am244
244.064280
—
β⁻
—
10.1 h

Pu243
243.061998
—
β⁻
—
4.956 h

U240
240.056587
—
β⁻
—
14.1 h

Cf246
246.068800
—
α
—
35.7 h

Cm243
243.061382
—
α (99.76)
ε (0.24)
29.1 y

Am243
243.061373
—
α
—
7380 y

Pu242
242.058737
—
α
—
373,300 y

Np240
240.056169
—
β⁻
—
61.9 m

U239
239.054289
—
β⁻
—
23.5 m

Cm242
242.058830
—
α
—
162.8 d

Am242
242.059542
—
β⁻ (82.7)
ε (17.3)
16.02 h

Pu241
241.056846
—
β⁻
—
14.35 y

Np239
239.052932
—
β⁻
—
2.355 d

U238
238.050785
99.2745
α
—
4.468·10⁹ y

Am241
241.056823
—
α
—
432.7 y

Pu240
240.053808
—
α
—
6563 y

Np238
238.050941
—
β⁻
—
2.117 d

U237
237.048725
—
β⁻
—
6.75 d

Pu239
239.052157
—
α
—
24.120 y

Np237
237.048168
—
α
—
2.14·10⁶ y

U236
236.045563
—
α
—
23.42·10⁶ y

Pu238
238.049555
—
α
—
87.74 y

U235
235.043924
0.7200
α
—
704·10⁶ y

U234
234.040947
0.0055
α
—
245,000 y

A10. MINERALS

Minerals form rocks, and rocks form planets. Minerals, therefore, are the primary constituents of the solid phase of planetary matter. About 3,500 minerals have been described and classified. Table A10 lists about 350 common minerals, their chemical compositions, and their physical properties. The type of mineral that forms not only depends upon the chemical environment, but also upon the physical environment—mainly temperature and pressure. These two parameters change from the surface of the Earth to the center of the core, as shown below. Minerals change phase accordingly. The names of minerals often have colorful histories. Anyone interested in learning the origin of some mineral name, should consult the book *Mineral Names*, by Richard Mitchell, published in 1979 by Van Nostrand Reinhold, New York (226 p.).

Layer		Depth (km)	Temp. (°C)	Pressure (atm)	Density (g/cm³)	Gravity (m/s²)	Velocity (km/s) P waves	Velocity (km/s) S waves	Mass 10²¹ kg	percent
		0	1			9.81				
Crust (lithosphere / oceanic / continental)		—	—	—	2.8	—	6	3.6	24	0.4
		12								
		35	500	2200/	2.9	9.84	7.2	4.3		
				9600	3.3		8.1	4.5		
		65								
		120	1300	38,200	3.4	9.87	8.0	4.4		
		170	—	—	3.45	9.89	7.8	4.3		
Mantle	upper								1206	20.2
		220	—	72,200	3.5	9.90	8.0	4.4		
		670	2400	242,300	4.4	10.01	10.5	5.9		
	lower	2000	—	—	5.1	—	12.8	6.9	2846	47.6
		2885	3800	1,372,000	5.5	10.69	13.7	7.2		
					9.9		8.0	0.0		
Core	outer		—	—	—	—	—	—	1768	29.6
		4720	6000	3,067,200	11.9	5.74	9.9	0.0		
	transition	5000	—	—	—	—	10.1	0.0	12	0.2
		5170	6300	3,341,800	12.7	4.36	10.8			
	inner	5750	—	—	12.9	—	—	—	120	2
		6371	6600	3,680,500	13.0	0	11.15			

Table A10. Minerals

Name	Composition	Crystal System[a]	G[b]	H[c]	n[d]	Remarks
Acanthite	Ag_2S	M, C	7.3	$2-2\frac{1}{2}$	—	—
Acmite	$NaFe^{3+}(SiO_3)_2$	M	3.5	$6-6\frac{1}{2}$	1.82	—
Actinolite	$Ca_2(Mg,Fe)_5Si_8O_{22}(OH)_2$	M	3.1–3.3	5–6	1.65	—
Adularia	$KAlSi_3O_8$	M				Translucent
Aegirine	—	—				Acmite
Agate	—	—				Concentrically layered chalcedony
Alabaster						Cryptocrystalline gypsum
Albite	$NaAlSi_3O_8$	Tc	2.62	6	1.53	Na-end member of plagioclase series
Alexandrite						Gem chrysoberyl
Alkali feldspar						Na or K feldspar
Allanite	$(Ce,Ca)_3(Fe^{2+},Fe^{3+})Al_2O(SiO_4)(Si_2O_7)(OH)$	M	3.5–4.2	$5\frac{1}{2}-6$	1.70–1.81	A garnet
Almandine	$Fe_3Al_2Si_3O_{12}$	C	4.32	7	1.83	Almandine
Alunite	$KAl_3(SO_4)_2(OH)_6$	T	2.6–2.8	4	1.57	
Amalgam						An alloy of Hg with Ag or Au
Amblygonite	$LiAlFPO_4$	Tc	3.0–3.1	6	1.60	
Amethyst	SiO_2					Purple quartz
Amphiboles	$Q_{2-3}R_5(Si,Al)_8O_{22}(OH)_2$					A group of minerals where $Q=$ Mg, Fe^{2+}, Ca, Na; $R=$ Mg, Fe^{2+} Fe^{3+}, Al
Analcime	$NaAlSi_2O_6\cdot H_2O$	C	2.27	$5-5\frac{1}{2}$	1.48–1.49	
Anatase	TiO_2	Tt	3.9	$5\frac{1}{2}-6$	2.6	
Andalusite	Al_2SiO_5	O	3.14–3.20	$7\frac{1}{2}$	1.64	
Andesine	$Ab_{70}An_{30}-Ab_{50}An_{50}$	Tc	2.69	6	1.55	A plagioclase; Ab = albite; An = anorthite
Andradite	$Ca_3Fe_2Si_3O_{12}$	C	3.86	7	1.89	A garnet

(cont.)

Table A10. Minerals (*continued*)

Name	Composition	Crystal System[a]	G^b	H^c	n^d	Remarks
Anglesite	$PbSO_4$	O	6.2–6.4	3	1.88	—
Anhydrite	$CaSO_4$	O	2.89–2.98	3–$3\frac{1}{2}$	1.58	—
Ankerite	$CaFe(CO_3)_2$	T	2.95–3	$3\frac{1}{2}$	1.70–1.75	—
Anorthite	$CaAl_2Si_2O_8$	Tc	2.76	6	1.58	Ca-end member of the plagioclase series
Anorthoclase	$(K,Na)AlSi_3O_8$–$NaAlSi_3O_8$	Tc	2.58	6	1.53	An alkali feldspar
Anthophyllite	$(Mg,Fe)_7Si_8O_{22}(OH)_2$	O	2.85–3.2	$5\frac{1}{2}$–6	1.61–1.71	An amphibole
Antigorite	$Mg_3Si_2O_5(OH)_4$	M	2.5–2.6	4	1.55	Platy serpentine
Antimony	Sb	R	6.7	3	—	—
Antlerite	$Cu_3SO_4(OH)_4$	O	3.9	$3\frac{1}{2}$–4	1.74	—
Apatite	$Ca_5(PO_4)_3(F,Cl,OH)$	H	3.15–3.20	5	1.63	—
Apophyllite	$KCa_4(Si_4O_{10})_2F \cdot 8H_2O$	Tt	2.3–2.4	$4\frac{1}{2}$–5	1.54	—
Aquamarine	—	—	—	—	—	Gem beryl
Aragonite	$CaCO_3$	O	2.95	$3\frac{1}{2}$–4	1.68	—
Argentite	—	—	—	—	—	Acanthite
Arsenic	As	T	5.7	$3\frac{1}{2}$	—	—
Arsenopyrite	$FeAsS$	M	6.07	$5\frac{1}{2}$–6	—	—
Asbestos	—	—	—	—	—	Commercial name for a group of fibrous silicates (see chrysotile, crocidolite)
Atacamite	$Cu_2Cl(OH)_3$	O	3.75–3.77	3–$3\frac{1}{2}$	1.86	—
Attapulgite	—	—	—	—	—	Palygorskite
Augite	$(Ca,Na)(Mg,Fe,Al)[(Si,Al)O_3]_2$	M	3.2–3.5	5–6	1.67–1.73	A pyroxene
Autunite	$Ca(UO_2)_2(PO_4)_2 \cdot 10$–$12H_2O$	Tt	3.1–3.2	2–$2\frac{1}{2}$	1.58	—
Azurite	$Cu_3(CO_3)_2(OH)_2$	M	3.77	$3\frac{1}{2}$–4	1.76	—
Baddeleyite	ZrO_2	M	5.4–6.0	6.5	2.13–2.20	—
Barite	$BaSO_4$	O	4.5	3–$3\frac{1}{2}$	1.64	—
Bauxite	—	—	—	—	—	A mixture of Al oxides and hydroxides

Name	Formula	System	G	H	n	Remarks
Beidellite	$(Ca,Na)_{0.3}Al_2(OH)_2(Al,Si)_4O_{10}\cdot(H_2O)_4$	M	2–3	1–2	—	Member of the montmorillonite group
Bentonite	—	—	—	—	—	Montmorillonite and colloidal silica produced by devitrification of volcanic ash
Beryl	$Be_3Al_2(Si_6O_{18})$	H	2.64–2.8	$7\frac{1}{2}$–8	1.57–1.61	—
Biotite	$K(Mg,Fe)_3(AlSi_3O_{10})(OH)_2$	M	2.8–3.2	$2\frac{1}{2}$–3	1.61–1.70	Black mica
Bismuth	Bi	T	9.8	2–$2\frac{1}{2}$	—	—
Bismuthinite	Bi_2S_3	O	6.78	2	—	—
Bloodstone	—	—	—	—	—	Heliotrope
Boehmite	$\gamma AlO(OH)$	O	3.1–3.6	$3\frac{1}{2}$–4	1.65	γ-phase of diaspore
Bog iron ore	—	—	—	—	—	Iron hydroxides, mainly limonite
Boracite	$Mg_3ClB_7O_{13}$	O	2.9–3.0	7	1.66	—
Borax	$Na_2B_4O_5(OH)_4\cdot8H_2O$	M	1.7	2–$2\frac{1}{2}$	1.47	—
Bornite	Cu_5FeS_4	Tt, C	5.06–5.08	3	—	—
Brochantite	$Cu_4SO_4(OH)_6$	M	3.9	$3\frac{1}{2}$–4	1.78	—
Bronzite	$(Mg,Fe)SiO_3$	O	3.3	$5\frac{1}{2}$–6	1.68	An orthopyroxene
Brucite	$Mg(OH)_2$	T	2.39	$2\frac{1}{2}$	1.57	—
Bytownite	$Ab_{30}An_{70}$–$Ab_{10}An_{90}$	Tc	2.74	6	1.57	A plagioclase
Calcite	$CaCO_3$	T	2.71	3	1.66	—
Cancrinite	$Na_6Ca(CO_3)(AlSiO_4)_6\cdot2H_2O$	H	2.45	5–6	1.52	A feldspathoid
Carnallite	$KMgCl_3\cdot6H_2O$	O	1.6	1	1.48	—
Carnelian	—	—	—	—	—	Improper spelling for cornelian
Carnotite	$K_2(UO_2)_2(VO_4)_2\cdot3H_2O$	M	4.7–5	—	1.93	Powdery incrustations
Cassiterite	SnO_2	Tt	6.8–7.1	6–7	2.00	—
Cat's eye	—	—	—	—	—	Gem variety of chrysoberyl
Celestite	$SrSO_4$	O	3.95–3.97	3–$3\frac{1}{2}$	1.62	—
Cerargyrite	$AgCl$	C	5.5	2–3	2.07	—
Cerussite	$PbCO_3$	O	6.55	3–$3\frac{1}{2}$	2.08	—
Chabazite	$Ca_2Al_2Si_4O_{12}\cdot6H_2O$	T	2.05–2.15	4–5	1.48	—

(cont.)

Table A10. Minerals (continued)

Name	Composition	Crystal System^a	G^b	H^c	n^d	Remarks
Chalcanthite	$CuSO_4 \cdot 5H_2O$	Tc	2.12–2.30	$2\frac{1}{2}$	1.54	—
Chalcedony	—	—	—	—	—	Semitranslucent microcrystalline quartz
Chalcocite	Cu_2S	O, H	5.5–5.8	$2\frac{1}{2}$–3	—	—
Chalcopyrite	$CuFeS_2$	Tt	4.1–4.3	$3\frac{1}{2}$–4	—	—
Chalcosiderite	$CuFe_6(PO_4)_4(OH)_8 \cdot 4H_2O$	Tc	3.22	$4\frac{1}{2}$	1.84	—
Chalk	—	—	—	—	—	An aggregate of small calcitic particles
Chert	SiO_2	—	2.65	7	1.54	Opaque, compact, microcrystalline quartz
Chiastolite	—	—	—	—	—	A variety of andalusite with carbonaceous impurities
Chlorapatite	$Ca_5(PO_4)_3Cl$	—	—	—	—	A Cl-rich apatite
Chlorargyrite	AgCl	C	5.5	2–3	2.07	—
Chlorite	$(Mg,Fe)_3(Si,Al)_4O_{10} \cdot (OH)_2(Mg,Fe)_3(OH)_6$	M, Tc	2.6–3.3	2–$2\frac{1}{2}$	1.57–1.67	—
Chloritoid	$(Fe,Mg)Al_4O_2(SiO_4)_2(OH)_4$	M, Tc	3.5–3.8	$6\frac{1}{2}$	1.72–1.73	—
Chondrodite	$Mg_5(SiO_4)_2(F,OH)_2$	M	3.1–3.2	6–$6\frac{1}{2}$	1.60–1.63	—
Chromite	$FeCr_2O_4$	C	4.6	$5\frac{1}{2}$	2.16	—
Chrysoberyl	$BeAl_2O_4$	O	3.65–3.8	$8\frac{1}{2}$	1.75	—
Chrysocolla	$Cu_2H_2(Si_2O_5)(OH)_4$	—	2.0–2.4	2–4	1.4	Cryptocrystalline or amorphous
Chrysolite	—	—	—	—	—	A variety of olivine with 10–30 mol percent of Fe_2SiO_4
Chrysotile	$Mg_3Si_2O_5(OH)_4$	M	2.5–2.6	4	1.55	Fibrous serpentine (an asbestos)
Cinnabar	HgS	T	8.10	$2\frac{1}{2}$	2.81	—
Citrine	SiO_2	—	—	—	—	Yellow quartz
Clinoenstatite	$MgSiO_3$	M	3.19	6	1.66	A clinopyroxene
Clinohypersthene	$(Mg,Fe)SiO_3$	M	3.4–3.5	5–6	1.68–1.72	A clinopyroxene

Name	Formula		S.G.	Hardness	R.I.	Remarks
Clinozoisite	$Ca_2Al_3O(SiO_4)Si_2O_7(OH)$	M	3.25–3.37	6–$6\frac{1}{2}$	1.67–1.72	—
Cobaltite	$(Co, Fe)AsS$	O	6.33	$5\frac{1}{2}$	—	—
Coesite	SiO_2	M	2.915	7	1.49	High-pressure phase of silica
Colemanite	$CaB_3O_4(OH)_3 \cdot H_2O$	M	2.42	4–$4\frac{1}{2}$	1.59	—
Collophane	—	—	—	—	—	Cryptocrystalline apatite
Columbite	$(Fe, Mn)Nb_2O_6$	O	5.2–7.3	6	—	—
Copper	Cu	C	8.9	$2\frac{1}{2}$–3	—	—
Cordierite	$(Mg, Fe)_2Al_4Si_5O_{18} \cdot nH_2O$	O	2.60–2.66	7–$7\frac{1}{2}$	1.53–1.57	—
Cornelian	—	—	—	—	—	Red chalcedony
Corundum	Al_2O_3	T	3.99	9	1.77	—
Coulsonite	FeV_2O_4	C	—	—	—	A spinel
Covellite	CuS	H	4.60–4.76	$5\frac{1}{2}$–$6\frac{1}{2}$, $1\frac{1}{2}$–2	—	—
Cristobalite	SiO_2	Tt, C	2.32	$6\frac{1}{2}$	1.48	High-temperature phase of silica
Crocidolite	$Na(Mg, Fe^{2+})_3Fe_2^{3+}Si_8O_{22}(OH)_2$	M	3.2–3.3	4	1.70	Blue amphibole asbestos
Crocoite	$PbCrO_4$	M	5.9–6.1	$2\frac{1}{2}$–3	2.36	—
Cryolite	Na_3AlF_6	M	2.95–3.0	$2\frac{1}{2}$	1.34	—
Cummingtonite	$(Mg, Fe)_7Si_8O_{22}(OH)_2$	M	3.1–3.3	$5\frac{1}{2}$–6	1.66–1.68	—
Cuprite	Cu_2O	C	6.1	$3\frac{1}{2}$–4	—	—
Dahllite	$Ca_5(PO_4, CO_3)_3(OH)$	H	3.2–3.3	5	—	Carbonate-hydroxylapatite; cf. francolite
Datolite	$CaB(SiO_4)(OH)$	M	2.8–3.0	5–$5\frac{1}{2}$	1.65	—
Diallage	—	—	—	—	—	A lamellar variety of augite or diopside
Diamond	C	C	3.51	10	2.42	High-pressure phase of graphite
Diaspore	$\alpha AlO(OH)$	O	3.35–3.45	$6\frac{1}{2}$–7	1.72	Cf. Boehmite
Diatomaceous earth	—	—	—	—	—	Diatomite
Diatomite	—	—	—	—	—	An aggregate of diatom frustules
Diopside	$CaMg(SiO_3)_2$	M	3.3	5–6	1.67	A clinopyroxene
Dioptase	$Cu_6(Si_6O_{18}) \cdot 6H_2O$	H	3.3	5	1.65	—

Table A10. Minerals (continued)

Name	Composition	Crystal System[a]	G[b]	H[c]	n[d]	Remarks
Dolomite	$CaMg(CO_3)_2$	T	2.87	$3\frac{1}{2}$–4	1.68	—
Electron	—	—	—	—	—	See electrum
Electrum	—	—	—	—	—	A natural alloy of Au (80%) and Ag (20%)
Emerald	—	—	—	—	—	Gem beryl
Emery	—	—	—	—	—	Corundum with Fe oxides
Enargite	Cu_3AsS_4	O	4.45	3	—	—
Enstatite	$MgSiO_3$	O	3.2	$5\frac{1}{2}$	1.65	—
Epidote	$Ca_2(Al,Fe)Al_2O(SiO_4)\cdot(Si_2O_7)(OH)$	M	3.25–3.50	6–7	1.72–1.78	—
Epsomite	$MgSO_4\cdot7H_2O$	O	1.75	$2-2\frac{1}{2}$	1.46	—
Epsom salt	—	—	—	—	—	See epsomite
Euclase	$BeAl(SiO_4)(OH)$	M	3.1	$7\frac{1}{2}$	1.66	—
Fayalite	Fe_2SiO_4	O	4.39	$6\frac{1}{2}$	1.86	Fe-end member of the olivine group
Feldspars	$QAl(Al,Si)Si_2O_8$	—				A group of minerals where Q = K, Na, Ca, Ba
Feldspathoids	—	—				A group of low-silica Na, K, Ca, Al-silicates, the most common being leucite, nepheline, cancrinite, sodalite
Fergusonite	$(REE,Fe)NbO_4$	Tt	5.8	$5\frac{1}{2}$–6	2.07	REE = rare-earth elements
Flint	SiO_2	—				Homogeneous microcrystalline quartz
Fluorapatite	$Ca_5(PO_4)_3F$	—	—	—	—	An F-rich apatite
Fluorite	CaF_2	C	3.18	4	1.43	—
Forsterite	Mg_2SiO_4	O	3.2	$6\frac{1}{2}$	1.63	Mg-end member of the olivine group

Francolite	$Ca_5(PO_4,CO_3)_3F$	H	3.1–3.2	5	—	Carbonate-fluorapatite; cf. dahllite
Franklinite	$(Zn,Fe,Mn)(Fe,Mn)_2O_4$	C	5.15	6	—	—
Gadolinite	$YFeBe_2(SiO_4)_2O_2$	M	4–4.5	$6\frac{1}{2}$–7	1.79	—
Gahnite	$ZnAl_2O_4$	C	4.55	$7\frac{1}{2}$–8	1.80	—
Galena	PbS	C	7.6	$2\frac{1}{2}$	—	—
Garnet	$Q_3R_2(SiO_4)_3$	C	3.5–4.3	$6\frac{1}{2}$–$7\frac{1}{2}$	1.71–1.88	A group of minerals where Q = Ca, Mg, Fe^{2+}, Mn^{2+}; R = Al, Fe^{3+}, Mn^{3+}, V^{3+}, Cr, Ti, Zr
Garnierite	$(Ni,Mg)_3Si_2O_5(OH)_4$	M	2.2–2.8	2–3	1.59	—
Gaylussite	$Na_2Ca(CO_3)_2 \cdot 5H_2O$	M	1.99	2–3	1.52	—
Geyserite	—	—	—	—	—	Opaline silica incrustations in hot springs
Gibbsite	$Al(OH)_3$	M	2.3–2.4	$2\frac{1}{2}$–$3\frac{1}{2}$	1.57	—
Glauconite	$(K,Na)(Al,Fe,Mg)_2(Al,Si)_4O_{10}(OH)_2$	M	2.4	2	1.62	—
Glaucophane	$Na_2(Mg,Fe)_3Al_2Si_8O_{22}(OH)_2$	M	3.1–3.3	6–$6\frac{1}{2}$	1.62–1.67	An amphibole
Goethite	$\alpha FeO(OH)$	O	4.37	5–$5\frac{1}{2}$	2.39	—
Gold	Au	C	19.32	$2\frac{1}{2}$–3	—	—
Goldmanite	$Ca_3V_2Si_3O_{12}$	C	—	—	—	A garnet
Graphite	C	H	2.27	1–2	—	—
Greenalite	$(Fe,Mg)_3Si_2O_5(OH)_4$	M	3.2	—	1.67	—
Greenockite	CdS	H	4.9	3–$3\frac{1}{2}$	—	—
Grossularite	$Ca_3Al_2Si_3O_{12}$	C	3.59	$6\frac{1}{2}$	1.73	A garnet
Grünerite	$Fe_7Si_8O_{22}(OH)_2$	M	3.6	6	1.71	—
Gypsum	$CaSO_4 \cdot 2H_2O$	M	2.32	2	1.52	—
Halite	$NaCl$	C	2.16	$2\frac{1}{2}$	1.54	—
Halloysite	$Al_2Si_2O_5(OH)_4$ and $Al_2Si_2O_5(OH)_4 \cdot 2H_2O$	M	2.0–2.2	1.2	1.54	A clay mineral
Harmotome	$Ba(Al_2Si_6O_{16}) \cdot 6H_2O$	M	2.45	$4\frac{1}{2}$	1.51	A zeolite
Hausmannite	Mn_3O_4	Tt	4.84	$5\frac{1}{2}$–6	2.15–2.46	—
Haüynite	$(Na,Ca)_{4-8}(AlSiO_4)_6(SO_4)_{1-2}$	C	2.4–2.5	$5\frac{1}{2}$–6	1.5	A feldspathoid
Hedenbergite	$CaFe(SiO_3)_2$	M	3.55	5–6	1.73	Ca-end member of the clinopyroxene group

(cont.)

Table A10. Minerals (continued)

Name	Composition	Crystal System[a]	G^b	H^c	n^d	Remarks
Heliotrope	—	—	—	—	—	Red and green chalcedony
Hematite	αFe_2O_3	R	5.27	$5\frac{1}{2}-6\frac{1}{2}$	—	—
Hemimorphite	$Zn_4(Si_2O_7)(OH)_2 \cdot H_2O$	O	3.4–3.5	$4\frac{1}{2}-5$	1.62	—
Hercynite	$FeAl_2O_4$	C	4.39	$7\frac{1}{2}-8$	1.80	Fe spinel
Hessite	Ag_2Te	C, M	8.4	$2\frac{1}{2}-3$	—	—
Heulandite	$(Na, Ca)_{2-3}Al_3(Al, Si)_2 Si_{13}O_{36} \cdot 12H_2O$	M	2.18–2.20	$3\frac{1}{2}-4$	1.48	—
Hornblende	$(Ca, Na)_{2-3}(Mg, Fe, Al)_5(Si, Al)_8O_{22}(OH)_2$	M	3.0	5–6	1.62–1.72	The commonest amphibole
Hyacinth	—	—	—	—	—	Gem zircon
Hyalite	—	—	—	—	—	Colorless opal
Hydroxylapatite	$Ca_5(PO_4)_3(OH)$	—	—	—	—	—
Hypersthene	$(Mg, Fe)SiO_3$	O	3.4–3.5	5–6	1.68–1.73	An orthopyroxene
Iceland spar	—	—	—	—	—	Pure and transparent calcite
Idocrase	—	—	—	—	—	See vesuvianite
Illite	$(K, H_3O)(Al, Mg, Fe)_2 \cdot (Si, Al)_4O_{10}[(OH)_2, H_2O]$	M	2.6–2.9	1–2	1.54–1.63	Predominant clay mineral in mid-latitudes
Ilmenite	$FeTiO_3$	T	4.8	$5\frac{1}{2}-6$	—	—
Indicolite	—	—	—	—	—	A blue, gem variety of tourmaline
Iridium	Ir	C	22.65	6–7	—	A platinum-group metal
Iron	Fe	C	7.3–7.9	$4\frac{1}{2}$	—	—
Jacinth	—	—	—	—	—	Improper spelling of hyacinth
Jade	—	—	—	—	—	Microcrystalline jadeite or nephrite
Jadeite	$NaAl(SiO_3)_2$	M	3.3	$6\frac{1}{2}-7$	—	Microcrystalline clinopyroxene
Jasper	—	—	—	—	—	Microcrystalline red quartz

Name	Formula		Density	Hardness	Refractive index	Description
Kainite	$MgSO_4 \cdot KCl \cdot 3H_2O$	M	2.1	3	1.51	—
Kamacite	$\alpha Fe, Ni$	C	7.3–7.9	4	—	Fe(93–95%)–Ni(5–7%) alloy occurring in meteorites
Kaolin	—	—	—	—	—	A mixture of kaolinite and other clay minerals
Kaolinite	$Al_4Si_4O_{10}(OH)_8$	Tc	2.6	2	1.55–1.57	A clay mineral dominant in the tropics
Kermesite	Sb_2S_2O	M	4.5–4.6	1–1.5	—	—
Kernite	$Na_2B_4O_6(OH)_2 \cdot 3H_2O$	M	1.95	3	1.47	—
K feldspars	—	—	—	—	—	Orthoclase, microcline, or sanidine
Kieserite	$MgSO_4 \cdot H_2O$	M	2.57	$3\frac{1}{2}$	1.53	—
Kyanite	Al_2SiO_5	Tc	3.67	5–7	1.72	—
Labradorite	$Ab_{50}An_{50}$–$Ab_{30}An_{70}$	Tc	2.71	6	1.56	A plagioclase
Lapis lazuli	—	—	—	—	—	A blue rock consisting mainly of lazurite
Laumontite	$Ca(Al_2Si_4O_{12}) \cdot 4H_2O$	M	2.28	4	1.52	A zeolite
Lawsonite	$CaAl_2(Si_2O_7)(OH)_2 \cdot H_2O$	O	3.09	8	1.67	—
Lazulite	$(Mg, Fe)Al_2(PO_4)_2(OH)_2$	M	3.0–3.1	$5–5\frac{1}{2}$	1.64	—
Lazurite	$(Na, Ca)_8(AlSiO_4)_6(SO_4, S, Cl)_2$	C	2.40–2.45	$5–5\frac{1}{2}$	1.5	—
Lechatelierite	SiO_2	A	2.2	6–7	1.46	Fused, glassy silica
Lepidocrocite	$\gamma FeO(OH)$	O	4.09	5	2.2	—
Lepidolite	$(K, Rb)(Li, Al)_3(Si, Al)_4O_{10}(OH, F)_2$	M	2.8–2.9	$2\frac{1}{2}–4$	1.55–1.59	A mica
Leucite	$KAlSi_2O_6$	Tt, C	2.47	$5\frac{1}{2}–6$	1.51	A feldspathoid
Limonite	$FeO(OH) \cdot nH_2O$	A	3.6–4.0	$5–5\frac{1}{2}$	—	—
Litharge	PbO	Tt	9.14	2	2.66	—
Lithiophilite	$Li(Mn, Fe)PO_4$	O	3.5	5	1.67	—
Lodestone	—	—	—	—	—	Naturally magnetized magnetite
Maghemite	γFe_2O_3	C	4.88	5	2.52–2.74	—
Magnesite	$MgCO_3$	T	3.0–3.2	$3\frac{1}{2}–5$	1.70	—
Magnetite	Fe_3O_4	C	5.20	6	—	—
Malachite	$Cu_2CO_3(OH)_2$	M	3.90–4.03	$3\frac{1}{2}–4$	1.88	—

(cont.)

645

Name	Composition	Crystal System[a]	G[b]	H[c]	n[d]	Remarks
Manganite	$MnO(OH)$	M	4.3	4	—	—
Marcasite	FeS_2	O	4.89	$6-6\frac{1}{2}$	—	—
Meerschaum	—	—	—	—	—	Massive sepiolite
Melilite	$Ca_2(Mg,Al)(Al,Si)_2O_7$	Tt	2.9-3.0	5-6	1.65	—
Mercury	Hg	—	13.6	—	—	—
Micas	$(K,Na,Ca)(Mg,Fe,Li,Al)_{2-3} \cdot (AlSi)_4O_{10}(OH,F)_2$	M	—	—	—	A group of phyllosilicates
Microcline	$KAlSi_3O_8$	Tc	2.54-2.57	6	1.53	Low-temperature K feldspar
Microlite	$Ca_2Ta_2O_6(O,OH,F)$	C	5.48-5.56	$5\frac{1}{2}$	1.92-1.99	—
Microperthite	—	—	—	—	—	A perthite with thin (5–100 μm) lamellae
Minium	Pb_3O_4	—	8.9-9.2	$2\frac{1}{2}$	2.42	—
Molybdenite	MoS_2	H	4.62-4.73	$1-1\frac{1}{2}$	—	—
Monazite	$(REE,Th)PO_4$	M	4.6-5.4	$5-5\frac{1}{2}$	1.79	—
Monticellite	$CaMgSiO_4$	O	3.2	5	1.65	—
Montmorillonite	$(Na,Ca)(Al,Mg)_4(Si_4O_{10})_3(OH)_6 \cdot nH_2O$	M	2.5	$1-1\frac{1}{2}$	1.50-1.64	A clay mineral
Moonstone		—				Translucent adularia
Mother of pearl	$CaCO_3$ (aragonite)	O	2.95	$3\frac{1}{2}-4$	1.68	Microcrystalline, lamellar aragonite
Mullite	$Al_6Si_2O_{13}$	O	3.23	6-7	1.67	—
Muscovite	$KAl_2(AlSi_3O_{10})(OH)_2$	M	2.76-2.88	$2-2\frac{1}{2}$	1.60	A mica
Nacrite	$Al_2Si_2O_5(OH)_4$	M	2.6	$2-2\frac{1}{2}$	1.56	—
Natrolite	$Na_2Al_2Si_3O_{10} \cdot 2H_2O$	O	2.25	$5-5\frac{1}{2}$	1.48	—
Natron	$Na_2CO_3 \cdot 10H_2O$	M	—	—	—	—
Nepheline	$(Na,K)AlSiO_4$	H	2.60-2.65	$5\frac{1}{2}-6$	1.54	—
Nephrite	—	—	—	—	—	Microcrystalline tremolite or actinolite
Niccolite	$NiAs$	H	7.78	$5-5\frac{1}{2}$	—	—
Nickeline	—	—	—	—	—	Niccolite
Niter	KNO_3	O	2.09-2.14	2	1.50	—

Name	Formula	Crystal system	Specific gravity	Hardness	Refractive index	Remarks
Nontronite	$Na_{0.33}Fe^{3+}_2(Al_{0.33}Si_{3.67})O_{10}(OH)_2 \cdot nH_2O$	M	2.5	$1-1\frac{1}{2}$	1.60	A clay mineral
Oligoclase	$Ab_{90}An_{10}-Ab_{70}An_{30}$	Tc	2.65	6	1.54	A plagioclase
Olivine	$(Mg,Fe)_2SiO_4$	O	3.27–4.37	$6\frac{1}{2}-7$	1.69	—
Omphacite	$(Ca,Na)(Mg,Fe,Al)(SiO_3)_2$	M	3.2–3.4	5.6	1.67–1.70	A clinopyroxene
Onyx	—	A	—	—	—	Banded chalcedony
Opal	$SiO_2 \cdot nH_2O$	A	2.0–2.25	5–6	1.44	—
Orpiment	As_2S_3	M	3.49	$1\frac{1}{2}-2$	2.8	—
Orthite	—	—	—	—	—	Allanite in slender crystals
Orthoclase	$KAlSi_3O_8$	M	2.55	6	1.52	An alkali feldspar
Orthoferrosilite	$FeSiO_3$	O	3.9	6	1.79	Fe-end member of orthopyroxenes
Palladium	Pd	C	11.9	$4\frac{1}{2}-5$	—	A platinum-group metal
Palygorskite	$Mg_2(Al,Fe)_2(Si_2O_5)_4(OH)_2 \cdot 4H_2O$	—	—	—	—	A fibrous clay mineral
Paragonite	$NaAl_2(AlSi_3O_{10})(OH)_2$	M	2.85	2	1.60	A mica
Patronite	VS_4	M	—	—	—	—
Pearl	$CaCO_3$ (aragonite)	M	—	—	—	Microcrystalline, lamellar aragonite
Pectolite	$NaCa_2Si_3O_8(OH)$	Tc	2.8	5	1.60	—
Pentlandite	$(Fe,Ni)_9S_8$	C	4.6–5.0	$3\frac{1}{2}-4$	—	—
Periclase	MgO	C	3.58	$5\frac{1}{2}$	1.73	—
Peridot	—	—	—	—	—	Gem olivine
Perovskite	$CaTiO_3$	O	4.03	$5\frac{1}{2}$	2.38	—
Perthite	—	—	—	—	—	Lamellar interspacing of microcline and albite
Petalite	$Li(AlSi_4O_{10})$	M	2.4	$6-6\frac{1}{2}$	1.51	—
Phenacite	Be_2SiO_4	R	2.97–3.0	$7\frac{1}{2}-8$	1.65	—
Phillipsite	$(K_2,Na_2,Ca)Al_2Si_4O_{12} \cdot 4-H_2O$	M	2.2	$4\frac{1}{2}-5$	1.50	A zeolite
Phlogopite	$KMg_3(AlSi_3O_{10})(OH)_2$	M	2.86	$2\frac{1}{2}-3$	1.56–1.64	—
Phosphorite	—	—	—	—	—	A sedimentary rock consisting mainly of phosphatic and calcitic minerals and bioclasts
Pigeonite	$(Ca,Mg,Fe^{2+})(Mg,Fe^{2+})(SiO_3)_2$	M	3.30–3.46	6	1.64–1.72	A clinopyroxene

(cont.)

647

Table A10. Minerals (continued)

Name	Composition	Crystal System[a]	G^b	H^c	n^d	Remarks
Pitchblende				—	—	Massive uraninite
Plagioclase	$Ab_{100}An_0$–Ab_0An_{100}	Te	2.62–2.76	6	1.53–1.59	Complete solid solution from albite ($NaAlSi_3O_8$) to anorthite ($CaAl_2Si_2O_8$)
Platinum	Pt	C	21.45	4–$4\frac{1}{2}$	—	
Pollucite	$CsAlSi_2O_6 \cdot H_2O$	C	2.9	$6\frac{1}{2}$	1.52	
Prehnite	$Ca_2Al_2Si_3O_{10}(OH)_2$	O	2.8–2.95	6–$6\frac{1}{2}$	1.63	
Proustite	Ag_3AsS_3	T	5.57	2–$2\frac{1}{2}$	3.09	
Pseudowollastonite	$CaSiO_3$	Te			—	High-temperature phase of wollastonite
Psilomelane	$BaMn^{2+}Mn_8^{4+}O_{16}(OH)_4$	O	3.7–4.7	5–6	—	
Pyrargyrite	Ag_3SbS_2	T	5.85	2–$2\frac{1}{2}$	3.08	
Pyrite	FeS_2	C	5.02	6–$6\frac{1}{2}$	—	
Pyrochlore	$(Ca,Na)_2(Nb,Ta)_2O_6(O,OH,F)$	C	4.3	5	—	
Pyrolusite	MnO_2	Tt	4.75	1–2	—	
Pyromorphite	$Pb_5(PO_4)_3Cl$	H	7.04	$3\frac{1}{2}$–4	2.06	
Pyrope	$Mg_3Al_2Si_3O_{12}$	C	3.58	7	1.71	A garnet
Pyrophyllite	$AlSi_4O_{10}(OH)_2$	M	2.8	1–2	1.59	
Pyroxenes	$QRSi_2O_6$				—	A group of Ca, Na, Mg, Fe silicates; Q = Ca, Na, Mg, Fe^{2+}; R = Mg, Fe^{2+}, Fe^{3+}, Fe, Cr, Mn, Al
Pyrrhotite	$F_{0.8-1}S$	M, H	4.58–4.65	4	—	
Quartz	SiO_2	T, H	2.65	7	1.54	
Realgar	AsS	M	3.48	$1\frac{1}{2}$–2	2.60	
Rhodochrosite	$MnCO_3$	T	3.5–3.7	$3\frac{1}{2}$–4	1.82	
Rhodonite	$MnSiO_3$	Te	3.4–3.7	$5\frac{1}{2}$–6	1.73–1.75	
Riebeckite	$Na_2(Mg,Fe^{2+})_3Fe_2^{3+}Si_8O_{22}(OH)_2$	M	3.4	5	1.66–1.71	
Rock crystal					—	Megacrystalline quartz
Rock salt					—	Halite

Mineral	Composition	Crystal system	G	H	n	Remarks
Rubellite	—	—	—	—	—	Red-to-pink tourmaline
Ruby	—	—	—	—	—	Red gem corundum
Ruby spinel	—	—	—	—	—	Red gem spinel
Rutile	TiO_2	Tt	4.25	6–$6\frac{1}{2}$	2.61	—
Saltpeter	—	—	—	—	—	Niter
Sanidine	$KAlSi_3O_8$	M	2.56–2.62	6	1.53	High-temperature K feldspar
Saponite	$(0.5Ca, Na)_{0.33}(Mg, Fe)_3(Si_{3.67}Al_{0.33})O_{10}(OH)_2 \cdot 4(H_2O)$	M	2.5	1–$1\frac{1}{2}$	1.52	A montmorillonitic clay
Sapphire	—	—	—	—	—	Blue gem corundum
Scapolite	$3NaAlSi_3O_8 \cdot NaCl$ to $3CaAl_2Si_2O_8 \cdot CaCO_3$	Tt	2.55–2.74	5–6	1.55–1.60	—
Scheelite	$CaWO_4$	Tt	5.9–6.1	$4\frac{1}{2}$–5	1.92	—
Schreibersite	$(Fe, Ni)_3P$	Tt	—	—	—	In iron meteorites
Scolecite	$CaAl_2Si_3O_{10} \cdot 3H_2O$	M	2.2	5–$5\frac{1}{2}$	1.52	A zeolite
Selenite	—	—	—	—	—	Megacrystalline gypsum
Sepiolite	$Mg_4(Si_2O_5)_3(OH)_2 \cdot 6H_2O$	O	2.0	2–$2\frac{1}{2}$	1.52	A fibrous clay mineral
Sericite	—	—	—	—	—	Fine-grained mica
Serpentine	$(Mg, Fe)_3Si_2O_5(OH)_4$	M, O	2.5–2.6	3–5	1.55	—
Siderite	$FeCO_3$	T	3.94	$3\frac{1}{2}$–4	1.88	—
Sillimanite	Al_2SiO_5	O	3.25	6–7	1.66	—
Silver	Ag	C	10.5	$2\frac{1}{2}$–3	—	—
Smithsonite	$ZnCO_3$	T	4.30–4.45	4–$4\frac{1}{2}$	1.85	—
Smoky quartz	—	—	—	—	—	Brown gem quartz
Soapstone	—	—	—	—	—	Steatite
Sodalite	$Na_8(AlSiO_4)_6Cl_2$	C	2.15–2.30	$5\frac{1}{2}$–6	1.48	—
Sperrylite	$PtAs_2$	C	10.50	6–7	—	—
Spessartine	$Mn_2Al_2Si_3O_{12}$	C	4.19	7	1.80	A garnet
Sphalerite	ZnS	C	3.9–4.1	$3\frac{1}{2}$–4	2.37	—
Sphene	$CaTiO(SiO_4)$	M	3.40–3.55	5–$5\frac{1}{2}$	1.91	—
Spinel	$MgAl_2O_4$	C	3.5–4.1	8	1.72	—
Spinel group	QR_2O_4	C	—	—	—	A group of minerals where $Q = Mg$, Fe^{2+}, Fe^{3+}, Mn, Zn; $R = Al$, Fe^{2+}, Fe^{3+}, Ti^{4+}, Cr, V

Table A10. Minerals (continued)

Name	Composition	Crystal System[a]	G[b]	H[c]	n[d]	Remarks
Spodumene	$LiAl(SiO_3)_2$	M	3.15–3.20	$6\frac{1}{2}$–7	1.67	—
Stannite	Cu_2FeSnS_4	Tt	4.3–4.5	4	—	—
Staurolite	$(Mg,Fe)_2Al_9Si_4O_{22}(O,OH)_2$	M	3.65–3.75	7–7½	1.75	Massive, compact talc
Steatite	—	—	—	—	—	
Stibnite	Sb_2S_3	O	4.52–4.62	2.0	1.50	—
Stilbite	$CaAl_2Si_7O_{18}\cdot 7H_2O$	M	2.1–2.2	$3\frac{1}{2}$–4	1.80	—
Stishovite	SiO_2	Tt	4.28	7	—	High-pressure phase of silica
Strontianite	$SrCO_3$	O	3.7	$3\frac{1}{2}$–4	1.67	—
Sulfur	S	O	2.07	$1\frac{1}{2}$–$2\frac{1}{2}$	2.04	—
Sylvanite	$(Au,Ag)Te_2$	M	8.0–8.2	$1\frac{1}{2}$–2	—	—
Sylvite	KCl	C	1.99	2	1.49	—
Taenite	$\gamma Fe,Ni$	C	7.8–8.2	5	—	Fe(35–70%)–Ni (30–65%) alloy occurring in meteorites
Talc	$Mg_3Si_4O_{10}(OH)_2$	M	2.7–2.8	1	1.59	—
Tantalite	$(Fe,Mn)(Ta,Nb)_2O_6$	O	6.5	6	—	—
Tanzanite	—	—	—	—	—	Blue gem zoisite
Tennantite	$Cu_{12}As_4S_{13}$	C	4.6–5.1	3–$4\frac{1}{2}$	—	—
Tenorite	CuO	Tc	6.5	3–4	—	—
Tetrahedrite	$Cu_{12}Sb_4S_{13}$	C	4.6–5.1	3–$4\frac{1}{2}$	—	—
Thorianite	ThO_2	C	9.7	$6\frac{1}{2}$	—	—
Thorite	$ThSiO_4$	Tt	5.3	5	1.8	—
Tiger eye	—	—	—	—	—	A gem variety of quartz
Tin	Sn	Tt	7.3	2	—	—
Titanaugite	$Ca(Mg,Fe,Ti)[(Si,Al)O_3]_2$	M	3.3	6	1.7	Cf. augite
Titanite	—	—	—	—	—	Sphene
Topaz	$Al_2SiO_4(OH,F)_2$	O	3.4–3.6	8	1.61–1.63	—
Tourmaline	$(Na,Ca)(Li,Mg,Al)\cdot(Al,Fe,Mn)_6(BO_3)_3Si_6O_{18}(OH)_4$	T	3.0–3.25	7–$7\frac{1}{2}$	1.64–1.68	—
Travertine	—	—	—	—	—	Hardened freshwater limestone

Mineral	Formula	Crystal system[a]	Density[b]	Hardness[c]	Refractive index[d]	Comment
Tremolite	$Ca_2(Mg,Fe)_5Si_8O_{22}(OH)_2$	M	3.0–3.2	5–6	1.61	—
Tridymite	SiO_2	M, O	2.26	7	1.47	High-temperature polymorph of quartz
Troilite	FeS	H	4.83	4	—	—
Trona	$Na_2CO_3 \cdot NaHCO_3 \cdot 2H_2O$	M	2.13	3	1.49	—
Tufa	—	—	—	—	—	Soft freshwater limestone
Turquoise	$CuAl_6(PO_4)_4(OH)_8 \cdot 4H_2O$	Tc	2.6–2.8	6	1.62	—
Ulexite	$NaCaB_5O_6(OH)_6 \cdot 5H_2O$	Tc	1.96	$1–2\frac{1}{2}$	1.50	—
Ulvöspinel	Fe_2TiO_4	C	4.78	$7\frac{1}{2}–8$	—	—
Uraninite	UO_2	C	7.5–9.7	$5\frac{1}{2}–$	—	—
Uvarovite	$Ca_3Cr_2Si_3O_{12}$	C	3.90	$7\frac{1}{2}$	1.87	A garnet
Vanadinite	$Pb_5(VO_4)_3Cl$	H	6.9	3	2.25–2.42	—
Vermiculite	$(Mg,Fe,Al)_3(Al,Si)_4O_{10}(OH)_2 \cdot 4H_2O$	M	2.4	$1\frac{1}{2}$	1.55–1.58	—
Vesuvianite	$Ca_{10}Mg_2Al_4(SiO_4)_5(Si_2O_7)_2(OH)_4$	Tt	3.35–3.45	$6\frac{1}{2}$	1.70–1.75	—
Wavellite	$Al_3(PO_4)_2(OH)_3 \cdot 5H_2O$	O	2.36	$3\frac{1}{2}–4$	1.54	—
Wernerite	—	—	—	—	—	Scapolite
Willemite	Zn_2SiO_4	T	3.9–4.2	$5\frac{1}{2}$	1.69	—
Witherite	$BaCO_3$	O	4.3	$3\frac{1}{2}$	1.68	—
Wolframite	$(Fe,Mn)WO_4$	M	7.0–7.5	$4–4\frac{1}{2}$	—	—
Wollastonite	$CaSiO_3$	Tc	2.8–2.9	$5–5\frac{1}{2}$	1.63	—
Wulfenite	$PbMoO_4$	Tt	6.8	3	2.40	—
Wurtzite	ZnS	H	3.98	4	2.35	—
Wüstite	FeO	—	—	—	—	—
Xanthophyllite	$Ca(Mg,Al)_3(Al_2Si_2O_{10})(OH)_2$	M	3–3.1	$3\frac{1}{2}$	1.65	—
Xenotime	YPO_4	Tt	4.3–4.7	4–5	1.72–1.83	—
Zeolites	—	—	—	—	—	Hydrous Na, K, Ca aluminosilicates
Zincite	ZnO	H	5.68	4	2.01	—
Zircon	$ZrSiO_4$	Tt	4.68	$7\frac{1}{2}$	1.92–1.96	—
Zoisite	$Ca_2Al_3Si_3O_{12}(OH)$	O	3.35	6	1.69	—

[a] Crystal systems: C, cubic; H, hexagonal; M, Monoclinic; O, orthorhombic; T, trigonal; Tc, triclinic; Tt, tetragonal; A, amorphous.

[b] Density relative to that of water at 3.98°C.

[c] Hardness (Mohs' scale).

[d] Refractive index.

Source: Emiliani (1987, pp 303–14).

A11. PERIODIC TABLE

The Periodic Table was constructed by the Russian chemist Dmitri Mendeléev in 1869. It contained only 61 elements (31 more were discovered later). Mendeléev lined up the elements in columns containing elements with similar properties, and in rows along which the properties showed a gradient. He left empty positions in which no known element would fit. He assumed that the positions would be filled as new elements were discovered. Five new elements had been discovered in the eight years preceding the publication of the Table, which made Mendeléev confident that new discoveries would be forthcoming. In fact, he was so confident that he predicted the properties of three of the missing elements—gallium, scandium, and germanium. When these three elements were discovered (in 1875, 1879, and 1885, respectively), their properties matched Mendeléev's predictions. Mendeléev quickly became the most famous chemist of his time.

In the Periodic Table, the columns are called *groups*, and the rows *periods*. The table shows the new notation, numbering the groups consecutively from 1 to 18. Below it the older notations of IUPAC (*International Union of Pure and Applied Chemistry*) and of CAS (*Chemical Abstract Service*) are shown. The number of electron shells increases downcolumn, and the number of electrons in the valence shells increases from left to right along the rows. The shells are shown in the column to the extreme right; after the top three rows in this column, only the three outer shells are shown. The "key to chart" explains the numbers and symbols contained in each square. Notice that the electron configuration is given only for the three outermost shells, the inner ones having been filled. The two bottom rows, separate from the rest of the table, include the lanthanides and actinides, all of which have -8-2, -9-2, or -10-2 electrons in their outer shells and therefore have very similar chemical properties.

The element symbols Unq, Unp, Unh, Uns, Uno, and Une mean *Unnilquadium, Unnilpentium, Unnilhexium, Unnilseptium, Unniloctium,* and *Unnilennium,* referring to the yet unnamed elements 104, 105, 106, 107, 108, and 109, respectively. These crazy symbols have been concocted by IUPAC as follows: *Un* (abbreviated from the Latin *unus*) refers to the number 1; *nil* (contracted from the Latin *nihil*) refers to the number 0; and the remaining letter is the initial for each the invented pseudo-Latin words *quadium* (for *quattuor*), *pentium* (from the Greek πέντε, *five*), *hexium* (from the Greek ἕξ, *six*), *septium* (from the Latin *septem, seven*), *octium* (from the Latin *octo* or the Greek ὀκτώ, *eight*), and *ennium* (from the Greek ἐννέα, *nine*). Thus, unnilquadium (abbreviated Unq) means 104, unnilpentium (Unp) means 105, and so forth. All of this may be found in the *Nomenclature of Inorganic Chemistry,* Recommendations 1990, page I-3.4, Section I-3.3.5, issued by IUPAC and published by Blackwell Scientific Publications, Oxford.

Table A11. Periodic Table of the Elements

KEY TO CHART

50	+2
Sn	+4
118.71	
18 18 4	

- Atomic Number → 50
- Oxidation States → +2 +4
- Symbol → Sn
- 1989 Atomic Weight → 118.71
- Electron Configuration → 18 18 4

New notation → Previous IUPAC form → CAS version

Group	1 IA	2 IIA	3 IIIA IIIB	4 IVA IVB	5 VA VB	6 VIA VIB	7 VIIA VIIB	8 VIIIA VIII	9 VIIIA VIII	10	11 IB	12 IIB	13 IIIB IIIA	14 IVB IVA	15 VB VA	16 VIB VIA	17 VIIB VIIA	18 VIIIA	Shell
	1 H +1 −1 1.00794 1																	2 He 0 4.0020602 2	K
	3 Li +1 6.941 2-1	4 Be +2 9.012182 2-2											5 B +3 10.811 2-3	6 C +2 +4 −4 12.011 2-4	7 N +1 +2 +3 +4 +5 −1 −2 −3 14.00674 2-5	8 O −2 15.9994 2-6	9 F −1 18.9984032 2-7	10 Ne 0 20.1797 2-8	K-L
	11 Na +1 22.989768 2-8-1	12 Mg +2 24.3050 2-8-2											13 Al +3 26.981539 2-8-3	14 Si +2 +4 −4 28.0855 2-8-4	15 P +3 +5 −3 30.97362 2-8-5	16 S +4 +6 −2 32.066 2-8-6	17 Cl +1 +5 +7 −1 35.4527 2-8-7	18 Ar 0 39.948 2-8-8	K-L-M
	19 K +1 39.0983 -8-8-1	20 Ca +2 40.078 -8-8-2	21 Sc +3 44.955910 -8-9-2	22 Ti +2 +3 +4 47.88 -8-10-2	23 V +2 +3 +4 +5 50.9415 -8-11-2	24 Cr +2 +3 +6 51.9961 -8-13-1	25 Mn +2 +3 +4 +6 +7 54.93085 -8-13-2	26 Fe +2 +3 55.847 -8-14-2	27 Co +2 +3 58.93320 -8-15-2	28 Ni +2 +3 58.6934 -8-16-2	29 Cu +1 +2 63.546 -8-18-1	30 Zn +2 65.39 -8-18-2	31 Ga +3 69.723 -8-18-3	32 Ge +2 +4 72.61 -8-18-4	33 As +3 +5 −3 74.92159 -8-18-5	34 Se +4 +6 −2 78.96 -8-18-6	35 Br +1 +5 −1 79.904 -8-18-7	36 Kr 0 83.80 -8-18-8	-L-M-N
	37 Rb +1 85.4678 -18-8-1	38 Sr +2 87.62 -18-8-2	39 Y +3 88.90585 -18-9-2	40 Zr +4 91.224 -18-10-2	41 Nb +3 +5 92.90638 -18-12-1	42 Mo +6 95.94 -18-13-1	43 Tc +4 +6 +7 (98) -18-13-2	44 Ru +3 101.07 -18-15-1	45 Rh +3 102.90550 -18-16-1	46 Pd +2 +4 106.42 -18-18-0	47 Ag +1 107.8682 -18-18-1	48 Cd +2 112.411 -18-18-2	49 In +3 114.82 -18-18-3	50 Sn +2 +4 118.710 -18-18-4	51 Sb +3 +5 −3 121.757 -18-18-5	52 Te +4 +6 −2 127.60 -18-18-6	53 I +1 +5 +7 −1 126.90447 -18-18-7	54 Xe 0 131.29 -18-18-8	-M-N-O
	55 Cs +1 132.90543 -18-8-1	56 Ba +2 137.327 -18-8-2	57* La +3 138.9055 -18-9-2	72 Hf +4 178.49 -32-10-2	73 Ta +5 180.9479 -32-11-2	74 W +6 183.85 -32-12-2	75 Re +4 +6 +7 186.207 -32-13-2	76 Os +3 +4 190.2 -32-14-2	77 Ir +3 +4 192.22 -32-15-2	78 Pt +2 +4 195.08 -32-16-2	79 Au +1 +3 196.96654 -32-18-1	80 Hg +1 +2 200.59 -32-18-2	81 Tl +1 +3 204.3833 -32-18-3	82 Pb +2 +4 207.2 -32-18-4	83 Bi +3 +5 208.98037 -32-18-5	84 Po +2 +4 (209) -32-18-6	85 At (210) -32-18-7	86 Rn 0 (222) -32-18-8	-N-O-P
	87 Fr +1 (223) -18-8-1	88 Ra +2 226.025 -18-8-2	89** Ac +3 227.028 -18-9-2	104 Unq +4 (261) -32-10-2	105 Unp (262) -32-11-2	106 Unh (263) -32-12-2	107 Uns (262) -32-13-2												O P Q

*Lanthanides

58 Ce +3 +4 140.115 -20-8-2	59 Pr +3 +4 140.90765 -21-8-2	60 Nd +3 144.24 -22-8-2	61 Pm +3 (145) -23-8-2	62 Sm +2 +3 150.36 -24-8-2	63 Eu +2 +3 151.965 -25-8-2	64 Gd +3 157.25 -25-9-2	65 Tb +3 158.92534 -27-8-2	66 Dy +3 162.50 -28-8-2	67 Ho +3 164.93032 -29-8-2	68 Er +3 167.26 -30-8-2	69 Tm +3 168.93421 -31-8-2	70 Yb +2 +3 173.04 -32-8-2	71 Lu +3 174.967 -32-9-2	N O P

**Actinides

90 Th +4 232.0381 -18-10-2	91 Pa +4 +5 231.03588 -20-9-2	92 U +3 +4 +5 +6 238.0289 -21-9-2	93 Np +3 +4 +5 +6 237.048 -22-9-2	94 Pu +3 +4 +5 +6 (244) -24-8-2	95 Am +3 +4 +5 +6 (243) -25-8-2	96 Cm +3 (247) -25-9-2	97 Bk +3 +4 (247) -27-8-2	98 Cf +3 (251) -28-8-2	99 Es +3 (252) -29-8-2	100 Fm +3 (257) -30-8-2	101 Md +2 +3 (258) -31-8-2	102 No +2 +3 (259) -32-8-2	103 Lr +2 +3 (260) -32-9-2	O P Q

The new IUPAC format numbers the groups from 1 to 18. The previous IUPAC numbering system and the system used by Chemical Abstracts Service (CAS) are also shown. For radioactive elements that do not occur in nature, the mass number of the most stable isotope is given in parentheses.

653

A12. PLANETS: PHYSICAL DATA

Table A12 includes basic data concerning the planets. Many interesting observations can be made from these data. For example:

The Earth is 3.3% closer to the Sun at perihelion (which occurs during the northern winter) than at aphelion.

One year for Mercury lasts 88 of our days, whereas Pluto's year is as long as 248.5 of our years.

Planetary orbits, except for Mercury and Pluto, have inclinations within 3.5° of each other.

The outer planets (Jupiter to Pluto) rotate faster than the inner planets (Mercury to Mars).

Except for Uranus and Pluto, the rotational axes of the planets are within 30° of the normal to their orbital plane.

Jupiter is 2.5 times more massive than all other planets combined.

The mean density of the inner planets is $5.32 \, \text{g/cm}^3$, while that of the outer planets is $1.2 \, \text{g/cm}^3$.

We would weigh 2.4 times more on Jupiter than on Earth.

Table A12. Planets: Physical Data

	Mercury	Venus	Earth	Mars	Jupiter	Saturn	Uranus	Neptune	Pluto
Mean distance from Sun (AU)	0.387099	0.723332	1.000000	1.523688	5.202561	9.554747	19.21814	30.10957	39.44
Perihelion distance (10^6 km)	46.0	107.4	147.1	206.6	740.5	1,349	2.738	4.463	4.443
Aphelion distance	69.9	108.8	152.1	249.1	815.8	1,504	3.002	4.537	7,375
Eccentricity	0.2056	0.0068	0.0167	0.0934	0.0485	0.0556	0.0472	0.0086	0.248
Sidereal period (tropical years)	0.24085	0.61521	1.00004	1.88089	11.8623	29.4577	84.0139	164.79	248.5
Sidereal period (d)	87.969	224.701	365.256	686.980	4,322.71	10,759.5	30,685	60,190	90,800
Inclination of orbit over ecliptic (degrees)	7.00	3.39	0.00	1.85	1.30	2.49	0.77	1.77	17.17
Sidereal rotation period	58.65 d	243.01 d[a]	23.9345 h	24.6229 h	9.841 h[b]	10.233 h[b]	17.24 h[a,b]	18.2 h[b]	6.3874 d
Inclination of equator to orbit (degrees)	0.0	3.4	23.44	23.98	3.08	29.00	97.92	28.8	88 to 112
Equatorial radius (km)	2,439	6,051.4	6,378.164	3,398	71,492	60,268	25,400	24,750	1,145
Equatorial radius (Earth = 1)	0.382	0.949	1.000	0.533	11.21	9.45	3.98	3.81	0.18
Oblateness	0.000	0.000	0.00334894	0.0059	0.0637	0.102	0.024	0.0266	?
Mean orbital velocity (km/s)	47.89	35.03	29.79	24.13	13.06	9.64	6.81	5.43	4.74
Mass (10^{24} kg)	0.3302	4.871	5.9737	0.6421	1,899.728	568.8	86.9	103.0	0.0130
Volume (10^{21} m³)	0.0603	0.929	1.0834	0.1617	1,460.841	851.832	66.238	62.263	0.0000063
Mean density (g/cm³)	5.48	5.243	5.515	3.970	1.33	0.67	1.31	1.65	2.06
Equatorial surface gravity[c] (m/s²)	3.78	8.60	9.78	3.72	23.12	9.05	7.77	11.00	0.59
Equatorial escape velocity (km/s)	4.3	10.3	11.2	5.0	59.5	35.6	21.2	23.6	1.33
Albedo	0.06	0.72	0.39	0.16	0.70	0.75	0.90	0.82	0.61
Number of satellites	0	0	1	2	16	17	15	8	1

Note: explanation of terms:

albedo: Reflectivity of a surface. It is usually expressed as percentage of incident energy. Examples: smooth, open ocean with vertical sun, 2–4%; forest, 10%; grassland, 20%; desert, 30%; cloud tops, 50–80%; fresh snow, 80%.

aphelion: The point along a planetary orbit that is farthest from the sun.

ecliptic: The intersection of the plane of the Earth's orbit with the celestial sphere.

oblateness: The degree of flattening of an ellipsoid of revolution. It is expressed as the ratio $(a-c)/a$, where a = semimajor axis, and c = semiminor axis.

perihelion: The point along a planetary orbit that is closest to the sun.

prograde: Counterclockwise orbital or rotational motion as seen from the north. All planets have prograde orbital motions; the Sun and all planets, except Venus, Uranus, and Pluto also have prograde rotational motions.

retrograde: Clockwise orbital or rotational motion as seen from the north.

sidereal: Pertaining to or referring to the stars.

tropical: Pertaining to the apparent motion of the Sun with respect to the Earth. The tropical year is the time interval between successive passages of the Sun through the vernal (spring) equinox.

[a] Retrograde.
[b] At equator.
[c] Including centrifugal terms.

Source: Emiliani (1987, p. 326), updated 1992.

A13. SYMBOLS AND ABBREVIATIONS

Innumerable symbols and abbreviations, from both the Latin and Greek alphabets, are used in science. Here we present a list of the more common ones. Notice that often a given symbol is used for more than one quantity. The meaning is generally clear from the context.

Table A13. Symbols and Abbreviations

Symbol	Name
α	activity
	alpha particle
	angular acceleration
	fine-structure constant
	isotopic fractionation factor
	right ascension
a	absorbance
	acceleration
	activity
	semimajor axis of an elliptical orbit
A	ampere
	annus (Latin for *year*)
A	area
	atomic mass number
	azimuth
Å	angstrom
a_0	Bohr radius
aA	abampere
ac	alternating current
aC	abcoulomb
A.C.	*Ante Christum*, Latin for *before Christ* (cf. A.D.)
A.D.	*Anno Domini*, Latin for *year of the Lord*, referring to any year after the birth of Christ, taken as having occurred at 0h 0m 0s on January 1, 1 C.E. The preceding year was 1 B.C.E. (there is no year 0).
aF	abfarad
aH	abhenry
AM	amplitude modulation
amp	ampere
amu	atomic mass unit, taken as being equal to 1/16 of the mass of the neutral atom ^{16}O (obsolete; see u)
atm	atmosphere (unit of pressure)
atto-	10^{-18}
AU	astronomical unit
Av	avogadro
av.	average
aV	abvolt
$a\Omega$	abohm
β	beta particle
β^+	beta-plus particle (= positron)
β^-	beta-minus particle (= electron)
b	bar
	barn
	semiminor axis of an elliptical orbit
B	magnetic flux density; magnetic induction
B.C.	before Christ (syn. A.C.; cf. A.D.)
B.C.E.	Before the Common Era ($\equiv$ B.C.)

Symbol	Name
BeV	billion electron volts (obsolete abbreviation, now replaced by GeV)
BIF	banded iron formation (referring to Proterozoic stromatolites)
b.p.	boiling point
B.P.	before the present (referring to the time before 1950, the year when W. F. Libby developed radiocarbon dating)
BTU	British Thermal Unit
c	centi-
	specific heat capacity
	speed of light in vacuo ($m\,s^{-1}$ in the SI; $cm\,s^{-1}$ in the cgs system)
	speed of sound
C	capacitance
	Celsius; centigrade
	coulomb
	heat capacity
C_p	heat capacity at constant pressure
C_v	heat capacity at constant volume
cal	calorie
cal_{IT}	International Table calorie
cc	cubic centimeter
cd	candela
C.E.	Common Era ($\equiv$ A.D.)
cf.	*confer*, Latin for *compare with*
cgs	centimeter-gram-second, a metric system of units now replaced by the SI
Ci	curie
centi-	10^{-2}
cm	centimeter
COP	coefficient of performance
cos	cosine
cosec	cosecant
cot	cotangent
covers	coversine
cP	centipoise
cps	cycles per second (cf. Hz)
CRT	cathode-ray tube
csc	cosecant
cSt	centistoke
ct	carat
δ	declination
∂	partial derivative
d	day
	diameter
d	dextrorotatory
d-	deci-
D	deuterium
	dextral chirality
D	diffusion coefficient
D	electric displacement
d_E	ephemeris day
da-	deca-, deka-
dam	decameter
dB	decibel
dc	direct current
deca-	10

(*cont.*)

Table A13.   **Symbols and Abbreviations** (*continued*)

Symbol	Name
deci-	10^{-1}
deka-	10
dm	decimeter
DNA	deoxyribonucleic acid
dpm	disintegrations per minute
dps	decays per second
dyn	dyne
ε	permittivity ($= \mathbf{D/E}$)
ε_0	permittivity constant
e	eccentricity
	electron charge
	$2.7182818284590\ldots$
$\mathbf{E}$	electrical field strength
E_k	kinetic energy
E_p	potential energy
EDTA	ethylenediaminetetraacetic acid
e.g.	*exempli gratia*, Latin for *for example*
emf	electromotive force
emu	electromagnetic unit
ERTS	Earth Resources Technology Satellite
ESCA	Electron Spectroscopy for Chemical Analysis
ESR	Electron Spin Resonance
esu	electrostatic unit
eV	electron volt
exp	exponential
η	viscosity
θ, ϑ	phase angle
φ	angular displacement
	latitude
	phase angle
f	focal length
	force
	frequency
F	Fahrenheit
	farad
	faraday ($=$ Faraday constant)
	Faraday constant ($=$ charge of 1 Av of electrons or protons $= 96{,}484.53$ coulombs)
	force
	formality
femto-	10^{-15}
FET	field-effect transistor
fm	femtometer
FM	frequency modulation
fps	foot-pound-second
ft	foot
γ	activity coefficient
	gamma ($= 10^{-5}$ oersted)
	photon
	surface tension

Symbol	Name
Γ	gamma ($= 10^{-5}$ oersted)
g	gram
g	gravitational acceleration at the Earth's surface
G	conductance
G	gauss
	gravitational constant
G-	giga-
g_0	standard gravitational acceleration at the Earth's surface ($= 980.665$ gal, exactly)
gal	galileo ($= 1\,\mathrm{cm\,s}^{-2}$)
g cal	gram-calorie
GeV	giga electron volt ($= 10^9\,\mathrm{eV}$)
giga-	10^9
GMAT	Greenwich Mean Astronomical Time
GMT	Greenwich Mean Time
GST	Greenwich Sidereal Time
g.u.	gravity unit ($= 10^{-6}\,\mathrm{ms}^{-2} = 10^{-4}\,\mathrm{gal}$)
Gy	gigayear ($= 10^9\,\mathrm{y}$)
h	celestial altitude
	hour
	Planck's constant
h-	hecto-
$\hbar$	h-bar ($= h/2\pi$)
H	enthalpy ($= U + pV$)
H	henry
H	magnetic field strength
H_0	Hubble constant
HI	neutral hydrogen (in space)
HII	ionized hydrogen (in space)
hecto-	100
HFU	heat-flow unit
hp	horsepower
hr	hour
Hz	hertz
i	$(-1)^{1/2}$
	electric-current density
	inclination
I	electric current
	ionic strength
	luminous intensity
	moment of inertia
IAT	International Atomic Time
IC	Integrated Circuit
i.e.	*id est*, Latin for *that is*
in.	inch
IR	infrared
	insoluble residue
j	$(-1)^{1/2}$
	electric-current density
	total-angular-momentum quantum number
j	electric-current density
J	electric-current density

(cont.)

Table A13. Symbols and Abbreviations (*continued*)

Symbol	Name
	total-angular-momentum quantum number
	joule
JD	Julian date
	Julian day
κ	electrical conductivity ($= \mathbf{J/E}$)
k	Boltzmann constant
	thermal conductivity
k-	kilo-
K	kelvin
	kinetic energy
K	equilibrium constant
kb	kilobar
kcal	kilocalorie ($= 1{,}000$ g cal)
keV	kilo electron volt ($= 10^3$ eV)
kg	kilogram
kilo-	10^3
km	kilometer
kt	karat
kV	kilovolt ($= 10^3$ V)
kWh	kilowatt-hour
λ	decay constant
	wavelength
l	length
	levorotatory
	orbital-angular-momentum quantum number
l	liter
L	angular momentum
	inductance
	left-handed chirality
	luminance
	luminosity
	self-inductance
	sinistral chirality
	lambert
lb	pound
lbf	pound force
LED	light-emitting diode
LIL	large-ion lithophyle element
lm	lumen
ln	logarithm (natural)
log	logarithm (common)
LST	local sidereal time
LT	local time
lx	lux
l.y.	light year
μ	chemical potential; ionic strength
	magnetic moment
	magneton
	micron
	permeability ($= \mathbf{B/H}$)
	proper motion

Symbol	Name
	viscosity (dynamic)
μ-	micro-
μ_B	Bohr magneton
μ_N	nuclear magneton
μ_0	permeability constant
μ_r	relative permeability ($= \mu/\mu_0$)
μF	microfarad
$\mu\mu$F	micromicrofarad
μb	microbar
μg	microgram
μm	micrometer
m	apparent magnitude
	mass
	meter
	minute
	molal concentration
m-	milli-
$\mathfrak{m}$	electromagnetic moment
m-	meta-
m_e	electron rest mass
m_l	magnetic orbital quantum number
m_n	neutron rest mass
m_0	rest mass
m_p	proton rest mass
m_s	magnetic-spin angular-momentum quantum number
M	absolute magnitude
	Messier number
	molar concentration
M	million
	mutual induction
M-	mega- ($= 10^6$)
M	magnetization
Ma	mega-annus (10^6 y)
MeV	million electron volts
mg	milligram
mgal	milligal
mi	mile
micro-	10^{-6}
milli-	10^{-3}
MKS	meter-kilogram-second, a metric system of units
MKSA	meter-kilogram-second-ampere, a metric system of units
MKSΩ	meter-kilogram-second-ohm, a metric system of units
ml	milliliter ($= 1$ cm^3)
MLW	mean low water
mm	millimeter
mmf	magnetomotive force
mmHg	millimeters of mercury
mol	mole
MORB	mid-oceanic-ridge basalt
MOS	metal-oxide semiconductor
MOSFET	metal-oxide-semiconductor field-effect transistor
Mpc	megaparsec
mRNA	messenger RNA
MSL	mean sea level

(*cont.*)

Table A13. Symbols and Abbreviations (*continued*)

Symbol	Name
Mt	megaton
MTL	mean tide level
mV	millivolt
mW	milliwatt
MW	megawatt
Mx	maxwell
m.y.	million years
ν	frequency
	neutrino
	viscosity (kinematic)
n-	nano-
n	index of refraction
	principal quantum number
n	neutron
N	newton
	north-seeking pole of magnetic dipole
N	normal concentration (1 Av of valences per liter)
N_A	Avogadro number (see Av)
nano-	10^{-9}
ng	nanogram ($= 10^{-9}$ g)
NGC	*New General Catalogue of Nebulae and Stars*
nm	nanometer ($= 10^{-9}$ m)
NMR	nuclear magnetic resonance
n.n.	*nomen nudum*, Latin for *naked name*
Np	neper
NRM	natural remanent magnetization
nt	nit
o-	ortho-
OD	ordinance datum
Oe	oersted
π	3.141592653589...
p	proton
p	momentum
	pressure
P	parity
	poise
	primary or pressure wave
P	osmotic pressure
	permeance
P	dielectric polarization
p-	pico
p-	para-
Pa	pascal
pc	parsec
pdl	poundal
pg	picogram
pH	p(otential of) H(ydrogen) $= -\log$ of H^+ ion concentration in aqueous solutions
pico-	10^{-12}
pK	$-\log$ of ionization constant K
Pl	poiseuille
pOH	$-\log$ of OH^- ion concentration in aqueous solutions

Symbol	Name
ppb	parts per billion
ppm	parts per million
ppt	parts per thousand
PSI	pounds per square inch
q	quintal
q	perihelion distance
Q	aphelion
Q	electric charge
	heat
	quality factor
QCD	quantum chromodynamics
QED	quantum electrodynamics
Q.E.D.	*quod erat demonstrandum*, Latin for *which was to be demonstrated*
QSO	quasi-stellar object (= quasar)
q.v.	*quod vide*, Latin for *which see*
ρ	density
	electric charge density (volumetric)
	resistivity
r	radius
R	cosmic scale factor
	gas constant
	radical
	resistance
R	roentgen
$\mathfrak{R}$	reluctance
R_∞	Rydberg constant
RA	right ascension
rad	radian
	radiation
RAM	random-access memory
REE	rare-earth elements
rem	roentgen-equivalent-man
rms	root-mean-square
RNA	ribonucleic acid
ROM	read-only memory
σ	electrical conductivity
	electric-charge density (surface)
	neutron-capture cross section
	Poisson ratio
	standard deviation
	Stefan–Boltzmann constant
Σ	summation
s	second
S	secondary or shear wave
	siemens
	south-seeking pole of magnetic dipole
S	entropy
s_A	atomic second
s_E	ephemeris second
sb	stilb
sec	secant

(cont.)

Table A13.   **Symbols and Abbreviations** (*continued*)

Symbol	Name
SEM	scanning electron microscope
sr	steradian
s.s.	*sensu stricto*, Latin for *in the strict sense*
St	stoke
statA	statampere
statC	statcoulomb
statF	statfarad
statH	stathenry
statV	statvolt
statΩ	statohm
STP	standard temperature and pressure (0°C, 1 atm)
τ	hour angle
	mean life
	shear stress
t	ton
t	temperature (Celsius)
	time
T	centuries from 1900.0 C.E.
	tesla
	tritium
T	temperature (absolute)
T-	tera-
$t_{1/2}$	half-life
t_a	atomic time
t_E	ephemeris time
T_{eff}	effective temperature
T_U	Universal Time
TAI	Temps Atomique International
tan	tangent
tera-	10^{12}
TL	thermoluminescence
torr	torricelli ($=1$ mmHg)
TRM	thermoremanent magnetization
tRNA	transfer RNA
u	atomic mass unit ($=1/12$ of mass of neutral atom of ^{12}C; cf. amu)
u	velocity
U	internal energy
UT	Universal Time
UV	ultraviolet
v	velocity
v_p	velocity of P waves
v_s	velocity of S waves
V	electric potential
	potential energy
	voltage
	volume
V	volt
vers	versine
VRM	viscous remanent magnetization
w	watt

Symbol	Name
W	energy, work
Wb	weber
Whr	watt-hour
χ	magnetic susceptibility
χ_e	electric susceptibility
X	reactance
X_C	capacitative reactance
X_{CL}	capacitative-inductive reactance
X_L	inductive reactance
XPS	x-ray photoelectron spectroscopy
y	year
Y	admittance $(=1/Z)$
	Young's modulus
yd	yard
ζ	zenith distance
z	redshift parameter
Z	atomic number
	valence
	Zulu
Z	ac impedance $(=R+iX)$
Zulu	Greenwich Mean Time
ω	angular frequency $(=2\pi f)$
	angular velocity $(=d\phi/dt)$
	circular frequency $(=2\pi v)$
Ω	ohm
	solid angle

Source: Emiliani (1987, pp. 339–45).

A14. TAXONOMY

Taxonomy derives from τάξις, which means *arrangement*, and νόμος, which means *law*. In other words, *taxonomy* means *legal arrangement*. Legal arrangements are, of course, a crashing bore, but taxonomy is necessary to see some order in the bewildering variety of life-forms on Earth. Unfortunately, taxonomy is in perpetual flux. In the good old times there were only two kingdoms—animals and plants. Now there are five, with some people pushing for a sixth—the Archaebacteria. The number of phyla has now reached over 100, vying in terms of growth rate with the number of elementary particles. Taxonomists, like particle physicists, age prematurely.

The universal language of taxonomy is Latin, and most of the Latin names of animals and plants have Greek origin or roots. Proper handling of these names requires a knowledge of Latin and Greek. Unfortunately, these languages are far more complicated than English. Both Latin and Greek have declensions, with six cases in Latin and five in Greek. Latin has 67 verbal forms (134 counting the passive), and Greek has 180. By comparison, English has 3 to 6 verbal forms. But not to worry. I have written an article to help people who do not know Latin or Greek to wade through the scientific nomenclature. You will find this article, entitled "Nomenclature and Grammar," in the *Proceedings of the Washington Academy of Sciences*, volume 42 (1952), pages 137–41.

The classification of living and fossil groups presented here is based on Margulis (1981, pp. 353–63). Two superkingdoms (Procaryota and Eucaryota) and five kingdoms (Monera, Protoctista, Fungi, Animalia, Plantae) are recognized. Exclusively fossil taxa are included. No stratigraphic range is given for taxa that left no verified fossil record.

Table A14. Taxonomy

Superkingdom Procaryota (Archean–Recent)
  Kingdom Monera (Achean–Recent)
    Phylum 1. Aphragmabacteria (unable to form cell walls; *Mycoplasma*)
           2. Chemoautotrophs
             Class 1. Sulfur-oxidizing bacteria (*Thiobacillus*)
                   2. Ammonia-oxidizing bacteria (*Nitrobacter*)
                   3. Iron-oxidizing bacteria (*Ferrobacillus*)
           3. Thiopneutes (anaerobic reducers of sulfate or sulfur to H_2S; *Desulfovibrio*)
           4. Methanocreatrices (methane synthesizers; *Methanobacterium*)
           5. Fermenting bacteria (unable to synthesize porphyrins; *Clostridium*)
           6. Spirochaetae (facultative or obligate anaerobes; *Spirochaeta*, *Treponema*)
           7. Thiorhodaceae (green and purple anaerobic protosynthesizers using bacteriochlorophyll and chlorobium chlorophyll; *Chlorobium*, *Chromatium*)
           8. Athiorhodaceae (nonsulfur anaerobic or facultative anaerobic photosynthesizers using bacteriochlorophyll and chlorobium chlorophyll, Archean–Recent; *Rhodospirillum*)
           9. Cyanophyta (blue–green algae, aerobic or facultative anaerobes; use chlorophyll a and b mixtures, Archean–Recent; *Nostoc*, *Oscillatoria*)
          10. Prochlorophyta (procaryotic green algae; *Prochloron*)
          11. Nitrogen-fixing aerobic bacteria (*Azobacter*)
          12. Pseudomonads (aerobic heterotrophs; *Pseudomonas*)
          13. Aeroendospora (aerobic endospore bacteria; *Bacillus*)
          14. Micrococci (aerobes with Krebs; *Paracoccus*, *Sarcina*)
          15. Omnibacteria (aerobic heterotrophs; *Acetobacter*, *Caulobacter*, *Escherichia*, *Leptothrix*, *Neisseria*, *Salmonella*, *Spirillum*)
          16. Actinobacteria (*Actinomyces*, *Streptomyces*)
          17. Myxobacteria (heterotrophic aerobic gliding bacteria; *Beggiatoa*, *Saprospira*)

Superkingdom Eucaryota (Proterozoic–Recent)
  Kingdom Protoctista (single-celled microorganisms and their immediate multicellular descendants; Proterozoic–Recent)
    Phylum 1. Caryoblastea (amitotic amoebae; *Pelomyxa*)
           2. Dinoflagellata (dinoflagellates, Triassic–Recent; *Gymnodinium*, *Notciluca*)
           3. Sarcodina and Rhizopoda (*Amoeba*, *Difflugia*)
           4. Chrysophyta (golden–yellow algae; *Dinobryon*, *Synura*)
           5. Euglenophyta (*Euglena*)
           6. Cryptophyta (cryptomonads; *Cryptomonas*)
           7. Zoomastigina (animal flagellates; *Diplomonas*, *Trichomonas*)
           8. Eumastigophyta (*Vischeria*)
           9. Bacillariophyta (diatoms, Cretaceous–Recent; *Coscinodiscus*, *Nitzschia*)
          10. Haptophyta (Coccolithophoridae, Triassic–Recent; *Coccolithus*, *Emiliania*)
          11. Actinopoda (heliozoans and radiolaria, Cambrian–Recent; *Lampocyrtis*, *Pterocanium*)
          12. Foraminifera (Cambrian–Recent; *Globigerina*, *Globorotalia*, *Cibicides*, *Nodosaria*)
          13. Gamophyta and desmids (*Spirogyra*)
          14. Ciliophora (ciliates; *Paramecium*)

15. Cnidosporidia (parasites; *Myxobolus, Nosema*)
16. Apicomplexa (parasites; *Plasmodium*)
17. Xanthophyta (yellow-green algae; *Botrydium, Vaucheria*)
18. Rhodophyta (red algae, Cambrian–Recent; *Archaeolithothamnium, Lithothamnium, Lithophyllum, Solenopora*)
19. Chlorophyta (green algae, Archean–Recent; *Botryococcus, Chara, Halimeda, Oedogonium, Penicillus, Sargassum, Spyrogyra, Ulotrix, Valonia*)
20. Phaeophyta (brown algae, Silurian–Recent; *Focus, Laminaria, Protaxites*)
21. Labyrinthulamycota (slime nets; *Labyrinthula*)
22. Acrasiomycota (cellular slime molds; *Acrasia*)
23. Myxomycota (plasmodial noncellular slime molds; *Dictyostelium, Polyspondylium*)
24. Plasmodiophoromycota (*Polimyxa*)
25. Hyphochytridiomycota (*Rhyzidiomyces*)
26. Chytridiomycota (*Olpidium*)
27. Oomycota (*Saprolegnia*)

Kingdom Fungi
  Phylum 1. Zygomycota (zygomycetes; *Phycomyces*)
       2. Ascomycota (sac fungi, including yeasts, molds, truffles, etc.; *Ascobolus, Aspergillus, Neurospora, Penicillium, Tuber*, and other truffles)
       3. Basidiomycota (club fungi, including rusts, smuts, mushrooms, etc.; *Agaricus, Amanita* and other mushrooms, *Puccinia*)
       4. Deuteromycota (fungi imperfecti; *Candida, Monilia*)
       5. Mycophycophyta [lichens, consisting of a fungus + a blue-green (often *Nostoc*) or a green alga (often *Trebouxia* or *Pseudotrebouxia*)]

Kingdom Animalia
 Subkingdom Parazoa
  Phylum 1. Placozoa (no polarity or bilateral symmetry; *Trichoplax*)
       2. Porifera (sponges, Cambrian–Recent)
    Class 1. Heteractinida (calcareous spicules, Cambrian–Permian)
         2. Calcarea (calcareous spicules, Carboniferous–Recent; *Clathrina, Eudea, Leucosolenia*)
         3. Demospongiae (spongin with or without siliceous spicules, Cambrian–Recent; *Cliona*, boring sponge; *Euspongia*, horny sponge; *Hippospongia*, bath sponge)
         4. Sclerospongiae (aragonitic skeleton, Ordovician–Recent)
      Order 1. Stromatoporida (Ordovician–Devonian; *Stromatopora*)
           2. Ceratoporellida (Caribbean; *Asterosclera, Ceratoporella*)
           3. Tabulofungida (Pacific; *Stromatospongia*)
         5. Hexactinellida (triaxial siliceous spicules, Cambrian–Recent; *Hexactinella, Hydnoceras, Ventriculites*)
       3. Archaeocyatha (Early–Middle Cambrian)
    Class 1. Monocyatha (single-walled skeleton, Early–Middle Cambrian; *Monocyathus*)
         2. Archaeocyatha (double-walled skeleton, Early–Middle Cambrian; *Archaeocyathellus*)
 Subkingdom Eumetazoa
  Branch Radiata
  Phylum 4. Cnidaria (coelenterates, Ediacaran–Recent)
    Class 1. Hydrozoa (Ediacaran–Recent)
      Order 1. Hydroida (Ediacaran–Recent; *Hydra, Obelia*)
           2. Hydrocorallina (Triassic–Recent; *Millepora, Stylaster*)
           3. Trachylina (Ediacaran–Recent; *Cyclomedusa, Gonionemus, Kirklandia, Olindias*)
           4. Siphonophora (Cambrian–Recent; *Physalia, Velella*)
         2. Scyphozoa (jellyfish, Ediacaran–Recent; *Aurelia, Conomedusites*)
         3. Conulata (Ediacaran–Triassic; *Conomedusites, Conularia*)
         4. Anthozoa (corals and sea anemones, Ediacaran–Recent)

(cont.)

Table A14. Taxonomy (*continued*)

Subclass 1. Tabulata (tabulate corals, Ordovician–Permian; *Favosites, Halysites, Syringopora, Tubipora*)
2. Rugosa (tetracorals, Ordovician–Permian; *Lithostrotion, Zaphrentis*)
3. Schizocorallia (Ordovician–Jurassic; *Tetradium*)
4. Zoantharia (hexacorals, Triassic–Recent)
    Order 1. Actinaria (sea anemones, Cambrian–Recent; *Actinia, Edwardsia, Mackenzia*)
        2. Scleractinia (stony corals, Triassic–Recent; *Acropora, Fungia Montastrea, Porites*)
        3. Zoanthidea (some resemblance to extinct Rugosa in the arrangement of septa, *Zoanthus*)
        4. Antipatharia (black corals; *Antipathes*)
        5. Ceriantharia (*Cerianthus*)
5. Alcyonaria (octocorals Ediacaran–Recent)
    Order 1. Stolonifera (*Tubipora*)
        2. Telestacea (*Telesto*)
        3. Alcyonacea (the soft corals, Cretaceous–Recent; *Alcyonium, Xenia*)
        4. Coenothecalia (Cretaceous–Recent; *Heliopora*)
        5. Gorgonacea (the horny corals, Cretaceous–Recent; *Corallium*, the red corals; *Gorgonia*, the sea fan; *Plexaura*, the sea whip)
        6. Pennatulacea (sea pens, Ediacaran–Recent; *Pennatula*)
Phylum 5. Ctenophora (comb jellies; *Cestum, Folia*)
Branch Bilateria
Grade Acoelomata (lack coelom)
Phylum 6. Mesozoa (small, parasitic, wormlike organisms; *Rhopalura*)
7. Platyhelminthes (flatworms)
    Class 1. Turbellaria (planarians; *Dugesia*)
        2. Trematoda (flukes; *Fasciola*)
        3. Cestoda (tapeworm; *Taenia*)
8. Nemertina (ribbon worms, Recent; *Lineus*)
9. Gnathostomulida [small (< 3.5 mm) marine worms; *Gnathostomula*]
Grade Pseudocoelomata
Phylum 10. Gastrotricha (microscopic pseudocoelomates; *Chaetonotus*)
11. Kinorhyncha (tiny marine animals with segmented cuticle; *Echinoderes*)
12. Acanthocephala (spiny-headed worms; *Echinorhynchus*)
13. Nematoda (roundworms; *Ascaris, Trichina*)
14. Nematomorpha (Gordiacea) (horsehair worms; *Gordius*)
15. Entoprocta (endoproct bryozoids; *Urnatella*)
Grade Coelomata (with mesodermal coelom)
Phylum 16. Ectoprocta (Bryozoa) (Late Cambrian–Recent; *Bugula*)
17. Phoronida (tubular mud-dweller, marine; Cambrian?–Recent; *Phoronis*)
18. Brachiopoda (Cambrian–Recent)
    Class 1. Inarticulata (Cambrian–Recent; *Lingula*)
        2. Articulata (Cambrian–Recent; *Productus, Spirifer, Terebratula*)
19. Mollusca (Cambrian–Recent)
    Class 1. Aplacophora (without shell; *Neomenia*)
        2. Monoplacophora (Cambrian–Recent; *Neopilina*)
        3. Polyplacophora (chitons, Cambrian–Recent; *Chiton*)
        4. Scaphopoda (Ordovician–Recent; *Dentalium*)
        5. Hyolitha (Cambrian–Permian; *Ceratotheca, Hyolithes*)
        6. Rostroconchia (Cambrian–Permian; *Conocardium, Eopteria, Hippocardia, Ribeiria, Technophorus*)
        7. Gastropoda (Cambrian–Recent; *Batillaria, Cerithium, Haliotis, Helix, Littorina, Murex, Natica, Oliva, Patella, Purpura, Strombus*)

8. Bivalvia (pelecypods, Ordovician–Recent; *Arca, Anomia, Astarte, Codakia, Cyprina, Lucina, Macoma, Mactra, Mercenaria, Mya, Mytilus, Natica, Pecten, Solen, Tellina*)
9. Cephalopoda (Cambrian–Recent)
    Order 1. Nautiloidea (nautiloids, Cambrian–Recent; *Nautilus*)
          2. Ammonoidea (ammonoids, Ordovician–Cretaceous; *Ceratites, Turrilites*)
          3. Belemnoidea (belemnoids, Mississippian–Cretaceous; *Belemnites*)
          4. Sepiodiea (cuttlefishes, Jurassic–Recent; *Belosepia, Sepia, Spirula, Spirulirostra*)
          5. Teuthoidea (squids, Jurassic–Recent; *Loligo*)
          6. Octopoda (octopi, Cretaceous–Recent; *Octopus*)
20. Priapulida (wormlike marine animals; *Priapulus*)
21. Siphunculida (peanut worms; *Dendrostoma*)
22. Echiurida (sea cucumbers, Ediacaran–Recent; *Echiurus*)
23. Annelida (worms and wormlike animals; Ediacaran–Recent)
    Class 1. Oligochaeta (oligochaete worms, Silurian?–Recent; *Lumbricus*)
          2. Polychaeta (polychaete worms, Ediacaran–Recent; *Dickinsonia, Nereis, Serpula, Spirorbis, Spriggina*)
          3. Clitellata (Hirudinea) (leeches; *Hirudo*)
24. Pentastomida (wormlike parasites; *Linguatula*)
25. Tardigrada (microscopic, bilaterally symmetrical animals; *Macrobiotus*)
26. Onychophora (with features of both Annelida and Arthropoda, Cambrian–Recent; *Peripatus*)
27. Arthropoda (anthropods)
    Subphylum 1. Chelicerata (Cambrian–Recent)
        Class 1. Trilobita (trilobites; Cambrian–Permian)
            Order 1. Eodiscida (Lower–Middle Cambrian; *Eodiscus*)
               2. Agnostida (Lower Cambrian–Ordovician; *Agnostus*)
               3. Olenellida (Lower Cambrian; *Holmia, Olenellus*)
               4. Proparia (Lower Ordovician–Devonian; *Calymene, Phacops*)
               5. Opisthoparia (Lower Cambrian–Permian; *Bumastus, Scutellum*)
          2. Merostomata
            Order 1. Aglaspida (Cambrian–Ordovician; *Aglaspella, Strabops*)
               2. Eurypterida (Ordovician–Permian; *Eurypterus*)
               3. Xiphosura (Ordovician–Recent; *Limulus*)
          3. Arachnida (spiders, scorpions, ticks, Silurian–Recent)
          4. Pycnogonida (sea spiders, Devonian–Recent; *Palaeopantopus*)
        2. Mandibulata (Cambrian–Recent)
        Class 1. Crustacea (Cambrian–Recent)
            Order 1. Branchiopoda (small, mostly fresh water, Devonian–Recent; *Artemia, Daphnia*)
               2. Ostracoda (ostracods, Cambrian–Recent; *Cypris*)
               3. Cirripedia (barnacles, Silurian–Recent; *Balanus*)
               4. Malacostraca (crabs, crayfishes, lobsters, shrimps, Carboniferous–Recent)
               5. Copepoda (copepods; *Calanus*)
          2. Myriapoda (centipedes, millipedes)
          3. Insecta (insects, Devonian–Recent)
28. Pogonophora (body in three segments, each with separate coelom; no mouth, digestive canal, anus; marine; Ediacaran–Recent; *Lamellisabella*)
29. Echinodermata (Cambrian–Recent)
    Subphylum Crinozoa
    Class 1. Cyamoidea (Middle Cambrian; *Peridionites*)
          2. Cycloidea (Middle Cambrian; *Cymbionites*)
          3. Cystoidea (Ordovician–Permian; *Aristocystites*)
          4. Blastoidea (Ordovician–Permian; *Pentremites*)
          5. Eocrinoidea (Cambrian–Ordovician; *Cryptocrinus, Macrocystella*)

(cont.)

Table A14. Taxonomy (*continued*)

Class 6. Paracrinoidea (Ordovician; *Amygdalocystites, Canadocystis, Comarocystites*)
      7. Crinoidea (sea lilies, Ordovician–Recent; *Antedon, Pentacrinus*)
Subphylum Echinozoa
  Class 1. Edrioasteroidea (Cambrian–Pennsylvanian; *Edrioaster, Stromatocystites*)
      2. Carpoidea (Cambrian–Devonian; *Dendrocystis*)
      3. Machaeridia (Ordovician–Devonian; *Lepidocoleus, Turrilepas*)
      4. Somasteroidea (earliest Ordovician, *Villebrunaster*)
      5. Asteroidea (starfishes, Ordovician–Recent; *Asterias*)
      6. Auluroidea (Ordovician–Mississippian; *Lysophiura, Streptophiura*)
      7. Ophiuroidea (brittle stars, Mississippian–Recent; *Amphipholis*)
      8. Holothuroidea (sea cucumbers, Cambrian–Recent; *Holothuria*)
      9. Echinoidea (sea urchins, sand dollars, Ordovician–Recent; *Cidaris, Echinus, Strongylocentrotus*)
    30. Chaetognatha (arrowworms, marine; *Sagitta*)
    31. Conodonta (eel-like animals known mainly from their teeth, Ordovician–Permian; *Belodus, Falcodus, Paltodus*)
    32. Hemichordata
    Class 1. Enteropneusta (*Balanoglossus*)
      2. Pterobranchia
        Order 1. Rhabdopleurida (Cretaceous–Recent; *Rhabdopleura*)
          2. Cephalodiscida (Ordovician?–Recent; *Cephalodiscus*)
      3. Graptolithina (Cambrian–Mississippian; *Dendrograptus, Dictyonema, Monographus, Tetragraptus*)
    33. Chordata
    Subphylum 1. Urochordata (tunicates, ascidians; *Ciona*)
      2. Cephalochordata (*Branchiostoma*, formerly *Amphioxus*)
      3. Craniata (Vertebrata) Ordovician–Recent)
      Class 1. Agnatha [jawless fishes, Ordovician–Recent; *Cephalaspis, Myxine* (hagfish), *Pteromyzon* (lamprey)]
        2. Acanthodii (spiny fishes with jaws, Silurian–Permian; *Acanthodes*)
        3. Placodermi (jawed, often armored fishes, Silurian–Mississippian; *Coccosteus*)
        4. Chodrichthyes (cartilaginous fish, Devonian–Recent; *Squalus, Raja*)
        5. Osteichthyes (bony fishes, Devonian–Recent)
        Subclass 1. Actinopterygii [ray-finned fishes, Devonian–Recent; *Acipenser* (sturgeon), *Perca, Salmo*]
          2. Sarcopterygii (air-breathing, lobe-finned fishes)
          Order 1. Crossopterygii (ancestors to amphibians, Devonian–Recent; *Latimeria*)
            2. Dipnoi (lungfishes, Devonian–Recent, *Protopterus*)
        6. Amphibia (amphibians, Mississippian–Recent; *Bufo, Rana, Seymouria*)
        7. Reptilia (reptiles, Pennsylvanian–Recent; *Brontosaurus, Diplodocus, Ichthyosaurus, Pteranodon, Pterodactylus, Rhamphorhynchus, Sphenodon, Stegosaurus, Testudo, Tyrannosaurus*)
        8. Aves (birds, Jurassic–Recent; *Archaeopteryx, Dinornis, Hesperonis, Ichthyornis, Aquila, Columba, Gallus*)
        9. Mammalia (mammals, Triassic–Recent)
        Subclass 1. Protheria
          Order 1. Monotremata (egg-laying mammals, *Ornithorhynchus, Tachyglossus*)
          2. Allotheria
          Order 1. Multituberculata (Jurassic–Paleocene; *Bolodon, Psalodon*)
            2. Pantotheria (Jurassic; *Docodon, Melanodon*)

3. Metatheria
    Order 1. Marsupialia [marsupials, Cretaceous–Recent; *Didelphis* (opossum), *Macropus* (kangaroo)]
4. Eutheria [placental mammals, Cretaceous–Recent; *Bos* (ox), *Canis* (dog), *Cebus* (New World monkeys), *Cercopithecus* (Old World monkeys), *Cervus* (stag), *Dasypus* (armadillo), *Delphinus* (dolphin), *Elephas* (Indian Elephant), *Equus* (horse), *Felis* (cat, lynx, ocelot, puma), *Gorilla*, *Homo*, *Lemur*, *Loxodonta* (African elephant), *Mus* (mouse), *Myotis*, (bat), *Orcinus* (killer whale), *Panthera* (leopard, lion, tiger), *Phoca* (seal), *Phocaena* (porpoise), *Rattus* (rat), *Sorex* (shrew), *Talpa* (mole), *Tarsius*, *Tupaia* (tree shrew), *Tursiops* (bottlenose dolphin)]

Kingdom Plantae
  Grade Bryophyta (no true roots, stem, leaves)
    Phylum 1. Bryophyta (Carboniferous–Recent)
        Class 1. Anthocerotae (hornworts)
             2. Hepaticae (liverworts)
             3. Musci (mosses)
        2. Psylophyta (Devonian; *Psylophyton*)
  Grade Tracheophyta (vascular plants)
    Phylum 3. Lycopodiophyta (Devonian–Recent; *Lepidodendron*, *Sigillaria*, *Lycopodium*, *Selaginella*)
        4. Sphenophyta (Equisetophyta) (Devonian–Recent, horsetails; *Calamites*, *Equisetum*)
        5. Polyodiophyta (ferns, Devonian–Recent)
        7. Cycadophyta
        Class 1. Lyginopteridales (seed ferns, Carboniferous–Recent; *Glossopteris*)
             2. Cycadales (cycads, Permian–Recent; *Cycas*)
             3. Bennettitales (Permian–Oligocene; *Zamites*)
        8. Gingkophyta (maidenhair tree, Triassic–Recent; *Gingko*)
        9. Coniferophyta(Devonian–Recent)
        Class 1. Cordaitales (Devonian–Jurassic; *Cordaites*)
             2. Pinales (conifers, Jurassic–Recent; *Abies*, *Cedrus*, *Cupressus*, *Larix*, *Picea*, *Pinus*, *Sequoia*, *Tsuga*)
             3. Taxales (yews, Jurassic–Recent; *Taxus*)
             4. Gneticae (some climbing shrubs and small tropical trees)
             5. Caytonicae (ancestral angiosperms; Early Jurassic–Early Cretaceous)
     10. Angiospermophyta (angiosperms, the flowering plants, Cretaceous–Recent)
        Class 1. Liliatae (Monocotyledonae) (grasses, sedges, lilies, palms, orchids, Cretaceous–Recent)
             2. Magnoliatae (Dicotyledonae) (oaks, elms, sycamores, poplars, birch trees, roses, legumes, cactuses, Cretaceous–Recent)

A15. Units of Measurement

Table A15 lists the more important units in alphabetical order, together with their symbols and the quantities to which they refer. If you find a unit with which you are not familiar, Table A15 will tell you to which quantity it refers. Once you know the quantity, you will find the definition in Tables A16 and A2.

Table A15. Units of Measurement (an asterisk identifies a fundamental SI unit)

Unit	Symbol	Quantity
aeon	Ae	time
ampere*	A	electric current
ampere-turn	A	magnetomotive force
are	a	area
atmosphere	atm	pressure
atomic mass unit	u	mass
avogadro	Av	number of items
bar	b	pressure
barn	b	area
barrel	bbl	volume
barye	—	pressure
becquerel	Bq	radioactivity
bel	b, B	ratio of two powers
bushel	bu	volume
calorie	cal	energy
candela*	cd	luminous intensity
carat	ct	mass
centimeter	cm	length
coulomb	C	electric charge
curie	Ci	radioactivity
dalton	d	mass
day	d	time
debye	D	molecular dipole moment
decameter	da	length
decibel	db, dB	ratio of two powers
decimeter	dm	length
dyne	dyn	force
electron volt	eV	energy
ephemeris second	s_E	time
erg	—	energy
farad	F	capacitance
faraday	F	charge
femtometer	fm	length
fermi	fm	length
foot	ft	length
galileo	gal	acceleration
gamma	γ	magnetic-field intensity
gauss	G	magnetic-flux density
gilbert	Gb	magnetomotive force
grain	gr	mass
gram	g	mass
gram-calorie	g-cal	energy
hectare	ha	area
hectogram	hg	mass
hectoliter	hL	volume
hectometer	hm	length
henry	H	inductance and permeance
hertz	Hz	frequency
hour	h	time
inch	in	length
joule	J	energy
karat	kt	1/24 gold in alloy

Unit	Symbol	Quantity
kayser	—	number of waves per centimeter
kelvin*	K	temperature
kilogram*	kg	mass
kilometer	km	length
kiloton	kt	energy
kilowatt	kW	power
knot	kn	velocity
lambert	L	luminance
light-year	l.y.	distance
line	—	magnetic flux
liter	L	volume or capacity
lumen	lm	luminous flux
lux	lx	illuminance
maxwell	Mx	magnetic flux
megaton	Mt	energy
meter*	m	length
mho (see siemens)		
microfarad	μF	capacitance
microgram	μg	mass
micrometer	μm	length
microsecond	μs	time
milligal	mgal	acceleration
milligram	mg	mass
milliliter	mL	volume or capacity
millimeter	mm	length
million electron volts	MeV	energy
millisecond	ms	time
millimeter of mercury	mmHg	pressure
mole*	mol	number of items
minute	m	time
month	mo	time
nanogram	ng	mass
nanometer	nm	length
nanosecond	ns	time
newton	N	force
nit	nt	luminance
oersted	Oe	magnetic-field intensity
ohm	Ω	resistance
parsec	pc	distance
pascal	Pa	pressure
peck	pk	volume
picogram	pg	mass
picometer	pm	length
picosecond	ps	time
poise	P	viscosity (dynamic)
poiseuille	Pl	viscosity (dynamic)
quintal	q	mass
radian*	rad	angle, plane
radiation	rad	radiation absorbed
roentgen	r	radiation absorbed
second*	s	time
siemens (formerly mho)	S	conductance ($=1/$ohm)
slug	—	mass

(*cont.*)

Table A15. Units of Measurement (*continued*)

Unit	Symbol	Quantity
stere	s	volume
steradian*	sr	angle, solid
stilb	sb	luminance
stoke	St	viscosity (kinematic)
sverdrup	sv	volume per second
tesla	T	magnetic-flux density
ton	t	mass
torricelli	torr	pressure
volt	V	electric potential
watt	W	power
watt-hour	W-h	energy
weber	Wb	magnetic flux
yard	yd	length
year (sidereal)	—	time
year (tropical)	y	time

Source: From Emiliani (1988, pp. 270–1).

A16. Units and Their Definitions

Table A16 lists in alphabetical order the quantities encountered in the physical world, and, for each quantity, it gives the name, symbol, and definition of the pertinent unit (or units).

Table A16. Units and Their Definitions (an asterisk identifies a fundamental SI unit)

Quantity	Unit name	Symbol	Definition
acceleration	—	—	$m\,s^{-2}$
	galileo	gal	$cm\,s^{-2}$
angle			
plane	*radian	rad	angle, with vertex at the center of a circle, that subtends a segment on the circumference equal to the radius
solid	*steradian	sr	solid angle, with vertex at the center of a sphere, that subtends an area on the surface equal to the square of the radius
angular acceleration	—	—	$rad\,s^{-2}$
angular momentum	—	—	$kg\,m^2\,s^{-1}$
angular velocity	—	—	$rad\,s^{-1}$
area	—	—	m^2
	are	a	$100\,m^2$
	barn	b	$10^{-28}\,m^2$
	hectare	ha	$10^4\,m^2$
capacitance	farad	F	$C\,V^{-1}$
	microfarad	μF	$10^{-6}\,F$
charge (see electric charge)			

Quantity	Unit name	Symbol	Definition
conductance	siemens	S	$\Omega^{-1} = A\,V^{-1}$
conductivity			
electrical	—	σ	$A\,V^{-1}\,m^{-1}$
thermal	—	k	$J\,m^{-1}\,s^{-1}\,K^{-1}$
density	—	ρ	$kg\,m^{-3}$
			$g\,cm^{-3}$(more practical than $kg\,m^{-3}$)
electric charge	coulomb	C	$A\,s$
	faraday	F	96,485,309 C
electric current	*ampere	A	current that, if maintained along two straight, parallel conductors of negligible cross section and infinite length, placed 1 m apart in vacuo, will cause each to produce on the other a force of 10^{-7} N per meter of length
electric-current density	—	j	$A\,m^{-2}$
electric-field intensity	—	**E**	$V\,m^{-1}, N\,C^{-1}$
electric potential	volt	V	$J\,C^{-1}$
electromotive force	volt	V	$J\,C^{-1}$
energy	joule	J	$N\,m$
	erg	—	$10^{-7}\,J$
	electron volt	eV	$1.6021773 \cdot 10^{-19}\,J$
	million electron volts	MeV	$1.6021773 \cdot 10^{-13}\,J$
	calorie (International Table)	cal_{IT}	4.1868 J (exactly)
	atomic mass unit	u	931.49432 MeV
	gram	g	$5.609586 \cdot 10^{26}$ MeV
	kiloton	kt	10^{12} cal (exactly) $= 4.1898 \cdot 10^{12}\,J$ (exactly)
	megaton	Mt	10^{15} cal (exactly) $= 4.1898 \cdot 10^{15}\,J$ (exactly)
enthalpy	—	H	J
entropy	—	S	$J\,K^{-1}$
force	newton	N	$kg\,m\,s^{-2}$
	dyne	dyn	$10^{-5}\,N$
frequency	hertz	Hz	s^{-1}
gravitational acceleration	—	—	$m\,s^{-2}$
	galileo	gal	$cm\,s^{-2}$
illuminance	lux	lx	$lm\,m^{-2}$
impulse	—	—	$kg\,m\,s^{-1}$
inductance	henry	H	$V\,s\,A^{-1} = Wb\,A^{-2}$
irradiance	—	—	$W\,m^{-2}$
length	angstrom	Å	$10^{-10}\,m$
	astronomical unit	AU	mean distance from Earth to Sun = 149,597.870.7 km
	attometer	am	$10^{-18}\,m$
	centimeter	cm	$10^{-2}\,m$
	decameter	da	$10\,m$
	decimeter	dm	$10^{-1}\,m$
	femtometer	fm	$10^{-15}\,m$
	hectometer	hm	$100\,m$
	kilometer	km	$10^3\,m$
	light year	l.y.	distance traveled by light in one tropical year $= 9.4605284 \cdot 10^{12}$ km

(cont.)

Table A16.   Units and Their Definitions (*continued*)

Quantity	Unit name	Symbol	Definition
	megaparsec	Mpc	10^6 parsecs $= 3.261633 \cdot 10^6$ l.y.
	*meter	m	distance traveled by light in vacuo in 1/299,792,458 s (exactly).
	micrometer	μm	10^{-6} m
	millimeter	mm	10^{-3} m
	nanometer	nm	10^{-9} m
	picometer	pm	10^{-12} m
	parsec	pc	distance at which 1 AU subtends 1 second of arc $= 3.261633$ l.y.
luminance	nit	nt	cd m^{-2}
	lambert	L	cd cm^{-2}
luminosity	*candela	cd	luminous intensity of 1/683 W sr^{-1} emitted by a monochromatic source radiating at a frequency of $540 \cdot 10^{12}$ Hz ($\lambda = 555.171218$ nm)
luminous flux	lumen	lm	cd sr^{-1}
magnetic-field intensity	—	H	A m^{-1}
	oersted	Oe	$10^3/4\pi$ A m^{-1}
	gamma	γ	10^{-5} Oe
magnetic flux	weber	Wb	V s
	maxwell	Mx	10^{-8} Wb
magnetic-flux density	tesla	T	Wb m^{-2}
magnetic induction (see magnetic-flux density)			
magnetomotive force	ampere-turn	A	A
	gilbert	Gb	10 A$/4\pi$
mass	atomic mass unit	u	$1.660540 \cdot 10^{-27}$ kg
	attogram	ag	10^{-8} g
	carat	ct	0.2 g
	dalton	—	1 atomic mass unit
	gram	g	10^{-3} kg
	femtogram	fg	10^{-15} g
	*kilogram	kg	mass of Pt-Ir International Prototype Kilogram kept at Sèvres, S-et-O, France
	microgram	μg	10^{-6} g
	milligram	mg	10^{-3} g
	nanogram	ng	10^{-9} g
	picogram	pg	10^{-12} g
moment of inertia	—	—	kg m^2
momentum	—	—	kg m s^{-1}
number of items	avogadro	Av	$6.022137 \cdot 10^{23}$ items
	*mole	mol	one Avogadro number of items
permeability	—	μ	H m^{-1}
permittivity	—	ε	F m^{-1}
potential (see electric potential)			
power	watt	W	J s^{-1}
pressure	pascal	Pa	N m^{-2}
	atmosphere	atm	76 cmHg
	bar	b	Nm^{-2}
	torricelli	torr	1 mmHg
radiant energy	joule	J	N m
radiant flux	watt	W	J s^{-1}

Quantity	Unit name	Symbol	Definition
radiant-flux density (see irradiance)			
radiation	radiation	rad	radiation energy absorption of $10^{-5}\,\mathrm{J\,g^{-1}}$
radioactivity	curie	Ci	quantity of a radioactive substance that produces $3.7\cdot10^{7}$ decays per second
resistance	ohm	Ω	$\mathrm{V\,A^{-1}}$
resistivity	—	ρ	$\Omega\,\mathrm{m}$
surface tension	—	γ	$\mathrm{N\,m^{-1}}$
temperature	*kelvin	K	temperature interval equal to 1/273.16 of the absolute temperature of the triple point of pure water
time	aeon	ae	$10^{9}\,\mathrm{y}$
	ephemeris second	s_E	1/31,556,925.9747 of tropical year 1900
	attosecond	as	$10^{-18}\,\mathrm{s}$
	femtosecond	fs	$10^{-15}\,\mathrm{s}$
	microsecond	μm	$10^{-6}\,\mathrm{s}$
	millisecond	ms	$10^{-3}\,\mathrm{s}$
	nanosecond	ns	$10^{-9}\,\mathrm{s}$
	picosecond	ps	$10^{-12}\,\mathrm{s}$
	*second	s	time interval equal to the duration of 9,192,631,770 periods of the radiation associated with the transition between the two hyperfine levels of the ground state of ^{133}Cs
	sidereal year	—	time required for the longitude of a distant star to increase by $360° =$ 31,558,149.99 s (1987–2015 C.E.)
	tropical year 1900	—	31,556,925.9747 s
	tropical year	y	31,556,925.5 s (1987–1988 C.E.)
velocity	—	—	$\mathrm{m\,s^{-1}}$
	knot	kn	1 nautical mile per hour
viscosity			
dynamic	poiseuille	Pl	$\mathrm{kg\,m^{-1}\,s^{-1}}$
	poise	P	$\mathrm{g\,cm^{-1}\,s^{-1}}$
kinematic	—	—	$\mathrm{m^2\,s^{-1}}$
	stoke	St	$\mathrm{cm^2\,s^{-1}}$
voltage	volt	v	$\mathrm{J\,C^{-1}}$
volume	barrel	bbl	42 gallons (U.S., liquid)
	liter	L	$1{,}000\,\mathrm{cm^3}$
	hectoliter	hL	100 L
	gigaliter	GL	$10^{9}\,\mathrm{L}$
	stere	—	$\mathrm{m^3}$
volume per second	sverdrup	sv	$10^{6}\,\mathrm{m^3\,s^{-1}} = 1\,\mathrm{GL\,s^{-1}}$
volumetric rate flow	—	—	$\mathrm{m^3\,s^{-1}}$
volumetric heat release	—	—	$\mathrm{W\,m^{-3}}$
work (see energy)			

Source: Emiliani (1988, pp. 272–6).

CHEMICAL FORMULARY

How many times do we hear a chemical name that is familiar, but we do not know or do not remember the formula? Table B1 lists the compounds frequently encountered here and there, together with their chemical formulas and some comments (if needed). I have often omitted such data as freezing and boiling points, solubilities, and so forth, because there are many manuals for that. See, for instance, the *CRC Handbook of Chemistry and Physics*, published yearly by the CRC Press, Boca Raton, Florida.

Table B1.   Chemical Formulary

Name	Formula	Synonyms, data, and notes
acetaldehyde	CH_3CHO	—
acetic acid	CH_3COOH	—
acetone	CH_3COCH_3	—
acetylene	$CHCH$	—
acetylsalicylic acid	$CH_3CO_2C_6H_4COOH$	aspirin
actinium*	Ac	a.m. = 227.021750
adenine	$C_5H_5N_5$	a nucleic acid base
adenosine	$C_{10}H_{13}N_5O_4$	—
monophosphate	$C_{10}H_{12}O_4(PO_2OH)H$	AMP
diphosphate	$C_{10}H_{12}O_4(PO_2OH)_2H$	ADP
triphosphate	$C_{10}H_{12}O_4(PO_2OH)_3H$	ATP
ADP	—	see adenosine diphosphate
alanine	$NH_2CHCOOH[CH_3]$	a natural amino acid; m.m. = 89.094
alloleucine	$NH_2CHCOOH[CH_3CHCH_2CH_3]$	a natural amino acid; m.m. = 131.174
allyl alcohol	CH_2CHCH_2OH	—
aluminum	Al	a.m. = 26.981539
oxide	Al_2O_3	corundum (α-alumina)
oxide	Al_2O_3	alumina (γ-alumina)
hydroxide	$AlO(OH)$	boehmite
hydroxide	$Al(OH)_3$	bayerite
hydroxide	$Al(OH)_3$	gibbsite
silicate	Al_2SiO_5	sillimanite
silicate	Al_2SiO_5	andalusite
silicate	Al_2SiO_5	kyanite
americium*	Am	a.m. = 243.061375
ammonia	NH_3	b.p. = − 33.35
hydroxide	NH_4OH	—
nitrate	NH_4NO_3	—
nitrite	NH_4NO_2	—
phosphate	$(NH_4)_2HPO_4$	—

(cont.)

Table B1.   **Chemical Formulary** (*continued*)

Name	Formula	Synonyms, data, and notes
ammonia (*cont.*)		
sulfate	$(NH_4)_2SO_4$	mascagnite
sulfide	$(NH_4)_2S$	—
AMP	—	see adenosine monophosphate
amylum	—	see starch
aniline	$C_6H_5NH_2$	—
anthracene	$C_{14}H_{10}$	—
antimony	Sb	a.m. = 121.757
chloride	$SbCl_3$	—
sesquioxide	Sb_2O_3	senarmontite
sesquioxide	Sb_2O_3	valentinite
sulfide	Sb_2S_3	stibnite
aqua regia	50% HCl + 50% HNO₃	dissolves gold and platinum
arginine	$NH_2CHCOOH[(CH_2)_3NHCNHNH_2]$	a natural amino acid; m.m. = 174.203
argon	Ar	a.m. = 39.9477
arsenic (gray)	As	a.m. = 74.921594
arsenic (black)	As	a.m. = 74.921594
arsenic (yellow)	As_4	m.m. = 299.686376
hydride	AsH_3	arsine
sesquioxide	As_2O_3	arsenolite
sesquioxide	As_2O_3	claudetite
sulfide	AsS	realgar
sesquisulfide	As_2S_3	orpiment
ascorbic acid	$C_6H_8O_6$	vitamin C
asparagine	$NH_2CHOOH[CH_2CONH_2]$	a natural amino acid; m.m. = 132.119
aspartic acid	$NH_2CHCOOH[CH_2COOH]$	a natural amino acid; m.m. = 133.104
aspirin	—	see acetylsalicylic acid
astatine*	At	a.m. = 219.0113
ATP	—	see adenosine triphosphate
barium	Ba	a.m. = 137.327
carbonate	$BaCO_3$	witherite
chloride	$BaCl_2$	—
nitrate	$Ba(NO_3)_2$	nitrobarite
sulfate	$BaSO_4$	barite
benzaldehyde	C_6H_5CHO	—
benzene	C_6H_6	—
benzil	—	see diphenyl glioxal
benzoic acid	C_6H_5COOH	—
benzyl alcohol	$C_6H_5CH_2OH$	—
berkelium*	Bk	a.m. = 247.070300
beryllium	Be	a.m. = 9.012182
aluminate	$BeAl_2O_4$	chrysoberyl
aluminum silicate	$Be_3Al_2(Si_6O_{18})$	beryl
oxide	Beo	bromellite
silicate	Be_2SiO_4	phenacite
bismuth	Bi	a.m. = 208.980374
sesquioxide	Bi_2O_3	—
sulfide	Bi_2S_3	bismuthinite
boric acid	H_3BO_3	sassolite
boron	B	a.m. = 10.811
oxide	B_2O_3	—
bromine	Br	a.m. = 79.904
bromoform	—	see methane, tribromo-
butane	$CH_3CH_2CH_2CH_3$	—

Name	Formula	Synonyms, data, and notes
butanoic acid	$Ch_3CH_2CH_2COOH$	butyric acid
butanol	$CH_3CH_2CH_2CH_2OH$	butyl alcohol
butyl alcohol	—	see butanol
butylene (iso-)	$(CH_3)_2C:CH_2$	—
butyric acid	—	see butanoic acid
cadmium	Cd	a.m. = 112.41
chloride	$CdCl_2$	—
cyanide	$Cd(CN)_2$	—
nitrate	$Cd(NO_3)_2$	—
oxide	CdO	—
sulfide	CdS	greenockite
calcium	Ca	a.m. = 40.078
aluminosilicate	$CaAl_2Si_2O_8$	anorthite
carbonate	$CaCO_3$	aragonite, calcite, vaterite
chloride	$CaCl_2$	—
chlorite	$Ca(ClO_2)_2$	—
fluoride	CaF	fluorite
hydroxide	$Ca(OH)_2$	portlandite
hypochlorite	$Ca(ClO)_2$	—
magnesium carbonate	$CaMg(CO_3)_2$	dolomite
magnesium silicate	$CaMgSiO_4$	monticellite
magnesium silicate	$CaMg(SiO_3)_2$	diopside
magnesium silicate	$Ca_2MgSi_2O_7$	akermanite
magnesium silicate	$Ca_3MgSi_2O_8$	merwinite
oxide	CaO	—
phosphate	$Ca_3(PO_4)_2$	whitlockite
phosphate	$Ca_5(PO_4)_3F$	apatite, fluorapatite
phosphate	$Ca_5(PO_4)_3(OH)$	hydroxylapatite
phosphate	$CaHPO_4 \cdot 2H_2O$	brushite
silicate	Ca_2SiO_4	—
silicate	$CaSiO_3$	wollastonite
sulfate	$CaSO_4$	anhydrite
sulfate	$CaSO_4 \cdot 2H_2O$	gypsum
sulfate	$CaSO_4 \cdot \frac{1}{2}H_2O$	plaster of Paris
californium*	Cf	a.m. = 251.079581
carbon	C	a.m. = 12.000000
carbon	C	diamond; m.p. = 3,550; b.p. = 4,827
carbon	C	graphite; m.p. = 3,652; b.p. = 4,827
carbon	C	carbon black; m.p. = 3,652; b.p. = 4,827
dioxide	CO_2	subl. at − 78.44
hydride	CH_4	methane
monoxide	CO	—
tetrachloride	CCl_4	—
carbonic acid	H_2CO_3	—
carbonyl chloride	$COCL_2$	phosgene
carotene	$C_{40}H_{56}$	—
cellosolve	$C_2H_5O(CH_2)_2OH$	—
acetate	$CH_3CO_2C_4H_9O$	—
cellulose	$(C_6H_{10}O_5)_n$	—
cerium	Ce	a.m. = 140.115
chloride	$CeCl_3$	—
oxide	CeO_2	ceria
phosphate	$CePO_4$	monazite

(cont.)

Table B1. **Chemical Formulary** (*continued*)

Name	Formula	Synonyms, data, and notes
cesium	Cs	a.m. = 132.905429; m.p. = 28.40
oxide	Cs_2O	—
sulfate	Cs_2SO_4	—
chloral	Cl_3CCHO	—
chlorine	Cl	a.m. = 35.453
chlorine	Cl_2	m.m. = 70.906
chloroform	$CHCl_3$	—
chlorophyll a	$MgN_4O_5C_{55}H_{72}$	m.m. = 893.502
chlorophyll b	$MgN_4O_6C_{55}H_{70}$	m.m. = 907.486
chromium	Cr	a.m. = 51.9961
carbide	Cr_3C_2	—
chloride	$CrCl_2$	—
chloride	$CrCl_3$	—
sesquioxide	Cr_2O_3	—
trioxide	CrO_3	chromic acid
citric acid	$HOC(CH_2COOH)_2COOH$	—
cobalt	Co	a.m. = 58.933198
sulfate	$CoSO_4$	—
sulfate, heptahydrate	$CoSO_4 \cdot 7H_2O$	bieberite
sulfide	CoS	sycoporite
(tri-) sulfide	Co_3S_4	linneite
copper	Cu	a.m. = 63.546
acetate	$Cu(C_2H_3O_2)_2 \cdot H_2O$	neutral verdigris
carbonate	Cu_2CO_3	—
carbonate, basic	$CuCO_3 \cdot Cu(OH)_2$	malachite
carbonate, basic	$2CuCO_3 \cdot Cu(OH)_2$	azurite
chloride	CuCl	nantokite
chloride	$CuCl_2$	—
chloride, basic	$CuCl_2 \cdot Cu(OH)_2$	—
cyanide	CuCN	—
fluoride	CuF	—
iodide	CuI	marshite
nitrate, trihydrate	$Cu(NO_3)_2 \cdot 3H_2O$	—
oxide	Cu_2O	cuprite
oxide	CuO	tenorite
oxychloride	$Cu_2Cl(OH)_3$	atacamite
sulfate	Cu_2SO_4	—
sulfate	$CuSO_4 \cdot 3Cu(OH)_2$	brochantite
sulfate	$CuSO_4 \cdot 5H_2O$	chalcantite
sulfide	Cu_2S	chalcocite
sulfide	CuS	covellite
telluride	Cu_4Te_3	rickardite
corrin	$HN_4C_{19}H_{21}$	m.m. = 264.389; tetracyclic ring in cobalamins
curium*	Cm	a.m. = 247.070347
cyanamide	NH_2CN	—
cyanic (isocyanic) acid	HOCN	—
cyanocobalamin	$CoN_4PO_4N_{10}O_{10}C_{63}H_{88}$	vitamin B_{12}; m.m. = 1,355.381
cyanogen	NCCN	—
cysteine	$NH_2CHCOOH[CH_2SH]$	a natural amino acid; m.m. = 121.158
cystine	$[NH_2CHCOOH(CH_2S)]_2$	a natural amino acid; m.m. = 240.301
cytochrome c	$FeN_4C_8(CH)_4(CH_3)_4(CH_2CH_2$ $COOH)_4 \cdot (CH_3CH)_2 \cdot protein$	m.m. = 502 + protein
cytosine	$C_4H_5N_3O$	a nucleic acid base; m.m. = 111.055

Name	Formula	Synonyms, data, and notes
deuterium	^2H	a.m. = 2.01410177
deuterium	D_2	m.m. = 4.02820354
deuterium dioxide	D_2O	heavy water
dextrin	$(C_6H_{12}O_5)_n$	starch
dextrose	$C_6H_{12}O_6$	glucose, D-
dieldrin	$C_{12}H_8OCl_6$	octalox
digitoxin	$C_{41}H_{64}O_{13}$	—
dimethyl sulfoxide	$(CH_3)_2SO$	—
DMSO	—	see dimethyl sulfoxide
diphenyl glioxal	$C_6H_5COCOC_6H_5$	benzil
diphenyl methane	$(C_6H_5)_2CH_2$	ditan
ditan	—	see diphenyl methane
dysprosium	Dy	a.m. = 162.498
einsteinium*	Es	a.m. = 254.088024
ephedrine	$C_6H_5CH(OH)CH(CH_3)NHCH_3$	—
erbium	Er	a.m. = 167.26
ethane	CH_3CH_3	—
ethanol	CH_3CH_2OH	ethyl alcohol
ether	—	see ethyl ether
ethyl ether	$CH_3CH_2OCH_2CH_3$	—
ethylene	$CH_2:CH_2$	—
tetrachloro	$CCl_2:CCl_2$	—
trichloro	$CHCl:CCl_2$	—
europium	Eu	a.m. = 151.965
oxide	Eu_2O_3	—
fermium*	Fm	a.m. = 257.095100
fluorine	F	a.m. = 18.998403
folic acid	$C_{19}N_{19}N_7O_6$	Vitamin B complex
formaldehyde	$HCH:O$	methanal
formic acid	HCOOH	methanoic acid
francium*	Fr	a.m. = 223.019733
Freon 12	—	see methane, chlorodifluoro-
Freon 13	—	see methane, chlorotrifluoro-
Freon 21	—	see methane, dichlorofluoro-
Freon 22	—	see methane, chlorodifluoro-
fumaric acid	COOHCH:CHCOOH	
gadolinium	Gd	a.m. = 157.252
galactose	$C_6H_{12}O_6$	—
galena	PbS	the principal Pb mineral
gallium	Ga	a.m. = 69.273
germanium	Ge	a.m. = 72.61
glucose	—	see dextrose
glutamic acid	$NH_2CHCOOH[CH_2CH_2COOH]$	a natural amino acid; m.m. = 147.131
glutamine	$NH_2CHCOOH[CH_2CH_2CONH_2]$	a natural amino acid; m.m. = 146.146
glycerin	$HOCH_2CH(OH)CH_2OH$	glycerol
glycerol	—	see glycerin
glycine	$NH_2CHCOOH[H]$	a natural amino acid; m.m. = 75.067
glycogen	$(C_6H_{10}O_5)_n$	animal starch
gold	Au	a.m. = 196.966543
chloride	AuCl	—
telluride	$AuTe_2$	krennerite

(cont.)

Table B1.   **Chemical Formulary** (*continued*)

Name	Formula	Synonyms, data, and notes
guanine	$C_5H_5ON_5$	a nucleic acid base; m.m. = 151.068
guanidine	$NH:C(NH_2)_2$	
hafnium	Hf	a.m. = 178.49
carbide	HfC	m.p. ~ 3,890
oxide	HfO_2	hafnia
heavy water	—	see deuterium dioxide
helium	He	a.m. = 4.002602
hematin	$Fe(OH)N_4C_{16}(CH)_4(CH_3)_4(CH:CH_2)_2\cdot$	m.m. = 633.505
	$(CH_2CH_2COOH)_2$	m.m. = 633.505
hematite	Fe_2O_3	the principal Fe mineral
hemocyanin (average)	$(Cu\cdot S_2\cdot N_{203}O_{208}C_{738}H_{1166})_n$	m.m. = $(16,338.230)_n$
hemoglobin (average)	$(Fe\cdot S_2\cdot N_{203}O_{208}C_{738}H_{1166})_4$	m.m. = 65,322.124
heptane	$CH_3(CH_2)_5CH_3$	—
heptanol	$CH_3(CH_2)_5CH_2OH$	—
hexachlorophene	$(OH\cdot Cl_3C_6H)_2CH_2$	—
hexadecanoic acid	$CH_3(CH_2)_{14}COOH$	palmitic acid
hexane	$CH_3(CH_2)_4CH_3$	—
hexanol	$CH_3(CH_2)_4CH_2OH$	—
histamine	$NH_2(CH_2)_2C(CH)_2NNH$	—
histidine	$NH_2CHCH_2C(CH)_2NNH$	a natural amino acid; m.m. = 155.1567
holmium	Ho	a.m. = 164.930319
oxide	Ho_2O_3	holmia
hydrazine	NH_2NH_2	—
hydrogen	H	a.m. = 1.00794
hydrogen	H_2	m.m. = 2.01588
antimonide	H_3Sb	stibine
arsenide	H_3As	arsine
bromide	HBr	hydrobromic acid
chloride	HCl	hydrochloric acid
cyanide	HCN	hydrocyanic acid
fluoride	HF	hydrofluoric acid
iodide	HI	—
oxide	H_2O	water
peroxide	H_2O_2	—
phosphide	H_3P	phosphine
sulfide	H_2S	—
hydrobenzoic acid	—	see salicylic acid
hydroquinone	$C_6H_4(OH)_2$	—
indium	In	a.m. = 114.82
inositol	$C_6H_{12}O_6$	—
iodine	I	a.m. = 126.904473
iodine	I_2	m.m. = 253.808946
idoform	—	see methane, triiodo-
iridium	Ir	a.m. = 192.22
iron	Fe	a.m. = 55.847
carbide	F_3C	—
carbonate	$FeCO_3$	siderite
chloride	$FeCl_2$	lawrencite
hydroxide	$FeO(OH)$	goethite
metasilicate	$FeSiO_3$	—
orthosilicate	Fe_2SiO_4	fayalite
oxide	FeO	wüstite
oxide	Fe_2O_3	hematite

Name	Formula	Synonyms, data, and notes
oxide	Fe_2O_3	maghemite
oxide	Fe_3O_4	magnetite
sulfide	FeS_2	marcasite
sulfide	FeS_2	pyrite
sulfide	FeS	troilite
tungstate	$FeWO_4$	ferberite
isoleucine	$NH_2CHCOOH[CHCH_3CH_2CH_3]$	a natural amino acid; m.m. = 131.175
krypton	Kr	a.m. = 83.80
lactic acid	$CH_3CH(OH)COOH$	—
lactose	$C_{12}H_{22}O_{11}$	—
lanthanum	La	a.m. = 138.9055
oxide	La_2O_3	lanthana
lawrencium*	Lw	a.m. = 260.105320
lead	Pb	a.m. = 207.2
acetate	$Pb(CH_3COO)_2$	—
carbonate	$PbCO_3$	cerussite
chloride	$PbCl_2$	cotunite
chromate	$PbCrO_4$	crocoite
fluoride	PbF_2	—
fluorochloride	$PbFCl$	matlockite
hydroxide	$Pb(OH)_2$	—
molybdate	$PbMoO_4$	wulfenite
nitrate	$Pb(NO_3)_2$	—
oxide, di-	PbO_2	plattnerite
oxide, mono-	PbO	litharge
oxide, red	Pb_3O_4	minium
oxide, sesqui-	Pb_2O_3	—
oxychloride	$PbCl_2 \cdot 2PbO$	mendipite
oxychloride	$PbCl_2 \cdot 3PbO$	—
oxichloride	$PbCl_2 \cdot 7PbO$	Cassel yellow
oxychloride	$PbCl_2 \cdot PbO \cdot H_2O$	paralaurionite
oxychloride	$2PbCl_2 \cdot PbO \cdot H_2O$	fiedlerite
oxychloride	$PbCl_2 \cdot Pb(OH)_2$	laurionite
phosphate, ortho-	$Pb_3(PO_4)_2$	—
phosphate, pyro-	$Pb_2P_2O_7$	—
selenide	$PbSe$	clausthalite
silicate, meta-	$PbSiO_3$	alamosite
silicate, diortho-	$Pb_2Si_2O_7$	barysilite
sulfate	$PbSO_4$	anglesite
sulfate	$PbSO_4 \cdot PbO$	lanarkite
sulfide	PbS	galena
telluride	$PbTe$	altaite
tungstate	$PbWO_4$	raspite
tungstate	$PbWO_4$	stolzite
leucine	$NH_2CHCOOH[CH_2CH(CH_3)_2]$	a natural amino acid; m.m. = 131.175
levulose	—	see fructose
lindane	$C_6H_6Cl_6$	—
lithium	Li	a.m. = 6.941
bromide	$LiBr$	—
carbonate	Li_2CO_3	—
chloride	$LiCl$	—
hydride	LiH	—
hydroxide	$LiOH$	—

(cont.)

Table B1.   **Chemical Formulary** (*continued*)

Name	Formula	Synonyms, data, and notes
lithium (*cont.*)		
oxide	Li_2O	—
silicate, meta-	Li_2SiO_3	—
silicate, ortho-	Li_4SiO_4	—
sulfate	Li_2SO_4	—
lutetium	Lu	a.m. = 174.967
oxide	Lu_2O_3	—
lysergic acid	$C_{16}H_{16}N_2O_2$	—
lysine	$NH_2CHCOOH[(CH_2)_4NH_2]$	a natural amino acid; m.m. = 146.190
magnesium	Mg	a.m. = 24.3050
aluminate	$MgAl_2O_4$	spinel
borate, di-	$Mg_2B_2O_5 \cdot H_2O$	ascharite
carbonate	$MgCO_3$	magnesite
carbonate	$MgCO_3 \cdot Mg(OH)_2 \cdot 3H_2O$	artinite
chloride	$MgCl_2$	—
chloride	$MgCl_2 \cdot 6H_2O$	bischofite
hydroxide	$Mg(OH)_2$	brucite
oxide	MgO	periclase
phosphate, ortho-	$Mg_3(PO_4)_2$	—
phosphate, pyro-	$Mg_2P_2O_7$	—
silicate, meta-	$MgSiO_3$	clinoenstatite
silicate, ortho-	Mg_2SiO_4	forsterite
sulfate	$MgSO_4$	—
heptahydrate	$MgSO_4 \cdot 7H_2O$	epsomite
monohydrate	$MgSO_4 \cdot H_2O$	kieserite
malathion	$C_{10}H_{19}PS_2O_6$	—
malonic acid	$COOH \cdot CH_2 \cdot COOH$	—
maltose	$C_{12}H_{22}O_{11}$	—
manganese	Mn	a.m. = 54.938047
chloride, di-	$MnCl_2$	scacchite
oxide	Mn_3O_4	hausmannite
oxide, di-	MnO_2	pyrolusite
oxide, mono-	MnO	manganosite
oxide, sesqui-	Mn_2O_3	braunite
silicate, meta-	$MnSiO_3$	rhodonite
sulfate	$MnSO_4$	—
sulfide, mono-	MnS	alabandite
sulfide, di-	MnS_2	hauerite
titanate	$MnTiO_3$	pyrophanite
mannose	$C_6H_{12}O_6$	—
mendelevium*	Md	a.m. = 258.0985
menthol	$C_{10}H_{20}O$	—
mercurochrome	$C_{20}H_7O_5Br_2Na_2HgOH \cdot 3H_2O$	—
mercury	Hg	a.m. = 200.59
chloride	Hg_2Cl_2	calomel
fulminate	$Hg(CNO)_2$	—
oxide	HgO	montroydite
oxide	Hg_2O	—
sulfide	Hg_2S	—
sulfide	HgS	cinnabar
sulfide	HgS	metacinnabar
methanal	—	see formaldehyde
methane	CH_4	b.p. = − 164

Name	Formula	Synonyms, data, and notes
chlorodifluoro-	$ClCHF_2$	Freon 22
chlorotrifluoro-	$ClCF_3$	Freon 23
dichlorodifluoro-	Cl_2CF_2	Freon 12
dichlorofluoro-	Cl_2CHF	—
tetrachloro-	CCl_4	—
tetrafluoro-	CF_4	—
tribromo-	$CHBr_3$	bromoform
trichloro-	$CHCl_3$	chloroform
trifluoro-	CHF_3	fluoroform
triiodo-	CHI_3	iodoform
trinitro-	$CH(NO_2)_3$	nitroform
methanoic acid	—	see formic acid
methanol	CH_3OH	methyl alcohol
methionine	$NH_2CHCOOH[(CH_2)_2SCH_3]$	a natural amino acid; m.m. = 149.212
methyl acetate	—	see acetic acid
alcohol	—	see methanol
amine	CH_3NH_2	—
methylene blue	$C_{16}H_{18}N_3ClS$	—
molybdenum	Mo	a.m. = 95.94
carbide	Mo_2C	m.p. = 2,687
carbide	MoC	m.p. = 2,692
oxide, tri-	MoO_3	molybdite
sulfate, di-	MoS_2	molybdenite
mustard gas	$ClCH_2CH_2SCH_2CH_2Cl$	yperite
naphthalene	$C_{10}H_8$	—
neodymium	Nd	a.m. = 144.242
oxide	Nd_2O_3	neodymia
neon	Ne	a.m. = 20.1795
neptunium*	Np	a.m. = 237.048168
niacin	—	see nicotinic acid
nickel	Ni	a.m. = 58.6934
antimonide	NiSb	breithauptite
arsenide	NiAs	niccolite
oxide	NiO	bunsenite
silicide	Ni_2Si	—
sulfate	$NiSO_4$	—
nicotine	$C_{10}H_{14}N_2$	—
nicotinic acid	C_5H_4NCOOH	niacin
niobium	Nb	a.m. = 92.906377
nitric acid	HNO_3	—
nitrocellulose	$(NO_2)_3C_6H_7O_5$	—
nitroform	—	see methane, trinitro-
nitrogen	N	a.m. = 14.0067; b.p. = −195.8
nitrogen	N_2	m.m = 28.0134
oxide, nitric	NO	—
oxide, nitrous	N_2O	laughing gas
pentoxide	N_2O_5	—
peroxide	NO_2	—
nitroglycerin	$NO_2OCH_2CHO(NO_2)CH_2ONO_2$	glycerol trinitrate
nobelium*	No	a.m. = 255.093260
octadecane	$CH_3(CH_2)_{16}CH_3$	—
octadecanoic acid	$CH_3(CH_2)_{16}COOH$	stearic acid

(cont.)

Table B1. **Chemical Formulary** (*continued*)

Name	Formula	Synonyms, data, and notes
octadecanol	$CH_3(CH_2)_{16}CH_2OH$	—
octadecenoic acid	$CH_3(CH_2)_7CH:CH(CH_2)_7COOH$	oleic acid
octane	$CH_3(CH_2)_6CH_3$	—
osmium	Os	a.m. = 190.2
oxalic acid	$COOH \cdot COOH$	—
oxygen	O	a.m. = 15.9994; b.p. = $-$182.962
oxygen	O_2	m.m. = 31.9988
ozone	O_3	—
palladium	Pd	a.m. = 106.42
palmitic acid	—	see hexadecanoic acid
parathion	$C_{10}H_{14}NO_5PS$	—
pentane	$CH_3(CH_2)_3CH_3$	—
perchloric acid	$HClO_4$	—
phenol	C_6H_5OH	—
phenolphthalein	$C_{20}H_{14}H_4$	—
phenylalanine	$NH_2CHCOOH[CH_2C_6H_5]$	a natural amino acid; m.m = 165.192
phosgene	$COCl_2$	carbonyl chloride
phosphoric acid, ortho-	H_3PO_4	—
phosphorus	P	a.m. = 30.973762
phosphorus, black	P_4	m.m. = 123.895048
phosphorus, red	P_4	m.m. = 123.895048
phosphorus, violet	P_4	m.m. = 123.895048
phosphorus, yellow	P_4	m.m. = 123.895048
hydride, tri-	PH_3	phosphine
oxide, penta-	P_2O_5	—
phytane	$C_{20}H_{42}$	—
phytol	$C_{20}H_{39}OH$	—
platinum	Pt	a.m. = 195.08
arsenide	$PtAs_2$	sperrylite
plutonium*	Pu	a.m. = 244.064200
fluoride, hexa-	PuF_6	—
polonium*	Po	a.m. = 208.982404
porphin	$N_2(NH)_2C_8(CH)_{12}$	—
potassium	K	a.m. = 39.0983
aluminosilicate	$KAlSi_3O_8$	orthoclase
aluminosilicate	$KAlSi_3O_8$	microcline
aluminosilicate	$KAlSi_3O_8$	sanidine
aluminosilicate	$KAl_2(AlSi_3O_{10}) \cdot (OH)_2$	muscovite
carbonate	K_2CO_3	—
chlorate	$KClO_3$	—
chlorate, per-	$KClO_4$	—
chloride	KCl	sylvite
chromate	K_2CrO_4	tarapacaite
chromate, di-	$K_2Cr_2O_7$	—
cobaltinitrite	$K_3[Co(NO_2)_6]$	Fischer's salt
cyanide	KCN	—
fluosilicate	K_2SiF_6	hieratite
hydroxide	KOH	—
magnesium chloride	$KCl \cdot MgCl_2 \cdot 6H_2O$	carnallite
magnesium chloride sulfate	$KCl \cdot MgSO_4 \cdot 3H_2O$	kainite
magnesium sulfate	$K_2SO_4 \cdot 2MgSO_4$	langbeinite
magnesium sulfate	$K_2SO_4 \cdot MgSO_4 \cdot 4H_2O$	leonite

Name	Formula	Synonyms, data, and notes
manganate	K_2MnO_4	—
manganate, per-	$KMnO_4$	—
nitrate	KNO_3	saltpeter, niter
nitrite	KNO_2	—
oxide	K_2O	potassa
sulfate	K_2SO_4	arcanite
praseodymium	Pr	a.m. = 140.907647
proline	$NHCHCOOH[(CH_2)_3]$	a natural amino acid; m.m. = 115.132
promethium*	Pm	a.m. = 140.907647; no stable isotopes
propane	$CH_3CH_2CH_3$	—
propanol	$CH_3CH_2CH_2OH$	—
propanone	CH_3COCH_3	acetone
propene	$CH_3CH:CH_2$	propylene
propylene	—	see propene
protactinium*	Pa	a.m. = 231.03588
purine	$C_5H_4N_4$	—
pyridine	C_6H_5N	—
pyrimidine	$C_4H_4N_2$	—
pyrrole	$NH(CH)_4$	—
pyruvic acid	$CH_3COCOOH$	—
quinine	$C_{20}H_{24}N_2O_2 \cdot 3H_2O$	—
radium*	Ra	a.m. = 226.025403
radon*	Rn	a.m. = 222.017571
rhenium	Re	a.m. = 186.207
rhodium	Rh	a.m. = 102.905500
riboflavin	$C_{17}H_{20}N_4O_6$	vitamin B_2
ribose	$CH_2OH(CHOH)_3CHO$	—
rubidium	Rb	a.m. = 85.4678
ruthenium	Ru	a.m. = 101.07
salicylic acid	HOC_6H_4COOH	hydrobenzoic acid
samarium	Sm	a.m. = 150.36
scandium	Sc	a.m. = 44.955910
selenium	Se	a.m. = 78.96
serine	$NH_2CHCOOH[CH_2OH]$	a natural amino acid; m.m. = 105.094
silicon	Si	a.m. = 28.0855
carbide	SiC	—
hydride	SiH_4	silane
oxide, di-	SiO_2	silica
oxide, di-	SiO_2	cristobalite
oxide, di-	SiO_2	lechatelierite
oxide, di-	SiO_2	quartz
oxide, di-	SiO_2	tridymite
oxide, di-	$SiO_2 \cdot nH_2O$	opal
silver	Ag	a.m. = 107.8682
acetate	$AgC_2H_3O_2$	—
bromide	AgBr	bromyrite
chloride	AgCl	cerargyrite
iodide	AgI	iodyrite
nitrate	$AgNO_3$	—
nitrite	$AgNO_2$	—
sulfate	Ag_2SO_4	—

(*cont.*)

Table B1.   **Chemical Formulary** (*continued*)

Name	Formula	Synonyms, data, and notes
silver (*cont.*)		
sulfide	Ag_2S	acanthite
sulfide	Ag_2S	argentite
telluride	Ag_2Te	hessite
thioantimonite	Ag_3SbS_3	pyrargyrite
thioarsenite	Ag_3AsS_3	proustite
sodium	Na	a.m. = 22.989768
aluminum trisilicate	$NaAlSi_3O_8$	albite
aluminum metasilicate	$NaAl(SiO_3)_2$	jadeite
orthosilicate	$NaAlSiO_4$	nepheline
carbonate	Na_2CO_3	—
carbonate	$Na_2CO_3 \cdot 10 H_2O$	natron
carbonate, bi-	$NaHCO_3$	—
chloride	NaCl	halite
chlorite	$NaClO_2$	—
chlorite, hypo-	NaOCl	—
cyanide	NaCN	—
fluoride	NaF	villiaumite
hydroxide	NaOH	—
magnesium sulfate	$Na_2Mg(SO_4)_2 \cdot 4H_2O$	bloedite
nitrate	$NaNO_3$	—
nitrite	$NaNO_2$	—
nitrite, hypo-	$Na_2N_2O_2$	—
pentothal	$C_{11}H_{17}N_2O_2NaS$	—
sulfate	Na_2SO_4	mirabilite
starch	$(C_6H_{10}O_5)_n$	amylum
stearic acid	—	see octadecanoic acid
strontium	Sr	a.m. = 87.62
carbonate	$SrCO_3$	strontianite
sulfate	$SrSO_4$	celestite
styrene	$C_6H_5CH{:}CH_2$	—
sucrose	$C_{12}H_{22}O_{11}$	saccharose
sulfonal	$(CH_3)_2C(C_2H_5SO_2)_2$	—
sulfur	S	a.m. = 32.066
sulfur	S_8	m.m = 256.528
sulfur	S_8	m.m. = 256.528
oxide, di-	SO_2	—
oxide, tri-	SO_3	—
sulfuric acid	H_2SO_4	—
tannic acid	$C_{76}H_{52}O_{46}$	tannin
tannin	—	see tannic acid
tantalum	Ta	a.m. = 180.9479; m.p. = 2,996; b.p. = 5,425
carbide	TaC	m.p. = 3,880; b.p. = 5,500.
tartaric acid	$COOH \cdot CHOHCHOH \cdot COOH$	—
technetium*	Tc	a.m. = 97.907215; no stable isotopes
tellurium	Te	a.m. = 127.60
oxide	TeO_2	tellurite
terbium	Tb	a.m. = 158.925342
oxide	Tb_2O_3	terbia
thallium	Tl	a.m. = 204.3833
thorium*	Th	a.m. = 232.038051
carbide	ThC_2	m.p. = 2,655; b.p. = ca. 5,000

Name	Formula	Synonyms, data, and notes
oxide	ThO_2	thorianite; m.p. = 3,220; b.p. = 4,400
threonine	$NH_2CHCOOH[CHCH_3OH]$	a natural amino acid; m.m. = 119.120
thulium	Tm	a.m. = 168.934212
thymine	$C_5H_6N_2O_2$	a nucleic acid (DNA) base; m.m. = 126.115
thymol	$(CH_3)(C_3H_7)C_6H_3OH$	—
tin, white	Sn	a.m. = 118.710
tin, gray	Sn	a.m. = 118.710
chloride	$SnCl_2$	—
fluoride	SnF_2	—
oxide	SnO_2	cassiterite
titanium	Ti	a.m. = 47.88
carbide	TiC	m.p. = 3,140; b.p. = 4,820
oxide, di-	TiO_2	brookite
oxide, di-	TiO_2	octahedrite, anatase
oxide, di-	TiO_2	rutile
TNT	—	see toluene, trinitro-
tocopherol	$C_{29}H_{50}O_2$	vitamin E
toluene	$C_6H_5CH_3$	—
trinitro-	$(NO_2)_3C_6H_2CH_3$	explosive
tryptophan	$NH_2 \cdot CH \cdot COOH[CH_2CCH(C_6H_4)NH]$	a natural amino acid; m.m. = 204.229
tungsten (wolfram)	W	a.m. = 183.85
carbide	WC	m.p. = 2,870; b.p. = ca. 6,000
carbide	W_2C	m.p. = 2,860; b.p. = ca. 6,000
oxide	WO_2	—
tyrosine	$NH_2CHCOOH[CH_2C_6H_4OH]$	a natural amino acid; m.m. = 182.199
uracil	$C_4H_4N_2O_2$	a nucleic acid (RNA) base; m.m. = 112.088
uranium*	U	a.m. = 238.0289
carbide, di-	UC_2	m.p. = 2350; b.p. = 4,370
fluoride, hexa-	UF_6	subl. at 56.2
oxide, di-	UO_2	m.p. = 2,878
urea	$NH_2 \cdot CO \cdot NH_2$	carbamide
urethane	$NH_2 \cdot CO_2 \cdot C_2H_5$	—
valine	$NH_2CHCOOH[CH(CH_3)_2]$	a natural amino acid; m.m. = 117.148
vanadium	V	a.m. = 50.9462
carbide	VC	m.p. = 2,810; b.p. = 3,900
chloride, tetra-	VCl_4	—
oxide	VO_2	—
vanillin	$(CH_3O)C_6H_3(OH) \cdot CHO$	—
vinyl acetate	$CH_3CO_2CH:CH_2$	—
chloride	$CH_2:CHCl$	—
water	H_2O	—
wolfram	—	see tungsten
xanthophyll	$C_{40}H_{56}O_2$	lutein
xenon	Xe	a.m. = 131.29
xylene	$(CH_3)_2C_6H_4$	—
yperite	—	see mustard gas
ytterbium	Yb	a.m. = 173.04
yttrium	Y	a.m. = 88.905849

(*cont.*)

Table B1. **Chemical Formulary** (*continued*)

Name	Formula	Synonyms, data, and notes
zinc	Zn	a.m. = 65.39
aluminate	$ZnAl_2O_4$	gahnite
carbonate	$ZnCO_3$	smithsonite
chloride	$ZnCl$	—
hydroxide	$Zn(OH)_2$	—
oxide	ZnO	zincite
silicate, ortho-	Zn_2SiO_4	willemite
sulfate	$ZnSO_4$	zinkosite
sulfide	ZnS	wurtzite
sulfide	ZnS	sphalerite
zirconium	Zr	a.m. = 91.224
carbide	ZrC	m.p. = 3,540; b.p. = 5,100
oxide	ZrO_2	beddeleyite
oxide	ZrO_2	zirconia
silicate, ortho-	$ZrSiO_4$	zircon

Note: a.m., atomic mass in atomic mass units; m.m., molecular mass in atomic mass units; m.p., melting point; b.p., boiling point; subl., sublimates. Temperatures given in degrees Celsius. Asterisks identify radioactive elements. Data for short-lived unstable elements refer to the longest-lived isotope. Many compounds occur in nature as minerals. Their names are given in the third column of this table. The more common ones are included in Table A.11.

SOURCES FOR THE ILLUSTRATIONS
AND THE TABLES

Sources for the text illustrations and tables and the tables that appear in Appendix A are given in the standard abbreviated form below each pertinent item. The full reference is given here. If no source is given, the material has been compiled, developed, calculated, or otherwise derived by the present author. Reproduction of materials from published sources is with the written consent of the original authors and publishers, to whom the present author wishes to express his appreciation and gratitude.

Abell G. O., Morrison D., and Wolff S. C. 1987. *Exploration of the Universe*. Saunders, Philadelphia, 755 p.

Alexander R. M. 1989. *Dynamics of Dinosaurs and Other Extinct Giants*. Columbia University Press, New York, 167 p.

Algermissen S. T. and Perkins D. M. 1976. *A Probabilistic Estimate of Maximum Acceleration in Rock in the Contiguous United States*. U.S. Geological Survey, Open File Report 76–416, 45 p.

Anderson R. N. 1986. *Marine Geology*. Wiley, New York, 328 p.

Anderson T. F. and Wollman E. L. 1957. *Annales de l'Institut Pasteur* (*Paris*), 93:450.

Andreatta C. 1943. *Mineralogia*. Università di Bologna, Bologna, Italy, 710 p.

Ayala F. J. 1982. *Population and Evolutionary Genetics*. Benjamin/Cummings, Menlo Park, Calif., 268 p.

Bakker R. T. 1987. The return of the dancing dinosaurs. In: Czerkas S. J. and Olson E. C. (eds.), *Dinosaurs Past and Present*, vol. 1, pp. 39–69. University of Washington Press, Seattle.

Barazangi M. and Dorman J. 1969. World seismicity maps compiled from ESSA, Coast and Geodetic Survey, epicenter data, 1961–1967. *Seismological Society of America Bulletin* 59:369–80.

Bard E., Hamelin B., Fairbanks R. G., and Zindler A. 1990. Calibration of the ^{14}C time scale over the past 30,000 years using mass spectrometric U-Th ages from Barbados corals. *Nature*, 345:405–9.

Barry R. G. and Chorley R. J. 1987. *Atmosphere, Weather, and Climate*. Methuen, London, 460 p.

Beatty J. K. and Chaikin A. (eds.). 1990. *The New Solar System*. Cambridge University Press, 326 p.

Berry A. 1961. *A Short History of Astronomy*. Dover, New York (reprint of original edition by John Murray, 1898), 440 p.

Berry L. G., Mason B., and Dietrich R. V. 1983. *Mineralogy*. Freeman, New York, 561 p.

Bloss F. D. 1971. *Crystallography and Crystal Chemistry*. Holt, Rinehart, and Winston, New York, 545 p.

Boardman R. S., Cheetham A. H., and Rowell H. A. (eds.). 1987. *Fossil Invertebrates*. Blackwell Scientific, Oxford, 713 p.

Borradaile L. A. and Potts F. A. 1961. *The Invertebrata.* Cambridge University Press, 820 p.

Bowers R. L. and Deeming T. 1984. *Astrophysics*, 2 vols. Jones & Bartlett, Boston, Mass., 619 p.

Bragg W. L. 1955. *Atomic Structure of Minerals*, 3 vols. Cornell University Press, Ithaca, N.Y.

Brookins D. G. 1981. *Earth Resources, Energy, and the Environment.* Charles E. Merrill Publishing, Columbus, Ohio, 190 p.

Brusca R. C. and Brusca G. J. 1990. *Invertebrates.* Sinauer Associates, Sunderland, Mass., 922 p.

Buettner-Janusch J. 1966. *Origins of Man.* Wiley, New York, 674 p.

Burke K. 1976. Development of graben associated with the initial ruptures of the Atlantic Ocean. *Tectonophysics*, 36: 83–112.

Calvin M. 1969. *Chemical Evolution.* Oxford University Press, 278 p.

Carr N. G. and Whitton B. A. 1973. *The Biology of Blue-Green Algae.* University of California Press, Berkeley, 676 p.

Clarkson E. N. K. 1986. *Invertebrate Paleontology and Evolution.* Allen & Unwin, London, 382 p.

Clausen C. J., Cohen A. D., Emiliani C., Holman J. A., and Stipp J. J. 1976. Little Salt Spring, Florida: an underwater site preserving earliest wooden artifacts in North America. *Science*, 203: 609–14.

Clemens D. P. 1985. Massachusetts–Stony Book galactic plane co-survey: the galactic disk rotation curve. *Astrophysical Journal* 295: 422–36.

Colbert E. H. 1968. *Men and Dinosaurs.* E. P. Dutton, New York, 283 p.

Coleman J. M. 1968. Deltaic evolution. In Fairbridge R. W. (ed.), *Encyclopedia of Geomorphology*, pp. 255–61. Reinhold, New York, 1,295 p.

Curtis H. and Barnes N. D. 1989. *Biology.* Worth Publishers, New York, 1,280 p.

Czerkas S. A. 1987. A reevaluation of the plate arrangement on *Stegosaurus stenops.* In: Czerkas S. J. and Olson E. C. (eds.), *Dinosaurs Past and Present*, vol. 2. pp. 83–99. University of Washington Press, Seattle, 164 p.

Dahlstron D. C. A. 1969. Balanced cross sections. *Canadian Journal of Earth Sciences*, 6: 746.

Dana J. D. 1896. *Manual of Geology.* American Book Co., New York, 1,087 p.

Darragh P. J., Gaskin A. J., and Sanders J. V. 1976. Opals. *Scientific American*, vol. 234, no. 4, pp. 84–95.

Diem K. and Lentner G. 1970. *Scientific Tables.* Geigy Pharmaceuticals, Ardley, N. Y., 810 p.

Dietrich G., Kalle K., Krauss W., and Siedler G. 1975. *Allgemeine Meereskunde.* Gebrüder Borntraeger, Berlin, 593 p.

Donaldson J. A. 1963. *Stromatolites of the Denault Formation, Marion Lake, Coast of Labrador, Newfoundland.* Geological Society of Canada, Bulletin No. 102.

Donn W. L. 1965. *Meteorology.* McGraw-Hill, New York, 484 p.

Ehlers E. G. and Blatt H. 1982. *Petrology—Igneous, Sedimentary, and Metamorphic.* Freeman, New York, 732 p.

Eisenberg D. and Kauzman W. 1969. *The Structure and Properties of Water.* Oxford University Press, 296 p.

Elders W. A., Rex R. W., Meidav T., Robinson P. T., and Biehler S. 1972. Crustal spreading in Southern California. *Science*, 178: 15–24.

Emery K. O. and Uchupi E. 1972. *Western North Atlantic Ocean.* American Association of Petroleum Geologists, Tulsa, Okla., 516 p.

Emiliani C. 1955. Pleistocene temperatures. *Journal of Geology*, 63: 538–78.

Emiliani C. 1978. The cause of the ice ages. *Earth and Planetary Science Letters*, 37: 349–52.

Emiliani C. 1980. Death and renovation and the end of the Mesozoic. *EOS*, 61: 502–6.

Emiliani C. 1987. *Dictionary of the Physical Sciences.* Oxford University Press, 365 p.

Emiliani C. 1988. *The Scientific Companion.* Wiley, New York, 287 p.

Encyclopaedia Britannica. 1952. Encyclopaedia Britannica, 14th ed., Chicago, 24 vols.

Fakhry A. 1961. *The Pyramids.* University of Chicago Press, 272 p.

Faure G. 1986. *Principles of Isotope Geology*. Wiley, New York, 589 p.

Finkelnburg W. 1964. *Structure of Matter*. Springer-Verlag, Berlin, and Academic Press, Orlando, Fla., 511 p.

Flint R. F. 1952. Glacial Epoch. *Encyclopaedia Britannica*, 14th ed., Chicago, vol. 10, pp. 374–81.

Freifelder D. 1987. *Molecular Biology*. Jones & Bartlett, Boston, 834 p.

Frye K. 1974. *Modern Mineralogy*. Prentice-Hall, Englewood Cliffs, N. J., 325 p.

Gamow G. 1958. *Matter, Earth, and Sky*. Prentice-Hall, Englewood Cliffs, N. J., 593 p.

Glaessner M. 1984. *The Dawn of Animal Life*. Cambridge University Press, 244 p.

Gornitz V. S., Lebedeff S., and Hansen J. 1982. Global sealevel trend in the past century. *Science*, 215: 1611–14.

Goudie A. 1989. *The Nature of the Environment*. Basil Blackwell, Oxford, 370 p.

Gribbin J. 1988. *The Omega Point*. Bantam Books, New York, 246 p.

Gross M. G. 1977. *Oceanography*. Prentice-Hall, Englewood Cliffs, N. J., 497 p.

Gross M. G. 1987. *Oceanography*. Prentice-Hall, Englewood Cliffs, N. J., 406 p.

Hansen J. and Lebedeff S. 1988. Global surface air temperature: update through 1987. *Geophysical Research Letters*, 15: 323–6.

Haq B. U. 1978. Calcareous nannoplankton. In: Haq B. U. and Boersma A. (eds.), *Introduction to Marine Micropaleontology*. Elsevier, New York, pp. 79–107.

Haller J. 1971. *Geology of the East Greenland Caledonides*. Wiley Interscience, New York, 413 p.

Harland W. B., Cox A. V., Llewellyn P. G., Pickton C. A. G., Smith A. G., and Walters R. 1982. *A Geologic Time Scale*. Cambridge University Press, 263 p.

Hess H. H. 1946. Drowned ancient islands of the Pacific Basin. *American Journal of Science*, 244: 772–91.

Hofmann H. J. and Schopf J. W. 1983. Early Proterozoic microfossils. In: Schopf J. W. (ed.), *The Earth's Earliest Biosphere*. Princeton University Press, Princeton, N. J., 543 p.

Holland C. H. 1971. *Cambrian of the New World*. Wiley, New York, 456 p.

Holland C. H. 1974. *Cambrian of the British Isles, Norden, and Spitzbergen*. Wiley, New York, 300 p.

Holland H. D. 1984. *The Chemical Evolution of the Atmosphere and Ocean*. Princeton University Press, Princeton, N. J., 582 p.

Holmes A. 1965. *Principles of Physical Geology*. Ronald Press, New York, 1,288 p.

Jaramayan A. and Cohen L. H. 1970. Phase diagrams in high-pressure research. In Alper A. M. (ed.), *Phase Diagrams*, vol. 1, pp. 245–93. Academic Press, Orlando, Fla., 358 p.

Kaufmann W. J. III. 1988. *Universe*. Freeman, New York, 634 p.

Kennett J. P. 1982. *Marine Geology*. Prentice-Hall, Englewood Cliffs, N. J., 813 p.

King H. C. 1955. *The History of the Telescope*. Charles Griffin & Co., London (Dover reprint, New York, 1979), 456 p.

King P. B. 1965. *The Geologic Map of North America*. Geological Society of America, Boulder, Colo.

King R. C. and Stansfield W. D. 1985. *A Dictionary of Genetics*. Oxford University Press, 480 p.

Krauskopf K. B. 1979. *Introduction to Geochemistry*. McGraw-Hill, New York, 617 p.

Krumbein W. C. and Sloss L. L. 1963. *Stratigraphy and Sedimentation*. Freeman, New York, 660 p.

Kummel B. 1970. *History of the Earth*. Freeman, New York, 707 p.

Larson R. and Pitman W. C. III. 1972. World-wide correlation of Mesozoic magnetic anomalies and its implications. *Geological Society of America, Bulletin*, 83: 3645–62.

Larson R. and Pitman W. C. III. 1988. *Bedrock Ages of the World*. Freeman, New York.

Leet L. D. 1950. *Earth Waves*. Harvard University Press, Cambridge, Mass.

Lehmann U. and Hillmer G. 1983. *Fossil Invertebrates*. Cambridge University Press, 350 p.

Littman M. 1988. *Planets Beyond*. Wiley, New York, 286 p.

Lorius C., Jouzel J., Ritz C., Merlivat L., Barkov N. I., Korotkevich Y. S., and Kotlyakov V. M. 1985. A 150,000-year climatic record from Antarctic ice. *Nature*, 316: 591–6.

Luterbacher H. P. and Premoli-Silva I. 1964. Biostratigrafia del limite Cretaceo-Terziario nell' Appennino Centrale. *Rivista Italiana di Paleontologia*, 70: 67–128.

Manchester R. N. and Taylor J. H. 1977. *Pulsars*. Freeman, San Francisco, 281 p.

Margulis L. and Schwartz K. V. 1988. *Five Kingdoms*. Freeman, New York, 376 p.

Martin P. S. and Wright H. E. Jr. 1967. *Pleistocene Extinctions*. Yale University Press, New Haven, Conn., 453 p.

Mason B. and Moore C. B. 1982. *Principles of Geochemistry*. Wiley, New York, 344 p.

McConnell R. B. 1972. Geological development of the rift system of East Africa. *Geological Society of America, Bulletin*, 83: 2549–72 (Fig. 1).

McGraw-Hill Encyclopedia of Science and Technology. 1987. McGraw-Hill, New York, 20 vols.

Mitchell R. S. 1979. *Mineral Names*. Van Nostrand Reinhold, New York, 229 p.

Moment G. B. and Habermann H. M. 1973. *Biology—A Full Spectrum*. Williams & Wilkins, Baltimore, 756 p.

Moore R. C. 1958. *Introduction to Historical Geology*. McGraw-Hill, New York, 656 p.

Moore W. J. 1950. *Physical Chemistry*. Prentice-Hall, Englewood Cliffs, N. J., 592 p.

Moret L. 1940. *Manuel de Paléontologie Animale*. Masson, Paris, 675 p.

Moret L. 1943. *Manuel de Paléontologie Végétale*. Masson, Paris, 216 p.

Moullade M. 1978. The Ligurian Sea and adjacent areas. In: Nairn A. E. M., Kanmes W. H., and Stehli F.G. (eds.), *The Ocean Basins and Margins*, vol. IV-B, *The Western Mediterranean*, pp. 67–148. Plenum Press, New York, 447 p.

Newell N. A. 1956. Geological reconnaissance of Raroia (Kon Tiki) atoll, Tuamotu archipelago. *American Museum of Natural History, Bulletin*, 109: 317–72.

Paul G. S. 1987. The science and art of restoring the life appearance of dinosaurs and their relatives: a rigorous how-to guide. In: Czerkas S. J. and Olson E. C. (eds.), *Dinosaurs Past and Present*, vol. 2, pp. 5–49. University of Washington Press, Seattle, Wash., 164 p.

Paul G. S. 1988. *Predatory Dinosaurs of the World*. Simon & Schuster, New York, 464 p.

Price R. A. and Mountjoy E. W. 1970. *Geologic Structure of the Canadian Rocky Mountains between Bow and Athabasca Rivers—Progress Report*. Geological Association of Canada, Special Paper 6, pp. 7–25.

Reineck H.-E. and Singh I. B. 1973. *Depositional Sedimentary Environments*. Springer-Verlag, Berlin, 439 p.

Rigutti M. 1984. *A Hundred Billion Stars*. MIT Press, Cambridge, Mass., 285 p.

Robinson A. S. and Coruh C. 1988. *Basic Exploration Geophysics*. Wiley, New York, 562 p.

Romer A. S. 1966. *Vertebrate Paleontology*. University of Chicago Press, 468 p.

Schopf J. W. (ed.), 1983. *The Earth's Earliest Biosphere*. Princeton University Press, Princeton, N. J., 543 p.

Schwemmler W. 1984. *Reconstruction of Cell Evolution*. CRC Press, Boca Raton, Fla., 248 p.

Seldner M., Siebers B., Groth E. J. and Peebles P. J. E. 1977. New reduction of the Lick catalog of galaxies. *Astronomical Journal*, 82: 249–56 (Plates 1 and 2).

Sellers W. D. 1966. *Physical Climatology*. University of Chicago Press, 272 p.

Shackleton N. J. and Kennett J. P. 1975. Paleotemperature history of the Cenozoic and the initiation of Antarctic glaciation: oxygen and carbon isotope analysis of DSDP sites 277, 279, and 281. *Initial Reports of the Deep-Sea Drilling Project*, vol. 29, pp. 743–55. U. S. Government Printing Office, Washington, D. C., 1197 p.

Shackleton N. J. and Opdyke N. D. 1977. Oxygen isotope and palaeomagnetic evidence of early Northern Hemisphere glaciation. *Nature*, 270: 216–19.

Shinn E. A. 1963. Spur and groove formation on the Florida reef tract. *Jounal of Sedimentary Petrology*, 33: 291–303.

Shinn E. A. 1978. Atlantis: Bimini hoax. *Sea Frontiers*, 24: 130–41.

Shrock R. R. and Twenhofel W. H. 1943. *Principles of Invertebrate Paleontology*. McGraw-Hill, New York, 816 p.

Siever R. 1983. The dynamic Earth. *Scientific American*, 249(3): 46–55.

Simpson G. G., Pittendrigh C. S., and Tiffany L. H. 1957. *Life*. Harcourt Brace Jovanovich, Orlando, Fla., 845 p.

Skinner B. J. 1969. *Earth Resources*. Prentice-Hall, Englewood Cliffs, N.J., 150 p.

Spencer E. W. 1965. *Geology*. Thomas Y. Crowell Co., New York, 653 p.

Spencer E. W. 1988. *Introduction to the Structure of the Earth*. McGraw-Hill, New York, 551 p.

Stacey F. D. 1977. *Physics of the Earth*. Wiley, New York, 414 p.

Stainer R. Y., Doudoroff M., and Adelberg E. A. 1986. *The Microbial World*. Prentice-Hall, Englewood Cliffs, N.J., 960 p.

Stewart 1983. *Paleobotany and the Evolution of Plants*. Cambridge University Press, 405 p.

Stitt J. H. 1977. *Late Cambrian and Earliest Ordovician Trilobites, Wichita Mountains Area, Oklahoma*. Oklahoma Geological Survey, Norman, Okla. Bulletin 124, 79 p.

Stowe K. 1987. *Essentials of Ocean Science*. Wiley, New York, 353 p.

Strahler A. N. 1971. *The Earth Sciences*. Harper & Row, New York, 824 p.

Sverdrup H. U., Johnson M. W., and Fleming R. H. 1942. *The Oceans*. Prentice-Hall, Englewood Cliffs, N.J., 1,087 p.

Tasch P. 1973. *Paleobiology of the Invertebrates*. Wiley, New York, 946 p.

Taylor T. N. 1981. *Paleobotany*. McGraw-Hill, New York, 589 p.

Tijo J. H. and Puck T. T. 1958. The somatic chromosomes of man. *National Academy of Sciences, Proceedings*, 44:1229–37.

U. S. Standard Atmosphere 1976. National Oceanographic and Atmospheric Administration, Washington, D. C., 227 p.

Vail P. R., Mitchum R. M., and Thompson S. III. 1977. Seismic stratigraphy and global changes of sea level. Part 4: Global cycles of relative changes of sea level. In: Payton C. E. (ed.), *Seismic Stratigraphy—Applications to Hydrocarbon Exploration*. American Association of Petroleum Geologists, Tulsa, Oklahoma Memoir 26, 516 p.

von Prahl H., and Erhardt H. 1985. *Colombia—Corales y Arrecifes Coralinos*. Universidad del Valle, Colombia, 295 p.

Weller J. M. 1969. *The Course of Evolution*. McGraw-Hill, New York, 696 p.

Williams G. L. 1978. Dinoflagellates, acritarchs, and tasmanitids. In: Haq B. U. and Boersma A. (eds.), *Introduction to Marine Micropaleontology*. Elsevier, New York, pp. 293–326.

Windley B. F. (ed.), 1976. *The Early History of the Earth*. Wiley, New York, 619 p.

Windley B. F. 1984. *The Evolving Continents*. Wiley, New York, 399 p.

Young G. M. 1974. Stratigraphy, paleocurrents, and stromatolites of the Hadrynian (Upper Cambrian) rocks, Victoria Island, Arctic Archipelago, Canada. *Precambrian Research*, 74:13–41.

FOR REFERENCE AND FOR LEARNING

FOR REFERENCE

American Institute of Physics Handbook. McGraw-Hill, New York.

Asimov I. 1982. *Asimov's Biographical Encyclopedia of Science and Technology.* Doubleday, New York, 1982, 941 p.

CRC Handbook of Chemistry and Physics. CRC Press, Boca Raton, Florida.

Emiliani C. 1987. *Dictionary of the Physical Sciences.* Oxford University Press, 1987, 365 p.

Encyclopaedia Britannica. Encyclopaedia Britannica, Chicago.

Geigy Scientific Tables. CIBA-GEIGY, Basel, Switzerland.

General Electric Company. *Nuclides and Isotopes—Chart of the Nuclides.* General Electric, San Jose, California.

Millar D., Millar I., Millar J., and Millar M. 1989. *Concise Dictionary of Scientists.* Chambers/Cambridge University Press, 1989, 461 p.

The McGraw-Hill Encyclopedia of Science and Technology, 7th ed., 20 vols. McGraw-Hill, New York, 1992.

Van Nostrand's Scientific Encyclopedia. Van Nostrand Reinhold, New York.

FOR LEARNING

Physics and Chemistry

Atkins P. W. 1990. *Physical Chemistry.* Freeman, New York, 857 p.

Davies P. 1984. *Quantum Mechanics.* Routledge & Kegan Paul, London, 139 p.

Davies P. 1986. *The Forces of Nature.* Cambridge University Press, 175 p.

Davies P. (ed.). 1989. *The New Physics.* Cambridge University Press, 516 p.

Feynman R. P. 1985. *QED.* Princeton University Press, 158 p.

Feynman R., Leighton R., and Sands M. 1963–1965. *Lectures on Physics.* Addison-Wesley, Reading, Mass. 3 vols.

Krane K. 1983. *Modern Physics.* Wiley, New York, 512 p.

Pauling L. 1988. *General Chemistry.* Dover Publications, New York, 959 p.

Polkinghorne J. C. 1984. *The Quantum World.* Princeton University Press, 100 p.

Cosmology

Beatty J. K. and Chaikin A. 1990. *The New Solar System.* Cambridge University Press, 326 p.

Glass B. P. 1982. *Introduction to Planetary Geology.* Cambridge University Press, 469 p.

Harrison E. R. 1988. *Cosmology.* Cambridge University Press, 430 p.

Hamblin W. K. and Christiansen E. H. 1990. *Exploring the Planets.* Macmillan, New York, 451 p.

Hawking S. W. 1988. *A Brief History of Time.* Bantam Books, New York, 198 p.

Kaufmann W. J. III 1987. *Universe.* Freeman, New York, 634 p.

Riordan M. and Schramm D. N. 1991. *The Shadows of Creation.* Freeman, New York, 278 p.

Silk J. 1989. *The Big Bang.* Freeman, New York, 485 p.

Weinberg S. 1988. *The First Three Minutes*. Basic Books, New York, 198 p.

Zeilik M. and Smith E. v. P. 1987. *Introductory Astronomy and Astrophysics*. Saunders, Philadelphia, 503 p.

Geology

Broecker W. S. 1985. *How to Build a Habitable Planet*. Lamont-Doherty Geological Observatory, Columbia University, Palisades, New York, 291 p.

Cloud P. 1988. *Oasis in Space*. Norton, New York, 508 p.

Condie K. C. 1989. *Plate Tectonics and Crustal Evolution*. Pergamon Press, Oxford, 476 p.

Ehlers E. G. and Blatt H. 1982. *Petrology*. Freeman, New York, 732 p.

Fowler C. M. R. 1990. *The Solid Earth*. Cambridge University Press, 472 p.

Lemon R. R. 1990. *Principles of Stratigraphy*. Merrill Publishing, Columbus, Ohio, 559 p.

Press F. and Siever R. 1986. *Earth*. Freeman, New York, 656 p.

Stanely S. 1989. *Earth and Life Through Time*. Freeman, New York, 690 p.

Van Andel T. H. 1985. *New Views on an Old Planet*. Cambridge University Press, 324 p.

Biology and Paleontology

Brusca R. C. and Brusca G. J. 1990. *Invertebrates*. Sinauer Associates, Sunderland, Mass., 922 p.

Carrol R. L. 1988. *Vertebrate Paleontology and Evolution*. Freeman, New York, 698 p.

Clarkson E. 1986. *Invertebrate Paleontology and Evolution*. Allen & Unwin, London, 382 p.

Curtis H. and Barnes N. S. 1989. *Biology*. Worth Publishers, New York, 1280 p.

Freifelder D. 1987. *Molecular Biology*. Jones & Bartlett, Boston, 834 p.

Loomis W. F. 1988. *Four Billion Years*. Sinauer Associates, Sunderland, Mass., 286 p.

Margulis L. 1981. *Symbiosis in Cell Evolution*. Freeman, New York, 419 p.

Margulis L. and Schwartz K. V. 1988. *Five Kingdoms*. Freeman, New York, 376 p.

Mayr E. 1942. *Systematics and the Origin of Species*. Columbia University Press, New York, 334 p. (Reprinted by Dover Publications, New York, 1964.)

SUBJECT INDEX

Note: Italics identify illustrations or tables.

EXPLANATIONS FOR THE ILLUSTRATIONS ON THE FRONT AND BACK COVERS

Front: The Earth photographed from a distance of 40,300 km by the crew of the *Apollo 17* mission on their way to the Moon (December 1972).

Back: The plaque, 15.2 × 22.9 cm, that is traveling beyond the solar system with a message from Earth. The plaque is on board *Pioneer 10*, a spacecraft that was launched on March 1972, passed Jupiter on December 3, 1973, and left the solar system on June 14, 1983. On the upper left corner is an outline of two hydrogen atoms, the left one with the proton and electron spin axes pointing in opposite directions, a state of lower energy, the right one with the two spins aligned in the same direction, a state of higher energy. The energy difference is $5.873 \cdot 10^{-6}$ eV, corresponding to a wavelength of 21.11 cm and a frequency of $1.427 \cdot 10^9$ Hz. The number 1 just beneath the line connecting the two atoms, indicates that this frequency is to be taken as unit of frequency (i.e., $1.427 \cdot 10^9 = 1$) and the wavelength is to be taken as unit of length ($21.11 = 1$). The rays radiating from the point at center left indicate the direction of 14 pulsars as seen from Earth. Along the rays are numbers in binary code (with dashes instead of zeros) giving the period of each pulsar in frequency units. Because pulsars slow down with time (see Section 6.4), the illustration will reveal to an intelligent alien the location of the Earth and the time and spacecraft was launched. The line of circles along the bottom shows the Sun (extreme left) and the planets, with their relative distances from the Sun in binary notation. Also shown is the path of *Pioneer 10* through the solar system and out. Center right are two human figures superimposed, to scale, on an outline of the spacecraft. To the right of the female figure is the number 8 in binary notation (1000), giving her height in units of length (the height is thus 21.11 cm × 8 = 1.69 m).

Binary notation expresses any number as the sum of powers of the number 2. If a power of 2 is missing, the number 0 is entered. Thus, 1 ($= 2^0$) is written as 1; 2 is written 10 ($2^1 + 0 \cdot 2^0$); 3 is written as 11 ($2^1 + 2^0$); 4 is written as 100 ($2^2 + 0 \cdot 2^1 + 0 \cdot 2^0$); 10 is written 1010 ($= 2^3 + 0 \cdot 2^2 + 2^1 + 0 \cdot 2^0$); etc. Notice that the exponent is one less than indicated by the position from the right. This is so because the first position on the right is taken by the exponent 0. To read a number in the binary notation, skip the zeros and add the powers of 2 corresponding to the positions of the 1's from the right (less 1). To write a number in binary notation, one fits decreasing powers of 2 into that number. For instance, 238, the atomic mass of ^{238}U, is equal to 2^7 (128) + 110; 110 = 2^6 (64) + 46; 46 = 2^5 (32) + 14; 14 = 2^3 (8) + 6; 6 = 2^2 (4) + 2; 2 = 2^1 with zero left. The binary notation of 238 is, therefore 11101110. In binary notation one would write ^{11101110}U. Conversely, $11101110 = 2^7 + 2^6 + 2^5 + 0 \cdot 2^4 + 2^3 + 2^2 + 2^1 + 0 \cdot 2^0 = 238$. Numbers with decimals are converted the same way. For instance, $\pi = 3.1416$ becomes $2^1 + 2^0 + 0.1416$; $0.1416 = 2^{-3} + 0.0166$; $0.0166 = 2^{-6} + 0.000975$; $0.000975 = 2^{-11} + 0.000486\ldots$ In binary notation, π is therefore $11.001000100001\ldots$

Carl Sagan (b. 11110001110), the noted U.S. astronomer who devised the plaque and is now professor at Cornell, expressed the planetary distances from the Sun by arbitrarily assigning the number 10 to the distance Sun–Mercury and scaling the other planetary distances accordingly. In order to signify to the alien reader that the planetary distances were given in units different from the 21.11 wavelength, he used the symbol I instead of 1 (as for the pulsars and the lady). He did not use the 21.11 cm wavelength as a unit of length, because the distance Sun–Mercury, 274250951970.1563 units, would translate into a binary number too long to fit on the plate. The distances to the other planets would be expressed by even longer binary numbers. Can you figure out a way to express these large numbers in a way that would preserve the binary notation, fit the plaque, and be comprehensible to intelligent aliens?

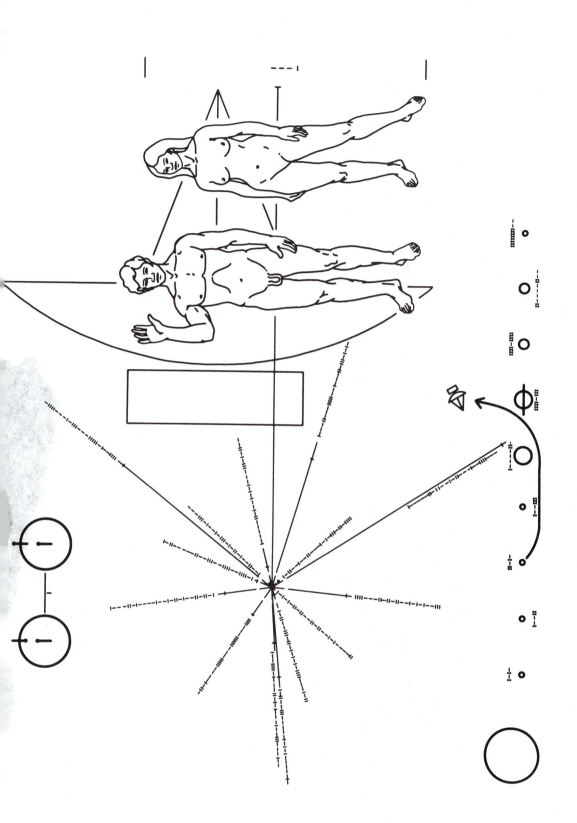